T0180872

Lecture Notes in Computer Science 13884

Founding Editors

Gerhard Goos
Juris Hartmanis

Editorial Board Members

Elisa Bertino, *Purdue University, West Lafayette, IN, USA*
Wen Gao, *Peking University, Beijing, China*
Bernhard Steffen, *TU Dortmund University, Dortmund, Germany*
Moti Yung, *Columbia University, New York, NY, USA*

The series Lecture Notes in Computer Science (LNCS), including its subseries Lecture Notes in Artificial Intelligence (LNAI) and Lecture Notes in Bioinformatics (LNBI), has established itself as a medium for the publication of new developments in computer science and information technology research, teaching, and education.

LNCS enjoys close cooperation with the computer science R & D community, the series counts many renowned academics among its volume editors and paper authors, and collaborates with prestigious societies. Its mission is to serve this international community by providing an invaluable service, mainly focused on the publication of conference and workshop proceedings and postproceedings. LNCS commenced publication in 1973.

Andre A. Cire
Editor

Integration of Constraint Programming, Artificial Intelligence, and Operations Research

20th International Conference, CPAIOR 2023 Nice, France, May 29 – June 1, 2023 Proceedings

Editor
Andre A. Cire
Department of Management, University of Toronto
Scarborough and Rotman School of Management
University of Toronto
Toronto, ON, Canada

ISSN 0302-9743 ISSN 1611-3349 (electronic)
Lecture Notes in Computer Science
ISBN 978-3-031-33270-8 ISBN 978-3-031-33271-5 (eBook)
https://doi.org/10.1007/978-3-031-33271-5

© The Editor(s) (if applicable) and The Author(s), under exclusive license
to Springer Nature Switzerland AG 2023
This work is subject to copyright. All rights are reserved by the Publisher, whether the whole or part of the material is concerned, specifically the rights of translation, reprinting, reuse of illustrations, recitation, broadcasting, reproduction on microfilms or in any other physical way, and transmission or information storage and retrieval, electronic adaptation, computer software, or by similar or dissimilar methodology now known or hereafter developed.
The use of general descriptive names, registered names, trademarks, service marks, etc. in this publication does not imply, even in the absence of a specific statement, that such names are exempt from the relevant protective laws and regulations and therefore free for general use.
The publisher, the authors, and the editors are safe to assume that the advice and information in this book are believed to be true and accurate at the date of publication. Neither the publisher nor the authors or the editors give a warranty, expressed or implied, with respect to the material contained herein or for any errors or omissions that may have been made. The publisher remains neutral with regard to jurisdictional claims in published maps and institutional affiliations.

This Springer imprint is published by the registered company Springer Nature Switzerland AG
The registered company address is: Gewerbestrasse 11, 6330 Cham, Switzerland

Preface

This volume contains the papers that were presented at the 20th International Conference on the Integration of Constraint Programming, Artificial Intelligence, and Operations Research (CPAIOR 2023). The conference was held as an in-person event in Nice, France, at the Université Côte d'Azur, campus of Saint Jean d'Angely.

The conference received a total of 95 submissions, including 71 regular papers and 24 extended abstracts. The regular papers reflect original unpublished work, whereas the extended abstracts contain either original unpublished work or a summary of work that was published in another venue. Each regular paper was reviewed by at least three Program Committee members in a single-blind process. The reviewing phase was followed by an author response period and an extensive discussion period carried out by the Program Committee. The extended abstracts were reviewed for appropriateness for the conference. At the end of the review period, 32 regular papers were accepted for presentation during the conference and publication in this volume, and 11 abstracts were accepted for a short presentation at the conference.

In addition to the regular papers and extended abstracts, three invited talks were given by John Paul Dickerson (University of Maryland, Arthur), Ivana Ljubić, (ESSEC Business School), and Rodrigo Acuna Agost (Amadeus). The conference program also included a Master Class on the topic of *Transportation: New Frontiers in Practice and Theory*, organized by Jean-Charles Régin. The Master Class included invited talks by Nicolas Isoart (Zeloce), Willem-Jan van Hoeve (Carnegie Mellon University), Arthur Finkelstein (Instant System France), Steven Gay (Google Paris), and Andre A. Cire (University of Toronto).

Of the regular papers accepted to the conference, the paper "Objective-Based Counterfactual Explanations for Linear Discrete Optimization" by Anton Korikov and J. Christopher Beck was selected for the Best Paper Award, and the paper "Column Elimination for Capacitated Vehicle Routing Problems" by Anthony Karahalios and Willem-Jan van Hoeve was selected for the Best Student Paper Award. The selection process was based on ranking and voting by the Program Committee members, paper scores, and extensive consultation with reviewers.

We acknowledge the local organizer, Jean-Charles Régin, and the generous support of our sponsors, which were at the time of publication the Artificial Intelligence Journal (AIJ), Springer, Université Côte d'Azur, Laboratory I3S, Gurobi Optimization, Association for Constraint Programming (ACP), COPT GmbH, Groupe de Recherche Raisonnement, Apprentissage et Décision en Intelligence Artificielle (GRD RADIA), and Groupe de Recherche – Recherche Opérationnelle et Décision (GRD ROD).

April 2023 Andre A. Cire

Organization

Program Chair

Andre A. Cire	University of Toronto, Canada

Conference/Master Class Chair

Jean-Charles Régin	Université Côte d'Azur, France

Program Committee

Brandon Amos	Facebook AI, USA
Marleen Balvert	Tilburg University, The Netherlands
Beste Basciftci	University of Iowa, USA
J. Christopher Beck	University of Toronto, Canada
Nicolas Beldiceanu	IMT Atlantique (LS2N), France
David Bergman	University of Connecticut, USA
Armin Biere	University of Freiburg, Germany
Christian Blum	Spanish National Research Council (CSIC), Spain
Merve Bodur	University of Toronto, Canada
Hadrien Cambazard	Université Grenoble Alpes, France
Quentin Cappart	Polytechnique Montréal, Canada
Carlos H. Cardonha	University of Connecticut, USA
Mats Carlsson	RISE Research Institutes of Sweden, Sweden
Margarida Carvalho	University of Montreal, Canada
Margarita Castro	Pontifica Universidad Católica de Chile, Chile
Simon de Givry	INRA - MIAT, France
Emir Demirović	Delft University of Technology, The Netherlands
Guillaume Derval	Université Catholique de Louvain, France
Maria Andreina Francisco Rodriguez	Uppsala University, Sweden
Maria Garcia de la Banda	Monash University, Australia
Tias Guns	KU Leuven, Belgium
Emmanuel Hebrard	Université de Toulouse, France
John Hooker	Carnegie Mellon University, USA
Serdar Kadioglu	Brown University, USA

Roger Kameugne	University of Maroua, Cameroon
George Katsirelos	AgroParisTech, France
Elias Khalil	University of Toronto, Canada
Joris Kinable	Amazon, USA
Zeynep Kiziltan	University of Bologna, Italy
Christophe Lecoutre	University of Artois, France
Jiaoyang Li	Carnegie Mellon University, USA
Michele Lombardi	University of Bologna, Italy
Pierre Lopez	Université de Toulouse, France
Arnaud Malapert	Université Côte d'Azur, France
Ciaran McCreesh	University of Glasgow, UK
Laurent Michel	University of Connecticut, USA
Nysret Musliu	TU Wien, Austria
Nina Narodytska	VMware Research, USA
Justin Pearson	Uppsala University, Sweden
Marie Pelleau	Université Côte d'Azur, France
Laurent Perron	Google France, France
Gilles Pesant	Polytechnique Montréal, Canada
Claude-Guy Quimper	Laval University, Canada
Jean-Charles Régin	University Nice-Sophia Antipolis, France
Michael Römer	Bielefeld University, Germany
Louis-Martin Rousseau	Polytechnique Montréal, Canada
Domenico Salvagnin	University of Padova, Italy
Pierre Schaus	Université Catholique de Louvain, Belgium
Thomas Schiex	INRAE, France
Paul Shaw	IBM, France
Mohamed Siala	INSA Toulouse, France
Helmut Simonis	University College Cork, Ireland
Christine Solnon	INSA Lyon, France
Christian Tjandraatmadja	Google, USA
Willem-Jan van Hoeve	Carnegie Mellon University, USA
Hélène Verhaeghe	Polytechnique Montréal, Canada
Petr Vilím	CoEnzyme, Czechia
Mark Wallace	Monash University, Australia
Roland Yap	National University of Singapore, Singapore

Contents

Efficiently Approximating High-Dimensional Pareto Frontiers for Tree-Structured Networks Using Expansion and Compression

Yiwei Bai[1(✉)], Qinru Shi[2], Marc Grimson[1], Alexander Flecker[3], and Carla P. Gomes[1]

[1] Department of Computer Science, Cornell University, Ithaca, USA
{yb263,mg2425}@cornell.edu, gomes@cs.cornell.edu

[2] Center for Applied Mathematics, Cornell University, Ithaca, USA
qs63@cornell.edu

[3] Department of Ecology and Evolutionary Biology, Cornell University, Ithaca, USA
asf3@cornell.edu

Abstract. Real-world decision-making often involves working with many distinct objectives. However, as we consider a larger number of objectives, performance degrades rapidly and many instances become intractable. Our goal is to approximate higher-dimensional Pareto frontiers within a reasonable amount of time. Our work is motivated by a problem in computational sustainability that evaluates the trade-offs between various ecological impacts of hydropower dam proliferation in the Amazon river basin. The current state-of-the-art algorithm finds a good approximation of the Pareto frontier within hours for three-objective problems, but a six-objective problem cannot be solved in a reasonable amount of time. To tackle this problem, we developed two different approaches: an *expansion method*, which assembles Pareto-frontiers optimized with respect to subsets of the original set of criteria, and a *compression method*, which assembles Pareto-frontiers optimized with respect to compressed criteria, which are a weighted sum of multiple original criteria. Our experimental results show that the aggregation of the different methods can reliably provide good approximations of the true Pareto-frontiers in practice. Source code and data are available at https://github.com/gomes-lab/Dam-Portfolio-Selection-Expansion-and-Compression-CPAIOR.

Keywords: Multi-objective Optimization · Approximation DP

1 Introduction

Multi-objective optimization (MOO) is of vital importance in many real-world problems in computational sustainability [7,13,30], which often involve balancing various environmental, economic, and social objectives, as captured e.g., in

Y. Bai and Q. Shi—Equal contribution.

© The Author(s), under exclusive license to Springer Nature Switzerland AG 2023
A. A. Cire (Ed.): CPAIOR 2023, LNCS 13884, pp. 1–17, 2023.
https://doi.org/10.1007/978-3-031-33271-5_1

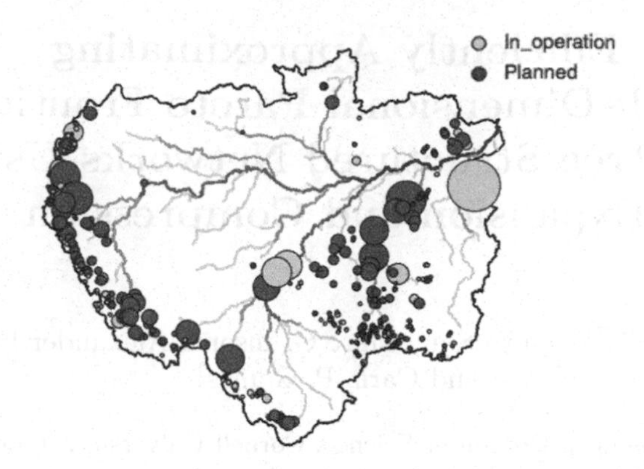

Fig. 1. Amazon hydropower dam portfolio selection problem. Green circles refer to potential dam sites while yellow circles represent already built dams. The sizes of the circles reflect the sizes of the dams in terms of energy output. (Color figure online)

the Sustainable Development Goals [28], that are all crucial to consider when designing solutions to these problems in alignment with human values. Such multi-objective optimization problems often have a large number of competing objectives that must be simultaneously optimized. However, most multi-objective algorithms only work efficiently for 2 or 3 objectives due to the curse of dimensionality [4,17,29]. Thus, finding methods to adapt state-of-the-art MOO algorithms to higher-dimensional problems is a topic of great interest. We propose two effective methods for efficiently approximating higher-dimensional Pareto Frontiers on tree-structured networks, using a state-of-the-art approximation algorithm, which works well in practice on lower-dimensional multi-objective problems.

Our main motivation comes from the real-world problem of strategic planning of hydropower dam expansion [16,32] in the Amazon basin (see Fig. 1), which has a lasting impact on a multitude of ecosystem services provided by the river network such as fish habitat and migration routes, sediment transportation, and fish biodiversity [2,8,34]. Finding optimal portfolios of hydropower dams while balancing the trade-offs between various ecological, social, and economic goals is a good example of a challenging combinatorial multiobjective optimization problem with a relatively large number of objectives. The current state-of-the-art algorithm [14,31] exploits the underlying tree-structure of the river networks and uses a dynamic programming scheme to approximate the Pareto frontier with provable guarantees, within an arbitrary small ϵ factor, and a runtime that is polynomial in the size of the instance and $\frac{1}{\epsilon}$. However, the runtime of this algorithm is still exponential with respect to the number of objectives. For large river networks such as the Amazon basin, while the algorithm is able to solve three or four-objective optimization problems efficiently with a small approximation factor, its performance drops off dramatically once we reach five objectives.

To encompass the complexity of balancing hydropower generation with ecosystem service impacts in the Amazon, higher numbers of objectives need to be considered. Here we address a set of six objectives associated with the proliferation of hydropower dams in the Amazon: **hydropower generation**, the main benefit provided by dams; **River connectivity index**, an indicator of the amount of habitat accessible to migratory fish; **sediment transportation**, the amount of sediment and nutrients transported by the river to the main stem and is essential for flood plain agriculture and fish habitat; **biodiversity impact**, which indicates the overall impact of dams on local biodiversity; **degree of regulation**, which represents the change of river flow regimes caused by dams and has a lasting influence on fish populations; and **greenhouse gases emissions**, which is an estimate of the total amount of greenhouse gases emitted by the construction and operation of dams, such as methane emissions due to the anaerobic decomposition of organic matter from areas flooded by the dams. Failing to consider any one of these six objectives leads to a less comprehensive representation of overall dam impacts. Thus, we aim to approximate the higher-dimensional (e.g., 6 criteria) Pareto frontier with the state-of-the-art algorithm that works efficiently on lower-dimensional (e.g., 3 criteria) problems.

High-dimensional real-world data are often shown to dwell on low-dimensional manifolds [15,18]. Similarly, we make the assumption that, for multi-objective optimization problems on river networks, the six-objective Pareto frontier might approximately lie on a lower-dimensional manifold. Current state-of-the-art works are able to solve this type of MOO problem with three or four objectives efficiently and with a guaranteed approximation factor. Given that these solutions are very likely to be on the Pareto frontier for more objectives, we conjecture that the aggregated solutions from Pareto frontiers optimized for all combinations of three or four-element subsets of the six objectives may form a good approximation of various local regions of the six-objective Pareto frontier. This naturally leads to two questions. First, for a specific n-objective optimization problem, can the true Pareto frontier be approximated by the Pareto frontiers defined by combinations of $k < n$ objectives? Complementary to the first question, can we reduce $k' > k$ objectives to k objectives via different linear combinations of criteria and still approximate the true k' dimensional Pareto frontier?

In answering these questions, we provide two major contributions to greatly improve the approximation of the Pareto frontier for 6 objectives for the Amazon river basin: (1) An ***expansion*** method, which computes the Pareto frontier with respect to different combinations of subsets of the original n criteria, aggregating the resulting non-dominated solutions with respect to all original criteria. (2) A ***compression*** method, complementary to the *expansion* method, which computes the Pareto frontier with respect to the original criteria compressed into fewer criteria via linear combinations, aggregating the resulting non-dominated solutions with respect to all original criteria. (3) We show our approaches produce high-quality Pareto frontiers, in a reasonable amount time, and demonstrate their effectiveness for three different sub-basins within the Amazon and for the entire Amazon basin.

Related Work. Our work leverages a state-of-the-art dynamic programming (DP) algorithm for computing the exact or approximation-guaranteed Pareto frontier for tree-structured networks, referred to as tree-DP [10,14,31]. Typically, the size of the Pareto frontier increases dramatically when the number of criteria increases and tree-DP's running time is proportional to it. Moreover, tree-DP considers all the criteria at the same time so they cannot run in parallel, which is not computationally efficient. Our methods approximate the Pareto frontier from subsets of all the criteria and they can naturally be computed in parallel. Parallel DP [9] may also be employed to boost its speed. Moreover, Genetic Algorithms (GA) have been widely applied to approximate Pareto frontiers and solve multiobjective optimization problems. Many well-established multiobjective GA methods have been developed over the past 40 years, including, but not limited to, vector evaluated GA (VEGA) [25], Multi-objective GA (MOGA) [11], Non-dominated Sorting Genetic Algorithm and its iterations (NSGA, NSGA-II, and NAGA-III) [5,6,27], and multiobjective evolutionary algorithm based on decomposition (MOEA/D) [33]. Nevertheless, GA approaches are not competitive with the current state-of-the-art algorithm [14,31], which exploits the underlying tree-structure of the river networks and uses a dynamic programming scheme to be able to approximate the Pareto frontier with provable guarantees with a runtime that is polynomial in the number of nodes in the network. Other methods, for instance, decision diagrams [3], propositional logic [26] and ray-based methods [19–21,23] are also be used for multiobjective optimization problems, but they cannot scale for the dam portfolio selection problem.

2 Problem Formulation

In this paper, we consider a multi-objective optimization problem with n ($n \geq 3$) objective functions $z^1, z^2, \cdots, z^n$, where the values of these functions are determined by a solution π (also referred to as a policy). Without loss of generality, we assume that all these objectives are to be **maximized**. For any solution π, we define the value vector of π to be

$$v(\pi) = (z^1(\pi), \cdots, z^n(\pi)).$$

Pareto Dominance: For two solutions π and π', if $z^i(\pi) \geq z^i(\pi')$ for all $i = 1, 2, \cdots, n$ and $z^i(\pi) > z^i(\pi')$ holds for at least one $i = 1, 2, \cdots, n$, then we say that the solution π dominates the solution π'

Pareto Frontier: If a solution π is not dominated by any other feasible solution, we say that π is a Pareto-optimal solution. The set of all Pareto-optimal solutions is called the Pareto frontier (denoted as P).

ϵ-approximations for multi-objective solutions: for two Pareto frontiers P_1, P_2, we say P_1 is ϵ-approximated by P_2 if and only if for any $\pi_1 \in P_1$, there exists a $\pi_2 \in P_2$, we have $\pi_1 \geq (1-\epsilon)\pi_2$ for all objectives.

Hydropower Dam Portfolio Selection Problem: Hydropower dams generate hydroelectricity, which accounts for 16.6% of the world's total electricity and 70% of all renewable electricity [1]. However, the construction of a hydropower dam can cause significant adverse environmental impacts, e.g., disruption of fish migration routes, alteration of river flow regimes, and greenhouse gas emissions.

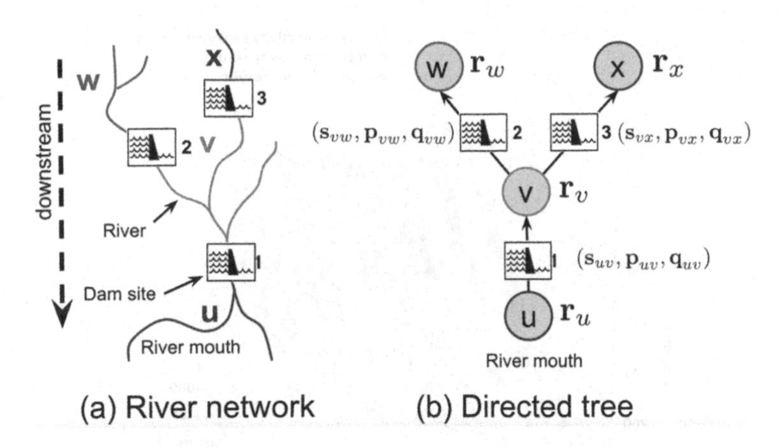

Fig. 2. An example of converting a river network (a) to a tree-structure (b). A node in the tree is a contiguous section of river uninterrupted by dam sites. Edges in the tree are dam sites that connect upstream and downstream segments. The mouth of the river (labelled u in this example) becomes the root of the tree. The tree-DP algorithm leverages this tree structure to be an efficient approximation algorithm.

[2,12]. So the selection of which potential dam sites to build is of vital importance for balancing energy production with ecosystem impacts. The hydropower dam portfolio selection problem is to generate an (approximated) Pareto frontier (the portfolio) of deciding what dams should be built (or selected) from a candidate pool of dam locations proposed by experts with respect to the six important criteria mentioned in the introduction. One solution in the portfolio is a subset of the dam candidate pool to be built.

The Off-the-Shelf Algorithm: Our methods leverages an algorithm that can compute low-dimensional Pareto frontiers efficiently for tree-structured problems. In this paper, we use the state-of-the-art tree dynamic programming (tree-DP) based approximation algorithm [14,31]. It can compute the exact solution given enough time or compute an ϵ approximated solution. The tree-DP algorithm models the entire river system as a tree structure (directed tree). Each dam site represents an edge and two vertices of that edge are the upstream river region and downstream river region respectively, where the river region is a contiguous part of the river, i.e., the streams of that region are connected and not blocked by any potential dam position (see Fig. 2). A bottom-up DP process can be done to compute the Pareto frontier. The running complexity of the DP algorithm is proportional to the number of solutions considered at each node, it can round the value of each criterion to the multiplicative of a small value (the approximation factor) to merge many similar solutions into one solution.

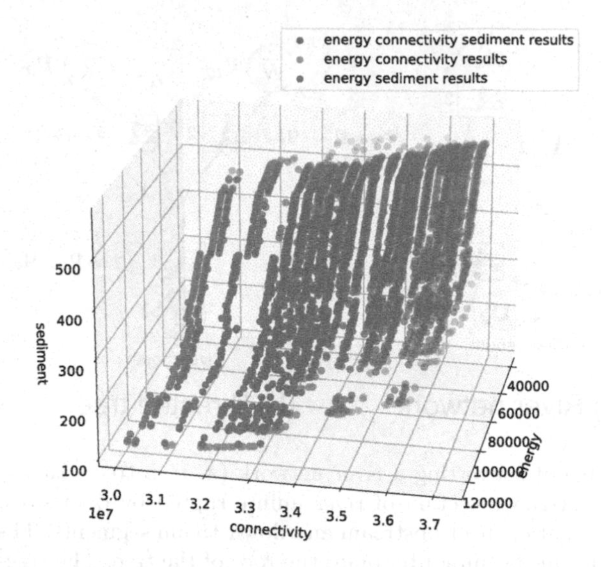

Fig. 3. An example of trying to approximate a three-criteria Pareto frontier with two-criteria optimization results. Each dot represents the values of the three criteria of one solution. For the two criteria solutions, we compute the value of the remaining criterion based on the dams built in that solution. We can observe that the two-criteria solutions only cover the edges of the Pareto optimal sets formed by the three-criteria optimization results.

3 The Expansion Method

We denote the actual n-objective Pareto frontier as P_n and define $V_n = \{v(\pi)|\pi \in P_n\}$. Given a positive integer $2 \leq k < n$, for all $1 \leq i_1 < i_2 < \cdots < i_k \leq n$, i.e., all the possible sized k combinations of n criteria, we compute an ϵ-approximate Pareto frontier $\tilde{P}_{i_1,\cdots,i_k}$ w.r.t. $z^{i_1}, z^{i_2}, \cdots, z^{i_k}$. We define the union for these k-objective Pareto frontiers to be

$$\tilde{P}_k = \bigcup_{1 \leq i_1 < \cdots < i_k \leq n} \tilde{P}_{i_1,\cdots,i_k},$$

and

$$\tilde{V}_k = \{v(\pi)|\pi \in \tilde{P}_k\}.$$

$\tilde{P}_k$ is the output of the ***Expansion method*** (see Fig. 4). In this paper we study the following proposition: for some real-world problems, $\tilde{V}_k$ forms a sufficiently good coverage of V_n with appropriate choices of k and ϵ.

The ***Expansion method*** method might look counter-intuitive at first because when we consider the smallest possible cases, two-criteria optimization solutions usually only cover the edges of a three-criteria Pareto frontier (see Fig. 3). However, in practice, we are able to approximate higher-dimensional Pareto frontiers using lower-dimensional Pareto frontiers. We will show an example after introducing the compression method.

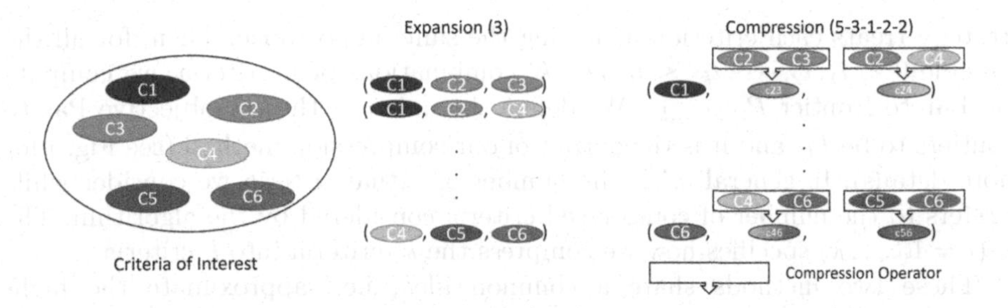

Fig. 4. High-level depiction of the expansion and compression methods. The left circle contains all the criteria we are interested in. Both the expansion and compression methods use the off-the-shelf Pareto-frontier algorithm optimized with respect to the criteria in the parentheses. In the expansion method, we choose all possible combinations of three criteria, from all criteria, and merge their results to generate the final results (evaluated with respect to all criteria). The compression method reduces the number of criteria by compressing the original criteria into fewer criteria. For example, 5-3-1-2-2 denotes that five criteria are reduced into three by keeping the first one as is and compressing the last two pairs of criteria. The compression operator is defined in Eq. 1.

4 The Compression Method

The ***Expansion method*** is likely to miss some solutions since it only optimizes with respect to a subset of the full criteria. We, therefore, propose a ***compression method*** (see Fig. 4) to further complement the Pareto frontier computed by the expansion method. By compressing $k' > k$ criteria into k criteria, the off-the-shelf algorithm can compute the Pareto frontier for k criteria while implicitly considering k' criteria.

Formally, as defined before, P_n refers to the actual $n-$objective Pareto frontier and $V_n = \{v(\pi)|\pi \in P_n\}$. The compression configuration can be defined as $(k', k, a_1, a_2, \ldots, a_k$ (Fig. 4) where $0 < k < k' \leq n$ and $\sum_{i=1}^{k} a_i = k'$. The idea of this configuration is to compress k' criteria into k criteria and a_i $(i = 1, \cdots, k)$ describe what criteria should be merged. The compression operator is defined as follows: for all the possible sized k' combinations of n criteria: $0 < i_1, i_2, \cdots, i_{k'} \leq n$, the compressed criteria evaluation function c_i' can be computed as:

$$c_i' = \sum_{j=sum[i-1]+1}^{sum[i]} w_j * z^j \quad (1)$$

where $a_0 = 0$, $sum[i] = \sum_{j=1}^{i} a_j, sum[0] = 0$ and w_j is the scalar weight. Note, if any two criteria i_p, i_q are compressed into one criterion, then $p \neq q$. The scales of different criteria vary substantially so the selection of the weights w_i is vital to the performance. A straightforward selection strategy is to normalize the criteria into the same scale: denote z_j^{max} as the max j-th criterion value among all the dams/rivers and w_j can be set as $1 - \frac{z_j^{max}}{\sum_{i=1}^{n} z_i^{max}}$. This normalized strategy treats each criterion as having the same importance. Then, for all the possible $0 < i_1, i_2, \cdots, i_{k'} \leq n$, i.e., k' combinations of n criteria, we compute the Pareto frontier $\tilde{P}_{1',2',\ldots,k'}$. We define the union of these k-objective Pareto frontiers to be $\tilde{P}_{k'}$ and it is the output of our compression method (see Fig. 4 for more details). In general, k' is the number of actual criteria we consider while k refers to the number of compressed criteria considered by the algorithm. The a_i $(i = 1, \ldots, k)$ specifies how we compress the k' criteria into k criteria.

These two methods share a common idea, i.e., approximate the high-dimensional Pareto frontier using many low-dimensional Pareto frontiers. The difference is that the compression method implicitly considers more criteria by compressing multiple criteria into fewer criteria.

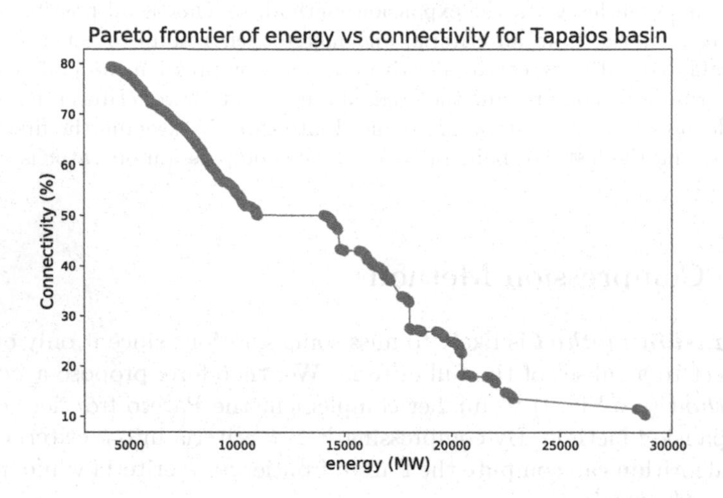

Fig. 5. Exact non-Convex Pareto frontier of energy-connectivity for the Tapajós basin.

5 Experiments

5.1 Experimental Setup

Our study focuses on the Amazon basin, where more than 350 large hydropower dams have been proposed. To show the generalizability of our methods and provide scalability insights, we also considered three sub-basins of the Amazon basin: Marañón, Tapajós and the West Amazon. We compute the Pareto frontier with respect to the six important criteria introduced in the introduction: hydropower generation, river connectivity index, sediment transportation, biodiversity impact, the degree of river regulation, and greenhouse gases emissions. For our underlying off-the-shelf algorithm and baseline, we use the state-of-the-art tree DP algorithm that computes the exact or approximate Pareto frontier, adopting the original papers' recommended configurations [14,31]. Our baseline is to directly consider all six criteria with the minimal approximation factor the runtime constraints allow.

5.2 Evaluation Method

To compare the optimization results of the various methods, we need a metric that can evaluate both the optimality and the coverage of the approximate Pareto frontiers. Note that the exact Pareto frontier we are trying to approximate can be non-convex (see Fig. 5). So, not only do we care about the overall shape of the Pareto frontier, but also the evaluation of the individual solutions. Therefore, we propose an evaluation method that divides the solution space into ϵ hypercubes following [24]'s approach.

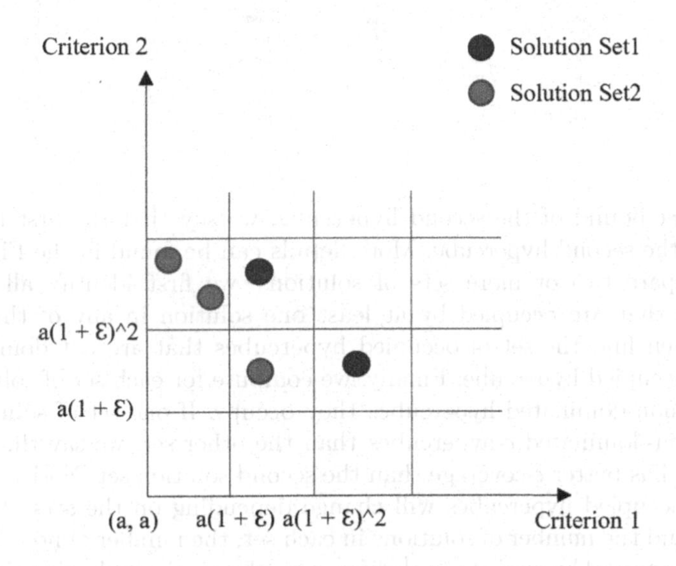

Fig. 6. We use two criteria as an example. The solution space is divided into several hypercubes. The upper bound of each cube is $1+\epsilon$ of its lower bound. The lower bound a is the minimum value of its criterion. The number of hypercubes one solution covers is a good metric. Consider two solutions sets 1 and 2. Set1 covers two hypercubes, while set2 covers two hypercubes. Note that these numbers are computed when we consider each solution set individually. When we compare them, we need to compute the new Pareto frontier after merging their solutions.

More specifically, for a n-objective optimization problem where, without loss of generality, every objective is to be maximized and the objective values are strictly non-negative, we define the solution space to be a n-dimensional space where each axis represents the value of one objective. We also make the assumption that the minimum possible value of each objective is non-negative. For a given error bound $\epsilon > 0$, we divide the solution space into hypercubes where the upper bound is $1 + \epsilon$ of the lower bound on each axis, with the smallest value of the lower bounds being the minimum possible value of the corresponding objective. Similar to the definition of Pareto-dominance, for two different hypercubes, if for each axis, the upper bound of the first hypercube is greater than or equal

Table 1. Non-dominated hypercubes occupied by the different methods. Note that the occupied non-dominated hypercubes are computed by merging and comparing tree-DP solutions, Expansion solutions, and Compression solutions. The approximation factors are shown in Table 2. The number in the parentheses is the number of criteria that are considered by the tree-DP algorithm. The epsilon is used in the hypercube computation. The Compression method further improves the performance. Expansion+Compression is a good approximation for all the basins, outperforming the baseline tree-DP, which provides a theoretical approximation guarantee.

Basin	epsilon	Tree DP (6)	Expansion-3 (3)	Expansion-4 (4)	Compression-3 (2)	Compression-4 (3)	Compression-5 (3)	Expansion + Compression
Marañón	0.01	12070	9	2	1344	14884	1620	17425
Marañón	0.05	35	1	0	747	771	2	816
Tapajós	0.01	0	681	382	1435	14878	0	17371
Tapajós	0.05	0	44	22	187	1057	0	1277
West Amazon	0.01	0	66	149	306	20851	0	21371
West Amazon	0.05	0	6	11	27	1160	0	1191
Amazon	0.01	0	6778	9	1623	1243	2397	12044
Amazon	0.05	0	485	1	216	85	75	847

to the upper bound of the second hypercube, we say that the first hypercube dominates the second hypercube. More details can be found in the Fig. 6.

To compare two or more sets of solutions, we first identify all of the ϵ-hypercubes that are occupied by at least one solution in any of the solution sets. We then find the set of occupied hypercubes that are not dominated by any other occupied hypercube. Finally, we compute for each set of solutions the number of non-dominated hypercubes they occupy. If one set of solutions covers more non-dominated ϵ-hypercubes than the other set, we say that the first solution set has better ϵ-coverage than the second solution set. Notice that since the set of occupied hypercubes will change depending on the sets of solutions compared and the number of solutions in each set, the number of non-dominated hypercubes covered by one set of solutions may change depending on the solution sets compared to, so the number of non-dominated hypercubes covered cannot be used as a universal metric of the quality of approximate Pareto frontiers. However, when comparing fixed sets of solutions, the metric provides a good comparison of the accuracy and coverage of the solution sets.

The visualized Pareto frontiers computed by different solutions can be more straightforward for comparison. However, it is difficult to visualize Pareto frontiers of dimensions higher than three. For the sake of clear visualization and easier comparison, we use the Uniform Manifold Approximation and Projection (UMAP) [22] method to project a high-dimensional Pareto-frontier onto a two-dimensional plane while preserving the general proximity relationships between the values of solutions. We merge the solutions generated by the tree-DP and our methods, then only save the non-dominated solutions, and finally, use UMAP to visualize these solutions. We have developed a website for visualizing the Pareto frontier.

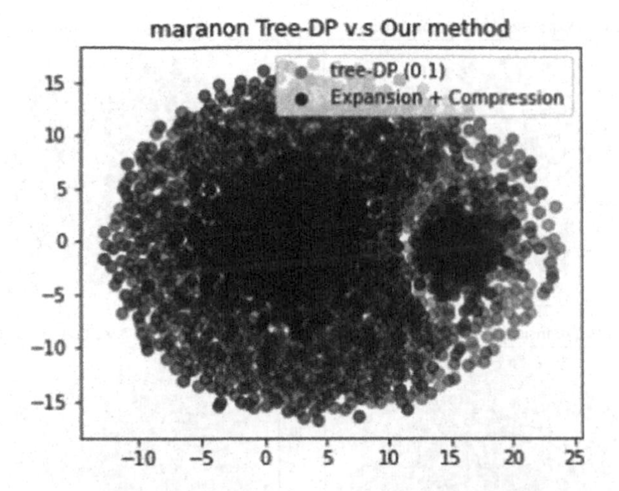

Fig. 7. UMAP results of the baseline approximate six-criteria Pareto frontier of the Marañón basin ($\epsilon = 0.1$) and the Expansion + Compression approximation results. We can observe that our method covers most solutions of the tree-DP algorithm. Note that the solutions fed into the UMAP results are all non-dominated solutions.

5.3 Experimental Results

We first show how well each of the ***Expansion methods*** and ***Compression methods*** can approximate the six criteria Pareto frontier for the full Amazon Basin and three sub-basins. Since the Pareto frontier may be non-convex (see Fig. 5), our metric is the number of non-dominated hypercubes occupied by the solutions computed by a given method. For methods' solution comparison, from both fine-grain and coarse perspectives, we used two hypercube error bounds: $\epsilon = 0.01$ and $\epsilon = 0.05$.

For all the experiments, we always include hydropower generation as a single criterion, otherwise, the optimal solution would be the trivial solution of building no dams as hydropower generation is the only criterion positively correlated with construction. Due to the scale of the problem, the state-of-the-art tree-DP algorithm can only compute or approximate the Pareto frontier in a reasonable amount of time for $k = 3, 4$ criteria. Thus, for both the expan-

Table 2. Approximation factors of the baseline, the expansion method, and the compression method. The number in the parentheses is the number of criteria considered by the tree-DP algorithm. Factors vary and are set so that every experiment is under the 80 hour limit, except for the baseline.

Basin	Baseline (6)	Expansion-3 (3)	Expansion-4 (4)	Compression-3 (2)	Compression-4 (3)	Compression-5 (3)
Marañón	0.1	0	0	0	0	0
Tapajós	1.0	0.1	0.4	0.3	0.3	0.3
West Amazon	1.0	0.2	0.2	0.2	0.2	0.2
Amazon	1.25	0.5	2.0	0.5	0.5	0.5

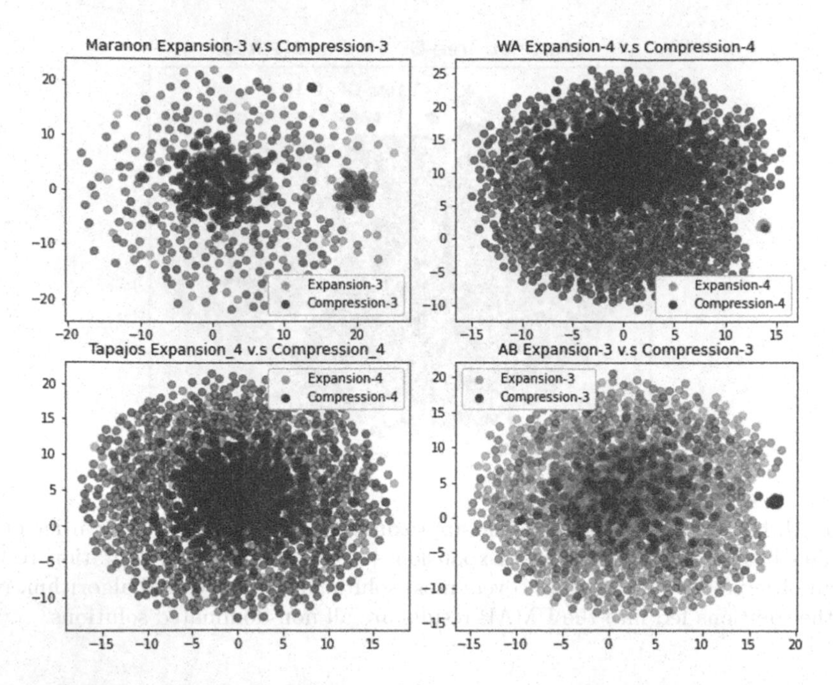

Fig. 8. UMAP results of four basins' Expansion and Compression solutions. We merge their sets of solutions and use all the non-dominated solutions to compute these UMAP results. The number in the names of the method refers to the actual criteria (k' of the Compression method) considered. We can observe that for all the basins, two methods capture different perspectives of the problem and this leads to very different solutions.

sion and compression methods, we consider combinations of 3 and 4 criteria. For the ***Expansion method*** we refer to the experiment that computes all the possible 3 criteria combinations ***Expansion-3*** and 4 criteria combinations as ***Expansion-4***, and they both consider $C^5_{2/3} = 10$ combinations (here we choose from 5 criteria instead of 6 since we always include energy as one criterion). For the compression method, we describe the configuration in the following format, assuming the criteria are sequentially numbered: $(k', k, a_1, \ldots, a_k)$, which denotes that we compress k' criteria into k criteria using the scheme $a_1, \ldots, a_k$. Each a_i denotes how many of the original criteria in the sequence are compressed to produce the final criterion i. The formal definition of compressing these a_i criteria can check Eq. 1. We consider three situations for the compression method: (1) $(3, 2, 1, 2)$ (denoted as ***Compression-3***): since the first target criterion must be the single (uncompressed) energy criterion, we have a total of $C^5_2 = 10$ combinations; (2) $(4, 3, 1, 2, 1)$ (denoted as ***Compression-4***): where we have a total of $C^5_2 \times C^5_1 = 50$ combinations; and (3) $(5, 3, 1, 2, 2)$ (denoted as ***Compression-5***): where we have a total of $C_2^{\frac{4*5}{2}} = 45$ combinations. To reduce the computational overhead, we assign each (compressed) criterion the same importance factor when reducing multiple criteria into one. We tune the

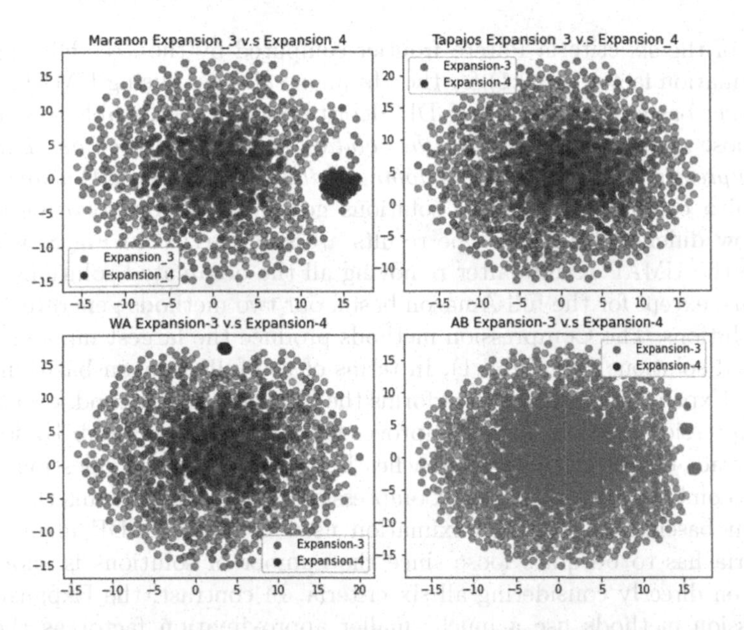

Fig. 9. UMAP results for the four basins' Expansion-3 and Expansion-4 solutions. We merge their sets of solutions and use all the non-dominated solutions to compute these UMAP results. For all the basins, except for the entire Amazon, the Expansion methods compute very different solutions, when considering different numbers of criteria. For the entire Amazon basin, due to the large approximation factor (2.0) used by the Expansion-4 method, it can only find a few solutions.

approximation factor of the tree-DP algorithm to ensure a single experiment (e.g. a combination of energy-connectivity-GHG (greenhouse gas emissions) using the expansion method) is finished within an 80 hours time budget running on a computation node that has 24 Intel(R) Xeon(R) CPU X5690 @ 3.47GHz. Note we do not set a time limit for the baseline method. The baseline runtime for all four basins is greater than 10 days.

The main results are summarized in Table 1 and the approximation factors of all methods are shown in Table 2. We compute the non-dominated hypercubes for each method, then combine these hypercubes and remove all dominated ones to form the hypercube Pareto frontier. Since the number of solutions can be quite large and there are many solutions that are quite similar to each other, we sort the solutions with respect to the number of dams they build and sample 3,000 solutions uniformly from each sub-experiment. For the baseline method, we either select all of its solutions or uniformly sample 1,000,000 solutions. For each method, we then count how many grids of its solution set belong to the hypercube Pareto frontier. Table 1 shows that even for the smallest sub-basin, the Marañón, which the tree-DP algorithm can finish with an $\epsilon = 0.1$ approximation factor, our method can get a better approximation (17425 v.s. 12070). In Fig. 7, we also show visually how our method can cover almost every solution of the six criteria Pareto frontier computed by the Tree-DP algorithm (approximation factor $\epsilon = 0.1$) for the Marañón sub-basin using UMAP. For the other larger basins, where the tree-DP algorithm alone is only able to scale with a very loose approximation factor, *the solution sets computed by our Expansion and Compression methods entirely dominate those of the tree-DP algorithm.*

We also compare the sets of solutions generated by our two methods to study how different they are. The results are summarized in Fig. 8, where we compare the UMAP results after removing all the dominated solutions. For all the basins, except for the full Amazon basin, our two methods generate very different solutions. The Compression methods produce the largest number of non-dominated solutions (see Table 1). In terms of the full Amazon basin, interestingly the Expansion method outperforms the Compression method, even though the Compression method further improves the Expansion method. Understanding the trade-offs of the two approaches is a future research question. In any case, the combination Expansion+Compression clearly outperforms the tree-DP algorithm baseline, as the approximation used by the tree-DP algorithm for six criteria has to be quite loose since the number of solutions is enormously large when directly considering all six criteria. In contrast, the Expansion and Compression methods use a much smaller approximation factor as the number of target criteria is small (2 or 3), efficiently handled by the tree-DP algorithm, which makes up for only optimizing with respect to subsets of the original criteria.

5.4 Ablation Study

We conducted experiments to study how the number of criteria affects the solution sets computed by the expansion method and the compression method. We merge the solution sets of all expansion methods and all compression methods separately, and then remove all the dominated solutions from them. We then run UMAP to project their non-dominated solutions to a 2-dimensional space to analyze the relative distances between the solutions. The results are shown in Fig. 9 (Expansion) and Fig. 10 (Compression). For all the basins, except for the full Amazon basin, Expansion-3 and Expansion-4 cover very different areas. For the entire Amazon basin, Expansion-3 dominates Expansion-4 since it is able to use a much smaller approximation factor (0.5 v.s. 2.0). For the Compression method, in general, solutions from Compression-4 and Compression-5 dominate Compression-3 solutions since the first two methods actually consider one more criterion Moreover, except for the full Amazon basin, Compression-4 and Compression-5 methods cover diverse areas. These results show that considering different numbers of criteria can provide different solution perspectives.

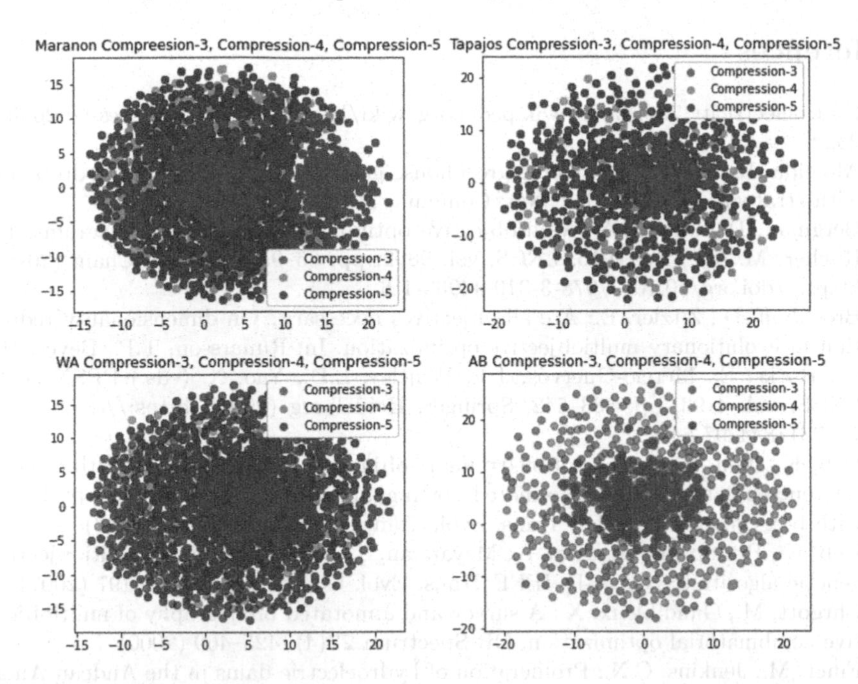

Fig. 10. UMAP results of four basins' Compression-3, Compression-4 and Compression-5 solutions. We merge their sets of solutions and use all the non-dominated solutions to compute these UMAP results. For all the basins, solution sets from the Compression-4 and Compression-5 experiments in general dominate the Compression-3 solutions since the first two methods actually consider one more criterion. Moreover, for all the basins except for the full Amazon basin, the solution sets of Compression-4 and Compression-5 cover very different solutions.

6 Conclusion

We propose the Expansion method to efficiently approximate an n (high)-dimension Pareto frontier by computing all the possible k (low)-dimension Pareto frontiers and merging their solutions together. Moreover, we also introduce a Compression method that compresses multiple criteria into fewer criteria, allowing the algorithm to consider more criteria implicitly, further improving the Expansion method. The combination of the Expansion and Compression methods provides a good Pareto frontier approximation for three Amazon sub-basins and the full Amazon basin for six criteria, in practice outperforming the baseline tree-DP approach, which provides a theoretical approximation guarantee. Understanding the trade-offs between the Expansion and Compression approaches is an interesting topic for further research. We hope this work inspires other approaches for efficiently approximating high-dimensional Pareto frontiers.

Acknowledgments. We thank the reviewers for all the constructive feedback. This research is supported in part by grants from the National Science Foundation, Air Force Office of Scientific Research, and Cornell Atkinson Center for Sustainability.

References

1. Hydroelectricity. https://en.wikipedia.org/wiki/Hydroelectricity. Accessed 26 Jan 2022
2. Almeida, R.M., et al.: Reducing greenhouse gas emissions of amazon hydropower with strategic dam planning. Nat. Commun. **10**(1), 1–9 (2019)
3. Bergman, D., Cire, A.A.: Multiobjective optimization by decision diagrams. In: Rueher, M. (ed.) CP 2016. LNCS, vol. 9892, pp. 86–95. Springer, Cham (2016). https://doi.org/10.1007/978-3-319-44953-1_6
4. Brockhoff, D., Zitzler, E.: Are all objectives necessary? On dimensionality reduction in evolutionary multiobjective optimization. In: Runarsson, T.P., Beyer, H.-G., Burke, E., Merelo-Guervós, J.J., Whitley, L.D., Yao, X. (eds.) PPSN 2006. LNCS, vol. 4193, pp. 533–542. Springer, Heidelberg (2006). https://doi.org/10.1007/11844297_54
5. Deb, K., Jain, H.: An evolutionary many-objective optimization algorithm using reference-point-based nondominated sorting approach, part i: solving problems with box constraints. IEEE Trans. Evol. Comput. **18**(4), 577–601 (2013)
6. Deb, K., Pratap, A., Agarwal, S., Meyarivan, T.: A fast and elitist multiobjective genetic algorithm: NSGA-II. IEEE Trans. Evol. Comput. **6**(2), 182–197 (2002)
7. Ehrgott, M., Gandibleux, X.: A survey and annotated bibliography of multiobjective combinatorial optimization. OR Spectrum **22**(4), 425–460 (2000)
8. Finer, M., Jenkins, C.N.: Proliferation of hydroelectric dams in the Andean Amazon and implications for Andes-Amazon connectivity. PLOS ONE **7**(4), 1–9 (2012). https://doi.org/10.1371/journal.pone.0035126
9. Fioretto, F., Pontelli, E., Yeoh, W., Dechter, R.: Accelerating exact and approximate inference for (distributed) discrete optimization with GPUs. Constraints **23**, 1–43 (2018)
10. Flecker, A.S., et al.: Reducing adverse impacts of amazon hydropower expansion. Science **375**(6582), 753–760 (2022)
11. Fonseca, C.M., Fleming, P.J., et al.: Genetic algorithms for multiobjective optimization: formulation discussion and generalization. In: ICGA, vol. 93, pp. 416–423 (1993)
12. Forsberg, B.R., et al.: The potential impact of new Andean dams on amazon fluvial ecosystems. Plos One **12**(8), 1–35 (2017). https://doi.org/10.1371/journal.pone.0182254
13. Gomes, C., et al.: Computational sustainability: computing for a better world and a sustainable future. Commun. ACM **62**(9), 56–65 (2019)
14. Gomes-Selman, J.M., Shi, Q., Xue, Y., García-Villacorta, R., Flecker, A.S., Gomes, C.P.: Boosting efficiency for computing the Pareto frontier on tree structured networks. In: van Hoeve, W.-J. (ed.) CPAIOR 2018. LNCS, vol. 10848, pp. 263–279. Springer, Cham (2018). https://doi.org/10.1007/978-3-319-93031-2_19
15. Huang, D., Yi, Z., Pu, X.: Manifold-based learning and synthesis. IEEE Trans. Syst. Man Cybern. Part B (Cybern.) **39**(3), 592–606 (2009). https://doi.org/10.1109/TSMCB.2008.2007499
16. Kareiva, P.M.: Dam choices: analyses for multiple needs. Proc. Natl. Acad. Sci. **109**(15), 5553–5554 (2012)
17. Khare, V., Yao, X., Deb, K.: Performance scaling of multi-objective evolutionary algorithms. In: Fonseca, C.M., Fleming, P.J., Zitzler, E., Thiele, L., Deb, K. (eds.) EMO 2003. LNCS, vol. 2632, pp. 376–390. Springer, Heidelberg (2003). https://doi.org/10.1007/3-540-36970-8_27

18. Li, B., Li, J., Tang, K., Yao, X.: Many-objective evolutionary algorithms: a survey. ACM Comput. Surv. **48**(1) (2015). https://doi.org/10.1145/2792984
19. Lin, X., Zhen, H.L., Li, Z., Zhang, Q., Kwong, S.: Pareto multi-task learning (2019). https://doi.org/10.48550/ARXIV.1912.12854
20. Ma, P., Du, T., Matusik, W.: Efficient continuous pareto exploration in multi-task learning (2020). https://doi.org/10.48550/ARXIV.2006.16434
21. Mahapatra, D., Rajan, V.: Exact pareto optimal search for multi-task learning: touring the pareto front (2021). https://doi.org/10.48550/ARXIV.2108.00597
22. McInnes, L., Healy, J., Melville, J.: UMAP: uniform manifold approximation and projection for dimension reduction. arXiv preprint arXiv:1802.03426 (2018)
23. Nowak, D., Küfer, K.H.: A ray tracing technique for the navigation on a non-convex pareto front (2020). https://doi.org/10.48550/ARXIV.2001.03634
24. Papadimitriou, C.H., Yannakakis, M.: On the approximability of trade-offs and optimal access of web sources. In: Proceedings 41st Annual Symposium on Foundations of Computer Science, pp. 86–92. IEEE (2000)
25. Schaffer, J.D.: Some experiments in machine learning using vector evaluated genetic algorithms (1985). https://www.osti.gov/biblio/5673304
26. Soh, T., Banbara, M., Tamura, N., Le Berre, D.: Solving multiobjective discrete optimization problems with propositional minimal model generation. In: Beck, J.C. (ed.) CP 2017. LNCS, vol. 10416, pp. 596–614. Springer, Cham (2017). https://doi.org/10.1007/978-3-319-66158-2_38
27. Srinivas, N., Deb, K.: Muiltiobjective optimization using nondominated sorting in genetic algorithms. Evol. Comput. **2**(3), 221–248 (1994)
28. United Nations General Assembly: Transforming our world: the 2030 agenda for sustainable development (2015). https://sdgs.un.org/2030agenda
29. Wagner, T., Beume, N., Naujoks, B.: Pareto-, aggregation-, and indicator-based methods in many-objective optimization. In: Obayashi, S., Deb, K., Poloni, C., Hiroyasu, T., Murata, T. (eds.) EMO 2007. LNCS, vol. 4403, pp. 742–756. Springer, Heidelberg (2007). https://doi.org/10.1007/978-3-540-70928-2_56
30. Wiecek, M.M., Ehrgott, M., Fadel, G., Figueira, J.R.: Multiple criteria decision making for engineering (2008)
31. Wu, X., et al.: Efficiently approximating the pareto frontier: hydropower dam placement in the amazon basin. In: Proceedings of the AAAI Conference on Artificial Intelligence, vol. 32 (2018)
32. Zarfl, C., Lumsdon, A.E., Berlekamp, J., Tydecks, L., Tockner, K.: A global boom in hydropower dam construction. Aquat. Sci. **77**(1), 161–170 (2015)
33. Zhang, Q., Li, H.: MOEA/D: a multiobjective evolutionary algorithm based on decomposition. IEEE Trans. Evol. Comput. **11**(6), 712–731 (2007)
34. Ziv, G., Baran, E., Nam, S., Rodríguez-Iturbe, I., Levin, S.A.: Trading-off fish biodiversity, food security, and hydropower in the Mekong river basin. Proc. Natl. Acad. Sci. **109**(15), 5609–5614 (2012). https://doi.org/10.1073/pnas.1201423109. https://www.pnas.org/content/109/15/5609

Objective-Based Counterfactual Explanations for Linear Discrete Optimization

Anton Korikov[(✉)] and J. Christopher Beck

Department of Mechanical and Industrial Engineering, University of Toronto, Toronto, Canada
{korikov,jcb}@mie.utoronto.ca

Abstract. Given a user who asks why an algorithmic decision did not satisfy some conditions, a counterfactual explanation takes the form of a minimally perturbed input that would have led to a decision satisfying the user's conditions. Building on recent work, this paper develops techniques to generate counterfactual explanations for linear discrete constrained optimization problems. These explanations take the form of a minimally perturbed objective vector that induces an optimal solution satisfying the newly stated user constraints. Drawing inspiration from the inverse combinatorial optimization literature, we introduce a novel non-convex quadratic programming algorithm to generate such explanations. Furthermore, we develop conditions for the existence of an explanation, addressing a limitation of past approaches. Finally, we discuss several future directions for explanations in discrete optimization such as actionable and sparse explanations.

1 Introduction

As the use of automated decision-making systems has increased, research has turned toward the question of providing explanations for the decisions that are made [12,13]. Such explanations enhance people's ability to interact with automated systems, improving performance of deployed systems [8] and facilitating better human oversight [7]. While much of explainability research has focused on deep learning (e.g., [13]), explainability is also important for model-based decision making systems, such as those studied in Constraint Programming, Operations Research, and Artificial Intelligence. Unlike deep learning models, declarative models are often decomposable into human-understandable symbols (e.g., costs, weights, priorities, etc.), yet decision algorithms are typically too complex or involve too many steps for a human to easily follow, making it difficult for people to contemplate relationships between modelling choices and algorithmic decisions [12]. Working in the context of AI planning, Smith [21] therefore proposed that explainability techniques are needed to support human reasoning about the effects of modeling choices on algorithmic decisions. Explainable AI Planning (XAIP) has since emerged as a rapidly growing research area [3], successful both in developing techniques specialized to AI planning [4,10]

© The Author(s), under exclusive license to Springer Nature Switzerland AG 2023
A. A. Cire (Ed.): CPAIOR 2023, LNCS 13884, pp. 18–34, 2023.
https://doi.org/10.1007/978-3-031-33271-5_2

as well as integrating broader explainable AI (XAI) research (e.g., contrastive explanations [17]).

In the field of optimization, an explainability literature similar in scope to XAIP has yet to emerge. Most explainability research in optimization has focused on explaining why a problem instance is infeasible, typically by identifying minimal sets of conflicting constraints [20], and there is little work on explaining feasible or optimal decisions [12]. Furthermore, integration between explainability research in optimization and XAI research at large has been limited.

Aiming to address this gap, our recent work [14,15] applied the XAI technique of counterfactual explanations [24] to optimal solutions of discrete optimization problems. Given an explainee[1] asking why an algorithmic decision was not different in a specific way, a counterfactual explanation presents minimally perturbed inputs to the algorithm that would have led to the decision being different in the way specified. In the framework introduced in our past work [14], an explainee asks why an optimal solution to a discrete optimization problem did not satisfy a previously unstated set of constraints. An explanation then takes the form of a minimally perturbed objective vector, so that, with the perturbed objective vector replacing the original one, an optimal solution would satisfy both the initial and new constraints.

Example 1. (Production Scheduling). A manager at a factory examines a monthly production schedule computed by an optimization system, the objective of which is determined by job priority levels and completion times. She asks: "Why were jobs 1 and 2 not completed in less than one week?". The explanation shows her the minimal change to the job priorities in the upcoming month so that jobs 1 and 2 would be completed in under one week in an optimal schedule. After receiving the explanation, the manager can either choose to keep the initial job priorities and accept the initial schedule or to produce a new schedule with the modified priority levels.

The task of generating such objective-based counterfactual explanations is a valuable research direction for several reasons. Objectives are the result of modeling choices, however, complex optimization algorithms make it difficult for a person to understand how a particular modeling choice leads to certain solution features [21]. For instance, in Example 1, the manager desires more information on how the job priority values (a modeling choice) impact whether jobs 1 and 2 are completed within a week (a solution feature). While such information can take many forms (see Sect. 7), one of the standard forms studied in XAI is a minimal change to a set of problem inputs that induces the solution feature in question [24]. The explanation in Example 1 takes this form. The goal of providing such information is to facilitate counterfactual reasoning, a fundamental human reasoning strategy [11], about an algorithm's inputs and outputs. Furthermore, when explainees are shown how decisions can be changed, they are empowered to contest or act to change decisions they believe are wrong [24].

[1] A person receiving an explanation.

Finally, research on counterfactual explanations in machine learning is developing rapidly, and many opportunities for cross-pollination exist between this literature and explainable optimization (see Sect. 7).

Furthermore, an *objective-based* explanation formulation allows connections to be made between explanation and inverse combinatorial optimization [6]. Given a possibly suboptimal solution, inverse optimization aims to find the minimal change to an optimization problems objectives such that the given solution becomes optimal. The use of inverse optimization as a methodological basis for objective-based explanation is discussed further in Sects. 2 and 4.

However, while our past work [14,15] defined an explanation framework for discrete optimization, the algorithms we used could only compute explanations for limited forms of questions. Specifically, an explainee could only ask why a subset of variables did not satisfy a partial assignment, or why a single variable did not satisfy a linear constraint (see Sect. 2.2). Their experiments also showed that for these restricted questions, no explanation existed for many natural problem instances.

Addressing these limitations, we introduce a novel, non-convex quadratic programming algorithm which can compute explanations for any linear discrete optimization problem when the user's question can be represented by linear or quadratic constraints. This new algorithm is inspired by a well-known algorithm in inverse combinatorial optimization [25], and capitalizes on recent advances in commercial solvers. Numerical simulations are performed to evaluate the new algorithm and demonstrate the explanation process for two combinatorial optimization problems. Additionally, we establish conditions for the existence of an explanation and apply them to design experimental problem instances. Finally, several future directions are identified, such as actionable and sparse explanations, and the limitations of the methods in this paper are discussed.

2 Background

2.1 Counterfactual Explanations

Given an algorithm which computes output p from input k, a counterfactual explanation responds to a *contrastive question* of the form: "Why p, and not some other output $q \in Q$?" [24]. Here, Q, called a *foil set*, is a set of alternative outputs, and each alternative output $q \in Q$ is called a *foil.* Given such a question, a counterfactual explanation shows the explainee an alternative input l which would have led to the output being in Q, with l typically selected such that it minimally perturbs k.[2]

2.2 Nearest Counterfactual Explanations

Given an objective vector $c \in \mathcal{D} \subseteq \mathbb{R}^z$, the purpose of a standard (or *forward*) optimization problem $\mathcal{FW}\langle c, f, X\rangle$ is to find values for a decision vector $x \in$

[2] *Counterfactual* means "contrary to the facts". The alternative input l and outputs $q \in Q$ are contrary to the initial input k and output p, respectively.

$X \subseteq \mathbb{R}^n$ which optimize an objective function $f : \mathcal{D} \times X \to \mathbb{R}$. In a minimization problem, the goal is to find an optimal x^* so that $f(c, x^*) = \min_x\{f(c, x) : x \in X\}$. If no optimization direction is specified, we assume minimization.

A counterfactual explanation process for an optimal solution x^* to a forward problem $\mathcal{FW}\langle c, f, X\rangle$ can be modeled using a Nearest Counterfactual Explanation (NCE) problem [14]. The explainee must first describe a set of alternative solutions $X_\psi \subset X$, and ask the contrastive question "Why x^* and not a solution $x^\psi \in X_\psi$?". As per Sect. 2.1, X_ψ is called the foil set and each solution $x^\psi \in X_\psi$ is called a foil. To define the foil set, the explainee must specify an additional set of constraints describing a feasible set $\psi \subset \mathbb{R}^n$, with $x^* \notin \psi$. These additional constraints are called *foil constraints*, and the foil set is defined as $X_\psi = X \cap \psi$.

Example 2. (Production Schedule - Contrastive Question). In Example 1, assume that a schedule for n jobs is generated by solving a $\mathcal{FW}\langle c, f, X\rangle$ with a decision vector $x \in X \subseteq \mathbb{N}_0^n$, where x_i represents the number of days before job i is completed. In the objective vector $c \in \mathcal{D} \subseteq \mathbb{N}_0^n$, c_i represents the priority of job i, and the objective is to minimize $f = c \cdot x$,[3] the sum of priority weighted completion times. Given $n = 4$ and $c = [4, 4, 3, 3]$, an optimal solution is $x^* = [8, 13, 1, 3]$, prompting the manager to ask why jobs 1 and 2 were not completed in under one week. In this case, the foil constraints are $x_1 \leq 7$ and $x_2 \leq 7$, and the contrastive question is "Why x^* and not an $x^\psi \in X_\psi$?", where $X_\psi = X \cap \psi$ and $\psi = \{x \in \mathbb{N}_0^n : x_1 \leq 7, x_2 \leq 7\}$.

The NCE addresses this type of question by searching for a counterfactual objective vector $d \in \mathcal{D} \subseteq \mathbb{R}^z$ that would lead to one of the foils $x^\psi \in X_\psi$ being optimal to the modified problem $\mathcal{FW}\langle d, f, X\rangle$, such that d is minimally perturbed from the initial objective vector c. This perturbation is measured by some norm $||\cdot||$, assumed to be L_1 if unspecified. If such a d is found, an explanation is: "A solution $x^\psi \in X_\psi$ would have been optimal if the objective vector had been d instead of c." Formally, assuming a minimization forward objective, the $\mathcal{NCE}\langle c, \mathcal{D}, f, \psi, x^*, X, ||\cdot||\rangle$ is

$$\min_{d \in \mathcal{D}} ||d - c|| \tag{1}$$

$$\text{s.t.} \quad \min_{x^\psi \in X_\psi} f(d, x^\psi) = \min_{x \in X} f(d, x). \tag{2}$$

If the optimization direction of the underlying forward problem is maximization, the minimization terms in constraint (2) are replaced with maximization terms.

Example 3. (Production Scheduling - Explanation). Assume the manager from Examples 1 & 2 is interested in explanations where job priorities can be adjusted to integers between 1 and 5, giving $\mathcal{D} = \{d \in \mathbb{N}^4 : 1 \leq d_i \leq 5,\ \forall i \in \{1, ..., 4\}\}$. If an optimal solution to the resulting $\mathcal{NCE}\langle c, \mathcal{D}, f, \psi, x^*, X\rangle$ is $d^* = [5, 5, 2, 2]$, the explanation is "Jobs 1 and 2 would have finished in under a week if their priorities were increased to the maximum level (5) and the priorities of the other two jobs were both one level lower (2)".

[3] Where clear from context, $c \cdot x$ is used as shorthand for $c^T x$.

NCEs provide a general way to model objective-based explanations of optimal forward solutions. However, previous solution methods could only solve NCEs for two restricted question types: questions about a single variable [14] and questions that ask why x^* did not satisfy a partial assignment [15]. Both methods, in addition, restrict some components of d from being perturbed. In fact, neither of these methods can solve the NCE described by Examples 1–3. Also, other than observing that it is necessary for $X_\psi \neq \emptyset$ in a feasible NCE, we previously did not formally study NCE feasibility conditions.

The main contribution of this paper is a novel quadratic programming algorithm which can solve any NCE where the forward problem is a discrete linear optimization problem, the foil constraints are linear or quadratic, and the norm is L_1. This new algorithm is inspired by inverse combinatorial optimization [25].

2.3 Inverse Combinatorial Optimization

Given a forward problem $\mathcal{FW}\langle c, f, X\rangle$ and a feasible target solution $x^d \in X$, the inverse optimization problem is to find a new objective vector $d \in \mathcal{D} \subseteq \mathbb{R}^z$, minimally perturbed from c, so that x^d becomes optimal. Given some norm $||\cdot||$, the inverse optimization problem $\mathcal{INV}\langle c, \mathcal{D}, f, x^d, X, ||\cdot||\rangle$ [6] is

$$\min_{d \in \mathcal{D}} ||d - c|| \tag{3}$$

$$\text{s.t. } f(d, x^d) = \min_{x \in X} f(d, x). \tag{4}$$

The inverse optimization problem can be interpreted as a special case of the NCE where the foil set is the singleton $X_\psi = \{x^d\}$. Though most inverse optimization algorithms have focused on continuous optimization [5], methods also exist for discrete optimization, with the standard technique being the *InvMILP* algorithm for inverse Mixed Integer Linear Programming (MILP) [25]. Our new algorithm is inspired by *InvMILP*.

A $\mathcal{MILP}\langle c, X\rangle$ is a forward problem $\mathcal{FW}\langle c, f, X\rangle$ with $c \in \mathcal{D} \subseteq \mathbb{R}^n$, $f = c \cdot x$, and $X = \{x \in \mathbb{R}^n_+ : Ax \leq b, x_I \in \mathbb{N}_0\}$ with $A \in \mathbb{R}^{v \times n}$, $b \in \mathbb{R}^v$, and $I \subseteq \{1, ..., n\}$. An inverse MILP, $\mathcal{INV}_{\mathcal{MILP}}\langle c, \mathcal{D}, x^d, X\rangle$, is an inverse problem where the forward problem is a $\mathcal{MILP}\langle c, X\rangle$, $\mathcal{D} \subseteq \mathbb{R}^n$, and the norm is L_1.

To solve such inverse MILPs, *InvMILP* (Algorithm 1) uses an iterative, two-level approach where a master problem $\mathcal{MP}_{\mathcal{INV}}$ (5)–(8) is initialized with a set $\mathcal{S}^0 \subseteq X$ of known extreme points of $conv(X)$, the convex hull of X. $\mathcal{MP}_{\mathcal{INV}}$ then searches for a d, minimizing $||d - c||_1$, such that x^d is at least as good of a solution to $\mathcal{MILP}\langle d, X\rangle$ as any point in $\mathcal{S}^0$. If such a d is found, a subproblem $\mathcal{MILP}\langle d, X\rangle$ is solved to optimality, returning an extreme point x^0. If x^0 gives a better objective value for $d \cdot x$ than x^d, then x^0 is added to $\mathcal{S}^0$ and the algorithm proceeds to the next iteration of $\mathcal{MP}_{\mathcal{INV}}$. *InvMILP* continues iterating either until the subproblem finds that x^d is optimal to $\mathcal{MILP}\langle d, X\rangle$, in which case d is an optimal solution to the inverse problem, or until the master problem is found infeasible, which will occur if the inverse problem is infeasible.

To formulate the master problem $\mathcal{MP}_{\mathcal{INV}}$ (5)–(8), the objective $||d - c||_1$ is first linearized using $g, h \in \mathbb{R}^n_+$, such that $c - d = g - h$: the magnitude of

Algorithm 1. *InvMILP* [25].

```
1  Inputs: INV_MILP⟨c, D, x^d, X⟩.
2  Output: d*.
3  Step 1: Initialize S^0 ← ∅.
4  Step 2: Solve MP_INV⟨c, D, x^d, X, S^0⟩.
5    If infeasible, return INFEAS.
6    Otherwise, get d^i = (c − g^i + h^i).
7  Step 3: Solve MILP⟨d^i, X⟩ to get x^0.
8    If d^{i,T} x^d ≤ d^{i,T} x^0, stop. Return d^i = d*.
9    Otherwise, update S^0 = S^0 ∪ {x^0} and return to Step 2.
```

the change to parameter c_j is represented by g_j if the change is negative and h_j if it is positive. Constraints (6) force x^d to be at least as good a solution to $\mathcal{MILP}\langle d, X\rangle$ as any point in $\mathcal{S}^0$. Finally, to avoid any d for which the forward problem is unbounded, Wang introduces the decision variable $u \in \mathbb{R}^v_+$ and adds the constraint $A^T u \geq d$ (7), ensuring that d results in a feasible dual problem. Thus, $\mathcal{MP}_{\mathcal{INV}}\langle c, \mathcal{D}, x^d, X, \mathcal{S}^0\rangle$, a linear program, is given by

$$\min_{u,g,h} g + h \tag{5}$$

$$\text{s.t } (c - g + h)^T x^d \leq (c - g + h)^T x^0 \qquad \forall x^0 \in \mathcal{S}^0 \tag{6}$$

$$A^T u \geq c - g + h \tag{7}$$

$$(c - g + h) \in \mathcal{D},\ g \in \mathbb{R}^n_+,\ h \in \mathbb{R}^n_+, u \in \mathbb{R}^v_+. \tag{8}$$

The complete *InvMILP* algorithm, which has been proven to terminate finitely [25], is given by Algorithm 1.

Noticeably absent from the discrete inverse optimization literature are algorithms capable of handling changes to constraint parameters. Such constraint parameter changes would add a degree of difficulty to the inverse optimization problem since they could induce the existence of multiple alternative feasible sets. The absence of such constraint-based inverse optimization methods is the reason that we choose to study objective-based explanations, as opposed to explanations based on changes to both objectives and constraints.

3 Problem Definition

This paper focuses on NCEs where the forward problem is a $\mathcal{MILP}\langle c, X\rangle$, the foil constraints defining X_ψ are linear or quadratic, and $||\cdot||$ is L_1. Such an NCE is denoted $\mathcal{NCE}_{\mathcal{MILP}}\langle c, \mathcal{D}, \psi, x^*, X\rangle$, and Examples 1–3 are examples of an $\mathcal{NCE}_{\mathcal{MILP}}$, given that the forward scheduling problem is a MILP.

Definition 1. *($\mathcal{NCE}_{\mathcal{MILP}}$). An $\mathcal{NCE}_{\mathcal{MILP}}\langle c, \mathcal{D}, \psi, x^*, X\rangle$ is an $\mathcal{NCE}\langle c, \mathcal{D}, f, \psi, x^*, X, ||\cdot||\rangle$ where the forward problem is a $\mathcal{MILP}\langle c, X\rangle$, ψ is defined by linear or quadratic constraints, $f = c^T x$, and $||\cdot||$ is L_1. A feasible $\mathcal{NCE}_{\mathcal{MILP}}$ solution, d, must not result in an unbounded $\mathcal{MILP}\langle d, X\rangle$.*

3.1 Existence of an Explanation

We now introduce conditions for the existence of a feasible, non-trivial solution to an $\mathcal{NCE}_{\mathcal{MILP}}$, defined as any $d \in \mathcal{D}$ feasible to (1)–(2) such that $d \neq \mathbf{0}$. While our past work [15] showed that NCE infeasibility can be an issue, no NCE feasibility conditions have been established other than the observation that it is necessary for $X_\psi \neq \emptyset$. We formalize a simple, necessary condition (Proposition 1) as well as a sufficient condition (Theorem 1) for $\mathcal{NCE}_{\mathcal{MILP}}$ feasibility, and Sect. 5 uses these conditions to design experimental instances.

Proposition 1. *For an $\mathcal{NCE}_{\mathcal{MILP}}\langle c, \mathcal{D}, \psi, x^*, X\rangle$ to have a non-trivial feasible solution, X_ψ cannot lie entirely in the interior region of $conv(X)$.*

Proof. If X_ψ lies entirely in the interior region of $conv(X)$, X_ψ cannot contain an optimal solution to $\mathcal{MILP}\langle d, X\rangle$ for any $d \neq \mathbf{0}$. □

Next, we present a sufficient feasibility condition for an $\mathcal{NCE}_{\mathcal{MILP}}$. Intuitively, assuming minimization, setting a single variable x_j to its minimal value in the feasible set will lead to an optimal $\mathcal{MILP}\langle d, X\rangle$ solution if d_j is greater than zero while all other components of d are zero. Formally, for all $i \in \{1, ..., n\}$, let $x_{i,\min} = \min_x\{x_i : x \in X \subseteq \mathbb{R}^n_+\}$ and $x_{i,\max} = \max_x\{x_i : x \in X \subseteq \mathbb{R}^n_+\}$. Also, let $\mathcal{D}^{f,+}_j = \{d \in \mathbb{R}^n_+ : 0 < d_j \leq d^{UB}_j, d_i = 0 \ \forall \ i \neq j\}$, where $d^{UB}_j = \max_d\{d_j : d \in \mathcal{D}\}$ and $j \in \{1, ..., n\}$.

Theorem 1. *An $\mathcal{NCE}_{\mathcal{MILP}}\langle c, \mathcal{D}, \psi, x^*, X\rangle$ has a non-trivial feasible solution if all following conditions hold:*

1. – *If the forward optimization direction is minimization, $\exists \ \tilde{x}^\psi \in X_\psi$ and $\exists \ j \in \{1, ..., n\}$ so that $\tilde{x}^\psi_j = x_{j,\min}$.*
   – *If the forward optimization direction is maximization, $\exists \ \tilde{x}^\psi \in X_\psi$ and $\exists \ j \in \{1, ..., n\}$ so that $\tilde{x}^\psi_j = x_{j,\max}$.*
2. *$\mathcal{D}^{f,+}_j \subseteq \mathcal{D}$.*
3. *$\exists M \in \mathbb{R}$ where $M > d^{UB}_j \tilde{x}^\psi_j$.*

Proof. For any $d^f \in \mathcal{D}^{f,+}_j$, the non-negative term $d^f_j x_j$ is the only component contributing to the forward objective value $(d^f)^T x$. If the forward objective is minimization, no minimization of $d^f_j x_j$ is possible below $d^f_j \tilde{x}^\psi_j$. Similarly, if the forward objective is maximization, no maximization of $d^f_j x_j$ is possible above $d^f_j \tilde{x}^\psi_j$. Condition (3) ensures the objective is bounded. Thus, $\tilde{x}^\psi$ is optimal to $\mathcal{MILP}\langle d, X\rangle$ for any $d^f \in \mathcal{D}^{f,+}_j$. □

An analogous theorem can be defined for negative d^f_j values that isolate a non-positive objective component $d^f_j x_j$, which we omit in the interests of space.

4 The *NCXplain* Algorithm

This section introduces *NCXplain*, a novel non-convex quadratic programming algorithm which optimally solves an $\mathcal{NCE}_{\mathcal{MILP}}$. Letting $\mathcal{S}$ be the set of all extreme points of $conv(X)$ and decision vector $x^\psi \in X_\psi$ be a foil, the $\mathcal{NCE}_{\mathcal{MILP}}\langle c, \mathcal{D}, \psi, x^*, X\rangle$ can be expressed as:

$$\min_{d,x^\psi,u} ||d-c||_1 \tag{9}$$

$$\text{s.t. } d \cdot x^\psi \le d \cdot x^0 \qquad \forall\, x^0 \in \mathcal{S} \tag{10}$$

$$A^T u \ge d \tag{11}$$

$$x^\psi \in X_\psi,\ d \in \mathcal{D},\ u \in \mathbb{R}^v_+. \tag{12}$$

Constraints (10) force a foil to have a forward objective no worse than any extreme point of $conv(X)$, and have a non-convex, quadratic left-hand side which is bilinear in d and x^ψ. Constraints (11) ensure that $\mathcal{MILP}\langle d, X\rangle$ is bounded by forcing its dual problem to be feasible.

NCXplain (Algorithm 2) follows a similar cutting plane approach to *InvMILP* (Algorithm 1), with the main difference being *NCXplain*'s quadratic master problem $\mathcal{MP}_{\mathcal{NCE}}$ (13)–(17) and stopping conditions. The $\mathcal{NCE}_{\mathcal{MILP}}$ objective is linearized using $d = c - g + h$, where $g, h \in \mathbb{R}^n_+$. Then, taking the $\mathcal{NCE}_{\mathcal{MILP}}$ (9)–(12) and relaxing constraints (10) by replacing $\mathcal{S}$ with a set of known extreme points $\mathcal{S}^0 \subseteq \mathcal{S}$ gives the $\mathcal{MP}_{\mathcal{NCE}}\ \langle c, \mathcal{D}, \psi, x^*, X, \mathcal{S}^0\rangle$:

$$\min_{g,h,x,u} g+h \tag{13}$$

$$\text{s.t. } (c-g+h)\cdot x \le (c-g+h)\cdot x^0, \forall\ x^0 \in \mathcal{S}^0 \tag{14}$$

$$A^T u \ge c-g+h \tag{15}$$

$$x \in X_\psi,\ (c-g+h) \in \mathcal{D} \tag{16}$$

$$g, h \in \mathbb{R}^n_+, u \in \mathbb{R}^v_+. \tag{17}$$

Given an optimal $\mathcal{MP}_{\mathcal{NCE}}$ solution $(d^i, x^{\psi,i}, u^i)$ at iteration i of *NCXplain*, solving a subproblem $\mathcal{MILP}\langle d^i, X\rangle$ to get an optimal extreme point $x^{0,i}$ allows *NCXplain* to either show d^i is an optimal solution to the $\mathcal{NCE}_{\mathcal{MILP}}$ or add a new extreme point of $conv(X)$ to $\mathcal{S}^0$. The complete *NCXplain* algorithm is given by Algorithm 2, and its properties are formalized by Lemmas 1–2 and Theorem 2. Both the master problem and the MILP subproblem can be modelled in Gurobi 9.0+ due to recent advances allowing non-convex quadratic constraints such as (14) to be expressed directly.

Lemma 1. *NCXplain only terminates in Step 3 if d^i is feasible for $\mathcal{NCE}_{\mathcal{MILP}}$.*

Proof. (Lemma 1). If $d^i \cdot x^{0,i} = d^i \cdot x^{\psi,i}$ (Case 1), then the foil $x^{\psi,i}$ is optimal to $\mathcal{MILP}\langle d^i, X\rangle$. If $d^i \cdot x^{0,i} < d^i \cdot x^{\psi,i}$ but $x^{0,i} \in X_\psi$ (Case 2), then $x^{0,i}$ is a foil and optimal to $\mathcal{MILP}\langle d^i, X\rangle$. □

Algorithm 2. *NCXplain.*

```
1  Inputs: NCE_MILP⟨c, D, ψ, x*, X⟩
2  Output: d*
3  Step 1: Initialize S^0 ← x*.
4  Step 2: Solve MP_NCE ⟨c, D, ψ, x*, X, S^0⟩.
5    If infeasible, return INFEAS.
6    Else get (d^i, x^{ψ,i}, u^i) with d^i = (c − g^i + h^i).
7  Step 3: Solve MILP⟨d^i, X⟩ to get x^{0,i}.
8    If d^i · x^{0,i} = d^i · x^{ψ,i} (Case 1)
9      Stop and return d* = d^i.
10   Elif d^i · x^{0,i} < d^i · x^{ψ,i} and x^{0,i} ∈ X_ψ (Case 2)
11     Stop and return d* = d^i.
12   Else (Case 3)
13     Update S^0 = S^0 ∪ {x^{0,i}}, go to Step 2.
```

Lemma 2. *Let $\mathcal{S}^{0,i}$ be $\mathcal{S}^0$ during iteration i of NCXplain. If in Step 3, $x^{0,i} \in \mathcal{S}^{0,i}$, NCXplain must terminate. If $x^{0,i} \notin \mathcal{S}^{0,i}$, then NCXplain either terminates or a new extreme point of $conv(X)$ is added to $\mathcal{S}^0$ before iteration $i+1$.*

Proof. (Lemma 2). If $x^{0,i} \in \mathcal{S}^{0,i}$, then due to constraint (14), $x^{\psi,i}$ must satisfy $d^i \cdot x^{\psi,i} \leq d^i \cdot x^{0,i}$, but due to the optimality of $x^{0,i}$ to $\mathcal{MILP}\langle d^i, X\rangle$, $d^i \cdot x^{\psi,i} \not< d^i \cdot x^{0,i}$. Thus, $d^i \cdot x^{\psi,i} = d^i \cdot x^{0,i}$ (Case 1), and *NCXplain* must terminate. If $x^{0,i} \notin \mathcal{S}^{0,i}$ and *NCXplain* does not terminate (Case 3), the new extreme point of $conv(X)$, $x^{0,i}$, is added to $\mathcal{S}^0$. □

Theorem 2. *NCXplain will optimally solve an $\mathcal{NCE}_{\mathcal{MILP}}\langle c, \mathcal{D}, \psi, x^*, X\rangle$ or prove it is infeasible in a finite number of iterations.*

Proof. (Theorem 2). Let $X_{\mathcal{MP}}$ and $X_{\mathcal{NCE}}$ represent the solution sets of $\mathcal{MP}_{\mathcal{NCE}}$ (13)–(17) and $\mathcal{NCE}_{\mathcal{MILP}}$ (9)–(12), respectively. Since the two problems have the same objective, differ only in constraints (10) and (14), and $\mathcal{S}^0 \subseteq \mathcal{S}$, then $X_{\mathcal{NCE}} \subseteq X_{\mathcal{MP}}$ and $\mathcal{MP}_{\mathcal{NCE}}$ is a relaxation of $\mathcal{NCE}_{\mathcal{MILP}}$. Thus, if an $\mathcal{MP}_{\mathcal{NCE}}$ is found infeasible in Step 2, then $\mathcal{NCE}_{\mathcal{MILP}}$ must also be infeasible. Similarly, if a solution $(d^i, x^{\psi,i}, u^i)$ is optimal to an $\mathcal{MP}_{\mathcal{NCE}}$ and d^i is feasible to $\mathcal{NCE}_{\mathcal{MILP}}$, then d^i must also be optimal to $\mathcal{NCE}_{\mathcal{MILP}}$.

By these observations and Lemmas 1 and 2, in any iteration, *NCXplain* either terminates having proven the $\mathcal{NCE}_{\mathcal{MILP}}$ is infeasible, terminates having found an optimal $\mathcal{NCE}_{\mathcal{MILP}}$ solution d^*, or continues after adding a new extreme point of $conv(X)$ to $\mathcal{S}^0$. Because $\mathcal{S}$ is a finite set, the number of iterations before $\mathcal{S}^0 = \mathcal{S}$ is finite, and when $\mathcal{S}^0 = \mathcal{S}$, Lemma 2 implies that *NCXplain* must terminate since any extreme point $x^{0,i}$ obtained in Step 3 is in $\mathcal{S}$. □

5 Experimental Method

Simulations demonstrating our explanation approach and testing *NCXplain* were carried out based on two forward MILP problems. These experiments were performed in three steps, focusing on the last:

1. Optimally solving a $\mathcal{MILP}\langle c, X\rangle$ instance to get x^*.
2. Simulating a contrastive question and creating a $\mathcal{NCE}_{\mathcal{MILP}}\langle c, \mathcal{D}, \psi, x^*, X\rangle$ instance.
3. Optimally solving the $\mathcal{NCE}_{\mathcal{MILP}}\langle c, \mathcal{D}, \psi, x^*, X\rangle$ with *NCXplain*.

We do not numerically compare *NCXplain* to alternatives because it is the first algorithm capable of solving an $\mathcal{NCE}_{\mathcal{MILP}}$.

5.1 Forward Problems

The two forward MILP problems were the 0-1 knapsack problem (KP) and the single machine scheduling with release dates problem, $1|r_j|\sum w_jC_j$. The KP was selected because it is NP-complete [19], has a simple structure, and is easy to understand. The scheduling problem was chosen because it matches a potential use case for $\mathcal{NCE}_{\mathcal{MILP}}$ based explanations (Examples 1–3) and is a relatively simple, though strongly NP-Hard [16] problem.

0-1 Knapsack Problem (KP). We are given a set of $n \in \mathbb{N}$ items, a profit vector $c \in \mathbb{N}_0^n$, a weight vector $w \in \mathbb{N}_0^n$, and a knapsack capacity $W \in \mathbb{N}_0$, with $W < \sum_{i=1}^n w_i$. A decision variable $x_i \in \{0,1\}$, $i \in \{1,...,n\}$, is assigned to 1 if item i is included in the knapsack and 0 otherwise, and the complete KP is $\max_x\{c \cdot x : x \in X\}$, $X = \{x \in \{0,1\}^n : w \cdot x \le W\}$. Problem instances of sizes $n \in \{250, 500, 1000\}$ were generated using independent random uniform distributions $c_i \in [1, R]$ and $w_i \in [1, R]$ with $R = 1000$, where $W = \max\{\lfloor P\sum_{i=1}^n w_i\rfloor, R\}$ with $P = 0.5$.

Scheduling Problem ($1|r_j|\sum w_jC_j$). There are $n \in \mathbb{N}$ jobs with each job $i \in \{1,...,n\}$ having a processing time $q_i \in \mathbb{N}$, a weight[4] $c_i \in \mathbb{N}$, a release date $r_i \in \mathbb{N}_0$, and a completion time $t_i^c \in \mathbb{N}_0$. The objective is to minimize the weighted sum of all completion times, $c \cdot t^c$, given that no job can start before its release date or be interrupted and no two jobs can be processed at the same time. Letting $x_{i,t} \in \{0,1\}$ be a decision variable which is 1 if job i starts at time t and 0 otherwise, and T be an upper bound on latest completion time of any job, a MILP model for $1|r_j|\sum w_jC_j$ is

$$\min_x \sum_{i=1}^{n}\sum_{t=0}^{T-q_i} c_i(t+q_i)x_{i,t} \tag{18}$$

[4] Though w is typically used for job weights, we use c instead to keep notation consistent throughout the paper.

$$\text{s.t.} \sum_{t=0}^{T-q_i} x_{i,t} = 1, \quad \forall\, i = 1, ..., n \tag{19}$$

$$\sum_{i=1}^{n} \sum_{s=\max(0,t-q_i+1)}^{t} x_{i,s} \leq 1, \; \forall\, t = 0, ..., T-1 \tag{20}$$

$$\sum_{t=0}^{r_i-1} x_{i,t} = 0, \quad \forall\, i \in \{1, ..., n\} \tag{21}$$

$$x_{i,t} \in \{0,1\}^{n \times (T-1)}. \tag{22}$$

Constraints (19) force each job to start exactly once. Constraints (20) ensure no two jobs are processed at the same time, and constraints (21) enforce the release dates. Problem instances of sizes $n \in \{6, 9, 12\}$ were generated using random uniform distributions $q_i \in [1, 10]$, $c_i \in [1, 10]$, and $r_i \in [0, \lfloor \alpha Q \rfloor]$, where $\alpha = 0.3$ and $Q = \sum_{i=1}^{n} q_i$, and the time horizon T was calculated as $T = \lfloor \alpha Q \rfloor + Q$.

5.2 $\mathcal{NCE}_{\mathcal{MILP}}$ Instances

Knapsack Questions. Given a subset of m items $\mathcal{S}_\psi \subseteq \{1, ..., n\}$, $|\mathcal{S}_\psi| = m$, $\mathcal{S}_\psi \neq \emptyset$, and a parameter $\beta_\psi \in (0, 1]$, the simulated question asked "Why were at least $\beta_\psi m$ items from $\mathcal{S}_\psi$ not included in the knapsack?". The foil set corresponding to this question is $X_\psi = \{x \in X : \sum_{j \in \mathcal{S}_\psi} x_j \geq \beta_\psi m\}$. Questions were simulated with $\beta = 0.75$ by randomly selecting m items to form $\mathcal{S}_\psi$ such that $x^* \notin X_\psi$.

Scheduling Questions. The simulated question asked why m randomly selected jobs $\mathcal{M} \subseteq \{1, ..., n\}$ were not scheduled earlier, as in Examples 1–3. Letting $t^* \in [0, T]^n$ denote job start times in x^*, a $t^\psi \in [0, T]^m$ was created with t_j^ψ representing the maximal counterfactual start time of job $j \in \mathcal{M}$ such that $r_j \leq t_j^\psi < t_j^*$. Then, the question asked "Why was each job $j \in \mathcal{M}$ not completed by $(t_j^\psi + q_j)$, respectively?". This question is represented with the foil set $X_\psi = \{x \in X : \sum_{t=0}^{T-q_j} t x_{j,t} \leq t_j^\psi \quad \forall\, j \in \mathcal{M}\}$. For each job $j \in \mathcal{M}$, the maximal counterfactual start time t_j^ψ was randomly selected from the interval $[t_j^{\psi,LB}, t_j^{\psi,UB}]$, where $t_j^{\psi,UB} = t_j^* - 1, t_j^{\psi,LB} = \lceil r_j + \theta(t_j^* - 1 - r_j) \rceil$, and $\theta = 0.5$.

Non-Empty Foil Sets. For both problems, after a foil set was generated, it was checked whether X_ψ was non-empty. If X_ψ was empty, the question data was re-randomized until a non-empty foil set was found, though such cases were rare.

Counterfactual Objectives. The set of feasible counterfactual objectives was set to $\mathcal{D} = \{d \in \mathbb{N}_0^n : 0 \leq d_i \leq c_i^{UB}\ \forall i \in \{1, ..., n\}\}$, where $c_i^{UB} \in \mathbb{N}$ is the

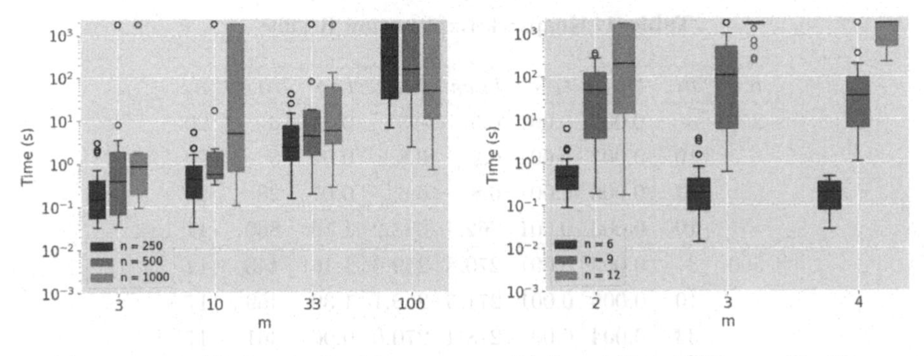

(a) Knapsack $\mathcal{NCE}_{\mathcal{MILP}}$ Problems (b) Scheduling $\mathcal{NCE}_{\mathcal{MILP}}$ Problems

Fig. 1. *NCXplain* Runtime Distributions.

maximum value for c_i in a forward instance ($c_i^{UB} = 1000$ for KP, $c_i^{UB} = 10$ for $1|r_j|\sum w_j C_j$). Given that X is finite for both forward problems, it is impossible for any $d \in \mathcal{D}$ to result in an unbounded $\mathcal{MILP}\langle d, X\rangle$, so constraints (15) were omitted in these simulations.

$\mathcal{NCE}_{\mathcal{MILP}}$ Feasibility. Any $\mathcal{NCE}_{\mathcal{MILP}}$ in these experiments meets Conditions (1)–(3) of Theorem 1, and thus has a non-trivial feasible solution $d^f \in \mathcal{D}_j^{f,+}$. Intuitively, any $d^f \in \mathcal{D}_j^{f,+}$ implies that in the KP, there is no benefit from including any items other than item j, while in the scheduling problem, there is no benefit from achieving an earlier completion time for any jobs other than job j. Specifically, Condition (2) is satisfied since $\mathcal{D}_j^{f,+} \subseteq \mathcal{D}$ for any $j \in \{1, ..., n\}$. Condition (1) is met by the KP instances since the maximal value of any x_j is 1, and any foil $x^\psi \in X_\psi$ must contain at least one component $x_j^\psi = 1$. For the scheduling problem, taking any schedule $x^\psi \in X_\psi$ and left-shifting it causes the first job in the schedule, which we will call job j, to start at its release date r_j. Condition (1) is thus satisfied since there exists a schedule in the foil set where job j is completed at its minimal possible time, $t_{j,\min}^c = r_j + q_j$. Finally, Condition (3) is met since $d_j^f x_{j,\max}$ is bounded from above by c^{UB} for KP, while for scheduling, $d_j^f t_{j,\min}^c$ is bounded from above by $c^{UB}(r_j + q_j)$.

5.3 Computational Details

Python 3.9.7 and Gurobi 9.5 were used to implement *NCXplain* for $\mathcal{NCE}_{\mathcal{MILP}}$ instances, as well as to solve the initial $\mathcal{MILP}$ instances. Twenty instances were tested for each value of (n, m) reported in Sect. 6, using a single core of a 2.6 GHz Intel Core i7-10750H CPU. A time limit of 30 min was used for *NCXplain*, and if an $\mathcal{NCE}_{\mathcal{MILP}}$ was not solved before this time limit, its runtime was recorded as 30 min. Thus, the *NCXplain* runtimes should be interpreted as lower bounds on the true runtimes.

Table 1. Knapsack Explanation Results

n	m	$t_{F,\mu}$	$t_{F,\sigma}$	$t_{\mathcal{NCE}}$	$t_{\mathcal{MP}}$	$t_{\mathcal{SP}}$	n_{ITR}	n_S
250	3	0.003	0.001	0.5	0.5	0.01	6	20
	10	0.003	0.001	0.9	0.8	0.02	8	20
	33	0.003	0.001	6.8	6.6	0.07	28	20
	100	0.003	0.001	862.4	853.2	1.16	860	11
500	3	0.004	0.001	270.8	259.6	1.16	443	17
	10	0.005	0.001	271.5	259.1	1.39	469	17
	33	0.004	0.001	278.4	270.0	0.96	361	17
	100	0.004	0.001	778.0	770.5	1.18	355	12
1000	3	0.008	0.003	360.7	343.4	2.11	376	16
	10	0.011	0.003	721.3	648.5	10.14	1129	12
	33	0.008	0.003	371.3	351.6	2.25	420	16
	100	0.008	0.003	655.7	640.2	3.00	345	13

6 Experimental Results

Figure 1 illustrates the *NCXplain* runtime distributions. For *NCXplain*, Tables 1 and 2 report the mean runtime ($t_{\mathcal{NCE}}$), mean number of iterations (n_{ITR}), the number of instances solved optimally before the time limit (n_S), as well as the mean cumulative time in the subproblem ($t_{\mathcal{SP}}$) versus the master problem ($t_{\mathcal{MP}}$). For the initial forward problem $\mathcal{MILP}\langle c, X\rangle$, these tables show the runtime mean ($t_{F,\mu}$) and standard deviation ($t_{F,\sigma}$). All runtimes are in seconds.

No instances were infeasible, demonstrating the successful use of Theorem 1. The number of instances solved (n_S) shows that most $\mathcal{NCE}_{\mathcal{MILP}}$ instances were solved in under 30 min, though the solution times for the initial forward problem ($t_{F,\mu}$) were much faster.

As indicated by the values of $t_{\mathcal{MP}}$ versus $t_{\mathcal{SP}}$, *NCXplain* spends 90%-99.9% of its runtime solving master problems (Step 2, Algorithm 2), showing that these non-convex, quadratic problems are significantly harder than the $\mathcal{MILP}\langle d, X\rangle$ subproblems (Step 3, Algorithm 2). That is, it is computationally cheaper to add a new point to $\mathcal{S}^0$ than to solve $\mathcal{MP}_{\mathcal{NCE}}$. A direction for future work may thus be to reduce the number of master problem iterations with a variation of *NCXplain* which adds multiple points to $\mathcal{S}^0$ for each iteration of $\mathcal{MP}_{\mathcal{NCE}}$.

While a larger n almost always resulted in longer $\mathcal{NCE}_{\mathcal{MILP}}$ solve times, the effects of m (the number of items or jobs in a user question), as well as compound effects of n and m, are difficult to observe from our data. Future work should study these effects more rigorously, including investigating whether any phase transitions exist. The existence of phase transitions may explain why the scheduling $\mathcal{NCE}_{\mathcal{MILP}}$s with $n = 6$ became easier as m increased, while those with $n = 12$ became harder (Fig. 1b, Table 2).

Table 2. Scheduling Explanation Results

n	m	$t_{F,\mu}$	$t_{F,\sigma}$	$t_{\mathcal{NCE}}$	$t_{\mathcal{MP}}$	$t_{\mathcal{SP}}$	n_{ITR}	n_S
6	2	0.006	0.002	0.9	0.9	0.01	4	20
	3	0.005	0.002	0.6	0.6	0.02	4	20
	4	0.005	0.002	0.2	0.2	0.02	5	20
9	2	0.010	0.002	89.5	89.3	0.08	7	20
	3	0.011	0.005	467.0	466.6	0.15	13	16
	4	0.012	0.008	150.8	150.3	0.19	15	19
12	2	0.025	0.022	694.8	694.5	0.13	7	15
	3	0.016	0.005	1499.0	1498.4	0.24	13	5
	4	0.021	0.013	1365.4	1364.6	0.33	17	6

7 Limitations and Future Work

Algorithmic Improvements. Currently, *NCXplain* can only explain small $\mathcal{MILP}$ problems. However, Bodur *et al.* [1] recently showed that *InvMILP* can be sped up by modifying Step 3 of Algorithm 1 to add non-extreme point solutions of $\mathcal{MILP}\langle c, X\rangle$ to $\mathcal{S}^0$ after finding these points using early stopping criteria and trust regions. Additionally, Duan and Wang [9] extend *InvMILP* with a heuristic to parallelize cut generation and compute feasible solutions as upper bounds to the inverse MILP problem. These extensions to *InvMILP* can likely be adapted to *NCXplain* to improve performance and produce feasible solutions to $\mathcal{NCE}_{\mathcal{MILP}}$ before the problem is solved optimally.

Minimizing Decision Perturbation. A limitation of the $\mathcal{NCE}_{\mathcal{MILP}}$ is that an optimal solution to $\mathcal{MILP}\langle d^*, X\rangle$ may be arbitrarily far from x^*, the decision being explained. Given an arbitrarily large change to x^*, an explainee may find it difficult to evaluate the effects of perturbing c on the decision, especially if multiple iterations of explanation and objective modification are performed. A future modification to the $\mathcal{NCE}_{\mathcal{MILP}}$ and *NCXplain* could add a term $|x^* - x^\psi|$ to the objective (9), thus optimizing for smaller perturbations to both c and x^*.

Actionability and Sparsity. The concepts of actionability [22] and sparsity [18] could be adapted from machine learning (ML) to $\mathcal{NCE}_{\mathcal{MILP}}$ explanations. Since some objective components c_i may be easier to change, or more actionable, than others, a weighted L_1 norm $w^T||d-c||_1$, $w \in \mathbb{R}^n_+$, could replace objective (9), with weight w_i representing the ease of changing parameter c_i. To induce sparsity, an L_0 term measuring the number of perturbed objective components could be added to objective (9), since an explainee may prefer explanations perturbing fewer components of c.

Meaningful Objectives. A fundamental assumption in an $\mathcal{NCE}_{\mathcal{MILP}}$ is that the objective parameters represented by c and d are meaningful to the explainee. Otherwise, the explainee requires an additional explanation of what these parameters mean before the $\mathcal{NCE}_{\mathcal{MILP}}$ explanation is useful.

8 Related Work

Our past work [14,15] discussed in Sect. 2, introduced the NCE (1)–(2) and solved two restricted versions of it. The only other work we are aware of which uses counterfactual explanations for a model-based optimization problem is that of Brandao *et al.* [2], in which inverse optimization is applied in its classical form to explain a path planning problem. As mentioned in Sect. 2.3, the inverse optimization problem can be interpreted as a special case of an NCE where the explainee is interested in exactly one alternative solution x^d. Our approach is more general since we enable an explainee to define a set of alternative solutions using linear or quadratic constraints.

In the inverse optimization literature, Wang [26] formulates a variant of inverse optimization which is similar to the $\mathcal{NCE}_{\mathcal{MILP}}$. However, this variant assumes d is continuous, while the $\mathcal{NCE}_{\mathcal{MILP}}$ allows the domain of $d \in \mathcal{D}$ to be an integer or mixed-integer set, such as the set of integer job priorities in the scheduling experiments. Additionally, other than the equivalent of bilinear constraints (10), all constraints in Wang's problem must be linear, while the $\mathcal{NCE}_{\mathcal{MILP}}$ allows constraints (12) defining X_ψ and $\mathcal{D}$ to be quadratic. Most notably, while providing interesting theoretical contributions, Wang's work does not connect inverse optimization with explanation, the focus of our paper.

Finally, an emerging literature on counterfactual explanations exists in ML [23], providing potential for cross-polination with explainability research in declarative optimization, such as the actionability and sparsity extensions proposed in Sect. 7.

9 Conclusion

This paper presents techniques to respond to users asking why an optimal solution x^* to a linear discrete optimization problem $\mathcal{MILP}\langle c, X\rangle$ did not satisfy some previously unstated constraints. We address such questions by formulating the $\mathcal{NCE}_{\mathcal{MILP}}$ (9)–(12), the solution to which is a counterfactual explanation d: an alternative objective vector minimally perturbed from c so that an optimal solution to $\mathcal{MILP}\langle d, X\rangle$ satisfies the additional user constraints. After establishing feasibility conditions for the $\mathcal{NCE}_{\mathcal{MILP}}$, we introduce *NCXplain*, a non-convex, quadratic cutting-plane algorithm which solves the $\mathcal{NCE}_{\mathcal{MILP}}$. Experiments are performed to simulate explanations for two discrete optimization problems, evaluating *NCXplain* and identifying next steps for improving it. Finally, we discuss future directions for counterfactual explanations in optimization such as actionability, sparsity, and minimizing decision perturbation.

References

1. Bodur, M., Chan, T.C., Zhu, I.Y.: Inverse mixed integer optimization: Polyhedral insights and trust region methods. INFORMS J. Comput. (2022)

2. Brandao, M., Coles, A., Magazzeni, D.: Explaining path plan optimality: fast explanation methods for navigation meshes using full and incremental inverse optimization. In: Proceedings of the International Conference on Automated Planning and Scheduling, vol. 31, pp. 56–64 (2021)
3. Chakraborti, T., Sreedharan, S., Kambhampati, S.: The emerging landscape of explainable automated planning & decision making. In: IJCAI, pp. 4803–4811 (2020)
4. Chakraborti, T., Sreedharan, S., Zhang, Y., Kambhampati, S.: Plan explanations as model reconciliation: moving beyond explanation as soliloquy. In: IJCAI (2017)
5. Chan, T.C., Mahmood, R., Zhu, I.Y.: Inverse optimization: theory and applications. arXiv preprint arXiv:2109.03920 (2021)
6. Demange, M., Monnot, J.: An introduction to inverse combinatorial problems. In: Paradigms of Combinatorial Optimization: Problems and New Approaches, pp. 547–586 (2014)
7. Doshi-Velez, F., Kortz, M.: Accountability of AI under the law: the role of explanation. Technical report, Berkman Klein Center Working Group on Explanation and the Law, Berkman Klein Center for Internet and Society (2017)
8. Doshi-Velez, F., Kim, B.: Towards a rigorous science of interpretable machine learning. arXiv preprint arXiv:1702.08608 (2017)
9. Duan, Z., Wang, L.: Heuristic algorithms for the inverse mixed integer linear programming problem. J. Global Optim. **51**(3), 463–471 (2011)
10. Eiffer, R., Cashmore, M., Hoffmann, J., Magazzeni, D., Steinmetz, M.: A new approach to plan-space explanation: analyzing plan-property dependencies in oversubscription planning. In: AAAI (2020)
11. Epstude, K., Roese, N.J.: The functional theory of counterfactual thinking. Pers. Soc. Psychol. Rev. **12**(2), 168–192 (2008)
12. Freuder, E.: Explaining ourselves: human-aware constraint reasoning. In: AAAI (2017)
13. Kim, B., Wattenberg, M., Gilmer, J., Cai, C., Wexler, J., Viegas, F., et al.: Interpretability beyond feature attribution: quantitative testing with concept activation vectors (TCAV). In: International Conference on Machine Learning, pp. 2668–2677. PMLR (2018)
14. Korikov, A., Shleyfman, A., Beck, J.C.: Counterfactual explanations for optimization-based decisions in the context of the GDPR. In: International Joint Conferences on Artificial Intelligence (IJCAI) (2021)
15. Korikov, A., Beck, J.C.: Counterfactual explanations via inverse constraint programming. In: Michel, L.D. (ed.) 27th International Conference on Principles and Practice of Constraint Programming (CP 2021). Leibniz International Proceedings in Informatics (LIPIcs), vol. 210, pp. 35:1–35:16. Schloss Dagstuhl - Leibniz-Zentrum für Informatik, Dagstuhl, Germany (2021). https://doi.org/10.4230/LIPIcs.CP.2021.35. https://drops.dagstuhl.de/opus/volltexte/2021/15326
16. Lenstra, J.K., Kan, A.R., Brucker, P.: Complexity of machine scheduling problems. In: Annals of Discrete Mathematics, vol. 1, pp. 343–362. Elsevier (1977)
17. Miller, T.: Explanation in artificial intelligence: insights from the social sciences. Artif. Intell. **267**, 1–38 (2019)
18. Mothilal, R.K., Sharma, A., Tan, C.: Explaining machine learning classifiers through diverse counterfactual explanations. In: Proceedings of the 2020 Conference on Fairness, Accountability, and Transparency, pp. 607–617 (2020)
19. Pisinger, D., Kellerer, H., Pferschy, U.: Knapsack problems. In: Handbook of Combinatorial Optimization, p. 299 (2013)

20. Senthooran, I., et al.: Human-centred feasibility restoration. In: Michel, L.D. (ed.) 27th International Conference on Principles and Practice of Constraint Programming (CP 2021). Leibniz International Proceedings in Informatics (LIPIcs), vol. 210, pp. 49:1–49:18. Schloss Dagstuhl - Leibniz-Zentrum für Informatik, Dagstuhl, Germany (2021). https://doi.org/10.4230/LIPIcs.CP.2021.49. https://drops.dagstuhl.de/opus/volltexte/2021/15340
21. Smith, D.E.: Planning as an iterative process. In: Twenty-Sixth AAAI Conference on Artificial Intelligence (2012)
22. Ustun, B., Spangher, A., Liu, Y.: Actionable recourse in linear classification. In: Proceedings of the Conference on Fairness, Accountability, and Transparency, pp. 10–19 (2019)
23. Verma, S., Dickerson, J., Hines, K.: Counterfactual explanations for machine learning: a review. In: NeurIPS Workshop on ML Retrospectives, Surveys and Meta-Analyses (2020)
24. Wachter, S., Mittelstadt, B., Russell, C.: Counterfactual explanations without opening the black box: automated decisions and the GDPR. Harv. JL & Tech. **31**, 841 (2017)
25. Wang, L.: Cutting plane algorithms for the inverse mixed integer linear programming problem. Oper. Res. Lett. **37**(2), 114–116 (2009)
26. Wang, L.: Branch-and-bound algorithms for the partial inverse mixed integer linear programming problem. J. Global Optim. **55**(3), 491–506 (2013)

Column Elimination for Capacitated Vehicle Routing Problems

Anthony Karahalios(✉) and Willem-Jan van Hoeve

Carnegie Mellon University, Pittsburgh, PA 15213, USA
{akarahal,vanhoeve}@andrew.cmu.edu

Abstract. We introduce a column elimination procedure for the capacitated vehicle routing problem. Our procedure maintains a decision diagram to represent a relaxation of the set of feasible routes, over which we define a constrained network flow. The optimal solution corresponds to a collection of paths in the decision diagram and yields a dual bound. The column elimination process iteratively removes infeasible paths from the diagram to strengthen the relaxation. The network flow model can be solved as a linear program with a conventional solver or via a Lagrangian relaxation. To solve the Lagrangian subproblem more efficiently, we implement a special successive shortest paths algorithm. We introduce several cutting planes to strengthen the dual bound, including a new type of clique cut that exploits the structure of the decision diagram. We experimentally compare the bounds from column elimination with those from column generation for capacitated vehicle routing problems.

1 Introduction

The capacitated vehicle routing problem (CVRP) can be stated as follows [29]. Given a set of locations each with a specified weight and a fleet of trucks each with a specified capacity, the problem asks to design a route for each truck such that each location is visited by a truck, for each truck the total weight of its visited locations does not exceed the capacity, and the sum of the truck route lengths is minimized. It is a central problem in logistics and has become increasingly important over the last decade due to the rise of last-mile delivery applications. The CVRP is among the most studied NP-hard combinatorial optimization problems and finding provably optimal solutions remains a challenge in practice. Current state-of-the-art exact methods can solve up to around 200 locations optimally within a reasonable of time, with branch-cut-and-price (BCP) methods performing particularly well [5,12,23–25].

BCP is an effective method for solving generic large-scale integer programming models [7]. It relies on column generation to solve the linear programming relaxation: working with a restricted set of variables (or columns), column generation iteratively adds new variables to the model until an optimal basis is found. Despite its successes, column generation has some weaknesses. For example, it may take many iterations to converge to the optimal solution due to dual degeneracy of the intermediate solutions. Furthermore, branching decisions or cutting

© The Author(s), under exclusive license to Springer Nature Switzerland AG 2023
A. A. Cire (Ed.): CPAIOR 2023, LNCS 13884, pp. 35–51, 2023.
https://doi.org/10.1007/978-3-031-33271-5_3

planes that strengthen the relaxation may complicate the pricing problem that finds new variables.

We study an alternative approach that does not rely on a pricing problem, thereby avoiding the potential drawbacks of column generation mentioned above. Instead of using a restricted set of columns, column elimination works with a *relaxed* set of columns, from which infeasible ones are iteratively eliminated. As the total number of columns can be exponentially large, we use relaxed decision diagrams to compactly represent and manipulate the set of columns. This method was first introduced for the graph coloring problem in [15,31,33], then applied to the traveling salesperson problem with a drone [27,28], and later termed 'column elimination' [32].

The main focus of this work is to develop strong *dual bounds* for the CVRP using column elimination. As will be formalized later, column elimination and column generation will produce the same dual bound if they work from the same underlying route relaxation. Column elimination can potentially produce stronger bounds than the initial route relaxation as it can remove infeasible columns beyond those that are excluded by the initial route relaxation (cf. [26]). Moreover, column elimination allows a more liberal use of cutting planes to strengthen the relaxation. We show how existing cuts from the column generation literature can be expressed directly into the column elimination model, while in addition the decision diagram representation of the columns permits us to develop new cuts. The novel contributions include introducing cuts to column elimination, developing an efficient solution method via a Lagrangian reformulation, and showing how column elimination can produce bounds competitive with state-of-the-art solvers for the CVRP.

The paper is organized as follows. In Sect. 2 we present the column formulation of the CVRP. Section 3 describes the decision diagram-based constrained network flow formulation. The column elimination procedure is presented in Sect. 4. Section 5 present our Lagrangian relaxation. In Sect. 6 we describe how cutting planes can be added to strengthen the model. Section 7 presents a reduced cost-based arc fixing procedure to reduce the model size. We conduct experimental results in Sect. 8 and conclude in Sect. 9.

2 Column Formulation for CVRP

We first give a formal definition of the CVRP [29]. Let $G = (V, A)$ be a complete directed graph with vertex set $V = \{0, 1, \ldots, n\}$ and arc set $A = \{(i, j) \mid i, j \in V, i \neq j\}$. Vertex 0 represents the depot and vertices $\{1, \ldots, n\}$ represent the locations to be visited. We will interchangeably use vertices and locations. Each vertex $i \in V$ has a demand $q_i \geq 0$ and each arc $a \in A$ has a length $l_a \geq 0$. Let K be the number of (homogeneous) vehicles, each with capacity Q. A *route* is a sequence of vertices $[v_1, v_2, \ldots, v_k]$ starting and ending at the depot with total demand at most Q. The *distance* of a route is the sum of its arc lengths, i.e., $\sum_{i=1}^{k-1} l_{(v_i, v_{i+1})}$. The CVRP consists in finding K routes such that each vertex except for the depot belongs to exactly one route and the sum of the route distances is minimized.

The column formulation for the CVRP is based on the set R of all feasible elementary routes [6]. We let d_r denote the distance of route $r \in R$. We define an $n \times |R|$ matrix M such that $M_{ir} = 1$ if vertex $i \in \{1, 2, \ldots, n\}$ belongs to route $r \in R$, and $M_{ir} = 0$ otherwise. That is, each column vector in M corresponds to a route. Lastly, we define a binary decision variable x_r for each $r \in R$. The column formulation of the CVRP is:

$$
\begin{aligned}
\min \sum_{r \in R} d_r x_r & \\
\text{s.t.} \sum_{r \in R} M_{ir} x_r = 1 & \quad \forall i \in \{1, 2, \ldots, n\} \\
\sum_{r \in R} x_r = K & \\
x_r \in \{0, 1\} & \quad \forall r \in R.
\end{aligned}
\tag{1}
$$

This model is also known as the *set partitioning* formulation. In practice the set of routes R often has exponential size, which restricts the direct application of the set partitioning model to very small instances. Branch-and-price [7] provides a more scalable approach by using a column generation procedure to solve the continuous linear programming relaxation of (1).

Column generation starts by solving the linear programming relaxation of the set partitioning model defined on a (small) subset of variables, known as the *restricted master problem.* Using the dual variables of the optimal solution it then solves a *pricing problem* to find a new variable with a negative reduced cost. This process continues until no more improving variables exist and the restricted master has a provably optimal basis. To ensure integer feasibility, column generation is embedded into a systematic search.

Solving the pricing problem for the CVRP is not straightforward, because it corresponds to the NP-hard elementary shortest path problem with resource constraints [13]. It can be solved with dynamic programming, which is however limited by the exponential state space size. A computationally efficient alternative is to relax the pricing problem to find a shortest path that is not necessarily elementary, i.e., certain locations can be visited more than once [20]. Recent examples include the q-route relaxation [12] and the ng-route relaxation [5]. The linear programming model from route relaxations can be further strengthened by adding cutting planes to the restricted master problem [23].

3 Decision Diagram Formulation for CVRP

The key ingredient of the column elimination procedure is to compactly represent the set of routes R as a decision diagram. The CVRP can then be formulated as a constrained integer network flow over the decision diagram following the methodology in [27,33].

3.1 From Dynamic Programming to Decision Diagrams

For our purposes, a *decision diagram* is a layered acyclic weighted directed graph $D = (\mathcal{N}, \mathcal{A})$ with node set $\mathcal{N}$ and arc set $\mathcal{A}$. Each arc $a \in \mathcal{A}$ has an associated cost c_a and arc label ℓ_a. Graph D has a single root node r and a single terminal node t. While there are different methods to compile decision diagrams, we employ a generic approach that constructs a decision diagram from a dynamic programming formulation [8]. It requires a state definition, an (implied) set of states $\mathcal{S}$, a set of labels $\mathcal{L}$, a state transition function $f : (\mathcal{S} \times \mathcal{L}) \rightarrow \mathcal{S}$ and a transition cost function $g : (\mathcal{S} \times \mathcal{L}) \rightarrow \mathbb{R}$.

For the CVRP, we can use the dynamic programming formulation for the elementary shortest path problem with resource constraints [13], which we will refer to as DP_{ESPRC}. We define each state as a tuple (S, w, e) where $S \subseteq V$ represents the set of visited locations, $w \geq 0$ represents the accumulated 'weight', and $e \in V$ represents the last visited location. The initial state is defined as $(\varnothing, 0, 0)$. The set of labels is $\mathcal{L} = V$. Given a state $s = (S, w, e)$ and control (or label) $i \in V$ such that $i \notin S$ and $w + q_i \leq Q$, we define the transition function $f(s, i)$ as

$$f(s, i) = (S \cup \{i\}, w + q_i, i)$$

with associated transition cost function $g(s, i) = l_{(e,i)}$.

The decision diagram is defined similar to the state-transition graph of the dynamic programming model: the nodes in $\mathcal{N}$ correspond to the states and the arcs in $\mathcal{A}$ correspond to the transitions. That is, the root node r corresponds to the initial state $(\varnothing, 0, 0)$. For each transition $f(s_1, i) = s_2$ from state s_1 to state s_2 we define an arc $(u, v) \in \mathcal{A}$ where u corresponds to s_1 and v to s_2. The arc (u, v) has associated label $\ell_{(u,v)} = i$ and arc cost $c_{(u,v)} = g(s, i)$. We define the terminal node t as the collection of all states (S, w, e) with $|S| \geq 1$ and $e = 0$, i.e., t is the endpoint of all transitions that take label $i = 0$ to finish the route at the depot.

3.2 Dynamic Programming for Route Relaxations

The two most-used route relaxations for the CVRP in the column generation literature are the q-route relaxation [12] and the ng-route relaxation [5]. Both are based on the DP_{ESPRC} dynamic programming formulation, but relax the set of visited locations S.

The *q-route relaxation* maintains the last q visited locations. The dynamic program has state definition (SQ, w) where w is defined as above, and $SQ = [i_1, \ldots, i_q]$ is a sequence of locations. The initial state is $([\text{-}, \ldots, \text{-}], 0)$. Given a state $s = (SQ, w)$ and label $i \in V$ such that $i \notin SQ$ and $w + q_i \leq Q$, we define the transition function as

$$f^{\text{SQ}}(s, i) = ([i_2, \ldots, i_q, i], w + q_i)$$

with associated transition cost function $g^{\text{SQ}}(s, i) = l_{(i_q, i)}$. We denote the resulting dynamic programming model as DP_{SQ_q}.

For the *ng-route relaxation*, we assume that a set $N_i \subseteq V$ of size g exists for each $i \in \{1, \ldots, n\}$. The set N_i must include i and typically represents the g locations closest to i. As state definition, we use (NG, w, e) where the 'no-good' set $NG \subseteq V$ is a subset of visited locations, and w and e are as above. The initial state is $(\varnothing, 0, 0)$. Given a state $s = (NG, w, e)$ and label $i \in V$ such that $i \notin NG$ and $w + q_i \leq Q$, we define the transition function as

$$f^{\text{NG}}(s, i) = ((NG \cup \{i\}) \cap N_i, w + q_i, i)$$

with associated transition cost function $g^{\text{NG}}(s, i) = l_{(e,i)}$. We denote the resulting dynamic programming model as DP_{NG_g}. Observe that DP_{SQ_q} and DP_{NG_g} forbid cycles of length at most q and g, respectively.

3.3 Exact and Relaxed Decision Diagrams

We next specify the concepts of exact and relaxed decision diagrams [8] in the context of the CVRP. Given a decision diagram D, we let P_D denote the set of arc-label specified r-t paths in D. We slightly abuse notation and let c_p denote the sum of the arc costs of path $p \in P_D$. Recall that d_r represents the distance of route $r \in R$.

Definition 1. *Let R be a set of routes for the CVRP and let D be a decision diagram. We say that D is an* exact *diagram w.r.t. R if $P_D = R$ and $c_p = d_r$ for all $p \in P_D$, where r is the route representation of p. We say that D is a* relaxed *diagram w.r.t. R if $P_D \supseteq R$ and $c_p \leq d_r$ for all $p \in P_D$.*

Theorem 1. *The decision diagram derived from DP_{ESPRC} is exact w.r.t. R. The decision diagrams derived from DP_{SQ_q} and DP_{NG_g} are both relaxed w.r.t. R.*

Proof. Because DP_{ESPRC} encodes elementary paths and represents all possible feasible routes and their associated length, the resulting decision diagram is exact. Both DP_{SQ_q} and DP_{NG_g} encode a relaxation that contains elementary paths, and therefore represent a superset of all possible feasible routes. Because they both maintain the last visited location their cost functions are not relaxed, i.e., $c_p = d_r$ for each path p and associate route r (which is not necessarily elementary). Hence, DP_{SQ_q} and DP_{NG_g} yield a relaxed decision diagram. □

3.4 Constrained Network Flow Formulation

We next reformulate the set partitioning model (1) as a constrained integer network flow model over a given decision diagram $D = (\mathcal{N}, \mathcal{A})$. We introduce a 'flow' variable $y_a \geq 0$ for each $a \in \mathcal{A}$. The set of incoming arcs of a node u is denoted by $\delta^-(u)$. Likewise $\delta^+(u)$ denotes the set of outgoing arcs of u. We denote the set of arcs in $\mathcal{A}$ with label i by $\mathcal{A}^i$. The model is as follows:

$$F(D): \quad \min \sum_{a \in \mathcal{A}} c_a y_a \tag{2}$$

$$\text{s.t.} \sum_{a \in \delta^-(u)} y_a - \sum_{a \in \delta^+(u)} y_a = 0 \qquad \forall u \in \mathcal{N} \backslash \{r, t\} \tag{3}$$

$$\sum_{a \in \mathcal{A}^i} y_a = 1 \qquad \forall i \in V \backslash \{0\} \tag{4}$$

$$\sum_{a \in \delta^+(r)} y_a = K \tag{5}$$

$$y_a \in \{0, 1\} \qquad \forall a \in \mathcal{A}. \tag{6}$$

The objective function (2) minimizes the sum of all arc costs. The 'flow conservation' constraints (3) ensure that the solution is a collection of labeled r-t paths. Constraints (4) ensure that all locations are visited once. Constraint (5) enforces that exactly K units of flow originate from r. The binary constraints (6) complete the formulation.

Theorem 2. *Let D be an exact decision diagram w.r.t. the set of routes R. Model $F(D)$ is an exact formulation of the CVRP.*

The proof relies on the fact that the dynamic programming model represents all possible routes, that each solution of the network flow model consists of exactly K r-t paths, and that each r-t path corresponds to a feasible route.

Corollary 1. *Let D be a relaxed decision diagram w.r.t. the set of routes R. Model $F(D)$ yields a dual bound for the CVRP.*

In the remainder of this paper, we will use the continuous linear programming relaxation of model $F(D)$, referred to as $LP(F(D))$, which is obtained by replacing the integrality constraints (6) by $0 \leq y_a \leq 1$ for all $a \in \mathcal{A}$.

4 Column Elimination Procedure

We present a schematic representation of column elimination in Fig. 1. Starting with an initial relaxed decision diagram D, the column elimination procedure iteratively 1) solves the constrained network flow model $F(D)$, 2) decomposes the solution into paths (routes), 3) identifies infeasible paths and removes them from D, and repeats. The process terminates when no infeasible paths are detected in which case $F(D)$ is solved to optimality. It can also terminate earlier when the dual bound matches a given (or heuristically generated) primal bound, or when a different stopping criterion such as a time or memory limit is met. The procedure can utilize either the integer model $F(D)$ or its continuous relaxation $LP(F(D))$; using $LP(F(D))$ would solve the continuous linear programming relaxation of (1), but could be embedded in branch-and-bound to solve the full problem.

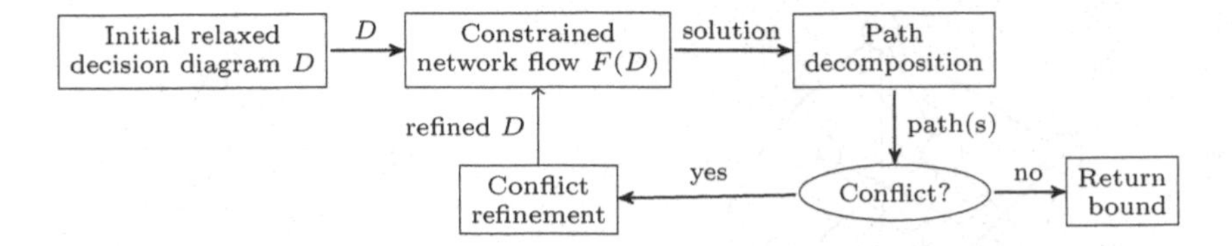

Fig. 1. Overview of the column elimination framework, adapted from [27].

Locations $V = \{0, 1, 2, 3, 4\}$
Depot $= 0$
Demands $q_1 = q_2 = q_3 = 1$, $q_4 = 2$
Number of trucks $K = 2$
Vehicle capacity $Q = 3$

l_{ij}	0	1	2	3	4
0	0	5	10	5	10
1	5	0	10	10	15
2	10	10	0	10	15
3	5	10	10	0	10
4	10	15	15	10	0

Fig. 2. Input data for the CVRP instance in Example 1.

Any (existing) route relaxation for the CVRP can be applied to construct the initial relaxed decision diagram. Recall that model DP_{ESPRC} has state definition (S, w, e), and each of these three elements can potentially be relaxed to define a relaxed decision diagram. The q-route and ng-route relaxations only relax the elementarity constraint, i.e., the set S. This means that conflicts will only come in the form of repeated labels; each path respects the truck capacity constraint and the route costs are exact. For a decision diagram D derived from such a relaxation, $F(D)$ is an exact formulation for the CVRP. In practice, we prefer using a relaxation that is relatively small and provides a 'good' starting point in terms of bound quality from $LP(F(D))$. In our experiments, we therefore use DP_{SQ_q} with $q = 1$ and DP_{NG_g} with $g = 2$ to initialize the relaxed decision diagram, with DP_{NG_2} performing best.

Given the initial relaxed decision diagram D, we solve the associated model $LP(F(D))$, apply a path decomposition of the solution, and inspect the paths for any conflicts. For our choice of route relaxations, the only conflicts arise from repetition of locations along a path. To remove a conflict, we follow the (partial) path elimination process outlined in [33]: it essentially separates the path by introducing a new node at each layer, and removing the arc associated with the repeated label. During this process, we will update the state information of the nodes along the separated path. We illustrate conflict separation in the next example, and refer to [33] for more details.

Example 1. Consider the CVRP instance with the problem data given in Fig. 2. The integer optimal solution uses routes $[0, 1, 2, 0]$ and $[0, 3, 4, 0]$ with total distance 50. The relaxed decision diagram based on DP_{SQ_q} with $q = 1$ is presented in Fig. 3(a). Each node in the diagram is associated with its SQ state, i.e., the last visited location. The weights are omitted from the states; instead nodes with the same cumulative weight are represented in the same layer. For clarity, we also

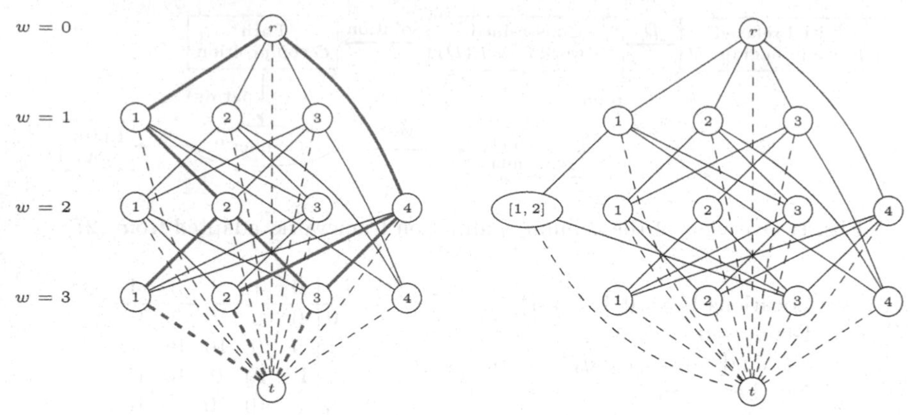

Fig. 3. Decision diagrams for the CVRP instance in Example 1. Figure (a) depicts the relaxed decision diagram obtained from the q-route relaxation. The optimal solution to model $LP(F(D))$ is indicated by thick blue arcs. Figure (b) represents the refined decision diagram after eliminating the partial path $[1, 2, 1]$ that contains a conflict.

omit the arc labels and arc costs. Arcs into t correspond to terminating a route and are dashed. The optimal solution to the linear programming relaxation of $F(D)$ yields dual bound 48.333 and uses the following arc-label specified paths: path $(1, 2, 1, 0)$ with flow value $\frac{1}{3}$, path $(1, 2, 3, 0)$ with flow value $\frac{1}{3}$, path $(4, 2, 0)$ with flow value $\frac{1}{3}$, and path $(4, 3, 0)$ with flow value $\frac{2}{3}$.

The first path contains a conflict: label 1 is repeated. We separate this conflict by rerouting the path to a new node with state $SQ = [1, 2]$ memorizing location 1 in addition to 2. As a result, we eliminate the arc with label 1 from the new state. The refined decision diagram is depicted in Fig. 3(b). It yields a dual bound of value 50, which is optimal.

5 Lagrangian Relaxation

Because the decision diagrams can grow large in size, solving the constrained network flow model can become the computational bottleneck of our method, even when we consider the continuous linear programming relaxation. To potentially solve the model more efficiently, we consider solving a Lagrangian relaxation, similar to [27,28], that has optimal bound equivalent to $LP(F(D))$. We obtain our Lagrangian relaxation of the constrained network flow model by dualizing constraints (4) that require that each location is visited once. We introduce a Lagrangian multiplier λ_i for each $i \in V$ ($\lambda_0 = 0$ is only introduced for notational ease), and define the Lagrangian relaxation as

$$L(D, \lambda): \quad \min \sum_{a \in \mathcal{A}} c_a y_a + \sum_{i \in V \setminus \{0\}} \lambda_i (1 - \sum_{a \in \mathcal{A}^i} y_a) \tag{7}$$

$$\text{s.t.} \sum_{a\in\delta^-(u)} y_a - \sum_{a\in\delta^+(u)} y_a = 0 \qquad \forall u \in \mathcal{N}\backslash\{r,t\} \quad (8)$$

$$\sum_{a\in\delta^+(r)} y_a = K \quad (9)$$

$$y_a \in \{0,1\} \qquad \forall a \in \mathcal{A}. \quad (10)$$

The objective function (7) can be rewritten as

$$\min \sum_{a\in\mathcal{A}} c_a y_a - \sum_{i\in V\backslash\{0\}} \lambda_i \sum_{a\in\mathcal{A}^i} y_a + \sum_{i\in V\backslash\{0\}} \lambda_i =$$

$$\min \sum_{a\in\mathcal{A}} (c_a - \lambda_{\ell_a}) y_a + \sum_{i\in V\backslash\{0\}} \lambda_i.$$

As a consequence, for fixed λ, the Lagrangian relaxation can be solved as a (continuous) minimum-cost network flow problem over the decision diagram, using $c_a - \lambda_{\ell_a}$ as the cost for arc $a \in \mathcal{A}$, yielding an integer optimal solution. In fact, given constraints (9) and the unit capacity constraints on the arcs, each solution consists of K arc-disjoint r-t paths. By applying the successive shortest paths (SSP) algorithm [1] to solve $L(D,\lambda)$ we obtain the following result:

Lemma 1. *Given a decision diagram $D = (\mathcal{N},\mathcal{A})$ and fixed λ, the Lagrangian relaxation $L(D,\lambda)$ can be solved in $O(K(|\mathcal{N}|\log(|\mathcal{N}|) + |\mathcal{A}|))$ time.*

We also implemented a dedicated algorithm, based on the 'minimum update Successive Shortest Paths' (muSSP) algorithm that was developed for specific directed acyclic graphs in the content of multi-object tracking in computer vision [34]. Although graphs with a slightly different structure are considered in [34], the algorithm generalizes to our case: weighted directed acyclic graphs with one source (the root), one sink (the terminal), and unit capacities. The muSSP algorithm leverages the fact that most updates to the shortest path tree through Dijkstra's algorithm are not useful, and it aims instead to make minimal updates to the shortest path tree. While it has the same theoretical worst-case time complexity as the SSP, in practice the muSSP algorithm can be an order of magnitude more efficient than the standard SSP algorithm.

The Lagrangian 'dual' subproblem $\max_\lambda L(D,\lambda)$ finds the multipliers that provide the best Lagrangian bound. Because the objective in $L(D,\lambda)$ is concave and piecewise linear, the dual can be solved via a subgradient method. At each iteration k of the subgradient method, one choice for a subgradient that we will use is γ^k such that $\gamma_i^k = (1 - \sum_{a\in\mathcal{A}^i} y_a^k)$ where y_a^k is the solution to $L(D,\lambda^k)$. Then we update the dual multipliers for the next iteration as $\lambda^{k+1} = \lambda^k + \alpha^k\gamma^k$, where we use an estimated Polyak step size α^k [10]. Note that the initial choice of multipliers λ^0 can be important for solving the dual quickly [9].

We remark that the optimal Lagrangian dual bound is equal to the optimal linear programming bound from model $LP(F(D))$, when both apply the same decision diagram. Moreover, when the column elimination process uses model $LP(F(D))$ or $L(D,\lambda)$, its bound at termination is equal to the column generation bound of the set partitioning model (1), assuming that all methods use the same underlying dynamic programming formulation, as was observed in [27]. That is, the decision diagram applies the same dynamic programming formulation in its construction as column generation uses in the pricing problem.

Lastly, we note that in each iteration of the subgradient method for solving the Lagrangian dual the solution can potentially be used to identify and separate conflicts. Similar to [28], we separate these conflicts in batches of size 100, after which we restart the Lagrangian process.

6 Cutting Planes

Results from the literature show that the LP relaxation of the set partitioning formulation for CVRP, solved via column generation, frequently has a 1–4% optimality gap. To further strengthen the LP relaxation several classes of valid inequalities can be added. According to the literature, the most effective are rounded capacity cuts, strengthened comb inequalities, and subset-row cuts [12, 18,23]. The first two types of cuts are called *robust* in the column generation literature because they do not affect the runtime of the pricing problem, while the subset-row cuts are not robust. We next show how rounded capacity cuts and strengthened comb inequalities can be implemented in our decision diagram-based model $LP(F(D))$, as well as a generalization of subset-row cuts as a type of clique cut.

Rounded capacity cuts ensure that a subset of locations S is visited by a sufficient number of trucks to meet its aggregate demand. In column generation these cuts can be added to model (1) so long as the underlying routes are stored for each $r \in R$. Let p_r^S be the number of times route r uses an edge between S and $V \backslash S$, and let $k(S) = \lceil \frac{1}{Q} \sum_{i \in S} q_i \rceil$. The cut added to the restricted master problem is $\sum_{r \in R} p_r^S x_r \geq 2k(S)$, and the associated dual variable is added to controls in the dynamic program for the pricing problem that correspond to a route traversing an edge between S and $V \backslash S$. To add this cut in column elimination, let A^S be the set of arcs $a \in A$ such that $\ell(a) \in S$ and the node u that is the head of a has state with last visited location $i \in V \backslash S$, or the other way around with $\ell(a) \in V \backslash S$ and $i \in S$. A rounded capacity cut for set S can be modeled by adding to $LP(F(D))$ the following inequality: $\sum_{a \in A^S} y_a \geq 2k(S)$. Note that when solving $LP(F(D))$ using the Lagrangean formulation, this constraint can be dualized.

Strengthened comb inequalities are a generalization of comb inequalities that have been proven highly useful for solving the Traveling Salesman Problem [17]. A strengthened comb inequality is defined by a handle set of locations H and teeth sets of locations T_t for $t \in \{1, ..., T\}$. Let $S(H, T_1, ..., T_T)$ be the appropriately defined right hand side for the inequality [17]. In column generation, this cut also requires storing the underlying routes and can be added to the restricted master problem as $\sum_{r \in R} p_r^H x_r + \sum_{t \in T} \sum_{r \in R} p_r^{T_t} x_r \geq S(H, T_1, ..., T_T)$. The associated dual variable is then added to controls in the dynamic program for the pricing problem that correspond to traversing edges with one endpoint in H or one of T_i and the other endpoint not in that set. In column elimination, a strengthened comb inequality with handle H and teeth T_t can be modeled by adding to $LP(F(D))$ the following inequality: $\sum_{a\in A^H} y_a + \sum_{t\in\{1,...,T\}} \sum_{a\in A^{T_t}} y_a \geq S(H, T_1, ..., T_T)$. This constraint can also be dualized when using the Lagrangean formulation to solve $LP(F(D))$.

Subset row cuts are non-robust cuts that have been successfully applied to the CVRP. In particular, the limited memory subset row cuts are an important part of the success of the column generation method in [23]. Since the decision diagram representation does not have a matrix view of the set of routes, subset row cuts do not directly translate to the $LP(F(D))$ model. However, they can be generalized by a class of clique cuts on a specific conflict graph [3]. These cuts are non-robust and have been used in [4] but not until the problem size has been reduced. The structure of our decision diagram allows column elimination to implement a specific version of these cuts. Let $D = (\mathcal{N}, \mathcal{A})$ be a decision diagram as defined above. The conflict graph $G^C = (\mathcal{N}, A^C)$ is defined on node set $\mathcal{N}$. Its arc set A^C contains all arcs (i, j) such that 1) the set of visited locations in the states associated to nodes i and j have a non-empty intersection, and 2) nodes i and j never appear on the same directed path in D. A *clique cut* states that the flow through nodes in a clique of G^C must be at most 1:

Theorem 3. *Let $\mathcal{C}$ be a clique in the conflict graph G^C derived from a decision diagram D. The associated* clique cut $\sum_{i\in\mathcal{C}} \sum_{a\in\delta^-(i)} y_a \leq 1$ *is a valid inequality for model $LP(F(D))$.*

Proof. By construction of G^C, each pair of nodes $i, j \in \mathcal{C}$ has at least one common visited location (say u) in their associate states, and there is no directed path between i to j in D. Suppose that for an integral optimal solution we have $\sum_{a\in\delta^-(i)\cup\delta^-(j)} y_a > 1$. This means that location u is visited twice, which cannot occur in an optimal solution: a contradiction. □

Given G^C and a set of cliques in G^C, clique cuts can be easily separated for $LP(F(D))$ by evaluating whether a given fractional solution violates a cut. Because a solution to the Lagrangian model $L(D, \lambda)$ is integral, we cannot directly use it to separate any cuts. In [2] it is shown that a weighted average of the subproblem solutions converges to an optimal primal solution and we apply this method to identify valid inequalities.

7 Reduced Cost-Based Arc Fixing

Variable fixing based on reduced costs is often applied to reduce the problem size of integer programs [21], including the CVRP [14,23]. It uses a feasible dual solution and suitably small optimality gap to set the value of a primal variable equal to 0 [1,11,16]. We develop an arc fixing method for the $LP(F(D))$ model, using similar arguments as [23].

Let $D = (\mathcal{N}, \mathcal{A})$ be a decision diagram that is exact w.r.t. some set of routes $R' \subseteq R$. Consider a feasible dual solution (ν, κ) to the LP relaxation of the set partitioning model (1) over R', where ν correspond to the 'set partitioning' constraints and κ to the 'number of trucks' constraint. For each arc $a \in \mathcal{A}$ we define a 'reduced cost distance' $rc(a) = l_a - \nu_{\ell_a}$. For each node $u \in \mathcal{N}$, we define $sp_u^{\downarrow}$ as the shortest r-u path in D with respect to the reduced cost distances, and similarly define $sp_u^{\uparrow}$ to be the shortest u-t path in D.

Theorem 4. *Consider arc $a = (v_1, v_2) \in \mathcal{A}$. Let $v(\nu, \kappa)$ be the dual solution value, and let UB an upper bound on (1). If $v(\nu, \kappa) + sp_{v_1}^{\downarrow} + sp_{v_2}^{\uparrow} + rc(a) - \kappa > UB$, then arc a can be fixed to have flow 0 in $F(D)$ and accordingly in $LP(F(D))$.*

Proof. Given (ν, κ), each route $\overline{r} \in R'$ in the LP relaxation of (1) has reduced cost $rc(\overline{r}) = d_{\overline{r}} - \sum_{i=1}^{n} M_{i\overline{r}}\nu_i - \kappa$. Each $\overline{r}$ corresponds to a path $p = \{a_1, ..., a_l\}$ in D, so $rc(\overline{r})$ can be decomposed into $rc(\overline{r}) = \sum_{i=1}^{l} rc(a_i) - \kappa$. For all p that contain arc a, let p' be the path that corresponds to the route r' with lowest reduced cost. Denote $rc(r') = sp_{v_1}^{\downarrow} + sp_{v_2}^{\uparrow} + rc(a) - \kappa$. Now for sake of contradiction assume an optimal solution to $F(D)$ has $y_a = 1$. Then some path p'' in D that contains arc a will have flow of 1, so we can consider this as some $x_{r''} = 1$ in an optimal solution to (1). To construct the remainder of an optimal solution to the LP relaxation of (1) we can solve this LP relaxation with constraints for locations in r'' removed and only requiring $K-1$ trucks. Because (ν, κ) remains feasible to the dual of this updated problem and has value $v(\nu, \kappa) - \sum_{i=1}^{n} M_{ir''}\nu_i - \kappa$, it gives a valid lower bound on (1) that contradicts UB, namely $v(\nu, \kappa) - \sum_{i=1}^{n} M_{ir''}\nu_i - \kappa + d_{r''} = v(\nu, \kappa) + rc(r'') \geq v(\nu, \kappa) + rc(r') \geq v(\nu, \kappa) + sp_{v_1}^{\downarrow} + sp_{v_2}^{\uparrow} + rc(a) - \kappa > UB$. □

Note that while Theorem 4 relies on the set partitioning model (1) to build the reduced cost argument, we can use the optimal dual solution to $LP(F(D))$ in the application of the theorem. When solving $LP(F(D))$ with a standard linear programming solver, we can use the feasible dual from the previous iteration – which remains feasible even with cuts and separations – to fix arcs. One important note is that these fixed arcs are reintroduced if separation happens before the next iteration, as the change in the decision diagram structure may disrupt previous arc fixing arguments. When solving $LP(F(D))$ via its Lagrangian relaxation $L(D, \lambda)$, we must ensure that we work with a feasible dual solution. In addition, we include a dual variable for constraint (10) and set it to its maximum value while ensuring dual feasibility.

8 Experimental Results

We use the benchmark set of CVRP instances from http://vrp.atd-lab.inf.puc-rio.br/index.php/en/, including the new challenge set of instances from [30]. All experiments are run on an Intel(R) Xeon(R) Gold 6248R CPU @ 3.00 GHz. We use CPLEX version 22.1 [19] as a linear programming solver and change (4) to $\geq$ to help find an initial feasible solution. We use the package CVRPSEP [18] to heuristically find rounded capacity cuts and strengthened comb inequalities when given a fractional primal solution. At each iteration we add at most 10 robust capacity cuts and 5 strengthened comb inequalities, using the most violated ones

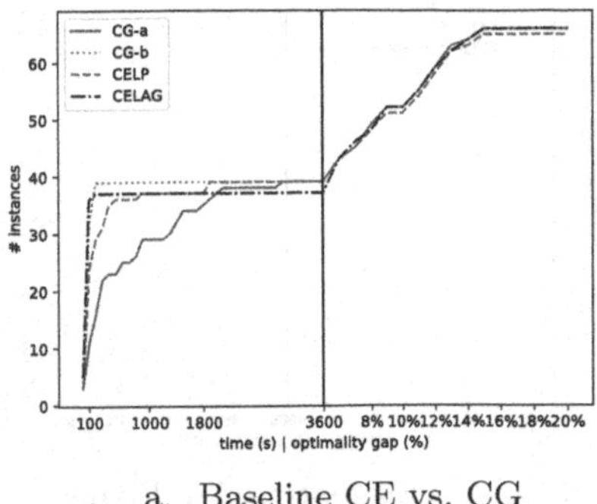

a. Baseline CE vs. CG

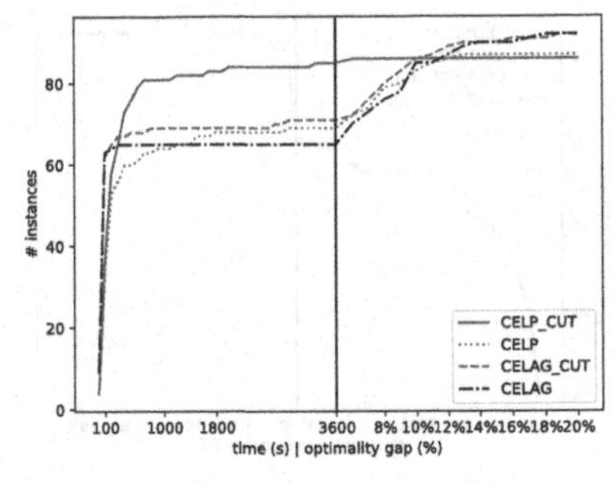

b. Adding Cuts to CE

Fig. 4. a) Comparing column generation over DP_{Q_2} with column elimination starting from DP_{Q_1} and using different methods to solve $LP(F(D))$ without any added cuts. b) Performance plot for adding cuts to CELP and CELAG.

possible. We use Cliquer [22] to heuristically find large weighted cliques in the conflict graph used to derive clique cuts.

Comparing Column Elimination and Column Generation. We compare column generation over DP_{Q_2} with column elimination starting from the DP_{Q_1} route relaxation and eliminating cycles of size 2. Doing so, the final bounds are the same, which allows us to compare how quickly column generation and column elimination reach the optimal bound. We implement a vanilla version of column generation that starts with a small set of greedily chosen routes and solves the pricing problem as shortest paths through the pre-compiled decision diagram for DP_{Q_2}. We compare column generation not including and including time to compile the decision diagram (CG-a, CG-b), column elimination using CPLEX (CELP), and column elimination using a subgradient method over the Lagrangian dual (CELAG). We run each method for 3,600 s over benchmark sets A, B, E, F, M, P. We remove instances when the decision diagram for DP_{Q_2} does not finish compiling. Arc fixing uses the best known solution as an upper bound and is used in CELP but not in CELAG. Lower bounds for column generation are computed before termination as in [35]. Figure 4.a is a performance plot of the number of instances solved to within a 5% optimality gap in a given amount of time, extended by the number of instances solved to larger optimality gaps. Over the given relaxation, it is evident that column elimination with the different methods can work appropriately and be competitive with column generation.

Evaluating the Impact of Cuts. We compare solving column elimination using CPLEX with and without cuts (CELP_CUT, CELP) and using the Lagrangian method with and without cuts (CELAG_CUT, CELAG). Figure 4.b is a performance plot for solving the instances up to 5% as in the last experiment. Figure 4.b shows how cuts greatly improve column elimination when using CPLEX as the LP solver, and benefit when solving the Lagrangian reformulation but not as much.

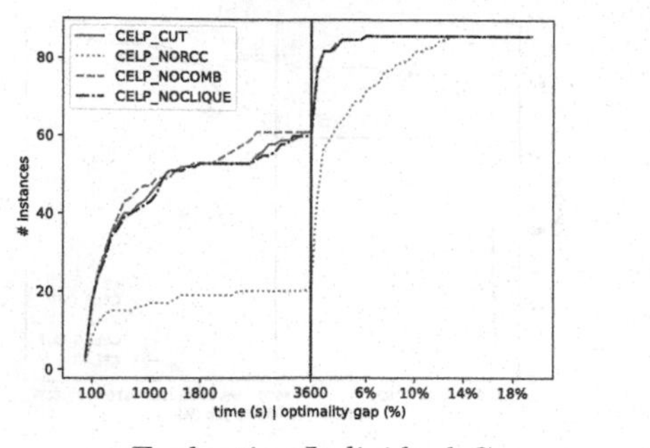

a. Evaluating Individual Cuts

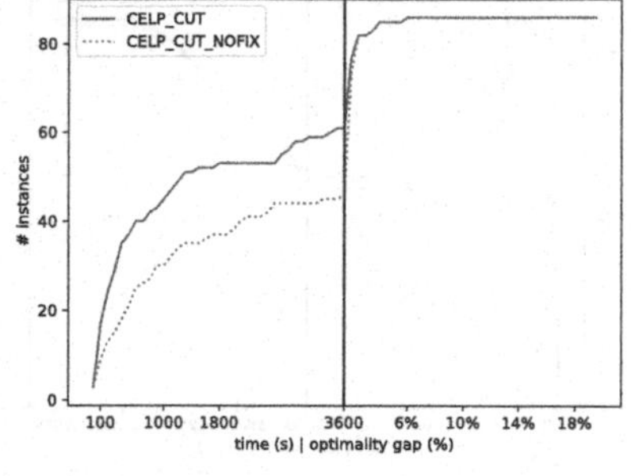

b. Evaluating Arc Fixing

Fig. 5. a) Evaluating the performance of individual cuts on CELP_CUT. b) Performance plot of CELP_CUT with and without arc fixing

We then compare the performance of CELP_CUT removing one class of cuts at a time: without the rounded capacity cuts (CELP_NORCC), without the strengthened comb inequalities (CELP_NOCOMB), and without the clique inequalities (CELP_NOCLIQUE). Figure 5.a is a performance plot of the number of instances solved to within a 1% optimality gap. Rounded capacity cuts provide the most benefit, the overhead of strengthened comb inequalities sometimes outweigh their benefit but not entirely if we more closely examine the bounds achieved for each instance, and clique inequalities can provide some benefit later in the method when it is able to be distinguished from separations and other cuts.

Evaluating the Impact of Arc Fixing. We consider the impact of arc fixing by removing the feature from CELP_CUT to get CELP_CUT_NOFIX. Figure 5.b is a performance plot using 1% optimality gap. Arc fixing speeds up column elimination to find stronger bounds in less time.

Evaluating the Impact of muSSP. We evaluate the impact of using the muSSP algorithm to solve the subproblem in CELAG by removing it in CELAG_-NOMUSSP. The performance plot using 5% optimality gap is for 64 large X instances and shows that there is a significant speedup. We chose to use the X class here because the speedup is more pronounced on large instances.

Comparison to State-of-the-Art. Figure 6.b compares the state-of-the-art BCP method's root node lower bounds (Pecin) with the best column elimination method settings that we chose through experimentation (CE). For each class of problems the table gives the number of problems in the class (NP) and the average optimality gap found at the root node over all instances. Pecin takes less than 3600 s to compute its bounds for all instances except the X class where it can take several hours. CE gaps are computed based on 3600 s second runs for all classes except M, F, and X which are given 7200 s. The decision diagram did not compile for 12 X instances, so we leave these out of the analysis. One F instance with large capacity resulted in a large diagram and 27% gap that can

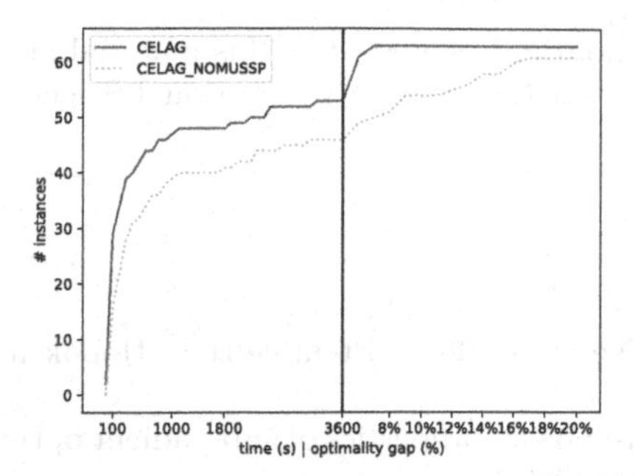

a. Evaluating muSSP for CELAG

Class	NP	Pecin Gap (%)	CE Gap (%)
A	22	0.36	0.66
B	20	0.14	0.61
E-M	12	0.33	2.60
F	3	0.00	16.41
P	24	0.42	0.85
X	100	0.44	2.13

b. Comparing root node bounds

Fig. 6. (a) A comparison of column elimination using the Lagrangean reformulation with and without the muSSP algorithm. (b) Comparing the root node lower bounds from Pecin et al. [23] (Pecin) and the lower bounds from column elimination (CE); both methods include cuts.

be reduced with more runtime. The better of the CELAG and CELP results is used; most small instances use CELP while large instances like almost all of the X class use CELAG. We also remove two E class instances with unconventional demand formatting.

9 Conclusion

We introduced a column elimination procedure for the capacitated vehicle routing problem (CVRP). Our methods works with a relaxed set of routes that are compactly represented in a decision diagram, and from which infeasible routes are iterative removed. We showed how we can use existing route relaxations for the CVRP, such as the q-route and ng-route relaxation, to compile good initial relaxed decision diagrams. When the decision diagram is exact, and only contains all feasible routes, we showed that a solution to the CVRP can be found by solving a constrained network flow problem over the diagram. When the diagram is relaxed, this model yields a dual bound. To strengthen the linear programming relaxation of our model we added valid inequalities; in particular, we showed how a class of clique cuts can be derived from the structure of the diagram. To solve the model more efficiently, we considered solving a Lagrangian dual formulation for which we implemented a specialized successive shortest paths algorithm. In our experimental results, we demonstrated that column elimination is a viable alternative to column generation for the CVRP, although the best known dual bounds from the literature, obtained by column generation with cutting planes, are generally stronger.

Acknowledgements. This work is partially supported by Office of Naval Research Grant No. N00014-21-1-2240 and National Science Foundation Award #1918102. This material is also based upon work supported by the National Science Foundation Graduate Research Fellowship Program under Grant No. DGE1745016, DGE2140739. Any opinions, findings, and conclusions or recommendations expressed in this material are those of the author(s) and do not necessarily reflect the views of the National Science Foundation.

References

1. Ahuja, R.K., Magnanti, T.L., Orlin, J.B.: Network Flows. Prentice Hall, Hoboken (1993)
2. Anstreicher, K.M., Wolsey, L.A.: Two "well-known" properties of subgradient optimization. Math. Program. **120**(1), 213–220 (2009)
3. Balas, E., Ho, A.: Set covering algorithms using cutting planes, heuristics, and subgradient optimization: a computational study. In: Padberg, M.W. (ed.) Combinatorial Optimization. Mathematical Programming Studies, vol. 12, pp. 37–60. Springer, Heidelberg (1980). https://doi.org/10.1007/BFb0120886
4. Baldacci, R., Christofides, N., Mingozzi, A.: An exact algorithm for the vehicle routing problem based on the set partitioning formulation with additional cuts. Math. Program. **115**(2), 351–385 (2008)
5. Baldacci, R., Mingozzi, A., Roberti, R.: New route relaxation and pricing strategies for the vehicle routing problem. Oper. Res. **59**(5), 1269–1283 (2011)
6. Balinski, M.L., Quandt, R.E.: On an integer program for a delivery problem. Oper. Res. **12**(2), 300–304 (1964)
7. Barnhart, C., Johnson, E.L., Nemhauser, G.L., Savelsbergh, M.W.P., Vance, P.H.: Branch-and-price: column generation for solving huge integer programs. Oper. Res. **46**(3), 316–329 (1998)
8. Bergman, D., Cire, A.A., Van Hoeve, W.J., Hooker, J.: Decision Diagrams for Optimization, vol. 1. Springer, Heilderberg (2016)
9. Bertsekas, D.P.: Constrained optimization and Lagrange multiplier methods. Academic Press, Cambridge (2014)
10. Boyd, S., Xiao, L., Mutapcic, A.: Subgradient methods. Lect. Notes EE392o, **2004**, 2004–2005 (2003)
11. Fischetti, M., Toth, P.: An additive bounding procedure for combinatorial optimization problems. Oper. Res. **37**(2), 319–328 (1989)
12. Fukasawa, R.: Robust branch-and-cut-and-price for the capacitated vehicle routing problem. Math. Program. **106**(3), 491–511 (2006)
13. Irnich, S., Desaulniers, G.: Shortest path problems with resource constraints. In: Desaulniers, G., Desrosiers, J., Solomon, M.M. (eds.) Column Generation, pp. 33–65. Springer, Boston (2005). https://doi.org/10.1007/0-387-25486-2_2
14. Irnich, S., Desaulniers, G., Desrosiers, J., Hadjar, A.: Path-reduced costs for eliminating arcs in routing and scheduling. INFORMS J. Comput. **22**(2), 297–313 (2010)
15. Karahalios, A., van Hoeve, W.-J.: Variable ordering for decision diagrams: a portfolio approach. Constraints **27**(1), 116–133 (2022)
16. Lodi, A., Milano, M., Rousseau, L.-M.: Discrepancy-based additive bounding for the AllDifferent constraint. In: Rossi, F. (ed.) CP 2003. LNCS, vol. 2833, pp. 510–524. Springer, Heidelberg (2003). https://doi.org/10.1007/978-3-540-45193-8_35
17. Lysgaard, J., Letchford, A., Eglese, R.: A new branch-and-cut algorithm for the capacitated vehicle routing problem. Math. Program. Ser. A **100**, 423–445 (2004)
18. Lysgaard, J.: CVRPSEP: a package of separation routines for the capacitated vehicle routing problem (2003). http://www.hha.dk/~lys/CVRPSEP.html

19. CPLEX User's Manual: Ibm ilog cplex optimization studio. Version 12(1987–2018): 1 (1987)
20. Christofides, N., Mingozzi, A., Toth, P.: Exact algorithms for the vehicle routing problem, based on spanning tree and shortest path relaxations. Math. Program. **20**(1), 255–282 (1981)
21. Nemhauser, G., Wolsey, L.: Integer and Combinatorial Optimization. Wiley, Hoboken (1988)
22. Östergård, P.R.J.: A fast algorithm for the maximum clique problem. Discret. Appl. Math. **120**(1–3), 197–207 (2002)
23. Pecin, D., Pessoa, A., Poggi, M., Uchoa, E.: Improved branch-cut-and-price for capacitated vehicle routing. Math. Program. Comput. **9**(1), 61–100 (2017)
24. Pessoa, A., Sadykov, R., Uchoa, E., Vanderbeck, F.: Automation and combination of linear-programming based stabilization techniques in column generation. INFORMS J. Comput. **30**(2), 339–360 (2018)
25. Pessoa, A., Sadykov, R., Uchoa, E., Vanderbeck, F.: A generic exact solver for vehicle routing and related problems. Mathematical Programming **1**, 483–523 (2020). https://doi.org/10.1007/s10107-020-01523-z
26. Roberti, R., Mingozzi, A.: Dynamic ng-path relaxation for the delivery man problem. Transp. Sci. **48**(3), 413–424 (2014)
27. Tang, Z., van Hoeve, W.-J.: Dual bounds from decision diagram-based route relaxations: an application to truck-drone routing. Optim. Online (2022)
28. Tang, Z.: Theoretical and Computational Methods for Network Design and Routing. PhD thesis, Carnegie Mellon University (2021)
29. Toth, P., Vigo, D.: Vehicle Routing: Problems, Methods, and Applications. SIAM, 2 edition (2014)
30. Uchoa, E., Pecin, D., Pessoa, A., Poggi, M., Vidal, T., Subramanian, A.: New benchmark instances for the capacitated vehicle routing problem. Eur. J. Oper. Res. **257**(3), 845–858 (2017)
31. Hoeve, W.-J.: Graph coloring lower bounds from decision diagrams. In: Bienstock, D., Zambelli, G. (eds.) IPCO 2020. LNCS, vol. 12125, pp. 405–418. Springer, Cham (2020). https://doi.org/10.1007/978-3-030-45771-6_31
32. van Hoeve, W.-J., Tang, Z.: Column "Elimination": dual bounds from decision diagram-based route relaxations. In: INFORMS Computing Society Conference (2022)
33. van Hoeve, W.-J.: Graph coloring with decision diagrams. Math. Program. **192**(1), 631–674 (2022)
34. Wang, C., Wang, Y., Wang, Y., Wu, C.T., Yu, G.: muSSP: Efficient min-cost flow algorithm for multi-object tracking. In: Advances in Neural Information Processing Systems, vol. 32 (2019)
35. Wolsey, L.A.: Integer Programming. Wiley, Hoboken (2020)

Cutting Plane Selection with Analytic Centers and Multiregression

Mark Turner[1,2](✉), Timo Berthold[1,3], Mathieu Besançon[2], and Thorsten Koch[1,2]

[1] Institute of Mathematics, Technische Universität Berlin, Berlin, Germany
[2] Zuse Institute Berlin, Berlin, Germany
{turner,koch,besancon}@zib.de
[3] Fair Isaac Deutschland GmbH, Berlin, Germany
timoberthold@fico.com

Abstract. Cutting planes are a crucial component of state-of-the-art mixed-integer programming solvers, with the choice of which subset of cuts to add being vital for solver performance. We propose new distance-based measures to qualify the value of a cut by quantifying the extent to which it separates relevant parts of the relaxed feasible set. For this purpose, we use the analytic centers of the relaxation polytope or of its optimal face, as well as alternative optimal solutions of the linear programming relaxation. We assess the impact of the choice of distance measure on root node performance and throughout the whole branch-and-bound tree, comparing our measures against those prevalent in the literature. Finally, by a multi-output regression, we predict the relative performance of each measure, using static features readily available before the separation process. Our results indicate that analytic center-based methods help to significantly reduce the number of branch-and-bound nodes needed to explore the search space and that our multiregression approach can further improve on any individual method.

1 Introduction

Branch-and-cut is the algorithm at the core of most Mixed-Integer Programming (MIP) solvers. A key component of branch-and-cut is the resolution of Linear Programming (LP) relaxations of the original problem over partitions of the variable domains. *Cutting planes* – or cuts – tighten those relaxations around integer-feasible points. Given a MIP:

$$\underset{\mathbf{x}}{\operatorname{argmin}}\{\mathbf{c}^\intercal\mathbf{x} \mid \mathbf{A}\mathbf{x} \le \mathbf{b},\ \mathbf{l} \le \mathbf{x} \le \mathbf{u},\ \mathbf{x} \in \mathbb{Z}^{|\mathcal{J}|} \times \mathbb{R}^{n-|\mathcal{J}|}\} \tag{P}$$

a cut is an inequality $\boldsymbol{\alpha}^\intercal\mathbf{x} \le \beta$ that is violated by at least one solution of the LP relaxation but that does not increase the optimal value of the problem when added i.e. it is valid for (P). Thereby, the inequality added as a constraint to (P) tightens the relaxation, potentially increasing the relaxation's optimal value. The use of cutting planes is one of the crucial aspects to solving MIPs efficiently [3].

© The Author(s), under exclusive license to Springer Nature Switzerland AG 2023
A. A. Cire (Ed.): CPAIOR 2023, LNCS 13884, pp. 52–68, 2023.
https://doi.org/10.1007/978-3-031-33271-5_4

A well-designed cutting plane separation procedure often helps to reduce the branch-and-bound tree size while accelerating the overall solving process.

In MIP solvers, two key algorithms related to cuts are their generation and their selection. Cut generation is the problem of computing a set of cuts that tighten the relaxation and separate the current continuous relaxation solution from the feasible MIP solutions. Modern MIP solvers implement various general-purpose and specialised cutting plane generation algorithms. Since the generation of cuts is, in general, far less expensive than solving the LP relaxation, many cuts are generated from the same relaxation. The cut selection algorithm takes the set of all cut candidates generated so far and selects a subset that is actually added to the LP relaxation. This two-step process of generation and selection constitutes a single *separation round.*

At the root node, the MIP solver interleaves separation rounds with solving the enhanced LP relaxation until the branch-and-bound search is started. At other search tree nodes, the solver often only performs a limited cut loop, if at all. We focus in this paper on globally-valid cuts, i.e. cuts that are valid for the original problem, as opposed to *local cuts*, which are generated with additional local bounds at a node.

Cut selection is a classical trade-off problem: too little cutting leads to large enumeration trees; too much cutting to a small node throughput and numerical instability. However, carefully selected cuts can help improve both the dual and the primal bound simultaneously by bringing the relaxation closer to the convex hull of feasible solutions. Since proximity to the convex hull is inherently hard to measure, cut selection methods often try to approximate it by various cheap measures, e.g., efficacy.

Efficacy, or *cutoff distance*, is used in commercial MIP solvers as one of the main criteria for whether to add a cut or not [1,9][1]. Efficacy measures the shortest distance between the LP optimal solution $\mathbf{x}^{LP}$ and the cut hyperplane $\boldsymbol{\alpha}^\intercal \mathbf{x} \leq \beta$. The function `eff` that maps a cut, and the LP solution $\mathbf{x}^{LP}$ to the efficacy is defined as:

$$\mathtt{eff}(\boldsymbol{\alpha}, \beta, \mathbf{x}^{LP}) := \frac{\boldsymbol{\alpha}^\intercal \mathbf{x}^{LP} - \beta}{||\boldsymbol{\alpha}||}$$

Introduced in [14], *directed cutoff distance* is the signed distance between the LP solution $\mathbf{x}^{LP}$ and the cut hyperplane in the direction of a primal solution, $\hat{\mathbf{x}}$. The measure has the property that the directed projection of $\mathbf{x}^{LP}$ onto the cut hyperplane is inside of the feasible region, and by using the best primal solution available, aims to cut in the direction of the optimal solution. We define the directed cutoff distance, with the function `dcd` as follows:

$$\mathtt{dcd}(\boldsymbol{\alpha}, \beta, \mathbf{x}^{LP}, \hat{\mathbf{x}}) := \frac{\boldsymbol{\alpha}^\intercal \mathbf{x}^{LP} - \beta}{|\boldsymbol{\alpha}^\intercal \mathbf{y}|}, \text{ where } \mathbf{y} = \frac{\hat{\mathbf{x}} - \mathbf{x}^{LP}}{||\hat{\mathbf{x}} - \mathbf{x}^{LP}||}$$

The last measure we consider, although not based on a distance, is *expected improvement*, see [27], which corresponds to the difference in objective between

[1] Confirmed as a main criterion in FICO Xpress 8.14.

$\mathbf{x}^{LP}$ and its orthogonal projection onto the hyperplane of a cut. We denote the measure as `exp-improv`, and define it as:

$$\texttt{exp-improv}(\boldsymbol{\alpha}, \beta, \mathbf{c}, \mathbf{x}^{LP}) := ||\mathbf{c}|| \cdot \frac{\boldsymbol{\alpha}^{\intercal}\mathbf{c}}{||\boldsymbol{\alpha}||||\mathbf{c}||} \cdot \texttt{eff}(\boldsymbol{\alpha}, \beta, \mathbf{x}^{LP})$$

In this paper, we propose new distance-based measures for the quality of cuts. These measures are designed to retain soundness properties in common cases that hinder the applicability, and to reduce the size of the search space, without focusing on runtime improvement. To this end, we perform extensive computational experiments to analyse the effects of our newly introduced measures and those that exist in the literature. Unlike previous beliefs, the choice of distance measure significantly impacts solver performance both in time and number of nodes, with different measures performing better on different groups of instances. Motivated by this observation, we design a multi-output regression model, which predicts the relative performance of each measure using static features readily available before the separation process. The scope of this paper is to establish a model that aims at reducing search space size rather than runtime.

2 Related Work

There is a prevailing sentiment in the MIP community, supported by a set of older computational studies, see [1,4,27], that inexpensive heuristics are sufficient for cut selection. Specifically, these studies suggest that cheap ranking metrics, predominantly efficacy, are sufficiently effective when combined with filtering mechanisms that ensure no two overly parallel cuts are added. The studies [1,27] argue that a weighted sum of different metrics is most effective for ranking cuts as opposed to any single metric.

Recently, the research focus for cuts in MIP has been on using deep learning to either calculate scores directly from a set of measures or to predict parameter values in scoring functions, see [6,7,17,21,25,26]. In [7] and [21] a neural network is trained to predict the objective value improvement of cuts when added. In [25], a neural network is trained using evolutionary strategies to select Gomory cuts. A neural network is also trained in [17], this time using multiple instance learning, to map a set of aggregate cut features to a scoring function. In [6], the cut selection parameter space is shown to be partitionable into a set of regions, such that all parameter choices within a region select the same set of cuts. Finally, for cut scores based on weighted criteria, [26] provides an illustrative example of worst-case scenario for parameter grid search in the cut selection parameter space, and phrases learning cut selection parameters as a reinforcement learning problem.

Closest to our work are papers that introduced cut selection measures other than efficacy, which were still based on a notion of measuring distances. *Directed cutoff distance* was introduced in SCIP 6.0 [14], other measures such as *rotated cutoff distance* and *distance with bounds* were explored in [27], and *depth* was introduced in [23]. Measures based on non-distance arguments are also prevalent in cut selection, albeit often as smaller weighted complements to a distance measure, see *objective parallelism* and *integer support* in [1,27], and enumeration of lattice points in [18].

In the presented distance measures of Sect. 1, an LP-optimal solution is used as a reference point. Note, however, that this optimum is not necessarily unique, with the set of minimisers frequently being a higher-dimensional face of the LP-feasible region due to dual degeneracy [13]. The optimal solution used in these degenerate cases inevitably has a large impact on cut generation [29], and therefore cut selection. This has been noted in previous research, with work such as [11] using multiple LP solutions from different LP random seeds to generate different sets of cuts. Additionally, the patent [2], proposes using a second LP optimal solution at the same cutting round to filter cuts derived from the original LP solution. They provide an example, where a second LP solution that prioritises integrality is found, which then filters all cuts that do not separate it. Dual degeneracy is one major aspect motivating some of the newly-proposed measures in our work.

Finally, analytic centers have been used in other aspects of MIP solving, namely, presolving [8], cut generation [12], branching [8], and heuristics [5,20], motivating the measures introduced in this paper.

3 Contributions and Methodology

The contribution of the present paper is threefold. First, we introduce new distance-derived cut quality measures, the most important of which utilise analytic centers, and analyse properties of interest for these measures in cases of dual degeneracy and infeasible projections. Secondly, we present an extensive set of computational experiments on the effectiveness of our new measures and those commonly found in practice, showing that the choice of cut selection measure does have a strong influence on root-node and tree-wide performance. Thirdly, we introduce a multi-output regression model that predicts a ranking of distance measures per instance from a set of root node features.

3.1 Analytic Center-Based Methods

We propose two new methods for measuring cut quality: *analytic efficacy*, and *analytic directed cutoff distance*. They are based on the analytic center of the polytope and of the optimal face, respectively. For a given bounded MIP formulation (P), the analytic center is unique and in the relative interior of the feasible region.

When a constraint $\mathbf{a}^\intercal \mathbf{x} \leq b$ is tight for any feasible solution or in the presence of equality constraints, the analytic center is not well-defined due to a log-barrier term being $+\infty$. In practice, algorithms relax all log-barriers with a fixed slack constant as long as constraints are imposed on solutions.

While the analytic center is invariant under affine transformations of Problem (P), it can change with reformulations. E.g., the analytic center can be shifted by the presence of redundant inequalities. For our work, we assume that the formulation has already been presolved by the MIP solver. Presolving includes tightening both variable bounds and constraints, and removing both redundant constraints and variables. It thereby limits the extent of such edge cases.

Analytic Efficacy. As opposed to using the $\mathbf{x}^{LP}$ extreme point returned by the MIP solver for `eff` calculations, we propose a new measure that uses the analytic center of the optimal face of the LP, $\mathbf{x}^F$. We define $\mathbf{x}^F$ by the following using the notation from Problem (P), where $\mathbf{A}_i$, $\mathbf{b}_i$, $\mathbf{u}_i$, and $\mathbf{l}_i$ are the i-th row or i-th entry of their respective matrix or vector:

$$\mathbf{x}^F := \underset{\mathbf{x}\,:\,\mathbf{c}^\intercal\mathbf{x}=\mathbf{c}^\intercal\mathbf{x}^{LP}}{\operatorname{argmin}} \{-\sum_{i=1}^{m}\log(\mathbf{b}_i - \mathbf{A}_i\mathbf{x}) - (\sum_{i=1}^{n}\log(\mathbf{x}_i - \mathbf{l}_i) + \log(\mathbf{u}_i - \mathbf{x}_i))\}$$

In practice, we find $\mathbf{x}^F$ using the barrier (or interior point) algorithm on the LP relaxation without crossover, see [8]. This algorithm is available in all modern MIP solvers, and is often run concurrently to the simplex at the root node, making our cut selection algorithms of practical interest for MIP solving.

The obtained center $\mathbf{x}^F$ is different from $\mathbf{x}^{LP}$ only if the current problem presents dual degeneracy. The purpose of evaluating cuts with respect to how much they separate the center of the optimal face is to favour those which cut a greater part or potentially all of the optimal face, thereby more likely favouring an improvement in the dual bound. Compare Figs. 1a and 1b for the intuition behind analytic efficacy.

Analytic Directed Cutoff Distance. This measure is inspired by directed cutoff distance, and uses the analytic center of the feasible region, $\mathbf{x}^C$, as opposed to the best incumbent solution $\hat{\mathbf{x}}$. Using the same notation as in Sect. 3.1, we define the analytic center of the feasible region, $\mathbf{x}^C$, as:

$$\mathbf{x}^C := \underset{\mathbf{x}}{\operatorname{argmin}}\{-\sum_{i=1}^{m}\log(\mathbf{b}_i - \mathbf{A}_i\mathbf{x}) - (\sum_{i=1}^{n}\log(\mathbf{x}_i - \mathbf{l}_i) + \log(\mathbf{u}_i - \mathbf{x}_i))\}$$

$\mathbf{x}^C$ is often interpreted as the "central-most" point of the polytope and can be computed efficiently by dropping the objective function $\mathbf{c}^\intercal\mathbf{x}$ and solving the resulting LP relaxation using the barrier algorithm without crossover. The motivation is that the incumbent is not necessarily representative of the feasible set of the MIP since it can be any point of the feasible set generated by heuristics. Furthermore, it introduces an additional source of variability in the cut selection since the best primal solution is prone to updates, particularly during root node cutting when many primal heuristics are employed. By contrast, $\mathbf{x}^C$ is unique and deterministically determined for the LP relaxation. Compare Figs. 1b and 1d for the intuition behind analytic directed cutoff distance.

For computational efficiency, we further introduce *approximate analytic directed cutoff distance*, which re-uses the analytic center, $\mathbf{x}^C$, from the previous separation round provided it is still LP-feasible. This is motivated by the intuition that the analytic center, as the "central-most" point, is rarely separated and remains close to the new analytic center after cuts have been added.

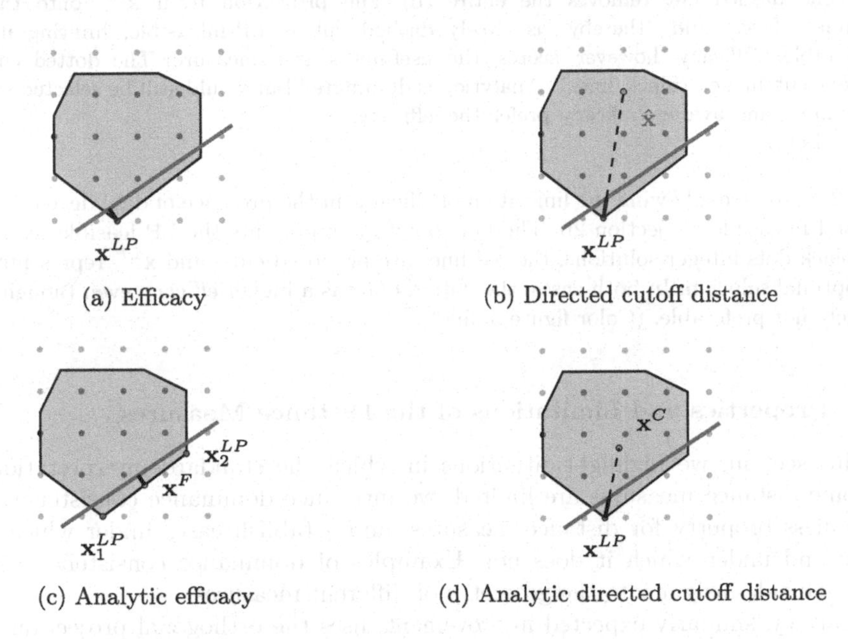

(a) Efficacy (b) Directed cutoff distance

(c) Analytic efficacy (d) Analytic directed cutoff distance

Fig. 1. A visualisation of distance measures. Note that in Fig. 1c we see that there are two alternative optimal vertices.

3.2 Multiple LP Solutions

We also introduce two distance measures, which mitigate cases of dual degeneracy and do not rely on analytic centers, but rather rely on multiple LP-optimal vertices. Let $\mathcal{X}^{LP}$ be a set of LP optimal solutions. We propose the measures *average efficacy*, denoted `avgeff`, and *minimum efficacy*, denoted `mineff`, which respectively take the average and minimum efficacy over all LP solutions in $\mathcal{X}^{LP}$.

$$\texttt{avgeff}(\boldsymbol{\alpha}, \beta, \mathcal{X}^{LP}) := \sum_{\mathbf{x}^{LP} \in \mathcal{X}^{LP}} \frac{\texttt{eff}(\boldsymbol{\alpha}, \beta, \mathbf{x}^{LP})}{|\mathcal{X}^{LP}|}$$

$$\texttt{mineff}(\boldsymbol{\alpha}, \beta, \mathcal{X}^{LP}) := \min\{\texttt{eff}(\boldsymbol{\alpha}, \beta, \mathbf{x}^{LP}) \mid \mathbf{x}^{LP} \in \mathcal{X}^{LP}\}$$

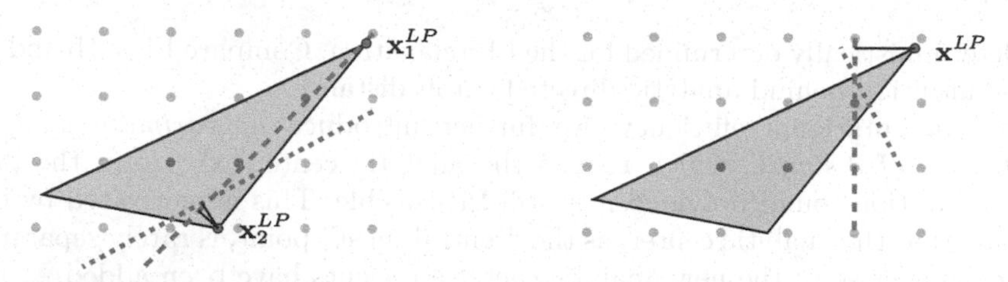

(a) The dashed cut removes the entire optimal face, and thereby is likely preferable. Efficacy however scores the dotted cut higher (black lines). Analytic, minimum, and average efficacy prefer the dashed cut.

(b) The projection from $\mathbf{x}^{LP}$ onto the dashed cut is LP-infeasible, limiting its usefulness as a measure. The dotted cut is dominated but would still be selected by efficacy.

Fig. 2. Two cases showing the limitation of efficacy, in the presence of dual degeneracy 2a, and infeasible projection 2b. The blue polytope represents the LP feasible region, the black dots integer solutions, the red lines are proposed cuts, and $\mathbf{x}^{LP}$ represents a LP optimal solution. In both cases, the dotted cut has a higher efficacy even though it is likely not preferable. (Color figure online)

3.3 Properties and Limitations of the Distance Measures

In this section, we highlight situations in which the standard interpretations of some distance measures are limited; we introduce dominance consistency, a soundness property for distance measures, and establish cases under which it holds and under which it does not. Examples of dominance consistency also offer valuable insight into the geometry of different measures.

Efficacy, similarly expected improvement, uses the orthogonal projection of $\mathbf{x}^{LP}$ onto the cut to measure distance. Unlike measures such as directed cutoff distance, the projection point might not be LP feasible, potentially making efficacy non-representative of the strength of the cut. A larger efficacy does not necessarily correspond to a larger part of the polyhedron being cut off nor to a better improvement in dual bound, see Fig. 2a. Note that minimum and analytic efficacy would not assign the dotted cut a positive score. Efficacy would prefer the dotted cut even though it does not improve the dual bound while the dashed cut does. Figure 2b visualises a basic example of when the orthogonal projection used for efficacy is LP-infeasible and thus not necessarily a good proxy for cut quality. Analytic efficacy, minimum efficacy, and average efficacy help overcome some limitations of efficacy, namely the dependence on the vertex returned by the LP solve. They are, however, equivalent when the current relaxation is not dual-degenerate since they then compute the distance to the cut using the same, unique optimal vertex.

Directed cutoff distance heavily depends on the incumbent solution, which is typically obtained from a primal heuristic.[2] This primal solution may be suboptimal and near a corner of the feasible region, biasing cuts in an unfavourable direction. Additionally, the primal solution may be LP-infeasible for local relaxations of the branch-and-bound tree, reducing the applicability of directed cutoff distance.

Ultimately, one would like a distance measure to be a surrogate for solver efficiency, in our case: the size of the search space in terms of branch and bound nodes. Such a measure is impossible to quantify however, with solution fractionality being insufficient, and the closest analogue, 'strong cutting', which measures the dual bound improvement of an added cut, see [21], being computationally intractable. The holy grail of cut selection is to either identify a computationally tractable measure that is a reasonably good proxy for solver performance, or to learn to adaptively select a suitable measure based on the input instance.

In the following, we will formalise these considerations. Therefore, we first recall the basic concepts of cut dominance and feasible rays. A cut $(A) = (\boldsymbol{\alpha}_A, \beta_A)$ *dominates* another cut $(B) = (\boldsymbol{\alpha}_B, \beta_B)$ if all points of the polytope cut by (A) are cut by (B) and there exists a point cut by (A) not cut by (B). We highlight that this definition of dominance is more general than, e.g., [28, Definition 9.2.1] since it only requires dominance to hold in the polytope and not in the whole space or the positive orthant. We define a *feasible ray* as a ray $\mathbf{r}$ starting from an LP-feasible point $\mathbf{x}$ for which there exists $\lambda > 0$ such that $\mathbf{x} + \lambda \mathbf{r}$ is LP-feasible. Such a ray always exists by convexity if the polyhedron is not a single point.

Definition 1 (Dominance consistency). *Given a MIP* (P) *and a relaxation point to cut off* $\mathbf{x}$, *a distance measure noted* $d(\mathbf{x}, \boldsymbol{\alpha}_X, \beta_X)$ *is* dominance-consistent *w.r.t. a set of cuts iff for any cut* (A) *and* (B) *in the set,* $d(\mathbf{x}, \boldsymbol{\alpha}_A, \beta_A) > d(\mathbf{x}, \boldsymbol{\alpha}_B, \beta_B)$ *implies that* (A) *is not dominated by* (B).

Note that with this definition, for a given MIP and LP solution $\mathbf{x}$, one set of cuts (e.g., from one separation round) can imply dominance consistency for a measure while another set of cuts (e.g., from another separation round) might not. We will see that for some measures, dominance consistency applies for all sets of cuts, all relaxation points, and all MIPs. Dominance consistency is a desirable property for a cut selection measure since a fully dominated cut will systematically be inferior to the dominating cut from the cut strength perspective.[3]

Proposition 1 (Consistency of Euclidean distance measures). *All measures consisting of the Euclidean distance of a given point* $\mathbf{x}$ *to the cut hyperplane are dominance-consistent with respect to any set of cuts if, for any two cuts in the set, the cut with the smallest distance measure cuts off* $\mathbf{x}$ *and the projection of* $\mathbf{x}$ *onto its hyperplane is LP-feasible.*

[2] At the root node, in particular, the incumbent – if existing – will always come from a heuristic, otherwise there would be no more cut rounds.

[3] Note that we disregard other cut properties such as density, numerical stability, or orthogonality here.

Proof. Let (A) and (B) be the two cuts of the set. The proof directly applies to more cuts by induction. We assume w.l.o.g. that $d(\mathbf{x}, \boldsymbol{\alpha}_A, \beta_A) > d(\mathbf{x}, \boldsymbol{\alpha}_B, \beta_B)$. The set of cuts with distance measure $d(\mathbf{x}, \cdot, \cdot) = d(\mathbf{x}, \boldsymbol{\alpha}_A, \beta_A)$ is a subset of the hyperplanes tangent to the sphere of radius $d(\mathbf{x}, \boldsymbol{\alpha}_A, \beta_A)$ centered at $\mathbf{x}$. The cut (B) does not separate the projection of $\mathbf{x}$ onto its own hyperplane, which is by assumption LP-feasible. The projected point lies on the sphere of radius $d(\mathbf{x}, \boldsymbol{\alpha}_B, \beta_B)$ centered at $\mathbf{x}$, and is contained in the open ball of radius $d(\mathbf{x}, \boldsymbol{\alpha}_A, \beta_A)$ centered at $\mathbf{x}$. As a tangent hyperplane to the sphere of radius $d(\mathbf{x}, \boldsymbol{\alpha}_A, \beta_A)$, (A) therefore cuts off the projected point by at least $d(\mathbf{x}, \boldsymbol{\alpha}_A, \beta_A) - d(\mathbf{x}, \boldsymbol{\alpha}_B, \beta_B)$ and can therefore not be dominated. □

Proposition 1 directly applies to efficacy and analytic efficacy, with the key restriction that the projection of the relaxation point onto the cut must be LP-feasible, which excludes a majority of real instances. Furthermore, since `eff` is a linear function of $\mathbf{x}^{LP}$, the property also applies to `avgeff` as a distance measure, with the point to project $\mathbf{x}$being the average of the multiple LP solutions. Finally, we can construct counter-examples where dominance-consistency does not hold for measures based on the Euclidean projection of a point onto the cut hyperplane in cases where the projection is not LP-feasible, as shown in Fig. 2b.

Proposition 2 (Consistency of the minimum efficacy). *Given the set of LP solutions* $\mathcal{X}^{LP}$*, we define the active solutions for a cut* $(\boldsymbol{\alpha}, \beta)$ *as the subset* $\arg\min_{\mathbf{x} \in \mathcal{X}^{LP}} \texttt{eff}(\boldsymbol{\alpha}, \beta, \mathbf{x})$. `mineff` *is dominance-consistent with respect to a set of cuts if for any two cuts* (A) *and* (B) *such that* (B) *has a strictly lower* `mineff`*, there exists an active solution* $\mathbf{x}_0$ *separated by* (B) *such that its projection onto the hyperplane of* (B) *is LP-feasible.*

Proof. Similarly to Proposition 1, the cut (B) forms a tangent hyperplane to the sphere centered at $\mathbf{x}_0$ and of radius equal to the score of (B). That point is not separated by (B) itself but (A) has to separate it by at least the difference in score between (A) and (B). □

Proposition 3 (Consistency of directed distance measures). *All measures based on the distance of a point* $\mathbf{x}$ *to the cut in the direction of an LP-feasible point* $\hat{\mathbf{x}}$ *are dominance-consistent w.r.t any set of cuts which all separate* $\mathbf{x}$.

Proof. All points on the segment $[\mathbf{x}, \hat{\mathbf{x}}]$ are LP-feasible. Let (A) and (B) be two cuts of the set such that (A) has a strictly greater measure value. Since the measure corresponds to the length of the segment cut off by cuts, some points are separated by (A) only, (B) cannot dominate (A). □

Proposition 3 notably applies to directed cutoff distance and (approximate) analytic directed cutoff distance. We note that dominance-consistency does not extend to `exp-improv`, even with feasible LP projections, with a counterexample visualised in Fig. 3.

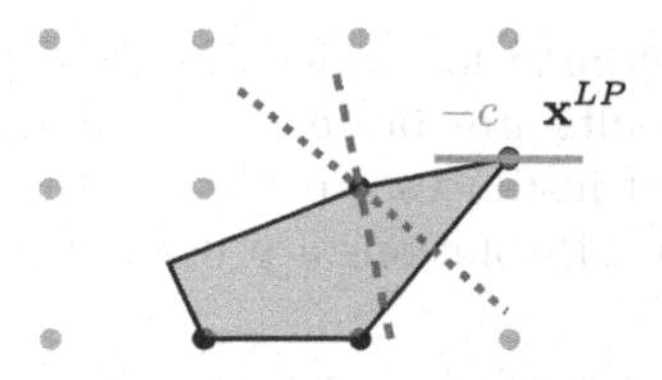

Fig. 3. Two cuts with LP-feasible orthogonal projections of $\mathbf{x}^{LP}$. The dotted cut is dominated, but has a better `exp-improv` score.

3.4 Multi-output Regression

Machine Learning for MIP has mainly focused on classification tasks, e.g. should an algorithm be run with option A or B, or run at all. In our case, the output of interest is the (relative) advantage of distance measures in terms of some performance criterion. For some instances, different measures may result in (near-)identical performance, e.g. if the same, "obvious" subset of cuts is selected by all methods. Classifying these (near-)ties with one distance measure or choosing an arbitrary threshold to classify a measure as well-performing could prevent the model from fitting well on the important data points where the selection method is significant. We therefore pose our learning task as a multi-output regression that predicts the relative performance of each method. We aim at an interpretable model to predict the preferred measure, with the goal of outperforming individual distance measures.

The feature space used as input to our model consists of: *dual degeneracy* (fraction of non-basic variables with zero-reduced cost), *primal degeneracy* (fraction of basic variables at their respective bounds), *solution fractionality* (fraction of integer variables with fractional LP values), *thinness* (fraction of equality constraints), and *density* (fraction of non-zero entries in constraint matrix). All features are obtainable at the root node before the separation process begins, are relevant to the separation process, and are easy to retrieve.

4 Experiments

We perform experiments on the MIPLIB 2017 collection set[4] [15], which we will now simply refer to as MIPLIB. We define a run as an instance random-seed pair for which we enforce exactly 50 separation rounds at the root node, with a maximum of 10 cuts to be added per round. We use default separators, but increase the amount of cuts that can be generated. Additionally, in order to reduce the variability of the solving process, restarts are disabled, no cuts are allowed to be added after the root node, and the best available MIPLIB solution is provided. All other aspects of the solver are untouched, with our experiments only replacing the cut scoring function in SCIP, not the selection algorithm itself, see [26]. The primary motivation is to assess the performance of different distance-based

[4] MIPLIB 2017 – The Mixed Integer Programming Library https://miplib.zib.de/.

cut measures, and to determine for which instance characteristics a distance measure is effective. All results are obtained by averaging results over the SCIP random seeds $\{1, 2, 3\}$, and instances are filtered subject to Table 1, with 162 instances remaining. Three LP solutions are used for `mineff` and `avgeff`.

Table 1. Criteria for which we removed instances from the MIPLIB collection and percentages of instances affected by each criterion

Criteria	% of instances removed
Tags: *feasibility, numerics, infeasible, no solution*	4.5%, 17.5%, 2.8%, 0.9%
Unbounded objective, MIPLIB solution unavailable	0.9%, 2.6%
Root optimal (any measure)	13.3%
Root node with separation rounds longer than 600s (any measure)	13.4%
No optional cuts generated (all measures)	2.3%
Numerical issues (any measure)	1.2%
Failed to prove optimality in branch and cut within 7200 s s (all measures)	21.7%

For all experiments, SCIP 8.0.2 [9] is used, with PySCIPOpt [19] as the API, and Xpress 8.14 [10] as the LP solver. All experiments are run on a cluster equipped with Intel Xeon Gold 5122 CPUs with 3.60 GHz and 96 GB main memory. The code used for all experiments is available and open-source[5]. The structure of this section is as follows. In Subsect. 4.1, we present results of our distance-based cut measures on root node restricted runs. In Subsect. 4.2 we present results of our distance measures generalised to branch and cut. Finally, in Subsect. 4.3 we present the performance of our support vector regression model on selecting distance measures.

4.1 Root Node Results

Table 2. Summary of all distance measures

Function	Measure	Description
`eff`	Efficacy	See Sect. 1
`dcd`	Directed cutoff distance	See Sect. 1
`a-eff`	Analytic Efficacy	See Subsect. 3.1
`a-dcd`	Analytic directed cutoff distance	See Subsect. 3.1
`app-a-dcd`	Approximate analytic directed cutoff distance	See Subsect. 3.1
`avgeff`	Average efficacy	See Subsect. 3.2
`mineff`	Minimum efficacy	See Subsect. 3.2
`exp-improv`	Expected Improvement	See Sect. 1

[5] https://github.com/Opt-Mucca/Analytic-Center-Cut-Selection.

Table 2 provides a summary of all cut selection measures we evaluated. We compare head-to-head results on the primal-dual difference after 50 separation rounds. We say that a scoring measure has outperformed another for an instance, if it is at least as good over all random seeds, and strictly better for at least one. Curiously, we observe a clear hierarchy of distance-based cut measures for root-restricted dual bound performance over MIPLIB. That is, `a-dcd` $\geq$ `app-a-dcd` $\geq$ `a-eff` $\geq$ `mineff` $\geq$ `avgeff` $\geq$ `dcd` $\geq$ `eff` $\geq$ `exp-improv`.

Table 3. Entry coordinate (i, j) is a tuple of win/loss percentage over all instances for dual bound improvement of measure i over measure j. A win is defined by at least as good performance over all seeds, and better performance for at least one seed.

	a-dcd	app-a-dcd	a-eff	mineff	avgeff	eff	dcd	exp-improv
a-dcd	-	**0.29/0.16**	**0.32/0.3**	0.29/0.29	**0.32/0.26**	**0.38/0.22**	**0.4/0.21**	**0.47/0.22**
app-a-dcd	*0.16/0.29*	-	**0.3/0.3**	**0.3/0.27**	**0.3/0.27**	**0.38/0.22**	**0.37/0.23**	**0.43/0.25**
a-eff	*0.3/0.32*	*0.3/0.3*	-	**0.25/0.23**	**0.22/0.2**	**0.31/0.14**	**0.37/0.26**	**0.41/0.23**
mineff	0.29/0.29	*0.27/0.3*	*0.23/0.25*	-	**0.14/0.13**	**0.28/0.12**	**0.32/0.26**	**0.36/0.24**
avgeff	*0.26/0.32*	*0.27/0.3*	*0.2/0.22*	*0.13/0.14*	-	**0.27/0.12**	**0.31/0.25**	**0.38/0.24**
eff	*0.22/0.38*	*0.22/0.38*	*0.14/0.31*	*0.12/0.28*	*0.12/0.27*	-	**0.26/0.25**	**0.32/0.25**
dcd	*0.21/0.4*	*0.23/0.37*	*0.26/0.37*	*0.26/0.32*	*0.25/0.31*	*0.25/0.26*	-	**0.36/0.26**
exp-improv	*0.22/0.47*	*0.25/0.43*	*0.23/0.41*	*0.24/0.36*	*0.24/0.38*	*0.25/0.32*	*0.26/0.36*	-

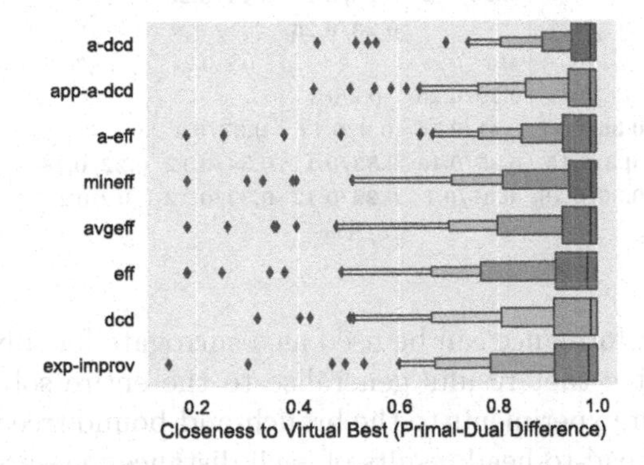

Fig. 4. Boxenplots of distance measures root-node performance. Each instance compared to the virtual best and averaged over random seeds.

A head-to-head comparison, while helpful for ranking measures, contains limited information on the distribution of performance. For this reason, we also visualise the primal-dual difference results using boxenplots in Fig. 4, see [16] for a description. Figure 4 shows a comparison of each method against the so-called *virtual best*. Therefore, we divide, instance by instance, the average gap (over all seeds) by the best average gap among the eight methods. We observe similar results to Table 3 in that `a-dcd`, `app-a-dcd`, and `a-eff` are superior to other methods in terms of dual bound improvement. It should be noted, however, that on average `a-eff` takes 32% of the root node processing time and `a-dcd` takes

25%. This is in contrast to eff, which only takes 0.8%. We conclude that using an analytic center for cut selection is beneficial for closing the primal-dual gap during root node cutting.

4.2 Branch and Bound Generalisation

Table 4. Entry coordinate (i, j) is a tuple of win/loss percentage over all instances for measure i over measure j. A win is defined by at least as good performance over all seeds, and better performance for at least one seed.

	a-dcd	app-a-dcd	a-eff	mineff	avgeff	eff	dcd	exp-improv
a-dcd	-	**0.17/0.16**	**0.29/0.17**	**0.25/0.17**	**0.23/0.19**	**0.27/0.17**	**0.23/0.17**	**0.28/0.15**
app-a-dcd	*0.16/0.17*	-	*0.2/0.22*	**0.23/0.2**	*0.19/0.27*	**0.22/0.19**	*0.21/0.22*	**0.23/0.18**
a-eff	*0.17/0.29*	**0.22/0.2**	-	0.15/0.15	0.2/0.2	0.17/0.17	**0.2/0.19**	**0.18/0.17**
mineff	*0.17/0.25*	*0.2/0.23*	0.15/0.15	-	*0.11/0.17*	**0.17/0.1**	*0.17/0.18*	**0.18/0.15**
avgeff	*0.19/0.23*	**0.27/0.19**	0.2/0.2	**0.17/0.11**	-	**0.17/0.14**	**0.21/0.19**	**0.23/0.22**
eff	*0.17/0.27*	*0.19/0.22*	0.17/0.17	*0.1/0.17*	*0.14/0.17*	-	*0.16/0.2*	*0.19/0.2*
dcd	*0.17/0.23*	**0.22/0.21**	*0.19/0.2*	**0.18/0.17**	*0.19/0.21*	**0.2/0.16**	-	**0.19/0.17**
exp-improv	*0.15/0.28*	*0.18/0.23*	*0.17/0.18*	*0.15/0.18*	*0.22/0.23*	**0.2/0.19**	*0.17/0.19*	-

(a) Number of nodes

	a-dcd	app-a-dcd	a-eff	mineff	avgeff	eff	dcd	exp-improv
a-dcd	-	**0.37/0.14**	**0.33/0.16**	**0.28/0.15**	**0.28/0.17**	*0.22/0.3*	*0.16/0.3*	*0.19/0.29*
app-a-dcd	*0.14/0.37*	-	**0.23/0.2**	**0.25/0.23**	**0.24/0.23**	*0.17/0.35*	*0.15/0.3*	*0.16/0.36*
a-eff	*0.16/0.33*	*0.2/0.23*	-	**0.23/0.23**	*0.26/0.29*	*0.14/0.41*	*0.15/0.35*	*0.1/0.37*
mineff	*0.15/0.28*	*0.23/0.25*	*0.23/0.23*	-	0.2/0.2	*0.12/0.4*	*0.11/0.33*	*0.12/0.29*
avgeff	*0.17/0.28*	*0.23/0.24*	**0.29/0.26**	0.2/0.2	-	*0.1/0.37*	*0.12/0.34*	*0.12/0.32*
eff	**0.3/0.22**	**0.35/0.17**	**0.41/0.14**	**0.4/0.12**	**0.37/0.1**	-	*0.18/0.22*	0.2/0.2
dcd	**0.3/0.16**	**0.3/0.15**	**0.35/0.15**	**0.33/0.11**	**0.34/0.12**	**0.22/0.18**	-	**0.21/0.2**
exp-improv	**0.29/0.19**	**0.36/0.16**	**0.37/0.1**	**0.29/0.12**	**0.32/0.12**	0.2/0.2	*0.2/0.21*	-

(b) Time

While root node performance can be used as a surrogate for solver performance, there is no guarantee that results generalise to the entire solving process. We therefore extend our experiments to the branch-and-bound tree with a time limit of two hours. The head-to-head results of each distance measure for the number of nodes and solve time are displayed in Table 4. For the number of nodes, note that we removed all instances where a measure timed out. We observe in the number of nodes comparison that a-dcd remains the superior method, however the ordering of methods is now less clear. Most interesting is the drop in performance of app-a-dcd compared to the root node results, which suggests that the analytic center from previous separation rounds is often not a good direction for distance measures. This is supported by the fact that 26.5% of the analytic centers from previous rounds are LP infeasible. In the solve time comparison, we see that the 'cheaper' measures, eff, dcd, exp-improv, which require no additional LP solver calls, are superior over the more 'expensive' measures. This suggests that while our introduced methods, especially a-dcd, can reduce the number of nodes and have better root node performance, the total solve time is not similarly improved. We note that while dcd is superior in

Table 4b, we believe that our experimental design is overly favourable since we start with an optimal MIPLIB solution (or best known, for unsolved instances).

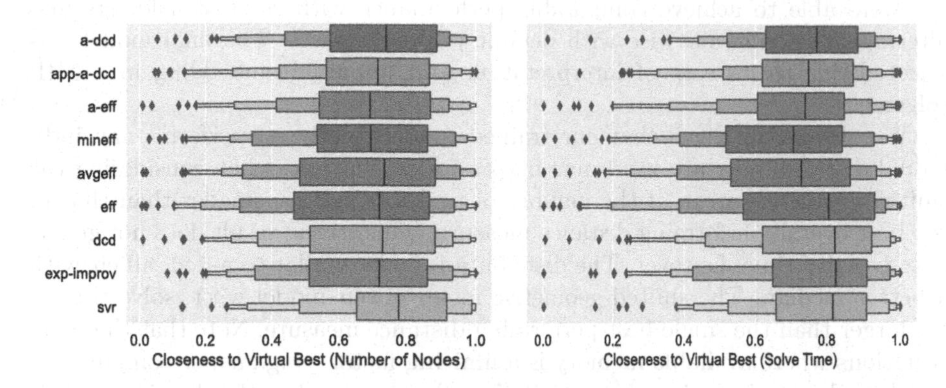

Fig. 5. Boxenplots of measures' tree performance. Nodes (left), time (right).

Similarly to Subsect. 4.1, we visualise an instance-wide comparison to the virtual best of all measures for number of nodes and solve time in Fig. 5. The results confirm our conclusion from Table 4a that a-dcd is the best performing measure, and that eff is the worst one w.r.t. number of nodes. All other measures, however, have similar distributions, making stronger conclusions difficult. We note that 90.4% of cuts have infeasible projections when scoring by eff, and that previous studies, see [13], identify an 87.5% occurrence rate of some level of dual degeneracy of the final root node LP in standard benchmark instances, confirming the practical geometric limitations of efficacy presented in Sect. 3.3. For solve time, we observe the improved performance of 'cheaper' methods through their relatively high median values. We also observe that a-dcd has the smallest performance variability of all measures, while the standard eff has the largest performance variability. This implies that using an analytic center for cut selection can help to reduce performance variability, an interesting observation by its own.

4.3 Regression Model Results

We have thus far observed that our newly introduced measures, especially a-dcd, have superior root node dual bound performance than traditional measures, and often result in smaller branch-and-bound trees. No single measure is however dominant, as seen in Tables 4a and 4b, with no single measure ever having less than 10% of instances as wins in the head-to-head contest. This motivates the need for an adaptive method, which decides on a distance measure at the start of the solving process that will best perform on the instance.

We use support vector regression (SVR) with a cubic kernel function, see [24], implemented in scikit-learn [22] with default parameters. We train on instance-seed pairs, with the virtual best number of nodes for each pair divided by the number of nodes under the distance measure as output. Our model was trained using 5-fold cross-validation, with 10% of pairs retained for validation. We were able to achieve comparable performance with regression forests and alternative kernels, likewise with default parameter sets. The final model was selected due to its ease of interpretation and potential embedding in a MIP solver.

We observe in Fig. 5 that our trained model clearly outperforms any individual distance measure w.r.t. number of nodes. Further, when considering the shifted geometric mean of the number of nodes, it is 12% smaller than that of the best overall performing distance measure. This strong result does not generalise to solve time, however. The distribution looks similar to a-dcd, albeit with a better median. The shifted geometric mean of our model w.r.t. solve time is 8% larger than the single best-performing distance measure. Note that there are situations where available memory is a limiting factor – e.g., super-computing – which makes node savings important. Finally, we visualise the decision boundaries of the trained model over the two first principal components from a PCA of the original features, maintaining 71% of the explained variance. We determine decision boundaries with the largest regression value over all distances and visualise the result in Fig. 6, with the component equations printed below.

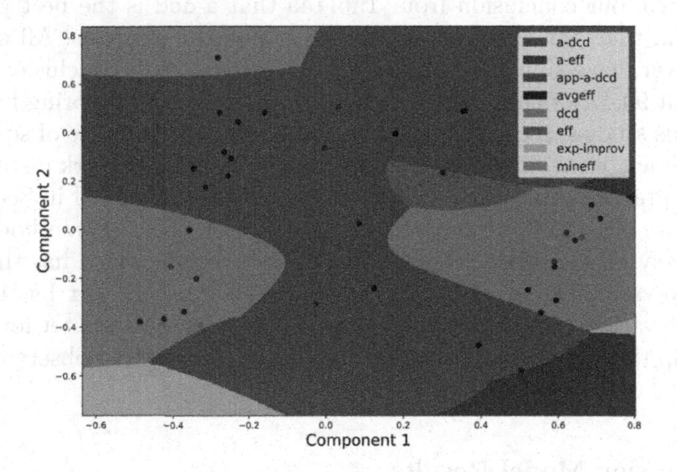

Fig. 6. Decision regions in transformed feature space. Dots are validation instances, with their opacity the relative performance of the predicted measure. Component 1: 0.947 dual_deg − 0.205 primal_deg + 0.22 frac − 0.089 thin − 0.063 density Component 2: − 0.27 dual_deg − 0.733 primal_deg + 0.467 frac − 0.256 thin + 0.326 density

5 Conclusion

In this paper, we reassessed the question of cut selection through the lens of distance measures. Motivated by geometric properties of polyhedra encountered in MIPs, we defined measures based on analytic centers and multiple LP solutions. We showed their performance, and more importantly, that the relative performance of distance measures can be learned for new instances with an interpretable and implementable model. We found that the introduced measures help to reduce root node gap, size of the branch and bound tree and performance variability. The focus of our work is on the improved evaluation of individual cuts; future directions will build upon these measures to enhance the whole separation process, incorporating combinations of cut measures, and generalising the promising node reductions to improved runtime.

Acknowledgements. The work for this article has been conducted in the Research Campus MODAL funded by the German Federal Ministry of Education and Research (BMBF) (fund numbers 05M14ZAM, 05M20ZBM). The described research activities are funded by the Federal Ministry for Economic Affairs and Energy within the project UNSEEN (ID: 03EI1004-C).

References

1. Achterberg, T.: Constraint integer programming. Ph.D. thesis, TU Berlin (2007)
2. Achterberg, T.: LP relaxation modification and cut selection in a MIP solver. US Patent US8463729B2 (2013). https://patents.google.com/patent/US8463729B2/en
3. Achterberg, T., Wunderling, R.: Mixed integer programming: analyzing 12 years of progress. In: Jünger, M., Reinelt, G. (eds.) Facets of Combinatorial Optimization, pp. 449–481. Springer, Cham (2013). https://doi.org/10.1007/978-3-642-38189-8_18
4. Andreello, G., Caprara, A., Fischetti, M.: Embedding {0, 1/2}-cuts in a branch-and-cut framework: a computational study. Informs J. Comput. **19**(2), 229–238 (2007)
5. Baena, D., Castro, J.: Using the analytic center in the feasibility pump. Oper. Res. Lett. **39**(5), 310–317 (2011). https://doi.org/10.1016/j.orl.2011.07.005. https://www.sciencedirect.com/science/article/pii/S0167637711000824
6. Balcan, M.F.F., Prasad, S., Sandholm, T., Vitercik, E.: Sample complexity of tree search configuration: cutting planes and beyond. In: Advances in Neural Information Processing Systems, vol. 34 (2021)
7. Baltean-Lugojan, R., Bonami, P., Misener, R., Tramontani, A.: Scoring positive semidefinite cutting planes for quadratic optimization via trained neural networks. Optimization-online preprint 2018/11/6943 (2019)
8. Berthold, T., Perregaard, M., Mészáros, C.: Four good reasons to use an interior point solver within a MIP solver. In: Kliewer, N., Ehmke, J.F., Borndörfer, R. (eds.) Operations Research Proceedings 2017. ORP, pp. 159–164. Springer, Cham (2018). https://doi.org/10.1007/978-3-319-89920-6_22
9. Bestuzheva, K., et al.: The SCIP Optimization Suite 8.0 (2021)
10. FICO Xpress Optimization. https://www.fico.com/en/products/fico-xpress-optimization. Accessed 10 Nov 2022
11. Fischetti, M., Lodi, A., Monaci, M., Salvagnin, D., Tramontani, A.: Improving branch-and-cut performance by random sampling. Math. Program. Comput. **8**(1), 113–132 (2016)

12. Fischetti, M., Salvagnin, D.: Yoyo search: a bisection cutting-plane method (2009)
13. Gamrath, G., Berthold, T., Salvagnin, D.: An exploratory computational analysis of dual degeneracy in mixed-integer programming. EURO J. Comput. Optim. **8**(3–4), 241–261 (2020)
14. Gleixner, A., et al.: The SCIP Optimization Suite 6.0. Technical report. 18–26, ZIB, Takustr. 7, 14195 Berlin (2018)
15. Gleixner, A., et al.: MIPLIB 2017: data-driven compilation of the 6th mixed-integer programming library. Math. Program. Comput. 1–48 (2021)
16. Hofmann, H., Kafadar, K., Wickham, H.: Letter-value plots: boxplots for large data. Technical report, had.co.nz (2011)
17. Huang, Z., et al.: Learning to select cuts for efficient mixed-integer programming. arXiv preprint arXiv:2105.13645 (2021)
18. Lodi, A., Pesant, G., Rousseau, L.-M.: On counting lattice points and Chvátal-Gomory cutting planes. In: Achterberg, T., Beck, J.C. (eds.) CPAIOR 2011. LNCS, vol. 6697, pp. 131–136. Springer, Heidelberg (2011). https://doi.org/10.1007/978-3-642-21311-3_13
19. Maher, S., Miltenberger, M., Pedroso, J.P., Rehfeldt, D., Schwarz, R., Serrano, F.: PySCIPOpt: mathematical programming in Python with the SCIP optimization suite. In: Greuel, G.-M., Koch, T., Paule, P., Sommese, A. (eds.) ICMS 2016. LNCS, vol. 9725, pp. 301–307. Springer, Cham (2016). https://doi.org/10.1007/978-3-319-42432-3_37
20. Naoum-Sawaya, J.: Recursive central rounding for mixed integer programs. Comput. Oper. Res. **43**, 191–200 (2014)
21. Paulus, M.B., Zarpellon, G., Krause, A., Charlin, L., Maddison, C.: Learning to cut by looking ahead: cutting plane selection via imitation learning. In: International Conference on Machine Learning, pp. 17584–17600. PMLR (2022)
22. Pedregosa, F., et al.: Scikit-learn: machine learning in Python. J. Mach. Learn. Res. **12**, 2825–2830 (2011)
23. Poirrier, L., Yu, J.: On the depth of cutting planes. arXiv preprint arXiv:1903.05304 (2019)
24. Smola, A.J., Schölkopf, B.: A tutorial on support vector regression. Stat. Comput. **14**(3), 199–222 (2004)
25. Tang, Y., Agrawal, S., Faenza, Y.: Reinforcement learning for integer programming: learning to cut. In: International Conference on Machine Learning, pp. 9367–9376. PMLR (2020)
26. Turner, M., Koch, T., Serrano, F., Winkler, M.: Adaptive cut selection in mixed-integer linear programming. arXiv preprint arXiv:2202.10962 (2022)
27. Wesselmann, F., Stuhl, U.: Implementing cutting plane management and selection techniques. Technical report, University of Paderborn (2012)
28. Wolsey, L.A.: Integer Programming. Wiley, Hoboken (2020)
29. Zanette, A., Fischetti, M., Balas, E.: Can pure cutting plane algorithms work? In: Lodi, A., Panconesi, A., Rinaldi, G. (eds.) IPCO 2008. LNCS, vol. 5035, pp. 416–434. Springer, Heidelberg (2008). https://doi.org/10.1007/978-3-540-68891-4_29

Handling Symmetries in Mixed-Integer Semidefinite Programs

Christopher Hojny[1(✉)] and Marc E. Pfetsch[2]

[1] Department of Mathematics and Computer Science, Eindhoven University of Technology, P.O. Box 513, 5600MB Eindhoven, The Netherlands
c.hojny@tue.nl

[2] Department of Mathematics, TU Darmstadt, Dolivostr. 15, 64293 Darmstadt, Germany
pfetsch@mathematik.tu-darmstadt.de

Abstract. Symmetry handling is a key technique for reducing the running time of branch-and-bound methods for solving mixed-integer linear programs. In this paper, we generalize the notion of (permutation) symmetries to mixed-integer semidefinite programs (MISDPs). We first discuss how symmetries of MISDPs can be automatically detected by finding automorphisms of a suitably colored auxiliary graph. Then known symmetry handling techniques can be applied. We demonstrate the effect of symmetry handling on different types of MISDPs. To this end, our symmetry detection routine is implemented in the state-of-the-art MISDP solver SCIP-SDP. We obtain speed-ups similar to the mixed-integer linear case.

Keywords: mixed-integer semidefinite programming · symmetry handling · branch-and-bound

1 Introduction

In this paper, we consider solving general Mixed-Integer Semidefinite Programs (MISDP) of the following form:

$$
\begin{aligned}
\inf \quad & b^\top y \\
\text{s.t.} \quad & \sum_{k=1}^{m} A^k y_k - A^0 \succeq 0, \\
& \ell_i \leq y_i \leq u_i \qquad && \forall i \in [m], \\
& y_i \in \mathbb{Z} && \forall i \in I,
\end{aligned}
\tag{1}
$$

with symmetric matrices $A^k \in \mathbb{R}^{n \times n}$ for $k \in [m]_0 := \{0, \dots, m\}$, $b \in \mathbb{R}^m$, $\ell_i \in \mathbb{R} \cup \{-\infty\}$, $u_i \in \mathbb{R} \cup \{\infty\}$ for all $i \in [m] := \{1, \dots, m\}$. The set of indices of integer variables is given by $I \subseteq [m]$. The notation $M \succeq 0$ indicates that a matrix M is positive semidefinite. Throughout this paper, we use the notation $A(y) := \sum_{k=1}^{m} A^k y_k - A^0$ for $y \in \mathbb{R}^m$.

© The Author(s), under exclusive license to Springer Nature Switzerland AG 2023
A. A. Cire (Ed.): CPAIOR 2023, LNCS 13884, pp. 69–78, 2023.
https://doi.org/10.1007/978-3-031-33271-5_5

One way to solve (1) is by SDP-based branch-and-bound, a special case of nonlinear branch-and-bound, see Dakin [5]. Here, branching on the integer variables creates a search tree and in each node a semidefinite program (SDP) is solved, which arises from the relaxation of the integrality requirements of that node. For more details on this approach see, e.g., [9,20].

Optimization problems (1) are quite general and, in particular, contain mixed-integer linear programs (MIPs) as a special case, where one uses only the diagonal entries of the matrices A^k, $k \in [m]_0$. MISDPs also have numerous applications, e.g., robust truss topology optimization with discrete bar diameters [19] and cardinality least squares [8,24].

The challenges for solving MIPs are inherited for the solution of MISDPs. This includes the presence of symmetries, which are defined as follows. Let

$$X = \{y \in \mathbb{R}^m : A(y) \succeq 0,\ \ell \leq y \leq u,\ y_i \in \mathbb{Z}\ \forall i \in I\}$$

be the feasible region of (1). A *symmetry* of (1) is a bijection $\pi\colon \mathbb{R}^m \to \mathbb{R}^m$ such that $X = \pi(X) := \{\pi(x) : x \in X\}$ and $b^\top \pi(x) = b^\top x$ for every $x \in X$. Thus, π maps feasible solutions of (1) to feasible solutions with the same objective value. The symmetries of (1) form the so-called *symmetry group*.

The presence of symmetries results in an unnecessarily large search tree, since many symmetric solutions have to be treated although they do not contain new information. This effect is well-known for MIPs, and many different techniques for handling symmetries in MIPs have been developed, see, e.g., [12,18,23] for an overview. Many of these methods are implemented in the solver SCIP [2,10].

As far as we know, symmetry handling for mixed-integer semidefinite programming has not been considered so far. For SDPs without integer variables, however, model reformulating techniques can be used to handle symmetries. The main idea is to aggregate a set of symmetric variables to a common variable that represents the average value of all symmetric variables, see, e.g., [1,6,11,13]. Moreover, symmetries in mixed-integer conic programming has been investigated very recently in [27]; note that the SDP cone is not mentioned there.

The goal of this paper is to generalize the techniques for MIPs to MISDPs and to investigate their computational impact. We first discuss how symmetries can be computed. In Sect. 2 permutations of the variables and so-called formulation symmetries of MISDPs are defined. In Sect. 3, we discuss how such symmetries can be computed through graph automorphisms. We conclude the paper in Sect. 4 with a numerical study of the impact of handling symmetries in MISDPs.

2 Computing Symmetries

Note that the definition of a symmetry above is based on the feasible region X, which is hard to handle in general (e.g., it is NP-hard to decide if $X = \varnothing$). In practice, one therefore often only considers permutations of variables that leave the description of X invariant. In the following, we generalize the corresponding definition for MIPs, see, e.g., Margot [18], to MISDPs.

Denote the (full) *symmetric group*, i.e., the set of all permutations of $[m]$, by $\mathcal{S}_m$. Then a permutation $\pi \in \mathcal{S}_m$ acts on $x \in \mathbb{R}^m$ by permuting its components, i.e., $\pi(x)_i = x_{\pi^{-1}(i)}$ for all $i \in [m]$. Thus, $\pi(x) := (x_{\pi^{-1}(1)}, \dots, x_{\pi^{-1}(m)})^\top$. Let $\sigma \in \mathcal{S}_n$ act on a matrix $A \in \mathbb{R}^{n \times n}$ as follows:

$$\sigma(A)_{ij} = A_{\sigma^{-1}(i),\sigma^{-1}(j)} \quad \forall i, j \in [n].$$

Definition 1. *A permutation $\pi \in \mathcal{S}_m$ of variables is a* formulation symmetry *of* (1) *if there exists a permutation $\sigma \in \mathcal{S}_n$ such that*

(P1) $\pi(I) = I$, $\pi(\ell) = \ell$, $\pi(u) = u$, and $\pi(b) = b$ (i.e., π leaves integer variables, variable bounds, and the objective coefficients invariant),
(P2) $\sigma(A^0) = A^0$ and, for all $i \in [m]$, $\sigma(A^i) = A^{\pi^{-1}(i)}$.

Thus, the variables are permuted by π and the matrices by σ.

Lemma 1. *Every formulation symmetry of* (1) *is a symmetry.*

Proof. Let $y \in \mathbb{R}^m$ and let $\pi \in \mathcal{S}_m$ be a formulation symmetry with corresponding matrix permutation $\sigma \in \mathcal{S}_n$. Since $\pi(b) = b$ and permutations are orthogonal maps, we find $b^\top \pi(y) = \pi^{-1}(b)^\top y = b^\top y$. Thus, π leaves the objective invariant. It remains to show that π also maps feasible solutions onto feasible solutions.

Note that

$$\sigma(A)(\pi(y)) = \sum_{k=1}^{m} \sigma(A^k)\pi(y)_k - \sigma(A^0) = \sum_{k=1}^{m} A^{\pi^{-1}(k)} y_{\pi^{-1}(k)} - A^0 \stackrel{(\star)}{=} \sum_{k'=1}^{m} A^{k'} y_{k'} - A^0 = A(y),$$

where $(\star)$ follows from π being a permutation of $[m]$. Consequently, since formulation symmetry π maps integer variables onto integer variables and respects the bounds of variables, y is feasible for (1) if and only if $\pi(y)$ is feasible. □

We note that we currently only treat symmetries in the above sense, but do not reduce symmetries in the SDP-formulations as described in the introduction, see, e.g., [1,6,13] for details; we leave this to future research.

3 Symmetry Detection

A common strategy for detecting formulation symmetries of MIPs is to construct a suitably colored graph whose color-preserving automorphisms correspond to formulation symmetries, see, e.g., [23,26]. We follow this line of research and present a colored graph to detect formulation symmetries of MISDPs.

Recall that each formulation symmetry π admits a permutation $\sigma \in \mathcal{S}_n$ with $\sigma(A^i) = A^{\pi^{-1}(i)}$ for all $i \in [m]$ as well as $\sigma(A^0) = A^0$. To model this matrix invariant, we first introduce some notation. For each matrix A^k, $k \in [m]_0$, let $N_k = \{(i,j)^k \in [n]\times[n] : A^k_{ij} \neq 0\}$ be the set of its non-zero entries, where the superscript at $(i,j)^k$ is used to distinguish non-zero entries of different matrices. For $p = (i,j)^k \in N_k$, let $r^k(p) = i$ be its row index and $c^k(p) = j$ be its column index. We define the symmetry detection graph $\mathcal{G} = (\mathcal{V}, \mathcal{E})$ as described next.

The graph needs to capture both the permutation of variables $\pi \in \mathcal{S}_m$ and the permutation of the matrix entries $\sigma \in \mathcal{S}_n$. For this reason, we consider the node set $\mathcal{V} = V \cup D \cup \bigcup_{k=0}^m N_k$, where $V := \{y_1, \dots, y_m\}$ and $D := [n]$ represent the variables and the "dimensions" of the matrices of the MISDP, respectively. Permutation π will then correspond to a permutation of the variable nodes V and σ will correspond to a permutation of the dimension nodes in D. The nodes in $N_0, \dots, N_k$ will make sure that both π and σ are correctly linked, which is achieved by adding appropriate edges.

The edge set $\mathcal{E}$ is partitioned into $E_V \cup E_R \cup E_C \cup E_P$, where

$$\begin{aligned}
E_V &= \{\{y_k, p\} : p \in N_k,\ k \in [m]\},\\
E_R &= \left\{\{p, r^k(p)\} : p \in N_k,\ k \in [m]_0\right\},\\
E_C &= \left\{\{p, c^k(p)\} : p \in N_k,\ k \in [m]_0\right\},\\
E_P &= \left\{\{(i,j)^k, (j,i)^k\} : (i,j)^k \in N_k,\ k \in [m]_0\right\}.
\end{aligned}$$

The edges in E_P are not necessary to encode formulation symmetries, however, we think that graph automorphism codes benefit from them since they allow to more easily recognize dependencies between different nodes. With this graph, permuting dimension nodes requires to also permute nodes corresponding to non-zero entries, which in turn might trigger a permutation of variable nodes and vice versa. To make sure that only nodes corresponding to the same problem information are mapped onto each other, we color the nodes and edges.

The variables are partitioned into groups, each with the same objective coefficient, lower & upper bound, and variable type (continuous or integer). Each of the groups is assigned a unique color and all variables within this group receive this color. Similarly, we define a unique color for the sets D and $\bigcup_{k=0}^m N_k$, and assign all nodes within such a set the corresponding color. That is, all nodes modeling a dimension of a matrix receive the same color and all nodes corresponding to matrix entries receive the same color. Finally, to also distinguish the entries of the matrices $A^0, \dots, A^m$, we color the edges in E_R according to their matrix coefficient, i.e., edge $\{p, r^k(p)\}$ gets color A^k_p. Similarly, we color the edges in E_C. To distinguish "row colors" from "column colors", we refer to the color of $\{p, c^k(p)\}$ as $\bar{A}^k_p$. All other edges remain uncolored.

A bijective map $\varphi\colon \mathcal{V} \to \mathcal{V}$ is an *automorphism* of $\mathcal{G}$ if it preserves adjacency, i.e., $\{u,v\} \in \mathcal{E}$ if and only if $\{\varphi(u), \varphi(v)\} \in \mathcal{E}$. We say that φ is *color-preserving* if, for every $v \in \mathcal{V}$, nodes $\varphi(v)$ and v have the same color and, for every $\{u,v\} \in \mathcal{E}$, edges $\{u,v\}$ and $\{\varphi(u), \varphi(v)\}$ have the same color.

Example 1. Consider the MISDP

$$\inf \left\{ y_1 + y_2 : \begin{pmatrix} 0&1&0\\ 1&0&0\\ 0&0&0 \end{pmatrix} y_1 + \begin{pmatrix} 0&0&0\\ 0&0&1\\ 0&1&0 \end{pmatrix} y_2 \succeq 0,\ 0 \le y_1,\ y_2 \le 1,\ y_1, y_2 \in \mathbb{Z} \right\}.$$

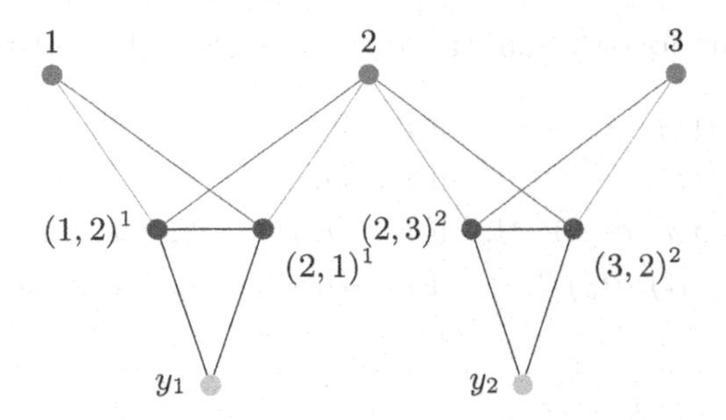

Fig. 1. Illustration of symmetry detection graph.

The corresponding symmetry detection graph is given in Fig. 1, where uncolored edges are drawn in black and $(i,j)^k$ denotes entry (i,j) of A^k. The only non-trivial color-preserving automorphism of the graph exchanges $y_1 \leftrightarrow y_2$, $(1,2)^1 \leftrightarrow (3,2)^2$, $(2,1)^1 \leftrightarrow (2,3)^2$, $1 \leftrightarrow 3$, and keeps node 2 fixed. This leads to the variable permutation π, which exchanges y_1 and y_2, and the matrix permutation σ, which exchanges 1 and 3.

We show that $\mathcal{G}$ captures all information about formulation symmetries. The *restriction* of φ to a set $B \subseteq \mathcal{V}$ is denoted $\varphi|_B$.

Proposition 1. *Let $\mathcal{G} = (\mathcal{V}, \mathcal{E})$ be constructed as described above.*

- *For each color-preserving automorphism φ of $\mathcal{G}$, $\varphi|_V$ is a formulation symmetry of* (1).
- *For every formulation symmetry π of* (1)*, there is a color-preserving automorphism φ such that $\varphi|_V = \pi$.*

Proof. For the first part, let φ be a color-preserving automorphism of $\mathcal{G}$. Define $\pi := \varphi|_V$ and $\sigma := \varphi|_D$. We claim that π is a formulation symmetry with corresponding matrix permutation σ. Since φ is a color-preserving automorphism, the image of π is V and the image of σ is D. Moreover, by the choice of colors, π can only map variables of the same type (integer/continuous, objective coefficient, upper and lower bounds) onto each other. Thus, π satisfies (P1). Moreover, for all $k \in [m]$ and $(i,j)^k \in N_k$, we have $\varphi((i,j)^k) = (\sigma(i), \sigma(j))^{\pi(k)}$ because φ preserves adjacency:

- If one element of N_k is mapped to $N_{k'}$, all elements from N_k need to be mapped to $N_{k'}$ by the edges in E_V; in particular, $k' = \pi(k)$.
- If $(i,j)^k$ is mapped to $(i',j')^{k'}$, the edges in E_R (resp. E_C) ensure $i' = \sigma(i)$ (resp. $j' = \sigma(j)$); in particular, $A^k_{i,j} = A^{k'}_{i',j'}$ as φ preserves edge colors.

Since $\varphi((i,j)^k) = (\sigma(i), \sigma(j)^{\pi(k)})$ describes the action of σ on matrix A^k, the second part of (P2) holds. The first part $\sigma(A^0) = A^0$ follows by the same argument, because φ cannot map $(i,j)^0$ to a node $(i',j')^k$ for $k \neq 0$ since the nodes in N_0 are the only matrix-entry nodes not connected to a variable node.

The second part follows from the above discussion by setting

$$\varphi(v) = \begin{cases} \pi(v), & \text{if } v \in V, \\ \sigma(v), & \text{if } v \in D, \\ (\sigma(i), \sigma(j))^{\pi(k)}, & \text{if } v = (i,j)^k \in N_k \text{ for some } k \in [m], \\ (\sigma(i), \sigma(j))^0, & \text{if } v = (i,j)^0 \in N_0 \text{ for some } k \in [m], \end{cases}$$

for each $v \in \mathcal{V}$. □

Remark 1. Graph automorphism codes like `bliss` [15] or `nauty` [21] cannot handle edge colors. Modifying $\mathcal{G}$ by replacing colored edges $\{u, v\}$ by a node with the same color that is connected to u and v allows to use these codes.

4 Computational Results

We implemented the symmetry detection method described in Sect. 3 in SCIP-SDP 4.1.0. SCIP-SDP is a framework for solving MISDPs and is available at https://wwwopt.mathematik.tu-darmstadt.de/scipsdp/. We compiled SCIP-SDP with a developer version of SCIP 8.0.2 (githash `878b1c5`) and used Mosek 9.2.40 for solving SDP-relaxations. All tests were performed on a Linux cluster with 3.5 GHz Intel Xeon E5-1620 Quad-Core CPUs, having 32 GB main memory and 10 MB cache. All computations were run single-threaded and with a time limit of one hour. Detailed results per instance can be found at [25].

To handle symmetries, we use the variant of the state-of-the-art method orbital fixing [17,22] as described in [23] and as implemented in SCIP. The idea of orbital fixing is to derive symmetry-based fixings of binary variables that are derived from the branching decisions and already fixed variables.

Table 1. Symmetries in the 21 symmetric instances from [20], where $\mathcal{S}_k$ refers to the full symmetric group on k elements and $\mathcal{D}_k$ is the dihedral group.

instance	symmetry group
0+-115305C_MISDPld000010	$\mathcal{S}_2$
0+-115305C_MISDPrd000010	$\mathcal{S}_2$
band60605D_MISDPld000010	$\mathcal{S}_2 \times \mathcal{S}_2 \times \mathcal{S}_2 \times \mathcal{S}_2 \times \mathcal{S}_2 \times \mathcal{S}_2 \times \mathcal{S}_{10} \times \mathcal{S}_3 \times \mathcal{S}_4$
band60605D_MISDPrd000010	$\mathcal{S}_2 \times \mathcal{S}_2 \times \mathcal{S}_2 \times \mathcal{S}_2 \times \mathcal{S}_2 \times \mathcal{S}_2 \times \mathcal{S}_{10} \times \mathcal{S}_3 \times \mathcal{S}_4$
band70704A_MISDPld000010	$\mathcal{S}_2 \times \mathcal{S}_2 \times \mathcal{S}_2 \times \mathcal{S}_3 \times \mathcal{S}_3$
band70704A_MISDPrd000010	$\mathcal{S}_2 \times \mathcal{S}_2 \times \mathcal{S}_2 \times \mathcal{S}_3 \times \mathcal{S}_3$
clique_60_k10_6_6, clique_60_k15_4_4, clique_60_k20_3_3, clique_60_k4_15_15, clique_60_k5_12_12, clique_60_k6_10_10, clique_60_k7_8_9, clique_60_k8_7_8, clique_60_k9_6_7, clique_70_k3_23_24	$\mathcal{S}_2$
diw_34	$\mathcal{S}_2 \times \mathcal{S}_2 \times \mathcal{S}_2 \times \mathcal{S}_2 \times \mathcal{D}_4 \times \mathcal{S}_4 \times \mathcal{S}_4$
diw_37	$\mathcal{S}_2 \times \mathcal{S}_4 \times \mathcal{S}_3 \times \mathcal{S}_4$
diw_38	$\mathcal{S}_2 \times \mathcal{S}_2 \times \mathcal{S}_2 \times \mathcal{S}_3$
diw_43	$\mathcal{S}_3$
diw_44	$\mathcal{S}_3$

Table 2. Results for MISDP instances from [20].

variant	all (184)			all optimal (168)		only symmetric (21)
	time (s)	symtime (s)	# gens	time (s)	#nodes	time (s)
no symmetry	130.6	–	–	95.0	778.3	45.07
orbital fixing	125.3	0.44	99	90.8	760.6	29.84

In the first experiment, we use the same 185 instances as in [20] from a variety of applications; the instances can be downloaded from [25]. Table 1 shows the detected symmetries of all 21 instances that contain symmetry. We can see that most instances admit symmetries of the full symmetric group $\mathcal{S}_k$, i.e., k variables can be permuted arbitrarily. As $|\mathcal{S}_k| = k!$, the symmetry groups become rather large, which indicates that symmetry handling might be beneficial for solving these instances. Note that the action of the symmetries might be nontrivial, i.e., several variables might be moved simultaneously like permuting the rows or columns of a matrix.

Table 2 shows a comparison of the default without symmetry handling and the new version with symmetry handling. Here, we excluded one numerically difficult instance for which both variants computed a wrong solution. The columns represent the shifted geometric mean of the CPU time in seconds (with a shift of 1s), the average time for symmetry handling (including detection), and the number of generators. The next two columns display the shifted geometric mean of the CPU time and number of nodes (with a shift of 100) for the 168 instances that could be solved by both variants. The last column shows the CPU time, but only for those instances for which at least one generator has been found.

One can see a 4% speed-up in CPU time for all instances and about 34% for the 21 instances that contain symmetry. Note that symmetry handling does not help to solve more instances (168), but to speed up computation. The time for handling and computing symmetry over all instances is quite small. Similarly, the number 99 of found generators is quite small. This fits well with Table 1 in which the symmetry groups admit a small set of generators.

In a second experiment, we consider the problem of finding a maximum stable set in an unweighted undirected graph $G = (V, E)$. Based on [3,16], we derive the MISDP formulation

$$\begin{aligned} \sup \quad & \sum_{v \in V} x_v \\ \text{s.t.} \quad & \begin{pmatrix} 1 & x^\top \\ x & X \end{pmatrix} \succeq 0, \\ & X_{uv} \leq x_u,\ X_{uv} \leq x_v,\ x_u + x_v \leq 1 + X_{uv} \quad \forall \{u, v\} \in \binom{V}{2}, \\ & X_{uv} = 0, \quad \forall \{u, v\} \in E, \\ & x \in \{0,1\}^V,\ X \in \{0,1\}^{V \times V}. \end{aligned} \tag{2}$$

Table 3. Results for stable set MISDP instances on Color02 graphs.

variant	all (54)			all optimal (47)		only symmetric (51)
	time (s)	symtime (s)	# gens	time (s)	#nodes	time (s)
no symmetry	89.2	–	–	51.1	11.0	85.35
orbital fixing	81.4	0.32	1028	46.4	7.2	77.54

Table 4. Results for the stable set MISDP instances on flower snark graphs.

variant	all (20)			all optimal (14)	
	time (s)	symtime (s)	# gens	time (s)	#nodes
no symmetry	310.8	–	–	108.3	10.2
orbital fixing	211.7	0.23	80	67.3	7.2

Model (2) can be easily transformed into a MISDP of type (1). It indeed models the stable set problem, where $x_v = 1$ if and only if v is contained in a stable set. The linear constraints model that $X_{uv} = x_u \cdot x_v$ and that not both endpoints of an edge can be contained in a stable set. The SDP constraint arises from the observation that $xx^\top \succeq 0$ for every incidence vector x of a stable set.

We conducted experiments for two sets of graph instances. The `Color02` test sets consists of the 55 smallest graphs from the Color02 symposium [4]; the `Snark` test sets consists of so-called flower snark graphs [7,14] with $4n$ nodes, where $n \in \{11, \dots, 49\} \cap (1 + 2\mathbb{Z})$ nodes. The latter instances are of particular interest, because they admit a large dihedral symmetry group.

Table 3 provides results for the `Color02` graphs. Here we excluded one instance for which the computations produced a wrong result. There is a speed-up of about 9% each for all instances, the ones the have been solved by all variants, and the ones in which a least one symmetry has been found. Note that the symmetry of the formulation only arises from the symmetry of the graph.

Table 4 shows the result for the flower snark graphs, which all contain symmetries. With symmetry handling one solves 17 instances, compared to 14 for the default. Moreover, there is a speed-up of about 32% and 38% on all instances and the ones solved to optimality by both variants, respectively.

Conclusion. Our numerical experiments indicate that symmetry handling is an important tool for reducing the running time of solving MISDPs. Our graph automorphism based approach allows to quickly compute symmetries such that the additional time needed to detect symmetries is negligible. Then, using state-of-the-art methods such as orbital fixing, already allows to substantially improve the solution time on a variety of test sets. An interesting question is whether new symmetry handling methods that are tailored for MISDPs allow to improve the running time even further. For instance, one could try to combine handling permutation symmetries and using model reformulations as mentioned in the introduction. We leave this for future research.

References

1. Bai, Y., de Klerk, E., Pasechnik, D., Sotirov, R.: Exploiting group symmetry in truss topology optimization. Optim. Eng. **10**(3), 331–349 (2009)
2. Bestuzheva, K., et al.: The SCIP Optimization Suite 8.0. Technical report, Optimization Online (2021). http://www.optimization-online.org/DB_HTML/2021/12/8728.html
3. Burer, S., Monteiro, R.D., Zhang, Y.: Maximum stable set formulations and heuristics based on continuous optimization. Math. Program. **94**, 137–166 (2022). https://doi.org/10.1007/s10107-002-0356-4
4. Color02 - computational symposium: graph coloring and its generalizations (2002). http://mat.gsia.cmu.edu/COLOR02
5. Dakin, R.J.: A tree-search algorithm for mixed integer programming problems. Comput. J. **8**(3), 250–255 (1965). https://doi.org/10.1093/comjnl/8.3.250
6. de Klerk, E., Sotirov, R.: Exploiting group symmetry in semidefinite programming relaxations of the quadratic assignment problem. Math. Program. **122**(2), 225–246 (2010)
7. Fiorini, S., Wilson, R.J.: Edge-colourings of graphs. No. 16 in Research Notes in Mathematics, Pitman Publishing Limited (1977)
8. Gally, T.: Computational Mixed-Integer Semidefinite Programming. Dissertation, TU Darmstadt (2019)
9. Gally, T., Pfetsch, M.E., Ulbrich, S.: A framework for solving mixed-integer semidefinite programs. Optim. Methods Softw. **33**(3), 594–632 (2017). https://doi.org/10.1080/10556788.2017.1322081
10. Gamrath, G., et al.: The SCIP Optimization Suite 7.0. Technical report, Optimization Online (2020). http://www.optimization-online.org/DB_HTML/2020/03/7705.html
11. Gatermann, K., Parrilo, P.: Symmetry groups, semidefinite programs, and sums of squares. J. Pure Appl. Algebra **192**(1–3), 95–128 (2004)
12. Hojny, C., Pfetsch, M.E.: Polytopes associated with symmetry handling. Math. Program. **175**(1), 197–240 (2019). https://doi.org/10.1007/s10107-018-1239-7
13. Hu, H., Sotirov, R., Wolkowicz, H.: Facial reduction for symmetry reduced semidefinite and doubly nonnegative programs. Math. Program. (2022, to appear)
14. Isaacs, R.: Infinite families of nontrivial trivalent graphs which are not tait colorable. Am. Math. Mon. **82**(3), 221–239 (1975). https://doi.org/10.1080/00029890.1975.11993805
15. Junttila, T., Kaski, P.: Bliss: a tool for computing automorphism groups and canonical labelings of graphs. https://users.aalto.fi/tjunttil/bliss/
16. Lovász, L.: On the Shannon capacity of a graph. IEEE Trans. Inf. Theory **25**, 1–7 (1979)
17. Margot, F.: Exploiting orbits in symmetric ILP. Math. Program. **98**(1–3), 3–21 (2003). https://doi.org/10.1007/s10107-003-0394-6
18. Margot, F.: Symmetry in integer linear programming. In: Jünger, M., et al. (eds.) 50 Years of Integer Programming 1958-2008, pp. 647–686. Springer, Heidelberg (2010). https://doi.org/10.1007/978-3-540-68279-0_17
19. Mars, S.: Mixed-Integer Semidefinite Programming with an Application to Truss Topology Design. Dissertation, FAU Erlangen-Nürnberg (2013)
20. Matter, F., Pfetsch, M.E.: Presolving for mixed-integer semidefinite optimization. INFORMS J. Optim. (2022, to appear). https://doi.org/10.1287/ijoo.2022.0079

21. McKay, B.D., Piperno, A.: Practical graph isomorphism, II. J. Symb. Comput. **60**, 94–112 (2014). https://doi.org/10.1016/j.jsc.2013.09.003
22. Ostrowski, J., Linderoth, J., Rossi, F., Smriglio, S.: Orbital branching. Math. Program. **126**(1), 147–178 (2011). https://doi.org/10.1007/s10107-009-0273-x
23. Pfetsch, M.E., Rehn, T.: A computational comparison of symmetry handling methods for mixed integer programs. Math. Program. Comput. **11**(1), 37–93 (2018). https://doi.org/10.1007/s12532-018-0140-y
24. Pilanci, M., Wainwright, M.J., El Ghaoui, L.: Sparse learning via Boolean relaxations. Math. Program. **151**(1), 63–87 (2015). https://doi.org/10.1007/s10107-015-0894-1
25. Project website: instance data, supplementary material. https://www2.mathematik.tu-darmstadt.de/pfetsch/MISDP-symmetries.html
26. Salvagnin, D.: A dominance procedure for integer programming. Master's thesis, University of Padova, Padova, Italy (2005)
27. Wiese, S.: Symmetry detection in mixed-integer conic programming. Mosek Whitepaper (2022). https://docs.mosek.com/whitepapers/symmetry.pdf

A Mixed-Integer Linear Programming Reduction of Disjoint Bilinear Programs via Symbolic Variable Elimination

Jihwan Jeong[1,3(✉)], Scott Sanner[1,3], and Akshat Kumar[2]

[1] University of Toronto, Toronto, Canada
jihwan.jeong@mail.utoronto.ca, ssanner@mie.utoronto.ca
[2] Singapore Management University, Singapore, Singapore
akshatkumar@smu.edu.sg
[3] Vector Institute, Toronto, Canada

Abstract. A disjointly constrained bilinear program (DBLP) has various practical and industrial applications, e.g., in game theory, facility location, supply chain management, and multi-agent planning problems. Although earlier work has noted the equivalence of DBLP and mixed-integer linear programming (MILP) from an abstract theoretical perspective, a practical and exact closed-form reduction of a DBLP to a MILP has remained elusive. Such explicit reduction would allow us to leverage modern MILP solvers and techniques along with their solution optimality and anytime approximation guarantees. To this end, we provide the first constructive closed-form MILP reduction of a DBLP by extending the technique of symbolic variable elimination (SVE) to constrained optimization problems with bilinear forms. We apply our MILP reduction method to difficult DBLPs including XORs of linear constraints and show that we significantly outperform Gurobi. We also evaluate our method on a variety of synthetic instances to analyze the effects of DBLP problem size and sparsity w.r.t. MILP compilation size and solution efficiency.

Keywords: Bilinear programming · Symbolic variable elimination

1 Introduction

A disjointly constrained bilinear program (DBLP) is formally defined as follows

$$\begin{aligned}
\min_{\mathbf{x},\mathbf{y}} \quad & f(\mathbf{x},\mathbf{y}) = \mathbf{c}^\top\mathbf{x} + \mathbf{x}^\top Q\mathbf{y} + \mathbf{d}^\top\mathbf{y} && (1)\\
\text{s.t.} \quad & \mathbf{a}_i^\top\mathbf{x} \le a_i \ \forall i \in I, \quad \mathbf{b}_j^\top\mathbf{y} \le b_j \ \forall j \in J\\
& x_k \ge 0 \ \forall k \in K; \quad y_l \ge 0 \ \forall l \in L\\
& x_m \in \{0,1\} \ \forall m \in M; \quad y_n \in \{0,1\} \ \forall n \in N,
\end{aligned}$$

where I and J are the index sets of the linear constraints. K, L and M, N are those of continuous and binary variables, respectively. Let $n_x = |K| + |M|$ and

© The Author(s), under exclusive license to Springer Nature Switzerland AG 2023
A. A. Cire (Ed.): CPAIOR 2023, LNCS 13884, pp. 79–95, 2023.
https://doi.org/10.1007/978-3-031-33271-5_6

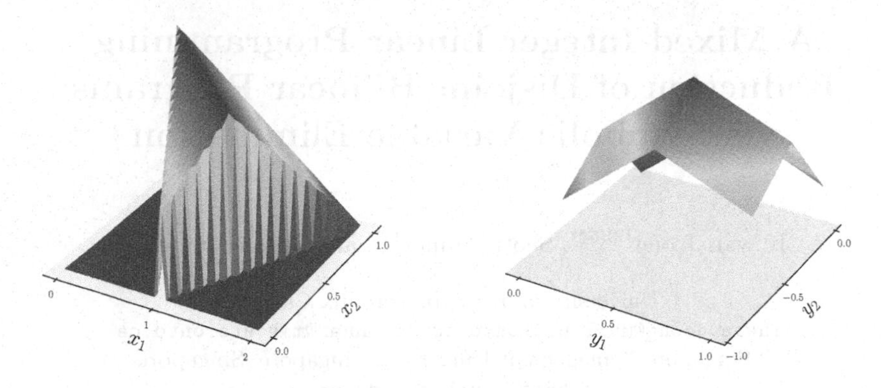

Fig. 1. The objective function of a DBLP instance constructed according to [15], evaluated on a range of values of $\mathbf{x}$ (left) and $\mathbf{y}$ (right). The piecewise linear structure hints at a MILP reduction. Details are provided in Sect. 5.

$n_y = |L| + |N|$, then we have $Q \in \mathbb{R}^{n_x \times n_y}$, $\mathbf{c}, \mathbf{a}_i \in \mathbb{R}^{n_x}$, and $\mathbf{d}, \mathbf{b}_j \in \mathbb{R}^{n_y}$. The disjointness property arises from the separation of linear constraints on $\mathbf{x}$ and $\mathbf{y}$. We define $\mathcal{X}$ and $\mathcal{Y}$ to be the feasible sets of $\mathbf{x}$ and $\mathbf{y}$ variables.

Historically, DBLPs have been used to formulate a variety of applications including uses in game theory, facility location, nonlinear multi-commodity network flows, dynamic assignment and production, risk management, and supply chain management [8,9,11,14]. In addition, DBLPs have found applications in multi-agent planning problems [10], particularly when the transitions of different agents are assumed to be independent, which leads to disjoint constraints.

While Gurobi [4] can directly solve DBLPs to optimality since version 9.0 (based on spatial branching and a locally valid McCormick-based LP relaxation), it can only solve small DBLP instances when they use complex logical constraints (e.g., XORs of linear constraints). Given that logical constraints can be naturally encoded in a MILP, we conjecture (and later empirically show) that Gurobi can better solve such DBLPs when transformed to a MILP formulation.

Earlier work has shown that a DBLP is a concave minimization problem with a piecewise linear objective and linear constraints over one set of variables, say, $\mathbf{x}$ [5,6]. To illustrate, Fig. 1 shows a DBLP objective from Sect. 5 evaluated on a range of $\mathbf{x}$ and $\mathbf{y}$ values, where we clearly observe piecewise linear structure. Formally, consider $\min_{\mathbf{x},\mathbf{y}} f(\mathbf{x}, \mathbf{y}) = \min_{\mathbf{x}} g(\mathbf{x})$ with

$$g(\mathbf{x}) := \min_{\mathbf{y} \in \mathcal{Y}} f(\mathbf{x}, \mathbf{y}) = \min_{\mathbf{y} \in V(\mathrm{Conv}(\mathcal{Y}))} f(\mathbf{x}, \mathbf{y}) = \mathbf{c}^\top \mathbf{x} + \min_{\mathbf{y} \in V(\mathrm{Conv}(\mathcal{Y}))} \{(\mathbf{d} + Q^\top \mathbf{x})^\top \mathbf{y}\},$$

where $V(\mathrm{Conv}(\mathcal{Y}))$ is the set of vertices of the convex hull of $\mathcal{Y}$. Theoretically, enumerating *all* vertices makes $g(\mathbf{x})$ piecewise linear and hence MILP-reducible, but a more compact and constructive MILP reduction has remained elusive.

In this work, we *extend symbolic variable elimination (SVE)* [12] *to bilinear expressions* and derive the *first* DBLP to MILP reduction that does not require enumeration of all vertices $V(\mathrm{Conv}(\mathcal{Y}))$. In addition to an investigation of the performance of our DBLP to MILP reduction on synthetic instances with varying size and sparsity, we demonstrate that *the Gurobi MILP solver applied to our DBLP reduction can outperform Gurobi's own bilinear solver for DBLPs.*

2 Reducing a DBLP to a MILP: A Worked Example

To foreshadow the general methodology that we explore in this paper, we first demonstrate how we can "deflate" a DBLP into a conditional DBLP by eliminating one variable from $\mathbf{y}$ at a time until the final result is a conditional LP, or a MILP. We proceed to show such deflation steps in close detail in Example 1.

Example 1. Consider the following simple DBLP (Fig. 2a):

$$\begin{aligned} \min_{x_1,y_1,y_2} \quad & -2x_1 + x_1(y_1+y_2) - y_1 - y_2 \\ \text{s.t.} \quad & -y_1 + 2y_2 \leq 2,\ y_1 \leq 2,\ y_2 \geq 1,\ 0 \leq x_1 \leq 2 \end{aligned} \tag{2}$$

Our goal is to symbolically minimize out y_1 and y_2 so that we can obtain a reduced form over just x_1. To do this, we can view the $\min_{x_1,y_1,y_2}$ from the perspective of symbolic variable elimination [12] where we can "min-out" y_1 first. Observe that when y_1 is minimized, x_1 and y_2 are considered free variables, allowing us to treat the bilinear objective as linear in y_1. The minimum, therefore, must occur at a boundary value of y_1. We can easily obtain symbolic bounds on y_1 if we isolate it in the linear constraints. In this example, $-y_1 + 2y_2 \leq 2$ and $y_1 \leq 2$ are equivalent to $y_1^{lb} \leq y_1 \leq y_1^{ub}$ with $y_1^{ub} = 2$ and $y_1^{lb} = 2y_2 - 2$.

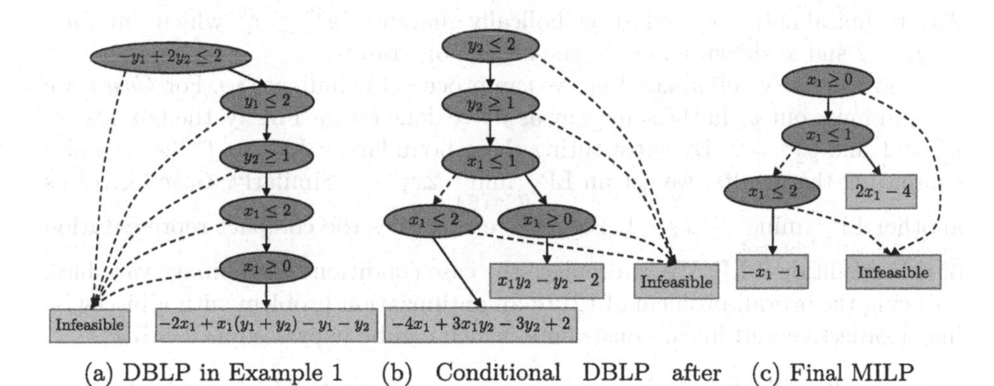

(a) DBLP in Example 1 (b) Conditional DBLP after eliminating y_1 (c) Final MILP

Fig. 2. Compact XADD [13] decision diagram representation of (2) in its (a) original form and after (b) y_1 and (c) y_2 are eliminated. Given values for x_1, y_1, and y_2, the XADD can be evaluated top-to-bottom. Oval constraints are decisions and the solid (dashed) edge is followed if the constraint evaluates to true (false). Leaf nodes provide the objective evaluation. In (c), once all $\mathbf{y}$ variables are symbolically eliminated, all constraints and leaves are linear leading to a conditional LP (=MILP).

We now plug the two bounds on y_1 into the objective and compare resulting values. To that end, let $f^{ub}(x_1, y_2)$ and $f^{lb}(x_1, y_2)$ be the objective values when the upper and lower bound of y_1 is substituted in, respectively. That is,

$$f^{ub}(x_1, y_2) = -2x_1 + x_1(2 + y_2) - 2 - y_2 = x_1 y_2 - y_2 - 2$$
$$f^{lb}(x_1, y_2) = -2x_1 + x_1[(2y_2 - 2) + y_2] - (2y_2 - 2) - y_2 = -4x_1 + 3x_1 y_2 - 3y_2 + 2$$

In order to determine which bound on y_1 minimizes the objective, we can check if the difference $f^{ub}(x_1, y_2) - f^{lb}(x_1, y_2)$ is positive or negative:

$$f^{ub}(x_1, y_2) - f^{lb}(x_1, y_2) = (y_1^{ub} - y_1^{lb})(x_1 - 1) = (4 - 2y_2)(x_1 - 1) \tag{3}$$

Crucially in (3), the terms in the objective that do not have y_1 *always* cancel out, while the ones multiplied to y_1 remain. Hence, when we substitute in the boundary values of y_1 into the objective, the difference always has two factors: one *linear* factor of $\mathbf{x}$ and one *linear* factor of $\mathbf{y}$ (see the discussion in Sect. 4). If (3) is positive (or negative), $f^{lb}(x_1, y_2)$ is smaller (or greater) than $f^{ub}(x_1, y_2)$. Fortunately since $y_1^{ub} - y_1^{lb}$ should be nonnegative, we need only check if *linear* factor $(x_1 - 1)$ is negative (positive) to determine if the upper (lower) bound substitution is minimal. Then we can write a reduced *conditional* DBLP form (Fig. 2b) with y_1 *eliminated*, *linear* conditions on $\mathbf{x}$, and a *bilinear* objective:

$$\begin{cases} (\mathit{Case1})\ x_1 - 1 \le 0: & \min_{x_1, y_2}\ f^{ub}(x_1, y_2) = x_1 y_2 - y_2 - 2 \\ & \text{s.t.}\quad 0 \le x_1 \le 1,\ 1 \le y_2 \le 2 \\ (\mathit{Case2})\ x_1 - 1 > 0: & \min_{x_1, y_2}\ f^{lb}(x_1, y_2) = -4x_1 + 3x_1 y_2 - 3y_2 + 2 \\ & \text{s.t.}\quad 1 < x_1 \le 2,\ 1 \le y_2 \le 2 \end{cases} \tag{4}$$

As a technical note, we need to symbolically guarantee $y_1^{ub} \ge y_1^{lb}$, which simplifies to $y_2 \le 2$ and is shown added to the above constraints.

Now that we've eliminated y_1, we can proceed to eliminate y_2. For *Case1*, we can minimize out y_2 in the same way as we've done for y_1. Firstly, the bounds are $y_2^{lb} = 1$ and $y_2^{ub} = 2$. By substituting these boundary values to $f^{ub}(x_1, y_2)$ and comparing the results, we get an LP $\min_{0 \le x_1 \le 1} 2x_1 - 4$. Similarly, *Case2* gives us another LP, $\min_{1 < x_1 \le 2} -x_1 - 1$. Figure 2c exemplifies the compact representation of this conditional LP. We can replace the case conditions with binary variables, reducing the overall problem of (2) to an optimization problem with a piecewise linear objective and linear constraints, which can be expressed as a MILP.

Example 1 illustrates that we can obtain a concrete MILP model by symbolically minimizing out one set of variables from a DBLP (e.g., $\mathbf{y}$) yielding a reduced MILP optimization problem over $\mathbf{x}$, which can be easily implemented and efficiently solved by off-the-shelf MILP solvers such as Gurobi. Substituting the optimal $\mathbf{x}$ in the original DBLP reduces to a MILP over $\mathbf{y}$ that is easily solved to obtain the corresponding $\mathbf{y}$. To move beyond this example and provide a fully automated reduction of an arbitrary DBLP to a MILP, we will need a general symbolic procedure to automate this reasoning, which we provide next.

3 Symbolic Calculus with Case Representation

Generalizing Example 1, we demonstrate the generic DBLP to MILP conversion using symbolic case representation and calculus [3,13] (Sect. 3) with a new extension for SVE [12] in continuous minimization operations with bilinear forms (Sect. 4). Finally, we present empirical analysis in Sect. 5.

3.1 Case Representation

We assume that all symbolic functions can be represented in *case form* [3,13]:

$$f = \begin{cases} \phi_1 : & f_1 \\ \vdots & \vdots \\ \phi_k : & f_k \end{cases} \tag{5}$$

Here, ϕ_i (a *partition*) are logical formulae, which can include arbitrary logical $(\wedge, \vee, \neg)$ combinations of linear inequalities. We assume that the set of conditions $\{\phi_1, \ldots, \phi_k\}$ disjointly and exhaustively partition the domain of the variables such that f is well-defined. We call ϕ_i "*disjointly linear*" if it consists only one of $\mathbf{x}$ or $\mathbf{y}$. We restrict f_i (a *function value*) to be linear or bilinear in $\mathbf{x}$ and $\mathbf{y}$. Further, we restrict ϕ_i to be disjointly linear if f has bilinear f_i. These restrictions ensure that we can represent an arbitrary DBLP in case form in Sect. 4.

Henceforth, we refer to functions with linear ϕ_i and f_i as linear piecewise linear (LPWL). Functions with disjointly linear ϕ_i and bilinear f_i are dubbed as disjointly linear piecewise bilinear (LPWB). Later, we discuss that in order for SVE of a DBLP to remain closed-form, it is critical that the procedural reduction of the original case function always produces an LPWB or LPWL function.

We remark that the DBLP in Example 1 can be easily rewritten in case form

$$f = \begin{cases} [-y_1 + 2y_2 \leq 2] \wedge [y_1 \leq 2] \wedge [y_2 \geq 1] \wedge [0 \leq x_1 \leq 2] : & -2x_1 + x_1(y_1+y_2) - y_1 - y_2 \\ \text{otherwise} : & \infty \end{cases}$$

where any finite value for f satisfying the first case (the feasible set) will always be chosen over ∞ (infeasibility), since we want $\min_{x_1,y_1,y_2} f$.

3.2 Basic Case Operators

One of the most simple case operations on f in (5) is a *unary operation* such as scalar multiplication $c \cdot f$ $(c \in \mathbb{R})$ or negation $-f$. This operation is simply applied to the function value f_i for every partition ϕ_i. We can also define *binary operations* between two case functions by taking the cross-product of the logical partitions from the two case statements and performing the operation on the resulting paired partitions.[1] For example, the "cross-sum" $\oplus$ of two cases is:

$$\begin{cases} \phi_1 : & f_1 \\ \phi_2 : & f_2 \end{cases} \oplus \begin{cases} \psi_1 : & g_1 \\ \psi_2 : & g_2 \end{cases} = \begin{cases} \phi_1 \wedge \psi_1 : & f_1 + g_1 \\ \phi_1 \wedge \psi_2 : & f_1 + g_2 \\ \phi_2 \wedge \psi_1 : & f_2 + g_1 \\ \phi_2 \wedge \psi_2 : & f_2 + g_2 \end{cases}$$

[1] Only the case operations that we actually use for SVE of a DBLP are introduced.

Likewise, we perform $\ominus$ by subtracting function values per each pair of partitions. Observe that LPWL and LPWB functions are closed under $\oplus$ and $\ominus$.

Next, we define symbolic *case min(max)* between two case functions as:

$$\text{casemin}\left(\begin{cases}\phi_1: & f_1\\ \phi_2: & f_2\end{cases}, \begin{cases}\psi_1: & g_1\\ \psi_2: & g_2\end{cases}\right) = \begin{cases}\phi_1 \wedge \psi_1 \wedge \boldsymbol{f_1 > g_1}: & g_1\\ \phi_1 \wedge \psi_1 \wedge \boldsymbol{f_1 \leq g_1}: & f_1\\ \phi_1 \wedge \psi_2 \wedge \boldsymbol{f_1 > g_2}: & g_2\\ \phi_1 \wedge \psi_2 \wedge \boldsymbol{f_1 \leq g_2}: & f_1\\ \vdots & \vdots\end{cases} \tag{6}$$

wherein the resulting partitions also include the comparison of associated function values f_i and g_j to determine $\min(f_i, g_j)$ (highlighted in bold). casemin of more than two case functions is straightforward since the operator is associative. Crucially, LPWL functions are closed under casemin (max), but LPWB functions are not because $f_i \leq g_j$ can be bilinear or jointly linear.

Another important symbolic operation is *symbolic substitution.* This operation takes a set σ of variables and their substitutions, e.g., $\sigma = \{y/(x_1 + x_2), z/(x_1 - x_2)\}$ where the LHS of '/' represents the substitution variable and the RHS of '/' is the expression being substituted in. Then, we write the substitution operation on f_i with σ as $f_i\sigma$. Then the operation follows:

$$f = \begin{cases}\phi_1: & f_1\\ \vdots & \vdots\\ \phi_k: & f_k\end{cases}, \quad f\sigma = \begin{cases}\phi_1\sigma: & f_1\sigma\\ \vdots & \vdots\\ \phi_k\sigma: & f_k\sigma\end{cases} \tag{7}$$

In this paper, we will only substitute linear expressions of $\{y_j\}_{j\neq i}$ variables into y_i, which clearly preserves the LPWL and LPWB properties.

In the next section, we show that the procedural reduction of a DBLP to a MILP only involves the application of the case operations that preserve an LPWB form, which eventually reduces to an LPWL form (equivalent to a MILP).

4 Symbolic Reduction of a DBLP to a MILP

Having introduced the case form and its basic operations in Sect. 3, we first note that the DBLP in (1) can be written in case form. That is, (1) is equivalent to $\min_{\mathbf{x},\mathbf{y}} f_{DBLP}(\mathbf{x}, \mathbf{y})$ where

$$f_{DBLP}(\mathbf{x}, \mathbf{y}) = \begin{cases}\phi(\mathbf{x}) \wedge \psi(\mathbf{y}): & \mathbf{c}^\top\mathbf{x} + \mathbf{x}^\top Q\mathbf{y} + \mathbf{d}^\top\mathbf{y}\\ \neg(\phi(\mathbf{x}) \wedge \psi(\mathbf{y})): & \infty\end{cases} \tag{8}$$

with $\phi(\mathbf{x}) := [\mathbf{x} \in \mathcal{X}]$, $\psi(\mathbf{y}) := [\mathbf{y} \in \mathcal{Y}]$. Note how the feasible set of the DBLP is encoded as a partition and the objective as its function value. Also, observe that $\phi(\mathbf{x}) \wedge \psi(\mathbf{y})$ is disjointly linear, so $f_{DBLP}(\mathbf{x}, \mathbf{y})$ is an LPWB function.

We have seen in Example 1 that we get a MILP out of a DBLP via symbolic minimization of $\mathbf{y}$ variables. In general, if the result of SVE of $\mathbf{y}$ from an arbitrary LPWB function can be shown to be equivalent to an LPWL function, we effectively reduce a DBLP to a MILP. However, existing symbolic min operators [17] fall short of dealing with LPWB functions, since none of them can handle bilinear function values. In the sequel, we show that we can always factorize the bilinear expressions appearing during the SVE of $\mathbf{y}$ variables into one factor in $\mathbf{x}$ and the other in $\mathbf{y}$. This in turn makes LPWB functions closed under the SVE operations. With this, we prove that a DBLP can be reduced to a MILP.

4.1 Symbolic Minimization of Linear Piecewise Linear Functions

To see why existing approaches fail to symbolically optimize variables in closed-form when it comes to LPWB functions, we first consider the symbolic min operator for LPWL functions [17].[2] This operator differs from casemin in that the former optimizes a symbolic function w.r.t. decision variables, whereas the latter compares multiple symbolic functions as in (6). Example 2 illustrates the application of the symbolic min operator to an LPWL function.

Example 2. Let $f(x_1, x_2)$ be a symbolic function of $x_1, x_2 \in [0, 10]^2$ as below:

$$f(x_1, x_2) = \begin{cases} x_1 + x_2 \geq 1: & 3x_1 + 2x_2 \\ x_1 + x_2 < 1: & -3x_1 + x_2 \end{cases} \tag{9}$$

As in Example 1, we can view the $\min_{x_1, x_2}$ from the perspective of symbolic variable elimination, and we write it as $\min_{x_2} \min_{x_1} f(x_1, x_2)$. When x_1 is being minimized out, we can treat x_2 as a symbolic free variable. Then,

$$\min_{x_2} \min_{x_1} f(x_1, x_2) = \min_{x_2} \left[\min_{x_1} \begin{cases} \phi_1(x_1, x_2): & f_1(x_1, x_2) \\ \phi_2(x_1, x_2): & f_2(x_1, x_2) \end{cases} \right]$$

$$= \min_{x_2} \left[\min_{x_1} \operatorname*{casemin}_{i=\{1,2\}} \begin{cases} \phi_i(x_1, x_2): & f_i(x_1, x_2) \\ \neg\phi_i(x_1, x_2): & \infty \end{cases} \right] \tag{10}$$

$$= \min_{x_2} \left[\operatorname*{casemin}_{i=\{1,2\}} \min_{x_1} \begin{cases} \phi_i(x_1, x_2): & f_i(x_1, x_2) \\ \neg\phi_i(x_1, x_2): & \infty \end{cases} \right] \tag{11}$$

where ϕ_i and f_i are defined as per (9). (10) follows since partitions are disjoint. The commutative property gives (11). As a result, $\min_{x_1} f(x_1, x_2)$ is equivalent to minimizing out x_1 from "$\{\phi_i : f_i$" for all i, followed by casemin of the results.

Now in order to compute $\min_{x_1} \{\phi_i(x_1, x_2) : f_i(x_1, x_2)$, we make three important observations: (a) a partition ϕ_i and domain bounds on x_1 prescribe the lower and upper bounds over the variable, $x_1^{lb,i}$ and $x_1^{ub,i}$ respectively; (b) since f_i is linear in x_1, either $x_1^{lb,i}$ or $x_1^{ub,i}$ will evaluate to the minimum (ties broken arbitrarily); and (c) if there is a subset of conditionals in ϕ_i that are independent of x_1, denoted as $\phi_i^{\perp x_1}$, it should still be satisfied after the min operation.

[2] This operator has been introduced firstly in [17] and later in more detail in [7]. However, we include the result here for completeness and to better illustrate our extension to handling bilinear function values in Sect. 4.2.

For example, from $\phi_1(x_1, x_2) = [x_1 + x_2 \geq 1]$,

$$x_1^{lb,1} = \text{casemax}(1 - x_2, 0) = \begin{cases} x_2 \geq 1: & 0 \\ x_2 < 1: & 1 - x_2 \end{cases} \tag{12}$$

In general, a domain bound (e.g., $x_1 \geq 0$) and each conditional (e.g., $[x_1 + x_2 \geq 1]$) of a partition can contribute at most one lower bound *candidate*, and $x_1^{lb,i}$ is the casemax among the candidates. Similarly, we get $x_1^{ub,i}$ as the casemin among candidates, which in this case is simply $x_1^{ub,1} = 10$. From these bounds, we additionally impose a set of constraints such that $x_1^{lb,i} \leq x_1^{ub,i}$ is ensured at all times, which are added to $\phi_i^{\perp\!\!\!\perp x_1}$. In this example, these are $[0 \leq 10]$ and $[1 - x_2 \leq 10]$, which trivially hold true, and so we set $\phi_1^{\perp\!\!\!\perp x_1} = true$.

With these bounds, it remains to determine the minimum value by substituting $x_1^{lb,i}$ and $x_1^{ub,i}$ into x_1 in f_1 and performing casemin. For $i = 1$, we have:[3]

$$\begin{aligned} \min_{x_1} \begin{cases} \phi_1(x_1, x_2): & f_1(x_1, x_2) \\ \neg\phi_1(x_1, x_2): & \infty \end{cases} &= \text{casemin}(f_1\sigma_1^{ub}, f_1\sigma_1^{lb}) \oplus \begin{cases} \phi_1^{\perp\!\!\!\perp x_1}: & 0 \\ \neg\phi_1^{\perp\!\!\!\perp x_1}: & \infty \end{cases} \\ &= \text{casemin}\left(30 + 2x_2, \begin{cases} x_2 \geq 1: & 2x_2 \\ x_2 < 1: & 3 - x_2 \end{cases}\right) \\ &= \begin{cases} x_2 \geq 1: & 2x_2 \\ x_2 < 1: & 3 - x_2 \end{cases} \end{aligned} \tag{13}$$

where $\sigma_1^{lb} = \{x_1/x_1^{lb,1}\}$ and $\sigma_1^{ub} = \{x_1/x_1^{ub,1}\}$.

If we follow the same procedure for $\phi_2(x_1, x_2)$ and $f_2(x_1, x_2)$, we get below:

$$\min_{x_1} \begin{cases} \phi_2(x_1, x_2): & f_2(x_1, x_2) \\ \neg\phi_2(x_1, x_2): & \infty \end{cases} = \begin{cases} x_2 \geq 1: & x_2 \\ x_2 < 1: & -3 + 4x_2 \end{cases} \tag{14}$$

Finally, we take casemin of (13) and (14), which becomes

$$g(x_2) := \min_{x_1} f(x_1, x_2) = \begin{cases} x_2 \geq 1: & x_2 \\ x_2 < 1: & -3 + 4x_2 \end{cases} \tag{15}$$

Note that x_1 has been *eliminated* from $f(x_1, x_2)$ in (15). The same procedure can be repeated for the elimination of x_2.

4.2 Symbolic Minimization of Disjointly Linear Piecewise Bilinear Functions

Example 2 highlights the key operations entailed in symbolic minimization of an LPWL function. However for DBLPs, the step in (13) would compare *bilinear*

[3] Note the way we enforce $\phi_1^{\perp\!\!\!\perp x_1}$ by the cross-sum operation.

expressions, leading to a case function with bilinear or jointly linear partitions, preventing naively applying the same symbolic manipulations. Despite these bilinear expressions, Proposition 1 affirms that we can still perform SVE of one set of variables from the DBLP, which eventually gives rise to an LPWL function. This in turn can be modeled as a MILP by introducing binary indicator variables.

Firstly, we formally define an LPWB function $f(\mathbf{x}, \mathbf{y})$ for $\mathbf{x} \in \mathbb{R}^{n_x}, \mathbf{y} \in \mathbb{R}^{n_y}$:

$$f(\mathbf{x}, \mathbf{y}) = \begin{cases} \phi_1(\mathbf{x}) \wedge \psi_1(\mathbf{y}): & f_1(\mathbf{x}, \mathbf{y}) = \mathbf{c}_1^\top \mathbf{x} + \mathbf{x}^\top Q_1 \mathbf{y} + \mathbf{d}_1^\top \mathbf{y} \\ \vdots & \vdots \\ \phi_n(\mathbf{x}) \wedge \psi_n(\mathbf{y}): & f_n(\mathbf{x}, \mathbf{y}) = \mathbf{c}_n^\top \mathbf{x} + \mathbf{x}^\top Q_n \mathbf{y} + \mathbf{d}_n^\top \mathbf{y} \end{cases} \tag{16}$$

where $\mathbf{c}_i \in \mathbb{R}^{n_x}$, $\mathbf{d}_i \in \mathbb{R}^{n_y}$, $Q_i \in \mathbb{R}^{n_x \times n_y}$, and $\phi_i(\mathbf{x})$ and $\psi_i(\mathbf{y})$ are conjunction of linear inequalities in $\mathbf{x}$ and $\mathbf{y}$.[4] Note that f_{DBLP} is a special case of (16). Proposition 1 establishes that LPWB functions are closed under symbolic min and eventually become LPWL, which uses the following result from Lemma 1.

Lemma 1. *Consider the symbolic substitution operations into bilinear* $f_i(\mathbf{x}, \mathbf{y})$ *with* $\sigma^j = \{y_1 / l^j(\mathbf{y}_{2:n_y})\}$, *where* $\mathbf{y}_{2:n_y} = \{y_2, \ldots, y_{n_y}\}$, $l^{ub}(\mathbf{y}_{2:n_y})$ *and* $l^{lb}(\mathbf{y}_{2:n_y})$ *are linear. Then,* $\text{casemin}(f_i(\mathbf{x}, \mathbf{y})\sigma^{ub}, f_i(\mathbf{x}, \mathbf{y})\sigma^{lb})$ *is an LPWB function.*

Proof. Define $h : \mathbb{R}^{n_x \times (n_y - 1)} \mapsto \mathbb{R}$ as $h(\mathbf{x}, \mathbf{y}_{2:n_y}) := f_i(\mathbf{x}, \mathbf{y})\sigma^{ub} - f_i(\mathbf{x}, \mathbf{y})\sigma^{lb}$. If $h \geq 0$, we select $f_i(\mathbf{x}, \mathbf{y})\sigma^{lb}$ as the casemin; otherwise, $f_i(\mathbf{x}, \mathbf{y})\sigma^{ub}$ is selected. In other words, we get a case function with bilinear partitions and bilinear values:

$$\text{casemin}(f_i(\mathbf{x}, \mathbf{y})\sigma^{ub}, f_i(\mathbf{x}, \mathbf{y})\sigma^{lb}) = \begin{cases} h(\mathbf{x}, \mathbf{y}_{2:n_y}) \geq 0: & f_i(\mathbf{x}, \mathbf{y})\sigma^{lb} \\ h(\mathbf{x}, \mathbf{y}_{2:n_y}) < 0: & f_i(\mathbf{x}, \mathbf{y})\sigma^{ub} \end{cases} \tag{17}$$

However, $h(\mathbf{x}, \mathbf{y}_{2:n_y})$ can always be factorized into two factors where each factor is linear in *either* $\mathbf{x}$ or $\mathbf{y}_{2:n_y}$. That is,

$$h(\mathbf{x}, \mathbf{y}_{2:n_y}) = \Big(l^{ub}(\mathbf{y}_{2:n_y}) - l^{lb}(\mathbf{y}_{2:n_y}) \Big) \Big[[\mathbf{d}_i]_1 + \sum_{r=1}^{n_x} x_r [Q_i]_{r,1} \Big] \geq 0 \tag{18}$$

since the terms in $f_i(\mathbf{x}, \mathbf{y})$ that do not include y_1 cancel out. Finally, we get

$$\begin{cases} [l^{ub}(\mathbf{y}_{2:n_y}) - l^{lb}(\mathbf{y}_{2:n_y}) \geq 0] \wedge [[\mathbf{d}_i]_1 + \sum_{r=1}^{n_x} x_r [Q_i]_{r,1} \geq 0]: & f_i(\mathbf{x}, \mathbf{y})\sigma^{lb} \\ [l^{ub}(\mathbf{y}_{2:n_y}) - l^{lb}(\mathbf{y}_{2:n_y}) < 0] \wedge [[\mathbf{d}_i]_1 + \sum_{r=1}^{n_x} x_r [Q_i]_{r,1} < 0]: & f_i(\mathbf{x}, \mathbf{y})\sigma^{lb} \\ [l^{ub}(\mathbf{y}_{2:n_y}) - l^{lb}(\mathbf{y}_{2:n_y}) \geq 0] \wedge [[\mathbf{d}_i]_1 + \sum_{r=1}^{n_x} x_r [Q_i]_{r,1} < 0]: & f_i(\mathbf{x}, \mathbf{y})\sigma^{ub} \\ [l^{ub}(\mathbf{y}_{2:n_y}) - l^{lb}(\mathbf{y}_{2:n_y}) < 0] \wedge [[\mathbf{d}_i]_1 + \sum_{r=1}^{n_x} x_r [Q_i]_{r,1} \geq 0]: & f_i(\mathbf{x}, \mathbf{y})\sigma^{ub} \end{cases} \tag{19}$$

which has disjointly linear partitions and bilinear values, hence an LPWB.

[4] A function value can be ∞, which implies that the corresponding partition is infeasible (see Fig. 2c).

Now, we present the main result in Proposition 1.

Proposition 1 (Symbolic minimization of LPWB functions). *Let $g(\mathbf{x})$ denote the result of symbolic minimization of $f(\mathbf{x}, \mathbf{y})$ over $\mathbf{y}$ variables, which we assume to be well-defined. That is,*

$$g(\mathbf{x}) := \min_{\mathbf{y}} f(\mathbf{x}, \mathbf{y}) \tag{20}$$

Then, it follows that $g(\mathbf{x})$ is an LPWL function of $\mathbf{x}$.

Proof. The proof relies on inductive reasoning as we show how each y_i can be eliminated in turn yielding an LPWB closed-form and ultimately a final LPWL form once all $\mathbf{y}$ have been eliminated.

Firstly, similar to (11), we note $\min_{\mathbf{y}} f(\mathbf{x}, \mathbf{y})$ is equivalent to the following:

$$\min_{y_{n_y}, \ldots, y_2} \left[\operatorname*{casemin}_{i=\{1,\ldots,n\}} \min_{y_1} \begin{cases} \phi_i(\mathbf{x}) \wedge \psi_i(\mathbf{y}) : & f_i(\mathbf{x}, \mathbf{y}) \\ \neg\phi_i(\mathbf{x}) \vee \neg\psi_i(\mathbf{y}) : & \infty \end{cases} \right] \tag{21}$$

For the ith partition, $\psi_i(\mathbf{y})$ and the generic domain bounds over y_1 specify the upper and lower bounds of y_1, denoted as $y_1^{ub,i}$ and $y_1^{lb,i}$, respectively. Notice that $y_1^{ub,i}$ and $y_1^{lb,i}$ are LPWL functions of $\mathbf{y}_{2:n_y}$. We now substitute the bounds in the place of y_1, followed by casemin to determine a smaller value, which gives:

$$\begin{aligned} g_i(\mathbf{x},\ \mathbf{y}_{2:n_y}) :=\ & \min_{y_1} \begin{cases} \phi_i(\mathbf{x}) \wedge \psi_i(\mathbf{y}) : & f_i(\mathbf{x}, \mathbf{y}) \\ \neg\phi_i(\mathbf{x}) \vee \neg\psi_i(\mathbf{y}) : & \infty \end{cases} \\ =\ & \operatorname{casemin}\left(f_i(\mathbf{x}, \mathbf{y})\sigma_i^{ub}, f_i(\mathbf{x}, \mathbf{y})\sigma_i^{lb} \right) \oplus \begin{cases} \phi_i(\mathbf{x}) \wedge \psi_i^{\perp y_1}(\mathbf{y}_{2:n_y}) : 0 \\ \neg(\phi_i(\mathbf{x}) \wedge \psi_i^{\perp y_1}(\mathbf{y}_{2:n_y})) : & \infty \end{cases} \end{aligned} \tag{22}$$

where $\sigma_i^{ub} = \{y_1 / y_1^{ub,i}\}$ and $\sigma_i^{lb} = \{y_1 / y_1^{lb,i}\}$.

The second term in (22) ensures that the conditionals *independent* of y_1 in $[\phi_i(\mathbf{x}) \wedge \psi_i(\mathbf{y})]$ hold true, which are not accounted for in $y_1^{ub,i}$ and $y_1^{lb,i}$. $\psi_i^{\perp y_1}(\mathbf{y}_{2:n_y})$ also includes a set of conditionals that require $y_1^{ub,i} \geq y_1^{lb,i}$ for all pairs of function values. Naturally, we use ∞ as the value of an infeasible partition such that it will be ignored in later steps since we are minimizing.

Now, we have that $g_i(\mathbf{x}, \mathbf{y}_{2:n_y})$ is an LPWB function. To see this, denote the casemin in (22) as $m(\mathbf{x}, \mathbf{y}_{2:n_y})$. A partition of $m(\mathbf{x}, \mathbf{y}_{2:n_y})$ is a conjunction of a partition from $y_1^{ub,i}$, say the jth, and another from $y_1^{lb,i}$, say the kth; the corresponding function value is $\operatorname{casemin}(f_i\{y_1/l_j^{ub}(\mathbf{y}_{2:n_y})\}, f_i\{y_1/l_k^{lb}(\mathbf{y}_{2:n_y})\})$, with l_j^{ub} and l_k^{lb} denoting the function values from $y_1^{ub,i}$ and $y_1^{lb,i}$, respectively. Then for this partition, we clearly see we get disjointly linear partitions and bilinear function values as per Lemma 1. This analysis can be extended to all other partitions and function values of $m(\mathbf{x}, \mathbf{y}_{2:n_y})$, and hence $g_i(\mathbf{x}, \mathbf{y}_{2:n_y})$ is LPWB $\forall i$.

Finally, we note (21) becomes

$$\min_{\mathbf{y}} f(\mathbf{x},\mathbf{y}) = \min_{y_{n_y},\dots,y_2} \left[\operatorname*{casemin}_{i=\{1,\dots,n\}} g_i(\mathbf{x},\mathbf{y}_{2:n_y}) \right]$$
$$= \operatorname*{casemin}_{i=\{1,\dots,n\}} \left[\min_{y_{n_y},\dots,y_3} \left(\min_{y_2} g_i(\mathbf{x},\mathbf{y}_{2:n_y}) \right) \right] \quad (23)$$

where (23) follows since min and casemin are commutative. Then, we see that the inner-most minimization is essentially SVE of y_2 of an LPWB function. Hence, we can repeat the elimination procedure until all $\mathbf{y}$ variables are minimized out, at which point we get a sequence of casemin applied to an LPWL function of $\mathbf{x}$. Since an LPWL function is closed under the casemin operator, we will get an LPWL function, $g(\mathbf{x})$ in closed-form.

Corollary 1. *The DBLP in* (1) *is equivalent to a MILP.*

Proof. The DBLP can be represented in case form as in (8), which is an LPWB function. Hence, $\min_{\mathbf{x},\mathbf{y}} f_{DBLP}(\mathbf{x},\mathbf{y})$ can be represented as $\min_{\mathbf{x}} g_{DBLP}(\mathbf{x})$ where $g_{DBLP}(\mathbf{x}) := \min_{\mathbf{y}} f_{DBLP}(\mathbf{x},\mathbf{y})$ is an LPWL function (Proposition 1). Therefore, the DBLP is equivalent to the minimization problem with piecewise linear objective and linear constraints, which is equivalent to a MILP.

We remark that maintaining a case representation of a DBLP or its LPWL equivalent with explicit partitions can be prohibitively expensive. Hence, in practice we use Extended Algebraic Decision Diagrams (XADDs) [13] (example in Fig. 2) to compactly represent the case statement and perform operations.

5 Empirical Analysis

In this section, we evaluate the proposed novel reduction of a DBLP to a MILP on various test problems. First, we present the problem constrained with XORs of linear constraints in which the proposed approach outperformed Gurobi (9.5.0). Then, we explore empirical characteristics of the MILP reduction on general DBLPs using a set of randomized test instances. Specifically, we analyze the effects of the problem size and sparsity on the MILP reduction and its solution efficiency. We use the XADD for practical implementation of case functions, and we ported the original XADD implementation in Java to our own in Python. Generated MILPs are then solved using Gurobi. All experiments were done on a Linux machine with a 2.90 GHz processor.[5]

[5] SVE runs on a single processor, but Gurobi made use of all 16 available cores.

Problems with XOR Conditional Constraints. Consider the following DBLP involving XOR ($\veebar$) combinations of constraints as motivated by [16]:

$$\begin{aligned}
\min \quad & \mathbf{c}^\top \mathbf{r} + \mathbf{r}^\top Q \mathbf{y} + \mathbf{d}^\top \mathbf{y} + c_z z, \quad \text{where} \qquad (24)\\
r_i = & \begin{cases} [[x_{3i-2} \geq x_{3i-1}] \veebar [x_{3i-1} \geq x_{3i}]] \wedge [z \geq 0] : & \max(x_{3i-1}, x_{3i}) - \min(x_{3i-1}, x_{3i}) \\ [[x_{3i-2} \geq x_{3i-1}] \veebar [x_{3i-1} \geq x_{3i}]] \wedge [z < 0] : & \min(x_{3i-1}, x_{3i}) - \max(x_{3i-1}, x_{3i}) \\ \neg[[x_{3i-2} \geq x_{3i-1}] \veebar [x_{3i-1} \geq x_{3i}]] \wedge [z \geq 0] : & \min(x_{3i-2}, x_{3i-1}) - \max(x_{3i-2}, x_{3i-1}) \\ \neg[[x_{3i-2} \geq x_{3i-1}] \veebar [x_{3i-1} \geq x_{3i}]] \wedge [z < 0] : & \max(x_{3i-2}, x_{3i-1}) - \min(x_{3i-2}, x_{3i-1}) \end{cases}\\
\text{s.t.} \quad & \mathbf{b}_j^\top \mathbf{y} \leq b_j, \quad \forall j = 1, \ldots, 15\\
& x_i \in [-10, 10], \; r_j \in [-20, 20], \; y_k \in [-10, 10]\\
& i = 1, \ldots, 3n, \quad j = 1, \ldots, n, \quad k = 1, \ldots, 15.
\end{aligned}$$

Here, $\mathbf{c}, \mathbf{r} \in \mathbb{R}^n$, $\mathbf{x} \in \mathbb{R}^{3n}$, $c_z, z \in \mathbb{R}$ and $\mathbf{b}_j, \mathbf{y} \in \mathbb{R}^{15}$. Observe that r_i in the objective is determined based on an XOR conditional expression involving $x_{3i-2}, x_{3i-1}, x_{3i}$ and a linear constraint of z $\forall i = 1, \ldots, n$. We randomly generate $(c_z, \mathbf{c}, \mathbf{d}, Q)$ and $(\mathbf{b}_j, b_j \; \forall j)$ to construct the objective and the feasible region over $\mathbf{y}$, respectively. We eliminate $\mathbf{x}$ from (24) and solve the resulting MILP using Gurobi for the remaining variables.

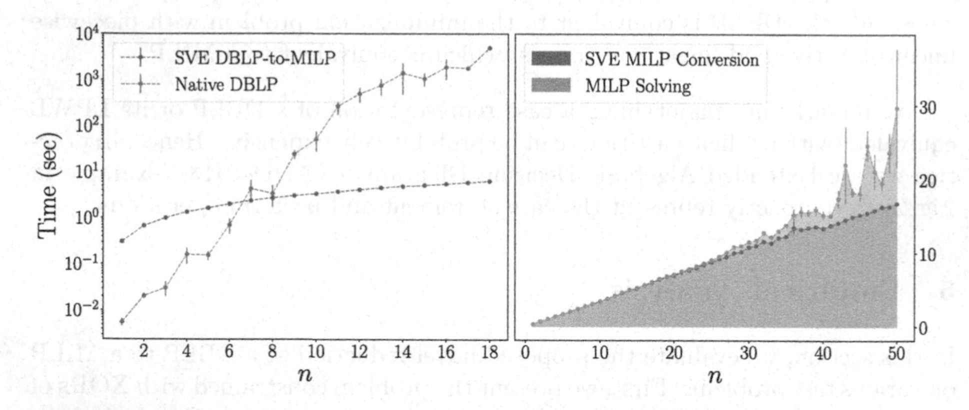

Fig. 3. Left: Runtime comparison of the Native DBLP form (using Gurobi's bilinear solver) and the SVE DBLP-to-MILP conversion (using Gurobi's MILP solver) vs. n (number of variables in XOR problem). Unlike Native DBLP whose time complexity appears exponential in n, SVE DBLP-to-MILP appears linear in n (nb. logarithmic y-axis). **Right**: Breakdown of total runtime of the SVE DBLP-to-MILP solution separated into SVE Conversion time and Gurobi MILP solve time. While SVE scales linearly in n, the MILP step takes a larger fraction of time as n increases (nb. linear y-axis and extended range of n on the x-axis, which only SVE DBLP-to-MILP can solve).

Note that this problem structure is particularly advantageous for the symbolic framework since each r_i can be compactly represented in XADD with only a small number of decision variables and the XOR constraints are sparse. In Fig. 3, we compare the runtime performance of our approach against that of Gurobi.

For each n, we generated 5 instances with different random seeds and plot the mean and its standard error. As the runtime grows *exponentially* for Gurobi, it quickly becomes impossible to solve problems with $n \geq 15$ ($n_x \geq 45$) within the given time limit of 5000 s. However, the solution time increases linearly in the number of variables for the symbolic approach, and we solve the problem with $n_x = 150$ within 30 s. In other words, we have effectively reformulated a DBLP that Gurobi cannot practically solve in its native form to the one that Gurobi can solve as a MILP!

Randomized Test Problems with Different Sizes and Sparsity. Now, we scrutinize the proposed approach on some general DBLP test problems. For the first set of experiments, we follow [15] for systematic generation of test problems with certain properties. In particular, they suggested a two-step method in which smaller DBLP problems are first constructed, which are then additively combined. Furthermore, the underlying structure of the problem is then concealed by random transformations on the decision variables using Householder matrices [1]. 5 instances with different random transformation matrices are constructed for each configuration (n_x, n_y) and we report the average and standard error.

In Table.1, we evaluate the impact of how balanced a problem is on computational complexity by fixing the total number of variables while altering (n_x, n_y) such that one instance has $n_x = n_y$ whereas $n_x > n_y$ for the other ($\mathbf{y}$ is eliminated). We have compared four sets of problem instances with varying total numbers of variables, i.e., $12, 16, 20, 24$. For each total number of variables, balanced and imbalanced instances are compared. We can see that it is in general much easier to solve imbalanced problems, which turn out to be more compact to encode as well. As the number of total variables increases, we observe that the discrepancy in the complexity between an imbalanced and its balanced counterpart widens.

Table 1. Time and space complexity for balanced and imbalanced problems. For every fixed number of total variables $(12, 16, 20, 24)$, the results for an imbalanced ($n_x > n_y$) and a balanced ($n_x = n_y$) are reported. Observe that imbalanced problems are easier to solve and more compact to encode than their balanced counterparts.

$n_x + n_y$	n_x	n_y	Time (Symbolic)	Time (MILP)	# XADD Nodes	# Cont var (MILP)	# Bin var (MILP)	# Constr (MILP)
12	8	4	4.35 ± 0.01	0.01 ± 0.00	44	35	16	55
	6	6	16.35 ± 0.17	0.04 ± 0.00	76	54	18	75
16	10	6	44.67 ± 0.17	0.04 ± 0.00	114	75	23	103
	8	8	121.45 ± 1.09	0.88 ± 0.02	214	140	24	168
20	12	8	391.87 ± 3.22	0.71 ± 0.01	318	185	30	221
	10	10	959.65 ± 66.15	28.84 ± 2.81	622	388	30	423
24	16	8	313.38 ± 1.65	0.18 ± 0.00	536	295	32	335
	12	12	5886.00 ± 142.75	356.23 ± 11.17	1840	1122	36	1164

Notably, the number of binary variables only rises at a moderate rate, whereas the numbers of continuous variables and constraints increase along with the size of the MILP reduction. This suggests that the case representation of the MILP equivalent of a given DBLP turns out to have a structure similar to a tree. For this type of problem, the computational gain attributed to using XADD can rather be small, and therefore we observe fast increases in complexity with the problem size. On the other hand, for types of problems we present in (24) and Fig. 4, the SVE step can be efficiently done even for larger problems. Finally, note also that regardless of n_y, the running times for the optimal MILP solution remain very small.

In order to better understand the solution efficiency with regard to the number of variables and the sparsity of the problem, we created other sets of random test problems. Concretely, the goals are to examine (i) whether the increase in the number of symbolically eliminated variables has greater impact than the increase in the total number of variables in solution efficiency and (ii) the effects of the sparsity of coefficients $(\mathbf{a}_i, \mathbf{b}_j, Q)$. For these problems, we generate feasible and bounded problems with 30 constraints ($n_a = n_b = 15$). 5 instances generated with different random seeds are used per each experiment configuration, and we plot the average and its standard error.

For (i), we symbolically eliminate $\mathbf{y}$ and compare two sets: one with $n_x = 8$ and n_y from 4 to 9, and the other with $n_y = 4$ and n_x increased from 8 to 13. This way, when we increment the total number of variables by 1, it is only for the first set that the number of symbolically eliminated variables increases. In Fig. 4, we see that the time requirements for solving problems with fixed n_y (solid) have virtually remained consistent regardless of the total number of variables. On the other hand, the runtimes for the symbolic solution with increasing n_y have seen a huge jump at $n_x + n_y = 16$ and they are generally on the increase along with the number of variables (dashed). On the contrary, the final sizes of the MILP reduction—in terms of the number of nodes in XADD, the number of binary

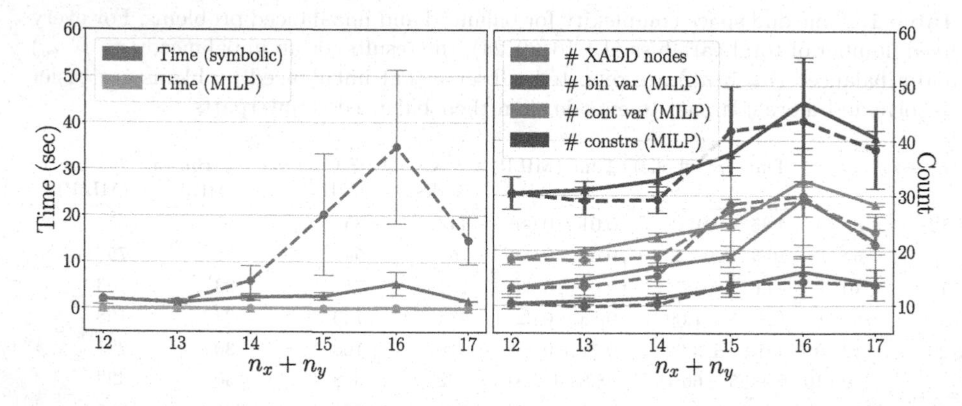

Fig. 4. Time and space complexity as the total number of variables increases. Here, $\mathbf{y}$ is symbolically minimized. The dashed lines correspond to the case of increasing n_y, whereas the solid lines represent the case of increasing n_x.

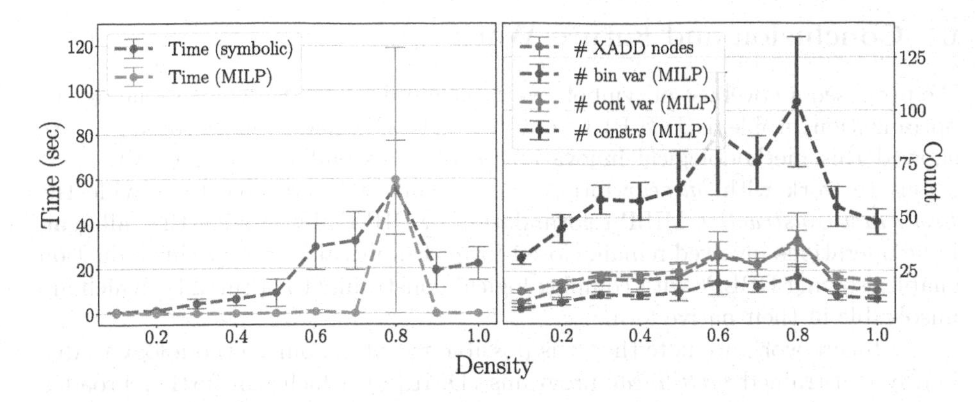

Fig. 5. Time and space complexity as the sparsity of $Q, \{\mathbf{a}_i\}_{i=1}^{n_a}, \{\mathbf{b}_j\}_{j=1}^{n_b}$ changes

and continuous variables, and the number of constraints—have shown only mild increasing patterns.

For (ii), we vary the density parameter used in the generation of the coefficient matrices ($\mathbf{a}, \mathbf{b}, Q$) from 0.1 to 1.0 (full matrices) and record the time and space complexity thereof. The numbers of variables are set to $(n_x, n_y) = (8, 4)$ and we eliminate $\mathbf{y}$ variables. Figure 5 shows a general trend where the MILP reduction becomes increasingly expensive as the density of the coefficient matrices rises. However, the complexity peaks at the density 0.8, and the instances with denser coefficients turn out to be easier to solve. Typically, instances that take longer symbolic compilation running times tend to result in XADDs with more nodes. Hence, it appears that sparse forms have few constraints leading to smaller encodings and solution times, while the highest density problems likely have redundant (implied) constraints that the XADD can eliminate also leading to smaller encodings and solution times.

To sum up, we have seen that there are types of DBLP problems that cannot be solved by Gurobi within a reasonable amount of time in their native form. We are able to solve such problems by solving the MILP equivalent of a DBLP which can be obtained via SVE. Using various test problems, we have also examined the efficiency of the proposed approach. In particular, we have observed that imbalanced problems are much easier to solve with SVE than their balanced counterparts with the same numbers of decision variables. Although it generally takes longer to solve a larger DBLP, there exists a set of problems with which we do not see much increase in solution time as the number of variables increases. These sorts of problems can benefit the most from our symbolic approach. Finally, we have seen that sparse instances can be more compactly represented via XADD, leading to smaller runtimes, while the densest form can be solved relatively easily as well.

6 Conclusion and Future Work

We proposed a novel use of symbolic variable elimination (SVE) for reducing one optimization problem (DBLP) to another (MILP) exactly in closed-form. We showed this methodological innovation involves extending existing SVE operations to work with *bilinear* forms. As a result, we were able to provide the *first exact constructive* MILP reformulation of DBLPs by proving that all symbolic operations involved remain closed-form. Empirically, we saw this reduction enables solving DBLPs with complex logical constraints to optimality, which are unsolvable in their native form.

As future work, we note that it is possible to extend our methodology to disjointly constrained *multilinear* programs (DMLPs), which will further broaden the applicability of our method to multi-agent decision-making problems [2].

Longer term, we hope that this work inspires the use of (and further research into) SVE as a technique for manipulating and reducing constrained optimization problems into alternative forms more amenable for use with highly efficient and optimal off-the-shelf solvers.

References

1. Audet, C., Hansen, P., Jaumard, B., Savard, G.: A symmetrical linear maxmin approach to disjoint bilinear programming. Math. Program. **85**(3), 573–592 (1999)
2. Becker, R., Zilberstein, S., Lesser, V., Goldman, C.V.: Transition-independent decentralized markov decision processes. In: Proceedings of the Second International Joint Conference on Autonomous Agents and Multiagent Systems, pp. 41–48. AAMAS 2003, Association for Computing Machinery, New York (2003). https://doi.org/10.1145/860575.860583
3. Boutilier, C., Reiter, R., Price, B.: Symbolic dynamic programming for first-order MDPs. In: IJCAI-01, pp. 690–697. Seattle (2001)
4. Gurobi Optimization, LLC: Gurobi Optimizer Reference Manual (2021). https://www.gurobi.com
5. Horst, R., Pardalos, P., Van Thoai, N.: Introduction to Global Optimization. Nonconvex Optimization and Its Applications. Springer, US (1995). https://books.google.ca/books?id=w6bRM8W-oTgC
6. Horst, R., Tuy, H.: Global Optimization: Deterministic Approaches. Springer, Heidelberg (2013)
7. Jeong, J., Jaggi, P., Sanner, S.: Symbolic dynamic programming for continuous state mdps with linear program transitions. In: Proceedings of the 30th International Joint Conference on Artificial Intelligence (IJCAI-21). Online (2021)
8. Konno, H.: A Bilinear Programming: Part II. Applications of Bilinear Programming, Technical report (1975)
9. Nahapetyan, A.G.: Bilinear Programming: Applications in the Supply Chain Management, pp. 282–288. Springer, Boston (2009)
10. Petrik, M., Zilberstein, S.: A bilinear programming approach for multiagent planning. J. Artif. Int. Res. **35**(1), 235–274 (2009)
11. Rebennack, S., Nahapetyan, A., Pardalos, P.M.: Bilinear modeling solution approach for fixed charge network flow problems. Optim. Lett. **3**(3), 347–355 (2009). https://doi.org/10.1007/s11590-009-0114-0, https://doi.org/10.1007/s11590-009-0114-0

12. Sanner, S., Abbasnejad, E.: Symbolic variable elimination for discrete and continuous graphical models. In: Proceedings of the AAAI Conference on Artificial Intelligence, vol. 26. no. 1, pp. 1954–1960 (2012)
13. Sanner, S., Delgado, K.V., de Barros, L.N.: Symbolic dynamic programming for discrete and continuous state MDPs. In: Proceedings of the 27th Conference on Uncertainty in AI (UAI-2011). Barcelona (2011)
14. Sherali, H.D., Alameddine, A.: A new reformulation-linearization technique for bilinear programming problems. J. Global Optim. **2**(4), 379–410 (1992)
15. Vicente, L.N., Calamai, P.H., Júdice, J.J.: Generation of disjointly constrained bilinear programming test problems. Comput. Optim. Appl. **1**, 299–306 (1992)
16. Ye, Z., Say, B., Sanner, S.: Symbolic bucket elimination for piecewise continuous constrained optimization. In: CPAIOR, pp. 585–594 (2018). https://doi.org/10.1007/978-3-319-93031-2_42
17. Zamani, Z., Sanner, S., Fang, C.: Symbolic dynamic programming for continuous state and action MDPs. In: Proceedings of the 26th AAAI Conference on Artificial Intelligence (AAAI-12). Toronto, Canada (2012)

Local Branching Relaxation Heuristics for Integer Linear Programs

Taoan Huang[1(✉)], Aaron Ferber[1], Yuandong Tian[2], Bistra Dilkina[1], and Benoit Steiner[2]

[1] University of Southern California, Los Angeles, USA
{taoanhua,aferber,dilkina}@usc.edu
[2] Meta AI (FAIR), Menlo Park, USA
yuandong@meta.com

Abstract. Large Neighborhood Search (LNS) is a popular heuristic algorithm for solving combinatorial optimization problems (COP). It starts with an initial solution to the problem and iteratively improves it by searching a large neighborhood around the current best solution. LNS relies on heuristics to select neighborhoods to search in. In this paper, we focus on designing effective and efficient heuristics in LNS for integer linear programs (ILP) since a wide range of COPs can be represented as ILPs. Local Branching (LB) is a heuristic that selects the neighborhood that leads to the largest improvement over the current solution in each iteration of LNS. LB is often slow since it needs to solve an ILP of the same size as input. Our proposed heuristics, LB-RELAX and its variants, use the linear programming relaxation of LB to select neighborhoods. Empirically, LB-RELAX and its variants compute as effective neighborhoods as LB but run faster. They achieve state-of-the-art anytime performance on several ILP benchmarks.

Keywords: Integer Linear Program · Large Neighborhood Search · Heuristic Search

1 Introduction

Combinatorial optimization problems (COP) concerns a wide variety of real-world applications, including vehicle routing [42], path planning [35] and resource allocation [34] problems. Many of them are difficult to solve with limited computational resources due to their NP-Hardness. Nonetheless, the widespread importance of COPs has inspired research in designing algorithms for solving them, including exact algorithms, approximation algorithms, heuristic algorithms and data-driven algorithms.

In this paper, we focus specifically on Integer Linear Programs (ILPs) since it is a powerful tool to model and solve a broad collection of COPs, including graph optimization [40], mechanism design [11], facility location [4,19] and network design [12,21] problems. Branch-and-Bound (BnB) is an optimal and complete

© The Author(s), under exclusive license to Springer Nature Switzerland AG 2023
A. A. Cire (Ed.): CPAIOR 2023, LNCS 13884, pp. 96–113, 2023.
https://doi.org/10.1007/978-3-031-33271-5_7

tree search algorithm and is one of the state-of-the-art algorithms for ILPs [27]. It is also the core of many ILP solvers such as SCIP [8] and Gurobi [17]. Huge research effort has been made to improve it over the past decades [2]. However, BnB still falls short of delivering practical impact due to scalability issues [14,24]. On the other hand, Large Neighborhood Search (LNS) is a powerful heuristic algorithm for hard COPs and has been recently applied to solve ILPs [40,41,43] in the machine learning (ML) community.

To solve ILPs, LNS starts with an initial solution, i.e., a feasible assignment of values to the variables. It then iteratively improves the best solution found so far (i.e., the *incumbent solution*), by applying *destroy heuristics* to select a subset of variables and solving a sub-ILP that optimizes only the selected variables while leaving others fixed. ML-based destroy heuristics are shown to be efficient and effective but they are often tailored for a specific problem domain and require extensive computational resources for learning. A few non-ML destroy heuristics have been studied, such as the randomized heuristics [40,41] and the Local Branching (LB) heuristic [13,41], but they are either less efficient or effective compared to the ML-based ones. The randomized heuristics select the neighborhood by quickly randomly sampling a subset of variables which is often of bad quality. LB computes the optimal solution across all possible search neighborhoods that differs from the current incumbent solutions on a limited number of variables; however, LB is computationally expensive since it requires solving an ILP that has the same size as the original problem.

To strike a balance between efficiency and effectiveness, we propose a simple yet effective destroy heuristic LB-RELAX that is based on the linear programming (LP) relaxation of LB. Instead of solving an ILP to find the neighborhood as LB does, LB-RELAX computes its LP relaxation. It then selects the variables greedily based on the difference between the values in the incumbent solution and the LP relaxation solution. We also propose two other variants, LB-RELAX-S and LB-RELAX-R, that deploy a sampling method and combine the randomized heuristic with LB-RELAX to help escape local optima more efficiently, respectively. In experiments, we compare LB-RELAX and its variants against LNS with baseline destroy heuristics and BnB on several ILP benchmarks and show that they achieve state-of-the-art anytime performance. We also show that LB-RELAX achieves competitive results with, sometimes even outperform, the ML-based destroy heuristics. We also test LB-RELAX and its variants on selected difficult MIPLIB instances [16] that encompass diverse problem domains, structures and sizes and show that they achieve best performance on at least 40% of the instances. We also empirically show that LB-RELAX and LB-RELAX-S find neighborhoods of similar quality but is much faster than LB. They sometimes even outperform LB due to LB being too slow to find good enough neighborhoods within a reasonable time cutoff.

2 Background

In this section, we first define ILP and introduce its LP relaxation. We then introduce LNS for ILP solving and the Local Branching (LB) heuristic.

Algorithm 1. LNS for ILPs

1: **Input:** An ILP.
2: $\boldsymbol{x}^0 \leftarrow$ Find an intial solution to the input ILP
3: $t \leftarrow 0$
4: **while** time limit not exceeded **do**
5: $\quad \mathcal{X}^t \leftarrow$ Select a subset of variables to destroy
6: $\quad \boldsymbol{x}^{t+1} \leftarrow$ Solve the ILP with additional constraints $\{x_i = x_i^t : x_i \notin \mathcal{X}^t\}$
7: $\quad t \leftarrow t + 1$
8: **return** $\boldsymbol{x}^t$

2.1 ILP and Its LP Relaxation

An *integer linear program (ILP)* is defined as

$$\min \mathbf{c}^\mathsf{T} \boldsymbol{x} \quad \text{s.t. } \boldsymbol{A}\boldsymbol{x} \leq \boldsymbol{b} \text{ and } \boldsymbol{x} \in \{0,1\}^n,$$

where $\boldsymbol{x} = (x_1, \ldots, x_n)^\mathsf{T}$ denotes the n binary variables to be optimized, $\mathbf{c} \in \mathbb{R}^n$ denotes the vector of objective coefficients and $\boldsymbol{A} \in \mathbb{R}^{m \times n}$ and $\boldsymbol{b} \in \mathbb{R}^m$ specify m linear constraints. A *solution* to the ILP is an feasible assignment of values to the variables.

The *linear programming (LP) relaxation* of an ILP is obtained by relaxing binary variables in the ILP to continuous variables between 0 and 1, i.e., by replacing the integer constraint $\boldsymbol{x} \in \{0,1\}^n$ with $\boldsymbol{x} \in [0,1]^n$.

Note that, in this paper, we focus on the formulation above that consists of only binary variables, but our methods can also be applied to mixed integer linear programs with continuous variables and/or non-binary integer variables.

2.2 LNS for ILP Solving

LNS is a heuristic algorithm that starts with an initial solution and then iteratively reoptimizes a part of the solution by applying the destroy and repair operations until a time limit is exceeded. Let $\boldsymbol{x}^0$ be the initial solution. In iteration $t \geq 0$ of the LNS, given the *incumbent solution* $\boldsymbol{x}^t$, defined as the best solution found so far, a destroy operation is done by a *destroy heuristic* where it selects a subset of k_t variables $\mathcal{X}^t = \{x_{i_1}, \ldots, x_{i_{k_t}}\}$. The repair operation is done by solving a sub-ILP with $\mathcal{X}^t$ being the variables while fixing the values of $x_j \notin \mathcal{X}^t$ to be the same as in $\boldsymbol{x}^t$. Compared to BnB, LNS is more effective in improving the objective value $\mathbf{c}^\mathsf{T} x$, or the primal bound, especially on difficult instances [40,41,43]. Compared to other local search methods, LNS explores a large neighborhood in each step and thus, is more effective in avoiding local minima. LNS for ILPs is summarized in Algorithm 1.

2.3 LB Heuristic

The LB Heuristic [13] is originally proposed as a primal heuristic in BnB but is also applicable in LNS for ILP solving [31,41]. Given the incumbent solution $\boldsymbol{x}^t$

in iteration t of LNS, the LB heuristic [13] aims to find the subset of variables to destroy $\mathcal{X}^t$ such that it leads to the optimal $\boldsymbol{x}^{t+1}$ that differs from $\boldsymbol{x}^t$ on at most k_t variables, i.e., it computes the optimal solution $\boldsymbol{x}^{t+1}$ that sits within a given Hamming ball of radius k_t centered around $\boldsymbol{x}^t$. To find $\boldsymbol{x}^{t+1}$, the LB heuristic solves the LB ILP that is exactly the same ILP from input but with one additional constraint that limits the distance between $\boldsymbol{x}^t$ and $\boldsymbol{x}^{t+1}$:

$$\sum_{i\in[n]:x_i^t=0} x_i^{t+1} + \sum_{i\in[n]:x_i^t=1} (1-x_i^{t+1}) \leq k_t.$$

The LB ILP is of the same size of the input ILP (i.e., it has the same number of variables and one more constraint), therefore, it is often slow to run in practice.

3 Related Work

In this section, we summarize related work on LNS for ILPs, LNS-based primal heuristics in BnB and LNS for other COPs.

3.1 LNS for ILPs

While a lot of effort has been made to improve BnB for ILPs in the past decades, LNS for ILPs has not been studied extensively in the past. Recently, Song et al. [40] show that even a randomized destroy heuristic in LNS can outperform state-of-the-art BnB in runtime. In the same paper, they show that an ML-guided decomposition-based LNS can achieve even better performance, where they apply reinforcement learning and imitation learning to learn the destroy heuristics. Since then, there have been a few more recent studies on ML-based LNS for ILPs. Sonnerat et al. [41] learn to select variables to destroy via imitating LB. Wu et al. [43] learn the same thing but they use reinforcement learning instead. The main difference between LB-RELAX and ML-based heuristics is that LB-RELAX does not require extra computational resource for learning and is agnostic to the underlying problem distributions. LB-RELAX also has a better balance between efficiency and effectiveness than those existing non-ML heuristics.

3.2 LNS-Based Primal Heuristics in BnB

LNS-based primal heuristics is one of the rich set of primal heuristics in BnB for ILPs and many techniques have been proposed in past decades. With the same purpose of improving primal bounds of the ILPs, the main differences between the LNS-based primal heuristics in BnB and LNS for ILPs are the following: (1) Since LNS-based primal heuristics are often more expensive to run than the others in BnB, they are executed periodically at different search tree nodes during the main search and the execution schedule is itself dynamic; (2) the destroy heuristics for LNS in BnB are often designed to use information, such as the dual bound and the LP relaxation at a search tree node, that is specific to BnB and not directly applicable in LNS for ILPs in our setting.

Next, we briefly summarize the destroy heuristics in LNS-based primal heuristics. The Crossover heuristics [37] destroy variables that have different values in a set of selected known solutions (typically two). The Mutation heuristics [37] destroys a random subset of variables. Relaxation Induced Neighborhood Search (RINS) [10] destroys variables whose values disagree in the solution of the LP relaxation at the current search tree node and the current incumbent solution. Relaxation Enforced Neighborhood Search (RENS) [7] restricts the neighborhood to be the feasible roundings of the LP relaxation at the current search tree node. Local Branching [13] restricts the neighborhood to a ball around the current incumbent solution. Distance Induced Neighborhood Search (DINS) [15] takes the intersection of the neighborhoods of the Crossover, LB and RINS heuristics. Graph-Induced Neighborhood Search (GINS) [33] destroys the breadth-first-search neighborhood of a variable in the bipartite graph representation of the ILP. An adaptive LNS primal heuristic that essentially solves a multi armed bandit problem has been proposed to combine the power of these heuristics [18].

LB-RELAX is closely related to RINS [10] since they both use LP relaxations to select neighborhoods. However, RINS is more suitable in BnB since it can adapt dynamically to the constraints added by branching. It uses the LP relaxation of the original problem, whereas LB-RELAX uses that of the LB ILP which takes into account the incumbent solutions that could change from iteration to iteration in LNS.

3.3 LNS for Other COPs

LNS has been applied to solve a wide range of COPs, such as the vehicle routing problem [5,36], the traveling salesman problem [39], scheduling problems [26,44] and path planning problems [23,28,29]. Recently, ML-based methods have been applied to improve LNS for those applications [9,20,22,30,32].

4 The Local Branching Relaxation Heuristic

Recently, designing effective destroy heuristics in LNS for ILPs has been a focus in the ML community [40,41,43]. However, it is difficult to apply ML-based destroy heuristics to general ILPs since they are often customized for ILPs from certain problem distributions, e.g., graph optimization problems from a given graph distribution or scheduling problems where resources and demands follow the distribution of historical data, and require extra computational resources for training. There has been a lack of study on destroy heuristics that are agnostic to the underlying distribution of the problem. Existing ones such as randomized heuristics are simple and fast but sometimes not effective [40,41]. LB are effective but not efficient [31,41] since it exhaustively solves an ILP the same size as input for the best improvement.

Algorithm 2. LB-RELAX (LB-RELAX-S)

1: **Input:** An ILP, incumbent solution $\boldsymbol{x}^t$ and neighborhood size k_t.
2: Construct the LB ILP given $\boldsymbol{x}^t$ and k_t
3: $\bar{\boldsymbol{x}}^{t+1} \leftarrow$ Solve the LP relaxation of the LB ILP
4: $\Delta_i \leftarrow |\bar{x}_i{}^{t+1} - x_i^t|$ for all $i \in [n]$
5: $\bar{\mathcal{X}}^t \leftarrow \{x_i : \Delta_i > 0, i \in [n]\}$
6: **if** $|\bar{\mathcal{X}}^t| \geq k_t$ **then**
7: $\quad\mathcal{X}^t \leftarrow$ Select k_t variables greedily with the largest Δ_i from $\bar{\mathcal{X}}^t$
$\quad(\mathcal{X}^t \leftarrow$ Select k_t variables uniformly at random from $\bar{\mathcal{X}}^t)$
8: **else**
9: $\quad\mathcal{X}' \leftarrow$ a random subset of $k_t - |\bar{\mathcal{X}}^t|$ variables from $\{x_i : \Delta_i = 0, i \in [n]\}$
10: $\quad\mathcal{X}^t \leftarrow \bar{\mathcal{X}}^t \cup \mathcal{X}'$
11: **return** $\mathcal{X}^t$

There are well-known approximation algorithms for NP-hard COPs based on LP relaxation [25]. Typically, they solve the LP relaxation of the ILP of the original problem and apply deterministic or randomized rounding afterwards to construct an integral solution. These algorithms often have theoretical guarantee on the effectiveness and are fast, since LP can be solved in polynomial time. Inspired by those algorithms, we propose destroy heuristic LB-RELAX that first solves the LP relaxation of the LB ILP and then constructs the neighborhood (selects variables $\mathcal{X}^t$ to destroy) based on the LP relaxation solution. Specifically, given an ILP and the incumbent solution $\boldsymbol{x}^t$ in iteration t, we construct the LB ILP with neighborhood size k_t and solve its LP relaxation. Let $\bar{\boldsymbol{x}}^{t+1}$ be the LP relaxation solution to the LB ILP. Also, let $\Delta_i = |\bar{x}_i{}^{t+1} - x_i^t|$ and $\bar{\mathcal{X}}^t = \{x_i : \Delta_i > 0, i \in [n]\}$. $\bar{\mathcal{X}}^t$ includes all the fractional variables in the LP relaxation solution and all integral variables that have different values from $\boldsymbol{x}^t$. In the following, we introduce (1) LB-RELAX, (2) LB-RELAX-S, a variant of LB-RELAX with randomized sampling and (3) LB-RELAX-R, another variant of LB-RELAX that combines a randomized destroy with LB-RELAX to help avoid local minima more effectively.

LB-RELAX first gets the LP relaxation solution $\bar{\boldsymbol{x}}^{t+1}$ of the LB ILP and then calculates Δ_i and $\bar{\mathcal{X}}^t$ from $\bar{\boldsymbol{x}}^{t+1}, \boldsymbol{x}^t$. To construct $\mathcal{X}^t$ (the set of variables to destroy), it then greedily selects k_t variables with the largest Δ_i and breaks ties uniformly at random. Intuitively, LB-RELAX greedily selects the variables whose values are more likely to change in the incumbent solution $\boldsymbol{x}^t$ after solving the LB ILP. LB-RELAX is summarized in Algorithm 2. Instead of using the LP relaxation of the LB ILP, one could argue that we alternatively use that of the original ILP similar to RINS [10]. However, the advantage of LB-RELAX over using the LP relaxation of the original problem is that, by approximating the solution to the LB ILP, LB-RELAX selects neighborhoods based on the incumbent solutions that change from iteration to iteration, whereas the original LP relaxation is a static and less informative feature that is pre-computed before the LNS procedure.

LB-RELAX-S is a variant of LB-RELAX with randomized sampling. To construct $\mathcal{X}^t$, instead of greedily choosing variables with the largest Δ_i, it selects k_t variables from $\bar{\mathcal{X}}^t$ uniformly at random. If $|\bar{\mathcal{X}}^t| < k_t$, it selects all variables from $\bar{\mathcal{X}}^t$ and $k_t - |\bar{\mathcal{X}}^t|$ variables from the remaining uniformly at random. LB-RELAX is summarized in Algorithm 2 where the parts in blue highlight the differences between LB-RELAX and LB-RELAX-S. Since $0 \leq \Delta_i \leq 1$, one could treat Δ_i as a probability distribution and sample k_t variables accordingly (see [41] for an example of how to normalize the distribution to sample k_t variables). However, this variant performs similarly to or slightly worse than LB-RELAX-S empirically and require extra hyperparameter tunings for the normalization. We therefore omit it and focus on the simpler variant in this paper.

LB-RELAX-R is another variant of LB-RELAX that leverages a randomized destroy to avoid local minima more effectively. Once LB-RELAX fails to find an improving solution in iteration t, if we let $k_{t+1} = k_t$, it will solve the exact same LP relaxation of the LB ILP again in the next iteration since the incumbent solution $\boldsymbol{x}^{t+1} = \boldsymbol{x}^t$ and the neighborhood size stay the same. Also, since LB-RELAX uses a greedy rule, it will select the same set of variables with the largest Δ_i's deterministically, except that it might need to break ties randomly in some cases when there are multiple variables with the same Δ_i. Therefore, it is susceptible to getting stuck at local minima. To tackle this issue, once LB-RELAX fails to find a new incumbent solution, we update k_{t+1} using the adaptive method described in the next paragraph. If it fails again in the next iteration, we switch to a randomized destroy heuristic that uniformly samples variables at random without replacement to construct the neighborhood. We switch back to LB-RELAX after running the randomized destroy heuristic for at least γ seconds and a new incumbent solution is found.

Next, we discuss an adaptive method to set the neighborhood size k_t for LB-RELAX and its variants. The initial neighborhood size k_0 is set to a constant or a fraction of the number of variables in the input ILP. In iteration t, if LNS finds a new incumbent solution, we let $k_{t+1} = k_t$. Otherwise, we increase k_t by a factor $\alpha > 1$. Also, we upper bound the neighborhood size k_t to a fraction $\beta < 1$ of the number of variables to make sure the sub-ILP in each iteration is not too difficult to solve, i.e., we let $k_{t+1} = \min\{\alpha \cdot k_t, \beta \cdot n\}$. This adaptive way of choosing k_t also helps address the issue of local minima by expanding the search neighborhood when LNS fails to improve the solution. It is applicable to not only LB-RELAX and its variants but also any destroy heuristics that require a given neighborhood size k_t.

5 Empirical Evaluation

In this section, we demonstrate the efficiency and effectiveness of LB-RELAX and its variants through extensive experiments on ILP benchmarks.

5.1 Setup

Instance Generation. We evaluate on four NP-hard problem benchmarks selected from previous work [38,40,43], which consist of synthetic minimum vertex cover (MVC), maximum independent set (MIS), set covering (SC) and multiple knapsack (MK) instances. MVC and MIS instances are generated according to the Barabasi-Albert random graph model [3], with 9,000 nodes and average degree 5 following [40]. SC instances are generated with 4,000 variables and 5,000 constraints following [43]. MK instances are generated with 400 items and 40 knapsacks following [38]. For each problem, we generate 100 instances.

Baselines. We compare LB-RELAX, LB-RELAX-R and LB-RELAX-S with the following baselines:

- BnB using SCIP (v8.0.1) as the solver with the aggressive mode turned on to focus on improving the primal bound;
- LB: LNS which selects the neighborhood with the LB heuristics;
- RANDOM: LNS which selects the neighborhood by uniformly sampling a subset of variables of a given neighborhood size k_t;
- GRAPH: LNS which selects the neighborhood based on the bipartite graph representation of the ILP similar to GINS [33]. A bipartite graph representation consists of nodes representing the variables and constraints on two sides, respectively, with an edge connecting a variable and a constraint if a variable has a non-zero coefficient in the constraint. It runs a breadth-first search starting from a random variable node in the bipartite graph and selects the first k_t variable nodes expanded.

Furthermore, we compare our approaches with state-of-the-art ML approaches:

- IL-LNS: LNS which selects the neighborhood using a GCN-based policy obtained by learning to imitate the LB heuristic [41]. We implement IL-LNS since the authors do not fully open source the code;
- RL-LNS: LNS which selects the neighborhood using a GCN-based policy obtained by reinforcement learning [43]. Note that this approach does not require a given neighborhood size k_t since the size is defined implicitly by how the trained policy is used. We use the code made available by the authors.

Hyperparameters. We conduct our experiments on 2.5 GHz Intel Xeon Platinum 8259CL CPUs with 32 GB RAM. All experiments use the hyperparameters described below unless stated otherwise. We use SCIP (v8.0.1) [8], the state-of-the-art open source ILP solver for the repair operations in LNS. To run LNS, we find an initial solution by running SCIP for 10 s for MVC, MIS and SC and 20 s for MK. We set the time limit to 60 min to solve each instance and 2 min for each repair operation in LNS. Except for LB, we set the time limit to 10 min for each repair operation since LB solves a larger ILP than other approaches in each iteration and typically requires a longer time limit. All approaches require a

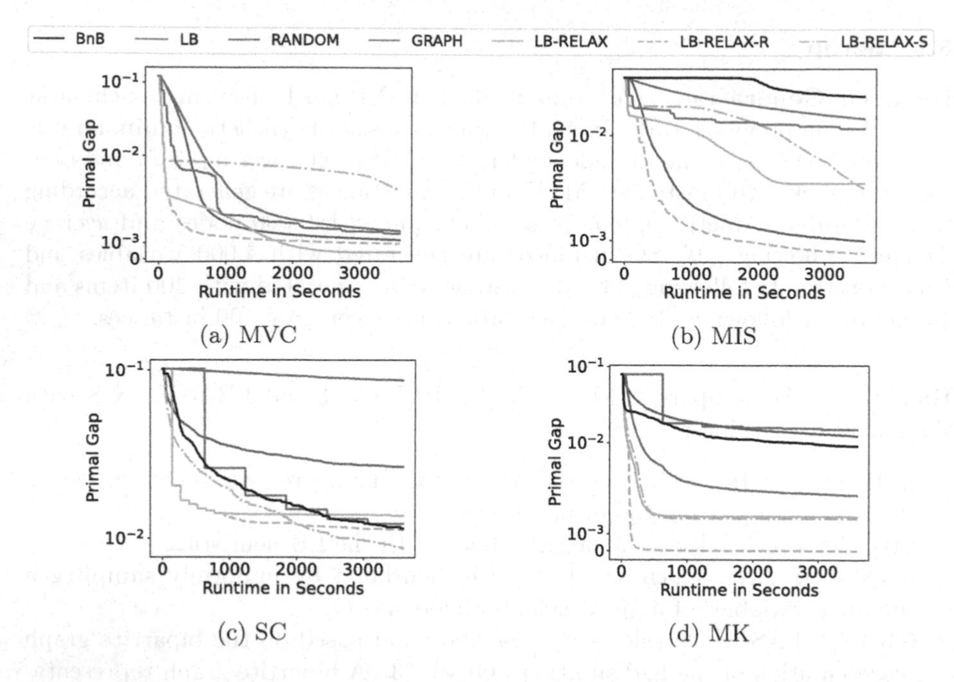

Fig. 1. Comparison with non-ML approaches: The primal gap as a function of time, averaged over 100 instances.

neighborhood size k_t in LNS, except for BnB and RL-LNS. The initial neighborhood size (k_0) is set to $k_0 = 400, 200, 150$ and 400 for MVC, MIS, SC and MK, respectively. For fair comparison, all baselines use adaptive neighborhood sizes with $\alpha = 1.02$ and $\beta = 0.5$, except for BnB and RL-LNS. For LB-RELAX-R, we set $\gamma = 30$ s. Additional details on tuning hyperparameters are included in Appendix[1].

Metrics. We use the following metrics to evaluate the efficiency and effectiveness of different approaches: (1) The *primal bound* is the objective value of the ILP. (2) The *primal gap* [6] is the normalized difference between the primal bound v and a precomputed best known objective value v^*, defined as $\frac{|v-v^*|}{\max(v,v^*,\epsilon)}$ if v exists and $v \cdot v^* \geq 0$, or 1 otherwise. We use $\epsilon = 10^{-8}$ to avoid division by zero and v^* is the best primal bound found within 60 min by any approach in the portfolio for comparison. (3) The *primal integral* [1] at time q is the integral on $[0, q]$ of the primal gap as a function of time. It captures the quality of and the speed at which solutions are found. (4) The *survival rate* to meet a certain primal gap threshold is the fraction of instances with the primal gap below the threshold [41]. Since BnB and LNS are both anytime algorithms, we

[1] Appendix is available in the full version of the paper: https://arxiv.org/abs/2212.08183.

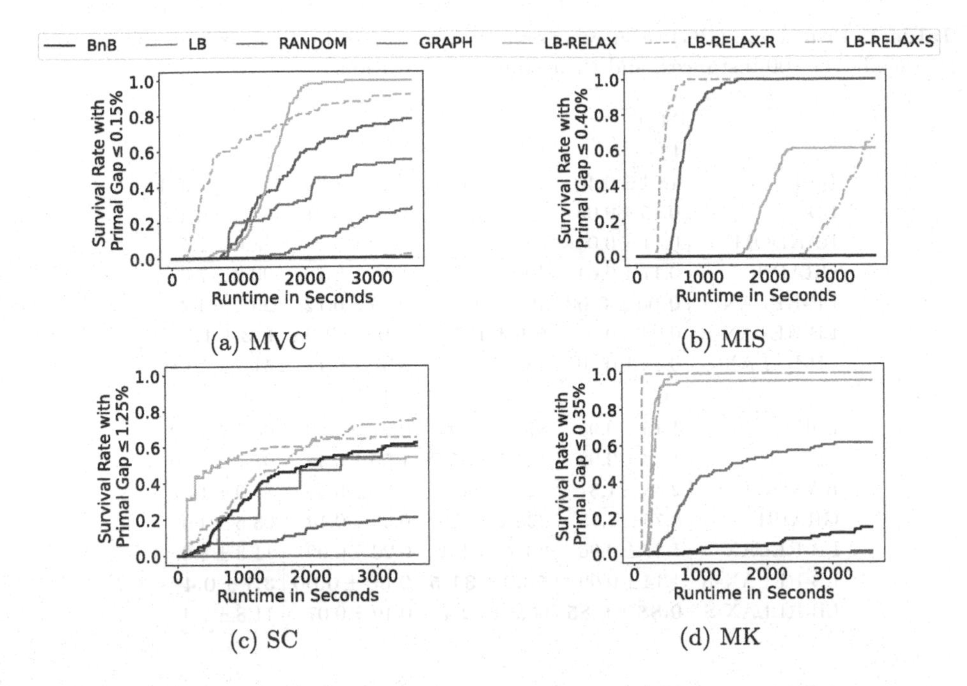

Fig. 2. Comparison with non-ML approaches: The survival rate over 100 instances as a function of time to meet a certain primal gap threshold. The primal gap threshold is chosen from Table 1 as the median of the average primal gaps at 60 min time cutoff over all approaches rounded to the nearest 0.05%.

show the metrics as a function of time or the number of iterations in LNS (when applicable) to demonstrate their anytime performance.

5.2 Results

Comparison with Non-ML Approaches. First, we compare LB-RELAX, LB-RELAX-R and LB-RELAX-S with non-ML approaches, namely BnB, LB, RANDOM and GRAPH. Figure 1 shows the primal gap as a function of time, averaged over 100 instances. The results show that LB-RELAX, LB-RELAX-R and LB-RELAX-S consistently improve the primal gap a lot faster than the baselines in the first few minutes of LNS. LB-RELAX improves the primal gap slightly faster than LB-RELAX-S in all cases. On average, LB-RELAX is always better than the baselines at any point of time on MK instances and LB-RELAX-S is always better than the baselines on SC and MK instances. However, both LB-RELAX and LB-RELAX-S could get stuck at some local minima. In those cases, they need some time to escape local minima by adjusting the neighborhood size and sometimes could be outperformed by some baselines with longer time on the MVC and MIS instances. By adding randomization to LB-RELAX, LB-RELAX-R escapes local minima more efficiently than LB-RELAX and LB-RELAX-S. On

Table 1. Primal gap (PG) (in percent) and primal integral (PI) at 60 min time cutoff, averaged over 100 instances, and their standard deviations.

	MVC		MIS	
	PG (%)	PI	PG (%)	PI
BnB	1.01 ± 0.46	128.6 ± 14.6	2.80 ± 1.36	144.0 ± 20.1
LB	0.15 ± 0.08	22.1 ± 3.6	1.20 ± 0.31	56.3 ± 9.4
RANDOM	0.11 ± 0.05	32.3 ± 2.3	0.10 ± 0.05	18.0 ± 2.5
GRAPH	0.17 ± 0.04	40.8 ± 2.5	1.56 ± 0.18	90.2 ± 7.6
LB-RELAX	**0.04 ± 0.03**	10.3 ± 1.7	0.39 ± 0.12	29.4 ± 4.3
LB-RELAX-R	0.09 ± 0.04	**9.6 ± 1.7**	**0.04 ± 0.04**	**9.3 ± 1.7**
LB-RELAX-S	0.42 ± 0.20	28.8 ± 8.1	0.37 ± 0.11	51.7 ± 10.1
	SC		MK	
BnB	1.15 ± 0.98	87.4 ± 38.6	0.91 ± 0.59	60.7 ± 17.9
LB	1.23 ± 0.98	114.1 ± 35.7	1.50 ± 0.48	97.7 ± 13.0
RANDOM	2.68 ± 1.31	124.4 ± 45.7	1.24 ± 0.36	68.9 ± 14.7
GRAPH	8.75 ± 2.15	338.2 ± 77.0	0.33 ± 0.14	23.6 ± 4.9
LB-RELAX	1.37 ± 0.96	63.9 ± 34.0	0.20 ± 0.09	11.3 ± 3.0
LB-RELAX-R	1.14 ± 0.90	**58.9 ± 31.5**	**0.00 ± 0.00**	**3.7 ± 0.4**
LB-RELAX-S	**0.88 ± 0.85**	63.8 ± 32.4	0.19 ± 0.07	11.8 ± 2.4

average, LB-RELAX-R is always better than the baselines at any point of time in the search on the MVC, MIS and MK instances.

Table 2. The time (in seconds) to improve the initial solution in one iteration and the improvement of the primal bound, averaged over 100 instances. The time for LB is the solving time of the LB ILP. The time for LB-RELAX and LB-RELAX-S is the sum of the solving times of the LB relaxation and the sub-ILP. The numbers in parentheses are the speed-ups. The improvement is computed by taking the difference between the initial solution and the new incumbent solution and the numbers in parentheses are the losses in quality in percent compared to LB. ↑ means higher is better, ↓ means lower is better.

		MVC	MIS	SC	MK
LB	Time↓	40.2	56.0	600.0	600.0
	Imp.↑	129.79	65.50	12.21	216.51
LB-RELAX	Time↓	12.1 (3.3x)	19.5 (2.9x)	125.3 (4.8x)	5.87 (102.2x)
	Imp.↑	129.41 (−0.3%)	65.19 (−0.5%)	15.77 (+29.2%)	141.10 (−34.8%)
LB-RELAX-S	Time↓	12.0 (3.4x)	19.5 (2.9x)	24.51 (24.5x)	5.12 (117.6x)
	Imp.↑	128.61 (−0.9%)	62.46 (−4.6%)	5.65 (−53.7%)	113.48 (−47.6%)

Table 1 presents the average primal gap and primal integral at 60 min time cutoff. (See results at 15, 30 and 45 min time cutoff in Appendix.) On MVC, SC and MK instances, all LB-RELAX, LB-RELAX-S and LB-RELAX-R have lower

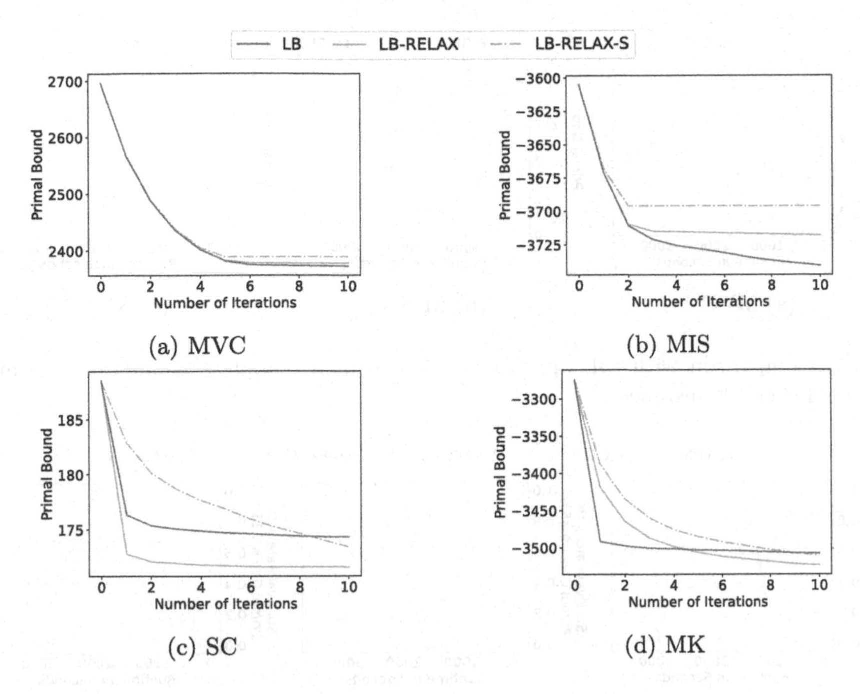

Fig. 3. Comparison with LB: The primal bound as a function of the number of iterations, averaged over 100 instances.

primal gaps and primal integrals on average than any baselines, demonstrating that they not only find higher quality solutions but also find them at a faster speed. On MIS and MK instances, LB-RELAX-R achieves the lowest primal gap and primal integral among all approaches. It also achieves the lowest primal integral on MVC and SC instances. Overall, LB-RELAX-R always comes up in the top 2 in both metrics on all problems.

Figure 2 shows the survival rate over 100 instances as a function of time to meet a certain primal gap threshold. On MVC instances, LB-RELAX and LB-RELAX-R achieve final survival rates above 0.9 while the best baseline RANDOM stays below 0.8. On MIS instances, both LB-RELAX-R and RANDOM achieve final survival rates of 1.0 but LB-RELAX-R uses shorter time. On SC instances, LB-RELAX-S and LB-RELAX-R consistently has a higher survival rate than the baselines. On MK instances, LB-RELAX and its variants achieve survival rates above 0.9 within 15 min while the best baseline GRAPH only gets to around 0.6 with 60 min.

One limitation of LB-RELAX and its variants is that they do not perform well on some problem domains, for example the maximum cut and combinatorial auction problems. Please see Appendix for more results.

Next, we run LB, LB-RELAX and LB-RELAX-S for 10 iterations to compare their effectiveness. We follow the same setup as earlier described, except that we do not use adaptive neighborhood sizes to make sure they have the same

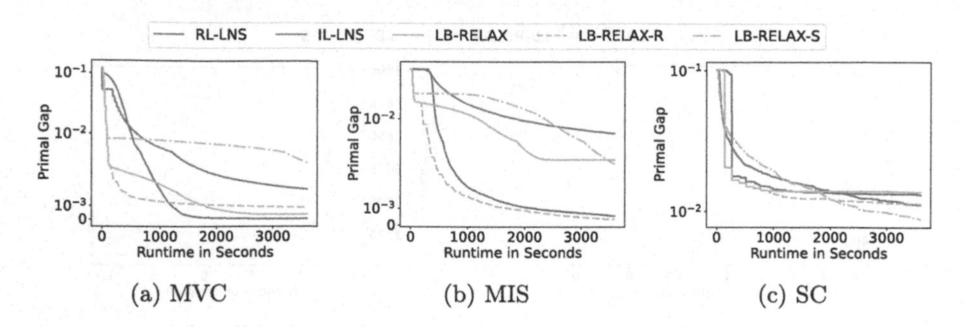

Fig. 4. Comparison with ML approaches: The primal bound as a function of time, averaged over 100 instances.

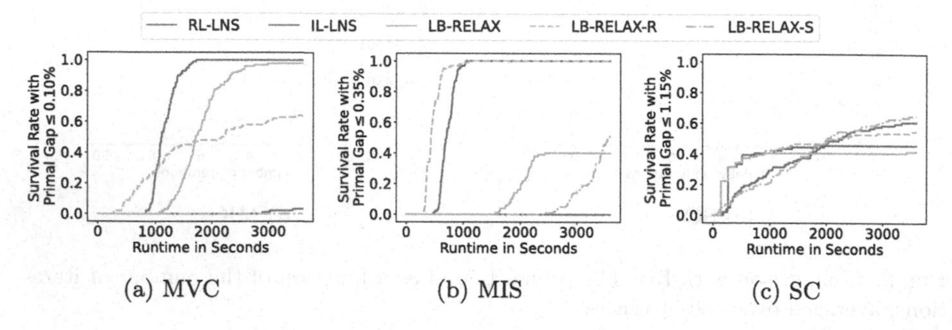

Fig. 5. Comparison with ML approaches: The survival rate over 100 instances as a function of time to meet a certain primal gap threshold. The primal gap thresholds are chosen in the same way as Fig. 2.

k_t in each iteration t. Note that the time limit for solving the sub-ILP in each iteration is set to 10 min for LB and 2 min for LB-RELAX and LB-RELAX-S. Table 2 shows the average time to improve the initial solutions and the average improvement of the primal bound in the first iteration of LNS. This allows us to compare how closely LB-RELAX and LB-RELAX-S approximate the quality of the neighborhood selected by LB and study the trade-off between quality and time. Compared to LB, LB-RELAX and LB-RELAX-S have 2.9x–117.6x speed-up but only lose at most 53.7% in quality. In particular, on MVC and MIS instances, both LB-RELAX and LB-RELAX-S lose 0.5% to 4.6% in quality but have at least 2.9x speed-up; on SC instances, LB-RELAX even gains 29.2% in quality and save 79.1% in time, due to LB cannot find a good enough neighborhood within its time limit (Fig. 5).

In Fig. 3, we show the primal bound as a function of the number of iterations. It allows comparing the effectiveness of different heuristics independently of their speed. On the MVC instances, both LB-RELAX and LB-RELAX-S perform similarly to but slightly worse than LB. On the SC and MK instances, LB-RELAX achieves better performance than LB, again due to scalability issues of LB, and LB-RELAX-S achieves competitive performance with LB after 10

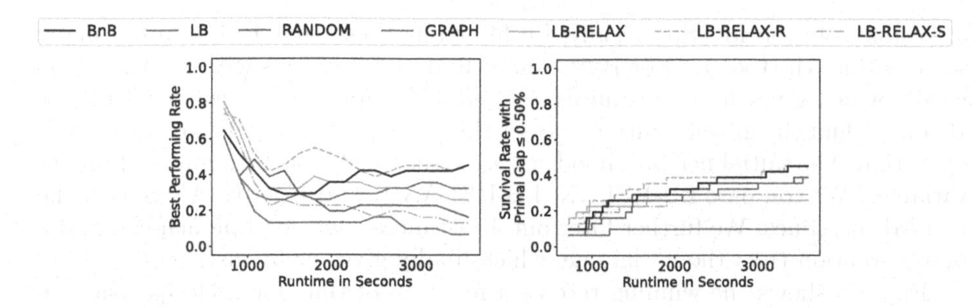

Fig. 6. Results on 31 selected MIPLIB instances: The best performing rate as a function of time (left) and the survival rate over 31 instances as a function of time to meet the primal gap threshold 0.50% (right).

iterations. However on the MIS instances, both LB-RELAX and LB-RELAX-S are able to quickly improve the primal bound in the first 2–3 iterations, but afterwards converge to local minima and the gaps between them and LB increase. To complete the first 10 iterations, both LB-RELAX and LB-RELAX-S take less than 21 min on SC instances and 3.3 min on the others, while LB takes at least 57 min and sometimes up to 100 min.

Comparison with ML Approaches. Then, we compare LB-RELAX, LB-RELAX-R and LB-RELAX-S on MVC, MIS and SC instances with ML approaches, namely IL-LNS and RL-LNS. Figure 4 shows the primal gap as a function of time averaged over 100 instances. The results show that LB-RELAX, LB-RELAX-R and LB-RELAX-S consistently improve the primal bound a lot faster than IL-LNS and RL-LNS in the first few minutes of LNS. On MVC instances, IL-LNS surpasses LB-RELAX-R with the smallest average primal gap best after 20 min and achieve (close-to-)zero gaps after 30 min. On MIS instances, LB-RELAX-R has a smaller gap than both IL-LNS and RL-LNS throughout the first 60 min. On SC instances, IL-LNS is very competitive with LB-RELAX and converges to a similar but slightly higher gap than LB-RELAX-R and LB-RELAX-S; RL-LNS converges to almost the same primal gap as LB-RELAX-R on average but is worse than the best performer LB-RELAX-S. Overall, LB-RELAX and its variants, that do not require extra computational resources for training, are competitive with and more often even better than state-of-the-art ML approaches, suggesting that they are agnostic to the distributions of the instances and easily applicable to different problem domains.

Results on Selected MIPLIB Instances. Finally, we examine how well LB-RELAX and its variants perform on ILPs that are diverse in structures and sizes. We test them on the MIPLIB dataset [16]. MIPLIB contains COPs from various real-world domains. We follow a procedure similar to [43] to filter out instances where we first filter to retain ILP instances with only binary variables. Among these, we select instances that are not too easy to solve but relatively easy to find a feasible solution for. Specifically, we filter out those that BnB can optimally solve within 3 h (too easy) or BnB cannot find any solutions within 10 min (too hard), which gives us 35 instances. For all LNS approaches, we run BnB for 10 min to find the initial solution and set the time limit to 10 min for each repair operation. The initial neighborhood size k_0 is set to 20% of the number of binary variables. We compare LB-RELAX, LB-RELAX-R and LB-RELAX-S with the non-ML baselines. We further filter out 4 instances that no approach can find a better solution than the initial one, which finally gives us 31 instances.

Figure 6 shows the winning rate as a function of time for each approach on the 31 instances. The best performing rate at a time q for an approach is the fraction of instances on which it achieves the best performance (including ties) compared to all approaches in the portfolio. LB-RELAX, LB-RELAX-R and LB-RELAX-S achieve the best performance with less than 1000 s seconds on 25, 23 and 24 instances out of 35, respectively. LB-RELAX-R has the highest best performing rates at different time cutoffs and ties with BnB at 14 instances at the 60-minute mark. Figure 6 also shows the survival rate over the 31 instances as a function of time to meet the primal gap threshold 0.50%. It demonstrates that RANDOM, GRAPH and BnB are competitive with our approaches but overall LB-RELAX-R has the highest survival rate over time. On some instances, LB-RELAX and its variants can significantly outperform the baselines and we show the anytime performance on those in Appendix.

6 Conclusion

In this paper, we focused on designing effective and efficient destroy heuristics to select neighborhoods in LNS for ILPs. LB is an effective destroy heuristic but is slow to run. We therefore proposed LB-RELAX, LB-RELAX-S and LB-RELAX-R to approximate LB's decisions by solving its LP relaxation that is a lot faster to run. Empirically, we showed that LB-RELAX, LB-RELAX-S and LB-RELAX-R efficiently selected almost as effective neighborhoods as LB and achieved state-of-the-art performance when compared against non-ML and ML approaches. One limitation of our approaches is that they do not work well on some problem domains, however we showed that they still outperformed the baselines on 14 to 25 (depending on the time cutoff) out of 31 difficult MIPLIB instances that are diverse in problem domains, structures and sizes. The other limitation is that they can get stuck at local minima. To address this issue, we proposed techniques to randomize the heuristics and adaptively adjust the neighborhood sizes. For future work, one could improve LB-RELAX and its variants to make them applicable on more problem domains. In addition, instead of using hard-coded rules for scheduling the randomized heuristic in LB-RELAX-R, one could use adaptive LNS to select destroy heuristics to run. It is also future work to develop theoretical claims to help support and explain the effectiveness of LB-RELAX, LB-RELAX-S, LB-RELAX-R and possibly their other variants.

Acknowledgements. This paper reports on research done while Taoan Huang and Aaron Ferber were interns at Meta AI (FAIR). The research at the University of Southern California was supported by the National Science Foundation (NSF) under grant number 2112533.

References

1. Achterberg, T., Berthold, T., Hendel, G.: Rounding and propagation heuristics for mixed integer programming. In: Klatte, D., Lüthi, H.J., Schmedders, K. (eds.) Operations Research Proceedings 2011. Operations Research Proceedings, pp. 71–76. Springer, Heidelberg (2012). https://doi.org/10.1007/978-3-642-29210-1_12
2. Achterberg, T., Wunderling, R.: Mixed integer programming: analyzing 12 years of progress. In: Jünger, M., Reinelt, G. (eds.) Facets of Combinatorial Optimization, pp. 449–481. Springer, Heidelberg (2013). https://doi.org/10.1007/978-3-642-38189-8_18
3. Albert, R., Barabási, A.L.: Statistical mechanics of complex networks. Rev. Mod. Phys. **74**(1), 47 (2002)
4. Amaral, A.R.: An exact approach to the one-dimensional facility layout problem. Oper. Res. **56**(4), 1026–1033 (2008)
5. Azi, N., Gendreau, M., Potvin, J.Y.: An adaptive large neighborhood search for a vehicle routing problem with multiple routes. Comput. Oper. Res. **41**, 167–173 (2014)
6. Berthold, T.: Primal heuristics for mixed integer programs. Ph.D. thesis, Zuse Institute Berlin (ZIB) (2006)
7. Berthold, T.: Rens. Math. Program. Comput. **6**(1), 33–54 (2014)
8. Bestuzheva, K., et al.: The SCIP optimization suite 8.0. Technical report, Optimization Online (2021). http://www.optimization-online.org/DB_HTML/2021/12/8728.html
9. Chen, X., Tian, Y.: Learning to perform local rewriting for combinatorial optimization. Adv. Neural Inf. Process. Syst. **32** (2019)
10. Danna, E., Rothberg, E., Pape, C.L.: Exploring relaxation induced neighborhoods to improve MIP solutions. Math. Program. **102**(1), 71–90 (2005)
11. De Vries, S., Vohra, R.V.: Combinatorial auctions: a survey. INFORMS J. Comput. **15**(3), 284–309 (2003)
12. Dilkina, B., Gomes, C.P.: Solving connected subgraph problems in wildlife conservation. In: Lodi, A., Milano, M., Toth, P. (eds.) CPAIOR 2010. LNCS, vol. 6140, pp. 102–116. Springer, Heidelberg (2010). https://doi.org/10.1007/978-3-642-13520-0_14
13. Fischetti, M., Lodi, A.: Local branching. Math. program. **98**(1), 23–47 (2003)
14. Gasse, M., Chételat, D., Ferroni, N., Charlin, L., Lodi, A.: Exact combinatorial optimization with graph convolutional neural networks. Adv. Neural Inf. Process. Syst. **32** (2019)
15. Ghosh, S.: DINS, a MIP improvement heuristic. In: Fischetti, M., Williamson, D.P. (eds.) IPCO 2007. LNCS, vol. 4513, pp. 310–323. Springer, Heidelberg (2007). https://doi.org/10.1007/978-3-540-72792-7_24
16. Gleixner, A., et al.: MIPLIB 2017: data-driven compilation of the 6th mixed-integer programming library. Math. Program. Comput. **13**(3), 443–490 (2021). https://doi.org/10.1007/s12532-020-00194-3
17. Gurobi Optimization, LLC: Gurobi Optimizer Reference Manual (2022). https://www.gurobi.com

18. Hendel, G.: Adaptive large neighborhood search for mixed integer programming. Math. Program. Comput. **14**(2), 185–221 (2022)
19. Heragu, S.S., Kusiak, A.: Efficient models for the facility layout problem. Eur. J. Oper. Res. **53**(1), 1–13 (1991)
20. Hottung, A., Tierney, K.: Neural large neighborhood search for the capacitated vehicle routing problem. In: ECAI 2020, pp. 443–450. IOS Press (2020)
21. Huang, T., Dilkina, B.: Enhancing seismic resilience of water pipe networks. In: Proceedings of the 3rd ACM SIGCAS Conference on Computing and Sustainable Societies, pp. 44–52 (2020)
22. Huang, T., Li, J., Koenig, S., Dilkina, B.: Anytime multi-agent path finding via machine learning-guided large neighborhood search. In: Proceedings of the AAAI Conference on Artificial Intelligence (AAAI), pp. 9368–9376 (2022)
23. Huang, T., et al.: Deadline-aware multi-agent tour planning. In: Proceedings of the International Conference on Automated Planning and Scheduling (ICAPS) (2023)
24. Khalil, E., Le Bodic, P., Song, L., Nemhauser, G., Dilkina, B.: Learning to branch in mixed integer programming. In: Proceedings of the AAAI Conference on Artificial Intelligence, vol. 30 (2016)
25. Kleinberg, J., Tardos, E.: Algorithm Design. Pearson Education India (2006)
26. Kovacs, A.A., Parragh, S.N., Doerner, K.F., Hartl, R.F.: Adaptive large neighborhood search for service technician routing and scheduling problems. J. Sched. **15**(5), 579–600 (2012)
27. Land, A.H., Doig, A.G.: An automatic method for solving discrete programming problems. In: Jünger, M., Liebling, T.M., Naddef, D., Nemhauser, G.L., Pulleyblank, W.R., Reinelt, G., Rinaldi, G., Wolsey, L.A. (eds.) 50 Years of Integer Programming 1958-2008, pp. 105–132. Springer, Heidelberg (2010). https://doi.org/10.1007/978-3-540-68279-0_5
28. Li, J., Chen, Z., Harabor, D., Stuckey, P.J., Koenig, S.: Anytime multi-agent path finding via large neighborhood search. In: Proceedings of the International Joint Conference on Artificial Intelligence (IJCAI), pp. 4127–4135 (2021)
29. Li, J., Chen, Z., Harabor, D., Stuckey, P.J., Koenig, S.: MAPF-LNS2: fast repairing for multi-agent path finding via large neighborhood search. In: Proceedings of the AAAI Conference on Artificial Intelligence (AAAI), pp. 10256–10265 (2022)
30. Li, S., Yan, Z., Wu, C.: Learning to delegate for large-scale vehicle routing. Adv. Neural. Inf. Process. Syst. **34**, 26198–26211 (2021)
31. Liu, D., Fischetti, M., Lodi, A.: Revisiting local branching with a machine learning lens
32. Lu, H., Zhang, X., Yang, S.: A learning-based iterative method for solving vehicle routing problems. In: International Conference on Learning Representations (2019)
33. Maher, S.J., et al.: The SCIP optimization suite 4.0 (2017)
34. Manne, A.S.: On the job-shop scheduling problem. Oper. Res. **8**(2), 219–223 (1960)
35. Pohl, I.: Heuristic search viewed as path finding in a graph. Artif. Intell. **1**(3–4), 193–204 (1970)
36. Ropke, S., Pisinger, D.: An adaptive large neighborhood search heuristic for the pickup and delivery problem with time windows. Transp. Sci. **40**(4), 455–472 (2006)
37. Rothberg, E.: An evolutionary algorithm for polishing mixed integer programming solutions. INFORMS J. Comput. **19**(4), 534–541 (2007)
38. Scavuzzo, L., et al.: Learning to branch with tree MDPS. arXiv preprint arXiv:2205.11107 (2022)
39. Smith, S.L., Imeson, F.: GLNS: an effective large neighborhood search heuristic for the generalized traveling salesman problem. Comput. Oper. Res. **87**, 1–19 (2017)

40. Song, J., Yue, Y., Dilkina, B., et al.: A general large neighborhood search framework for solving integer linear programs. Adv. Neural. Inf. Process. Syst. **33**, 20012–20023 (2020)
41. Sonnerat, N., Wang, P., Ktena, I., Bartunov, S., Nair, V.: Learning a large neighborhood search algorithm for mixed integer programs. arXiv preprint arXiv:2107.10201 (2021)
42. Toth, P., Vigo, D.: The Vehicle Routing Problem. SIAM (2002)
43. Wu, Y., Song, W., Cao, Z., Zhang, J.: Learning large neighborhood search policy for integer programming. Adv. Neural. Inf. Process. Syst. **34**, 30075–30087 (2021)
44. Žulj, I., Kramer, S., Schneider, M.: A hybrid of adaptive large neighborhood search and tabu search for the order-batching problem. Eur. J. Oper. Res. **264**(2), 653–664 (2018)

Online Learning for Scheduling MIP Heuristics

Antonia Chmiela[1(✉)], Ambros Gleixner[1,2], Pawel Lichocki[3], and Sebastian Pokutta[1,4]

[1] Zuse Institute Berlin, Berlin, Germany
{chmiela,gleixner,pokutta}@zib.de
[2] Hochschule für Technik und Wirtschaft Berlin, Berlin, Germany
[3] Google Research, Mountain View, USA
pawell@google.com
[4] Technische Universität Berlin, Berlin, Germany

Abstract. Mixed Integer Programming (MIP) is NP-hard, and yet modern solvers often solve large real-world problems within minutes. This success can partially be attributed to heuristics. Since their behavior is highly instance-dependent, relying on hard-coded rules derived from empirical testing on a large heterogeneous corpora of benchmark instances might lead to sub-optimal performance. In this work, we propose an online learning approach that adapts the application of heuristics towards the single instance at hand. We replace the commonly used static heuristic handling with an adaptive framework exploiting past observations about the heuristic's behavior to make future decisions. In particular, we model the problem of controlling Large Neighborhood Search and Diving – two broad and complex classes of heuristics – as a multi-armed bandit problem. Going beyond existing work in the literature, we control two different classes of heuristics simultaneously by a single learning agent. We verify our approach numerically and show consistent node reductions over the MIPLIB 2017 Benchmark set. For harder instances that take at least 1000 s to solve, we observe a speedup of 4%.

Keywords: Mixed Integer Programming · Machine Learning · Heuristics

1 Introduction

A multitude of problems arising from real-world applications can be modeled as *Mixed Integer Problems (MIPs)*. Because of that, there is high interest in

---

This work was partially funded by the Deutsche Forschungsgemeinschaft (DFG, German Research Foundation) under Germany's Excellence Strategy - The Berlin Mathematics Research Center MATH+ (EXC-2046/1, 390685689) and by the German Federal Ministry of Education and Research (BMBF) within the Research Campus MODAL (05M14ZAM, 05M20ZBM) and supported by a Google Research Award.

© The Author(s), under exclusive license to Springer Nature Switzerland AG 2023
A. A. Cire (Ed.): CPAIOR 2023, LNCS 13884, pp. 114–123, 2023.
https://doi.org/10.1007/978-3-031-33271-5_8

finding ways to solve MIPs efficiently. Generally, the *Branch-and-Bound (B&B)* framework [23] is used which decomposes the optimization problem in smaller subproblems that are then easier to handle. Since this approach involves a variety of decisions that significantly influence its behavior, the idea of using machine learning (ML) has gained interest: ML has been used to find good solver parameters [11,19,20], to improve node [14,32], variable [2,12,21,26,27,29], and cut selection [3,18,28,30,31], and to detect decomposable structures [22].

The objective of B&B is to solve MIPs to global optimality. However, it is often not feasible to wait until the optimum is found, thus finding good feasible solutions early on is important. Primal heuristics are crucial for this: In [4], the authors showed that heuristics improved the primal bound by 80% and the solving time by 30% on average. An overview of different primal heuristics and their impact can be found in [5,6,26].

Primal heuristics are powerful but can be very costly, thus it is important to be strategic about how they are applied in practice. Controlling their behavior by hard-coded rules derived from empirical tests on heterogeneous benchmark sets leads to strategies that work averagely well on a broad variety of instances. However, since the performance of heuristics is highly instance-dependent, this might lead to suboptimal behavior. For example, primal performance can be significantly improved by deriving problem-specific heuristic settings [10].

In this work, we present an online learning approach to control primal heuristics within B&B. We model heuristic selection as a multi-armed bandit problem and exploit past observations of heuristics' behavior to learn on-the-fly which heuristics are most likely to be successful. Our scheduler is, thus, capable to adapt to and leverage specific characteristic of the problem at hand. In particular, we control Large Neighborhood Search and Diving, two significantly different and complex classes of heuristics.

Contribution. To the best of our knowledge, this is the first time when two different classes of heuristics are treated simultaneously by a single learning agent. To summarize:

1. We propose an **online learning approach for heuristic scheduling** to replace more static heuristic handling (Sect. 3),
2. We support our findings by **extensive computational experiments** on a heterogeneous benchmark test set to numerically verify our approach (Sect. 4).

Related Work. Since heuristics have a large impact on solver performance, using ML to develop new strategies and to optimize their usage is a topic of ongoing research. For instance, [27] use neural networks to derive variable assignments to find primal solutions. The authors in [17] propose a bi-layer prediction model utilizing graph convolutional networks designed to help heuristics find solutions faster. To improve the usage of heuristics, the authors in [21] learn an oracle that aims to predict at which nodes a heuristic will be successful or not. Whereas in [10] a data-driven heuristic scheduling framework is proposed that learns problem-specific heuristic schedules to find many solutions at minimal cost.

In [15,16] adaptive heuristics are built that use bandit algorithms to decide which heuristics to additionally run. In particular, their ALNS heuristic [15] inspired the framework we present here: While ALNS was designed as another primal heuristic to be added in the pool of available heuristics, we extend it to a framework that aims to replace static heuristic handling and that can be easily extendable to handle any class of heuristics.

2 Background

Mixed Integer Problems. A MIP is an optimization problem of the form

$$\min_{x} \ c^T x, \ \text{s.t.} \ Ax \leq b, \ x_i \in \mathbb{Z}, i \in I, \tag{P}$$

with matrix $A \in \mathbb{R}^{m \times n}$, vectors $b, c \in \mathbb{R}^m$ and index set $I \subseteq [n]$. To solve (P), B&B partitions the feasible region, resulting in a tree structure with nodes correspond to the simpler subproblems.

Primal Heuristics. Heuristics aim to find feasible solutions for (P). Generally, a solver utilizes a variety of heuristics exploiting different ideas to find high-quality solutions. Two of the most complex and time consuming classes of heuristics are *diving* and *Large Neighborhood Search (LNS)*. Diving heuristics examine a single probing path by sub-sequentially fixing variables according to a specific rule. In contrast, LNS builds a neighborhood around a reference point by fixing a certain percentage of variables and then solving the resulting sub-MIP. Since no heuristic is guaranteed to be successful, the solver iterates over all available heuristics in a predefined order to hopefully find a new solution. Good heuristics, like diving and LNS, are typically computationally expensive. Thus, it is especially important to be strategic about controlling them.

Multi-Armed Bandit Problem. Given a set of actions $\mathcal{A}$, an agent aims to select a series of actions with maximal cumulative reward. In every iteration t, an action $a_t \in \mathcal{A}$ is selected for which a reward $r_t \in [0, 1]$ is observed. Since the agent only learns how the selected action behaves, a good strategy entails a balance between exploring unknown actions and exploiting the ones that performed well in the past. There are various approaches to finding a good strategy, see [9,24].

3 Scheduling Primal Heuristics Online

We present an online learning approach that models heuristic handling as a multi-armed bandit problem. Thereby, the set of actions $\mathcal{A}$ corresponds to the set of heuristics $\mathcal{H}$ we want to control. Two main challenges arise when modeling the scheduling of heuristics this way: (i) defining a suitable reward function and (ii) choosing the right bandit algorithm. After presenting our online scheduling framework, we describe how we tackle both in Sect. 3.2 and 3.3, respectively.

3.1 The Online Scheduling Framework

The scheduler controls a set of primal heuristics $\mathcal{H}$. Each heuristic has special working limits influencing its behavior. Whenever the scheduler is called, it selects and executes one heuristic $h \in \mathcal{H}$. Depending on how h performed, we dynamically adapt some of its working limits. This way, we not only tailor heuristic handling to the instance at hand but also reduce the number of user-defined parameters. To summarize, the scheduler executes the following steps:

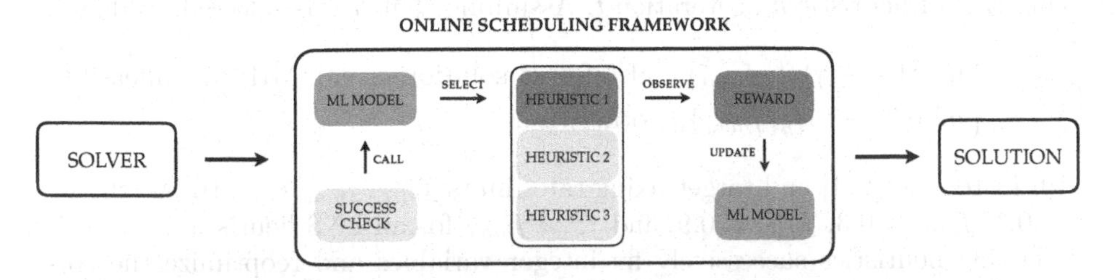

Fig. 1. Visualisation of the Online Scheduling Framework: When the solver decides to run heuristics, it is checked if the scheduler was successful enough in the past. If so, a bandit algorithm selects a heuristic which is executed with specific working limits. A reward is observed and then used to update the bandit as well as the working limits. A solution is returned to the solver if one was found.

1. **Select heuristic** h using a suitable bandit algorithm (introduced in Sect. 3.3).
2. **Execute heuristic** h using the current working limits.
3. **Observe reward** r after executing h (introduced in Sect. 3.2).
4. **Update bandit algorithm and working limits** of h using reward r.

An overview of the scheduling framework is shown in Fig. 1.

Often, a solver finds an optimal solution noticeably faster than it proves the solution's optimality [7]. Thus, always running heuristics with the same frequency is not necessarily the best strategy. To dynamically adapt how often the scheduler is executed, we track how often no solution was found. Whenever it is unsuccessful for too long, we skip a number of future calls to the scheduler: We skip $\lfloor \exp(\beta n_{\text{fail}}) \rfloor - 1$ calls, where n_{fail} counts consecutively failed calls and $\beta = 0.1$.

At the beginning of the solving process, when heuristics run for the first time, the scheduler does not have any information about the heuristic's behavior yet. Thus, any bandit algorithm would start by selecting heuristics at random. To avoid uninformed decisions, our framework uses expert knowledge to warmstart the bandit strategy: We execute all heuristics in their default order first and observe their rewards; only then the bandit algorithm takes over.

This is a general heuristic scheduling framework that can be applied to an arbitrary set $\mathcal{H}$. However, as mentioned in Sect. 2, we focus on LNS and diving since they cover the majority of the more complex heuristics. We control different types of working limits directly influencing the cost and success probability of the heuristics: For LNS, we impose a target fixing rate and for diving, we control the LP resolve frequency. We adapt both as follows.

The target fixing rate controls how many integer variable should be fixed in the sub-MIP. This directly influences the success rate as well as the costs of the heuristic. The more variables are fixed, the easier but also the more restrictive the resulting subproblem becomes. To dynamically adapt the fixing rate, we use the same approach as presented in [15]. Let us denote by $f_h^t \in [0, 1]$ the target fixing rate of heuristic h at iteration t. Assuming that h was selected, we have

$$f_h^{t+1} = \begin{cases} \max\{(1-\gamma)f_h^t, f_{min}\}, & \text{if } h \text{ found solution or sub-MIP was infeasible,} \\ \min\{(1+\gamma)f_h^t, f_{max}\}, & \text{otherwise,} \end{cases}$$

with factor $\gamma \in [0, 1]$ and target fixing rate limits $f_{min}, f_{max} \in [0, 1]$. We choose $\gamma = 0.1$, $f_{min} = 0.3$, $f_{max} = 0.9$, and $f_h^0 = f_{max}$ for all LNS heuristics.

Diving heuristics successively fix integer variables and reoptimize the corresponding LP relaxation in between. If the LP is solved more often, diving becomes more expensive, but also more successful: Fixings that led to infeasibility can be detected earlier and then be corrected by backtracking. To control how often the LP is solved, the fraction of variables is tracked that had their domains changed since the last LP solve. If this fraction exceeds a threshold parameter q, an LP solve is triggered; larger q results in less frequent LP solves.

We dynamically adjust this threshold in a similar fashion to the target fixing rate of LNS heuristics. Let us denote by $q_h^t \in [0, 1]$ the value for diving heuristic h at iteration t. If h was selected at t, then

$$q_h^{t+1} = \begin{cases} \max\{(1-\eta)q_h^t, q_{min}\}, & \text{if } h \text{ did not find an incumbent at iteration } t \\ \min\{(1+\eta)q_h^t, q_{max}\}, & \text{otherwise} \end{cases}$$

for factor $\eta \in [0, 1]$ and the bounds $q_{min}, q_{max} \in [0, 1]$. Thus, if h was not successful, we reduce q_h^t to increase the success probability of h in the future. Otherwise, we increase the value to reduce the cost of executing h. We choose $\eta = 0.1$, $q_{min} = 0.05$, $q_{max} = 0.3$, and $q_h^0 = q_{min}$ for all diving heuristics.

3.2 Choosing a Reward Function

The simplest choice to reward a heuristic $h \in \mathcal{H}$ would be the binary function

$$r_{\text{sol}}(h, t) = \begin{cases} 1, & \text{if } h \text{ found an incumbent at iteration } t \\ 0, & \text{otherwise.} \end{cases}$$

However, heuristics find improving solutions rather rarely: For instance, on the test set we consider in our experiments, the default settings of the solver found on average only 12 incumbents. Thus, using r_{sol} as the only reward signal might not give enough feedback to the agent. Furthermore, r_{sol} lacks a lot of important information. For example, a heuristic that fails fast is preferable over one that needs more time to terminate without a solution. Furthermore, if a solution is found, its quality should also be considered. Besides the obvious preference for better solutions, considering the current stage of the solving process is vital: At the beginning, it is much easier to find a new incumbent than at a more advanced stage. Another problem is that r_{sol} implicitly assumes the only objective of heuristics is finding solutions. This is not always true, for instance, diving heuristics can also generate conflict constraints [1], which profits the solver.

Thus, besides r_{sol}, we consider three additional metrics to reward h:

1. r_{gap} to reward the quality of the new incumbent if h was successful,
2. r_{eff} to punish the effort it took to execute h,
3. r_{conf} to reward the number of conflict constraints h found.

The overall reward function r is then

$$r(h,t) = \lambda_1 r_{\text{sol}}(h,t) + \lambda_2 r_{\text{gap}}(h,t) + \lambda_3 r_{\text{eff}}(h,t) + \lambda_4 r_{\text{conf}}(h,t),$$

with $\lambda_i \in [0,1]$. We choose $\lambda_1 = \lambda_2 = 0.3$ and $\lambda_3 = \lambda_4 = 0.2$.

The functions r_{gap}, r_{eff}, and r_{conf} are defined as follows. Assuming that h was successful, let us denote by x_{new} and x_{old} the new and old solution, respectively. Furthermore, let x_{LP} be the solution of the current linear relaxation. Then, we measure the quality of x_{new} relative to the current solving stage with

$$r_{\text{gap}}(h,t) = \begin{cases} 0, & \text{if } h \text{ did not find an incumbent at iteration } t \\ 1, & \text{if } h \text{ found the first incumbent at iteration } t \\ \frac{c^T x_{old} - c^T x_{new}}{c^T x_{old} - c^T x_{LP}}, & \text{otherwise.} \end{cases}$$

To define r_{eff}, let n_h^t be the number of nodes used by h, and n_{max} an upper bound on the maximal number of nodes used. For LNS, n_h^t refers to the number of nodes solved in the sub-MIP; for diving, it refers to the number of nodes visited during the partial search. Finally, we define $r_{\text{eff}}(h,t) = 1 - \frac{n_h^t}{n_{max}}$ and $r_{\text{conf}}(h,t) = \frac{v_h^t}{v_{max}}$ where v_h^t is the number of conflict constraints h found and v_{max} the maximal number of conflict constraints found by any heuristic in the past. The reward function r is an extension of the reward used in [15], which only uses r_{gap} and variants of r_{sol} and r_{eff}.

3.3 Choosing a Bandit Algorithm

As mentioned before, to solve the multi-armed bandit problem successfully, we need to balance exploitation and exploration carefully. In our case, this raises the following question: Should we prioritize heuristics that have not been executed (that often) or heuristics that have performed well in the past? Our experimental results suggests that for primal heuristics, exploitation is the better choice. Typically, a heuristic that performs bad at the beginning, will also be rather unsuccessful later on, since it only gets harder to find improving solutions.

Algorithm 1. Modified ϵ-greedy bandit algorithm

Input: Set of heuristics $\mathcal{H}$, reward function r, probability $\epsilon \in [0,1]$
$w(h,0) \leftarrow \frac{1}{|\mathcal{H}|}$
$t \leftarrow 0$
while not stopped **do**
  $t \leftarrow t+1$
  $\epsilon_t \leftarrow \epsilon \cdot \sqrt{\frac{|\mathcal{H}|}{t}}$
  Draw $\rho_t \sim \mathbb{U}([0,1])$
  if $\rho_t > \epsilon_t$ **then**
    $h_t \leftarrow \operatorname{argmax}_{h \in \mathcal{H}} w(h, t-1)$
  else
    Draw $h_t \sim w(\cdot, t-1)$
  end if
  Observe reward $r(h_t, t)$
  if h_t was selected for the first time **then**
    $w(h_t, t) \leftarrow r(h_t, t)$
  else
    $w(h_t, t) \leftarrow$ update average weight with $r(h_t, t)$
  end if
end while

That is why we propose to use a variant of the ϵ-greedy bandit algorithm. The ϵ-greedy, or follow-the-leader, algorithm pursues a simple strategy: Given an $\epsilon \in [0,1]$, the best action seen so far is chosen with probability $1-\epsilon$. Otherwise, an action is randomly selected following a uniform distribution. To characterize the best action at iteration t, we associated a weight $w(h,t)$ with every heuristic h. The weights w are equal to the average reward of h observed so far, that is, $w(h,t) = \sum_{\tilde{t} \in \mathcal{T}_h^t} r(h,\tilde{t})/|\mathcal{T}_h^t|$ with $\mathcal{T}_h^t \subseteq [t]$ being the subset of calls at which h was selected up to time t.

In the modified ϵ-greedy algorithm we consider, instead of selecting a heuristic uniformly at random, we draw it following the distribution imposed by the weights w. This variant allows for more exploitation; it is described in Algorithm 1. In our experiments, our online scheduling approach performed best with this bandit algorithm. We use $\epsilon = 0.7$.

To put more focus on heuristics that performed well in the *recent* past, we also tried another modification: Instead of looking at the average reward as w, we examined using an aggregation of the observed rewards where older observations contribute exponentially less. This performed considerably worse, suggesting that it is preferable to consider all past behavior to make future decisions.

4 Computational Results

To study the performance of our approach, we used the state-of-the-art open-source MIP solver SCIP 8.0 with SoPlex 6.0 [8]. We ran all experiments on a Linux cluster of Intel Xeon CPU E5-2630 v3 2.40GHz with 64GB RAM. The time

Table 1. Summary of results for B&B experiments. Rows labeled $[t, 7200]$ consist of instances solved with at least one settings taking at least t seconds. *heurtime* refers to time spent in heuristics controlled by the scheduler, relative to DEFAULT. Shifted geometric means are used.

subset	instances	DEFAULT			SCHEDULER			relative		
		solved	time	nodes	solved	time	nodes	time	nodes	heurtime
all	892	472	1157.44	4238	464	1189.03	4056	1.03	0.95	0.94
$[0, 7200]$	485	472	249.02	2522	464	261.70	2426	1.05	0.96	1.20
$[1, 7200]$	481	468	259.89	2593	460	273.23	2494	1.05	0.96	1.20
$[10, 7200]$	441	428	373.99	3439	420	394.12	3298	1.05	0.96	1.21
$[100, 7200]$	330	317	839.49	8231	309	862.90	7759	1.03	0.94	1.05
$[1000, 7200]$	175	162	2312.56	20769	154	2217.32	19627	0.96	0.95	0.68
all-optimal	451	451	199.60	2294	451	209.61	2168	1.05	0.95	1.25

limit was set two hours and the test set consists of the benchmark instances of the MIPLIB 2017 [13]. Since our framework aims to improve primal performance, we removed all infeasible instances and problems with a zero objective function. This leaves us with 226 instances. To filter out the effects of performance variability [25], all experiments are run with four random seeds.

We compare two settings: DEFAULT refers to the default settings of SCIP and SCHEDULER refers to the proposed online scheduling framework. In the latter, we deactivated all LNS and diving heuristics that are controlled by the scheduler, as well as the two adaptive heuristics presented in [15,16]. The scheduler is called at every node, right after cheaper heuristics like rounding.

Table 1 shows that SCHEDULER consistently reduces the size of the B&B tree by 4–6%. Unfortunately, this improvement does not directly translate into improving solving time. However, we perform the better the harder the instances get: On instances taking at least 1000 seconds to solve, the scheduling framework outperforms DEFAULT by about 4%. Even though SCHEDULER solves 13 instances that cannot be solved by DEFAULT, it fails to solve 21 instances solved by DEFAULT. One reason for this behavior could be the fact that for harder instances, the scheduler tends to spend less time then default in the heuristics controlled by it: On $[1000, 7200]$, SCHEDULER reduces time spent in heuristics by over 30%.

The heuristics' behavior shows that our scheduling framework succeeds in detecting more successful heuristics: The scheduler finds 88% more incumbents while increasing the probability of a heuristic finding a new solution by 57%. On average, the heuristics controlled by SCHEDULER find 3.80 solutions per instance as opposed to 2.01; with a success probability of 4.49% instead of 2.86%.

To conclude, the computational results show that our scheduling framework can improve the performance of a solver. However, for easier instances, it seems that there is not much potential for improvement by using an online learning approach; there we compete against the default parameters that have been tuned on the test set over a long period of time. This could be attributed to a lack of observations: When an instance is solved fast, a learning approach might not have enough time to gather meaningful information about the heuristics' behavior. Furthermore, the results also suggest that our approach might be too conservative for harder instances, since it reduces the time spent in heuristics considerably.

Hence, we believe that there is further room for improvement, also since we have not spent a large amount of effort on tuning any hyperparameters of our method in order to obtain the current results. As next steps, we need to combine the good performance of the static heuristic handling with our online scheduling approach and better detect when to apply heuristics more aggressively and when to rely on well-working default parameters.

References

1. Achterberg, T.: Conflict analysis in mixed integer programming. Discret. Optim. **4**(1), 4–20 (2007)
2. Balcan, M.F., Dick, T., Sandholm, T., Vitercik, E.: Learning to branch. In: International Conference on Machine Learning, pp. 344–353. PMLR (2018)
3. Baltean-Lugojan, R., Bonami, P., Misener, R., Tramontani, A.: Scoring positive semidefinite cutting planes for quadratic optimization via trained neural networks (2019). https://optimization-online.org/?p=17362
4. Berthold, T.: Measuring the impact of primal heuristics. Oper. Res. Lett. **41**(6), 611–614 (2013)
5. Berthold, T.: Primal MINLP heuristics in a nutshell. In: International Conference on Operations Research (2013)
6. Berthold, T.: A computational study of primal heuristics inside an MI(NL)P solver. J. Glob. Optim. **70**, 189–206 (2018)
7. Berthold, T., Hendel, G., Koch, T.: From feasibility to improvement to proof: three phases of solving mixed-integer programs. Optim. Methods Softw. **33**, 1–19 (2017)
8. Bestuzheva, K., et al.: The SCIP Optimization Suite 8.0. ZIB-Report 21-41, Zuse Institute Berlin (2021). https://nbn-resolving.de/urn:nbn:de:0297-zib-85309
9. Bubeck, S., Nicoló, C.B.: Regret analysis of stochastic and nonstochastic multi-armed bandit problems. Found. Trends Mach. Learn. **5**(1), 1–122 (2012)
10. Chmiela, A., Khalil, E., Gleixner, A., Lodi, A., Pokutta, S.: Learning to schedule heuristics in branch and bound. In: Advances in Neural Information Processing Systems, vol. 34 (2021)
11. Iommazzo, G., D'Ambrosio, C., Frangioni, A., Liberti, L.: A learning-based mathematical programming formulation for the automatic configuration of optimization solvers. In: Nicosia, G., et al. (eds.) LOD 2020. LNCS, vol. 12565, pp. 700–712. Springer, Cham (2020). https://doi.org/10.1007/978-3-030-64583-0_61
12. Etheve, M., Alès, Z., Bissuel, C., Juan, O., Kedad-Sidhoum, S.: Reinforcement learning for variable selection in a branch and bound algorithm. arXiv:2005.10026 (2020)
13. Gleixner, A., et al.: MIPLIB 2017: data-driven compilation of the 6th mixed-integer programming library. Math. Program. Comput. **13**(3), 443–490 (2021)
14. He, H., Daume III, H., Eisner, J.M.: Learning to search in branch and bound algorithms. In: Advances in Neural Information Processing Systems, vol. 27, pp. 3293–3301 (2014)

15. Hendel, G.: Adaptive large neighborhood search for mixed integer programming. Math. Program. Comput. **14**(2), 185–221 (2022)
16. Hendel, G., Miltenberger, M., Witzig, J.: Adaptive algorithmic behavior for solving mixed integer programs using bandit algorithms. In: International Conference on Operations Research (2018)
17. Huang, L., et al.: Improving primal heuristics for mixed integer programming problems based on problem reduction: a learning-based approach. arXiv:2209.13217 (2022)
18. Huang, Z., et al.: Learning to select cuts for efficient mixed-integer programming. Pattern Recognit. **123**, 108353 (2022)
19. Hutter, F., Hoos, H., Leyton-Brown, K., Stützle, T.: Paramils: an automatic algorithm configuration framework. J. Artif. Intell. Res. (JAIR) **36**, 267–306 (2009)
20. Hutter, F., Hoos, H.H., Leyton-Brown, K.: Sequential model-based optimization for general algorithm configuration. In: Learning and Intelligent Optimization, pp. 507–523 (2011)
21. Khalil, E.B., Bodic, P.L., Song, L., Nemhauser, G., Dilkina, B.: Learning to branch in mixed integer programming. In: Proceedings of the 30th AAAI Conference on Artificial Intelligence (2016)
22. Kruber, M., Lübbecke, M., Parmentier, A.: Learning when to use a decomposition. In: International Conference on AI and OR Techniques in Constraint Programming for Combinatorial Optimization Problems, pp. 202–210 (2017)
23. Land, A.H., Doig, A.G.: An automatic method of solving discrete programming problems. Econometrica **28**(3), 497–520 (1960)
24. Lattimore, T., Szepesvári, C.: Bandit Algorithms. Cambridge University Press, Cambridge (2020)
25. Lodi, A., Tramontani, A.: Performance variability in mixed-integer programming. Tutor. Oper. Res. **10**, 1–12 (2013)
26. Lodi, A., Zarpellon, G.: On learning and branching: a survey. TOP **25**(2), 207–236 (2017). https://doi.org/10.1007/s11750-017-0451-6
27. Nair, V., et al.: Solving mixed integer programs using neural networks. arXiv preprint: arXiv:2012.13349 (2020)
28. Paulus, M.B., Zarpellon, G., Krause, A., Charlin, L., Maddison, C.: Learning to cut by looking ahead: cutting plane selection via imitation learning. In: Proceedings of the 39th International Conference on Machine Learning, vol. 162, pp. 17584–17600 (2022)
29. Scavuzzo, L., et al.: Learning to branch with tree MDPs. arXiv:2205.11107 (2022)
30. Tang, Y., Agrawal, S., Faenza, Y.: Reinforcement learning for integer programming: learning to cut. In: Proceedings of the 37th International Conference on Machine Learning, vol. 119, pp. 9367–9376 (2020)
31. Turner, M., Koch, T., Serrano, F., Winkler, M.: Adaptive cut selection in mixed-integer linear programming. arXiv:2202.10962 (2022)
32. Yilmaz, K., Yorke-Smith, N.: A study of learning search approximation in mixed integer branch and bound: node selection in SCIP. AI **2**, 150–178 (2021)

Contextual Robust Optimisation with Uncertainty Quantification

Egon Peršak(✉) and Miguel F. Anjos

University of Edinburgh, Edinburgh, UK
E.Persak@sms.ed.ac.uk

Abstract. We propose two pipelines for convex optimisation problems with uncertain parameters that aim to improve decision robustness by addressing the sensitivity of optimisation to parameter estimation. This is achieved by integrating uncertainty quantification (UQ) methods for supervised learning into the ambiguity sets for distributionally robust optimisation (DRO). The pipelines leverage learning to produce contextual/conditional ambiguity sets from side-information. The two pipelines correspond to different UQ approaches: i) explicitly predicting the conditional covariance matrix using deep ensembles (DEs) and Gaussian processes (GPs), and ii) sampling using Monte Carlo dropout, DEs, and GPs. We use i) to construct an ambiguity set by defining an uncertainty around the estimated moments to achieve robustness with respect to the prediction model. UQ ii) is used as an empirical reference distribution of a Wasserstein ball to enhance out of sample performance. DRO problems constrained with either ambiguity set are tractable for a range of convex optimisation problems. We propose data-driven ways of setting DRO robustness parameters motivated by either coverage or out of sample performance. These parameters provide a useful yardstick in comparing the quality of UQ between prediction models. The pipelines are computationally evaluated and compared with deterministic and unconditional approaches on simulated and real-world portfolio optimisation problems.

Keywords: Prediction and Optimisation · Prescriptive Analytics · Uncertainty Quantification · Distributionally Robust Optimisation

1 Introduction

Real world decision problems are seldom deterministic. The perennial operational risk of mathematical programming is the sensitivity of the problem to its parameterisation. Small differences in the parameters governing the objective or the constraints can render solutions highly suboptimal or infeasible. Optimisation under uncertainty is a mature field which has devised a number of tractable approaches that derive robust or expectation optimal solutions, thus managing the uncertainty. Since the true underlying distributions are never known in practice, the most successful approaches exploit available samples of parameters

© The Author(s), under exclusive license to Springer Nature Switzerland AG 2023
A. A. Cire (Ed.): CPAIOR 2023, LNCS 13884, pp. 124–132, 2023.
https://doi.org/10.1007/978-3-031-33271-5_9

with statistically valid constructs of uncertainty. When contextual information exists, using what amounts to an unsupervised approach should lead to overly conservative decisions. Contexts of the problem for which existing samples are information-poor may result in overly confident decisions. Problem parameters can be estimated based on available contextual information using supervised learning. This is commonly referred to as predict-then-optimise and is how prescriptive analytics is often performed. This is a form of contextual optimisation, but prediction models tend to be overly confident and in their vanilla form do not quantify the certainty of their predictions. Using point parameter estimates preserves the operational risk, which may be exacerbated due to inconsistent out-of-sample performance. Given the ubiquity of predict-then-optimise decision making, improving its reliability and out-of-sample performance will result in tangible impact.

In this work we propose an approach to adapt predictive models with uncertainty quantification (UQ) to a robust optimisation setup. We mainly focus on distributionally robust models (DRO), forming the pipeline UQ-DRO. Similar logic can be applied in robust ways for safety-critical situations. Section 2 introduces the concept of robust prediction and optimisation. Section 3 presents implemented predictive methods with UQ. The ambiguity sets used with UQ are defined in Sect. 4. Section 5 sets out data-driven algorithms for robustness parameter specification and the objectives of the two pipelines. Section 6 computationally evaluates the approach on a simulated and a real data portfolio optimisation problem.

2 Robust Predict-then-Optimise

Prediction models provide an estimate of the conditional expected value of the target. Predictive uncertainty can be decomposed into epistemic, and aleatoric uncertainty. Epistemic uncertainty is a result of a lack of information about the true data generating process (DGP). Aleatoric uncertainty is the underlying stochasticity of the process and is irreducible. Even in the best-case supervised scenario, the persisting aleatoric uncertainty presents an operational risk, thus motivating the use of robust approaches. A well-tuned robust predict-then-optimise approach would provide a meaningful scoring criterion for predictive models used in optimisation, reflecting their worst-case outcome. The key driver of decision quality in such a system would be the level of epistemic uncertainty, which would determine the needed level of conservativeness. As such, the predictive model should be highly expressive, trained delicately, as phenomena such as overfitting may increase out-of-sample epistemic uncertainty, and capable of reasonably capturing the uncertainty of its predictions. The prediction task in this case is more difficult as we are typically interested in the values of many parameters, encouraging the use of multiple-output models to better capture interdependence.

Existing approaches for conditional optimisation have utilised local nonparametric regression methods such as K-nearest neighbours with DRO [2,3,15] to provide a conditional sample for a variety uncertain optimisation problems. Building on this [8] propose a method based on distribution trimmings and cast it as a partial mass transportation problem to hedge against the limitations of inferring conditional distributions with limited samples, providing an outer layer of robustness. On the other hand, a growing body of literature has looked at fused approaches wherein the prediction function is optimised with respect to decision loss. Often referred to as Predict-and-Optimise, it involves differentiating across a solver which has been achieved either with surrogate gradients [1,17] or differentiating the optimality conditions [7,16] of a potentially relaxed problem. These models implicitly learn how to deal with conditional uncertainty.

In contrast our work uses global supervised learning methods. This is motivated by the idea that global models have the potential to cross-learn about different contexts through shared patterns in the data. This improves the model's ability to infer about contexts which are information-poor, a key case of which are out-of sample contexts. We do not make any assumptions on the DGP, instead relying on a cross-validation type approach to determine robustness parameters.

3 Predictive Models with Uncertainty Quantification

We propose the use of three predictive approaches which have high expressive power and capacity for UQ. The approaches cover both main directions in UQ, namely ensemble, and Bayesian techniques. The ensemble approach is a deep ensemble (DE) [11], an ensembling technique which treats ensemble members as mixture model components. Constituent models are neural networks designed to predict both a mean vector and a covariance matrix. They are trained using a form of gradient descent to minimise the negative log likelihood given a parametric assumption about the DGP's uncertainty, usually heteroskedastic Gaussian, though Laplacian likelihood may be more appropriate for heavy tails. Denote the available contextual information as $x \in \mathbb{R}^n$, and the model parameters as θ. The model maps $\mathbb{R}^n \mapsto \mathbb{R}^p \times \mathbb{R}^{p\times p}$ or from contextual information x to a mean vector $\mu(x) \in \mathbb{R}^p$ and covariance matrix $\Sigma(x) \in \mathbb{R}^{p\times p}$. The optimisation problem with a Gaussian maximum likelihood objective is:

$$\min_{\theta} \mathcal{L}(\theta) = \sum_{(\mathbf{x},y)\in\mathcal{D}} \frac{1}{2}(\mathbf{y} - \mu(\mathbf{x};\theta))^T \Sigma(\mathbf{x};\theta)^{-1}(\mathbf{y} - \mu(\mathbf{x};\theta)) + \frac{1}{2}\ln(|\Sigma(\mathbf{x};\theta)|) \quad (1)$$

Note that the covariance matrix is symmetric, so only $p + \frac{p(p+1)}{2}$ outputs need to be predicted. The structural concern is that Σ should be positive semidefinite (PSD). We follow [14] who encourage a PSD estimate by using an exp activation function for variance terms and a tanh activation function for predicting correlation coefficients from which they construct covariances. The size of the estimated matrix grows quadratically, so this approach is unlikely to scale to large problems. In practice, the exponential activation and subsequent matrix construction can lead to numerical difficulties with the determinant. Since we are interested in the log of the determinant we can reformulate this part of the loss function into a sum of the log of its eigenvalues. If an odd number of eigenvalues are negative, we clip the value at a small positive ϵ. While this means that the resulting matrix may not be PSD in intermediate steps, it enables more informative gradients and tends to predict PSD matrices after training. This procedure is fully differentiable and was key to stabilising training in addition to standardising variables.

The second approach is Monte-Carlo dropout (MCD), which is an ensemble technique that became popular after a Bayesian analysis showed that it can be cast as approximate inference in deep Gaussian processes [9]. MCD relies on a regularisation technique for deep learning called dropout. Dropout deactivates neurons in the network randomly according to some prior parameterisation (usually Bernoulli), thus limiting the gradient information during that pass to the active units. Dropout regularisation can be thought of as training an implicit ensemble of models within the network, but is deactivated at test time. MCD retains dropout at test time and uses it as an empirical sampling technique to estimate the posterior uncertainty of predictions given contextual information x. This approach should scale better, but tends to be worse at UQ.

Finally, we propose the use of a Bayesian non-parametric regression approach. The most commonly used such approach is a Gaussian Process (GP) which is assumed to be the distribution across functions. Any set of observations about the function value is assumed to have a multivariate Gaussian joint distribution parameterised by a mean, and kernel function which measures the similarity of contextual information. Predictions are made by marginalising the probability distribution of a new point given its contextual information. The choice of kernel is key for modelling (scale and Matern in our case) and kernels are often parametric, thus allowing for some optimisation, typically by optimising the marginal likelihood. Given that we are interested in multi-task prediction, we follow the setup of [5], which models interdependence with a task-similarity kernel.

4 Conditional Ambiguity Sets

We incorporate UQ in various forms of DRO. DRO seeks to obtain a solution that has the least worst expected value across all distributions in an ambiguity set. To exemplify, say the uncertain parameter $\xi \sim \mathcal{P}$ is only in the objective $h(x, \xi)$. The DRO formulation is:

$$\min_{x \in S} \sup_{\mathcal{P} \in \mathcal{D}} \mathbb{E}_{\mathcal{P}}(h(x, \xi)) \tag{2}$$

It is less conservative than robust optimisation approaches and does not suffer from the optimiser's curse (overly optimistic out-of-sample) like stochastic programming. The two prevailing ways of defining ambiguity sets are by using moments or disturbance metrics. Moment-based sets are typically convex sets constructed in reference to a stated or estimated moment, usually using conic formulations. Disturbance sets are defined as all distributions within a certain disturbance metric. Even though these problems are semi-infinite, they often admit tractable reformulations by exploiting duality.

We propose the use of ambiguity sets that incorporate the output of UQ. DE and GP quantify their uncertainty with predicted covariances. We employ the approach from [6], which defines ambiguity sets in terms of moment-uncertainty. The ambiguity set $\mathcal{D}(\hat{\mu}, \hat{\Sigma}, \gamma_1, \gamma_2)$ is defined as:

$$\mathcal{D}(\hat{\mu}, \hat{\Sigma}, \gamma_1, \gamma_2) = \left\{ \mathbb{P} \in \mathcal{P}(\mathbb{S}) \;\middle|\; \begin{array}{l} \xi \sim \mathbb{P} \\ (\mathbb{E}(\xi) - \hat{\mu})^{\mathrm{T}} \hat{\Sigma}^{-1} (\mathbb{E}(\xi) - \hat{\mu}) \leq \gamma_1 \\ \mathbb{E}((\xi - \hat{\mu})^{\mathrm{T}} (\xi - \hat{\mu})) \preceq \gamma_2 \hat{\Sigma} \end{array} \right\}, \quad (3)$$

where $\mathbb{S} \subseteq \mathbb{R}^p$ is the support of the set. The set defines all distributions for which the expected value lies within a scaled ellipsoid uncertainty set centred on the model prediction and shaped by the UQ, and for which the true covariance lies within a PSD cone defined by scaled UQ. This ambiguity set does not assume that the model is correct or that it captures its conditional uncertainty well. It enables us to parametrically define a space of distributions around our model's predictions within which we can guarantee a worst case expectation. Under mild convexity assumptions about the objective function [6], this ambiguity set has a tractable semidefinite programming (SDP) robust counterpart.

For sampling-based UQ we propose the use of ambiguity sets defined by the Wasserstein metric [12]. The ambiguity set is defined as the set of distributions that are within a ball from the empirical reference distribution, which in our case is the n conditionally generated samples $\hat{\mathbb{P}}_n(x_i)$. The ambiguity set is defined as $\mathcal{D}_w(\hat{\mathbb{P}}_n, \phi) = \{\mathbb{Q} \in \mathcal{P}(\mathbb{S}) | d_{W,p}(\mathbb{Q}, \hat{\mathbb{P}}_n) \leq \phi\}$, where $d_{W,p}$ is the p-norm Wasserstein metric $d_{W,p}(\hat{\mathbb{P}}_n, \mathbb{Q}) = \inf_{\Pi}\{\int_{\mathbb{S}\times\mathbb{S}} ||\hat{p} - q||^p d\Pi(p,q)\}$, and Π denotes the joint distribution of p, q whose marginals are $\hat{\mathbb{P}}_n, \mathbb{Q}$ respectively. Wasserstein ambiguity sets are generally less tractable, but robust counterparts exist in a number of settings. In the context of predict-then-optimise, this allows us to account for sampling error and bias. Posterior sampling from a GP provides a conditional reference distribution based on our structural beliefs about the DGP. MCD is a black box, but provides a more centred form of sampling.

5 Data-driven Robustness Parameter Specification

We see two ways of leveraging the UQ-DRO pipeline depending on how robustness parameters are set. We can either set them to probabilistically cover potential outcomes, or induce limited robustness. The drawback of the former is that robustness tends to come with a cost to performance on average as it is overly conservative. In turn, limited robustness may increase average performance by reducing the sensitivity of decision quality to mild parameter uncertainty.

We aim to achieve coverage with the UQ and uncertain moments pipeline. We achieve this by finding the smallest values γ_1, γ_2 such that the defined uncertainty sets are likely to contain the true moments. We use a holdout set $\mathbb{V}$ as a proxy for the problem. Lower values of these parameters indicate that a model is better tuned for estimating its uncertainty in the context of the optimisation model.

We cast the setting of these parameters as optimisation problems. We want to find the smallest $\hat{\gamma}_1$ such that the (unknown) conditional mean is within the ellipsoidal uncertainty set defined by the predicted mean and variance:

$$\arg\min_{\gamma_1}\{\gamma_1|\mathbb{E}[(\xi-\hat{\mu}(x))^{\mathrm{T}}\hat{\Sigma}^{-1}(x)(\xi-\hat{\mu}(x))\leq\gamma_1|x]\}. \tag{4}$$

We use $(x,\xi)\in\mathbb{V}$ as a proxy for this problem, by calculating the distance for each point and then picking the median. This is motivated by an assumption that the true conditional distributions are symmetric on average, so realisations are more distant from the predicted mean than the true mean approximately half of the time. Setting γ_2 is slightly more difficult as the associated constraint is defined using a Loewner order. We solve the following SDP problem for each point in the holdout set for each $(x_i,\xi_i)\in\mathbb{V}$:

$$\hat{\gamma}_{2,i}=\arg\min_{\gamma_2}\{\gamma_2|\gamma_2\hat{\Sigma}(x_i)-Z\succeq 0,\gamma_2\geq 0\}, \tag{5}$$

where $Z=\frac{1}{|\mathbb{V}|}\sum_{i\in\mathbb{V}}((\xi_i-\hat{\mu}(x_i))^{\mathrm{T}}(\xi_i-\hat{\mu}(x_i)))$ and set $\hat{\gamma}_2(\alpha_2)$ as the $1-\alpha_2$ quantile of the obtained $\hat{\gamma}_{2,i}$. The linear matrix inequality constraint in this problem is simple so it should not be a computational bottleneck. We use $\alpha_2=0.1$ to encourage a 90% coverage of covariances, but this can be tinkered with.

We use the sampling-Wasserstein pipeline to achieve limited robustness. We want to obtain a reference holdout ϕ and then scale it by multiplying it with some constant $k\leq 1$. We obtain the reference ϕ by computing the p-Wasserstein distance between every distribution realisation ξ_i and the predicted sample $\hat{\mathbb{P}}_n(x_i)=\frac{1}{n}\sum_{j=1}^{n}\delta_{\hat{\xi}_{i,j}}$ which we treat as a mixture of Dirac delta distributions. Since both are discrete, this is equivalent to calculating the earth mover's distance (EMD) between the two. The p-Wasserstein distance between the two can be cast as a linear optimisation problem:

$$d_{W,p}(\hat{\mathbb{P}}_n(x_i),\delta_{\xi_i})=\min_{T}\{\langle T,M\rangle|T\mathbf{1}=\mathbf{p}_{\mathbb{P}},T^T\mathbf{1}=\mathbf{p}_{\xi}\}, \tag{6}$$

where T is the optimal transport matrix, $M\in\mathbb{R}^{n\times 1}$ is the moving cost, which is calculated as the point-wise p-norm between the sample $\hat{\mathbb{P}}_n(x_i)$ and ξ_i, and $\mathbf{p}_{\mathbb{P}},\mathbf{p}_{\xi}$ are the discrete densities (in this case equally weighted). Since T is only a vector, the optimisation is trivial: the optimal transport plan is $t_k=\frac{1}{n}$ for all k. The EMD for holdout entry i is therefore $\frac{1}{n}\sum_{j=1}^{n}|\hat{\xi}_{i,j}-\xi_i|_p^p$ and we set the reference ϕ as the 0.9 quantile of these values. However, the true distribution of ξ is very unlikely to be a discrete point and such a large ambiguity set will likely lead to overly conservative solutions, which is why we scale it down with k.

6 Computational Evaluation and Discussion

Our approach was evaluated on a common prediction and optimisation problem, namely portfolio optimisation. The code is publicly available[1]. The uncertain

[1] https://github.com/EgoPer/Contextual-Robust-Optimisation-with-UQ.

parameters are the asset returns $\mathbf{p}$. Asset returns are famously difficult to predict due to a high noise to signal ratio and concept drift. We followed the problem setup from [4,8], which uses a linear reformulation of CVAR [13] in the objective. The DRO optimisation problem for ϵ-CVAR is:

$$\begin{aligned} \min_{\mathbf{x},\beta} \inf_{\mathbb{P}} \quad & \mathbb{E}_{\mathbb{P}}[\beta + \tfrac{1}{\epsilon}(-\mathbf{p}^T\mathbf{x} - \beta)^+ - \lambda \mathbf{p}^T\mathbf{x}] \\ \text{s.t.} \quad & \mathbf{e}^T\mathbf{x} = 1, \mathbf{x} \geq 0 \end{aligned} \tag{7}$$

where λ governs the trade off between tail risk and returns, and $(a)^+ = \max(0, a)$. We set $\epsilon = 0.1$ (expected value of the 10% worst cases), and λ at 1. We use 25 samples for each conditional Wasserstein approach, set $k = 0.1$ on the simulated problem, and $k = 0.02$ on the real data problem.

6.1 Simulated Problem

The simulated version of this problem is based on a problem used in two existing papers [4,8] about DRO with side/contextual information, but we introduce significant non-linearity and heteroskedasticity. Three independent inputs are simulated as standard normal variables $x_{1,2,3} \sim \mathcal{N}(0,1)$. The conditional joint distribution of the simulated returns is $\mathcal{N}(\mu(\mathbf{x}), \Sigma(\mathbf{x}))$, where $\mu(\mathbf{x}) = \bar{\mu} + y(\mathbf{x})^2, \Sigma(\mathbf{x}) = [(\frac{1}{5}\tanh(x_1) + 1) \cdot \bar{\Sigma}^{\frac{1}{2}}]^2$ ($\bar{\mu}, \bar{\Sigma}$ as in [4]).

We generate five datasets at five training set sizes (20% holdout) and train models five times due to the stochastic nature of their training. The test set is the same across experiments and is deliberately generated out-of-sample (100 samples of $x_{1,3} \sim \mathcal{N}(2,1), x_2 \sim \mathcal{N}(-2,1)$, same DGP). Given that we have access to the true DGP, we can approximately evaluate the performance of each solution (in our case using a 10^4 sample Monte Carlo simulation). We construct a deterministic equivalent, which gives us the true optimal value, allowing us to measure regret. We run three unconditioned models that derive their uncertainty inputs from the whole training set, an uncertain moments (UM) model, a Wasserstein (WASS) model, and a sample average approximation (SAA) model. We run a conditional SAA model using the moment outputs of DE to sample a normal distribution. We run five of our contextual models: DE-UM, GP-UM, DE-Wasserstein (DE-WASS), GP-WASS, and MCD-WASS.

Figure 1 displays boxplots of the mean out-of-sample regret for each approach. The DE-WASS approach dominates across data sizes improving in performance with a growing size of the dataset, outdoing a robustly performing unconditional Wasserstein. MCD-WASS is close in performance to the unconditional case, while the GP version lags slightly. The prediction models were trained in an out-of-the-box manner, so it is conceivable that they would outperform using hyperparameter optimisation or in richer data environments. The conditional UM methods outperformed the unconditional case, though the gap

[2] The outputs of $y(\mathbf{x})$ are defined as:
$y_1 = 30\tanh(x_2 \exp(\frac{1}{2}x_1 - 2))], y_2 = 50\tanh(x_1)\sin(3x_3), y_3 = 10\ln(|x_1x_2x_3|)$, $y_4 = \sin(x_2) + x_1^2 - x_1x_2, y_5 = 20(\sin(x_1) + \sin(\frac{x_2}{10x_3})), y_6 = y_1 - y_2$.

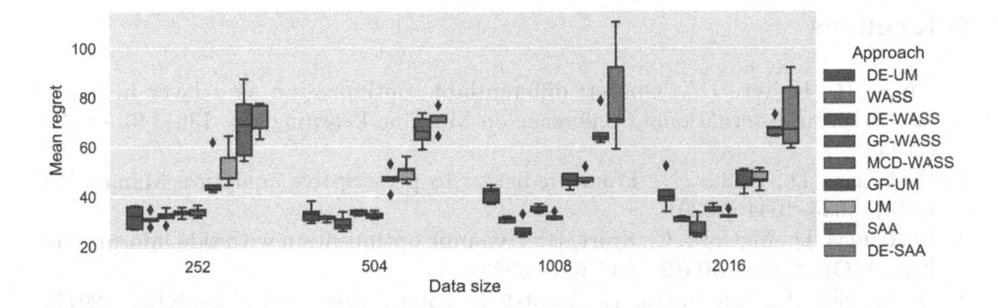

Fig. 1. Mean performance from simulated experiments.

between DE and the unconditional UM got smaller with an increase in the size of data, possibly reflecting overfitting as the models were trained for the same number of epochs in each case. The results support the use of UQ for constructing contextual DRO approaches.

For the real problem we use a slightly reduced version of the dataset from [10], which contains returns on five large US indices alongside 106 contextual covariates such as technical and economic indicators. We cannot directly evaluate the objective function in 7 so we approximate the CVAR and the expected return using a two year testing set training each model five times.

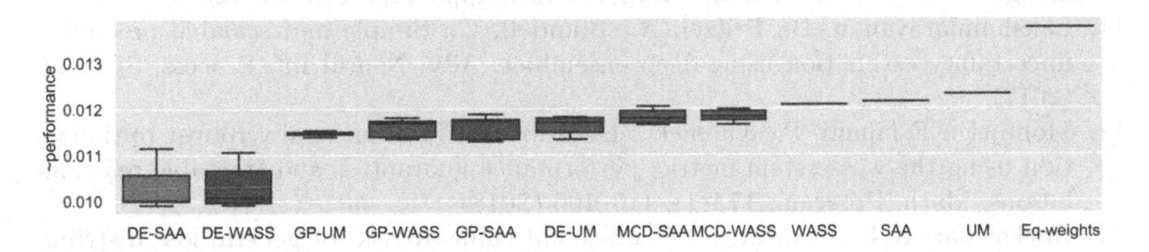

Fig. 2. Approximate performance from real-data experiment.

Figure 2 presents the approximate performance of these methods. The contextual approaches outperform all unconditioned approaches. The Wasserstein approaches are very competitive with non-robust contextual SAA equivalents and are less sensitive to model training, illustrating the positive trade-off of limited robustness. Notably, the DE-WASS approach is best performing in both experiments.

The pipelines offer an effective way of introducing robustness to prediction and optimisation problems by leveraging established methods for UQ. They can also be used a means of achieving contextual DRO, though analysis is needed to establish the desired convergence properties that established DRO techniques have. We see much potential for refining these pipelines with regularisation, predictive model architecture, contextual setting of robustness parameters, and end-to-end learning.

References

1. Amos, B., Kolter, J.Z.: Optnet: differentiable optimization as a layer in neural networks. In: International Conference on Machine Learning, pp. 136–145. PMLR (2017)
2. Bertsimas, D., Kallus, N.: From predictive to prescriptive analytics. Manag. Sci. **66**(3), 1025–1044 (2020)
3. Bertsimas, D., McCord, C., Sturt, B.: Dynamic optimization with side information. Eur. J. Oper. Res. **304**(2), 634–651 (2023)
4. Bertsimas, D., Van Parys, B.: Bootstrap robust prescriptive analytics (2017). https://arxiv.org/abs/1711.09974v1
5. Bonilla, E.V., Chai, K., Williams, C.: Multi-task gaussian process prediction. Adv. Neural Inf. Process. Syst. **20** (2007)
6. Delage, E., Ye, Y.: Distributionally robust optimization under moment uncertainty with application to data-driven problems. Oper. Res. **58**(3), 595–612 (2010)
7. Elmachtoub, A.N., Grigas, P.: Smart predict, then optimize. Manag. Sci. **68**(1), 9–26 (2022)
8. Esteban-Pérez, A., Morales, J.M.: Distributionally robust stochastic programs with side information based on trimmings. Math. Program. 1–37 (2021). https://doi.org/10.1007/s10107-021-01724-0
9. Gal, Y., Ghahramani, Z.: Dropout as a Bayesian approximation: representing model uncertainty in deep learning. In: International Conference on Machine Learning, pp. 1050–1059. PMLR (2016)
10. Hoseinzade, E., Haratizadeh, S.: CNNpred: CNN-based stock market prediction using a diverse set of variables. Expert Syst. Appl. **129**, 273–285 (2019)
11. Lakshminarayanan, B., Pritzel, A., Blundell, C.: Simple and scalable predictive uncertainty estimation using deep ensembles. Adv. Neural Inf. Process. Syst. **30** (2017)
12. Mohajerin Esfahani, P., Kuhn, D.: Data-driven distributionally robust optimization using the wasserstein metric: performance guarantees and tractable reformulations. Math. Program. **171**(1), 115–166 (2018)
13. Rockafellar, R.T., Uryasev, S.: Conditional value-at-risk for general loss distributions. J. Bank. Financ. **26**(7), 1443–1471 (2002)
14. Russell, R.L., Reale, C.: Multivariate uncertainty in deep learning. IEEE Trans. Neural Netw. Learn. Syst. **33**, 7937–7943 (2021)
15. Srivastava, P.R., Wang, Y., Hanasusanto, G.A., Ho, C.P.: On data-driven prescriptive analytics with side information: a regularized nadaraya-watson approach (2021). https://arxiv.org/abs/2110.04855
16. Vlastelica, M., Paulus, A., Musil, V., Martius, G., Rolínek, M.: Differentiation of blackbox combinatorial solvers. arXiv preprint arXiv:1912.02175 (2019)
17. Wilder, B., Dilkina, B., Tambe, M.: Melding the data-decisions pipeline: decision-focused learning for combinatorial optimization. In: Proceedings of the AAAI Conference on Artificial Intelligence, vol. 33, pp. 1658–1665 (2019)

Breaking Symmetries with High Dimensional Graph Invariants and Their Combination

Avraham Itzhakov(✉) and Michael Codish

Department of Computer Science, Ben-Gurion University of the Negev, Beer-Sheva, Israel
{itzhakoa,mcodish}@cs.bgu.ac.il

Abstract. This paper illustrates the application of graph invariants to break symmetries for graph search problems. The paper makes two contributions: (1) the use of higher dimensional graph invariants in symmetry breaking constraints; and (2) a novel technique to obtain symmetry breaking constraints by combining graph invariants. Experimentation demonstrates that the proposed approach applies to provide new results for the generation of a particular class of cubic graphs.

1 Introduction

Graph search problems are about finding simple graphs with desired structural properties. Such problems arise in many real-world applications and are fundamental in graph theory. Solving graph search problems is typically hard due to the enormous search space and the large number of symmetries in graph representation. For graph search problems, any graph obtained by permuting the vertices of a solution (or a non-solution) is also a solution (or a non-solution), which is isomorphic, or "symmetric". When solving graph search problems, the presence of symmetries often causes redundant search effort by revisiting symmetric objects. To optimize the search we aim to restrict it to focus on one "canonical" graph from each isomorphism class.

A standard approach to eliminate symmetries is to add *symmetry breaking constraints* which are satisfied by at least one member of each isomorphism class [8,22,23]. A symmetry breaking constraint is called *complete* if it is satisfied by exactly one member of each isomorphism class and *partial* otherwise. We say that a symmetry breaking constraint is of polynomial size, if it has a representation in propositional logic which is polynomial in size. There is no known polynomial size complete symmetry breaking constraint for graph search problems. Therefore, in practice, one typically applies partial symmetry breaking constraints [5–7] which are polynomial in size.

Over the past decade, there has been little progress in the research of partial symmetry breaking constraints for graph search problems. Codish *et al.* [6,7] introduced a polynomial sized partial symmetry breaking constraint, denoted

© The Author(s), under exclusive license to Springer Nature Switzerland AG 2023
A. A. Cire (Ed.): CPAIOR 2023, LNCS 13884, pp. 133–149, 2023.
https://doi.org/10.1007/978-3-031-33271-5_10

here as sb_{lex}, which restricts the search space to graphs with lexicographically minimal adjacency matrices with respect to permutations that swap two vertices. This constraint turns out to work well in practice, despite eliminating only a small portion of the symmetries. However, when dealing with hard instances of graph search problems, this constraint does not suffice.

Graph invariants [13] are properties of graphs (typically expressed numerically) which are preserved under isomorphism. Graph invariants have been extensively researched in various disciplines such as, chemistry [1], physics [9], and also in the context of graph isomorphism tools [15]. In this paper we use the following terminology. Graph invariants which relate to individual vertices, such as the degree of a vertex, are one dimensional and called "vertex invariants". Graph invariants which relate to pairs of vertices are two dimensional and called "pair invariants". Graph invariants which relate to sets of vertices (with at least two elements) are called "high dimensional". Previous approaches that consider structural information to improve on sb_{lex} apply one dimensional graph invariants in combination with the lexicographic order. For example, in [5,18], the authors combine lexicographic order with information about vertex degrees.

This paper explores the application of higher dimensional graph invariants to break symmetries. We focus on one and two dimensional invariants. However, all the techniques demonstrated apply to invariants of any dimension. We study two techniques to combine graph invariants. First, we introduce the "chain" symmetry breaking constraint which generalizes the standard approach for breaking symmetries with graph invariants. The chain constraint combines a given series of graph invariants to break symmetries such that each invariant refines its predecessors. We then introduce the "product" symmetry breaking constraint which combines graph invariants by interleaving them. We demonstrate the advantage of this approach over the chain constraint. Finally, we demonstrate the application of high dimensional graph invariants to generate connected claw-free cubic graphs of order $n \leq 36$ vertices where existing symmetry breaking methods do not suffice. The results for 32, 34, and 36 vertices are new.

The computations detailed throughout this paper are performed using the finite-domain constraint compiler BEE [17] which compiles constraints to a CNF and solves it applying an underlying SAT solver. We use Clasp 3.1.3 [12] as the underlying SAT solver. All experiments run on an Intel Xeon E5-2660 with CPU's clocked at 2 GHz, Each instance is run on a single thread.

2 Preliminaries and Notation

Throughout this paper we consider simple graphs, i.e. undirected graphs with no self loops. The vertex set of a graph $G = (V, E)$ of order n, is denoted $V(G)$ and assumed to be $V = \{1, \dots, n\}$. The edge set of G is denoted $E(G) \subseteq V \times V$. The adjacency matrix of G is an $n \times n$ Boolean matrix which, in abuse of notation, is also denoted G. The element at row i and column j is denoted $G_{i,j}$ and is *true* if and only if (i, j) is an edge in G. The set of neighbors of an edge $v \in V$ is denoted $N_G(v)$. The degree of a vertex $v \in V$ is the number of its neighbors, and is denoted $deg_G(v)$. The set of simple graphs on n vertices is denoted $\mathcal{G}_n$. An *unknown graph* of order n is represented as an $n \times n$ adjacency matrix of Boolean variables which is symmetric and has the values *false* (denoted by 0) on the diagonal. A graph H is a *subgraph* of G if $V(H) \subseteq V(G)$ and $E(H) \subseteq E(G)$. A graph H is called an *induced subgraph* of G if H is a subgraph of G and every edge in G that connects vertices from $V(H)$ also appears in $E(H)$. In other words, the graph H is an induced subgraph of G if H and G have the same edges between the vertices of H.

The group of all permutations on $\{1 \dots n\}$ is denoted S_n. We represent a permutation $\pi \in S_n$ as a sequence of length n where the ith element indicates the value of $\pi(i)$. For example, the permutation $[2, 3, 1] \in S_3$ maps as follows: $\{1 \mapsto 2, 2 \mapsto 3, 3 \mapsto 1\}$. A *transposition* is a permutation which swaps two elements and is the identity for all other elements. The set of all transpositions on $\{1 \dots n\}$ is denoted T_n. The transposition which swaps i and j is denoted $\pi_{i,j}$. For example, the transposition $\pi_{1,3} \in T_4$ maps as follows: $\{1 \mapsto 3, 2 \mapsto 2, 3 \mapsto 1, 4 \mapsto 4\}$. Permutations act on graphs and on unknown graphs in the natural way. For a graph $G \in \mathcal{G}_n$ and also for an unknown graph G, viewing G as an adjacency matrix, given a permutation $\pi \in S_n$, then $\pi(G)$ is the adjacency matrix obtained by mapping each element $G_{i,j}$ to $G_{\pi(i),\pi(j)}$ (for $1 \leq i, j \leq n$). The permutation, $\pi(G)$ of G, can equivalently be described as the adjacency matrix obtained by permuting both rows and columns of G using π. Two graphs $G, H \in \mathcal{G}_n$ are *isomorphic* if there exists a permutation $\pi \in S_n$ such that $G = \pi(H)$.

The standard lexicographic order on strings is denoted $\leq_{\text{lex}}$. We consider also lexicographic orders between integer and Boolean matrices, always comparing matrices of the same type, dimension and order. In our context, matrices of dimension $k > 1$ are always symmetric and have fixed values on the diagonal. We define the lexicographic ordering of two such matrices M_1 and M_2 as follows: $M_1 \leq_{\text{lex}} M_2$ if and only if $vec(M_1) \leq_{\text{lex}} vec(M_2)$ where $vec(M)$ is a string defined by concatenating the rows of M. For a matrix M with dimension $k = 1$, M is a vector and $vec(M)$ is the string of its elements. When M is of dimension $k = 2$, because of symmetry and fixed values on the diagonal, $vec(M)$ can be viewed as the concatenation of the rows of the upper triangle of M [4]. For higher dimensions, $k > 2$, the definition extends in the natural way.

In particular for graphs $G, H \in \mathcal{G}_n$, $G \leq_{\text{lex}} H$ defines a lexicographic ordering on graphs. When G, H are unknown graphs, represented as adjacency matrices of Boolean variables, then the lexicographic ordering, $G \leq_{\text{lex}} H$, can be viewed as specifying a *lexicographic order constraint* over these variables. This constraint is true with respect to an assignment for the variables of G, H if $G \leq_{\text{lex}} H$ under this assignment. We call such a constraint a *"lex-constraint"*. The case for $M_1 \leq_{\text{lex}} M_2$ where M_1, M_2 are matrices of integer variables is similar.

Example 1. Figure 1 depicts an unknown, order 5, graph G and its permutation $\pi(G)$, for $\pi = [2, 1, 3, 5, 4]$, both represented as adjacency matrices of Boolean variables. Note, for example, that the variable x_2 occurs at position $(1, 3)$ in G and at position $(\pi(1), \pi(3)) = (2, 3)$ in $\pi(G)$. The lex-constraint $G \leq_{\text{lex}} \pi(G)$ is

$$[x_1, x_2, x_3, x_4, x_5, x_6, x_7, x_8, x_9, x_{10}] \leq_{\text{lex}} [x_1, x_5, x_7, x_6, x_2, x_4, x_3, x_9, x_8, x_{10}]$$

$$\mathbf{G} = \begin{bmatrix} 0 & x_1 & x_2 & x_3 & x_4 \\ x_1 & 0 & x_5 & x_6 & x_7 \\ x_2 & x_5 & 0 & x_8 & x_9 \\ x_3 & x_6 & x_8 & 0 & x_{10} \\ x_4 & x_7 & x_9 & x_{10} & 0 \end{bmatrix} \qquad \pi(\mathbf{G}) = \begin{bmatrix} 0 & x_1 & x_5 & x_7 & x_6 \\ x_1 & 0 & x_2 & x_4 & x_3 \\ x_5 & x_2 & 0 & x_9 & x_8 \\ x_7 & x_4 & x_9 & 0 & x_{10} \\ x_6 & x_3 & x_8 & x_{10} & 0 \end{bmatrix}$$

Fig. 1. An unknown graph G and its permutation $\pi(G)$ for $\pi = [2,1,3,5,4]$.

where the sequences on the left and on the right are obtained by concatenating the rows of the upper triangles of the corresponding graphs. This constraint can be simplified as described by Frisch *et al.* [11] to

$$[x_2, x_3, x_4, x_8] \leq_{\text{lex}} [x_5, x_7, x_6, x_9]$$

□

An order n graph search problem is a predicate, $\varphi(G)$, on an unknown, order n graph G, which is closed under isomorphism. A solution to $\varphi(G)$ is a satisfying assignment for the variables of G. Given a (non-)solution for a graph search problem, each permutation of its vertices yields a symmetric (non-)solution. One common way to break symmetries in graph search problems is to define a symmetry breaking predicate which is satisfied only by the minimal representatives of each isomorphism class with respect to some total order $\preceq$.

Theorem 1. *Let G be an unknown order n graph and let $\preceq$ be a total order on graphs. Then,*

$$\textsc{Can}_{\preceq}(G) = \bigwedge_{\pi \in S_n} G \preceq \pi(G)$$

is a complete symmetry breaking constraint.

Proof. Since $\preceq$ is a total order, every isomorphism class $\mathcal{I}$ of graphs contains a unique minimal member G with respect to $\preceq$. By definition, G satisfies the constraint $\textsc{Can}_{\preceq}(G)$. Suppose that $H \in \mathcal{I}$ also satisfies $\textsc{Can}_{\preceq}$. Because $H \in \mathcal{I}$ it follows that $G = \pi(H)$ for some $\pi \in S_n$. Because H satisfies $\textsc{Can}_{\preceq}$ then $H \preceq G$. Because G is minimal $G \preceq H$. Hence $G = H$. □

The complete symmetry breaking constraint $\textsc{Can}_{\preceq}$ is impractical as it is composed of a super-exponential number of constraints, one for each permutation of the vertices. Hence, in practice, one often applies a partial symmetry breaking constraint defined in terms of a polynomial sized subset of the $\textsc{Can}_{\preceq}$ constraints.

Example 2. A classic example of a total order for graphs is the $\leq_{\text{lex}}$ order. The corresponding complete symmetry breaking constraint $\textsc{Can}_{\leq_{\text{lex}}}$ is often referred to as the lex-leader constraint [20]. Codish *et al.* [6,7] introduced a partial symmetry breaking constraint which is equivalent to taking the subset of the lex-leader constraints corresponding to all transpositions (permutations which swap two values), as specified below.

$$sb_{\text{lex}}(G) = \bigwedge_{\pi \in T_n} G \leq_{\text{lex}} \pi(G) \tag{1}$$

The sb_{lex} constraint is composed of a quadratic number of lex-constraints. It is compact and turns out to be effective when solving a wide range of graph search problems. □

Observation 1. *if $\preceq$ is a weak order on graphs, instead of a total order, then* $\text{CAN}_{\preceq}$ *is a partial symmetry breaking constraint. Moreover, any constraint defined as a subset of the constraints in* $\text{CAN}_{\preceq}$ *is also a partial symmetry breaking constraint.*

Proof. For the first claim, the proof is similar to the proof of Theorem 1. However, there is no guarantee that the minimum is unique. Hence, the corresponding symmetry breaking constraint is partial. For the second claim, weakening a partial symmetry breaking constraint results in a partial symmetry breaking constraint. □

3 Graph Invariants and Their Induced Graph Orderings

In this section, we recall the notion of graph invariants and in particular, high dimensional graph invariants. We propose a constraint-based representation for invariants of unknown graphs which is an essential component when defining symmetry breaking constraints. Finally, we introduce an ordering on graphs based on their corresponding graph invariant values.

A *k-dimensional graph invariant* is a function f which maps a graph G and a set $S = \{v_1, \ldots, v_k\} \subseteq V(G)$ of k vertices to a value which is invariant under graph isomorphism. Namely, for every permutation π of the vertex set $V(G)$ it holds that $f(G, S) = f(\pi(G), \pi(S))$. When the graph G is fixed, we denote the function $f^G(S) = f(G, S)$. In this paper, we focus primarily on the special cases of 1-dimensional and 2-dimensional graph invariants which we call *vertex invariants* and *pair invariants*, respectively.

Figure 2 introduces several graph invariants that we refer to in the remainder of the paper. Let G be a graph. The degree invariant assigns each vertex to its degree. The common neighbors invariant assigns each pair of vertices to the number of their common neighbors. The min (max) degree invariant assigns each pair of vertices to their minimal (maximal) degree. The triangles invariant assigns each vertex to the number of triangles (cycles of length 3) in which it occurs. In the figure, (u, v) denotes a pair of distinct vertices and the pair invariants are not defined when $u = v$. The *inverse* of a k-dimensional invariant f, denoted $-f$, is also a k-dimensional invariant which maps every input to the minus of the corresponding value of f. For instance, the invariant $-f^G_{\text{common}}$ specifies the minus of the number of common neighbors for each pair of vertices in G. Namely for vertices u and v, $(-f^G_{\text{common}})(u, v) = -(f^G_{\text{common}}(u, v))$.

For a fixed graph G, a k-dimensional graph invariant f^G can be viewed as a k-dimensional matrix. For a set of vertices, $S = \{v_1, \ldots v_k\}$, the element at *degree invariant*

$f^G_{\text{deg}} : V \to \mathbb{N}$
$f^G_{\text{deg}}(v) = deg_G(v)$

common neighbors invariant

$f^G_{\text{common}} : V \times V \to \mathbb{N}$
$f^G_{\text{common}}(u, v) = |N(u) \cap N(v)|$

triangles invariant

$f^G_{\text{triangles}} : V \to \mathbb{N}$
$f^G_{\text{triangles}}(v) = |\{\ (u, w) \in E(G) \mid v \in N(u) \cap N(w)\ \}|$

min degree invariant

$f^G_{\text{min}} : V \times V \to \mathbb{N}$
$f^G_{\text{min}}(u, v) = \min(deg_G(u), deg_G(v))$

max degree invariant

$f^G_{\text{max}} : V \times V \to \mathbb{N}$
$f^G_{\text{max}}(u, v) = \max(deg_G(u), deg_G(v))$

Fig. 2. Several example graph invariants.

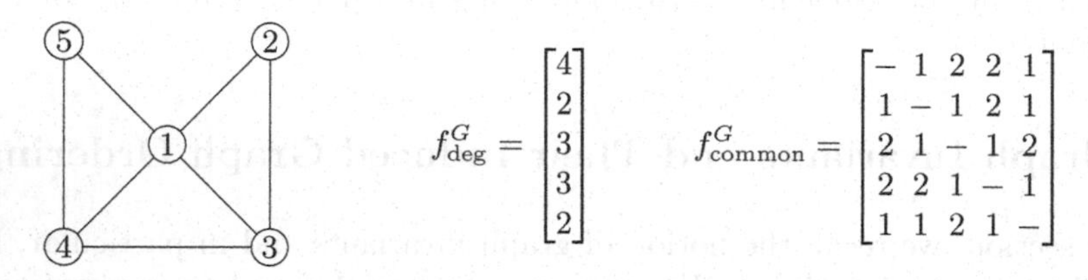

5
2
1
4
3
$f^G_{\text{deg}} = \begin{bmatrix} 4 \\ 2 \\ 3 \\ 3 \\ 2 \end{bmatrix}$
$f^G_{\text{common}} = \begin{bmatrix} - & 1 & 2 & 2 & 1 \\ 1 & - & 1 & 2 & 1 \\ 2 & 1 & - & 1 & 2 \\ 2 & 2 & 1 & - & 1 \\ 1 & 1 & 2 & 1 & - \end{bmatrix}$

Fig. 3. A graph G with matrix representation for f^G_{deg} and f^G_{common}.

position $\langle v_1, \ldots v_k \rangle$ in the matrix specifies the integer value $f^G(S)$. The following example illustrates this representation.

Example 3. Figure 3 details the matrix representation for graph invariants f^G_{deg} and f^G_{common} for the graph G depicted on the left. The i-th entry in the vector f^G_{deg} specifies the degree of vertex i in G. For instance, the degree of vertex 1 is 4. The (i, j) entry in the matrix f^G_{common} specifies the number of common neighbors of vertices i and j. For instance, the value in the entry $(1, 3)$ is 2 because vertices 1 and 3 share two neighbors (vertices 2 and 4). Notice that the values on the diagonal are not defined.

When G is an unknown graph, The invariant f^G can be viewed as a k-dimensional matrix of integer variables, together with a constraint μ which links the Boolean variables in G and the integer variables in f^G. Solutions of μ instantiate G to a graph and f^G to corresponding invariant values.

$$\mathbf{G} = \begin{bmatrix} 0 & x_1 & x_2 & x_3 & x_4 \\ x_1 & 0 & x_5 & x_6 & x_7 \\ x_2 & x_5 & 0 & x_8 & x_9 \\ x_3 & x_6 & x_8 & 0 & x_{10} \\ x_4 & x_7 & x_9 & x_{10} & 0 \end{bmatrix} \qquad \mathbf{f}^{\mathbf{G}}_{\mathbf{deg}} = \begin{bmatrix} d_1 \\ d_2 \\ d_3 \\ d_4 \\ d_5 \end{bmatrix} \qquad \mu = \begin{bmatrix} d_1 = x_1 + x_2 + x_3 + x_4 \ \wedge \\ d_2 = x_1 + x_5 + x_6 + x_7 \ \wedge \\ d_3 = x_2 + x_5 + x_8 + x_9 \ \wedge \\ d_4 = x_3 + x_6 + x_8 + x_{10} \ \wedge \\ d_5 = x_4 + x_7 + x_9 + x_{10} \ \wedge \end{bmatrix}$$

Example 4. Figure 3 details an unknown graph G of order 5 and the corresponding matrix representation of f^G_{deg}. The constraints in μ specify the relationship between the Boolean variables in G and the integer variables in f^G. The integer variable d_i in f^G represents the degree of the ith vertex in G.

We observe that a (possibly unknown) graph can also be viewed as a two dimensional graph invariant which specifies the adjacency relation. Let G be a graph. Then,

$$f^G_{\mathrm{adj}}(u,v) = \begin{cases} 1 & \text{if } (u,v) \in E(G) \\ 0 & \text{else} \end{cases}$$

The matrix representation of f^G_{adj} is identical to the adjacency matrix of G, except that integer values (one and zero) occur instead of Boolean values (true and false) and the diagonal calls are undefined instead of false.

An essential component to define symmetry breaking constraints based on graph invariants is a notion of graph ordering with respect to an invariant f.

Definition 1 *(invariant induced graph ordering). Let $G, H \in \mathcal{G}_n$ and let f be a graph invariant. Recall that vec is a flattening of the (upper triangle of the) matrix into a string of values. Then, $G \preceq_f H \Leftrightarrow vec(f^G) \leq_{lex} vec(f^H)$. We write $G =_f H$ if $G \preceq_f H$ and $H \preceq_f G$.*

In general, depending on the specific invariant f, $\preceq_f$ is possibly a weak order as distinct graphs may admit the same values for the invariant f. The following example demonstrates that $\preceq_{f_{\mathrm{deg}}}$ is a weak order.

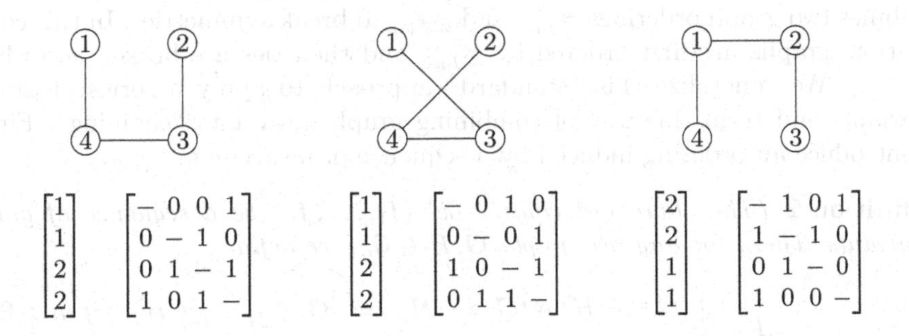

$$\begin{bmatrix}1\\1\\2\\2\end{bmatrix} \begin{bmatrix}- & 0 & 0 & 1\\0 & - & 1 & 0\\0 & 1 & - & 1\\1 & 0 & 1 & -\end{bmatrix} \quad \begin{bmatrix}1\\1\\2\\2\end{bmatrix} \begin{bmatrix}- & 0 & 1 & 0\\0 & - & 0 & 1\\1 & 0 & - & 1\\0 & 1 & 1 & -\end{bmatrix} \quad \begin{bmatrix}2\\2\\1\\1\end{bmatrix} \begin{bmatrix}- & 1 & 0 & 1\\1 & - & 1 & 0\\0 & 1 & - & 0\\1 & 0 & 0 & -\end{bmatrix}$$

Fig. 4. isomorphic representations of P_4 and their $f_{\mathrm{deg}}, f_{\mathrm{adj}}$ values.

Example 5. Consider three isomorphic representations of P_4 (path on four vertices), as depicted in Fig. 4. The minimal graph amongst them with respect to the total order $\preceq_{f_{\mathrm{adj}}}$ is the leftmost graph. The leftmost and the center graphs are both minimal with respect to the weak order $\preceq_{f_{\mathrm{deg}}}$. □

4 Symmetry Breaking Constraints with Graph Invariants

A classic way to refine the partial symmetry breaking constraint sb_{lex} presented in Eq. (1) is to specify a partition of the vertices with respect to a graph invariant, and to post a lex-constraint for the subset of transpositions which preserve the partition. In [5], the authors refine sb_{lex} with respect to a partition based on the degree invariant. This symmetry breaking constraint, which we denote here $sb_{\text{lex}}^{deg}(G)$, is defined as follows where G is an order n unknown graph:

$$\underbrace{\bigwedge_{1\leq i<n} f_{\text{deg}}^{G}(i) \leq f_{\text{deg}}^{G}(i+1)}_{(a)} \wedge \underbrace{\bigwedge_{1\leq i<j\leq n} f_{\text{deg}}^{G}(i) = f_{\text{deg}}^{G}(j) \implies G \leq_{\text{lex}} \pi_{i,j}(G)}_{(b)} \quad (2)$$

The left conjunct (a) constrains the degrees of the vertices of G to be sorted in non-decreasing order. This induces a vertex partition where vertices with equal degree are in the same part of the partition. The right conjunct (b) enforces G to be minimal with respect to all transpositions which preserve the vertex partition. Equation (2) can be rewritten using the invariant based graph ordering from Definition 1, as follows.

$$\underbrace{\bigwedge_{\pi\in T_n} G \preceq_{f_{\text{deg}}} \pi(G)}_{(a')} \wedge \underbrace{\bigwedge_{\pi\in T_n} G =_{f_{\text{deg}}} \pi(G) \implies G \preceq_{f_{\text{adj}}} \pi(G)}_{(b')} \quad (3)$$

The left and right parts (a') and (b') of Eq. (3) are equivalent respectively to parts (a) and (b) of Eq. (2). The formulation of $sb_{\text{lex}}^{deg}(G)$ as specified in Eq. (3) combines two graph orderings $\preceq_{f_{\text{deg}}}$ and $\preceq_{f_{\text{adj}}}$ to break symmetries. In this combination, graphs are first ordered by $\preceq_{f_{\text{deg}}}$ and then ties are broken according to $\preceq_{f_{\text{adj}}}$. We generalize this "standard" approach to apply a series of graph invariants and term this way of combining graph invariants "chaining". First, we introduce an ordering induced by a sequence of invariants.

Definition 2 *(The chain ordering). Let $\langle f_1, \ldots, f_n\rangle$ be a sequence of graph invariants. Then, for any two graphs $G, H \in \mathcal{G}_n$, we define*

$$G \preceq_{\langle f_1,\ldots,f_m\rangle} H = \begin{cases} (G \preceq_{f_1} H) \wedge (G =_{f_1} H \implies G \preceq_{\langle f_2,\ldots,f_m\rangle} H) & \text{if } m > 0 \\ \text{true} & \text{otherwise} \end{cases}$$

The chain ordering induces a chain symmetry breaking constraint, as specified in the following definition.

Definition 3 *(The chain constraint). Let G be an unknown graph of order n and let $\langle f_1, \ldots, f_m\rangle$ be a sequence of graph invariants. Then, the chain symmetry breaking constraint induced by $\langle f_1, \ldots, f_m\rangle$ is*

$$sb_{chain}^{f_1,\ldots,f_m}(G) = \bigwedge_{\pi\in T_n} G \preceq_{\langle f_1,\ldots,f_m\rangle} \pi(G)$$

One can check that, in general, the chain ordering is a weak order on graphs. Hence, by Observation 1, the chain symmetry breaking constraint induced by a sequence $\langle f_1, \ldots . f_n \rangle$ is a partial symmetry breaking constraint. Observe also that $sb^{deg}_{\text{lex}}(G)$ as expressed in Eq. (3) is a special case of Definition 3 and is equivalent to the chain symmetry breaking constraint induced by $\langle f_{deg}, f_{adj} \rangle$.

As illustrated in the following example, the chain constraint can be alternatively expressed as a conjunction of lex-constraints. Each constraint of the form $G \preceq_{\langle f_1,\ldots,f_{\text{m}} \rangle} \pi(G)$ is equivalent to the lex-constraint

$$vec(f_1^G, \ldots, f_m^G) \leq_{\text{lex}} vec(f_1^{\pi(G)}, \ldots, f_m^{\pi(G)})$$

where $vec(f_1^G, \ldots, f_m^G)$ is obtained by concatenating $vec(f_1^G), \ldots, vec(f_m^G)$ and similarly for $vec(f_1^{\pi(G)}, \ldots, f_m^{\pi(G)})$.

Example 6. Consider the unknown graph G of order 4, the invariant f^G_{common} and its constraints μ which are detailed below.

$$G = \begin{bmatrix} 0 & x_1 & x_2 & x_3 \\ x_1 & 0 & x_4 & x_5 \\ x_2 & x_4 & 0 & x_6 \\ x_3 & x_5 & x_6 & 0 \end{bmatrix} \quad f^G_{\text{common}} = \begin{bmatrix} - & y_1 & y_2 & y_3 \\ y_1 & - & y_4 & y_5 \\ y_2 & y_4 & - & y_6 \\ y_3 & y_5 & y_6 & - \end{bmatrix} \quad \mu = \begin{bmatrix} y_1 = x_2 * x_4 + x_3 * x_5 \wedge \\ y_2 = x_1 * x_4 + x_3 * x_6 \wedge \\ y_3 = x_1 * x_5 + x_2 * x_6 \wedge \\ y_4 = x_1 * x_2 + x_5 * x_6 \wedge \\ y_5 = x_1 * x_3 + x_4 * x_6 \wedge \\ y_6 = x_2 * x_3 + x_4 * x_5 \wedge \end{bmatrix}$$

The chain symmetry breaking constraint induced by $\langle f_{\text{common}}, f_{\text{adj}} \rangle$ consists of 6 constraints of the form $G \preceq_{\langle f_1,\ldots,f_m \rangle} \pi_{i,j}(G)$, one for each transposition $\pi_{i,j}$. Each of these can be expressed as a lex-constraint. The lex-constraint corresponding to $\pi_{1,2}$ is

$$[y_1, \ldots, y_6, x_1, \ldots, x_6] \leq_{\text{lex}} [y_1, y_4, y_5, y_2, y_3, y_6, x_1, x_4, x_5, x_2, x_3, x_6]$$

The vector on the left of the constraint consists of the variables from the invariant matrix followed by the variables from the adjacency matrix. The vector on the right, consists of the variables of the corresponding matrices obtained by swapping rows 1 and 2 as well as columns 1 and 2. Both vectors involve y variables first (from the graph invariant), followed by x variables (from the adjacency matrix). This constraint further simplifies to:

$$[y_2, y_3, x_2, x_3] \leq_{\text{lex}} [y_4, y_5, x_4, x_5]$$

□

When combining a sequence, $\langle f_1, \ldots f_m \rangle$, of graph invariants as a chain constraint, not every sequence "makes sense". Each invariant f_i in the sequence should "refine" those preceding it. We say that f_i refines $f_1, \ldots, f_{i-1}$ if the set of graphs which satisfy $sb^{f_1,\ldots,f_i}_{\text{chain}}$ is a strict subset of the set of graphs which satisfy $sb^{f_1,\ldots,f_{i-1}}_{\text{chain}}$. Adding an invariant which does not refine those preceding it does not make sense as it adds no precision. For example, the order induced by $\langle f_{\text{adj}}, f_{\text{deg}}\rangle$ is equivalent to the order induced by $\langle f_{\text{adj}}\rangle$. This is because if $G =_{f_{\text{adj}}} H$ holds, then also $G =_{f_{\text{deg}}} H$ holds. In practice, if f_{adj} occurs in a sequence combined as a chain constraint, then it should always be the last invariant in the sequence as it is the "most refined".

Table 1. Generating all order n graphs with various symmetry breaking constraints.

method	order						
	5	6	7	8	9	10	11
part 1: base							
exact	34	156	1,044	12,346	274,668	12,005,168	1,018,997,864
sb_{lex}	43 0.00 s	276 0.00 s	3,158 0.01 s	66,595 0.20 s	2,587,488 6.49 s	184,192,329 8.20 m	23,963,012,033 20.91 h
part 2: chain							
$f_{\text{deg}}, f_{\text{adj}}$	34 0.00 s	158 0.00 s	1,143 0.03 s	14,937 0.61 s	363,373 30.69 s	16,773,384 19.66 h	T.O
$f_{\text{common}}, f_{\text{adj}}$	43 0.00 s	231 0.02 s	1,933 0.37 s	28,184 7.32 s	748,727 32.08 m	T.O	T.O
$f_{\text{deg}}, f_{\text{common}}, f_{\text{adj}}$	34 0.00 s	156 0.02 s	1,075 0.28 s	13,223 5.62 s	305,189 54.33 m	T.O	T.O
part 3: product							
$f_{\text{common}}, f_{\text{adj}}$	43 0.00 s	226 0.00 s	1,852 0.11 s	26,030 1.24 s	673,069 47.25 s	32,881,227 12.01 h	T.O
$f_{\text{adj}}, f_{\text{common}}$	42 0.00 s	231 0.00 s	1,949 0.06 s	27,620 0.84 s	715,804 22.38 s	35,060,107 78.43 m	T.O
$f_{\text{adj}}, f_{\min}, f_{\max}$	43 0.00 s	215 0.01 s	1,669 0.07 s	22,464 0.86 s	562,234 24.98 s	26,480,344 95.02 m	T.O
$f_{\text{adj}}, f_{\min}, f_{\max}, f_{\text{common}}$	42 0.00 s	210 0.01 s	1,553 0.19 s	19,209 1.95 s	437,794 63.83 s	19,188,298 16.59 h	T.O

Table 1 illustrates the impact of various symmetry breaking constraints based on combinations of graph invariants. To this end, we compute all order $5 \leq n \leq 11$ graphs using various symmetry breaking constraints. Cells in the table which detail computations performed in this paper consist of two numbers: the number of solutions (above) and the computation time (below). All times are CPU running times specified in an appropriate unit: (s) seconds, (m) minutes, or (h) hours. A timeout (TO) of 24 h is applied. The rows of the table are divided into three parts titled: "base", "chain" and "product".

The first part of Table 1 (titled "base"): provides the base for comparison and consists of two rows. First, the "exact" number of order n graphs (modulo isomorphism) [19] (sequence A000088 of the OEIS). This is the base comparison for precision. For other computations, the closer the number of computed graphs is to these values, the more precise the result. The second row details the number of graphs computed using the sb_{lex} constraint introduced in [6,7]. Applying this constraint, the number of graphs generated is up to 20 times larger than the actual number of graphs modulo isomorphism. This is the least precise configuration described in the table but the only one that can find all solutions for $n = 11$ with the specified timeout.

The second part of Table 1 (titled "chain"): consists of three rows which detail the computation of all graphs using the chain symmetry breaking constraint combining various sequences of graph invariants. Note that the computations are more precise than with sb_{lex} but considerably slower. In particular, when combining three invariants (row three of part two), the computation becomes slightly more precise but considerably slower than when combining two.

A possible explanation for the inefficiency when chaining invariants is that they allow less propagations on the variables of the unknown graph. Generally speaking, for a lex-constraint of the form $a_1, \ldots, a_n \leq_{\text{lex}} b_1, \ldots, b_n$ between strings of variables, propagation on the domain of a variable a_i or b_i, depends on changes to the domains of the variables to the left: $a_1, \ldots, a_{i-1}$ and $b_1, \ldots, b_{i-1}$.

To better understand, we focus in the following example on the comparison of sb_{lex} and $sb_{\text{chain}}^{\langle f_{common}, f_{adj} \rangle}$ viewing both as conjunctions of lex-constraints.

Example 7. Consider the unknown graph G of order 4 and the invariant f^G_{common} detailed in Example 6. The following are the lex-constraints (after simplification) deriving from the transposition $\pi_{1,2}$. The first is from sb_{lex} and the second is from $sb_{\text{chain}}^{\langle f_{common}, f_{adj} \rangle}$:

$$[x_2, x_3] \leq_{\text{lex}} [x_4, x_5] \tag{4}$$

$$[y_2, y_3, x_2, x_3] \leq_{\text{lex}} [y_4, y_5, x_4, x_5] \tag{5}$$

In Eq. (4) all prefixes (of the sequences in the comparison) relate only to x variables, from the unknown adjacency matrix. In Eq. (5) all prefixes involve y variables from the graph invariant. Assignments for the y variables do not necessarily restrict the possible consistent values for the x variables because each y variable is defined in terms of a set of the x variables (see Example 6). □

The third part of Table 1 will be described later in the paper. First, we seek new ways to combine graph invariants that result in symmetry breaking constraints that improve on sb_{lex} in both efficiency and precision.

Let us first clarify notation. Let $\langle f_1, \ldots, f_m \rangle$ be a sequence of k-dimensional graph invariants. Recall that for a given graph G and each invariant f_i, f_i^G is a function which maps sets of k vertices to integer values. The Cartesian product, $f_1^G \times \ldots \times f_m^G$ of these functions maps each set S of k vertices to a tuple of integers, $\langle f_1^G(S), \ldots f_m^G(S) \rangle$. As demonstrated in Example 8, the product, $f_1^G \times \ldots \times f_m^G$, can also be viewed as a k dimensional matrix of tuples.

Definition 4 *(The product ordering). Let $\langle f_1, \ldots, f_m \rangle$ be a sequence of graph invariants of dimension k. Then, for any two graphs $G, H \in \mathcal{G}_n$, we say that $G \preceq_{f_1 \times \ldots \times f_m} H$ if and only if $vec(f_1^G \times \ldots \times f_m^G) \leq_{lex} vec(f_1^H \times \ldots \times f_m^H)$.*

Definition 5 *(The product constraint). Let G be an unknown graph of order n and let $\langle f_1, \ldots, f_m \rangle$ be a sequence of k-dimensional graph invariants. Then, the product symmetry breaking constraint induced by $\langle f_1, \ldots, f_m \rangle$ is*

$$sb_{prod}^{f_1,\ldots,f_m}(G) = \bigwedge_{\pi \in T_n} G \preceq_{f_1 \times \ldots \times f_m} \pi(G)$$

One can check that, in general, the product ordering is a weak order on graphs. Hence, by Observation 1, $sb_{prod}^{f_1,\ldots,f_m}(G)$ is a partial symmetry breaking constraint.

The following example demonstrates the construction of the product symmetry breaking constraint for the sequence of invariants $\langle f_{\text{adj}}, f_{\text{common}} \rangle$.

Example 8. Consider the unknown graph G of order 4 and the invariant f^G_{common} as detailed in Example 6. Recall that the x variables are from the adjacency matrix, and the y variables are from the graph invariant. Then,

$$f^G_{\text{adj}} \times f^G_{\text{common}} = \begin{bmatrix} - & \langle x_1, y_1 \rangle & \langle x_2, y_2 \rangle & \langle x_3, y_3 \rangle \\ \langle x_1, y_1 \rangle & - & \langle x_4, y_4 \rangle & \langle x_5, y_5 \rangle \\ \langle x_2, y_2 \rangle & \langle x_4, y_4 \rangle & - & \langle x_6, y_6 \rangle \\ \langle x_3, y_3 \rangle & \langle x_5, y_5 \rangle & \langle x_6, y_6 \rangle & - \end{bmatrix}$$

The product symmetry breaking constraint induced from $\langle f_{\text{adj}}, f_{\text{common}} \rangle$ consists of 6 constraints of the form $G \preceq_{f_{\text{adj}} \times f_{\text{common}}} \pi_{i,j}(G)$, one for each transposition $\pi_{i,j}$. Each of these can be expressed as a lex-constraint. The lex-constraint corresponding to $\pi_{1,2}$ is:

$$[x_1, y_1, \ldots, x_6, y_6] \leq_{\text{lex}} [x_1, y_1, x_4, y_4, x_5, y_5, x_2, y_2, x_3, y_3, x_6, y_6]$$

The vector on the left of the constraint consists of the variables of the matrix representing the product of the two invariants. The vector on the right consists of the variables from the permuted matrix obtained by swapping rows 1 and 2 as well as columns 1 and 2. This constraint further simplifies to:

$$[x_2, y_2, x_3, y_3] \leq_{\text{lex}} [x_4, y_4, x_5, y_5]$$

Note that the variable order in this constraint interleaves the x variables from the adjacency matrix and the y variables from the invariant whilst in the chain constraint (see Example 7) the adjacency matrix variables occur at the end of the vector. This is a property of the product constraint, that the variables of each invariant gets a "fair" place in the vectors occurring in the lex-constraints. We conjecture that interleaving allows for better propagation. The third part of Table 1 supports this conjecture, at least in the sence that computations are considerably more efficient than with the chain constraint. □

The third part of Table 1 (titled "product"): details the computation of graphs applying symmetry breaking constraints based on the product constraint. This part consists of four rows, each row describes the computation using the specified sequence of invariants as a product. Overall, the product constraints are more efficient than those using the chain constraint. The product symmetry breaking constraint induced from $\langle f_{\text{common}}, f_{\text{adj}} \rangle$ is more precise and faster than the corresponding induced chain symmetry breaking constraint. The product symmetry breaking constraint induced from $\langle f_{\text{adj}}, f_{\text{common}} \rangle$ is slightly less precise than that induced from $\langle f_{\text{common}}, f_{\text{adj}} \rangle$ but it is much faster. For example, when $n = 9$, it is about 5% less precise but is about 80 times faster. Moreover, all of the illustrated product constraints allow to generate the order 10 graphs within the 24 h timeout.

The next section demonstrates the advantage of using the (product) combination of graph invariants when solving graph search problems related to specific classes where knowledge about the structure of the graphs can be exploited to select invariants.

5 An Application: Generation of Cubic Graphs

This section demonstrates the application of symmetry breaking constraints induced from graph invariants to generate a specific class of cubic graphs. Cubic graphs are such that each vertex has degree 3. The class of cubic graphs is well studied, and many papers address the problem of generating small cubic graphs [2,3,16]. Brinkmann *et al.* [3] introduce a generation method which allows to generate all non-isomorphic connected cubic graphs for up to 32 vertices. In [21], the authors compute the number of cubic graphs for graphs of orders $n \leq 40$. However, their technique is non-constructive. That is, it allows to count the graphs but not to generate them. The number of cubic graphs is humongous [19] (sequence A005638 of the OEIS). For example, there are 8,832,736,318,937,756,165 cubic graphs of order 40.

We consider the problem of generating all connected cubic graphs, which are also "claw-free". A graph is called claw-free if it contains no $K_{1,3}$ as an induced subgraph. For cubic graphs the condition for being claw-free is equivalent to the requirement that each vertex participates in a triangle [14]. The number of connected cubic claw-free graphs for order $n \leq 30$ is specified in the OEIS as sequence A084656 [19]. Using our constraint based approach with symmetry breaking constraints based on various combinations of graph invariants, we were able to extend this sequence for $n \leq 36$. It is important to note that one cannot generate the sets of order n connected cubic claw-free graphs simply by testing the corresponding sets of cubic graphs. While the latter have been generated for $n \leq 32$, their sheer number is humongous.

Figure 5 details the constraint model we apply to generate order n cubic claw-free connected graphs. The variables $G_{i,j}$ are the Boolean variables of the unknown order n graph G. Equation (5) constrains the degree of each vertex to be 3. Equation (5) constrains the graph to be claw-free (each vertex must occur in a triangle). Finally, Eq. (5) constrains the graph G to be connected, using an encoding of the Floyd-Warshall shortest paths algorithm [10]. The variables $p^k_{i,j}$ indicate whether there is a path between vertices i and j in which intermediate vertices are from the set $\{1 \ldots k\}$. The left conjunct specifies that $p^0_{i,j}$ is true if and only if there is an edge between i and j. The center conjunct specifies the variables $p^k_{i,j}$ for $1 \leq k \leq n$, encoding the recursive part of the Floyd Warshall algorithm. The right conjunct ensures that there is a path in the graph between every two vertices.

$$\bigwedge_{1 \leq i \leq n} \sum_{1 \leq j \leq n} G_{i,j} = 3 \tag{6}$$

$$\bigwedge_{1 \leq i \leq n} \bigvee_{1 \leq j,k \leq n} (G_{i,j} \wedge G_{i,k} \wedge G_{j,k}) \tag{7}$$

$$\bigwedge_{1 \leq i,j \leq n} (p^0_{i,j} \leftrightarrow G_{i,j}) \wedge \bigwedge_{1 \leq i,j,k \leq n} p^k_{i,j} \leftrightarrow (p^{k-1}_{i,j} \vee (p^{k-1}_{i,k} \wedge p^{k-1}_{k,j})) \wedge \bigwedge_{1 \leq i,j \leq n} p^n_{i,j} \tag{8}$$

Fig. 5. The constraint model for connected cubic claw-free graphs.

In our experimentation, we applied chain combinations involving f_{adj} with one additional graph invariant from the set $\{f_{\text{common}}, f_{\text{triangles}}\}$ (and their inverses). We applied product combinations of f_{adj} with f_{common} and its inverse. For each approach (chain and product), we report the results for the symmetry breaking constraint that exhibit the best (time) performance. For comparison, we also apply the state-of-the-art partial symmetry breaking constraint, sb_{lex}.

Table 2 details the computation of claw-free cubic graphs. The first column specifies the order of the graphs. The second column details the number of non-isomorphic solutions. The next three columns detail the solving time and number of solutions for each symmetry breaking method. All times reported are CPU running times and specified in an appropriate unit: (s) seconds, (m) minutes, or (h) hours where we apply a timeout (TO) of 24 h. The numbers of non isomorphic solutions as specified in the second column are obtained by filtering isomorphic representations from the set of solutions using nauty [15]. The numbers below the solid line for $n \geq 32$ are new.

The results in Table 2 clearly show that symmetry breaking based on the product constraint, $f_{\text{adj}} \times -f_{\text{common}}$, is superior in both computation time and precision to the other techniques. This approach allows us to generate all solutions up to order 36, thus extending the OEIS sequence A084656 [19] with three new values.

Table 2. Generating connected cubic claw-free graphs for orders $4 \leq n \leq 36$.

n	graphs	sb_{lex}		$\langle -f_{\text{common}}, f_{\text{adj}} \rangle$		$f_{\text{adj}} \times -f_{\text{common}}$	
		time	sols	time	sols	time	sols
4	1	0.00 s	1	0.00 s	1	0.00 s	1
6	1	0.00 s	1	0.00 s	1	0.00 s	1
8	1	0.00 s	2	0.05 s	4	0.00 s	1
10	1	0.02 s	7	0.40 s	3	0.03 s	4
12	3	0.09 s	24	1.25 s	10	0.07 s	3
14	3	0.52 s	188	3.23 s	17	0.15 s	10
16	5	1.87 s	1,134	12.60 s	58	0.28 s	28
18	11	14.58 s	7,293	26.08 s	100	0.72 s	44
20	15	3.39 m	61,391	37.17 s	280	2.11 s	132
22	27	2.28 h	546,409	2.29 m	716	3.86 s	307
24	54	T.O	–	5.66 m	1,551	10.96 s	660
26	94	T.O	–	5.25 m	4,384	45.37 s	1,835
28	181	T.O	–	50.76 m	10,883	2.57 m	4,372
30	369	T.O	–	46.00 m	26,778	6.60 m	10,567
32	731	T.O	–	1.70 h	75,303	24.28 m	29,069
34	1,502	T.O	–	T.O	–	1.12 h	72,501
36	3,187	T.O	–	T.O	–	8.98 h	188,495

6 Conclusion

This paper explores the application of high dimensional invariants to define symmetry breaking constraints. To the best of our knowledge, this is the first time graph invariants of dimension higher than one have been applied in symmetry breaking constraints. We introduce two techniques to obtain symmetry breaking constraints by combining graph invariants. First, we introduce the chain constraint which generalizes the standard approach for combining several properties when breaking symmetries. Then, after observing the poor performance of this technique, we introduce the product combination and demonstrate its superior performance. We demonstrate the application of the product constraint to extend the computation of cubic claw-free graphs for order $n \leq 36$ vertices.

While we focus on two dimensional invariants in examples and experiments, the same techniques apply for invariants of any dimension.

A Note for the Modeller: When solving a specific graph search problem, selecting which invariants to combine is nontrivial. Two points to consider are: (1) properties of the graphs the problem seeks to find; and (2) the complexity of the invariants when expressed as a CNF. For example, when seeking regular graphs, one would not consider f_{deg} as all vertices have the same degree. Alternatives such as f_{common} and $f_{\text{triangles}}$ encode "similar" information on pairs of vertices. However, their encodings differ in complexity. Each variable of $f_{\mathrm{triangles}}$ encodes the number of triangles that involve a vertex i. This expression is quadratic. In contrast, each f_{common} variable encodes the number of common neighbors of a pair, i and j. This expression is linear in size. Hence, one might prefer the latter.

Acknowledgement. We thank the anonymous reviewers of this paper for their constructive suggestions.

References

1. Balaban, A.T., Balaban, T.S.: New vertex invariants and topological indices of chemical graphs based on information on distances. J. Math. Chem. **8**(1), 383–397 (1991). https://doi.org/10.1007/BF01166951
2. Brinkmann, G.: Fast generation of cubic graphs. J. Graph Theory **23**(2), 139–149 (1996)
3. Brinkmann, G., Goedgebeur, J., McKay, B.D.: Generation of cubic graphs. Discret. Math. Theor. Comput. Sci. **13**, 69–80 (2011)
4. Cameron, R., Colbourn, C., Read, R., Wormald, N.C.: Cataloguing the graphs on 10 vertices. J. Graph Theory **9**(4), 551–562 (1985)
5. Codish, M., Gange, G., Itzhakov, A., Stuckey, P.J.: Breaking symmetries in graphs: the nauty way. In: Rueher, M. (ed.) CP 2016. LNCS, vol. 9892, pp. 157–172. Springer, Cham (2016). https://doi.org/10.1007/978-3-319-44953-1_11
6. Codish, M., Miller, A., Prosser, P., Stuckey, P.J.: Breaking symmetries in graph representation. In: Rossi, F. (ed.) Proceedings of the 23rd International Joint Conference on Artificial Intelligence, IJCAI 2013, Beijing, China, 3–9 August 2013, pp. 510–516. IJCAI/AAAI (2013). https://ijcai.org/proceedings/2013
7. Codish, M., Miller, A., Prosser, P., Stuckey, P.J.: Constraints for symmetry breaking in graph representation. Constraints **24**(1), 1–24 (2018). https://doi.org/10.1007/s10601-018-9294-5
8. Crawford, J.M., Ginsberg, M.L., Luks, E.M., Roy, A.: Symmetry-breaking predicates for search problems. In: Aiello, L.C., Doyle, J., Shapiro, S.C. (eds.) Proceedings of the Fifth International Conference on Principles of Knowledge Representation and Reasoning (KR 1996), Cambridge, Massachusetts, USA, 5–8 November 1996, pp. 148–159. Morgan Kaufmann (1996)
9. da F. Costa, L., Rodrigues, F.A., Travieso, G., Boas, P.R.V.: Characterization of complex networks: a survey of measurements. Adv. Phys. **56**(1), 167–242 (2007). https://doi.org/10.1080/00018730601170527
10. Floyd, R.W.: Algorithm 97: shortest path. Commun. ACM **5**(6), 345 (1962). https://doi.org/10.1145/367766.368168
11. Frisch, A.M., Harvey, W.: Constraints for breaking all row and column symmetries in a three-by-two matrix. In: Proceedings of SymCon 2003 (2003)
12. Gebser, M., Kaufmann, B., Schaub, T.: Conflict-driven answer set solving: from theory to practice. Artif. Intell. **187**, 52–89 (2012)
13. Harary, F.: Graph Theory. Addison-Wesley, Reading (1969)
14. Hong, Y., Liu, Q., Yu, N.: Edge decomposition of connected claw-free cubic graphs. Discret. Appl. Math. **284**, 246–250 (2020). https://doi.org/10.1016/j.dam.2020.03.040

15. McKay, B.D., Piperno, A.: Practical graph isomorphism, II. J. Symb. Comput. **60**, 94–112 (2014)
16. Meringer, M.: Fast generation of regular graphs and construction of cages. J. Graph Theory **30**(2), 137–146 (1999)
17. Metodi, A., Codish, M., Stuckey, P.J.: Boolean equi-propagation for concise and efficient SAT encodings of combinatorial problems. J. Artif. Intell. Res. (JAIR) **46**, 303–341 (2013)
18. Miller, A., Prosser, P.: Diamond-free degree sequences. CoRR abs/1208.0460 (2012). https://arxiv.org/abs/1208.0460
19. The on-line encyclopedia of integer sequences (OEIS) (2010). Published electronically at https://oeis.org
20. Read, R.C.: Every one a winner or how to avoid isomorphism search when cataloguing combinatorial configurations. Ann. Discret. Math. **2**, 107–120 (1978)
21. Robinson, R.W., Wormald, N.C.: Numbers of cubic graphs. J. Graph Theory **7**(4), 463–467 (1983). https://doi.org/10.1002/jgt.3190070412
22. Shlyakhter, I.: Generating effective symmetry-breaking predicates for search problems. Discret. Appl. Math. **155**(12), 1539–1548 (2007)
23. Walsh, T.: General symmetry breaking constraints. In: Benhamou, F. (ed.) CP 2006. LNCS, vol. 4204, pp. 650–664. Springer, Heidelberg (2006). https://doi.org/10.1007/11889205_46

Optimization Bounds from Decision Diagrams in Haddock

Rebecca Gentzel[1(✉)], Laurent Michel[1], and Willem-Jan van Hoeve[2]

[1] University of Connecticut, Storrs, CT 06269, USA
{rebecca.gentzel,laurent.michel}@uconn.edu
[2] Carnegie Mellon University, Pittsburgh, PA 15213, USA
vanhoeve@andrew.cmu.edu

Abstract. We study the automatic generation of primal and dual bounds from decision diagrams in constraint programming. In particular, we expand the functionality of the HADDOCK system to optimization problems by extending its specification language to include an objective function. We describe how restricted decision diagrams can be compiled in HADDOCK similar to the existing relaxed decision diagrams. Together, they provide primal and dual bounds on the objective function, which can be seamlessly integrated into the constraint programming search. The entire process is automatic and only requires a high-level user model specification. We evaluate our method on the sequential ordering problem and compare the performance of HADDOCK to a dedicated decision diagram approach. The results show that HADDOCK achieves comparable results in similar time, demonstrating the viability of our automated decision diagram procedures for constraint optimization problems.

Keywords: Decision Diagrams · Constraint Programming Systems · Optimization Bounds

1 Introduction

Constraint Programming (CP) traditionally focuses on feasibility solving for Constraint Satisfaction Problems (CSP). Namely, it focuses on finding either one or all feasible solutions to a CSP $\langle X, D, C\rangle$. The ability to tackle *optimization* problems is added by solving a sequence of CSPs, each one with an additional constraint that requires the production of a solution that improves upon the last incumbent solution. Naturally, the last problem in the sequence is infeasible and the entire search tree itself is the optimality certificate. While search strategies to explore this search tree vary, the most common choice is a depth first search on a search tree dynamically-defined with variable and value selection heuristics. Black-box searches [4,9,13,14] provide pre-defined variable and value selection heuristics and non-sequential strategies such as limited discrepancy search offer compositional solutions to consider alternative strategies that remain orthogonal to the objective function.

Laurent Michel—Synchrony Chair in Cybersecurity.

© The Author(s), under exclusive license to Springer Nature Switzerland AG 2023
A. A. Cire (Ed.): CPAIOR 2023, LNCS 13884, pp. 150–166, 2023.
https://doi.org/10.1007/978-3-031-33271-5_11

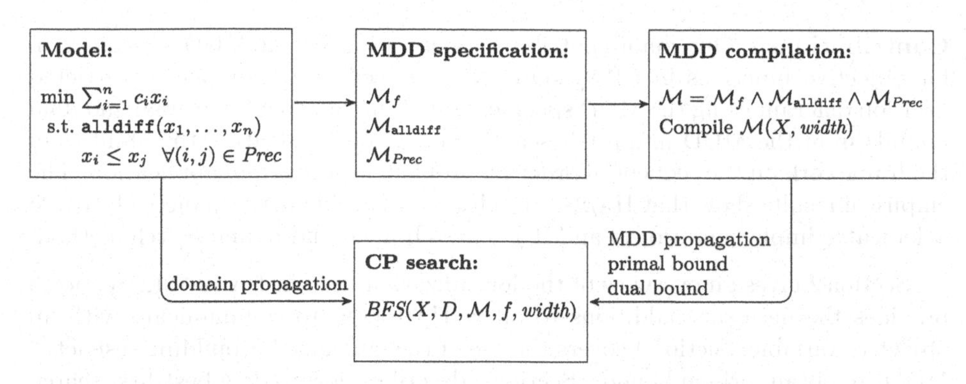

Fig. 1. Overview of automatic MDD-based constraint programming in HADDOCK on an example constraint optimization problem $\langle X, D, C, f \rangle$ with variables X, domains D, constraints C and objective function f. The constraint programming search employs a branch-and-bound best-first strategy (BFS). The MDD specification and compilation are derived automatically from the model declaration.

Mixed Integer Programming (MIP) solvers employ a different strategy. They rely on a *linear relaxation* of the MIP model that removes integrality constraints and leverage a linear programming solver to obtain a dual bound. MIP solvers then use branch-and-bound style techniques to organize and explore the frontier of nodes to be expanded. Like CP solvers, MIP solvers primarily rely on the search process to produce a sequence of improving incumbents (thus, tightening the primal bound), though techniques such as probing or the feasibility pump [5] offer additional mechanisms to tighten the primal bound. The dual bounds produced at each node by the relaxation give a mechanism to prioritize nodes in the frontier and explore the most promising options first.

Multi-valued decision diagrams (MDDs) were recently introduced as an effective tool to derive optimization bounds for discrete optimization problems, and embed these in a branch-and-bound search [2]. This paper explores the use of MDDs as a systematic mechanism to leverage both primal and dual bounds *within a CP solver* to enable MIP-style branch-and-bound search within the confines of a CP solver. HADDOCK was introduced as a generic architecture and language for MDD propagation in a CP framework [6]. To this end, we extend the existing HADDOCK framework to allow its use in optimization problems and derive both classes of bounds for problems that are expressible as an MDD in the HADDOCK language. A schematic overview is depicted in Fig. 1 on an example constraint optimization problem (COP), that has a weighted sum as objective function, an `alldiff` constraint, and precedence constraints (defined on a set *Prec*). In addition to automatically deriving an MDD specification for each constraint [6], we now also derive an MDD specification for the objective function. The specification language is compositional, which means that HADDOCK can take the conjunction of all MDD specifications to compile a single MDD. By adding the objective specification, the MDD can now be automatically used to derive primal and dual bounds during the CP search. For the primal bound, we assume that all constraints are represented in the MDD.

Contributions. Our main contributions are 1) a formal MDD specification for objective functions in CP systems, 2) a procedure to compile a restricted decision diagram using the MDD specification, 3) integrating the primal and dual bounds from the MDD into a CP search, and 4) demonstrating the abilities of the framework on the sequential ordering problem as a concrete application. The empirical results show that HADDOCK offers comparable performance relative to a dedicated implementation of an MDD-based branch-and-bound search method.

Section 2 gives an overview of the formalization used in HADDOCK. Section 3 provides the necessary additions to allow HADDOCK to communicate with an objective variable. Section 4 covers the use of the language for building restricted MDDs to obtain primal bounds. Section 5 describes how to do a best-first search using HADDOCK. Finally, Sect. 6 reports on the empirical results, and Sect. 7 concludes the paper.

2 Background

2.1 MDD as Layered Transition System

Following [6], we formally define an MDD as a labeled transition system [11]:

Definition 1. *A* labeled transition system *is a triplet* $\langle \mathcal{S}, \rightarrow, \Lambda \rangle$ *where* $\mathcal{S}$ *is a set of states,* $\rightarrow$ *is a relation of labeled transitions between states from* $\mathcal{S}$*, and* Λ *is a set of labels used to tag transitions.*

Definition 2. *Given an ordered set of variables* $X = \{x_1, \ldots, x_n\}$ *with domains* $D(x_1)$ *through* $D(x_n)$*, a* multi-valued decision diagram (MDD) *on* X *is a layered transition system* $\langle \mathcal{S}, \rightarrow, \Lambda \rangle$ *in which:*

- *the state set* $\mathcal{S}$ *is stratified in* $n+1$ *layers* $\mathcal{L}_0$ *through* $\mathcal{L}_n$ *with transitions from* $\rightarrow$ *connecting states between layers* i *and* $i+1$ *exclusively;*
- *the transition label set* Λ *is defined as* $\bigcup_{i \in 1..n} D(x_i)$*;*
- *a transition between two states* $a \in \mathcal{L}_{i-1}$ *and* $b \in \mathcal{L}_i$ *carries a label* $v \in D(x_i)$ $(i \in 1..n)$*;*
- *the layer* $\mathcal{L}_0$ *consists of a single* source *state* $s_\perp$*;*
- *the layer* $\mathcal{L}_n$ *consists of a single* sink *state* $s_\top$.

An MDD M can represent a constraint set with specific state definitions and transition functions. If each solution in the constraint set is represented by an $s_\perp$-$s_\top$ path in M, and vice-versa, M is *exact.* If M represents a superset of the solutions of the constraint set, it is *relaxed.* If M represents a subset of the solutions of the constraint set, it is *restricted.* In HADDOCK, states consist of integer-valued sets of *properties* to represent the constraints. We next describe how these are used to automatically compile the LTS, using the among constraint as an illustration. For a complete description, we refer to [6].

2.2 State Properties

Recall the definition of the among global constraint on an ordered set X of n variables [1]. It counts the number of occurrences of values taken from a given set Σ and ensures that the total number is between l and u, i.e.,

$$\texttt{among}(X, l, u, \Sigma) := l \leq \sum_{i=1}^{n} (x_i \in \Sigma) \leq u.$$

A state for $\texttt{among}(X, l, u, \Sigma)$ carries four properties, i.e., $\langle L^{\downarrow}, U^{\downarrow}, L^{\uparrow}, U^{\uparrow} \rangle$, for each node v in the MDD:

- $L^{\downarrow} \in \mathbb{Z}$: minimum number of times a value in Σ is taken from $s_{\perp}$ to v.
- $U^{\downarrow} \in \mathbb{Z}$: maximum number of times a value in Σ is taken from $s_{\perp}$ to v.
- $L^{\uparrow} \in \mathbb{Z}$: minimum number of times a value in Σ is taken from v to $s_{\top}$.
- $U^{\uparrow} \in \mathbb{Z}$: maximum number of times a value in Σ is taken from v to $s_{\top}$.

We initialize the state for the source $s_{\perp}$ as $\langle 0, 0, -, - \rangle$ and the sink $s_{\top}$ as $\langle -, -, 0, 0 \rangle$.

2.3 Transition Functions

The transition between a node $a \in \mathcal{L}_{i-1}$ and $b \in \mathcal{L}_i$ is an arc (a, b) labeled by a value $\ell \in D(x_i)$. We use *transition functions* $T^{\downarrow}(a, b, i, \ell)$ and $T^{\uparrow}(b, a, i, \ell)$ to derive the property values (the states) for b and a, respectively. For each individual property p, we use the function $f(s, p, \ell)$ for a given state s. For among, we apply $f(s, p, \ell) = p(s) + (\ell \in \Sigma)$ for each property p in $\langle L^{\downarrow}, U^{\downarrow}, L^{\uparrow}, U^{\uparrow} \rangle$. For example, we define $L^{\downarrow}(b) = f(a, L^{\downarrow}, \ell)$, i.e., $L^{\downarrow}(a) + (\ell \in \Sigma)$. We likewise define $L^{\uparrow}(a) = f(b, L^{\uparrow}, \ell)$, $U^{\downarrow}(b) = f(a, U^{\downarrow}, \ell)$ and $U^{\uparrow}(a) = f(b, U^{\uparrow}, \ell)$. The state-level transition functions $T^{\downarrow}$ and $T^{\uparrow}$ compute all the down or up properties of the next state as follows:

$$T^{\downarrow}(a, b, i, \ell) = \langle f(a, L^{\downarrow}, \ell), f(a, U^{\downarrow}, \ell), -, - \rangle$$
$$T^{\uparrow}(b, a, i, \ell) = \langle -, -, f(b, L^{\uparrow}, \ell), f(b, U^{\uparrow}, \ell) \rangle.$$

Note that slight variants of both functions that preserve the properties of states b and a, respectively, in the opposite directions are equally helpful. Those are:

$$T^{\downarrow}(a, b, i, \ell) = \langle f(a, L^{\downarrow}, \ell), f(a, U^{\downarrow}, \ell), L^{\uparrow}(b), U^{\uparrow}(b) \rangle$$
$$T^{\uparrow}(b, a, i, \ell) = \langle L^{\downarrow}(a), U^{\downarrow}(a), f(b, L^{\uparrow}, \ell), f(b, U^{\uparrow}, \ell) \rangle.$$

2.4 Transition Existence Function

The *transition existence function* $E_t(a, b, i, \ell)$ specifies whether an arc (a, b) with label $\ell \in D(x_i)$ exists in the LTS. For among, this function should ensure that the lower bound l is met and the upper bound u is not exceeded, i.e.:

$$U^{\downarrow}(a) + (\ell \in S) + U^{\uparrow}(b) \geq l \wedge L^{\downarrow}(a) + (\ell \in S) + L^{\uparrow}(b) \leq u.$$

2.5 Node Relaxation Functions

Two states a and b in the same layer $\mathcal{L}_i$ can be relaxed (merged) to produce a new state s' according to a *relaxation function* $R(a,b)$. For among, we can use:

$$R(a,b) = \langle\ \min\{L^{\downarrow}(a), L^{\downarrow}(b)\}, \max\{U^{\downarrow}(a), U^{\downarrow}(b)\}, \\ \min\{L^{\uparrow}(a), L^{\uparrow}(b)\}, \max\{U^{\uparrow}(a), U^{\uparrow}(b)\}\ \ \rangle.$$

State relaxation generalizes to an ordered set of states $\{s_0, s_1, \ldots, s_{k-1}\}$ as follows:

$$R(s_0, R(s_1, R(\ldots, R(s_{k-2}, s_{k-1})\ldots))).$$

For among, we maintain MDD-bounds consistency on this expression, i.e., we only maintain a lower and upper bound on the count to ensure feasibility and rely on the above relaxation function to merge nodes and bound the width of the MDD to at most w states. The usage of a relaxation is precisely why we maintain bounds (L and U) in both up and down directions. Note that full MDD consistency for among can be established in polynomial time by maintaining a set of exact counts [10].

2.6 MDD Language

All of the above are used to define an *MDD language* used to generate an MDD for propagation:

Definition 3 (MDD Language). *Given a constraint* $c(x_1, \ldots, x_n)$ *over an ordered set of variables* $X = \{x_1, \ldots, x_n\}$ *with domains* $D(x_1), \ldots, D(x_n)$ *the* MDD language *for* c *is a tuple* $\mathcal{M}_c = \langle X, \mathcal{P}, s_\perp, s_\top, T^{\downarrow}, T^{\uparrow}, U, E_t, E_s, R, H\rangle$ *where* $\mathcal{P}$ *is the set of properties used to model states,* $s_\perp$ *is the source state,* $s_\top$ *is the sink state,* $T^{\downarrow}$ *is the forward state transition function,* $T^{\uparrow}$ *is the reverse state transition function,* U *is the state update function [6],* E_t *is the transition existence function,* E_s *is the state existence function [6],* R *is the state relaxation function, and* H *is the trio of heuristics controlling the refinement process [7].*

3 MDDs for Optimization

Consider a COP $\langle X, D, C, f\rangle$ to be solved within the HADDOCK MDD framework. Without loss of generality, assume for now that for all constraints in C, HADDOCK has an MDD language for that constraint. Compiling the COP to solve it within HADDOCK requires one to *compose* the MDD languages for each constraint $c \in C$ as well as an MDD language for the objective function f. To carry out this compilation, one must rewrite the objective function $\{\min, \max\} f$ into an additional constraint of the form $z = f$ where z is an auxiliary variable, replace the objective with $\{\min, \max\}\ z$, and obtain an MDD language for this objective to be composed with the rest.

Some restrictions on f are needed. Since it is meant to model some form of transition costs over prefixes of the variable list, it is required to be separable (e.g., additive). Any inductive definition for f would meet this requirement. A simple example is $\sum_{i=1}^{n}(x_i \in \Sigma)$ which counts the number of variables taking their value from a prescribed set Σ. Likewise, a weighted sum $\sum_{i=1}^{n} w_i x_i$ that captures transition costs is acceptable.

Definition 4 (MDD Language for Objective Function). *Given an objective function* $\{\min, \max\} f(x_1, \ldots, x_n)$ *over an ordered set of variables* $X = \{x_1, \ldots, x_n\}$ *with domains* $D(x_1), \ldots, D(x_n)$ *let the auxiliary* z *be defined as* $z = f(x_1, \ldots, x_n)$ *and the concrete objective be* $\{\min, \max\}\ z$*. Then, the* MDD language *for the objective* $\{\min, \max\}\ f$ *is*

$$\mathcal{M}_f = \langle X, \mathcal{P}, s_\perp, s_\top, T^\downarrow, T^\uparrow, -, E_t, -, R, -, \{\min, \max\}\ z \rangle$$

where $\mathcal{P}$ *is the set of properties used to model states (for* $z = f(x_1, \ldots, x_n)$*),* $s_\perp$ *is the source state,* $s_\top$ *is the sink state,* $T^\downarrow$ *is the forward state transition function,* $T^\uparrow$ *is the reverse state transition function,* E_t *is the transition existence function, and* R *is the state relaxation function. Dashes denotes the absence of state update, state existence, and heuristic bundles.*

A few observations are in order. First, the auxiliary z is not a model variable and therefore does not occupy a layer in the MDD. Second, the auxiliary z is typically used within E_t to filter arcs that cannot produce solutions of the desired quality. Third, the source and sink states, respectively $s_\perp$ and $s_\top$, hold properties related to f (and therefore z) that pertain to all source-sink paths in the MDD and will be used to read both primal and dual bounds. Fourth, internal states of the MDD hold properties for f that are related to the source-sink paths going through that specific node.

Example 1 (Minimize a Weighted Sum Objective). For the objective function $\min \sum_{i=1}^{n} w_i \cdot x_i$, the auxiliary z is defined as $z = \sum_{i=1}^{n} w_i \cdot x_i$ and is associated to properties L and U giving the lower and upper bounds on f in both the up and down directions in the diagram. As a result, the values $L^\downarrow(s_\top)$ and $L^\uparrow(s_\perp)$ represent a lower bound[1] for z in a relaxed MDD while, for any internal state s, $L^\downarrow(s) + L^\uparrow(s)$ denotes z's bound for any internal state for all paths going through s. The transition functions are simply

$$T^\downarrow(a, b, i, \ell) = \langle L^\downarrow(a) + (w_i \cdot \ell), U^\downarrow(a) + (w_i \cdot \ell), L^\uparrow(b), U^\uparrow(b) \rangle$$
$$T^\uparrow(b, a, i, \ell) = \langle L^\downarrow(a), U^\downarrow(a), L^\uparrow(b) + (w_i \cdot \ell), U^\uparrow(b) + (w_i \cdot \ell) \rangle$$

while the relaxation of two states a and b is:

$$R(a, b) = \langle\ \min\{L^\downarrow(a), L^\downarrow(b)\}, \max\{U^\downarrow(a), U^\downarrow(b)\}, \min\{L^\uparrow(a), L^\uparrow(b)\}, \max\{U^\uparrow(a), U^\uparrow(b)\}\ \rangle$$

The arc existence function meant to test the viability of a value $\ell \in D(x_i)$ to connect states a and b is

$$E_t(a, b, i, \ell) = U^\downarrow(a) + w_i \cdot \ell + U^\uparrow(b) \geq \min(z) \wedge L^\downarrow(a) + w_i \cdot \ell + L^\uparrow(b) \leq \max(z)$$

[1] For a maximization, $U^\downarrow(s_\top)$ and $U^\uparrow(s_\perp)$ give the upper bound.

To derive the HADDOCK MDD language from a COP $\langle X, D, C, f\rangle$, it suffices to compile

$$\mathcal{M} = \bigwedge_{c \in C} \mathcal{M}_c \wedge \mathcal{M}_f$$

in which $\wedge$ is the MDD composition operator. To search for a global optimum over D using HADDOCK, one must instantiate propagators for $\mathcal{M}$. While it is often not tractable to maintain an *exact* MDD, it is natural to rely instead on *relaxed* and *restricted* MDD propagators to derive dual and primal bounds and carry out a branch-and-bound search. Given a maximum width w, we can obtain:

Relaxed MDD Let $\underline{M}$ be the relaxed MDD (to width w) where nodes are merged within each layer to never exceed width w;
Exact MDD Let M^* be the exact MDD;
Restricted MDD Let $\overline{M}$ be the restricted MDD (to width w) in which overflow nodes are discarded.

Note that the maximum width w together with the MDD language and a given COP instance will yield a unique relaxed or restricted diagram. (Exact diagrams are always unique for a given variable ordering.) This is because the MDD language also controls any heuristic compilation choices. Therefore, so long as the heuristics do not introduce any randomness, the relaxed or restricted MDD will be unique.

For a decision diagram M, let $\Psi(M)$ be the set of solutions ($s_\perp$-$s_\top$ paths) encoded by M. By construction, we can obtain a bound on z by reading $M.L^\downarrow(s_\top)$ from the sink state of M. We have the following results [2]:

Proposition 1. $\Psi(\overline{M}) \subseteq \Psi(M^*) \subseteq \Psi(\underline{M})$.

Proposition 2. $\underline{M}.L^\downarrow(s_\top) \leq M^*.L^\downarrow(s_\top) \leq \overline{M}.L^\downarrow(s_\top)$.

That is, the relaxed MDD $\underline{M}$ delivers a dual bound while the restricted MDD $\overline{M}$ delivers a primal bound.

Example 2 (COP). Consider the COP defined over $X = \{x_1, \ldots, x_4\}$, $z \in \{0, \ldots, 4\}$, and $D(x_i) \in \{0, 1\}$ for $i = 1, \ldots, 4$:

$$COP = \langle X, D, \{\texttt{among}(X, 1, 3, \{1\})\}, \min \sum_{i=1}^{4} (x_i \in \{1\})\rangle$$

With the auxiliary $z = \sum_{i=1}^{4} (x_i \in \{1\})$, the MDD language from $\mathcal{M} = \mathcal{M}_{\texttt{among}} \wedge \mathcal{M}_{\texttt{sum}} \wedge \mathcal{M}_{\min z}$ models the COP. It can be used to compile a relaxed, exact, and restricted diagram as shown in Fig. 2, where we impose a maximum width 2 on the relaxed and restricted MDDs. Each state is labeled with properties $(L^\downarrow, U^\downarrow, L^\uparrow, U^\uparrow)$.[2]

[2] With a slight abuse of notation as we do not repeat the bounds on z and among since those properties are identical.

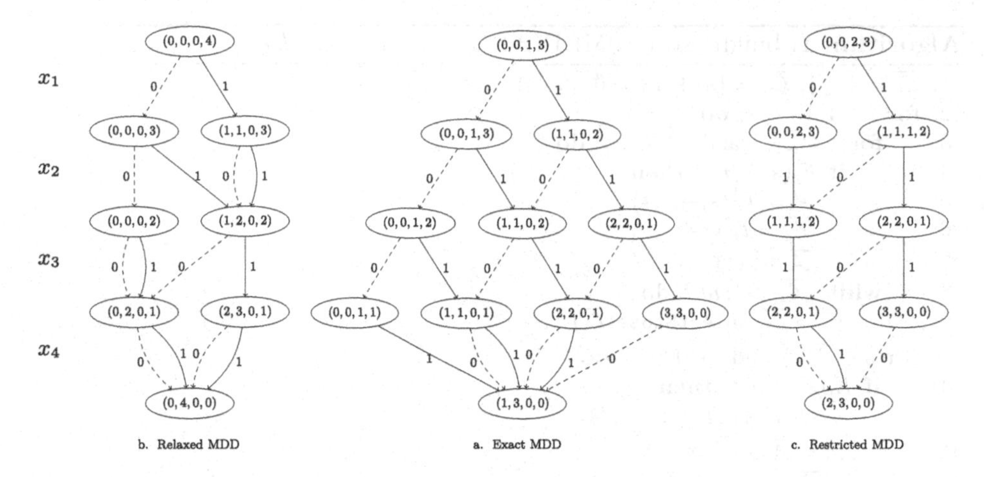

Fig. 2. MDDs for the COP $\langle X, D, \{\texttt{among}(X, 1, 3, \{1\})\}, \min \sum_{i=1}^{4} (x_i \in \{1\})\rangle$ of Example 2.

The CP solver maintains the relaxed and restricted variants within propagators and uses the bounds to drive the search, i.e., z is tightened using both $\underline{M}.L^{\downarrow}(s_{\top})$ and $\overline{M}.L^{\downarrow}(s_{\top})$. While the Exact MDD only contains paths with sums between 1 and 3, the Relaxed MDD includes paths of value 0 and 4, and the Restricted only contains paths of values 2 and 3. Observe that $\overline{M}.L^{\downarrow}(s_{\top})$ yields a primal bound of value 2 while $\underline{M}.L^{\downarrow}(s_{\top})$ delivers a dual bound of value 0.

4 Restricted Decision Diagrams

Reference [6] offers a way to compile a propagator for the relaxed diagram $\underline{M}$. This section adapts the mechanism to produce a propagator for the restricted diagram $\overline{M}$, meant to run at a higher priority, to compute primal bounds. When the propagator runs, if the restricted MDD is feasible, the best path through the restricted diagram from source to sink spells out a *witness solution* and its objective value which can be submitted to the solver as a new incumbent (and therefore trigger the usual addition of a global optimality cut based on this primal value). The restricted MDD construction is shown in Algorithm 1. The main loop (lines 2–9) constructs the layers sequentially. Each iteration starts with an empty layer and considers every node and outgoing arc from the prior layer (line 3). If the arc exists, then the transition $T^{\downarrow}$ produces a new state that is added to layer $\overline{\mathcal{L}}_i$. The loop on lines 8–9 trims layer i until it reaches the desired width, discarding the arcs chosen by the `selectState` heuristic introduced in [7]. Lines 10–13 conclude by connecting nodes of the penultimate layer to the sink and making use of the relaxation function R. Note that R is not used anywhere else, preferring instead to discard overflowing states. When creating a restricted MDD with top-down compilation, there are no bottom-up properties, hence the transition existence function must be updated to include this possibility. In place

Algorithm 1. buildRestrictedMDD($\mathcal{M}, [x_1, \ldots, x_n], width$)

1: $\overline{\mathcal{L}_0} = \{s_\perp\}, \overline{\mathcal{L}_n} = \{s_\top\}, \overline{\mathcal{L}_i} = \emptyset \ \forall i \in 1..n-1, \overline{\mathcal{A}} = \emptyset$
2: **for** $i \in 1..(n-1)$ **do**
3: **for** $s \in \overline{\mathcal{L}_{i-1}}$ and $\ell \in D(x_i)$ **do**
4: **if** $E_t(s, -, i, \ell)$ **then**
5: $s' = T^\downarrow(s, -, i, \ell)$
6: $\overline{\mathcal{L}_i} = \overline{\mathcal{L}_i} \cup s'$
7: $\overline{\mathcal{A}} = \overline{\mathcal{A}} \cup s \xrightarrow{\ell} s'$
8: **while** $|\overline{\mathcal{L}_i}| > width$ **do**
9: $\overline{\mathcal{L}_i} = \overline{\mathcal{L}_i} \backslash$`selectState`$(\overline{\mathcal{L}_i})$
10: **for** $s \in \overline{\mathcal{L}_{n-1}}$ and $\ell \in D(x_n)$ **do**
11: **if** $E_t(s, -, n, \ell)$ **then**
12: $s_\top = R(s_\top, T^\downarrow(s, -, i, \ell))$
13: $\overline{\mathcal{A}} = \overline{\mathcal{A}} \cup s \xrightarrow{\ell} s_\top$
14: `return` $\langle [\overline{\mathcal{L}_0}, \cdots, \overline{\mathcal{L}_n}], \overline{\mathcal{A}} \rangle$

Algorithm 2. Filter $\mathcal{F}_{\overline{M}}$ over variables X for MDD language $\mathcal{M}$ and width w

1: $\overline{M}$ = buildRestrictedMDD($\mathcal{M}, X, w$)
2: **for** $f \in$ *solver*.`onSolCallbacks` **do**
3: f(`bestPath`$(\overline{M}.s_\perp, \overline{M}.s_\top)$,$\overline{M}.L^\downarrow(s_\top)$)
4: **if** $\overline{M}$ is exact **then**
5: `failNow()`

of bottom-up properties, one can use the *rough relaxed bounds* introduced in [8]. Line 14 returns the produced restricted diagram.

4.1 Restricted MDDs in HADDOCK Propagation

Algorithm 2 gives the pseudocode of the restricted propagator. Line 1 builds the restricted diagram (these are not reused across invocations) for the MDD language $\mathcal{M}$ defined over variables in X. The loop on lines 2–3 iterates over the list of callbacks passing down the witness solution for the best path in $\overline{M}$ together with the primal bound for it, i.e., $\overline{M}.L^\downarrow(s_\top)$. As long as the callback tightens z's upper bound, the COP will be required to improve the incumbent for the remainder of the execution. Finally, line 4 determines whether the diagram is exact or not. If it is exact, then it contains the optimal solution and the search can stop.

4.2 Relaxed MDDs in HADDOCK Propagation

The propagator for the relaxed MDD $\underline{M}$ is unchanged from [6]. The only difference is that the propagator is accessible by the search procedure as an oracle capable of producing a dual bound on request (produced at its last fixpoint).

4.3 Restricted MDDs and Constraints External to the MDD

One of the strengths of HADDOCK is the ability to support additional constraints with their own propagators within a single solver. The algorithms presented here assume that all constraints are embedded in one MDD. This assumption means every $s_\perp$-$s_\top$ path in a restricted MDD corresponds to a feasible solution. If some constraints are external to the MDD, a solution sent to the solver on Line 3 of Algorithm 2 may violate one or more of those external constraints. Thankfully in this case, the callback f can invoke a sub-solver for all external constraints based on the binding imposed by the witness solution it receives to verify their feasibility. Note that this can entail a nested search [17]. For brevity's sake, this paper only considers models where the MDD contains all constraints.

5 Best-First Search

CP often uses a depth-first search and relies on optimality cuts to discard subtrees that cannot improve upon the incumbent. With both a primal and a dual bound, a best-first search strategy becomes feasible. Consider Algorithm 3 modeled after the DFS in miniCP [12]. Lines 1 and 2 specify the propagators for the restricted and the relaxed MDDs, respectively, of width w associated to $\mathcal{M}$. Line 3 creates a priority queue and populates it with an initial problem where the constraints and objective function f are embedded in M. A trivial upper bound for the primal is set to $+\infty$ (without loss of generality, we assume a minimization). Line 5 adds an anonymous function to be called each time an incumbent is produced. The purpose of this lambda is to tighten the primal bound. Lines 6–14 offer the main loop. Each iteration starts in line 7 with pulling the most promising node from the queue. Line 8 propagates this node fully with the filtering associated to all constraints to obtain the refined domains D'. Line 9 picks a variable to branch on (i.e., x_i). The loop spanning lines 10–14 considers each value v in turn. Line 11 queries layer i of the relaxed MDD to retrieve the state reachable via value v, and line 12 recovers the dual bound for that node. If the node appears viable (line 13), a problem is added to the queue with the revised domain, the binding of x_i to v, and the tightening of the objective.

A few observations are worth making:

- The propagator for $\mathcal{F}_{\overline{M}}$ should be scheduled at a higher priority than the propagator for $\mathcal{F}_{\underline{M}}$ since it is cheaper to compute and has the potential to end the search early. This is easy to achieve with any solver that has at least 2 priority lists.
- The BFS implementation outline above adopts a *lazy* technique by only enqueueing the specification of the new search node in *queue* on line 14 and propagating the effect of the branching constraint only where the node is de-queued on line 7. This lazy strategy dominates the eager when a node is propagated before being added to the queue. The rationale is that propagation is relatively expensive in the context of MDD solvers where a substantial computational effort is expanded when refining the MDD. BFS nodes that are ultimately fathomed do not have to carry this burden in the lazy approach.

Algorithm 3. $BFS(X = [x_1, \ldots, x_n], D, \mathcal{M}, f, w)$

1: $\mathcal{F}_{\overline{M}}$ = filtering function for the restricted MDD $\overline{M}$ of width w
2: $\mathcal{F}_{\underline{M}}$ = filtering function for the relaxed MDD $\underline{M}$ of width w
3: $queue = \{(\langle X, D, \{\overline{M}, \underline{M}\}\rangle, -\infty)\}$
4: $primal = +\infty$
5: $solver.$`onSolution`$(\lambda X.\lambda z \rightarrow primal = \min(primal, z))$
6: **while** $queue \neq \emptyset$ **do**
7: $\quad (\langle X, D, C\rangle, \_) = queue.$`extractBest()`
8: $\quad D' = \mathcal{F}_C(D)$
9: $\quad i = \min_{1\ldots n}\{i \mid x_i \text{ is not bound}\}$
10: $\quad$ **for** $v \in D(x_i)$ **do**
11: $\quad\quad s = \underline{M}.\mathcal{L}_i.$`stateWithIncomingArc`$(v)$
12: $\quad\quad dual = L^{\downarrow}(s) + L^{\uparrow}(s)$
13: $\quad\quad$ **if** $dual < primal$ **then**
14: $\quad\quad\quad queue = queue \cup (\langle X, D', C \cup \{x_i = v, f \geq dual\}\rangle, dual)$

- When branching on x_i, every variable sequentially before x_i is bound. This ensures that every earlier layer in the MDD consists of a single state with one outgoing arc. As a result, when obtaining the state reachable via value v (line 11), there can be only one state because the previous layer consists of a single node. If using a branching technique that does not ensure every previous variable is bound, then there may be multiple states in $\mathcal{L}_i$ reachable via v. In this case, the dual bound on layer 12 would instead be the minimum across all such states in $\mathcal{L}_i$.
- This search bears similarities to the MDD-based branch-and-bound proposed in [3]. In [3], the branching is done on MDD nodes from cutsets consisting exclusively of *exact* nodes, i.e. nodes whose states did not require merging, and required that every $s_{\perp}$-$s_{\top}$ path takes at least one node in the cutset. This paper intentionally applied a traditional CP style branching on variables. Yet, it is possible to adopt the same cutset branching provided that the necessary API is provided on an MDD propagator, a task reserved for future work.

6 Empirical Evaluation

HADDOCK is part of `MiniC++`, a C++ implementation of the MiniCP specification [12]. All benchmarks were executed on a Intel Xeon CPU E5-2640 v4 at 2.40 GHz with 32 GB.

Comparison to MDD-Based Branch and Bound. We compare constraint optimization in HADDOCK to the *dedicated* MDD-based branch and bound solver from [16]. We downloaded the provided source code for the dedicated solver[3], compiled, and ran it on the same machine as HADDOCK. We evaluate

[3] Source code located at https://github.com/IsaacRudich/PnB_SOP.

Table 1. Evaluating SOP instances. The maximum MDD width is $w = 64$.

		HADDOCK				HADDOCK w/ Priority			B&B		
Instance	n	# Nodes	Time (s)	Dual	Primal	Time (s)	Dual	Primal	Time (s)	Dual	Primal
esc07	9	2	**0.001**	2125	2125	**0.001**	2125	2125	**0.001**	2125	2125
esc11	13	7	**0.003**	2075	2075	**0.003**	2075	2075	0.26	2075	2075
esc12	14	109	0.102	1675	1675	**0.080**	1675	1675	1.89	1675	1675
esc25	27	1456	17.129	1681	1681	**16.786**	1681	1681	1002.01	1681	1681
esc47	49	77278	-	171	1441	-	326	**1427**	-	**335**	1542
esc63	65	46163	-	**21**	**62**	-	**21**	**62**	-	8	**62**
esc78	80	8602	-	2050	**19575**	-	2025	**19575**	-	**2230**	19800
br17.10	18	7995	39.387	55	55	**34.052**	55	55	270.92	55	55
br17.12	18	5765	25.456	55	55	**19.982**	55	55	146.50	55	55
ft53.1	54	18969	-	**2996**	**8198**	-	1625	**8198**	-	1785	8478
ft53.2	54	26150	-	**2322**	8840	-	1729	**8458**	-	1945	8927
ft53.3	54	36033	-	2018	11519	-	2138	**11707**	-	**2546**	12179
ft53.4	54	71566	-	3549	**14758**	-	3681	14776	-	**3773**	14811
ft70.1	71	7293	-	24428	41751	-	24556	**41647**	-	**25444**	41926
ft70.2	71	9546	-	24560	42294	-	24664	**41932**	-	**25237**	42805
ft70.3	71	15824	-	25263	**46497**	-	25220	47232	-	**25809**	48073
ft70.4	71	21663	-	28775	**56477**	-	**28928**	**56477**	-	28583	56644
kro124p.1	101	2800	-	**14667**	45025	-	9556	**44699**	-	10773	46158
kro124p.2	101	3801	-	**13901**	46802	-	10003	**46608**	-	11061	46930
kro124p.3	101	6407	-	10606	**55137**	-	10882	**55137**	-	**12110**	55991
kro124p.4	101	9854	-	**16524**	**84492**	-	15297	84685	-	13829	85533
p43.1	44	37428	-	375	**28785**	-	350	29090	-	**630**	29450
p43.2	44	68066	-	405	**28770**	-	370	29010	-	**440**	29000
p43.3	44	76591	-	505	**29530**	-	510	**29530**	-	**595**	**29530**
p43.4	44	121550	-	960	83800	-	1015	**83760**	-	**1370**	83900
prob.42	42	93554	-	90	271	-	**106**	**263**	-	99	289
prob.100	100	4232	-	166	**1673**	-	163	**1673**	-	**170**	1841
rbg048a	50	51446	-	55	**369**	-	60	**369**	-	**76**	379
rbg050c	52	59245	-	**70**	**500**	-	56	**500**	-	63	566
rbg109a	111	35077	-	**313**	**1127**	-	307	**1127**	-	91	1196
rbg150a	152	16265	-	**354**	**1863**	-	201	**1863**	-	63	1874
rbg174a	176	9447	-	**453**	**2156**	-	335	**2156**	-	118	2157
rbg253a	255	4408	-	**538**	**3178**	-	390	**3178**	-	112	3181
rbg323a	325	3492	-	**678**	3380	-	416	**3370**	-	89	3519
rbg341a	343	3321	-	**319**	**2968**	-	246	2970	-	68	3038
rbg358a	360	2074	-	**181**	**3202**	-	175	**3202**	-	69	3359
rbg378a	554	1789	-	**196**	**3402**	-	67	**3402**	-	52	3429
ry48p.1	49	33024	-	**6414**	16892	-	4668	**16763**	-	5198	17555
ry48p.2	49	46264	-	**6284**	17439	-	4908	**17410**	-	5290	18046
ry48p.3	49	46053	-	5772	**20890**	-	5793	20962	-	**6208**	21161
ry48p.4	49	41435	-	12443	33391	-	**14576**	**33261**	-	13598	34517

the implementations on the Sequential Ordering Problem (SOP) from [16]. This problem can be represented as an asymmetric traveling salesman problem with precedence constraints. Given n elements labeled $v_1, \ldots, v_n$ with asymmetric arcs connecting them, the objective is to find a minimum path from v_1 to v_n visiting each element once and respecting precedence constraints. The precedence constraints are defined as a precedence ordering of v_i before v_j, the index of v_i in the path must be before the index of v_j. The solvers were tested on the 41 SOP problems in TSPLIB [15].

HADDOCK represents the problem as the composition of an AllDifferent, a sum (for the TSP distances), and a global ordering (that encapsulates all precedence constraints) MDD languages. The language for the sum is a modified version from Sect. 3 to use the appropriate weight value in the transition functions. The language for the global ordering constraint is very simple, only requiring one forward property and one reverse property to track which elements have been selected. The solver uses n variables labeled x_1 to x_n with domains $D(x_i) = \{1, \dots n\}$ where the value of $x_i = v$ means element v is in position i of the sequence. Variables x_1 and x_n are restricted to be 1 and n, respectively. Following [7], the model uses heuristics to refine the MDD. First, the model uses equality for the equivalence function and prioritizes refinement to favor states with a smaller $L^{\downarrow}$. Second, in the initial refinement iteration, we make use of an approximate equivalence function to split nodes based on incoming arc values. We use a `maximum reboot distance` of 100.

All experiments use a 1-h timeout and record the primal and dual solutions as well as the time taken to terminate. Results appear in Table 1. Bold-faced entries report which solver terminates first (time) or with the best bounds (and thus best incumbent for the primal bound). The "HADDOCK" columns correspond to the default heuristics while "HADDOCK w/ Priority" refers to boosting the priority of the ordering constraint. The columns for "B&B" refer to the dedicated MDD-based branch-and-bound method from [16].

Out of the 41 instances, 6 terminate in under an hour. These terminate for Branch and Bound as well but with longer runtimes. This is most likely attributable to the impact of the heuristics used within the relaxed MDD propagator for merging MDD nodes. Without taking advantage of constraint priority, HADDOCK still obtained better times in the 6 terminating instances. Setting the ordering constraint at top priority, the bounds obtained by HADDOCK improve for several instances. For example, the dual bound for `esc47` increases from 128 to 336. However, we also observe a couple instances where this heuristic negatively impacts the dual bound. Most notably, `rbg150a` and `rbg341a` both fail to obtain a meaningful dual bound. A limited number of heuristics were tested in HADDOCK, which leads us to speculate that other heuristics may give tighter bounds within the same time frame. For benchmark instances that time-out after one hour, HADDOCK obtains competitive bounds compared to Branch and Bound. In most instances, HADDOCK has a better primal (incumbent) while the dual bound is often marginally weaker. Exceptions where the dual bound is better do exist, e.g., `esc63`, `ft70.4`, `kro124p.4`, `prob.42`, `rgb109a`. From a dual bound standpoint, it leads to the conclusion that neither solver dominates and the difference are most likely attributable to the differences in heuristics with the relaxed MDD propagator with the heuristic used in HADDOCK being either a better or worse fit depending on the benchmark structure.

Table 2. Time(s) and search nodes to reach target dual bound at different widths.

		$w = 32$		$w = 64$		$w = 128$		$w = 256$	
Instance	Target Dual	Time (s)	# Nodes	Time (s)	#Nodes	Time (s)	#Nodes	Time (s)	# Nodes
esc78	1800	365.457	1707	389.607	916	830.739	723	2210.184	701
ft70.4	28000	59.153	1186	121.803	1083	252.895	1005	530.146	947
prob.42	80	221.997	11870	438.860	10751	1031.687	10909	2124.905	10695
ry48p.2	5500	165.818	4471	73.132	690	210.550	690	685.847	690

Effects of Width. Table 2 shows how the performance of HADDOCK scales with the specified width on a subset of benchmarks from the various classes of instances. Since those are larger instances that time out at 1 h, to have a better comparison, the solvers were asked to stop once they reached a target value for the dual bound (reported in the second column). Note how, as observed before, there is a sweet spot for the width for which runtime is minimized. Also, the number of nodes for the branch-and-bound tree tends to reduce as width increases. Naturally, since the algorithm is not executed to its natural termination (with an optimality proof) the results should be interpreted conservatively.

Table 3. Impact of Restricted MDDs for the primal bound on BFS.

		Depth-First Search HADDOCK			Best-First Search HADDOCK		
Instance	n	Time (s)	Dual	Primal	Time (s)	Dual	Primal
esc07	9	0.002	2125	2125	0.001	2125	2125
esc11	13	0.079	2075	2075	0.003	2075	2075
esc12	14	0.737	1675	1675	0.102	1675	1675
esc25	27	605.560	1681	1681	17.129	1681	1681
esc47	49	-	-	7655	-	171	1441
esc63	65	-	-	170	-	21	62
esc78	80	-	-	29340	-	2050	19575
br17.10	18	55.099	-	55	39.387	55	55
br17.12	18	10.298	-	55	24.456	55	55

Comparison to Depth-First Search Without Restricted MDDs. Table 3 highlights the impact of using Best-First Search with restricted MDDs. DFS finds and proves optimality on the same instances that BFS did. Yet, in all but one of these cases, DFS takes longer. In the exception (br17.12), it appears that DFS gets 'lucky' and finds the optimal solution quickly with the search strategy alone. In other cases, DFS takes over a factor 10 longer, and when the instance takes over an hour, not only does DFS have a weaker incumbent solution, but it has no dual bound (effectively a dual bound of 0).

Comparison to Peel and Bound. We ran the Julia implementation of Peel & Bound on our hardware and share in Table 4 a qualitative comparison between HADDOCK and the results from [16]. First HADDOCK appears to remain competitive w.r.t. runtime. In addition, HADDOCK produces primal bounds within

Table 4. SOP Instances. Results from [16] at $w = 64$ for Peel & Bound.

		HADDOCK			Branch & Bound			Peel & Bound		
Instance	n	Time (s)	Dual	Primal	Time (s)	Dual	Primal	Time(s)	Dual	Primal
esc07	9	**0.001**	2125	2125	**0.001**	2125	2125	**0.001**	2125	2125
esc11	13	**0.003**	2075	2075	0.26	2075	2075	0.10	2075	2075
esc12	14	**0.102**	1675	1675	1.89	1675	1675	0.67	1675	1675
esc25	27	**17.129**	1681	1681	1002.01	1681	1681	319.47	1681	1681
esc47	49	-	171	**1441**	-	335	1542	-	**364**	1676
esc63	65	-	21	**62**	-	8	**62**	-	**44**	62
esc78	80	-	2050	**19575**	-	2230	19800	-	**4950**	20045
br17.10	18	39.387	55	55	270.92	55	55	**11.36**	55	55
br17.12	18	25.456	55	55	146.50	55	55	**25.26**	55	55
ft53.1	54	-	2996	**8198**	-	1785	8478	-	**3313**	8244
ft53.2	54	-	2232	8840	-	1945	8927	-	**3419**	**8815**
ft53.3	54	-	2018	**11519**	-	2546	12179	-	**4198**	12482
ft53.4	54	-	3549	**14758**	-	3773	14811	-	**6398**	14862
ft70.1	71	-	24428	41751	-	25444	41926	-	**31077**	**41607**
ft70.2	71	-	24560	**42294**	-	25237	42805	-	**31190**	42623
ft70.3	71	-	25263	**46497**	-	25809	48073	-	**31823**	47491
ft70.4	71	-	28775	**56477**	-	28583	56644	-	**35895**	56552
kro124p.1	101	-	14667	**45025**	-	10773	46158	-	**17541**	46158
kro124p.2	101	-	13901	**46802**	-	11061	46930	-	**17608**	46930
kro124p.3	101	-	10606	**55137**	-	12110	55991	-	**18542**	55991
kro124p.4	101	-	16524	**84492**	-	13829	85533	-	**24316**	85316
p43.1	44	-	375	**28785**	-	**630**	29450	-	380	29390
p43.2	44	-	405	**28770**	-	**440**	29000	-	420	29080
p43.3	44	-	505	**29530**	-	**595**	**29530**	-	480	**29530**
p43.4	44	-	960	**83800**	-	**1370**	83900	-	1010	83880
prob.42	42	-	90	**271**	-	**99**	289	-	94	289
prob.100	100	-	166	**1673**	-	170	1841	-	**174**	1841
rbg048a	50	-	55	**369**	-	**76**	379	-	45	380
rbg050c	52	-	70	**500**	-	63	566	-	**154**	512
rbg109a	111	-	313	**1127**	-	91	1196	-	**372**	1196
rbg150a	152	-	354	**1863**	-	63	1874	-	**563**	1865
rbg174a	176	-	453	**2156**	-	118	2157	-	**623**	**2156**
rbg253a	255	-	538	**3178**	-	112	3181	-	**707**	3181
rbg323a	325	-	**678**	**3380**	-	89	3519	-	281	3529
rbg341a	343	-	**319**	**2968**	-	68	3038	-	318	3064
rbg358a	360	-	**181**	**3202**	-	69	3359	-	72	3384
rbg378a	380	-	**196**	**3402**	-	52	3429	-	50	3429
ry48p.1	49	-	**6414**	**16892**	-	5198	17555	-	6140	17454
ry48p.2	49	-	6284	**17439**	-	5290	18046	-	**6442**	17970
ry48p.3	49	-	5772	**20890**	-	6208	21161	-	**6874**	21142
ry48p.4	49	-	12443	**33391**	-	13598	34517	-	**14171**	33804

the 1-h timeout that rival (and often exceeds) those produced by peel & bound. Finally, the dual bounds from peel & bound seem quite competitive, overtaking both HADDOCK and classic Branch & Bound with only a few exceptions.

7 Conclusion

This paper studied the automatic use of primal and dual bounds from Multi-valued Decision Diagrams (MDDs) in the context of branch-and-bound within a CP solver. The paper extended HADDOCK to support both relaxed and restricted diagrams for any constraints for which a labeled transition system can be specified. The paper described the derivation of the implementation and recognizes the possibility for extending this work to include branching directly on MDD nodes and supporting hybrid CP models that mix MDD propagators with conventional constraints. The empirical evaluation established that the generic implementation one derives is competitive with state of the art *dedicated* MDD branch-and-bound procedures including peel & bound.

Acknowledgements. Laurent Michel and Rebecca Gentzel were partially supported by Synchrony. Willem-Jan van Hoeve is partially supported by Office of Naval Research Grant No. N00014-21-1-2240 and National Science Foundation Award #1918102.

References

1. Beldiceanu, N., Contejean, E.: Introducing global constraints in CHIP. J. Math. Comput. Model. **20**(12), 97–123 (1994)
2. Bergman, D., Cire, A.A., Van Hoeve, W.-J., Hooker, J.: Decision Diagrams for Optimization, vol. 1. Springer, Cham (2016)
3. Bergman, D., Cire, A.A., van Hoeve, W.-J., Hooker, J.N.: Discrete optimization with decision diagrams. INFORMS J. Comput. **28**(1), 47–66 (2016)
4. Boussemart, F., Hemery, F., Lecoutre, C., Sais, L.: Boosting systematic search by weighting constraints. In: Proceedings of the 16th European Conference on Artificial Intelligence, ECAI 2004, pp. 146–150, NLD, August 2004. IOS Press (2004)
5. Fischetti, M., Glover, F. Lodi, A.: The feasibility pump. Math. Program. **104**(1), 91–104 (2005). https://doi.org/10.1007/s10107-004-0570-3
6. Gentzel, R., Michel, L., van Hoeve, W.-J.: HADDOCK: a language and architecture for decision diagram compilation. In: Simonis, H. (ed.) CP 2020. LNCS, vol. 12333, pp. 531–547. Springer, Cham (2020). https://doi.org/10.1007/978-3-030-58475-7_31
7. Gentzel, R., Michel, L., van Hoeve, W.-J.: Heuristics for MDD propagation in HADDOCK. In: 28th International Conference on Principles and Practice of Constraint Programming (CP 2022), vol. 235 of Leibniz International Proceedings in Informatics (LIPIcs), pp. 24:1–24:17. Schloss Dagstuhl - Leibniz-Zentrum für Informatik (2022)
8. Gillard, X., Coppé, V., Schaus, P., Cire, A.A.: Improving the filtering of branch-and-bound MDD solver. In: Stuckey, P.J. (ed.) CPAIOR 2021. LNCS, vol. 12735, pp. 231–247. Springer, Cham (2021). https://doi.org/10.1007/978-3-030-78230-6_15

9. Pesant, G., Quimper, C., Zanarini, A.: Counting-based search: branching heuristics for constraint satisfaction problems. J. Artif. Intell. Res. **43**, 173–210 (2012). https://doi.org/10.1613/jair.3463
10. Hoda, S., van Hoeve, W.-J., Hooker, J.N.: A systematic approach to MDD-based constraint programming. In: Cohen, D. (ed.) CP 2010. LNCS, vol. 6308, pp. 266–280. Springer, Heidelberg (2010). https://doi.org/10.1007/978-3-642-15396-9_23
11. Keller, R.M.: Formal verification of parallel programs. Commun. ACM **19**(7), 371–384 (1976)
12. Michel, L., Schaus, P., Van Hentenryck, P.: MiniCP: a lightweight solver for constraint programming. Math. Program. Comput. **13**, 133–184 (2021)
13. Michel, L., Van Hentenryck, P.: Activity-based search for black-box constraint programming solvers. In: Beldiceanu, N., Jussien, N., Pinson, É. (eds.) CPAIOR 2012. LNCS, vol. 7298, pp. 228–243. Springer, Heidelberg (2012). https://doi.org/10.1007/978-3-642-29828-8_15
14. Refalo, P.: Impact-based search strategies for constraint programming. In: Wallace, M. (ed.) CP 2004. LNCS, vol. 3258, pp. 557–571. Springer, Heidelberg (2004). https://doi.org/10.1007/978-3-540-30201-8_41
15. Reinelt, G.: TSPLIB-a traveling salesman problem library. ORSA J. Comput. **3**(4), 376–384 (1991)
16. Rudich, I., Cappart, Q., Rousseau, L.M.: Peel-and-bound: generating stronger relaxed bounds with multivalued decision diagrams. In: 28th International Conference on Principles and Practice of Constraint Programming (CP 2022), vol. 235 of Leibniz International Proceedings in Informatics (LIPIcs), pp. 35:1–35:20. Schloss Dagstuhl - Leibniz-Zentrum für Informatik (2022)
17. Van Hentenryck, P., Perron, L., Puget, J.F.: Search and Strategies in OPL. ACM Trans. Comput. Logic **1**(2), 1–36 (2000)

ZDD-Based Algorithmic Framework for Solving Shortest Reconfiguration Problems

Takehiro Ito[1], Jun Kawahara[2(✉)], Yu Nakahata[3], Takehide Soh[4], Akira Suzuki[1], Junichi Teruyama[5], and Takahisa Toda[6]

[1] Graduate School of Information Sciences, Tohoku University, Sendai, Japan
{takehiro,akira}@tohoku.ac.jp

[2] Graduate School of Informatics, Kyoto University, Kyoto, Japan
jkawahara@i.kyoto-u.ac.jp

[3] Graduate School of Science and Technology, Nara Institute of Science and Technology, Ikoma, Japan
yu.nakahata@is.naist.jp

[4] Information Infrastructure and Digital Transformation Initiatives Headquarters, Kobe University, Kobe, Japan
soh@lion.kobe-u.ac.jp

[5] Graduate School of Information Sciences, University of Hyogo, Kobe, Japan
junichi.teruyama@gsis.u-hyogo.ac.jp

[6] Graduate School of Informatics and Engineering, The University of Electro-Communications, Chofu, Japan
toda@disc.lab.uec.ac.jp

Abstract. This paper proposes an algorithmic framework for solving various combinatorial reconfiguration problems by using zero-suppressed binary decision diagrams (ZDDs), a data structure for representing families of sets. In general, a reconfiguration problem checks if there is a step-by-step transformation between two given feasible solutions (e.g., independent sets of an input graph) of a fixed search problem, such that all intermediate results are also feasible and each step obeys a fixed reconfiguration rule (e.g., adding/removing a single vertex to/from an independent set). The solution space formed by all feasible solutions can be exponential in the input size, and indeed, many reconfiguration problems are known to be PSPACE-complete. This paper shows that an algorithm in the proposed framework efficiently conducts breadth-first search by compressing the solution space using ZDDs, and that it finds a shortest transformation between two given feasible solutions if such a transformation exists. Moreover, the proposed framework provides rich information on the solution space, such as its connectivity and all feasible solutions that are reachable from a specified one. Finally, we demonstrate that the proposed framework can be applied to various reconfiguration problems, and experimentally evaluate its performance.

Partially supported by JSPS KAKENHI Grant Numbers JP18H04091, JP19H01103, JP19H04068, JP19K11814, JP20K11666, JP20K11748, JP20H05793 and JP20H05794, Japan.

© The Author(s), under exclusive license to Springer Nature Switzerland AG 2023
A. A. Cire (Ed.): CPAIOR 2023, LNCS 13884, pp. 167–183, 2023.
https://doi.org/10.1007/978-3-031-33271-5_12

Keywords: Combinatorial reconfiguration · Graph algorithms · Binary decision diagrams

1 Introduction

Combinatorial reconfiguration [11,13,23] is a family of problems that involve finding a procedure to change one solution of a combinatorial search problem into another solution while maintaining the conditions of the search problem, and has attracted much attention in recent years. Taking a change in the switch configuration of a power distribution network as an example, we can regard a switch configuration that satisfies all given electrical conditions as a solution for a search problem [12]. In the reconfiguration version of this search problem, the task is to find a change procedure from the current switch configuration to another (more desirable) configuration while maintaining required electrical conditions such as not causing power outages. In general, we can say that a combinatorial reconfiguration problem models a situation in which the goal is to change a current configuration to another one without requiring the system to stop.

Combinatorial reconfiguration problems have been actively studied in the theoretical algorithms research community in recent years. (See the surveys in [11,23].) In particular, combinatorial reconfiguration problems related to graph problems, such as independent set reconfiguration and graph coloring reconfiguration, have been well studied. Many of these studies have mainly followed a theoretical perspective, such as analysis of the computational complexity of the problem with respect to graph classes; in contrast, to the bast of our knowledge, there has been little research on the applied aspects of combinatorial reconfiguration. Since many reconfiguration problems, such as the independent set reconfiguration [15] and graph 4-coloring reconfiguration [2], are PSPACE-complete, it is hard to design an efficient algorithm. However, depending on the application, the number of vertices in the input graph may be at most tens or hundreds, and in such cases, we can expect the existence of an algorithm that runs within an acceptable time.

There are various promising methods for solving combinatorial optimization problems that appear in real applications, such as integer programming, SAT solvers, genetic algorithms, and metaheuristics. One approach that has attracted attention is the use of zero-suppressed binary decision diagrams (ZDDs) [4,20]. A ZDD is a data structure that compresses and compactly represents a family of sets. By representing the solution set of a combinatorial optimization problem as a ZDD and then performing set operations of ZDDs, it is possible to find an optimal solution by imposing constraints that integer programming methods and SAT solvers do not handle well.

In this study, we investigate the solution of combinatorial reconfiguration problems by using ZDDs. In particular, we leverage the fact that a ZDD preserves not just one solution, but *all* solutions. Hence, we propose an algorithm to obtain all solutions that are changeable from a given solution as a ZDD.

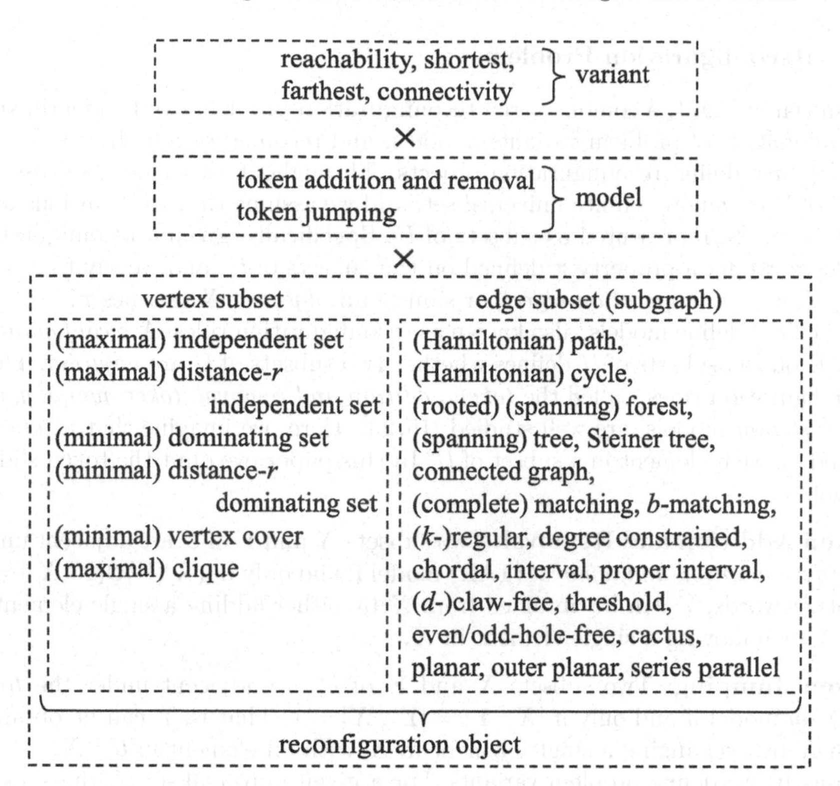

Fig. 1. Variants, models, and reconfiguration objects that our framework can handle.

This algorithm can be applied to various combinatorial reconfiguration problems whose solution sets can be represented as ZDDs. We give precise definitions in Sect. 2, but here, Fig. 1 shows combinatorial reconfiguration problems that our algorithm can handle. These problems can be identified by their combinations of problem variants, change rules (models), and solutions (reconfiguration objects). In addition, we demonstrate the effectiveness of the proposed algorithm through computer experiments.

The organization of the paper is as follows. Section 2 defines combinatorial reconfiguration problems and introduces ZDDs. We propose our algorithm using ZDDs in Sect. 3. Then, Sect. 4 shows that the proposed algorithm can solve various combinatorial reconfiguration problems. Finally, we describe the experimental results in Sect. 5, before concluding the paper in Sect. 6.

2 Preliminaries

Throughout this paper, we use the symbols G, V, and E to denote an input graph, its vertex set, and its edge set, respectively. For families $\mathcal{A}$, $\mathcal{B}$ of sets, we define $\mathcal{A} \bowtie \mathcal{B} = \{A \cup B \mid A \in \mathcal{A}, B \in \mathcal{B}\}$. In this paper, we sometimes simply call a family of sets a "family."

2.1 Reconfiguration Problems

As shown in Fig. 1, a combinatorial reconfiguration problem can be identified by a combination of problem variants, models, and reconfiguration objects.

We first define reconfiguration objects. Throughout this paper, we use the symbol U to denote a finite universal set, and we assume that the solutions for a change can be represented as subsets of U. Specifically, given a reconfiguration problem, we fix a property π defined on the subsets of U, and we say that a set $X \subseteq U$ is a *reconfiguration object*, or simply an *object*, if X satisfies π.

We next define models, also known as reconfiguration rules. A *reconfiguration rule* $\mathcal{R}$ on the subsets of U defines whether two subsets of U are *adjacent*. Three reconfiguration rules, called the *token addition and removal*, *token jumping*, and *token sliding* models, are well studied [15,23]. Here, we imagine that a *token* is placed on each element in a subset of U. In this paper, we omit the token sliding model.

Token Addition and Removal: Two subsets X and Y of U are adjacent under the *token addition and removal* (TAR) model if and only if $|(X \setminus Y) \cup (Y \setminus X)| = 1$. In other words, Y can be obtained from X by either adding a single element in $U \setminus X$ or removing a single element in X.

Token Jumping: Two subsets X and Y of U are adjacent under the *token jumping* model if and only if $|X \setminus Y| = |Y \setminus X| = 1$. That is, Y can be obtained from X by exchanging a single element in X with an element in $U \setminus X$.

Lastly, we define problem variants. For a given universal set U, the *solution space* under the property π and the reconfiguration rule $\mathcal{R}$ is a graph where each node corresponds to a reconfiguration object of U, and two nodes are joined by an edge if and only if their corresponding sets are adjacent under $\mathcal{R}$. Then, we can consider variants of reconfiguration problems on the solution space. The *reachability variant* asks whether the solution space contains a path connecting two given objects. The *shortest variant* asks to compute the shortest length (i.e., the minimum number of edges) of any path in the solution space that connects two given objects. The *farthest variant* asks to find an object farthest from a given object in the solution space (i.e., the shortest path between the two objects is the longest). Finally, the *connectivity variant* asks whether the solution space is connected.

2.2 Zero-Suppressed Decision Diagram (ZDD)

A ZDD, as defined below, is a data structure for efficiently representing a family of sets. In this section, we set $U = \{x_1, \dots, x_n\}$ and $x_1 < x_2 < \cdots < x_n$. A ZDD is a directed acyclic graph (DAG) $\mathcal{Z}$ that has the following properties. It has at most two nodes with outdegree zero, which are called *terminals* and denoted by $\bot$ and $\top$. Nodes other than the terminals are called non-terminal nodes. A non-terminal node ν has an element in U, which is called a *label* and denoted by $\mathsf{label}(\nu)$, and it has two arcs, called the *0-arc* and *1-arc*. If the 0-arc and 1-arc of a non-terminal node ν point at nodes ν_0, ν_1, we write $\nu = (\mathsf{label}(\nu), \nu_0, \nu_1)$, and we call ν_0 and ν_1 the *0-child* and *1-child*, respectively. Then, $\mathsf{label}(\nu) < \mathsf{label}(\nu_0)$ and $\mathsf{label}(\nu) < \mathsf{label}(\nu_1)$ must hold, where we guarantee that $x_i < \mathsf{label}(\bot)$ and $x_i < \mathsf{label}(\top)$ for all $i = 1, \ldots, n$. Lastly, a ZDD $\mathcal{Z}$ has exactly one node with indegree zero, called the *root*, and is denoted by $\mathsf{root}(\mathcal{Z})$.

A ZDD $\mathcal{Z}$ represents a family of sets whose universal set is U, as follows. We associate each node in $\mathcal{Z}$ with a family, denoted by $\mathcal{S}(\nu)$, in the following recursive manner. The terminal nodes $\bot$ and $\top$ are associated with $\emptyset$ and $\{\emptyset\}$, respectively; that is, $\mathcal{S}(\bot) = \emptyset$ and $\mathcal{S}(\top) = \{\emptyset\}$. Consider the case where ν is a non-terminal node. Let $\nu = (x, \nu_0, \nu_1)$, where $x \in U$ and ν_0, ν_1 are nodes of $\mathcal{Z}$. The node ν is associated with the union of the family that we associate with ν_0, and the family obtained by adding x to each set in the family that we associate with ν_1; that is, $\mathcal{S}(\nu) = \mathcal{S}(\nu_0) \cup (\{\{x\}\} \bowtie \mathcal{S}(\nu_1))$. (Note that $\{\{x\}\} \bowtie \mathcal{S}(\nu_1)$ is the family obtained by adding x to each set in $\mathcal{S}(\nu_1)$.) Observe that each of the sets in $\mathcal{S}(\nu_0)$ and $\mathcal{S}(\nu_1)$ does not contain x because of the ZDD property that $\mathsf{label}(\nu) < \mathsf{label}(\nu_0)$ and $\mathsf{label}(\nu) < \mathsf{label}(\nu_1)$ must hold. We interpret $\mathcal{Z}$ to represent the family that we associate with the root node. The family $\mathcal{S}(\mathcal{Z})$ represented by $\mathcal{Z}$ is defined by $\mathcal{S}(\mathcal{Z}) = \mathcal{S}(\mathsf{root}(\mathcal{Z}))$ (note that we use the same notation $\mathcal{S}$ for a node and a ZDD).

Every ZDD $\mathcal{Z}$ has the following recursive structure [3,20]. Let $\nu = \mathsf{root}(\mathcal{Z})$, and suppose that ν is represented by $\nu = (x, \nu_0, \nu_1)$. Then, for $i = 0, 1$, the DAG comprising the nodes and arcs reachable from ν_i can be considered a ZDD with root ν_i, and we denote it as $\mathsf{child}_i(\mathcal{Z})$ (see Fig. 2). The ZDD $\mathsf{child}_i(\mathcal{Z})$ represents $\mathcal{S}(\nu_i)$. For computation using a ZDD $\mathcal{Z}$, we often design a recursive algorithm $\mathsf{op}(\mathcal{Z})$, which calls $\mathsf{op}(\mathsf{child}_0(\mathcal{Z}))$ and $\mathsf{op}(\mathsf{child}_1(\mathcal{Z}))$ and manipulates the results of them. To clarify this notion, we describe the behavior of a recursive algorithm for the set union operation. Consider the construction of a ZDD $\mathcal{Z}$ for the union of $\mathcal{P} = \{\{x_1\}, \{x_2\}, \{x_1, x_2\}\}$ and $\mathcal{P}' = \{\{x_1, x_3\}, \{x_2, x_3\}\}$ from the ZDDs. First, we focus on sets in $\mathcal{P}$ and $\mathcal{P}'$ that do not include x_1, and we take the union over them. We do this by applying the union operation to the left children of the ZDDs for $\mathcal{P}$ and for $\mathcal{P}'$. The operation proceeds recursively: the base step occurs when one of the ZDDs is $\top$ or $\bot$, which is straightforward and thus omitted here. The induction step is described below. The same approach applies to the union of the other sets, i.e., the sets in $\mathcal{P}$ and $\mathcal{P}'$ that include x_1. Here, we have two ZDDs as a result of applying the recursive operation: one represents all sets in $\mathcal{P} \cup \mathcal{P}'$ that do not include x_1, and the other represents those sets in $\mathcal{P}'$ that include x_1; these ZDDs correspond to $\mathsf{child}_0(\mathcal{P}) \cup \mathsf{child}_0(\mathcal{P}')$ and $\mathsf{child}_1(\mathcal{P}) \cup \mathsf{child}_1(\mathcal{P}')$, respectively. We thus construct the final ZDD $\mathcal{Z}$ so that the 0-arc of x_1 points to the root of the ZDD $\mathsf{child}_0(\mathcal{P}) \cup \mathsf{child}_0(\mathcal{P}')$ and that of the ZDD $\mathsf{child}_1(\mathcal{P}) \cup \mathsf{child}_1(\mathcal{P}')$.

ZDDs have rich operations for manipulating families [3,20]. For example, given two ZDDs $\mathcal{Z}, \mathcal{Z}'$, we can efficiently compute ZDDs representing $\mathcal{S}(\mathcal{Z}) \cup \mathcal{S}(\mathcal{Z}')$, $\mathcal{S}(\mathcal{Z}) \cap \mathcal{S}(\mathcal{Z}')$, $\mathcal{S}(\mathcal{Z}) \setminus \mathcal{S}(\mathcal{Z}')$, and so on by the recursive process described above. For a binary operation $\circ \in \{\cup, \cap, \setminus, \ldots\}$, we denote the ZDD representing $\mathcal{S}(\mathcal{Z}) \circ \mathcal{S}(\mathcal{Z}')$ by $\mathcal{Z} \circ \mathcal{Z}'$. For more information on ZDDs, refer to [19].

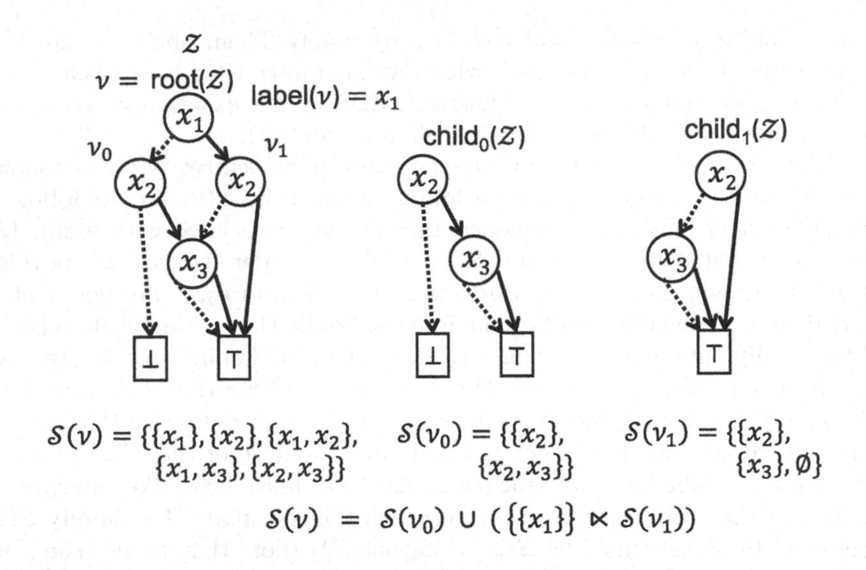

Fig. 2. Example of a ZDD and the recursive structure.

3 ZDD-Based Algorithmic Framework

3.1 Algorithmic Framework

We begin with the reachability variant under the TAR model for independent sets, where reconfiguration objects are independent sets in an input graph G whose cardinality is at least a threshold k given as an input. In this problem, we are given a graph $G = (V, E)$, an integer k, and two independent sets S, T. The task of the problem is to decide whether a reconfiguration sequence from S to T exists such that any set in the sequence is an independent set of G, its cardinality is at least k, and any set except for S is obtained by adding or removing a vertex to or from the previous set, respectively. For example, if $G = (V, E)$ with $V = \{1, 2, 3, 4, 5, 6\}$ and $E = \{\{1, 2\}, \{2, 3\}, \{3, 4\}, \{4, 5\}, \{5, 6\}\}$, $k = 2$, $S = \{1, 3\}$, and $T = \{1, 4, 6\}$, an example of a reconfiguration sequence is $\{1, 3\}, \{1, 3, 6\}, \{1, 6\}, \{1, 4, 6\}$, which is obtained by adding 6, removing 3, and adding 4.

Hayase et al. [10] proposed an algorithm that constructs a ZDD, say $\mathcal{Z}_{\text{ind}}$, that represents the family of all the independent sets of a given graph, where the universal set U is the vertex set of the graph and the elements (vertices) in U are ordered. The ZDD $\mathcal{Z}_{\text{ind}}$ is highly compressed if the given graph has a good structure, such as one with a small pathwidth. For example, an 8×250 grid graph has 3.07×10^{361} independent sets, but the ZDD representing them has just 49,989 nodes (with about 1MB of memory usage).

For the reconfiguration problem, we consider the use of a ZDD that represents a vast number of independent sets. Although $\mathcal{Z}_{\text{ind}}$ includes all the independent sets, it does not have information on their adjacency relations. Our goal is to obtain the family of independent sets that are adjacent to a given independent set, and more generally, the family of independent sets that are adjacent to any independent set in a given family. If we can obtain these by repeating the operation of obtaining the family of adjacent independent sets from the initial independent set, then we can obtain all independent sets that are reachable from the initial independent set and decide whether a reconfiguration sequence from that set to the target set exists.

The TAR model requires two operations: removal of a vertex from an independent set and addition of a vertex to an independent set. First, we consider the removal operation. Given a family $\mathcal{I}$ of independent sets, the removal operation entails removing each element from each independent set in $\mathcal{I}$, thus obtaining the family $\{I \setminus \{v\} \mid I \in \mathcal{I}, v \in I\}$. Given $\mathcal{I}$ as a ZDD, we propose an algorithm that constructs a ZDD representing $\{I \setminus \{v\} \mid I \in \mathcal{I}, v \in I\}$ without extracting elements from $\mathcal{I}$. As for the addition operation, we also propose an algorithm that constructs a ZDD representing $\{I \cup \{v\} \mid I \in \mathcal{I}, v \in U \setminus I\}$.

For later use, we describe the two operations in a somewhat general form. For a ZDD $\mathcal{Z}$ (whose universal set is U) and a set $R \subseteq U$, let $\mathsf{remove}(\mathcal{Z}, R)$ be the ZDD representing

$$\{I \setminus \{v\} \mid I \in \mathcal{S}(\mathcal{Z}), v \in I \cap R\},$$

which means that we remove an element only from R. For a ZDD $\mathcal{Z}$ and a set $A \subseteq U$, let $\mathsf{add}(\mathcal{Z}, A)$ be the ZDD representing

$$\{I \cup \{x\} \mid I \in \mathcal{S}(\mathcal{Z}), x \in A \setminus I\},$$

which means that we add an element only in A.

For the TAR model, we can solve the reachability variant of the independent set reconfiguration problem by using $\mathsf{remove}(\mathcal{Z}, R)$ and $\mathsf{add}(\mathcal{Z}, A)$ as follows. First, we construct a ZDD representing the solution space. Recall that a feasible independent set in the TAR model contains at least k vertices. It is easy to construct a ZDD, say $\mathcal{Z}_{\geq k}$, representing the family of all sets with cardinality at least k (i.e., $\{I \subseteq U \mid |I| \geq k\}$) [19]. The solution space ZDD $\mathcal{Z}_{\mathrm{sol}}$ is then obtained from the intersection operation of $\mathcal{Z}_{\mathrm{ind}}$ and $\mathcal{Z}_{\geq k}$, as mentioned in Sect. 2.2. In the above example, $\mathcal{S}(\mathcal{Z}_{\geq 2}) = \{\{1,2\},\{1,3\},\ldots,\{5,6\}\}$. $\mathcal{S}(\mathcal{Z}_{\mathrm{ind}})$ includes sets whose cardinality is less than k (for example, $\{1\}$ and $\{3\}$ are independent sets of G). By taking the union of $\mathcal{Z}_{\mathrm{ind}}$ and $\mathcal{Z}_{\geq k}$, we remove the sets whose cardinality is less than k from $\mathcal{S}(\mathcal{Z}_{\mathrm{ind}})$.

Next, for $i = 0, 1, \ldots$, let $\mathcal{Z}^i$ denote the ZDD representing the family of independent sets obtained by applying the reconfiguration rule (i.e., removing or adding a vertex) to S exactly i times, where $\mathcal{Z}^0$ is the ZDD such that $\mathcal{S}(\mathcal{Z}^0) = \{S\}$. The construction of $\mathcal{Z}^0$ is trivial. For $i = 1, 2, \ldots$, the ZDD $\mathcal{Z}^i$ can be constructed by

$$\mathcal{Z}^i \leftarrow \mathsf{op}(\mathcal{Z}^{i-1}) \cap \mathcal{Z}_{\mathrm{sol}}, \quad (1)$$

where $\mathsf{op}(\mathcal{Z}) = \mathsf{remove}(\mathcal{Z}, V) \cup \mathsf{add}(\mathcal{Z}, V)$ for ZDD $\mathcal{Z}$ (recall that V is the vertex set of a given graph). Note that $\cup$ and $\cap$ are ZDD operations mentioned in

Sect. 2.2. After constructing $\mathcal{Z}^i$, we decide whether $\mathcal{Z}^i = \bot$ (i.e., $\mathcal{S}(\mathcal{Z}^i) = \emptyset$) and whether $T \in \mathcal{S}(\mathcal{Z}^i)$ (both tasks are straightforward). If $\mathcal{Z}^i = \bot$, it means that a reconfiguration sequence from S to T does not exist, because $\mathcal{S}(\mathcal{Z}^j)$, $0 \leq j \leq i$, includes *all* independent sets that are reachable from S within j steps. We thus output NO and halt. On the other hand, if $T \in \mathcal{S}(\mathcal{Z}^i)$, then a reconfiguration sequence with length i from S to T exists: we output YES and halt. If neither case holds, we construct $\mathcal{Z}^{i+1}$.

Let us consider the above example. In the following description, we use the $\mathcal{S}$ notation, but actually, all operations are conducted as ZDD operations. We obtain $\mathcal{S}(\mathcal{Z}^0) = \{S\} = \{\{1,3\}\}$, $\mathcal{S}(\mathcal{Z}^1) = \{\{1,3,5\},\{1,3,6\}\}$, $\mathcal{S}(\mathcal{Z}^2) = \{\{1,3\},\{1,5\},\{1,6\},\{3,5\},\{3,6\}\}$, and $\mathcal{S}(\mathcal{Z}^3) = \{\{1,3,5\},\{1,3,6\},\{1,4,6\}\}$.

3.2 Removal and Addition Operations

Next, given a ZDD $\mathcal{Z}$ and two sets A and R, we describe how to construct the ZDDs $\mathsf{remove}(\mathcal{Z}, R)$ and $\mathsf{add}(\mathcal{Z}, A)$. We begin by designing an algorithm for $\mathsf{remove}(\mathcal{Z}, R)$ for a ZDD $\mathcal{Z}$, which is based on the recursive process described in Sect. 2.2. Let $\nu = \mathsf{root}(\mathcal{Z})$. Suppose that ν is a non-terminal node and $\nu = (x, \nu_0, \nu_1)$, where $x \in U$ and ν_i is the i-child of ν.

Here, we consider the case of $x \in R$. Letting $\mathcal{Z}^{\mathrm{rem}} = \mathsf{remove}(\mathcal{Z}, R)$, we observe the characteristics of $\mathcal{Z}^{\mathrm{rem}}$. First, $\mathsf{root}(\mathcal{Z}^{\mathrm{rem}}) = x$ because $\mathcal{S}(\mathcal{Z}^{\mathrm{rem}})$ contains a set that includes x and does not contain any set that includes an element smaller than x. Secondly, consider $\mathsf{child}_0(\mathcal{Z}^{\mathrm{rem}})$, which is a ZDD representing the family of sets in $\mathcal{S}(\mathcal{Z}^{\mathrm{rem}})$ that do not contain x. Each set in $\mathcal{S}(\mathsf{child}_0(\mathcal{Z}^{\mathrm{rem}}))$ is obtained in one of the following two ways: (i) we remove an element from a set in $\mathcal{S}(\mathcal{Z})$ that does not include x (i.e., a set in $\mathcal{S}(\mathsf{child}_0(\mathcal{Z}))$), or (ii) we remove x from a set in $\mathcal{S}(\mathcal{Z})$ that includes x (i.e., a set in $\{\{x\}\} \bowtie \mathcal{S}(\mathsf{child}_1(\mathcal{Z}))$). We collect all the sets obtained by (i), and we construct a ZDD representing them by recursively applying the remove operation to $\mathsf{child}_0(\mathcal{Z})$. In contrast, the ZDD for case (ii) is just $\mathsf{child}_1(\mathcal{Z})$. Hence, we obtain

$$\mathsf{child}_0(\mathcal{Z}^{\mathrm{rem}}) = \mathsf{remove}(\mathsf{child}_0(\mathcal{Z}), R \setminus \{x\}) \cup \mathsf{child}_1(\mathcal{Z}),$$

where '$\cup$' is the union operation of ZDDs described in Sect. 2.2.

Thirdly, we consider $\mathsf{child}_1(\mathcal{Z}^{\mathrm{rem}})$, which is the ZDD representing the family of sets each of which is obtained by removing x from a set in $\mathcal{S}(\mathcal{Z}^{\mathrm{rem}})$ that contains x. Each set in $\mathcal{S}(\mathsf{child}_1(\mathcal{Z}^{\mathrm{rem}}))$ is obtained by removing x from a set in $\mathcal{S}(\mathcal{Z}^{\mathrm{rem}})$ containing x. The ZDD is thus obtained by applying the remove operation to $\mathsf{child}_1(\mathcal{Z})$ as follows:

$$\mathsf{child}_1(\mathcal{Z}^{\mathrm{rem}}) = \mathsf{remove}(\mathsf{child}_1(\mathcal{Z}), R \setminus \{x\}).$$

Now, we consider the case of $x \notin R$, which means that we do not remove x from any independent set. We obtain

$$\mathsf{child}_0(\mathcal{Z}^{\mathrm{rem}}) = \mathsf{remove}(\mathsf{child}_0(\mathcal{Z}), R \setminus \{x\}),$$
$$\mathsf{child}_1(\mathcal{Z}^{\mathrm{rem}}) = \mathsf{remove}(\mathsf{child}_1(\mathcal{Z}), R \setminus \{x\}).$$

Our recursive algorithm for $\mathsf{remove}(\mathcal{Z}, U)$ is as follows: If $\mathcal{Z} = \bot$ or $\mathcal{Z} = \top$, return $\bot$. Otherwise, let $x = \mathsf{label}(\mathsf{root}(\mathcal{Z}))$, construct

$$\mathcal{Z}_0 \leftarrow \begin{cases} \mathsf{remove}(\mathsf{child}_0(\mathcal{Z}), R \setminus \{x\}) \cup \mathsf{child}_1(\mathcal{Z}) & \text{if } x \in R, \\ \mathsf{remove}(\mathsf{child}_0(\mathcal{Z}), R \setminus \{x\}) & \text{if } x \notin R, \end{cases}$$

$$\mathcal{Z}_1 \leftarrow \mathsf{remove}(\mathsf{child}_1(\mathcal{Z}), R \setminus \{x\}),$$

and then call and return $\mathsf{makenode}(x, \mathcal{Z}_0, \mathcal{Z}_1)$. Here, the $\mathsf{makenode}(x, \mathcal{Z}_0, \mathcal{Z}_1)$ function conducts the following procedure: if there is a node whose label is x and whose i-arc points at the root of $\mathcal{Z}_i$ for $i = 0, 1$, just return the node; otherwise, make a new node with label x, make its i-arc point at the root of $\mathcal{Z}_i$ for $i = 0, 1$, and return the new node.

Next, we design an algorithm for $\mathsf{add}(\mathcal{Z}, A)$ for any $A \subseteq U$. Note that there is a possibility that an element that appears in A but never in $\mathcal{Z}$ is added to a set. Let $x = \mathsf{label}(\mathsf{root}(\mathcal{Z}))$, and let y be the minimum element in A. First, we consider the case of $x \geq y$. Similarly to the remove operation, we call $\mathsf{makenode}(x, \mathcal{Z}_0, \mathcal{Z}_1)$, where

$$\mathcal{Z}_0 \leftarrow \mathsf{add}(\mathsf{child}_0(\mathcal{Z}), A \setminus \{x\}),$$

$$\mathcal{Z}_1 \leftarrow \begin{cases} \mathsf{add}(\mathsf{child}_1(\mathcal{Z}), A \setminus \{x\}) \cup \mathsf{child}_0(\mathcal{Z}), & \text{if } x \in A \\ \mathsf{add}(\mathsf{child}_1(\mathcal{Z}), A \setminus \{x\}) & \text{if } x \notin A. \end{cases}$$

Secondly, we consider the case of $x < y$, including the case where $\mathcal{Z} = \top$ and $A \neq \emptyset$, which means that the constructed family contains sets obtained by adding y to sets in $\mathcal{S}(\mathcal{Z})$. In this case, we consider a ZDD $\mathcal{Z}'$ that is equivalent to $\mathcal{Z}$ (i.e., $\mathcal{S}(\mathcal{Z}') = \mathcal{S}(\mathcal{Z})$), such that $\mathsf{label}(\mathsf{root}(\mathcal{Z}')) = y$. Such a $\mathcal{Z}'$ is constructed by calling $\mathsf{makenode}(y, \mathcal{Z}, \bot)$. We then call and return $\mathsf{add}(\mathcal{Z}', A)$ recursively.

At the end of the recursion, $\mathsf{add}(\bot, A) = \bot$ for any $A \subseteq U$, and $\mathsf{add}(\top, \emptyset) = \bot$ (the case of $\mathsf{add}(\top, A)$ for a non-emptyset A has already been described above).

4 Versatility of Proposed Algorithm

In this section, we show the versatility of the proposed algorithm in the following three directions. (i) By using $\mathcal{Z}^i$ (from Sect. 3.1), we can solve the variants introduced in Sect. 2.1 (discussed below in Sect. 4.1). (ii) By changing $\mathsf{op}(\mathcal{Z})$ in Eq. (1), we can solve certain models (Sect. 4.2). (iii) By constructing $\mathcal{Z}_{\mathrm{sol}}$, we can handle various reconfiguration objects and constraints (Sect. 4.3).

4.1 Shortest, Farthest, and Connectivity Variants

The ZDD $\mathcal{Z}^i$ represents the family of *all* independent sets that are reachable from the initial set S in i steps. Therefore, the smallest integer i such that $T \in \mathcal{S}(\mathcal{Z}^i)$ holds is the length of a shortest reconfiguration sequence from S to T. The proposed algorithm can solve not only the reachability variant but also the shortest variant.

The shortest sequence I_0 ($= S$), $\ldots$, I_h ($= T$) between S and T can be obtained by the following backtrack method, where h is the smallest integer such that $T \in \mathcal{S}(\mathcal{Z}^h)$. Here, we consider only the token jumping model; the other models are similar. Suppose that we have already obtained $I_p, \ldots, I_h$ ($2 \leq p \leq h$). Then, there are vertices $v \notin I_p$ and $w \in I_p$ such that $I_p \cup \{v\} \setminus \{w\} \in \mathcal{S}(\mathcal{Z}^{p-1})$ according to the construction of $\mathcal{Z}^p$. Thus, we let $I_{p-1} := I_p \cup \{v\} \setminus \{w\}$. By the above method, we obtain $I_1, \ldots, I_h$. Finally, $|I_0 \setminus I_1| = |I_1 \setminus I_0| = 1$ obviously holds according to the construction of $\mathcal{Z}^1$, which indicates that the sequence $I_0, \ldots, I_h$ is certainly the reconfiguration sequence between S and T. The computation time is as follows. We can test whether $I_p \cup \{v\} \setminus \{w\}$ is in $\mathcal{S}(\mathcal{Z}^{p-1})$ by a ZDD operation in $O(|V|)$ time. The number of candidates for I_{p-1} is $O(|V|^2)$. Therefore, the computation time to obtain the shortest sequence after constructing the ZDDs $\mathcal{Z}^0, \ldots, \mathcal{Z}^h$ is $O(h|V|^3)$.

Next, consider the farthest variant. We construct $\mathcal{Z}^0, \mathcal{Z}^1, \ldots$ without checking whether $T \in \mathcal{S}(\mathcal{Z}^i)$ in the algorithm until $\mathcal{Z}^i = \bot$ holds. Let h' be the smallest integer such that $\mathcal{Z}^{h'} = \bot$. Then, a set in $\mathcal{S}(\mathcal{Z}^{h'-1})$ is a farthest independent set from S.

As for the connectivity variant, we solve it by applying the following idea. If the solution space (graph) is connected, then all independent sets are reachable from any set S. Therefore, we randomly choose S from $\mathcal{S}(\mathcal{Z}_{\text{sol}})$ by a ZDD operation and construct $\mathcal{Z}^0, \mathcal{Z}^1, \ldots, \mathcal{Z}^{h'-1}$ in the same way as for the farthest variant. Then, by examining whether $\mathcal{Z}_{\text{sol}}$ is equivalent to $\bigcup_{i=0,\ldots,h'-1} \mathcal{Z}^i$, we obtain the answer. Note that checking the equivalency of two given ZDDs can be done in $O(1)$ time in many ZDD manipulation systems.

We conclude this subsection by pointing out that our algorithm can solve the reconfiguration problem with multiple starting sets $S_1, \ldots, S_s$ and goal sets $T_1, \ldots, T_t$, where the task is to decide whether a reconfiguration sequence between S_j and $T_{j'}$ exists for some j, j'. We simply let $\mathcal{Z}^0$ be the ZDD for $\{S_1, \ldots, S_s\}$ and decide whether $T_{j'} \in \mathcal{S}(\mathcal{Z}^i)$ for some j' instead of whether $T \in \mathcal{S}(\mathcal{Z}^i)$.

4.2 Token Jumping Model

We consider the token jumping model by designing $\mathsf{op}(\mathcal{Z})$ in Eq. (1).

A swap operation removes a vertex from and adds another vertex to an independent set. For a ZDD $\mathcal{Z}$ (whose universal set is U) and sets $A, R \subseteq U$, let $\mathsf{swap}(\mathcal{Z}, A, R)$ be the ZDD representing

$$\{I \cup \{v\} \setminus \{v'\} \mid I \in \mathcal{S}(\mathcal{Z}), v \in A \setminus I, v' \in I \cap R\},$$

which means that we add an element in A and remove an element in R. This can be represented by $\mathsf{add}(\mathsf{remove}(\mathcal{Z}^{i-1}, R), A) \setminus \mathcal{Z}^{i-1}$. The set subtraction of $\mathcal{Z}^{i-1}$ is needed because the family represented by $\mathsf{add}(\mathsf{remove}(\mathcal{Z}^{i-1}, R), A)$ includes sets obtained by removing and adding the same vertex.

We can design a more efficient algorithm. Here, we only show $\mathcal{Z}_0$ and $\mathcal{Z}_1$ when calling $\mathsf{makenode}(x, \mathcal{Z}_0, \mathcal{Z}_1)$ with $x = \mathsf{root}(\mathcal{Z})$. The other cases are similar to the addition operation. $\mathcal{Z}_0$ and $\mathcal{Z}_1$ are given as follows:

$$\mathcal{Z}_0 \leftarrow \begin{cases} \mathsf{swap}(\mathsf{child}_0(\mathcal{Z}), A \setminus \{x\}, R \setminus \{x\}) \cup \mathsf{add}(\mathsf{child}_1(\mathcal{Z}), A \setminus \{x\}), & x \in R, \\ \mathsf{swap}(\mathsf{child}_0(\mathcal{Z}), A \setminus \{x\}, R \setminus \{x\}) & x \notin R, \end{cases}$$

$$\mathcal{Z}_1 \leftarrow \begin{cases} \mathsf{swap}(\mathsf{child}_1(\mathcal{Z}), A \setminus \{x\}, R \setminus \{x\}) \cup \mathsf{remove}(\mathsf{child}_0(\mathcal{Z}), R \setminus \{x\}), & x \in A, \\ \mathsf{swap}(\mathsf{child}_1(\mathcal{Z}), A \setminus \{x\}, R \setminus \{x\}) & x \notin A. \end{cases}$$

This approach holds because $\mathcal{S}(\mathcal{Z}_0)$ includes the independent sets obtained by removing x from each set in $\{\{x\}\} \bowtie \mathcal{S}(\mathsf{child}_1(\mathcal{Z}))$ and then adding a vertex other than x if $x \in R$, and $\mathcal{S}(\mathcal{Z}_1)$ includes the independent sets obtained by adding x to each set in $\mathcal{S}(\mathsf{child}_0(\mathcal{Z}))$ and removing a vertex other than x if $x \in A$. At the end of the recursion, $\mathsf{swap}(\bot, A, R) = \mathsf{swap}(\top, A, R) = \bot$ holds for any A and R.

4.3 Reconfiguration Objects and Constraints

The proposed algorithm does not depend on the characteristics of independent sets, except for the construction of $\mathcal{Z}_{\mathrm{sol}}$. Therefore, to solve a certain reconfiguration problem, we can apply the algorithm by explaining how to construct $\mathcal{Z}_{\mathrm{sol}}$ for objects corresponding to the problem. Many researchers have proposed ZDD construction algorithms for various set families, some of which can be applied to ZDD construction for many reconfiguration objects. In this section, we overview the kinds of objects that we can handle.

Vertex Subsets. First, we show that we can construct many kinds of objects each of which is represented as a subset of the vertex set by set operations. Assume that the universal set U is V.

We begin with independent sets (although we mentioned in the previous section that there is a more efficient algorithm [10]). Let $\mathcal{X}_v$ and $\overline{\mathcal{X}}_v$ be respectively the families of all sets including v and the family of those not including v; that is, $\mathcal{X}_v = \{A \subseteq U \mid v \in A\}$ and $\mathcal{X}_v = \{A \subseteq U \mid v \notin A\}$. It is easy to construct ZDDs for $\mathcal{X}_v$ and $\overline{\mathcal{X}}_v$. For two vertices v and w, the family of all sets including at most one of v and w is $\overline{\mathcal{X}}_v \cup \overline{\mathcal{X}}_w$. Therefore, the family of all the independent sets is

$$\bigcap_{\{v,w\} \in E} (\overline{\mathcal{X}}_v \cup \overline{\mathcal{X}}_w),$$

and the ZDD for this family can simply be obtained by combining known ZDD operations [19]. Similarly, we can solve other reconfiguration objects via ZDD operations [19], and we list some of them in Table 1.

Table 1. Reconfiguration objects that can be represented as vertex subsets, and how to obtain them by set operations [19]. Let $N_k(v)$ be the set of vertices whose distance from v ranges from 1 to k.

Reconfiguration object	Set operations
Dominating set	$\bigcap_{v\in V}\left(\mathcal{X}_v \cup \left(\bigcup_{w\in N(v)} \mathcal{X}_w\right)\right)$
Vertex cover	$\bigcap_{\{v,w\}\in E}(\mathcal{X}_v \cup \mathcal{X}_w)$
Clique	$\bigcap_{\{v,w\}\notin E}(\overline{\mathcal{X}}_v \cup \overline{\mathcal{X}}_w)$
Distance-k independent set	$\bigcap_{v\in V}\bigcap_{w\in N_k(v)}(\overline{\mathcal{X}}_v \cup \overline{\mathcal{X}}_w)$
Distance-k dominating set	$\bigcap_{v\in V}\left(\mathcal{X}_v \cup \left(\bigcup_{w\in N_k(v)} \mathcal{X}_w\right)\right)$

Coudert [8] proposed algorithms that construct ZDDs representing the families of sets obtained by collecting only the maximal/minimal sets in a family given as a ZDD $\mathcal{Z}$; that is, $\mathsf{maximal}(\mathcal{Z})$ is the ZDD for $\{X \in \mathcal{S}(\mathcal{Z}) \mid \forall X' \in \mathcal{S}(\mathcal{Z}), X \subseteq X' \implies X = X'\}$ and $\mathsf{minimal}(\mathcal{Z})$ is the ZDD for $\{X \in \mathcal{S}(\mathcal{Z}) \mid \forall X' \in \mathcal{S}(\mathcal{Z}), X' \subseteq X \implies X = X'\}$. Using the maximal operation, we can solve (the token jumping model of) the maximal independent set reconfiguration problem [5], where every feasible solution of this problem is a maximal independent set. We can also solve certain maximal/minimal reconfiguration problems, such as minimal dominating set reconfiguration, minimal vertex cover reconfiguration, and maximal clique reconfiguration.

Subgraphs. Next, we consider a subgraph that can be represented by an edge set. For example, a path can be represented by the set of edges comprising the path. Formally, for an edge set $E' \subseteq E$, a subgraph is represented by (V', E'), where $V' = \bigcup_{\{v,w\}\in E'}(\{v\} \cup \{w\})$. Then, we set E as the universe set U. Note that this representation handle subgraphs that include isolated vertices.

Sekine et al. [24] proposed an algorithm that constructs a ZDD representing the family of all spanning trees. Knuth [19] proposed a similar algorithm that constructs a ZDD representing the family of all s-t paths. Kawahara et al. [16] generalized those algorithms to a framework that can handle various objects, including matchings, regular graphs, and Steiner trees. Moreover, their framework can impose constraints on parameters such as the degree of each vertex, the connectivity of vertices, the existence of a cycle, and the number of edges (equal to, less than, or more than a specified value) in any combination. Recent research on ZDD construction has enabled treatment of more complex graph classes such as degree constrained graphs [17], chordal graphs [18], interval graphs [18], and planar graphs [22]. All of these can be treated as reconfiguration objects and are shown in Fig. 1. For example, the proposed algorithms can solve Steiner tree reconfiguration [21], planar subgraph reconfiguration, and so on.

5 Experimental Results

To evaluate the performance of the proposed ZDD-based method, we conducted an experimental comparison using the 1st place solver that won CoRe Challenge 2022.

CoRe Challenge 2022. The 1st Combinatorial Reconfiguration Challenge (CoRe Challenge 2022)[1] was held in 2022. The goal of CoRe Challenge 2022 was practical exploration of combinatorial reconfiguration. This first competition targeted the token jumping model of the independent set reconfiguration problem [15]. It provided 369 instances, which included both instances that have a reconfiguration sequence (323 instances) and do not have any sequence (46 instances). The participating solvers included a state-of-the-art AI planner and BMC solver based methods [25], e.g., SymK [26] and NuSMV [7].

Experimental Conditions. We compared our system with recon (@telematik-tuhh) [27], which is based on a hybridization of the IDA* algorithm and breadth-first search. It found 280 shortest reconfiguration sequences and was the 1st-place solver in the overall solver track for the shortest reconfiguration metric[2]. Except for the queen benchmark series, the instances solved by recon included all instances solved by the 2nd- and 3rd-place solvers. In our experiment, we compared how many shortest reconfiguration sequences could be found by each system. We used a machine with a 2.30-GHz CPU and 2 TB of RAM. The proposed ZDD solver was written in C++ language with the SAPPOROBDD[3] and TdZdd [14] libraries, and it was compiled by g++ with the -O3 option. The variable of ZDDs were ordered by the heuristic described in [9]. We have published an implementation of the proposed algorithm and scripts to reproduce the experiments on GitHub[4].

Results. Table 2 summarizes the comparisons between the proposed algorithm and recon for each series of benchmark instances. In this experiment, the time limit was two hours. The first and second columns respectively indicate the name of the benchmark series and the number of instances included in each series. The third and fourth columns indicate the maximum numbers of vertices and edges in each series, respectively. The fifth column lists the longest shortest reconfiguration length known for each series. Finally, the sixth and seventh columns indicate the numbers of instances solved by the two systems in two hours. The benchmark series are sorted in order of the longest shortest reconfiguration length.

From this table, we can see the pros and cons of the proposed algorithm and recon. The proposed algorithm was good at solving instances having long reconfiguration sequences, but it was not so good for instances of large graphs. In contrast, the recon solver could adapt to instances of large graphs but could not solve many instances having long reconfiguration sequences. A major bottleneck

[1] https://core-challenge.github.io/2022/.
[2] https://core-challenge.github.io/2022result/.
[3] https://github.com/Shin-ichi-Minato/SAPPOROBDD.
[4] https://github.com/junkawahara/ddreconf-experiments2023.

in the proposed algorithm is the construction of the ZDD $\mathcal{Z}_{\text{ind}}$ representing all the independent sets. When a graph is large (e.g., more than 10,000 edges), it takes a very long time and large memory to construct $\mathcal{Z}_{\text{ind}}$. Table 3 summarizes the results when classified by the length of the shortest reconfiguration found. This table more clearly shows the characteristics of the two methods.

Next, Fig. 3 shows a plot of the relation between the length of the shortest reconfiguration sequence found and the CPU time in log scale for `sp` series. In this experiment, to examine the length of the reconfiguration sequences that the algorithms could find, we set the time limit to 200,000 s. From the plot, we can see that the CPU time for the proposed algorithm rose gently as the length increased, whereas the time for recon rose steeply. Figure 4 shows a similar plot, but for a comparison of memory usage. The proposed algorithm used more memory at the beginning for the ZDD library, but the situation was reversed when the reconfiguration sequence became longer. The proposed algorithm successfully computed the shortest sequence for the `sp019` instance with 247 vertices and 1,578 edges, with length 5,767,157, in 151,567 s. For `sp019`, the size (number of nodes) of the ZDDs $\mathcal{Z}^i$ in the proposed algorithm was at most tens of thousands for each i, but some $\mathcal{Z}^i$ contained more than 10^{10} independent sets. The proposed algorithm thus behaved as if it was searching the solution space like a breadth-first search while compressing the found solutions. This indicates that our algorithm is good at handling instances for which the solution space has a relatively small width, but is very long. The recon solver also conducts a breadth-first search, but it takes a long time because the solution space itself is enormous. Other solvers in the competition show the same trend as recon although we omit the detail.

Table 2. Experimental results for each benchmark series

Dataset	# Inst	Max $\|V\|$	Max $\|E\|$	Reconfig. len	# Solved ZDD	# Solved recon
`grid`	4	40000	79600	8	**2**	**2**
`handcrafted`	5	36	51	69	**5**	**5**
`color04`	202	10000	990000	112	76	**201**
`queen`	48	40000	13253400	159	8	**40**
`square`	17	204	303	1722	**17**	8
`power`	17	304	463	55139	**11**	6
`sp`	30	390	2502	90101	**15**	10

Table 3. Experimental results for each range of reconfiguration lengths.

Reconfig. len ℓ	# Instances	# Solved ZDD	# Solved recon
$1 \leq \ell \leq 10$	178	70	**178**
$10 < \ell \leq 100$	78	25	**73**
$100 < \ell \leq 1000$	21	**16**	13
$1000 < \ell \leq 10000$	11	**11**	6
$10000 < \ell \leq 100000$	7	**7**	2

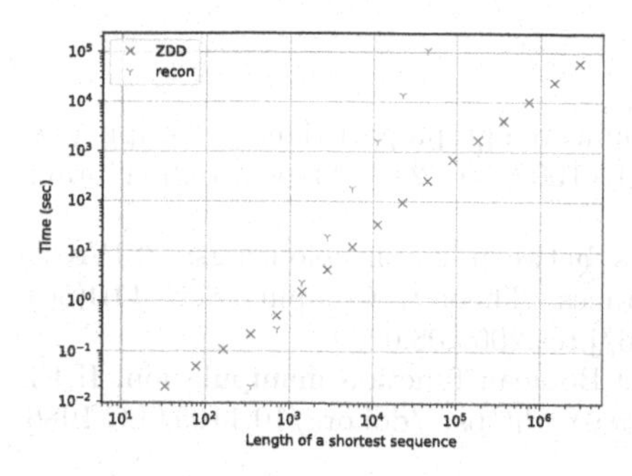

Fig. 3. CPU time for the **sp** series.

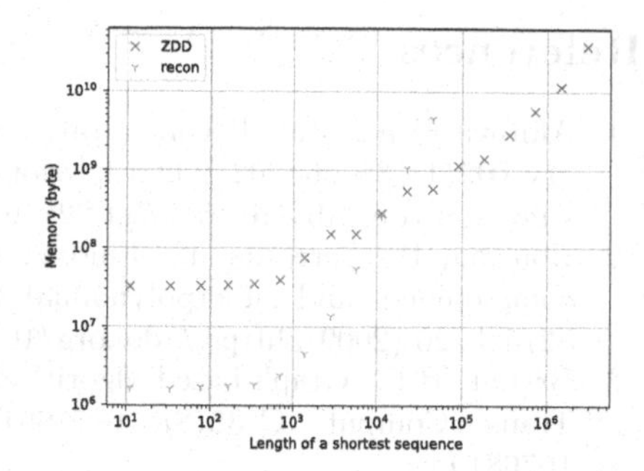

Fig. 4. Memory usage for the **sp** series.

6 Conclusion

We have proposed a ZDD-based framework for solving combinatorial reconfiguration problems, and we have demonstrated that the framework can handle various reconfiguration objects, as shown in Fig. 1. In particular, the framework can solve the TAR and the token jumping model on a graph. We have also shown that our framework can be used for analyzing the solution spaces of reconfiguration problems such as the reachability, shortest, farthest, and connectivity variants. We seek to implement all the features described in this paper, and some of them have already been published on Github[5]. Currently, our program can handle independent/dominating sets, matchings, (spanning/Steiner) trees, and forests as reconfiguration objects under the TAR and the token jumping model. We hope that these features will contribute to analysis of reconfiguration problems from both theoretical and practical viewpoints.

A power grid network described in the introduction is modelled as a graph where vertices correspond to substations (power suppliers) and houses (demand nodes), and edges correspond to power cables having switch gears. A switch on/off configuration is regarded as a rooted forest consisting of switch-on edges where roots correspond to substations. The solution space is the family of switch-on edge sets. Adding/removing an edge means turning on/off a switch. Therefore, the power grid reconfiguration problem described in the introduction can be solved by our framework. We expect that some power grid networks are (almost) planar [12] and that the ZDD that represents the solution space will be small.

Future work includes theoretical analysis of the complexity of the proposed algorithm, designing ZDD-based algorithms for problems in which the solution space cannot be directly represented as a set family, such as coloring reconfiguration problems [6] and graph partition reconfiguration problems [1], adopting relax/restrict DD techniques [4] and applying the algorithm to practical problems.

[5] https://github.com/junkawahara/ddreconf.

References

1. Akitaya, H.A., et al.: Reconfiguration of connected graph partitions. J. Graph Theory **102**(1), 35–66 (2023). https://doi.org/10.1002/jgt.22856. https://onlinelibrary.wiley.com/doi/abs/10.1002/jgt.22856
2. Bonsma, P., Cereceda, L.: Finding paths between graph colourings: PSPACE-completeness and superpolynomial distances. Theoret. Comput. Sci. **410**(50), 5215–5226 (2009). https://doi.org/10.1016/j.tcs.2009.08.023
3. Bryant, R.E.: Graph-based algorithms for Boolean function manipulation. IEEE Trans. Comput. **C-35**(8), 677–691 (1986). https://doi.org/10.1109/TC.1986.1676819
4. Castro, M.P., Cire, A.A., Beck, J.C.: Decision diagrams for discrete optimization: a survey of recent advances. INFORMS J. Comput. **34**(4), 2271–2295 (2022). https://doi.org/10.1287/ijoc.2022.1170
5. Censor-Hillel, K., Rabie, M.: Distributed reconfiguration of maximal independent sets. J. Comput. Syst. Sci. **112**, 85–96 (2020). https://doi.org/10.1016/j.jcss.2020.03.003. https://www.sciencedirect.com/science/article/pii/S0022000020300349
6. Cereceda, L., van den Heuvel, J., Johnson, M.: Connectedness of the graph of vertex-colourings. Discrete Math. **308**(5–6), 913–919 (2008). https://doi.org/10.1016/j.disc.2007.07.028
7. Cimatti, A., et al.: NuSMV 2: an OpenSource tool for symbolic model checking. In: Brinksma, E., Larsen, K.G. (eds.) CAV 2002. LNCS, vol. 2404, pp. 359–364. Springer, Heidelberg (2002). https://doi.org/10.1007/3-540-45657-0_29
8. Coudert, O.: Solving graph optimization problems with ZBDDs. In: Proceedings European Design and Test Conference, ED & TC 1997, pp. 224–228 (1997). https://doi.org/10.1109/EDTC.1997.582363
9. Fifield, B., Imai, K., Kawahara, J., Kenny, C.T.: The essential role of empirical validation in legislative redistricting simulation. Stat. Public Policy **7**(1), 52–68 (2020). https://doi.org/10.1080/2330443X.2020.1791773
10. Hayase, K., Sadakane, K., Tani, S.: Output-size sensitiveness of OBDD construction through maximal independent set problem. In: Du, D.-Z., Li, M. (eds.) COCOON 1995. LNCS, vol. 959, pp. 229–234. Springer, Heidelberg (1995). https://doi.org/10.1007/BFb0030837
11. van den Heuvel, J.: The complexity of change. In: Blackburn, S.R., Gerke, S., Wildon, M. (eds.) Surveys in Combinatorics 2013, London Mathematical Society Lecture No te Series, vol. 409, pp. 127–160. Cambridge University Press, Cambridge (2013). https://doi.org/10.1017/CBO9781139506748.005
12. Inoue, T., et al.: Distribution loss minimization with guaranteed error bound. IEEE Trans. Smart Grid **5**(1), 102–111 (2014). https://doi.org/10.1109/TSG.2013.2288976
13. Ito, T.: On the complexity of reconfiguration problems. Theoret. Comput. Sci. **412**(12–14), 1054–1065 (2011). https://doi.org/10.1016/j.tcs.2010.12.005
14. Iwashita, H., Minato, S.: Efficient top-down ZDD construction techniques using recursive specifications. TCS Technical reports TCS-TR-A-13-69 (2013)
15. Kamiński, M., Medvedev, P., Milanič, M.: Complexity of independent set reconfigurability problems. Theoret. Comput. Sci. **439**, 9–15 (2012). https://doi.org/10.1016/j.tcs.2012.03.004
16. Kawahara, J., Inoue, T., Iwashita, H., Minato, S.: Frontier-based search for enumerating all constrained subgraphs with compressed representation. IEICE Trans. Fund. Electron. Commun. Comput. Sci. **E100-A**(9), 1773–1784 (2017). https://doi.org/10.1587/transfun.E100.A.1773

17. Kawahara, J., Saitoh, T., Suzuki, H., Yoshinaka, R.: Solving the longest oneway-ticket problem and enumerating letter graphs by augmenting the two representative approaches with ZDDs. In: Proceedings of the Computational Intelligence in Information Systems Conference (CIIS 2016), vol. 532, pp. 294–305 (2016). https://doi.org/10.1007/978-3-319-48517-1_26
18. Kawahara, J., Saitoh, T., Suzuki, H., Yoshinaka, R.: Colorful frontier-based search: implicit enumeration of chordal and interval subgraphs. In: Kotsireas, I., Pardalos, P., Parsopoulos, K.E., Souravlias, D., Tsokas, A. (eds.) SEA 2019. LNCS, vol. 11544, pp. 125–141. Springer, Cham (2019). https://doi.org/10.1007/978-3-030-34029-2_9
19. Knuth, D.E.: The Art of Computer Programming, Volume 4A, Combinatorial Algorithms, Part 1, 1st edn. Addison-Wesley Professional (2011)
20. Minato, S.: Zero-suppressed BDDs for set manipulation in combinatorial problems. In: Proceedings of the 30th ACM/IEEE Design Automation Conference, pp. 272–277 (1993). https://doi.org/10.1145/157485.164890
21. Mizuta, H., Ito, T., Zhou, X.: Reconfiguration of steiner trees in an unweighted graph. IEICE Trans. Fundam. Electron. Commun. Comput. Sci. **100-A**(7), 1532–1540 (2017). https://doi.org/10.1587/transfun.E100.A.1532
22. Nakahata, Yu., Kawahara, J., Horiyama, T., Minato, S.: Implicit enumeration of topological-minor-embeddings and its application to planar subgraph enumeration. In: Rahman, M.S., Sadakane, K., Sung, W.-K. (eds.) WALCOM 2020. LNCS, vol. 12049, pp. 211–222. Springer, Cham (2020). https://doi.org/10.1007/978-3-030-39881-1_18
23. Nishimura, N.: Introduction to reconfiguration. Algorithms **11**(4), 52 (2018). https://doi.org/10.3390/a11040052
24. Sekine, K., Imai, H., Tani, S.: Computing the Tutte polynomial of a graph of moderate size. In: Proceedings of the 6th International Symposium on Algorithms and Computation, pp. 224–233 (1995). https://doi.org/10.1007/BFb0015427
25. Soh, T., Okamoto, Y., Ito, T.: Core challenge 2022: solver and graph descriptions. CoRR abs/2208.02495 (2022). https://doi.org/10.48550/arXiv.2208.02495
26. Speck, D., Mattmüller, R., Nebel, B.: Symbolic top-k planning. In: The Thirty-Fourth AAAI Conference on Artificial Intelligence, AAAI 2020, The Thirty-Second Innovative Applications of Artificial Intelligence Conference, IAAI 2020, The Tenth AAAI Symposium on Educational Advances in Artificial Intelligence, EAAI 2020, New York, NY, USA, 7–12 February 2020, pp. 9967–9974. AAAI Press (2020). https://ojs.aaai.org/index.php/AAAI/article/view/6552
27. Turau, V., Weyer, C.: Finding shortest reconfigurations sequences of independent sets. In: Core Challenge 2022: Solver and Graph Descriptions, pp. 3–14 (2022)

Neural Networks for Local Search and Crossover in Vehicle Routing: A Possible Overkill?

Ítalo Santana[1(✉)], Andrea Lodi[2], and Thibaut Vidal[1,3]

[1] Department of Computer Science, Pontifical Catholic University of Rio de Janeiro, Rio de Janeiro, Brazil
isantana@inf.puc-rio.br, thibaut.vidal@polymtl.ca

[2] Jacobs Technion-Cornell Institute, Cornell Tech and Technion - IIT, New York, USA
andrea.lodi@cornell.edu

[3] CIRRELT & SCALE-AI Chair in Data-Driven Supply Chains, Department of Mathematical and Industrial Engineering, Polytechnique Montréal, Montréal, Canada

Abstract. Extensive research has been conducted, over recent years, on various ways of enhancing heuristic search for combinatorial optimization problems with machine learning algorithms. In this study, we investigate the use of predictions from graph neural networks (GNNs) in the form of heatmaps to improve the Hybrid Genetic Search (HGS), a state-of-the-art algorithm for the Capacitated Vehicle Routing Problem (CVRP). The crossover and local-search components of HGS are instrumental in finding improved solutions, yet these components essentially rely on simple greedy or random choices. It seems intuitive to attempt to incorporate additional knowledge at these levels. Throughout a vast experimental campaign on more than 10,000 problem instances, we show that exploiting more sophisticated strategies using measures of node relatedness (heatmaps, or simply distance) within these algorithmic components can significantly enhance performance. However, contrary to initial expectations, we also observed that heatmaps did not present significant advantages over simpler distance measures for these purposes. Therefore, we faced a common —though rarely documented— situation of overkill: GNNs can indeed improve performance on an important optimization task, but an ablation analysis demonstrated that simpler alternatives perform equally well.

Keywords: Heuristic search · Vehicle routing problem · Graph neural networks

1 Introduction

Vehicle routing problems (VRP) represent one of the most studied classes of NP-hard problems due to their practical difficulty and ubiquity in real-life applications such as food distribution, parcel delivery, or waste collection, among others

© The Author(s), under exclusive license to Springer Nature Switzerland AG 2023
A. A. Cire (Ed.): CPAIOR 2023, LNCS 13884, pp. 184–199, 2023.
https://doi.org/10.1007/978-3-031-33271-5_13

[18,26]. Problems in this class generally seek to plan efficient itineraries for a fleet of vehicles to service a geographically-dispersed set of customers. The capacitated VRP (CVRP) is the most canonical variant among all existing routing problems. Its objective is to minimize the total distance traveled by the vehicles to service the customers, subject to a single constraint representing the vehicle capacities: the sum of customers' demands over a route should not exceed the vehicle capacity.

Over the years, there have been dramatic improvements in the heuristic and exact (i.e., provably optimal) solution of VRPs. To date, the best-performing exact algorithms rely on branch-cut-and-price strategies, with tailored cutting-plane algorithms and sophisticated column-generation routines [5,13]. With these methods, it is now possible to solve most existing instances with 200 or 300 customers. However, the time for an exact solution remains highly volatile at this scale, and most larger instances remain unsolved. Consequently, extensive research has been conducted on metaheuristics for this problem to find high-quality solutions in a shorter and more controlled time [24].

As it stands now, metaheuristics can consistently locate high-quality solutions for CVRPs with up to 1,000 customers in a matter of minutes [4,21]. Of all existing methods, the Hybrid Genetic Search (HGS) algorithm developed in [21,23,25] achieves the best-known solution quality consistently on most problems and instances of interest. During the 12th DIMACS implementation challenge on the CVRP organized in 2022 [2], it was used as the base algorithm for four of the five best methods. It was also adopted as a baseline for the EURO Meets NeurIPS 2022 Vehicle Routing Competition [8]. Moreover, very-large problem instances counting dozens of thousands of customers can also be solved using tailored data structures and decomposition strategies [1,15]. Within the latest generation of metaheuristics, including HGS, it is clear that two main operators —local search and crossover— are instrumental in finding improving solutions.

- **Local Search (LS)** consists in systematically exploring a neighborhood obtained by small changes over a current solution to identify improvements. This process is iterated until attaining a local optimum. Classical neighborhoods for the CVRP involve exchanges or relocations of client visits and edge reconnections. They typically include $\mathcal{O}(n^2)$ possible neighbors, where n represents the number of customers. Due to its iterative nature, LS typically takes the largest share of the computational time. Several techniques have been developed to reduce computational complexity. In particular, in [17], it is observed that the search could be limited to relocations and exchanges of customers that are *geographically related*. The resulting strategy, called granular search, limits classical neighborhoods to $\mathcal{O}(\Gamma n)$ moves, where Γ is a user-defined parameter. However, although very simple in design, a straight-forward distance-based relatedness criterion may hinder the search process, especially if optimal solutions require a few long edges.
- In contrast, **Crossover** operators focus on diversifying the search. They consist of recombining two existing (parent) solutions into a new (offspring)

solution that inherits promising characteristics from both. For the CVRP, crossover operators are not primarily designed for solution improvement, but instead used to create promising starting points for subsequent LS. Various crossovers have been used in previous works [10,21]. As shown in Sect. 2, the Ordered Crossover (OX) is widely used, and consists in juxtaposing a fragment of the first solution with the remaining client visits ordered as in the second solution. Doing so implicitly creates a re-connection point that is typically random.

Note that in both LS and Crossover, there is interest in using relatedness information between client vertices to (i) speed up the LS or (ii) identify a subset of more promising crossover operations. It is also noteworthy that the relatedness information used until now for the LS (and possibly used for the crossover) is a broader concept that goes beyond simple distance criteria and that could be possibly learned.

In recent years, graph neural networks (GNNs) have emerged as a tool to apply machine learning techniques to combinatorial optimization problems posed over graphs. To the best of our knowledge, the first attempt in this context was proposed by [7] for the Traveling Salesman Problem. Underpinned by enhancements in hardware and artificial intelligence research over the last years, the development of deep NNs made them relevant to a wide range of difficult combinatorial optimization problems, such as SAT, Minimum Vertex Cover, and Maximum Cut [6,28]. When applied to solve CVRPs, these networks are usually combined with reinforcement learning (RL) [3,11] or typically used for node classification or edge prediction [9,27]. Despite extensive research, GNNs for directly solving CVRPs remain limited to small problem instances (e.g., 100 customers) and generally do not compare favorably with classic optimization methods (exact or heuristic) regarding solution quality. This is possibly due to the fact that good solutions for combinatorial optimization problems result from tacit structural knowledge about the problem (learnable solution structures) along with a significant amount of trial-and-error to build the best possible solution satisfying almost perfectly the constraints at hand. Whereas better knowledge of solution structures can be learned to guide the search, avoiding some (explicit or implicit) enumeration of solutions without compromising solution quality is generally challenging.

Against this background, a promising path toward better solution methods for VRPs concerns the hybridization of learning-based and traditional solution methods. More specifically, given the importance of both LS and crossover operations in HGS, we are naturally led to question whether learned relatedness information can lead to substantial improvements in these components. To that end, we capitalize upon the work in [9], where a GNN is trained to predict the occurrence probabilities of edges in high-quality solutions (i.e., heatmap). We leverage the heatmaps as a source of relatedness information to define neighborhood restrictions in the LS and possible re-connection points in the crossover. To make this analysis possible, we introduce the following methodological contributions:

1. We introduce a framework for defining and exploiting *relatedness* information between pairs of customers in CVRP context. We show that relatedness measures can be exploited to steer the LS toward the most promising moves. Additionally, we use relatedness information to extend the classical OX crossover, trading some of its inherent randomness for better choices of re-connection points between the parents. Our approaches are generic and applicable to any relatedness measure. Specifically, we consider relatedness from two sources: geographical relatedness, given by the distance between two customers, and learnable relatedness (i.e., heatmap), obtained from a GNN.
2. We suggest a practical technique to exploit the output of a single GNN (heatmaps for fixed-size graphs) for problem instances of varying sizes. To this end, we decompose the original instance into a sequence of fixed-size subproblems and aggregate the resulting heatmap information. This approach has the benefit of only requiring a single trained model of moderate size.

Next, throughout an extensive experimental campaign, we evaluate how HGS performance varies with the proposed changes, for better or worse, on more than 10,000 different instances containing 100 to 1,000 customers. We, therefore, measure the enhancements achieved with the proposed techniques and analyze the impact of each change through an ablation analysis.

The first, positive result of our investigation is that incorporating relatedness information within the crossover and LS operators significantly benefit the search. This is maybe not totally surprising but, considering the quality of the baseline results, the significance of this improvement is remarkable. The second result gives us a more mixed message instead. Indeed, contrary to our initial intuition, and after closer analysis through the ablation study, we observe that the improvements we achieved are quite insensitive to the source of the relatedness information (geographical or learned). Essentially, there is potential to improve these operators by exploiting additional problem knowledge, but the learning strategies designed to do so did not perform much better than simple ad-hoc rules based on geographical relatedness. In other words, in this study, using GNN-based heatmaps seemed to be an overkill for the task at hand, although the door remains open to use them to capture possible relations in more complex problem variants.

2 Methodology

The CVRP is defined over a complete graph $G = (V, E)$, where the set of vertices $V = \{0, 1 \dots, n\}$ contains a vertex 0 representing the depot, and the remaining vertices represent customers. Each customer $i \in \{1, \dots, n\}$ is characterized by a demand d_i. Edges (i, j) model direct travel between vertices i and j for a distance d_{ij}. A solution to this problem is a set of routes originating and ending at the depot and visiting customers, such that (i) the total demand over each route does not exceed a vehicle-capacity limit Q, (ii) each customer is visited exactly once, and (iii) the total travel distance is minimized.

We additionally assume that we can calculate a *relatedness* metric $\phi(i, j)$ for each edge (i, j). This definition is general: in the simplest setting, relatedness could be the inverse of distance, i.e., $\phi(i, j) = 1/d_{ij}$. In a more informed setting, we can instead consider defining $\phi(i, j)$ as the output of a graph neural network (GNN) as seen in [9], predicting the probability of occurrence of an edge in a high-quality solution. Probabilities of this kind are typically called heatmaps. In the remainder of the paper, we will refer to $\phi_{\mathrm{D}}(i, j)$ for distance-relatedness, and $\phi_{\mathrm{N}}(i, j)$ for GNN-based relatedness. This information will now be used to refine the two most important HGS operators.

2.1 Hybrid Genetic Search

The Hybrid Genetic Search [21] relies on simple solution generation and improvement steps. The method starts by initializing a population of size μ with random solutions that are improved by local search. After this initialization phase, HGS iteratively generates new solutions by selecting two random solutions in the population, recombining them using an ordered crossover (OX), and applying local search for improvement. To promote exploration, solutions that exceed capacity limits are not directly rejected but instead penalized according to their amount of infeasibility. The penalty weights are adapted during the search to achieve a target percentage of feasible solutions, and infeasible solutions are maintained in a separate subpopulation. Whenever a solution is infeasible after the local search, an extra REPAIR step is applied, which simply consists of a classic local search with a temporarily (10×) higher penalty coefficient.

During the overall search process, the number of solutions in the feasible and infeasible populations is monitored. Whenever any population exceeds $\mu + \lambda$ solutions, a survivors' selection phase is triggered to retain only the best μ individuals, according to a ranking metric based on solution value and contribution to the population diversity. Finally, the algorithm restarts each time n_{IT} consecutive solution generations have been done without improvement of the best solution, and it terminates upon a time limit T_{MAX} by returning the best solution found over all the restarts.

2.2 Local Search Using Relatedness Measures

Local Search (LS) is a conceptually simple and efficient method to solve combinatorial optimization problems of the form $\min_{x \in X} c(x)$, where X is the space of all solutions and c is the objective function. A neighborhood is defined as a mapping $\mathcal{N} : X \to 2^X$ associating with any solution x a set of neighbors $\mathcal{N}(x) \subset X$. For the CVRP, $\mathcal{N}(x)$ is usually defined relative to a set of operations (i.e., moves) that can modify the current solution x. A move τ is a small modification that can be applied on x to obtain a neighbor $\tau(x) \in \mathcal{N}(x)$. HGS uses four main types of moves and some of their immediate extensions [21]:

Twenty customers most related to customer 63 according to ϕ_D and ϕ_N:

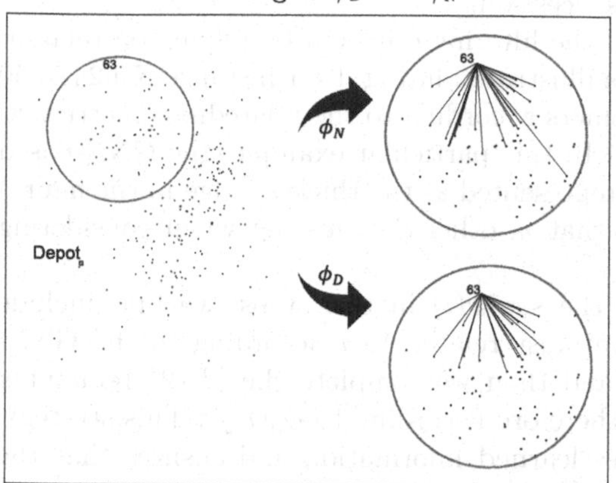

Customer 63 in an optimal solution:

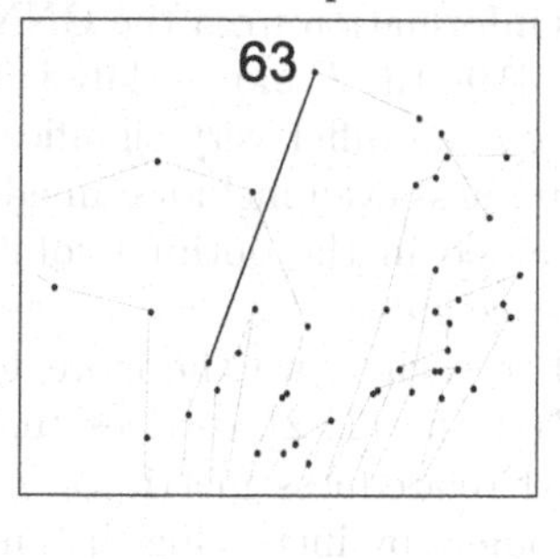

Fig. 1. Sets of related customers according to ϕ_D and ϕ_N, on instance X-n247-k50

- Relocate: Moves a visit to customer i immediately after a visit to a different customer j or the depot;
- Swap: Exchanges the visits of customers i and j;
- 2-Opt: Reverts a customer-visit sequence $(i, \ldots, j)$;
- 2-Opt*: Exchanges customers i and j and their succeeding visits.

From an incumbent solution represented as a set of routes, the moves are evaluated in a random order of the indices i and j (within the same or different routes), and any improvement is directly applied. This process is repeated until a local minimum is reached, i.e., a situation where no improving move exists for all the considered neighborhoods. Without further pruning, all these neighborhoods contain $O(n^2)$ solutions. Using incremental calculations (keeping track of partial load and distance over the routes), it is possible to conduct a complete evaluation of all neighborhoods in $O(n^2)$ time. Moreover, the number of complete neighborhood searches (i.e., loops) needed to converge is rarely greater than 10 in practice.

A quadratic complexity for the LS operator is adequate for small problems, but this can become a significant bottleneck otherwise due to its frequent use. Considering this, [17] introduced a "granular search" mechanism that consists in limiting the moves to customer pairs (i, j) that are geographically close, i.e., such that j belongs to a set $\Phi(i)$ formed of the Γ closest customers of i. Consequently, the total number of moves and the complexity of each LS loop reduce to $O(n\Gamma)$ time. Indeed, it rarely makes sense to relocate or exchange customer visits that are far away from each other. Moreover, this strategy ensures that each move creates at least one short edge [1,16,22].

Since its inception, granular search has been adapted to many VRP variants. Especially, to handle customer constraints on service-time windows, [22] extended the concept to filter node pairs (i, j) based on a compound metric that includes distance, unavoidable waiting times, and unavoidable time-window violations arising from this customer succession.

In this study, we instead extend the filtering criterion by relying on relatedness information from the GNN. As illustrated in Fig. 1 for instance X-n247-k50 from [19], the Γ most-related customers according to the relatedness metrics ϕ_{D} and ϕ_{N} can differ very significantly. In this particular example, the GNN-based relatedness even includes an edge (represented as the thickest edge in the figure) contained in the optimal solution that is otherwise missing when considering distance only.

For each i, we therefore form the set $\Phi(i)$ in two steps: we first include in $\Phi(i)$ the $\lfloor \Gamma/2 \rfloor$ vertices that are most related to i according to the GNN-based relatedness metric $\phi_{\mathrm{N}}(i,j)$, and then we complete the $\lceil \Gamma/2 \rceil$ remaining customers by increasing distance, therefore according to $\phi_{\mathrm{D}}(i,j)$. This strategy, named neural granular search, uses learned information and ensures that the $\Gamma/2$ closest customers are still considered in the moves.

2.3 Crossover Using Relatedness Measures

In HGS, each solution is represented as a single permutation of the customer's visits (i.e., a giant tour) during the crossover operation. The use of this representation is motivated by the fact that (i) one can simply represent any complete solution by concatenating the routes and omitting the visits to the depot, and (ii) reversely, given a sequence of customers visits, there exists a linear-time algorithm, called SPLIT, that optimally segments this giant tour into routes [20].

Based on this representation, HGS employs the ordered crossover (OX – [12]) illustrated in Fig. 2. OX works in two steps. First, a fragment F of the first parent defined by two randomly-selected cutting points is copied in place into an empty offspring. Next, the second parent is scanned from the position of the second cutting point to complete all missing customer visits circularly. This gives a new giant tour, which is then transformed into a complete CVRP solution using SPLIT.

As it stands, OX is completely dependent upon random choices. In particular, the second step tends to concatenate unrelated customers immediately after fragment F. This creates low-quality fragments of solution requiring many LS moves for improvement. To correct this issue, we suggest relying on the relatedness metric to modify the completion step. Let i be the last customer from fragment F. Instead of arbitrarily reconnecting F with the next customer from Parent 2 obtained by a circular sweep, we select a random related customer j among the Γ customers most related to i that are not part of F (or a random position if no such j exists) and then proceed to complete the offspring from this position following the order in Parent 2. This small but notable difference permits reconnecting visits that are more closely related among both parents, allowing for better solutions without sacrificing diversity. As previously, the choice of relatedness metric leads to different variants of the OX crossover. In the remainder of this paper, we will refer to the modified crossover using distance-relatedness as DOX, and to the modified crossover using GNN-relatedness as NOX.

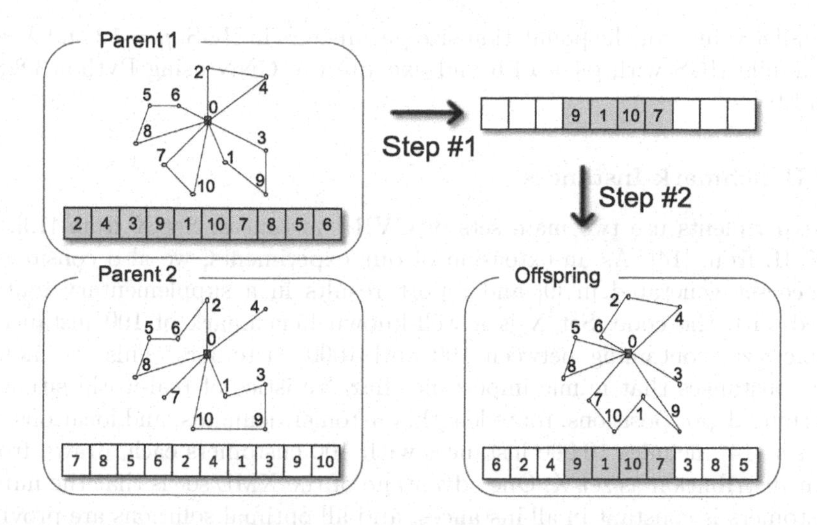

Fig. 2. Illustration of the ordered crossover (OX)

3 Experimental Analyses

This section presents extensive computational experiments designed to: (i) calibrate and evaluate the impact of the granular search parameter Γ, which governs the size of the LS neighborhoods; (ii) measure the impact of our enhancements on the LS and OX operators as well as the usefulness of different relatedness criteria; (iii) confront the characteristics, the computational effort, and the performance of heatmaps produced by different GNN configurations, and (iv) analyze the extension of GNNs originally trained on fixed-size graphs to instances of varying sizes. We address objective (i) in Sect. 3.4, whereas objectives (ii, iii, iv) are covered in Sects. 3.5 and 3.6.

We will analyze, in the following sections, the performance of HGS in its original form (baseline) along with five combinations of ϕ_{D} and ϕ_{N} for local search and crossover operators, which are listed below:

- **HGS-D-O (baseline):** HGS with granular search and OX;
- **HGS-D-D:** HGS with granular search and DOX;
- **HGS-D-N:** HGS with granular search and NOX;
- **HGS-N-O:** HGS with neural granular search and OX;
- **HGS-N-D:** HGS with neural granular search and DOX;
- **HGS-N-N:** HGS with neural granular search and NOX.

3.1 Computational Environment

All experiments are run on a single thread of an Intel Gold 6148 Skylake 2.4 GHz processor with 40 GB of RAM and NVIDIA Tesla P100 Pascal (12 G memory), running CentOS 7.8.2003. Unless otherwise stated, we use the original parameters defined for HGS in [21] and the GNN in [9]. To achieve fast convergence, we set smaller values for the population-size parameters in HGS: $\mu = 12$ and $\lambda = 20$. We compile HGS with g++ 9.1.0 and execute the GNN using Python 3.8.8 on Torch 1.9.1.

3.2 Benchmark Instances

Our experiments use two main sets of CVRP instances: Set X from [19], and Set XML from [14]. As an extension of our experiments, we also consider the instance set generated in [9] and report results in a supplementary material located with the code. Set X is a well-known benchmark of 100 instances of variable size, containing between 100 and 1000 customers. This set includes diverse instances that mimic important characteristics of real-world situations concerning depot positions, route length, customer demands, and locations. The XML set [14] includes 10,000 instances with 100 customers each, drawn from a similar distribution as set X. One advantage of the XML set is that the number of customers is constant in all instances, and all optimal solutions are provided. This permits comparisons with proven optima instead of best-known solutions (BKS) collected from all previous works. In contrast, many instances of set X are still unsolved to proven optimality. All instance sets, open-source codes, and scripts needed to run the experiments are provided at https://github.com/italogs/HGS-CVRP.

3.3 Parametrization and Training of the GNN

The GNN proposed by [9] is designed to be trained and applied for prediction over graphs (i.e., instances) of fixed size. The authors released their final model trained on 100-customers instances generated similarly to set X and XML. This model can therefore be directly applied for inference on the XML instances, which each contain 100 customers. In contrast, since the original Set X has instances with different numbers of customers, a different approach is needed to use the heatmaps. Due to these key differences, we will subdivide the presentation of our experiments into two parts, with results on XML instances in Sect. 3.5, and adaptations and results for Set X in Sect. 3.6.

In these experiments, we use the original trained GNN from [9] for heatmap generation, called ORIGINAL in the rest of this paper. However, although this model is already trained, it still takes around 0.85 s of inference time to produce the heatmap for a given instance. This is a similar order of magnitude as the time needed by HGS to solve the CVRP to near optimality (i.e., below 0.1% error) on instances containing 100 customers. Since we aim to compare CVRP solution algorithms under the same total CPU time budget (counting inference time and solution time), GNN-based methods would be at a disadvantage if a large share of the CPU time is invested in the inference step. Therefore, to estimate the performance of GNN-based algorithms in the most optimistic conditions (e.g., considering a hypothetical scenario where GPU inference is extremely fast), we will also report the results of the same method ignoring inference time. Additionally, we produce results (counting inference time) obtained with two lighter versions of the GNN, called MODEL #1 and MODEL #2, which were trained on the same examples as [9], with fewer internal nodes and internal layers. Table 1 summarizes the parameter setting of all the considered GNNs.

Table 1. GNN configurations

GNN	#NODES	#LAYERS	#EPOCHS	PRED-T(s)
ORIGINAL	300	30	1500	0.85
OPTIMISTIC	300	30	1500	IGNORED
MODEL #1	10	5	500	0.03
MODEL #2	10	5	1500	0.03

This table lists for each GNN the number of hidden layers (# LAYERS), nodes per layer (# NODES), and epochs (# EPOCHS) used for training. Finally, the last column reports the average inference time on an XML instance. The parameters of MODEL #1 and MODEL #2 were selected to achieve training and inference in a limited time. MODEL #1 (resp. #2) required 8 (resp. 24) hours of training time on our hardware.

3.4 Calibration of the Local Search

We focus here on the parameter Γ, which drives the exploration breadth of the LS (see Sect. 2.2) and significantly impacts the computational time of HGS. The aim of this experiment is to select a meaningful range of values for this parameter. Based on standard values used in previous works, we evaluate configurations $\Gamma \in \{5, 10, 15, 20, 30, 50, 100\}$ and analyze the sensitivity of the baseline method (i.e., HGS-D-O) to this parameter. To keep a simple experimental design, we focus on the performance of the LS by generating ten random initial solutions for each instance and applying a single LS to each of these solutions. We then report in Table 2 the quality of the best solution found as well as the computational time used by the ten LS runs.

Table 2. Impact of Γ on solution quality and CPU time

Γ	SET X		SET XML	
	GAP%	TIME (S)	GAP%	TIME (S)
5	4.664	0.157	2.976	0.024
10	4.018	0.160	2.365	0.026
15	3.817	0.162	2.194	0.027
20	3.667	0.165	2.137	0.029
30	3.630	0.197	2.087	0.034
50	3.696	0.238	2.089	0.043
100	3.690	0.375	2.087	0.057

In Table 2 and the rest of this paper, solution quality is expressed as a percentage error gap calculated as GAP(%) $= 100 \times (z - z_{\text{BKS}})/z_{\text{BKS}}$, where z represents the cost of the solution and z_{BKS} is the optimal or BKS cost value.

The results of this experiment indicate that solution quality generally improves with Γ, but with decreasing marginal returns. We cease to see notable solution quality improvements once Γ exceeds a value of 30, but CPU time dramatically increases. Given this, we set $\Gamma = 15$ in the remainder of our experiments, and additionally provide detailed results with $\Gamma \in \{20, 30, 50\}$ in the online material.

3.5 Experimental Results – Set XML

Having calibrated all the algorithmic components, we can now measure the impact of GNN-informed relatedness measures in the LS and crossover operator. We focus here on the instances of set XML. Given that HGS converges towards near-optimal solutions within seconds for these instances, we use a short termination criterion with $T_{\text{MAX}} = 5$ s per instance and report final results as well as convergence plots to measure the impact of the different versions of the LS and crossover.

Table 3 therefore reports the number of optimal solutions (#OPT) attained over the 10,000 instances and the average final GAP(%) for all of the methods, considering the four possible GNN configurations (ORIGINAL, OPTIMISTIC, MODEL #1, and MODEL #2). Best performance is indicated in boldface. Additionally, the convergence plots of Fig. 3 depict the progress of the average gap of the different methods over time for the OPTIMISTIC configuration of the GNN, and similar graphs are provided for the other GNN configurations in the online material.

Table 3. Results of all methods and GNN configurations for the instances of set XML

GNN	HGS-D-O		HGS-D-D		HGS-D-N†		HGS-N-O†		HGS-N-D†		HGS-N-N†	
	#OPT	GAP%	#OPT	GAP%	#OPT	GAP%	#OPT	GAP%	#OPT	GAP%	#OPT	GAP%
ORIGINAL	7715	0.030	**8105**	0.024	8086	**0.023**	7691	0.031	8046	**0.023**	8011	0.025
OPTIMISTIC	7715	0.030	8105	0.024	**8120**	**0.023**	7732	0.031	8062	**0.023**	8041	0.025
MODEL #1	7715	0.030	**8105**	0.024	8094	**0.023**	7697	0.031	8028	0.024	7994	0.025
MODEL #2	7715	0.030	**8105**	**0.024**	8102	**0.024**	7719	0.030	8067	**0.024**	8057	**0.024**

†: HGS VARIANTS WITH HEATMAPS EITHER IN THE LOCAL SEARCH OR CROSSOVER.

As seen in these results, all HGS versions progress (i.e., decrease the gap) smoothly within the time limit, and all approaches except HGS-N-O outperformed HGS-D-O (the baseline HGS algorithm) in terms of their number of optimal solutions and average gap. The significance of these improvements is confirmed by two-tailed paired-samples Wilcoxon tests between each method and HGS-D-O at a significance level of 0.05.

However, these experiments also show that HGS-N-O does not perform significantly better than HGS-D-O, even when ignoring the inference time (i.e.,

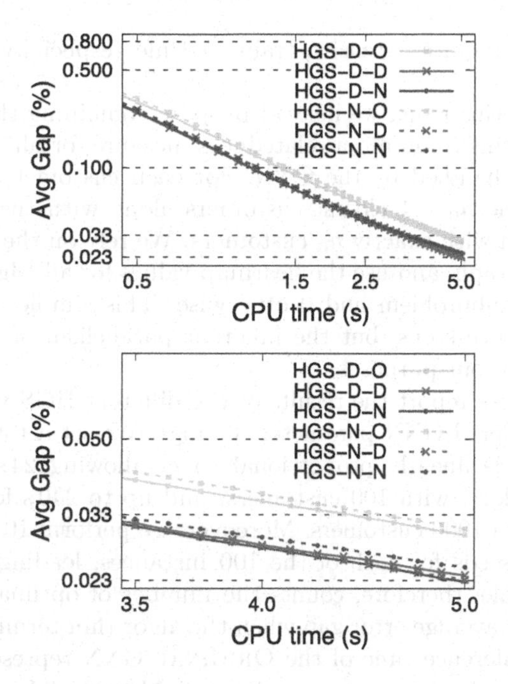

Fig. 3. Convergence plots for all HGS variants on set XML (upper graph = complete run, lower graph = last 1.5 s)

OPTIMISTIC evaluation of the GNN). This indicates that using the GNN-based relatedness criterion in the LS does not bring significant benefits. It is an open research question to determine if different GNN architectures may perform better in the task of filtering LS neighborhoods.

Now, a comparison of configurations HGS-D-O (baseline), HGS-D-D, and HGS-D-N permits us to assess the impact of our changes on the crossover operator. We remind that HGS-D-O refers to the original OX crossover, whereas HGS-D-D and HGS-D-N modify the reconnection step to integrate relatedness information. As seen in our experiments, HGS-D-D and HGS-D-N are much better than the baseline (final gaps of 0.024% compared to 0.030%), as confirmed by paired-samples Wilcoxon tests at 0.05 significance level. This is a notable breakthrough, given that it is uncommon to identify simple conceptual changes to HGS that significantly improve its state-of-the-art performance.

Finally, the choice of configuration for the GNN did not significantly affect the results, and our observations remain valid for the ORIGINAL, MODEL #1, and MODEL #2 configurations.

3.6 Experimental Results – Set X

The instances of set X include a different number of customers, but the GNN of [9] is designed to predict heatmaps only for fixed-size graphs. Moreover, training a model on an instance of maximal size (1000 customers) and relying on dummy nodes is likely to require extensive training time (especially with its original parametrization).

To circumvent this issue, we instead propose combining the heatmaps from different subproblems to obtain a relatedness measure for all customers. Let n_G be the graph size handled by the GNN. For each customer $i \in \{1, \ldots, n\}$, in turn, we collect the $n_G - 1$ closest customers along with the depot to form a CVRP subproblem with exactly n_G customers. We rely on the GNN to infer the heatmap for this graph, and use the heatmap values for all edges (i, j) such that j belongs to the subproblem and 0 otherwise. This simple approach requires n heatmaps inference steps, but the inherent parallelism of Pytorch makes it effective enough for our purposes.

As previously, we report the results of the different HGS variants in Table 4 for the four considered GNN parameter settings. We set a total computational time budget that is linearly proportional to n, allowing 24 s for the smallest instance (X-n101-k25) with 100 customers, and up to 240 s for the largest one (X-n1001-k43) with 1000 customers. Moreover, we perform 10 experiments with different random seeds for each of the 100 instances, leading to 1000 solution processes. The table, therefore, counts the number of optimal solutions out of 1000 as well as the average error gap when the algorithm terminates. With these time limits, the inference time of the ORIGINAL GNN represents 15.2% of the overall time budget, and the inference time of MODEL #1 and MODEL #2 is limited to 1.9% of the time budget. Convergence plots in the same format as before are additionally presented in Fig. 4.

Table 4. Results of all methods and GNN configurations for the instances of set X

GNN	HGS-D-O		HGS-D-D		HGS-D-N†		HGS-N-O†		HGS-N-D†		HGS-N-N†	
	#OPT	GAP%	#OPT	GAP%	#OPT	GAP%	#OPT	GAP%	#OPT	GAP%	#OPT	GAP%
ORIGINAL	**188**	0.368	187	**0.302**	184	0.317	177	0.395	169	0.325	172	0.317
OPTIMISTIC	**188**	0.368	187	0.302	187	**0.299**	185	0.365	177	0.307	182	**0.299**
MODEL #1	**188**	0.368	187	**0.302**	185	0.309	166	0.414	177	0.332	184	0.336
MODEL #2	**188**	0.368	187	0.302	186	**0.301**	169	0.391	175	0.314	170	0.316

†: HGS VARIANTS WITH HEATMAPS EITHER IN THE LOCAL SEARCH OR CROSSOVER.

These additional results on Set X confirm our previous observations: all HGS variants except HGS-N-O outperformed the HGS-D-O baseline. Additionally, the proposed modifications to the crossover operator (HGS-D-D and HGS-D-N) led to performance improvements that are even more expressive on that instance set, with final gaps of 0.302% and 0.299% compared to 0.368% for the original HGS. As previously, however, using learned information from the GNN instead of distance did not make a substantial difference in the crossover and even appeared to be detrimental in the context of the LS.

It is important to stress that, without a complete analysis involving HGS-D-D, a comparison of HGS-D-N versus HGS-D-O could have led to the conclusion that the GNN was responsible for the improvement. However, recommending the

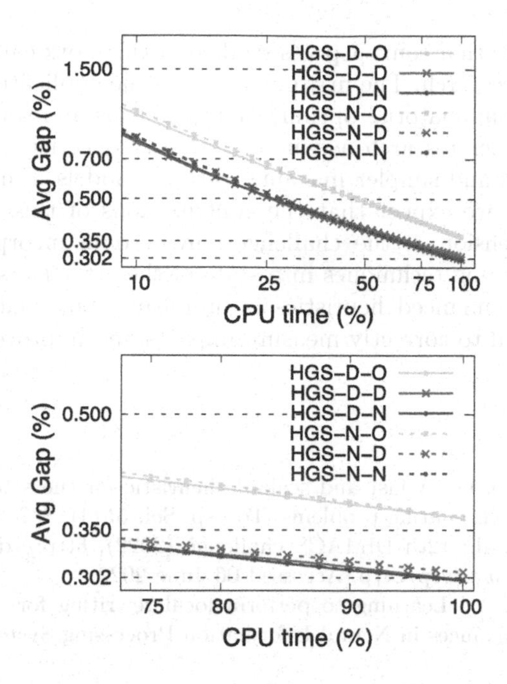

Fig. 4. Convergence plots for all HGS variants on set X (upper graph = complete run, lower graph = last 25% of CPU time)

use of this method in this context would have been an "overkill" since a simpler reconnection mechanism based on distance effectively produces the same gains.

4 Conclusions

In this work, we have shown that relatedness metrics can be broadly used to improve the performance of the HGS [21], a state-of-the-art solution algorithm for the CVRP. Relatedness has been exploited in two ways: to focus the LS on promising moves, and to steer the crossover operator towards meaningful reconnections. As relatedness is a fairly general concept, we can freely use geographical or learnable (i.e., GNN-based) information for that purpose. As seen in our experimental analyses, these adaptations lead to significant improvements on a large benchmark counting over 10,000 instances. Additionally, we show that a simple strategy to extend GNN heatmap predictions to instances of varying size is fairly effective, circumventing the limitation due to fixed-size training. Overall, exploiting heatmaps to boost HGS operators is very effective, but also not superior to a simpler application of distance-based relatedness for similar purposes, even considering subsets of the tested instances with different characteristics. This observation contrasts with the superiority claims of sophisticated learning mechanisms and ever-larger networks. Instead, it aligns with the "less-is-more" approach toward algorithmic design.

We acknowledge that some aspects studied in this work can be further investigated for future research. The first one refers to the applications of relatedness criteria to other combinatorial optimization problems and solvers (e.g., branch and bound). Another research avenue of interest concerns exploiting different relatedness sources and simpler machine learning models. Finally, from a more general viewpoint, we expect that the contributions of this work can lead to a better comprehension of the challenges involved in incorporating sophisticated machine-learning techniques into state-of-the-art solvers. We believe that research on GNN-enhanced heuristics is promising, but that careful ablation studies are essential to correctly measure impacts and improvements.

References

1. Accorsi, L., Vigo, D.: A fast and scalable heuristic for the solution of large-scale capacitated vehicle routing problems. Transp. Sci. **55**(4), 832–856 (2021)
2. Archetti, C., et al.: 12th DIMACS challenge (2022). http://dimacs.rutgers.edu/programs/challenge/vrp/cvrp. Accessed 06 June 2022
3. Chen, X., Tian, Y.: Learning to perform local rewriting for combinatorial optimization. In: Advances in Neural Information Processing Systems, pp. 6278–6289 (2019)
4. Christiaens, J., Vanden Berghe, G.: Slack induction by string removals for vehicle routing problems. Transp. Sci. **54**(2), 417–433 (2020)
5. Costa, L., Contardo, C., Desaulniers, G.: Exact branch-price-and-cut algorithms for vehicle routing. Transp. Sci. **53**(4), 946–985 (2019)
6. Dai, H., Khalil, E.B., Zhang, Y., Dilkina, B., Song, L.: Learning combinatorial optimization algorithms over graphs. In: Proceedings of the 31st International Conference on Neural Information Processing Systems, pp. 6351–6361 (2017)
7. Hopfield, J.J., Tank, D.W.: "Neural" computation of decisions in optimization problems. Biol. Cybern. **52**(3), 141–152 (1985)
8. Kool, W., et al.: The EURO meets NeurIPS 2022 vehicle routing competition. In: Proceedings of Machine Learning Research, NeurIPS 2022 Competitions (2023, in press)
9. Kool, W., van Hoof, H., Gromicho, J., Welling, M.: Deep policy dynamic programming for vehicle routing problems. In: Schaus, P. (ed.) CPAIOR 2022. LNCS, vol. 13292, pp. 190–213. Springer, Cham (2022). https://doi.org/10.1007/978-3-031-08011-1_14
10. Nagata, Y.: Edge assembly crossover for the capacitated vehicle routing problem. In: Cotta, C., van Hemert, J. (eds.) EvoCOP 2007. LNCS, vol. 4446, pp. 142–153. Springer, Heidelberg (2007). https://doi.org/10.1007/978-3-540-71615-0_13
11. Nazari, M., Oroojlooy, A., Takáč, M., Snyder, L.V.: Reinforcement learning for solving the vehicle routing problem. In: Proceedings of the 32nd International Conference on Neural Information Processing Systems, pp. 9861–9871 (2018)
12. Oliver, I.M., Smith, D.J., Holland, J.R.C.: A study of permutation crossover operators on the traveling salesman problem. In: Proceedings of the 2nd International Conference on Genetic Algorithms on Genetic Algorithms and Their Application, pp. 224–230 (1987)
13. Pessoa, A., Sadykov, R., Uchoa, E., Vanderbeck, F.: A generic exact solver for vehicle routing and related problems. Math. Program. **183**(1), 483–523 (2020)

14. Queiroga, E., Sadykov, R., Uchoa, E., Vidal, T.: 10,000 optimal CVRP solutions for testing machine learning based heuristics. In: AAAI-22 Workshop on Machine Learning for Operations Research (ML4OR) (2022)
15. Santini, A., Schneider, M., Vidal, T., Vigo, D.: Decomposition strategies for vehicle routing heuristics. INFORMS Journal on Computing, Articles in Advance (2023)
16. Schneider, M., Schwahn, F., Vigo, D.: Designing granular solution methods for routing problems with time windows. Eur. J. Oper. Res. **263**(2), 493–509 (2017)
17. Toth, P., Vigo, D.: The granular tabu search and its application to the vehicle-routing problem. Informs J. Comput. **15**(4), 333–346 (2003)
18. Toth, P., Vigo, D. (eds.): Vehicle Routing: Problems, Methods, and Applications, 2nd edn. Society for Industrial and Applied Mathematics, Philadelphia (2014)
19. Uchoa, E., Pecin, D., Pessoa, A., Poggi, M., Vidal, T., Subramanian, A.: New benchmark instances for the capacitated vehicle routing problem. Eur. J. Oper. Res. **257**(3), 845–858 (2017)
20. Vidal, T.: Split algorithm in O(n) for the capacitated vehicle routing problem. Comput. Oper. Res. **69**, 40–47 (2016)
21. Vidal, T.: Hybrid genetic search for the CVRP: open-source implementation and SWAP* neighborhood. Comput. Oper. Res. **140**, 105643 (2022)
22. Vidal, T., Crainic, T.G., Gendreau, M., Prins, C.: A hybrid genetic algorithm with adaptive diversity management for a large class of vehicle routing problems with time-windows. Comput. Oper. Res. **40**(1), 475–489 (2013)
23. Vidal, T., Crainic, T., Gendreau, M., Lahrichi, N., Rei, W.: A hybrid genetic algorithm for multidepot and periodic vehicle routing problems. Oper. Res. **60**(3), 611–624 (2012)
24. Vidal, T., Crainic, T., Gendreau, M., Prins, C.: Heuristics for multi-attribute vehicle routing problems: a survey and synthesis. Eur. J. Oper. Res. **231**(1), 1–21 (2013)
25. Vidal, T., Crainic, T., Gendreau, M., Prins, C.: A unified solution framework for multi-attribute vehicle routing problems. Eur. J. Oper. Res. **234**(3), 658–673 (2014)
26. Vidal, T., Laporte, G., Matl, P.: A concise guide to existing and emerging vehicle routing problem variants. Eur. J. Oper. Res. **286**, 401–416 (2020)
27. Xin, L., Song, W., Cao, Z., Zhang, J.: NeuroLKH: combining deep learning model with Lin-Kernighan-Helsgaun heuristic for solving the traveling salesman problem. In: Advances in Neural Information Processing Systems, vol. 34, pp. 7472–7483 (2021)
28. Yolcu, E., Poczos, B.: Learning local search heuristics for Boolean satisfiability. In: Advances in Neural Information Processing Systems, vol. 32, pp. 7992–8003 (2019)

Getting Away with More Network Pruning: From Sparsity to Geometry and Linear Regions

Junyang Cai[1], Khai-Nguyen Nguyen[1], Nishant Shrestha[1], Aidan Good[1], Ruisen Tu[1], Xin Yu[2], Shandian Zhe[2], and Thiago Serra[1(✉)]

[1] Bucknell University, Lewisburg, USA
{jc092,nkn002,ns037,wag011,rt024,thiago.serra}@bucknell.edu
[2] University of Utah, Salt Lake City, USA
xin.yu@utah.edu, zhe@cs.utah.edu

Abstract. One surprising trait of neural networks is the extent to which their connections can be pruned with little to no effect on accuracy. But when we cross a critical level of parameter sparsity, pruning any further leads to a sudden drop in accuracy. This drop plausibly reflects a loss in model complexity, which we aim to avoid. In this work, we explore how sparsity also affects the geometry of the linear regions defined by a neural network, and consequently reduces the expected maximum number of linear regions based on the architecture. We observe that pruning affects accuracy similarly to how sparsity affects the number of linear regions and our proposed bound for the maximum number. Conversely, we find out that selecting the sparsity across layers to maximize our bound very often improves accuracy in comparison to pruning as much with the same sparsity in all layers, thereby providing us guidance on where to prune.

Keywords: Model complexity · Network pruning · Solution counting

1 Introduction

In deep learning, there are often good results with little justification and good justifications with few results. Network pruning exemplifies the former: we can easily prune half or more of the connections of a neural network without affecting the resulting accuracy, but we may have difficulty explaining why we can do that. The theory of linear regions exemplifies the latter: we can theoretically design neural networks to express very nuanced functions, but we may end up obtaining much simpler ones in practice. In this paper, we posit that the mysteries of pruning and the wonders of linear regions can complement one another.

When it comes to pruning, we can reasonably argue that reducing the number of parameters improves generalization. While Denil et al. [12] show that the parameters of neural networks can be redundant, it is also known that the

J. Cai and K.-N. Nguyen—Equal contribution.

© The Author(s), under exclusive license to Springer Nature Switzerland AG 2023
A. A. Cire (Ed.): CPAIOR 2023, LNCS 13884, pp. 200–218, 2023.
https://doi.org/10.1007/978-3-031-33271-5_14

smoother loss landscape of larger neural networks leads to better training convergence [45,66]. Curiously, Jin et al. [36] argue that pruning also smooths the loss function, which consequently improves convergence during fine tuning—the additional training performed after pruning the network. However, it remains unclear to what extent we can prune without ultimately affecting accuracy, which is an important concern since a machine learning model with fewer parameters can be deployed more easily in environments with limited hardware.

The survey by Hoefler et al. [31] illustrates that a moderate amount of pruning typically improves accuracy while further pruning may lead to a substantial decrease in accuracy, whereas Liebenwein et al. [46] show that this tolerable amount of pruning depends on the task for which the network is trained. In terms of what to prune, another survey by Blalock et al. [6] observes that most approaches consist of either removing parameters with the smallest absolute value [14,16,22–24,28,35,44,48,54,67]; or removing parameters with smallest expected impact on the output [4,13,29,30,40,42,43,47,50,64,72,73,76,79,80, 82], to which we can add the special case of exact compression [18,60,63,65].

While most work on this topic has helped us prune more with a lesser impact on accuracy, fairness studies recently debuted by Hooker et al. [32] have focused instead on the impact of pruning on recall—the ability of a network to correctly identify samples as belonging to a certain class. Recall tends to be more severely affected by pruning in classes and features that are underrepresented in the dataset [32,33,56], which Tran et al. [70] attribute to differences across such groups in gradient norms and Hessian matrices of the loss function. In turn, Good et al. [20] showed that such recall distortions may also occur in balanced datasets, but in a more nuanced form: moderate pruning leads to comparable or better accuracy while reducing differences in recall, whereas excessive pruning leads to lower accuracy while increasing differences in recall. Hence, avoiding a significant loss in accuracy due to pruning is also relevant for fairness.

Overall, network pruning studies have been mainly driven by one question: **how can we get away with more network pruning?** Before we get there with our approach, let us consider the other side of the coin in our narrative.

When it comes to the theory of linear regions, we can reasonably argue that the number of linear regions may represent the expressiveness of a neural network—and therefore relate to its ability to classify more complex data. We have learned that a neural network can be a factored representation of functions that are substantially more complex than the activation function of each neuron. This theory is applicable to networks in which the neurons have piecewise linear activations, and consequently the networks represent a piecewise linear function in which the number of pieces—or linear regions—may grow polynomially on the width and exponentially on the depth of the network [52,57]. When the activation function is the Rectified Linear Unit (ReLU) [19,55], each linear region corresponds to a different configuration of active and inactive neurons. For geometric reasons that we discuss later, not every such configuration is feasible.

The study of linear regions bears some resemblance to universal approximation results, which have shown that most functions can be approximated to arbitrary precision with sufficiently wide neural networks [10,17,34]. These results were extended in [78] to the currently more popular ReLU activation and later focused on networks with limited width but arbitrarily large depth [27,49]. In comparison to universal approximation, the theory of linear regions tells us what piecewise linear functions are possible to represent—and thus what other functions can be approximated with them—in a context of limited resources translated as both the number of layers and the width of each layer.

Most of the literature is focused on fully-connected feedforward networks using the ReLU activation function, which will be our focus on this paper as well. Nevertheless, there are also adaptations and extensions of such results for convolutional networks by [77] and for maxout networks [21] by [52,53,62,71].

Several papers have shown that the right choice of parameters may lead to an astronomical number of linear regions [3,52,62,68], while other papers have shown that the maximum number of linear regions can be affected by narrow layers [51], the number of active neurons across different linear regions [62], and the parameters of the network [61]. Despite the exponential growth in depth, Serra et al. [62] observe that a shallow network may in some cases yield more linear regions among architectures with the same number of neurons. Whereas the number of linear regions among networks of similar architecture relates to the accuracy of the networks [62], Hanin and Rolnick [25,26] show that the typical initialization and subsequent training of neural networks is unlikely to yield the expressive number of linear regions that have been reported elsewhere.

These contrasting results lead to another question: **is the network complexity in terms of linear regions relevant to accuracy if trained models are typically much less expressive in practice?** Now that you have read both sides of our narrative, you may have guessed where we are heading.

We posit that these two topics—network pruning and the theory of linear regions—can be combined. Namely, that the latter can guide us on how to prune neural networks, since it can be a proxy to model complexity.

But we must first address the paradox in our second question. As observed by Hanin and Rolnick [25], perturbing the parameters of networks designed to maximize the number of linear regions, such as the one by Telgarsky [68], leads to a sudden drop on the number of linear regions. Our interpretation is that every architecture has a probability distribution for the number of linear regions. If by perturbing these especially designed constructions we obtain networks with much smaller numbers, we may infer that these constructions correspond to the tail of that distribution. However, if certain architectural choices lead to much larger numbers of linear regions at best, we may also conjecture that the entire distribution shifts accordingly, and thus that even the ordinary trained network might be more expressive if shaped with the potential number of linear regions in mind. Hence, we conjecture the architectural choices aimed at maximizing the number of linear regions may lead better performing networks.

That brings us to a gap in the literature: to the best of our understanding, there is no prior work on how network pruning affects the number of linear regions. We take the path that we believe would bring the most insight, which consists of revisiting—under the lenses of sparsity – the factors that may limit the maximum number of linear regions based on the neural network architecture.

In summary, this paper presents the following contributions:

(i) We prove an upper bound on the expected number of linear regions over the ways in which weight matrices might be pruned, which refines the bound in [62] to sparsified weight matrices (Sect. 3).
(ii) We introduce a network pruning technique based on choosing the density of each layer for increasing the potential number of linear regions (Sect. 4).
(iii) We propose a method based on Mixed-Integer Linear Programming (MILP) to count linear regions on input subspaces of arbitrary dimension, which generalizes the cases of unidimensional [25] and bidimensional [26] inputs; this MILP formulation includes a new constraint in comparison to [62] for correctly counting linear regions in general (Sect. 5).

2 Notation

In this paper, we study the linear regions defined by the fully-connected layers of feedforward networks. For simplicity, we assume that the entire network consists of such layers and that each neuron has a ReLU activation function, hence being denoted as a *rectifier network*. However, our results can be extended to the case in which the fully-connected layers are preceded by convolutional layers, and in fact our experiments show their applicability in that context. We also abstract the fact that fully-connected layers are often followed by a softmax layer.

We assume that the neural network has an input $\boldsymbol{x} = [x_1 \; x_2 \; \dots \; x_{n_0}]^T$ from a bounded domain $\mathbb{X}$ and corresponding output $\boldsymbol{y} = [y_1 \; y_2 \; \dots \; y_m]^T$, and each hidden layer $l \in \mathbb{L} = \{1, 2, \dots, L\}$ has output $\boldsymbol{h}^l = [h_1^l \; h_2^l \dots h_{n_l}^l]^T$ from neurons indexed by $i \in \mathbb{N}_l = \{1, 2, \dots, n_l\}$. Let $\boldsymbol{W}^l$ be the $n_l \times n_{l-1}$ matrix where each row corresponds to the weights of a neuron of layer l, $\boldsymbol{W}_i^l$ the i-th row of $\boldsymbol{W}^l$, and $\boldsymbol{b}^l$ the vector of biases associated with the units in layer l. With $\boldsymbol{h}^0$ for $\boldsymbol{x}$ and $\boldsymbol{h}^{L+1}$ for $\boldsymbol{y}$, the output of each unit i in layer l consists of an affine function $g_i^l = \boldsymbol{W}_i^l \boldsymbol{h}^{l-1} + \boldsymbol{b}_i^l$ followed by the ReLU activation $h_i^l = \max\{0, g_i^l\}$. We denote the neuron *active* when $h_i^l = g_i^l > 0$ and *inactive* when $h_i^l = 0$ and $g_i^l < 0$. We explain later in the paper how we consider the special case in which $h_i^l = g_i^l = 0$.

3 The Linear Regions of Pruned Neural Networks

In rectifier networks, small perturbations of a given input produce a linear change on the output before the softmax layer. This happens because the neurons that are active and inactive for the original input remain in the same state if the perturbation is sufficiently small. Hence, as long as the neurons remain in their current active or inactive states, the neural network acts as a linear function.

If we consider every configuration of active and inactive neurons that may be triggered by different inputs, then the network acts as a piecewise linear function. The theory of linear regions aims to understand what affects the achievable number of such pieces, which are also known as linear regions. In other words, we are interested in knowing how many different combinations of active and inactive neurons are possible, since they make the network behave differently for inputs that are sufficiently different from one another.

Many factors may affect such number of combinations. We consider below some building blocks leading to an upper bound for pruned networks.

(i) The Activation Hyperplane: Every neuron has an input space corresponding to the output of the neurons from the previous layer, or to the input of the network if the neuron is in the first layer. For the i-th neuron in layer l, that input space corresponds to $\boldsymbol{h}^{l-1}$. The hyperplane $\boldsymbol{W}_i^l \boldsymbol{h}^{l-1} + \boldsymbol{b}_i^l = 0$ defined by the parameters of the neuron separate the inputs in $\boldsymbol{h}^{l-1}$ into two half-spaces. Namely, the inputs that activate the neuron in one side ($\boldsymbol{W}_i^l \boldsymbol{h}^{l-1} + \boldsymbol{b}_i^l > 0$) from those that do not activate the neuron in the other side ($\boldsymbol{W}_i^l \boldsymbol{h}^{l-1} + \boldsymbol{b}_i^l < 0$). We discuss in (iii) how we regard inputs on the hyperplane ($\boldsymbol{W}_i^l \boldsymbol{h}^{l-1} + \boldsymbol{b}_i^l = 0$).

(ii) The Hyperplane Arrangement: With every neuron in layer l partitioning $\boldsymbol{h}^{l-1}$ into two half-spaces, our first guess could be that the intersections of these half-spaces would lead the neurons in layer l to partition $\boldsymbol{h}^{l-1}$ into a collection of 2^{n_l} regions [52]. In other words, that there would be one region corresponding to every possible combination of neurons being active or inactive in layer l. However, the maximum number of regions defined in such a way depends on the number of hyperplanes and the dimension of space containing those hyperplanes. Given the number of activation hyperplanes in layer l as n_l and assuming for now that the size of the input space $\boldsymbol{h}^{l-1}$ is n_{l-1}, then the number of linear regions defined by layer l, or N_l, is such that $N_l \leq \sum_{d=0}^{n_{l-1}} \binom{n_l}{d}$ [81]. Since $N_l \ll 2^{n_l}$ when $n_{l-1} \ll n_l$, we note that this bound can be much smaller than initially expected—and that does not cover the other factors discussed in (iv), (v), and (vi).

(iii) The Boundary: Before moving on, we note that the bound above counts the number of full-dimensional regions defined by a collection of hyperplanes in a given space. In other words, the activation hyperplanes define the boundaries of the linear regions and within each linear region the points are such that either $\boldsymbol{W}_i^l \boldsymbol{h}^{l-1} + \boldsymbol{b}_i^l > 0$ or $\boldsymbol{W}_i^l \boldsymbol{h}^{l-1} + \boldsymbol{b}_i^l < 0$ with respect to each neuron i in layer l. Hence, this bound ignores cases in which we would regard $\boldsymbol{W}_i^l \boldsymbol{h}^{l-1} + \boldsymbol{b}_i^l = 0$ as making the neuron inactive when $\boldsymbol{W}_i^l \boldsymbol{h}^{l-1} + \boldsymbol{b}_i^l \geq 0$ for any possible input in $\boldsymbol{h}^{l-1}$, and vice-versa when $\boldsymbol{W}_i^l \boldsymbol{h}^{l-1} + \boldsymbol{b}_i^l \leq 0$, since in either case the linear region defined with $\boldsymbol{W}_i^l \boldsymbol{h}^{l-1} + \boldsymbol{b}_i^l = 0$ would not be full-dimensional and would actually be entirely located on the boundary between other full-dimensional regions.

(iv) Bounding Across Layers: As we add depth to a neural network, every layer of the network breaks each linear region defined so far in even smaller pieces with respect to the input space $\boldsymbol{h}^0$ of the network. One possible bound would be the product of the bounds for each layer l by assuming the size of the input space to be n_{l-1} [58]. That comes with the assumption that every linear region defined by the first $l-1$ layers can be further partitioned by layer l in as many linear regions as possible. However, this partitioning is going to be more detailed in some linear regions than in others because their input space might be very different. The output of a linear region in layer l is defined by a linear transformation with rank at most n_l. The linear transformation would be

$\boldsymbol{h}^l = \boldsymbol{M}^l \boldsymbol{h}^{l-1} + \boldsymbol{d}^l$, where $\boldsymbol{M}^l_i = \boldsymbol{W}^l_i$ and $\boldsymbol{d}^l_i = \boldsymbol{b}^l_i$ if neuron i of layer l is active in the linear region and $\boldsymbol{M}^l_i = \boldsymbol{0}$ and $\boldsymbol{d}^l_i = 0$ otherwise. Hence, the output from a linear region is the composite of the linear transformations in each layer. If layer $l+1$ or any subsequent layer has more than n_l neurons, that would not imply that the dimension of the image from any linear region is greater than n_l since the output of any linear region after layer l is contained in a space with dimension at most $\min\{n_0, n_1, \ldots, n_l\}$ [51]. In fact, the dimension the of image is often much smaller if we consider that the rank of each matrix $\boldsymbol{M}^l_i$ is bound by how many neurons are active in the linear region, and that in only one linear region of a layer we would see all neurons being active [62].

(v) The Effect of Parameters: The value of the parameters may also interfere with the hyperplane arrangement. First, consider the case in which the rank of the weight matrix is smaller than the number of rows. For example, if all activation hyperplanes are parallel to one another and thus the rank of the weight matrix is 1. No matter how many dimensions the input space has, this situation is equivalent to drawing parallel lines in a plane. Hence, n_l neurons would not be able to partition the input space into more than $n_l + 1$ regions. In general, it is as if the dimension of the space being partitioned were equal to the rank of the weight matrix [62]. Second, consider the case in which a neuron is stable, meaning that this neuron is always active or always inactive for any valid input [69]. Not only that would affect the dimension of the image because a stably inactive neuron always outputs zero, but also the effective number of activation hyperplanes: since the activation hyperplane associated with a stable neuron has no inputs to one of its sides, it does not subdivide any linear region [61].

(vi) The Effect of Sparsity: When we start making parameters of the neural network equal to zero through network pruning, we may affect the number of linear regions due to many factors. First, some neurons may become stable. For example, neuron i in layer l becomes stable if $\boldsymbol{W}^l_i = 0$, i.e., if that row of parameters only has zeros, since the bias term alone ends up defining if the neuron is active ($\boldsymbol{b}^l_i > 0$) or inactive ($\boldsymbol{b}^l_i < 0$). That is also likely to happen if only a few parameters are left, such as when all the remaining weights and the bias are all either positive or negative, since the probability of all parameters having the same sign increases significantly as the number of parameters left decrease if we assume that parameters are equally likely to be positive or negative. Second, the rank of the weight matrix $\boldsymbol{W}^l$ may decrease with sparsity. For example, let us suppose that the weight matrix has n rows, n columns, and that there are only n nonzero parameters. Although it is still possible that those n parameters would all be located in distinct rows and columns to result in a full-rank matrix, that would only occur in $\frac{n!}{\binom{n^2}{n}}$ of the cases if we assume every possible arrangement for those n parameters in the n^2 different positions. Hence, we should expect some rank deficiency in the weight matrix even if we do not prune that much. Third, the rank of submatrices on the columns may decrease even if the weight matrix is full row rank. This could happen in the typical case where the number of columns exceeds the number of rows, such as when the number of neurons decreases from layer to layer, and in that case we could replace the number of active neurons with the rank of the submatrix on their columns for the dimension of the output from each linear region in order to obtain a tighter bound.

Based on the discussion above, we propose an expected upper bound on the number of linear regions over the possible sparsity patterns of the weight matrices. We use an expected bound rather than a deterministic one to avoid the unlikely scenarios in which the impact of sparsity is minimal, such as in the previous example with n parameters leading to matrix with rank n. This upper bound considers every possible sparsity pattern in the weight matrix as equally probable, which is an assumption that aligns with random pruning and does not seem to be too strict in our opinion. For simplicity, we assume that every weight of the network has a probability p of not being pruned; or, conversely, a probability $1 - p$ of being pruned. We denote p as the network *density*.

Moreover, we focus on the second effect of sparsity—through a decrease on the rank of the weight matrix—for two reasons: (1) it subsumes part of the first effect when an entire row becomes zero; and (2) we found it to be stronger than the third effect in preliminary comparisons with a bound based on it.

Theorem 1. *Let $R(l, d)$ be the expected maximum number of linear regions that can be defined from layer l to layer L with the dimension of the input to layer l being d; and let $P(k|R, C, S)$ be the probability that a weight matrix having rank k with R rows, C columns, and probability S of each element being nonzero. With p_l as the probability of each parameter in $\boldsymbol{W}^l$ from remaining in the network after pruning—the layer density, then $R(l, d)$ for $l = L$ is at most*

$$\sum_{k=0}^{n_L} P(k|R = n_L, C = n_{L-1}, S = p_L) \sum_{j=0}^{\min\{k,d\}} \binom{n_L}{j}$$

and $R(l, d)$ for $1 \leq l \leq L - 1$ is at most

$$\sum_{k=0}^{n_l} P(k|R = n_l, C = n_{l-1}, S = p_l) \sum_{j=0}^{\min\{k,d\}} \binom{n_l}{j} R(l + 1, \min\{n_l - j, d, k\}).$$

Proof. We begin with a recurrence on the number of linear regions similar to the one in [62]. Namely, let $R(l, d)$ be the maximum number of linear regions that can be defined from layer l to layer L with the dimension of the input to layer l being d, and let $N_{n_l,d,j}$ be the maximum number of regions from partitioning a space of dimension d with n_l activation hyperplanes such that j of the corresponding neurons are active in the resulting subspaces ($|S^l| = j$):

$$R(l, d) = \begin{cases} \sum\limits_{j=0}^{\min\{n_L,d\}} \binom{n_L}{j} & \text{if } l = L, \\ \sum\limits_{j=0}^{n_l} N_{n_l,d,j} R(l + 1, min\{j, d\}) & \text{if } 1 \leq l \leq L - 1 \end{cases} \quad (1)$$

Note that the base case of the recurrence directly uses what we know about the number of linear regions given the number of hyperplanes and the dimension of the space. That bound also applies to $\sum_{j=0}^{n_l} N_{n_l,d,j}$ in the other case from the recurrence. Based on Lemma 5 from [62], $\sum_{j=0}^{n_l} N_{n_l,d,j} \leq \sum_{j=0}^{\min\{n_l,d\}} \binom{n_l}{j}$. Some of these linear regions will have more neurons active than others. In fact, there are at most $\binom{n_l}{j}$ regions with $|S^l| = j$ for each j. In resemblance to BC, we can thus assume that the largest possible number of neurons is active in each linear region defined by layer l for the least impact on the input dimension of the following layers. Since $\binom{n_l}{j} = \binom{n_l}{n_l-j}$, we may conservatively assume that $\binom{n_l}{0}$ linear regions have n_l active neurons, $\binom{n_l}{1}$ linear regions have $n_l - 1$ active neurons, and so on. That implies the following refinement of the recurrence:

$$R(l,d) = \begin{cases} \sum_{j=0}^{\min\{n_L,d\}} \binom{n_L}{j} & \text{if } l = L, \\ \sum_{j=0}^{\min\{n_l,d\}} \binom{n_l}{j} R(l+1, min\{n_l - j, d\}) & \text{if } 1 \leq l \leq L-1 \end{cases} \quad (2)$$

Note that there is a slight change on the recurrence call, by which j is replaced with $n_l - j$, given that we are working backwards from the largest possible number of active neurons n_l with $n_l - j$.

Finally, we account for the rank of the weight matrix upon sparsification. For the base case of $l = L$, we replace n_L from the end of the summation range with the rank k of the weight matrix $\boldsymbol{W}^L$, and then we calculate the expected maximum number of linear regions using the probabilities of rank k having any value from 0 to n_L as

$$\sum_{k=0}^{n_L} P(k|R = n_L, C = n_{L-1}) \sum_{j=0}^{\min\{k,d\}} \binom{n_L}{j},$$

which corresponds to the first expression in the statement. For the case in which $l \in \{1, \dots, L-1\}$, we similarly replace n_l from the end of the summation range with the rank k of the weight matrix $\boldsymbol{W}^l$, and then we calculate the expected maximum number of linear regions using the probabilities of rank k having any value from 0 to n_l as

$$\sum_{k=0}^{n_l} P(k|R = n_l, C = n_{l-1}) \sum_{j=0}^{\min\{k,d\}} \binom{n_l}{j} R^H(l+1, \min\{n_l - j, d, k\}),$$

which corresponds to the second expression in the statement. ■

Please note that the probability of the rank of a sparse matrix is not uniform when the probability of the sparsity patterns is uniform. We discuss how to compute the former from the later as one of the items in Sect. 6.

4 Pruning Based on Linear Regions

Based on Theorem 1, we devise a network pruning strategy for maximizing the number of linear regions subject to the total number of parameters to be pruned. For a global density p reflecting how much should be pruned, we may thus choose a density p_l for each layer l, some of which above and some of which below p if we do not prune uniformly. We illustrate below the simpler case of pruning two hidden layers and not pruning the connections to the output layer, which is the setting used in our experiments. We focused on two layers because there is only one degree of freedom in that case: for any density p_1 that we choose, the density p_2 is implied by p_1 and by the global density p. When there are more layers involved, trying to optimize the upper bound becomes more challenging. If the effect is not as strong, it could be due to issues solving this nonlinear optimization problem rather than with the main idea in the paper.

When pruning two layers, the relevant dimensions for us are the input size n_0 and the layer widths n_1 and n_2. Assuming the typical setting in which $n_0 > n_1 = n_2$, the maximum rank of both weight matrices is limited by the number of rows (n_1 for $\boldsymbol{W}^1$ and n_2 for $\boldsymbol{W}^2$). However, the greater number of columns in $\boldsymbol{W}^1$ (n_0) implies that we should expect the rank of $\boldsymbol{W}^1$ to be greater if $p_1 = p_2$, whereas preserving more nonzero elements in $\boldsymbol{W}^2$ by pruning a little more from $\boldsymbol{W}^1$ may change the probabilities for $\boldsymbol{W}^2$ with little impact on those for $\boldsymbol{W}^1$. In some of our experiments, the second layer actually has more parameters than the first, meaning that we need to consider $p_1 > p_2$ instead of $p_1 < p_2$.

From preliminary experimentation, we indeed observed that (i) pruning more from the layer with more parameters tends to be more advantageous in terms of maximizing the upper bound; and also that (ii) the upper bound can be reasonably approximated by a quadratic function. Hence, we use the extremes consisting of pruning as much as possible from each of the two layers, say $\overline{p}_1$ and $\overline{p}_2$, in addition to the uniform density p in both layers to interpolate the upper bound. If that local maximum of the interpolation is not pruning more from the layer l with more parameters, we search for the density p_l that improves the upper bound the most by uniformly sampling densities from p all the way to $\overline{p}_l$.

5 Counting Linear Regions in Subspaces

Based on the characterization of linear regions in terms of which neurons are active and inactive, we can count the number of linear regions defined by a trained network with a Mixed-Integer Linear Programming (MILP) formulation [62]. Among other things, these formulations have also been used for network verification [9], embedding the relationship between inputs and outputs of a network into optimization problems [5,11,59], identifying stable neurons [69] to facilitate adversarial robustness verification [75] as well as network compression [60,63], and producing counterfactual explanations [37]. Moreover, several studies have analyzed and improved such formulations [1,2,8,15,61,63].

In these formulations, the parameters $\boldsymbol{W}^l$ and $\boldsymbol{b}^l$ of each layer $l \in \mathbb{L}$ are constant while the decision variables are the inputs of the network ($\boldsymbol{x} = \boldsymbol{h}^0 \in \mathbb{X}$), the outputs before and after activation of each feedforward layer ($\boldsymbol{g}^l \in \mathbb{R}^{n_l}$ and $\boldsymbol{h}^l \in \mathbb{R}_+^{n_l}$ for $l \in \mathbb{L}$), and the state of the neurons in each layer ($\boldsymbol{z}^l \in \{0,1\}^{n_l}$ for $l \in \mathbb{L}$). By mapping these variables according to the parameters of the network, we can characterize every possible combination of inputs, outputs, and activation states as distinct solutions of the MILP formulation. For each layer $l \in \mathbb{L}$ and neuron $i \in \mathbb{N}_l$, the following constraints associate the input $\boldsymbol{h}^l$ with the outputs $\boldsymbol{g}_i^l$ and $\boldsymbol{h}_i^l$ as well as with the neuron activation $\boldsymbol{z}_i^l$:

$$\boldsymbol{W}_i^l \boldsymbol{h}^{l-1} + \boldsymbol{b}_i^l = \boldsymbol{g}_i^l \tag{3}$$
$$(\boldsymbol{z}_i^l = 1) \rightarrow \boldsymbol{h}_i^l = \boldsymbol{g}_i^l \tag{4}$$
$$(\boldsymbol{z}_i^l = 0) \rightarrow \boldsymbol{g}_i^l \leq 0 \tag{5}$$
$$(\boldsymbol{z}_i^l = 0) \rightarrow \boldsymbol{h}_i^l = 0 \tag{6}$$
$$\boldsymbol{h}_i^l \geq 0 \tag{7}$$
$$\boldsymbol{z}_i^l \in \{0,1\} \tag{8}$$

The indicator constraints (4)–(6) can be converted to linear inequalities [7].

We can use such a formulation for counting the number of linear regions based on the number of distinct solutions on the binary vectors $\boldsymbol{z}^l$ for $l \in \mathbb{L}$. However, we must first address the implicit simplifying assumption allowing us to assume that a neuron can be either active ($\boldsymbol{z}_i^l = 1$) or inactive ($\boldsymbol{z}_i^l = 0$) when the preactivation output is zero ($\boldsymbol{g}_i^l = 0$) in (3)–(8). We can do so by maximizing the value of a continuous variable that is bounded by the preactivation output of every active neuron and the negated preactivation output of every inactive neuron. In other words, we count the number of solutions on the binary variables for the solutions with positive value for the following formulation:

$$\max \quad f \tag{9}$$
$$\text{s.t.} \quad (3) - (8) \qquad \forall l \in \mathbb{L}, i \in \mathbb{N}_l \tag{10}$$
$$(\boldsymbol{z}_i^l = 1) \rightarrow f \leq \boldsymbol{g}_i^l \qquad \forall l \in \mathbb{L}, i \in \mathbb{N}_l \tag{11}$$
$$(\boldsymbol{z}_i^l = 0) \rightarrow f \leq -\boldsymbol{g}_i^l \qquad \forall l \in \mathbb{L}, i \in \mathbb{N}_l \tag{12}$$
$$\boldsymbol{h}^0 \in \mathbb{X} \tag{13}$$

We note that constraint (12) has not been used in prior work, where it is assumed that the neuron is inactive when $\boldsymbol{g}_i^l = 0$ [61,62]. However, its absence makes the counting of linear regions incompatible with the theory used to bound the number of linear regions, which assumes that only full-dimensional linear regions are valid. Hence, this represents a small correction to count all the linear regions.

Finally, we extend this formulation for counting linear regions on a subspace of the input. This form of counting has been introduced by [25] for 1-dimensional inputs and later extended by [26] to 2-dimensional inputs. Although far from the upper bound, the number of linear regions can still be very large even for networks of modest size, which makes the case for analyzing how neural networks partition subspaces of the input. In prior work, 1 and 2-dimensional inputs have been considered as the affine combination of 1 and 2 samples with the origin, and a geometric algorithm is used for counting the number of linear regions defined. We present an alternative approach by adding the following constraint to the MILP formulation above in order to limit the inputs of the neural network:

$$\boldsymbol{h}^0 = \boldsymbol{p}^0 + \sum_{i=1}^{S} \alpha_i (\boldsymbol{p}^i - \boldsymbol{p}^0) \tag{14}$$

where $\{\boldsymbol{p}^i\}_{i=0}^{S}$ is a set of $S+1$ samples and $\{\alpha_i\}_{i=1}^{S}$ is a set of S continuous variables. One of these samples, say $\boldsymbol{p}^0$, could be chosen to the be origin.

6 Computational Experiments

We ran computational experiments aimed at assessing the following items:

(1) if accuracy after pruning and the number of linear regions are connected;
(2) if this connection also translates to the upper bound from Theorem 1; and
(3) if that bound can guide us on how much to prune from each layer.

Our experiments involved models trained on the datasets MNIST [41], Fashion [74], CIFAR-10 [38], and CIFAR-100 [38]. We used multilayer perceptrons having 20, 100, 200, and 400 neurons in each of their 2 fully-connected layers (denoted as 2×20, 2×100, 2×200, and 2×400), and adaptations of the LeNet [41] and AlexNet [39] architectures. For each choice of dataset and architecture used, we trained and pruned 30 models. Only the fully-connected layers were pruned. In the case of LeNet and AlexNet, we considered the output of the last convolutional layer as the input for upper bound calculations, as if their respective dimensions were $400 \times 128 \times 84$ and $1024 \times 4096 \times 4096$. We removed the weights with smallest absolute value (magnitude pruning), using either the same density p on each layer or choosing different densities while pruning the same number of parameters in total. We discuss other experimental details later. The source code is available at https://github.com/caidog1129/getting_away_with_network_pruning.

Experiment 1: We compared the mean accuracy of networks that are pruned uniformly according to their network density with the number of linear regions on subspaces defined by random samples from the datasets (Fig. 1) as well as with the upper bound with input dimensions matching those subspaces (Fig. 2). We used a simpler architecture (2×20) to keep the number of linear regions small enough to count and a simpler dataset (MNIST) to obtain models with good accuracy. In this experiment, we observe that indeed the number of linear regions drops with network density and consequently with accuracy. However, the most relevant finding is that the upper bound also drops in a similar way, even if its values are much larger. This finding is important because it is actionable: if we compare the upper bound resulting from different pruning strategies, then we may prefer a pruning strategy that leads to a smaller drop in the upper bound. Moreover, it is considerably cheaper to work with the upper bound since we do not need to train neural networks and neither count their linear regions.

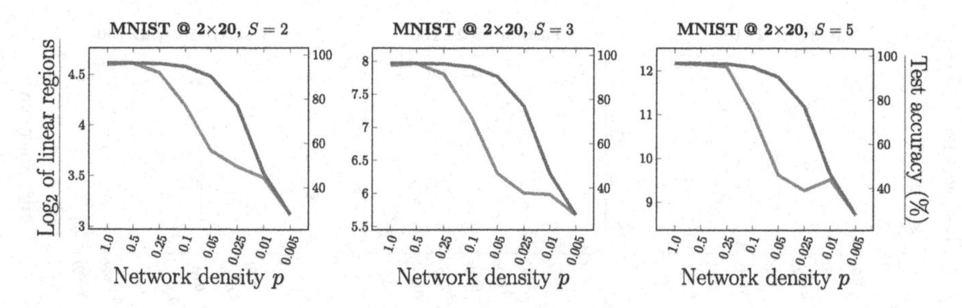

Fig. 1. Comparison between mean number of linear regions on the affine subspace defined by $S = 2$, 3, or 5 sample points (olive curve) and mean test accuracy (blue curve; right y axis) with the same density p used to prune both layers of the networks. (Color figure online)

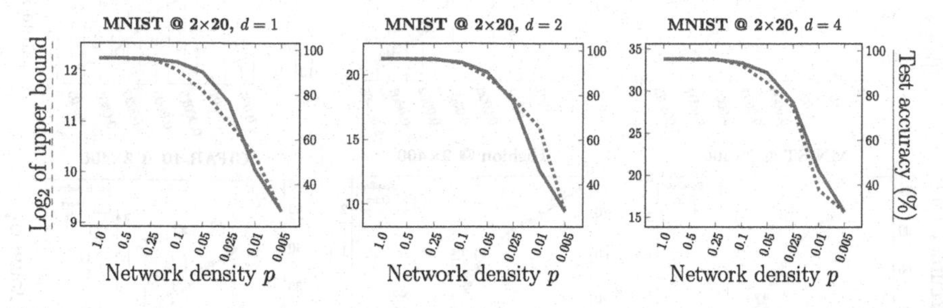

Fig. 2. Comparison between the upper bound from Theorem 1 (dashed blue curve) for input dimension $d = 1$, 2, and 4 (equivalent to $S = 2$, 3, and 5) and mean test accuracy (continuous blue curve; right y axis) for the same networks and densities from Fig. 1. (Color figure online)

Experiment 2: We compared using the same density p in each layer with using per layer densities as described in Sect. 4. We evaluated the simpler datasets (MNIST, Fashion, and CIFAR-10) on the simpler architectures (multilayer perceptrons and LeNet) in Fig. 3, where every combination of dataset and architecture is tested to compare accuracy gain across network sizes and datasets. We set aside the most complex architecture (AlexNet) and the most complex datasets (CIFAR-10 and CIFAR-100) in Fig. 4. In this experiment, we observe that pruning the fully-connected layers differently and oriented by the upper bound indeed leads to more accurate networks. The difference between the pruning strategies is noticeable once the network density starts impacting the network accuracy. We intentionally evaluated network densities leading to very different accuracies and all the way to a complete deterioration of network performance, and we notice that the gain is consistent across all of them. If the number of parameters is similar across fully-connected layers, such as in the case of 2×400, we notice that the gain is smaller because more uniform densities are better for the upper bound. Curiously, we also observe a relatively greater gain with our pruning strategy for CIFAR-10 on multilayer perceptrons.

Additional Details: Each network was trained for 15 epochs using stochastic gradient descent with batch size of 128 and learning rate of 0.01, pruned, and

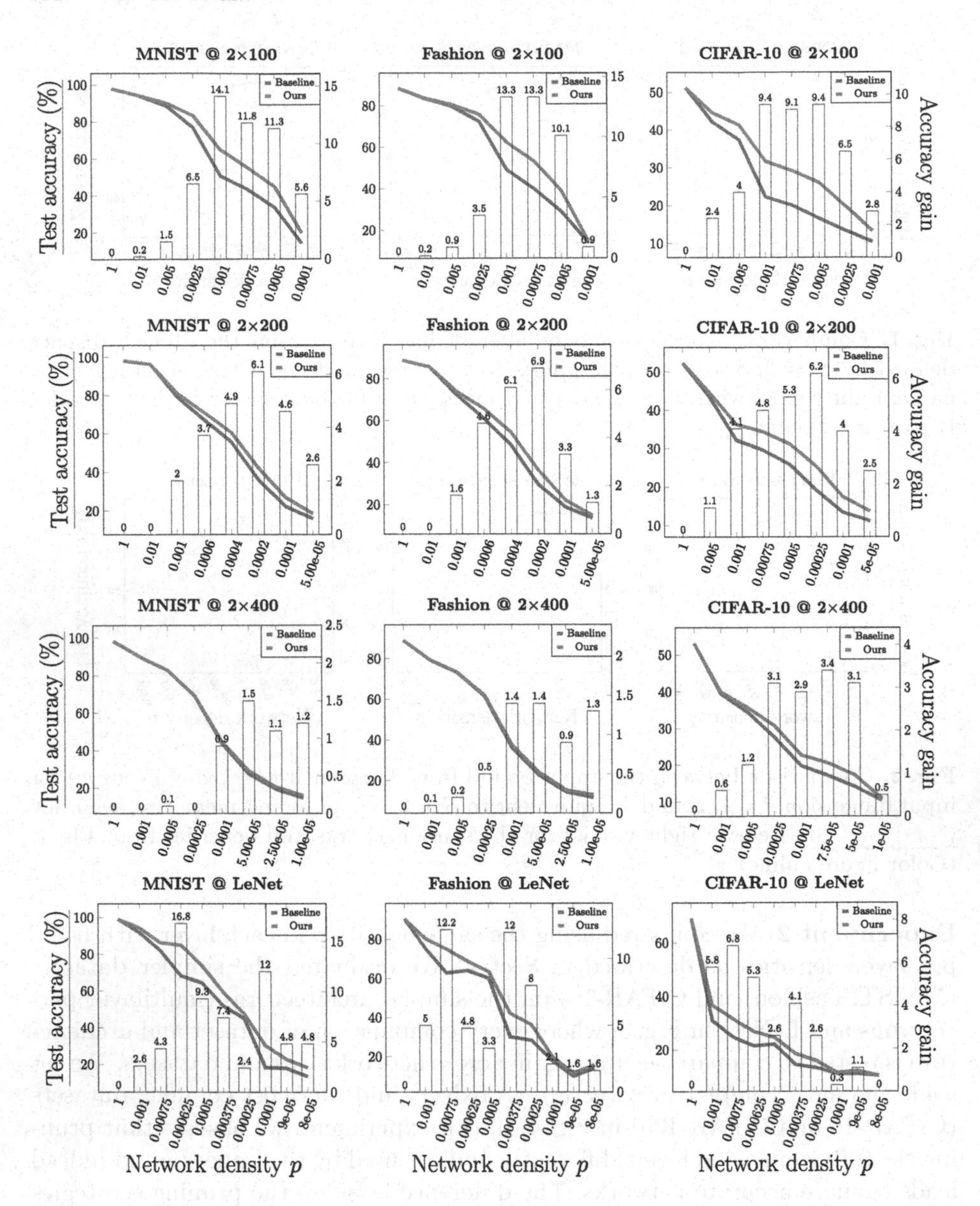

Fig. 3. Comparison between the mean test accuracy as fully-connected layers are pruned using the baseline method and our method with each network density p. In the **baseline** method, the same density is used in all layers (blue curve). In **our method**, layer densities are chosen to maximize the bound from Theorem 1 while pruning the same number of parameters (orange curve). The **accuracy gain** from using our method instead of the baseline is shown in the scaled columns (**maroon** bars; right y axis). Each column refers to a dataset among MNIST, Fashion, and CIFAR-10. Each row refers to an architecture among multilayer perceptrons (2×100, 2×200, and 2×400) and LeNet. We test every combination of dataset and architecture. (Color figure online)

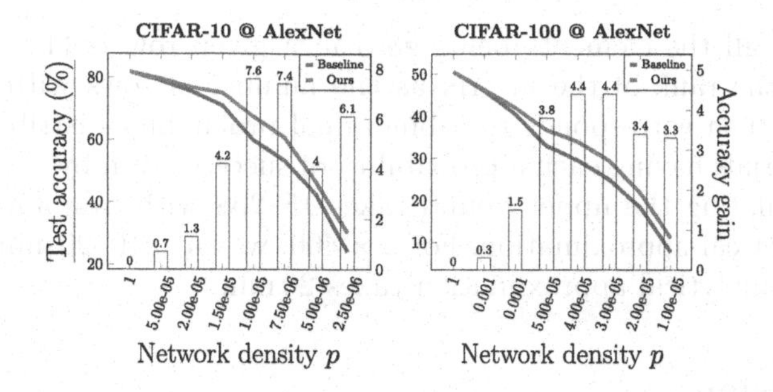

Fig. 4. Comparison between mean test accuracy for the same strategies as in Fig. 3 for the AlexNet architecture, in which we test the datasets CIFAR-10 and CIFAR-100.

then fine-tuned with the same hyperparameters for another 15 epochs. We have opted for magnitude-based pruning due to its simplicity, popularity, and frequent use as a component of more sophisticated pruning algorithms [6,16]. Our implementation is derived from the ShrinkBench framework [6]. In the baseline that we used, we opted for removing a fixed proportion of parameters from each layer (layerwise pruning) to avoid disconnecting the network, which we observed to happen under extreme sparsities if the parameters with smallest absolute value were mostly concentrated in one of the layers. We measured the mean network accuracy before pruning, which corresponds to network density $p = 1$, as well as for another seven values of p. In the experiments in Fig. 3, the choices of p were aimed at gradually degrading the accuracy toward random guessing, which corresponds to accuracy 10% accuracy in those datasets with 10 balanced classes (MNIST, Fashion, and CIFAR-10). In the experiments with AlexNet in Fig. 4, we aimed for a similar decay in performance.

Upper Bound Calculation: Estimating the probabilities $P(k|R, C, S)$ in Theorem 1 is critical to calculate the upper bound. For multilayer perceptrons and LeNet, we generated a sample of matrices with the same shape as the weight matrix for each layer and in which every element is randomly drawn from the normal distribution with mean 0 and standard deviation 1. These matrices were randomly pruned based on the density p, which may have been the same for every layer or may varied per layer as discussed later, and then their rank was calculated. We first generated 50 such matrices for each layer, kept track the minimum and maximum rank values obtained, $\min_r$ and $\max_r$, and then generated more matrices until the number of matrices generated was at least as large as $(\max_r - \min_r + 1) * 50$. For example, 50 matrices are generated if the rank is always the same, and 500 matrices are generated if the rank goes from 11 to 20. Finally, we calculated the probability of each possible rank based on how many times that value was observed in the samples. For example, if 10 out of 500 matrices have rank 11, then we assumed a probability of 2% for the rank of the matrix to be 11. For AlexNet, the time required for sampling is considerably longer. Hence, we resorted to an analytical approximation which is faster but possibly not as accurate. For an $m \times n$ matrix, $m \leq n$, with density p, the probability of all the elements being zero in a given row is $(1-p)^n$. We can overestimate the rank of the matrix as the number of rows with nonzero elements, which then corresponds to a binominal probability distribution with m independent trials having each a probability of success given by $1-(1-p)^n$. For 2×100, calculating the upper bound takes 15–20 s with sampling and 0.5–1 s with the analytical approximation. For 2×400, we have 10–20 min vs. 20 s. For AlexNet, the analytical approximation takes 20 min.

7 Conclusion

In this work, we studied how the theory of linear regions can help us identify how much to prune from each fully-connected feedforward layer of a neural network. First, we proposed an upper bound on the number of linear regions based on the density of the weight matrices when neural networks are pruned. We observe from Fig. 2 that the upper bound is reasonably aligned with the impact of pruning on network accuracy. Second, we proposed a method for counting the number of linear regions on subspaces of arbitrary dimension. In prior work, the counting of linear regions in subspaces is restricted to at most 3 samples and thus dimension 2 [26]. We observe from Fig. 1 to the number of linear regions is also aligned with the impact of pruning on network accuracy—although not as accurately as the upper bound. Third, and most importantly, we leverage this connection between the upper bound and network accuracy under pruning to decide how much to prune from each layer subject to an overall network density p. We observe from Fig. 3 that we obtain considerable gains in accuracy across varied datasets and architectures by pruning from each layer in a proportion that improves the upper bound on the number of linear regions rather than pruning uniformly. These gains are particularly more pronounced when the number of parameters differs across layers. Hence, the gains are understandably smaller when the width of the layers increases (from 100 to 200 and 400) but greater when the size of the input increases (from 784 for MNIST and Fashion to 3,072 for CIFAR-10 with a width of 400). We also obtain positive results with pruning fully connected layers of convolutional networks as illustrated with LeNet and AlexNet, and in future work we intend to investigate how to also make decisions about pruning convolutional filters. Althought we should not discard the possibility of a confounding factor affecting both accuracy and linear regions, our experiments indicate that the potential number of linear regions can guide us on pruning more from neural networks with less impact on accuracy.

Acknowledgments. We would like to thank Christian Tjandraatmadja, Anh Tran, Tung Tran, Srikumar Ramalingam, and the anonymous reviewers for their advice and constructive feedback. Junyang Cai, Khai-Nguyen Nguyen, Nishant Shrestha, Aidan Good, Ruisen Tu, and Thiago Serra were supported by the National Science Foundation (NSF) award IIS 2104583. Xin Yu and Shandian Zhe were supported by the NSF CAREER award IIS 2046295.

References

1. Anderson, R., Huchette, J., Ma, W., Tjandraatmadja, C., Vielma, J.P.: Strong mixed-integer programming formulations for trained neural networks. Math. Program. **183**(1), 3–39 (2020). https://doi.org/10.1007/s10107-020-01474-5
2. Anderson, R., Huchette, J., Tjandraatmadja, C., Vielma, J.P.: Strong mixed-integer programming formulations for trained neural networks. In: Lodi, A., Nagarajan, V. (eds.) IPCO 2019. LNCS, vol. 11480, pp. 27–42. Springer, Cham (2019). https://doi.org/10.1007/978-3-030-17953-3_3
3. Arora, R., Basu, A., Mianjy, P., Mukherjee, A.: Understanding deep neural networks with rectified linear units. In: ICLR (2018)
4. Baykal, C., Liebenwein, L., Gilitschenski, I., Feldman, D., Rus, D.: Data-dependent coresets for compressing neural networks with applications to generalization bounds. In: ICLR (2019)
5. Bergman, D., Huang, T., Brooks, P., Lodi, A., Raghunathan, A.: JANOS: an integrated predictive and prescriptive modeling framework. INFORMS J. Comput. **34**, 807–816 (2022)
6. Blalock, D., Ortiz, J., Frankle, J., Guttag, J.: What is the state of neural network pruning? In: MLSys (2020)
7. Bonami, P., Lodi, A., Tramontani, A., Wiese, S.: On mathematical programming with indicator constraints. Math. Program. **151**(1), 191–223 (2015). https://doi.org/10.1007/s10107-015-0891-4
8. Botoeva, E., Kouvaros, P., Kronqvist, J., Lomuscio, A., Misener, R.: Efficient verification of ReLU-based neural networks via dependency analysis. In: AAAI (2020)
9. Cheng, C.-H., Nührenberg, G., Ruess, H.: Maximum resilience of artificial neural networks. In: D'Souza, D., Narayan Kumar, K. (eds.) ATVA 2017. LNCS, vol. 10482, pp. 251–268. Springer, Cham (2017). https://doi.org/10.1007/978-3-319-68167-2_18
10. Cybenko, G.: Approximation by superpositions of a sigmoidal function. Math. Control Signals Syst. **2**, 303–314 (1989). https://doi.org/10.1007/BF02551274
11. Delarue, A., Anderson, R., Tjandraatmadja, C.: Reinforcement learning with combinatorial actions: an application to vehicle routing. In: NeurIPS (2020)
12. Denil, M., Shakibi, B., Dinh, L., Ranzato, M., Freitas, N.: Predicting parameters in deep learning. In: NeurIPS (2013)
13. Dong, X., Chen, S., Pan, S.: Learning to prune deep neural networks via layer-wise optimal brain surgeon. In: NeurIPS (2017)
14. Elesedy, B., Kanade, V., Teh, Y.W.: Lottery tickets in linear models: an analysis of iterative magnitude pruning (2020)
15. Fischetti, M., Jo, J.: Deep neural networks and mixed integer linear optimization. Constraints **23**(3), 296–309 (2018). https://doi.org/10.1007/s10601-018-9285-6
16. Frankle, J., Carbin, M.: The lottery ticket hypothesis: finding sparse, trainable neural networks. In: ICLR (2019)
17. Funahashi, K.I.: On the approximate realization of continuous mappings by neural networks. Neural Netw. **2**(3) (1989)
18. Ganev, I., Walters, R.: Model compression via symmetries of the parameter space (2022). https://openreview.net/forum?id=8MN_GH4Ckp4
19. Glorot, X., Bordes, A., Bengio, Y.: Deep sparse rectifier neural networks. In: AISTATS (2011)
20. Good, A., et al.: Recall distortion in neural network pruning and the undecayed pruning algorithm. In: NeurIPS (2022)

21. Goodfellow, I., Warde-Farley, D., Mirza, M., Courville, A., Bengio, Y.: Maxout networks. In: ICML (2013)
22. Gordon, M., Duh, K., Andrews, N.: Compressing BERT: studying the effects of weight pruning on transfer learning. In: Rep4NLP Workshop (2020)
23. Han, S., Mao, H., Dally, W.: Deep compression: compressing deep neural networks with pruning, trained quantization and Huffman coding. In: ICLR (2016)
24. Han, S., Pool, J., Tran, J., Dally, W.: Learning both weights and connections for efficient neural network. In: NeurIPS (2015)
25. Hanin, B., Rolnick, D.: Complexity of linear regions in deep networks. In: ICML (2019)
26. Hanin, B., Rolnick, D.: Deep ReLU networks have surprisingly few activation patterns. In: NeurIPS (2019)
27. Hanin, B., Sellke, M.: Approximating continuous functions by ReLU nets of minimal width. arXiv:1710.11278 (2017)
28. Hanson, S., Pratt, L.: Comparing biases for minimal network construction with back-propagation. In: NeurIPS (1988)
29. Hassibi, B., Stork, D.: Second order derivatives for network pruning: optimal Brain Surgeon. In: NeurIPS (1992)
30. Hassibi, B., Stork, D., Wolff, G.: Optimal brain surgeon and general network pruning. In: IEEE International Conference on Neural Networks (1993)
31. Hoefler, T., Alistarh, D., Ben-Nun, T., Dryden, N., Peste, A.: Sparsity in deep learning: pruning and growth for efficient inference and training in neural networks. arXiv:2102.00554 (2021)
32. Hooker, S., Courville, A., Clark, G., Dauphin, Y., Frome, A.: What do compressed deep neural networks forget? arXiv:1911.05248 (2019)
33. Hooker, S., Moorosi, N., Clark, G., Bengio, S., Denton, E.: Characterising bias in compressed models. arXiv:2010.03058 (2020)
34. Hornik, K., Stinchcombe, M., White, H.: Multilayer feedforward networks are universal approximators. Neural Netw. **2**(5) (1989)
35. Janowsky, S.: Pruning versus clipping in neural networks. Phys. Rev. A (1989)
36. Jin, T., Roy, D., Carbin, M., Frankle, J., Dziugaite, G.: On neural network pruning's effect on generalization. In: NeurIPS (2022)
37. Kanamori, K., Takagi, T., Kobayashi, K., Ike, Y., Uemura, K., Arimura, H.: Ordered counterfactual explanation by mixed-integer linear optimization. In: AAAI (2021)
38. Krizhevsky, A.: Learning multiple layers of features from tiny images. Technical report, University of Toronto (2009)
39. Krizhevsky, A., Sutskever, I., Hinton, G.E.: ImageNet classification with deep convolutional neural networks. Commun. ACM **60**(6), 84–90 (2017)
40. Lebedev, V., Lempitsky, V.: Fast ConvNets using group-wise brain damage. In: CVPR (2016)
41. LeCun, Y., Bottou, L., Bengio, Y., Haffner, P.: Gradient-based learning applied to document recognition. In: Proceedings of the IEEE (1998)
42. LeCun, Y., Denker, J., Solla, S.: Optimal brain damage. In: NeurIPS (1989)
43. Lee, N., Ajanthan, T., Torr, P.: SNIP: single-shot network pruning based on connection sensitivity. In: ICLR (2019)
44. Li, H., Kadav, A., Durdanovic, I., Samet, H., Graf, H.: Pruning filters for efficient convnets. In: ICLR (2017)
45. Li, H., Xu, Z., Taylor, G., Studer, C., Goldstein, T.: Visualizing the loss landscape of neural nets. In: NeurIPS (2018)

46. Liebenwein, L., Baykal, C., Carter, B., Gifford, D., Rus, D.: Lost in pruning: the effects of pruning neural networks beyond test accuracy. In: MLSys (2021)
47. Liebenwein, L., Baykal, C., Lang, H., Feldman, D., Rus, D.: Provable filter pruning for efficient neural networks. In: ICLR (2020)
48. Liu, S., et al.: Sparse training via boosting pruning plasticity with neuroregeneration. In: NeurIPS (2021)
49. Lu, Z., Pu, H., Wang, F., Hu, Z., Wang, L.: The expressive power of neural networks: a view from the width. In: NeurIPS (2017)
50. Molchanov, P., Tyree, S., Karras, T., Aila, T., Kautz, J.: Pruning convolutional neural networks for resource efficient inference. In: ICLR (2017)
51. Montúfar, G.: Notes on the number of linear regions of deep neural networks. In: SampTA (2017)
52. Montúfar, G., Pascanu, R., Cho, K., Bengio, Y.: On the number of linear regions of deep neural networks. In: NeurIPS (2014)
53. Montúfar, G., Ren, Y., Zhang, L.: Sharp bounds for the number of regions of maxout networks and vertices of Minkowski sums (2021)
54. Mozer, M., Smolensky, P.: Using relevance to reduce network size automatically. Connection Sci. (1989)
55. Nair, V., Hinton, G.: Rectified linear units improve restricted Boltzmann machines. In: ICML (2010)
56. Paganini, M.: Prune responsibly. arXiv:2009.09936 (2020)
57. Pascanu, R., Montúfar, G., Bengio, Y.: On the number of response regions of deep feedforward networks with piecewise linear activations. In: ICLR (2014)
58. Raghu, M., Poole, B., Kleinberg, J., Ganguli, S., Dickstein, J.: On the expressive power of deep neural networks. In: ICML (2017)
59. Say, B., Wu, G., Zhou, Y., Sanner, S.: Nonlinear hybrid planning with deep net learned transition models and mixed-integer linear programming. In: IJCAI (2017)
60. Serra, T., Kumar, A., Ramalingam, S.: Lossless compression of deep neural networks. In: Hebrard, E., Musliu, N. (eds.) CPAIOR 2020. LNCS, vol. 12296, pp. 417–430. Springer, Cham (2020). https://doi.org/10.1007/978-3-030-58942-4_27
61. Serra, T., Ramalingam, S.: Empirical bounds on linear regions of deep rectifier networks. In: AAAI (2020)
62. Serra, T., Tjandraatmadja, C., Ramalingam, S.: Bounding and counting linear regions of deep neural networks. In: ICML (2018)
63. Serra, T., Yu, X., Kumar, A., Ramalingam, S.: Scaling up exact neural network compression by ReLU stability. In: NeurIPS (2021)
64. Singh, S.P., Alistarh, D.: WoodFisher: efficient second-order approximation for neural network compression. In: NeurIPS (2020)
65. Sourek, G., Zelezny, F.: Lossless compression of structured convolutional models via lifting. In: ICLR (2021)
66. Sun, R., Li, D., Liang, S., Ding, T., Srikant, R.: The global landscape of neural networks: an overview. IEEE Signal Process. Mag. **37**(5), 95–108 (2020)
67. Tanaka, H., Kunin, D., Yamins, D., Ganguli, S.: Pruning neural networks without any data by iteratively conserving synaptic flow. In: NeurIPS (2020)
68. Telgarsky, M.: Representation benefits of deep feedforward networks (2015)
69. Tjeng, V., Xiao, K., Tedrake, R.: Evaluating robustness of neural networks with mixed integer programming. In: ICLR (2019)
70. Tran, C., Fioretto, F., Kim, J.E., Naidu, R.: Pruning has a disparate impact on model accuracy. In: NeurIPS (2022)
71. Tseran, H., Montúfar, G.: On the expected complexity of maxout networks. In: NeurIPS (2021)

72. Wang, C., Grosse, R., Fidler, S., Zhang, G.: EigenDamage: structured pruning in the Kronecker-factored eigenbasis. In: ICML (2019)
73. Wang, C., Zhang, G., Grosse, R.: Picking winning tickets before training by preserving gradient flow. In: ICLR (2020)
74. Xiao, H., Rasul, K., Vollgraf, R.: Fashion-MNIST: a novel image dataset for benchmarking machine learning algorithms. arXiv:1708.07747 (2017)
75. Xiao, K., Tjeng, V., Shafiullah, N., Madry, A.: Training for faster adversarial robustness verification via inducing ReLU stability. In: ICLR (2019)
76. Xing, X., Sha, L., Hong, P., Shang, Z., Liu, J.: Probabilistic connection importance inference and lossless compression of deep neural networks. In: ICLR (2020)
77. Xiong, H., Huang, L., Yu, M., Liu, L., Zhu, F., Shao, L.: On the number of linear regions of convolutional neural networks. In: ICML (2020)
78. Yarotsky, D.: Error bounds for approximations with deep ReLU networks. Neural Netw. **94** (2017)
79. Yu, R., et al.: NISP: pruning networks using neuron importance score propagation. In: CVPR (2018)
80. Yu, X., Serra, T., Ramalingam, S., Zhe, S.: The combinatorial brain surgeon: pruning weights that cancel one another in neural networks. In: ICML (2022)
81. Zaslavsky, T.: Facing up to arrangements: face-count formulas for partitions of space by hyperplanes. Am. Math. Soc. (1975)
82. Zeng, W., Urtasun, R.: MLPrune: multi-layer pruning for automated neural network compression (2018)

OAMIP: Optimizing ANN Architectures Using Mixed-Integer Programming

Mostafa ElAraby[1,3](✉), Guy Wolf[1,4], and Margarida Carvalho[2,3]

[1] Mila – Quebec AI institute, Montreal, Canada
moustafa.elarabi@Umontrea.ca
[2] CIRRELT, Montreal, Canada
[3] Department of Computer Science and Operations Research, Université de Montréal, Montreal, Canada
[4] Department of Mathematics and Statistics, Université de Montréal, Montreal, QC, Canada

Abstract. In this work, we concentrate on the problem of finding a set of neurons in a trained neural network whose pruning leads to a marginal loss in accuracy. To this end, we introduce *Optimizing ANN Architectures using Mixed-Integer Programming* (OAMIP) to identify critical neurons and prune non-critical ones. The proposed OAMIP uses a Mixed-Integer Program (MIP) to assign importance scores to each neuron in deep neural network architectures. The impact of simultaneous neuron pruning on the main learning tasks guides the neurons' scores. By carefully devising the objective function of the MIP, we drive the solver to minimize the number of critical neurons (i.e., with high importance score) that maintain the overall accuracy of the trained neural network. Our formulation identifies optimized sub-network architectures that generalize across different datasets, a phenomenon known as lottery ticket optimization. This optimized architecture not only performs well on a single dataset but also generalizes across multiple ones upon retraining of network weights. Additionally, we present a scalable implementation of our pruning methodology by decoupling the importance scores across layers using auxiliary networks. Finally, we validate our approach experimentally, showing its ability to generalize on different datasets and architectures.

Keywords: Pruning Neural Networks · Mixed Integer Programming · Neurons Ranking · Sparse Neural Networks

1 Introduction

Deep learning has proven its power to solve complex tasks and to achieve state-of-the-art results in various domains such as image classification, speech recognition, machine translation, robotics and control [6,24]. Over-parameterized artificial neural networks (ANN), which have more parameters than the training

G. Wolf and M. Carvalho—Equal contribution.

© The Author(s), under exclusive license to Springer Nature Switzerland AG 2023
A. A. Cire (Ed.): CPAIOR 2023, LNCS 13884, pp. 219–237, 2023.
https://doi.org/10.1007/978-3-031-33271-5_15

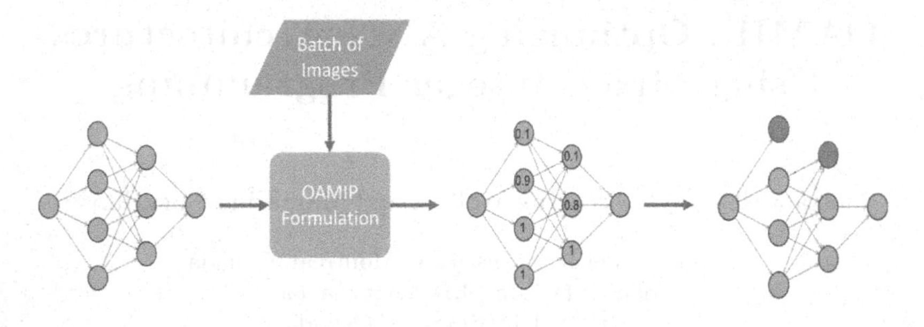

Fig. 1. The generic flow of OAMIP used to remove neurons with an importance score below a specific threshold.

samples, can be used to achieve state-of-the-art results in various tasks [39,57]. However, the large number of parameters comes at the expense of computational cost in terms of memory footprint, training time, and inference time on resource-limited devices.

In this context, the pruning of neurons in an over-parameterized neural model has been an active area of research, enabling the increase of computational efficiency and the uncovering of sub-networks with marginal (or even no) loss in the network's predictive capacity [1,9,17,28,41,42,45,50,51,56]. The typical sparsification procedure involves training a neural model to convergence, computing the parameters' importance, then pruning existing ones using specific criteria, and fine-tuning the neural model to regain its lost accuracy. Existing pruning and neuron ranking procedures [1,9,17,18,27,35,45,56] require iterations of fine-tuning on the sparsified model instead of pruning a pre-trained network directly. Moreover, the evaluation of the generalization of sparsified models across different datasets is under-explored in existing pruning and neuron ranking procedures [31], which is consistent with the lottery ticket hypothesis [13,34,37].

We remark that modern network architectures often use sparse neuron connectivity and, most notably, convolutional layers in image processing. Indeed, the limited size of the parameter space in such cases increases the effectiveness of network training and enables the learning of meaningful semantic features from the input images [15]. Inspired by the benefits of sparsity in such architecture designs, we aim to leverage the neuron sparsity achieved by our framework, Optimizing ANN Architectures using Mixed-Integer Programming (OAMIP) to obtain optimized neural architectures that can generalize well across different datasets. For this purpose, we create a sparse sub-network by optimizing on one dataset and then training the same architecture, i.e., masked, on another dataset. Our results indicate a promising direction of future research into the utilization of combinatorial optimization for effective automatic architecture tuning to augment handcrafted network architecture design.

Contributions and Paper Organization. In OAMIP, illustrated in Fig. 1, we formalize the notation of *neuron importance score* in a trained neural network and the associated dataset. The neuron importance score reflects how much activity decrease can be inflicted while controlling the loss on the neural network model accuracy. To this end, in Sect. 2, we begin by providing background on the constraints that serve as the basis for our Mixed-Integer Programming (MIP) formulation presented in Sect. 3. Concretely, we propose a MIP that allows the computation of the importance score for each fully connected neuron and convolutional feature map. The error propagation associated with pruning between different layers defines each neuron's importance score. In addition, we also discuss the extension of the MIP constraints for other layers besides ReLU-activated fully connected layers. Section 4 describes OAMIP in detail, namely the integration of the neuron importance scores on the pruning procedure. Here, we also propose a methodology to independently decouple the computation of neuron importance score per layer to represent deeper architectures and, thus, scale up our approach to models like VGG-16 [44]. Furthermore, in Sect. 5, we show OAMIP's robustness to various input data points besides its ability to parallelize the computation of importance score per class. Finally, we show that OAMIP's importance score generalizes well over various datasets complying with the lottery ticket hypothesis [13].

1.1 Related Work

Weight Pruning Methods. Early methods in weight pruning relied on the weight magnitude by disabling the lowest magnitude weights and re-training/fine-tuning the resulting sub-network [16,29,37]. Magnitude-based techniques rely on the intuition that large weight values are more critical during inference than smaller weight values. [36] devised a greedy criteria-based pruning with fine-tuning by back-propagation. The criteria devised are given by the absolute difference between dense and sparse neural model loss (ranker) to avoid a drop in the predictive capacity. [43] developed a framework that computes the neurons' importance at each layer through a single backward pass as an approximation to the interpretability of each neuron during inference. Other related techniques, using different objectives and interpretations of neuron importance, have been presented [1,3,19,20,22,54], and require either fine-tuning to recover the network's performance or dynamic re-training and pruning. Another line of research [10,33,42,49, 53,55] formulates an optimization model to select which neuron to disable without losing performance on the task at hand. With a less conservative perspective but also using an optimization-based model, OAMIP aims to quantify a generalizable per-neuron importance score for either a pre-trained network or at initialization without re-training or fine-tuning the network. Similarly, other pruning procedures aim to avoid the fine-tuning step by pruning the network during initialization. In particular, SNIP [28] and GraSP [51] focus on predicting critical weights during initialization via salience scores and then train the sub-network until convergence. SNIP [28] was the first to investigate the pruning of a network during initialization by computing the connection's sensitivity to an input batch of data through gradient back-propagation. OAMIP can be applied to the network at initialization or after training without requiring a long fine-tuning step.

Lottery Ticket. [13] introduced the lottery ticket theory that shows the existence of a lucky pruned sub-network, a *winning ticket.* The lucky pruned sub-network can be trained effectively with fewer parameters while achieving a marginal loss in accuracy. [37] proposed "one ticket to win them all" for sparsifying n over-parameterized trained neural models based on the lottery hypothesis. Searching for the winning ticket involves pruning the model and disabling some of its sub-networks. The pruned model can be trained on a different dataset using the same initialization (winning ticket), achieving good results. To this end, the dataset used for the pruning phase must be sufficiently large. The lucky sub-network is found by iteratively pruning the lowest magnitude weights and re-training. Another phenomenon discovered in [40,52] was the existence of smaller, high-accuracy models that reside in larger random networks. This phenomenon is called the strong lottery ticket hypothesis, which was proven [34] on ReLU fully connected layers. Furthermore, [51] proposed a technique to select the winning ticket at initialization (before training the ANN) by computing an importance score based on the gradient flow in each unit.

Mixed-Integer Programming. [12] presented a Mixed-Integer Linear Programming big-M formulation to represent trained ReLU neural networks. Later, [4] introduced the strongest possible tightening to the big-M formulation by adding strengthening separation constraints when needed, which reduced the solving time by several orders of magnitude. Recently, [48] presented efficient partitioning strategies that improved solving time. All the proposed formulations are designed to represent trained ReLU ANNs with fixed parameters. In our framework, we use the formulation from [12] since its performance was good due to our tight local variable bounds, and its polynomial number of constraints (while the models in [4,48] are non-compact). The interest of representing an ANN as a MIP lies in its use to evaluate robustness, carry out compression and create adversarial examples for trained ANNs. For instance, [21,47] used a big-M formulation to evaluate the robustness of neural models against adversarial attacks. [55] modeled an extension of the optimal brain surgeon [18], where the goal is to select and remove the weights that have the most negligible impact on the predictive capacity of the network as an Integer Quadratic Program. However, the optimal brain surgeon pruning criteria rely heavily on the weights scale. Moreover, the weights' scale will be sensitive to the architecture used; different normalization layers affect the scale and magnitude of weights in a different way [28]. [42] also used a MIP formulation to maximize the compression of a trained neural network without decreasing predictive accuracy. *Lossless compression* [42] relies on different compression methods, such as removing neurons and folding layers. However, the reported computational experiments lead only to the removal of inactive neurons. OAMIP can identify such neurons and quantify the importance of various neurons with respect to the predictive capacity while pruning neurons that are non-critical across different datasets. The latter means that the sub-networks found by our framework to a specific dataset generalize to others.

2 Preliminaries

Consider layer l of a trained ReLU neural network with $\boldsymbol{W}^l$ as the weight matrix, w_i^l as row i of $\boldsymbol{W}^l$, and b^l as the bias vector. For each input data point x, let h^l be a decision vector denoting the output value of layer l, i.e., $h^l = ReLU(\boldsymbol{W}^l h^{l-1} + b^l)$ for $l > 0$ and $h^0 = x$, and z_i^l be a binary variable taking value 1 if the unit i is active ($w_i^l h^{l-1} + b_i^l \geq 0$) and 0 otherwise. Finally, let L_i^l and U_i^l be constants indicating a valid lower and upper bound for the input of each neuron i in layer l. We discuss the computation of these bounds in Sect. 3.2. For now, we assume that L_i^l and U_i^l are sufficiently small and large numbers, respectively, i.e., the so-called big-M values. Next, we provide the representation of ReLU neural networks of [12]. Although [4] proposed an ideal MIP formulation with an exponential number of facet-defining constraints that can be separated efficiently, we use the formulation by [12], since it performed well in practice for our purpose. For the sake of simplicity, we describe the formulation for one layer l of the model at neuron i and one input data point x:

$$h_i^0 = x_i \tag{1a}$$

$$h_i^l \geq 0, \quad \text{for } l > 0 \tag{1b}$$

$$h_i^l + (1 - z_i^l) L_i^l \leq w_i^l h^{l-1} + b_i^l, \tag{1c}$$

$$h_i^l \leq z_i^l U_i^l, \tag{1d}$$

$$h_i^l \geq w_i^l h^{l-1} + b_i^l, \tag{1e}$$

$$z_i^l \in \{0, 1\}, h_i^l \in \mathbb{R}. \tag{1f}$$

In constraint (1a), the initial decision vector h^0 is forced to be equal to the input x of the first layer. When z_i^l is 0, constraints (1b) and (1d) force h_i^l to be zero, reflecting a non-active neuron. If an entry of z_i^l is 1, then constraints (1c) and (1e) enforce h_i^l to be equal to $w_i^l h^{l-1} + b_i^l$. After formulating the ReLU, if we relax the binary constraint (1f) on z_i^l to $[0, 1]$, we obtain a polyhedron, over which it is easier and faster to optimize. The *quality* (tightness) of such relaxation highly depends on the choice of tight upper and lower bounds, U_i^l, L_i^l. Indeed, the determination of tight bounds reduces the search space and hence, the solving time.

3 Neuron Importance Score

In what follows, we adapt constraints (1) to quantify neurons' importance, we describe the computation of the bounds L_i^l and U_i^l and we discuss the objective function for our MIP. Our goal is to compute importance scores for all layers in the model in an integrated fashion. In fact, [54] has shown that this integrated perspective leads to better predictive accuracy than layer by layer.

3.1 MIP Constraints

In ReLU-activated layers, we keep the previously introduced binary variables z_i^l and continuous variables h_i^l. Recall that these variables are linked to an input data point x, so if more than one data point is considered, copies of these variables must be created. Additionally, we create the continuous decision variables $s_i^l \in [0,1]$ representing neuron i importance score in layer l; contrarily to z_i^l and h_i^l, no copies of s_i^l are created for each input data point. In this way, we proceed to modify the ReLU constraints (1) by adding the neuron importance decision variable s_i^l to constraints (1c) and (1e):

$$h_i^l + (1 - z_i^l)L_i^l \leq w_i^l h^{l-1} + b_i^l - (1 - s_i^l)\max(U_i^l, 0), \tag{2a}$$

$$h_i^l \geq w_i^l h^{l-1} + b_i^l - (1 - s_i^l)\max(U_i^l, 0). \tag{2b}$$

Constraints (2) impose that when neuron i is activated due to the input h^{l-1}, i.e., $z_i^l = 1$, then h_i^l is equal to the right-hand-side of those constraints. This value can be directly decreased by reducing the neuron importance s_i^l. When neuron i is non-active, i.e., $z_i^l = 0$, constraint (2b) becomes irrelevant as its right-hand-side is negative. This fact together with constraints (1b) and (1d), imply that h_i^l is zero. Now, we claim that constraint (2a) allows s_i^l to be zero if that neuron is indeed non-important, i.e., for all possible input data points, neuron i is not activated. This claim can be shown through the following observations. Note that decisions h and z must be replicated for each input data point x as they represent the propagation of x over the neural network. On the other hand, s evaluates the importance of each neuron for the main learning task, and thus, it must be the same for all data input points. Thus, the key ingredients are the bounds L_i^l and U_i^l that are computed for each input data point, as explained in Sect. 3.2. In this way, if U_i^l is non-positive, s_i^l can be zero without interfering with constraints (2). The latter is driven by the objective function derived in Sect. 3.3. We designate a neuron as *critical* with respect to a trained ANN, if its importance score is higher than a predefined threshold, otherwise it is called *non-critical.*

We now discuss other architectures. Concerning convolutional feature maps, we convert them to toeplitz matrices and their input images to vectors. This allows us to use simple matrix multiplication which is computationally efficient and generates the full convolution output. For padded convolution we use only parts of the output of the full convolution, and for strided convolutions we use sum of 1 strided convolution as proposed by [7]. Moreover, we can represent the convolutional layer using the same formulation of fully connected layers presented in (2a). The importance score of convolutional layers is associated with each feature map [30,36].

We represent both max and average (avg) pooling on multi-input units in our MIP formulation. Pooling layers are used to reduce spatial representation of input images by applying an arithmetic operation on each feature map of the previous layer. Avg pooling layers compute the average operation on each feature map of the previous layer l having N^l as the number of neurons. This operation is linear and thus, it can directly be included in the MIP constraints:

$$h^{l+1} = \text{AvgPool}(h_1^l, \cdots, h_{N^l}^l) = \frac{1}{N^l}\sum_{i=1}^{N^l} h_i^l.$$

Max Pooling takes the maximum of each feature map of the previous layer:

$$h^{l+1} = \text{MaxPool}(h_1^l, \cdots, h_{N^l}^l) = \max\{h_1^l, \cdots, h_{N^l}^l\}.$$

This operation can be expressed by introducing a set of binary variables $m_1, \cdots, m_{N^l}$, where $m_i = 1$ implies $x = \text{MaxPool}(h_1^l, \cdots, h_{N^l}^l)$:

$$\sum_{i=1}^{N^l} m_i = 1$$

$$\left.\begin{aligned} x &\geq h_i^l, \\ x &\leq h_i^l m_i + U_i(1 - m_i) \\ m_i &\in \{0, 1\} \end{aligned}\right\} i = 1, \cdots, N^l.$$

3.2 Bound Propagation

In the previous section, we assumed a large upper bound U_i^l and a small lower bound L_i^l. However, using large bounds may lead to long computational times and a loss of freedom to reduce the importance score, as discussed above. In order to overcome these issues, we tailor these bounds accordingly with their respective input point x by considering small perturbations on its value:

$$L^0 = x - \epsilon \tag{3a}$$

$$U^0 = x + \epsilon \tag{3b}$$

$$L^l = \boldsymbol{W}^{(l-)} U^{l-1} + \boldsymbol{W}^{(l+)} L^{l-1} \tag{3c}$$

$$U^l = \boldsymbol{W}^{(l+)} U^{l-1} + \boldsymbol{W}^{(l-)} L^{l-1} \tag{3d}$$

$$\boldsymbol{W}^{(l-)} \triangleq \min(\boldsymbol{W}^{(l)}, 0) \tag{3e}$$

$$\boldsymbol{W}^{(l+)} \triangleq \max(\boldsymbol{W}^{(l)}, 0). \tag{3f}$$

Propagating the initial bounds of the input data points throughout the trained model will create the desired bound using a simple arithmetic interval. The obtained bounds are tight, narrowing the space of feasible solutions.

3.3 MIP Objective

Our framework aims at identifying non-critical neurons without significantly decreasing the predictive accuracy of the pruned ANN. To this end, we combine two optimization objectives.

Our first objective is to maximize the set of neurons sparsified from the trained ANN. Recall that N^l is the number of neurons at layer l, and let n be the number of layers, and $I^l = \sum_{i=1}^{N^l}(s_i^l - 2)$ be the sum of neuron importance scores at layer l with s_i^l scaled down to the range $[-2, -1]$.

In order to create a relation between neurons' importance score in different layers, our objective becomes the maximization of the number of neurons sparsified from the $n-1$ layers with higher score I^l. Hence, we denote $A = \{I^l : l = 1, \dots, n\}$ and formulate the sparsity loss as

$$\text{sparsity} = \frac{\max\limits_{A' \subset A, |A'|=(n-1)} \sum\limits_{I \in A'} I}{\sum_{l=1}^{n} |N^l|}. \tag{4}$$

Here, the goal is to maximize the number of non-critical neurons at each layer relative to the other layers of the trained neural model. Note that only the $n-1$ layers with the most significant importance score will weigh in the objective, allowing to reduce the pruning effort on some layers that will naturally have low scores. The total number of neurons then normalizes the sparsity quantification.

Our second objective is to minimize the loss of *important* information due to the sparsification of the trained neural model. Additionally, we aim for this minimization to be done without relying on the values of the logits, which are closely correlated with the neurons pruned at each layer. Otherwise, this would drive the MIP to simply give a total score of 1 to all neurons to keep the same output logit value. Instead, we formulate this optimization objective using the marginal softmax proposed in [14]. Using marginal softmax allows the solver to focus on minimizing the misclassification error without relying on logit values. Moreover, the scale of logits can be marginally different between the decision vector h^n computed by the MIP with some disabled neurons and the trained neural network predictions. To that end, in the proposed marginal softmax loss, the label with the highest logit value is optimized regardless of its value. Formally, we write the objective

$$\text{softmax} = \sum_{i=1}^{N^n} \log\left[\sum_{c} \exp(h_{i,c}^n)\right] - \sum_{i=1}^{N^n}\sum_{c} Y_{i,c} h_{i,c}^n, \tag{5}$$

where the index c stands for the class label. The softmax marginal objective retains the trained model's correct predictions for the batch of input images x having a one-hot encoded label Y without regard to the logit value. Finally, we combine the two objectives to formulate the loss

$$\text{loss} = \text{sparsity} + \lambda \cdot \text{softmax} \tag{6}$$

as a weighted sum of sparsification regularizer and marginal softmax.

4 OAMIP: Pruning Approach

Given a trained neural network and a dataset, our goal is to identify and prune non-critical neurons based on importance score s_i^l for neuron i at layer l. To this end, we formulated a neural network as a mixed-integer program, including the neuron importance score in its constraints and objective function. Algorithm 1 summarizes the integration of our formulation within a pruning procedure.

Algorithm 1: OAMIP: Optimizing ANN Architectures using a MIP

Require: Trained ANN, dataset D and a threshold.
Ensure: Sub-network selected from the trained ANN.
1: Select a per-class image $D' \subset D$ to be fed into the MIP.
2: Solve the MIP restricted to D' and save the neurons importance scores s.
3: Remove every neuron i from layer l with $s_i^l \leq$ threshold from the ANN.
4: Return pruned ANN (sub-network).

[58] highlights the phenomenon of neural collapse, where features of images from the same distribution in the training set collapse around a class mean and are maximally distant between different classes. Moreover, the neurons that are important for a specific class, as computed on an image, should not change drastically when another image from the same distribution as the training set is used. Besides, using all the training samples as input to the MIP solver is intractable. Hence, we use only a subset of the data points, each representing a class in the classification task for which we aim to approximate the neuron importance score (step 1). Then, OAMIP computes an estimation of the importance score of each neuron (step 2). With a small tuned threshold based on the network's architecture, we mask (prune) non-critical neurons with a score lower than the threshold (step 3). Finally, our proposed framework returns a pruned ANN (sub-network), achieving marginal loss in accuracy.

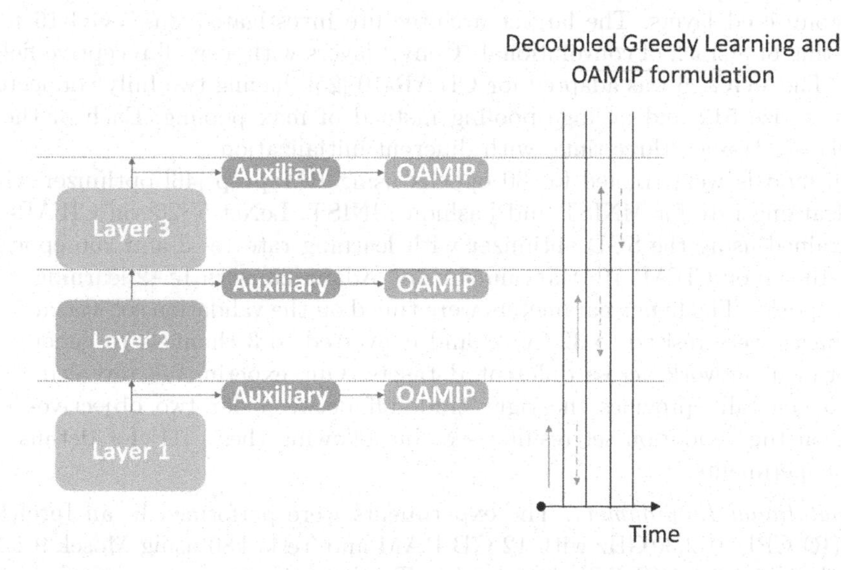

Fig. 2. Illustration of the auxiliary network attached to each sub-module along with the signal backpropagation during training as shown in [5].

The most time-sensitive step of OAMIP is the optimization of the MIP. The number of variables and constraints increases with the number of neurons and input data points. Indeed, if large and realistic ANNs are modeled with our MIP, the computation time for determining importance scores is expected to become very large, as observed in the problem tackled in [12]. To overcome the computational time issue, we propose independent computation of importance scores per layer using auxiliary networks [5]. In particular, we used decoupled greedy learning [5] to train each layer of VGG-16 [44] using a small auxiliary network, and, in this way, we computed the neuron importance score independently on each auxiliary network, as shown in Fig. 2. Then, we fine-tuned the generated masks for one epoch to propagate the errors across them resulting from the independent optimization. Decoupled training of each layer allowed us to represent deep models using the MIP formulation and to parallelize the computation per layer.

5 Empirical Results

This section shows experimentally that *(i)* our approach can efficiently find high-performance sub-networks from ANN architectures, *(ii)* the computed sub-networks generalize well to new datasets, and *(iii)* OAMIP outperforms the state-of-the-art approach SNIP with regards to generalization.

Experimental Setting. We used a simple fully connected 3-layer ANN (FC-3) model, with 300+100 hidden units, from [26], and another simple fully connected 4-layer ANN (FC-4) model, with 200+100+100 hidden units. In addition, we used the convolutional LeNet-5 [26] consisting of two sets of convolutional and average pooling layers, followed by a flattening convolutional layer, then two fully-connected layers. The largest architecture investigated was VGG-16 [44] consisting of a stack of convolutional (Conv.) layers with a small receptive field: 3×3. The VGG-16 was adapted for CIFAR-10 [25], having two fully connected layers of size 512 and average pooling instead of max pooling. Each of these models was trained three times with different initialization.

All models were trained for 30 epochs using RMSprop [46] optimizer with 1e-3 learning rate for MNIST and Fashion MNIST. LeNet-5 [26] on CIFAR-10 was trained using the SGD optimizer with learning rate 1e−2 and 256 epochs. VGG-16 [44] on CIFAR-10 was trained using Adam [23] with 1e−2 learning rate for 30 epochs. The hyper-parameters were tuned on the validation set's accuracy. All images were resized to 32 by 32 and converted to 3 channels to generalize the pruned network across different datasets. Our experiments revealed that $\lambda = 5$ generally provides the right trade-off between our two objectives (6) based on the validation set results; see the following thesis [11] for details on these experiments.

Computational Environment. The experiments were performed in an Intel(R) Xeon(R) CPU @ 2.30 GHz with 12 GB RAM and Tesla k80 using Mosek 9.1.11 [38] solver on top of CVXPY [2,8] and PyTorch 1.3.1[1].

[1] The code can be found here: https://github.com/chair-dsgt/mip-for-ann.

5.1 OAMIP Robustness

We examine the robustness of OAMIP against different batches of input images fed into the MIP, on the implementation of step 2 of OAMIP. Namely, we used 25 randomly sampled balanced images from the validation set. Figure 3 shows that changing the input images used by the MIP to compute neuron importance scores in step 2 resulted in marginal changes in the test accuracy between different batches. We remark that the input batches may contain images that were misclassified by the neural network. In this case, the MIP tries to use the score s to obtain the true label, which explains the variations in the pruning percentage. Furthermore, we show empirically that OAMIP is robust on different convergence levels of the trained neural network as shown in Fig. 4. Hence, we do not need to wait for the ANN to be trained to identify the target sub-network (strong lottery ticket hypothesis theory [34]).

Additionally, we experiment parallelizing per class neuron importance score computation using a balanced and imbalanced set of images per class. For those experiments, we sampled a random number of images per class (IMIDP), then

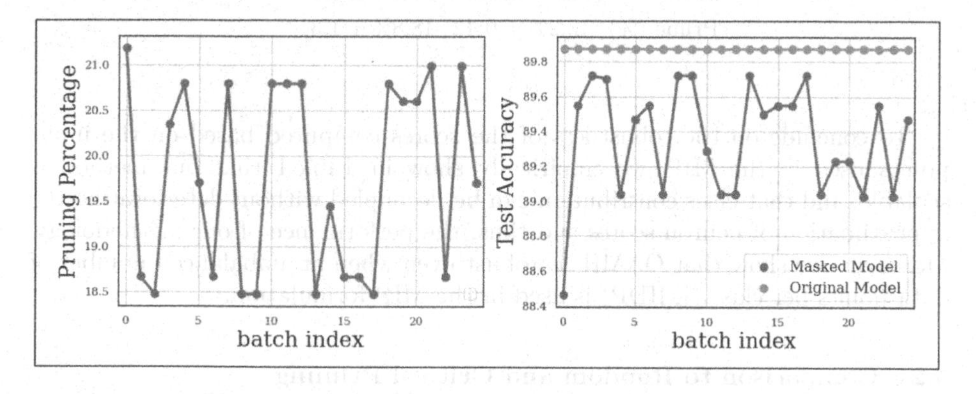

Fig. 3. Effect of changing validation set of input images.

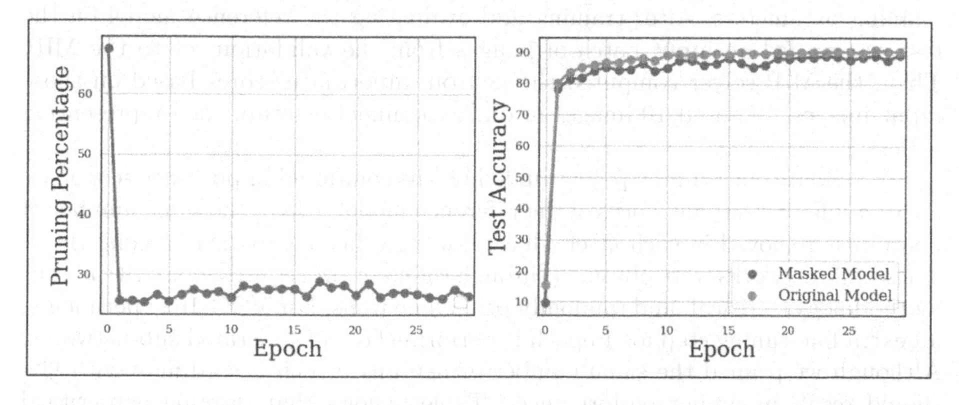

Fig. 4. Evolution of the computed masked sub-network during model training.

we took the average of the computed neuron importance scores from solving the MIP on each class. The obtained sub-networks were compared to solving the MIP with 1 image per class (IDP) and to solving the MIP with balanced images representing all classes (SIM). We achieved comparable results in terms of test accuracy and pruning percentage.

Table 1. Comparing test accuracy of Lenet-5 on imbalanced independent class by class (IMIDP.), balanced independent (IDP.) and simultaneously all classes (SIM) with 0.01 threshold, and $\lambda = 1$.

	MNIST	Fashion-MNIST
Ref	98.8% ± 0.09	89.5% ± 0.3
IDP.	98.6% ± 0.15	87.3% ± 0.3
Prune (%)	19.8% ± 0.18	21.8% ± 0.5
IMIDP.	98.6% ± 0.1	88% ± 0.1
Prune (%)	15% ± 0.1	18.1% ± 0.3
SIM.	98.4% ± 0.3	87.9% ± 0.1
Prune (%)	13.2% ± 0.42	18.8% ± 1.3

To conclude on the robustness of the scores computed based on the input points used in the MIP, we empirically show in Table 1 that our method is scalable, and that class contribution can be decoupled without deteriorating the approximation of neuron scores and thus, the performance of our methodology. Moreover, we show that OAMIP is robust even when an imbalanced number of data points per class (IMIDP) is used in the MIP formulation.

5.2 Comparison to Random and Critical Pruning

We started by training a reference model (REF.) using previously described training parameters. After training and evaluating the reference model on the test set, we fed an input batch of images from the validation set to the MIP. Then, the MIP solver computed the neuron importance scores based on those input images. We used 10 images in our experimental setup, each representing a class.

To validate our pruning policy guided by the computed importance scores, we created different sub-networks of the reference model, where the same number of neurons is removed in each layer, thus allowing a fair comparison among them. These sub-networks were obtained through different procedures: non-critical (our methodology), critical, and randomly pruned neurons. For VGG-16 experiments, an extra fine-tuning step for 1 epoch is performed on all generated sub-networks. Although we pruned the same number of neurons, which accordingly with [32] should result in similar performances, Table 2 shows that pruning non-critical neurons results in marginal loss and gives better performance. On the other hand, we observe a significant drop in the test accuracy when critical or a random set of neurons are removed compared with the reference model. If we fine-tune for just 1 epoch the sub-network obtained through our method, the model's accuracy can surpass the reference model. This is due to the fact that the MIP while computing neuron scores, is solving its marginal softmax (5) on true labels.

Table 2. Pruning results on fully connected (FC-3, FC-4) and convolutional (Lenet-5, VGG-16) network architectures using three different datasets. We compare the test accuracy between the unpruned reference network (REF.), randomly pruned model (RP.), model pruned based on critical neurons selected by the MIP (CP.) and our non-critical pruning approach with (OAMIP + FT) and without (OAMIP) fine-tuning for 1 epoch.

		REF.	RP.	CP.	OAMIP	OAMIP+FT	PRUNE (%)	RUNTIME (s)
MNIST	FC-3	98.1%	83.6%	44.5%	**95.9%**	**97.8%**	44.5%	12 s
		±0.1	±4.6	±7.2	±0.87	±0.2	±7.2	±0.7
	FC-4	97.9%	77.1%	50%	**96.6%**	**97.6%**	42.9%	9*s*
		±0.1	±4.8	±15.8	±0.4	±0.01	±4.5	±0.4
	LENET-5	98.9%	56.9%	38.6%	**98.7%**	**98.9%**	17.2%	1 s
		±0.1	±36.2	±40.8	±0.1	±0.04	±2.4	±0.6
FASHION-MNIST	FC-3	87.7%	35.3%	11.7%	**80%**	**88.1%**	68%	16 s
		±0.6	±6.9	±1.2	±2.7	±0.2	±1.4	±1
	FC-4	88.9%	38.3%	16.6%	**86.9%**	**88%**	60.8%	10 s
		±0.1	±4.7	±4.1	±0.7	±0.03	±3.2	±0.8
	LENET-5	89.7%	33%	28.6%	**87.7%**	**89.8%**	17.8%	10 s
		±0.2	±24.3	±26.3	±2.2	±0.4	±2.1	±1
CIFAR-10	LENET-5	72.2%	50.1%	27.5%	**67.7%**	**68.6%**	9.9%	6 s
		±0.2	±5.6	±1.7	±2.2	±1.4	±1.4	±0.5
	VGG-16	83.9%	85%	83.3%	N/A[a]	**85.3%**	36%	N/A[b]
		±0.4	±0.4	±0.3		±0.2	±1.1	

[a] A fine-tuning step is required to connect the results of independent layers.
[b] Computation was applied independently on each layer.

5.3 Generalization Between Different Datasets

Table 3. Cross-dataset generalization: sub-network masking is computed on source dataset (d_1) and then applied to target dataset (d_2) by re-training with the same early initialization. Test accuracies are presented for masked and unmasked (REF.) networks on d_2, as well as pruning percentage.

MODEL	SOURCE DATASET d_1	TARGET DATASET d_2	REF. ACC.	MASKED ACC.	PRUNING (%)
LENET-5	MNIST	FASHION MNIST	89.7% ± 0.3	89.2% ± 0.5	16.2% ± 0.2
		CIFAR-10	72.2% ± 0.2	68.1% ± 2.5	
VGG-16	CIFAR-10	MNIST	99.1% ± 0.1	99.4% ± 0.1	36% ± 1.1
		FASHION-MNIST	92.3% ± 0.4	92.1% ± 0.6	

In this experiment, we train the model on a dataset d_1 and create a masked neural model using our approach. After creating the masked model, we restart it to its original initialization. Finally, the new masked model is re-trained on another dataset d_2, and its generalization is analyzed.

Table 3 displays our experiments and respective results. When we compare generalization results to pruning using our approach on Fashion-MNIST and CIFAR-10, we discover that computing the critical sub-network for the LeNet-5 architecture on MNIST creates a more sparse sub-network. Moreover, this sub-network has a test accuracy better than zero-shot pruning without fine-tuning and comparable accuracy with the original ANN. This behavior occurs because the solver is optimizing on a batch of images that are classified correctly with high confidence from the trained model. Furthermore, computing the critical VGG-16 sub-network architecture on CIFAR-10 using decoupled greedy learning [5] generalizes well to Fashion-MNIST and MNIST.

5.4 Comparison to SNIP

OAMIP can be viewed as a compression technique of over-parameterized neural models. We compare it to SNIP [28].

SNIP computes connection sensitivities in a data-dependent way before the training. The sensitivity of a connection represents its importance based on the influence of the connection on the loss function. After computing the sensitivity, the connections below a predefined threshold are pruned before training (single shot).

In our methodology, we exclusively identify the importance of neurons and essentially prune all the connections of non-important ones. On the other hand, SNIP only focuses on pruning individual connections. Moreover, we highlight that SNIP can only compute connection sensitivity on the initialization of an ANN. Indeed, for a trained ANN, the magnitude of the derivatives concerning the loss function optimized during the training, makes SNIP keener to keep all the parameters. On the other hand, OAMIP can work on different convergence levels, as shown in Sect. 3.3. Furthermore, the connection sensitivity computed by SNIP is only network and dataset-specific; thus, the computed connection sensitivity for a single connection does not give a meaningful signal about its general importance for a given task. Rather, it needs to be compared to the sensitivity of other connections.

In order to bridge the differences between the two methods and provide a fair comparison in equivalent settings, we make a slight adjustment to our method. In step 2 of OAMIP, we compute neuron importance scores on the model's initialization[2]. We note that we used only 10 images as input to the MIP, corresponding to the 10 different classes, and 128 images as input to SNIP, following its original paper [28]. Our algorithm was able to prune neurons from fully connected and convolutional layers of LeNet-5. After creating the sparse

[2] Remark: we used $\lambda = 1$ and pruning threshold 0.2 and kept ratio 0.45 for SNIP. Training procedures as in Sect. 5.

networks using SNIP and our methodology, we trained them on the Fashion-MNIST dataset. The difference between SNIP (88.8% ± 0.6) and our approach (88.7% ± 0.5) was marginal in terms of test accuracy. SNIP pruned 55% of the ANN's parameters and OAMIP 58.4%.

Table 4. Cross-dataset generalization comparison between SNIP, with neurons having the lowest sum of connections' sensitivity pruned, and our framework (OAMIP), both applied on initialization, see Sect. 5.3 for the generalization experiment description.

Source dataset d_1	Target dataset d_2	Ref. Acc.	Method	Masked Acc.	Pruning (%)
Mnist	Fashion-MNIST	89.7% ± 0.3	SNIP	85.8% ± 1.1	53.5% ± 1.8
			OAMIP	**88.5% ± 0.3**	**59.1% ± 0.8**
	CIFAR-10	72.2% ± 0.2	SNIP	53.5% ± 3.3	53.5% ± 1.8
			OAMIP	**63.6% ± 1.4**	**59.1% ± 0.8**

Next, we compare SNIP and OAMIP in terms of generalization. In Table 4, we show that our framework outperforms SNIP in terms of generalization. We adjusted SNIP to prune entire neurons based on the value of the sum of its connections' sensitivity, and our framework was also applied to ANN's initialization. When our framework is applied on the initialization, more neurons are pruned as the marginal softmax part of the objective function discussed in Sect. 3.3 is weighing less ($\lambda = 1$), driving the optimization to focus on model sparsification.

Finally, we remark that the adjustments made to SNIP and OAMIP in the previous experiments are solely for comparison, while (unlike SNIP) the primary purpose of our method is to allow optimization at any stage – before, during, or after training. In the specific case of optimizing at initialization and discarding entire neurons based on aggregated connection sensitivity, the SNIP approach may have some advantages, notably in scalability for deep architectures. However, it also has some limitations, as previously discussed.

6 Discussion

We proposed a mixed integer program to compute neuron importance scores in ReLU-based deep neural networks. Our contributions focus on providing scalable computations of importance scores in fully connected and convolutional layers. We presented results showing that these scores can effectively prune unimportant parts of the network without significantly affecting its predictive capacity. Further, our results indicate that this approach allows the automatic construction of efficient sub-networks that can be transferred and retrained on different datasets. Knowing a neural network's critical components can further impact future work beyond the pruning applications presented here.

Acknowledgements. This work was partially funded by: IVADO (l'institut de valorisation des données) [*G.W.*, *M.C.*]; FRQ-IVADO Research Chair in Data Science for Combinatorial Game Theory, and NSERC grant 2019-04557 [*M.C.*] Canada CIFAR AI Chair, NIH grant R01GM135929 [*G.W.*].

References

1. Adamczewski, K., Park, M.: Dirichlet pruning for neural network compression. Proc. Mach. Learn. Res. **130** (2021)
2. Agrawal, A., Verschueren, R., Diamond, S., Boyd, S.: A rewriting system for convex optimization problems. J. Control Decis. **5**(1), 42–60 (2018)
3. Amjad, R.A., Liu, K., Geiger, B.C.: Understanding neural networks and individual neuron importance via information-ordered cumulative ablation. IEEE Trans. Neural Netw. Learn. Syst. **33**, 7842–7852 (2021)
4. Anderson, R., Huchette, J., Tjandraatmadja, C., Vielma, J.P.: Strong mixed-integer programming formulations for trained neural networks. In: International Conference on Integer Programming and Combinatorial Optimization, pp. 27–42. Springer (2019)
5. Belilovsky, E., Eickenberg, M., Oyallon, E.: Decoupled greedy learning of CNNs. In: Proceedings of the 37th International Conference on Machine Learning. Proceedings of Machine Learning Research, vol. 119, pp. 736–745. PMLR (2020)
6. Bengio, Y., Goodfellow, I., Courville, A.: Deep Learning, vol. 1. Citeseer (2017)
7. Brosch, T., Tam, R.: Efficient training of convolutional deep belief networks in the frequency domain for application to high-resolution 2D and 3D images. Neural Comput. **27**(1), 211–227 (2015)
8. Diamond, S., Boyd, S.: CVXPY: a Python-embedded modeling language for convex optimization. J. Mach. Learn. Res. **17**(83), 1–5 (2016)
9. Dong, X., Chen, S., Pan, S.: Learning to prune deep neural networks via layer-wise optimal brain surgeon. In: Advances in Neural Information Processing Systems, pp. 4857–4867 (2017)
10. Ebrahimi, A., Klabjan, D.: Neuron-based pruning of deep neural networks with better generalization using kronecker factored curvature approximation. arXiv preprint arXiv:2111.08577 (2021)
11. ElAraby, M.: Optimizing ANN architectures using mixed-integer programming. Master's dissertation, Université de Montréal (2020). http://hdl.handle.net/1866/24312
12. Fischetti, M., Jo, J.: Deep neural networks and mixed integer linear optimization. Constraints **23**(3), 296–309 (2018)
13. Frankle, J., Carbin, M.: The lottery ticket hypothesis: finding sparse, trainable neural networks. In: International Conference on Learning Representations (2019)
14. Gimpel, K., Smith, N.A.: Softmax-margin CRFs: training log-linear models with cost functions. In: The Annual Conference of the North American Chapter of the Association for Computational Linguistics, pp. 733–736. Association for Computational Linguistics (2010)
15. Goodfellow, I., Bengio, Y., Courville, A.: Deep Learning. MIT Press (2016)
16. Han, S., Mao, H., Dally, W.J.: Deep compression: Compressing deep neural networks with pruning, trained quantization and Huffman coding. arXiv preprint arXiv:1510.00149 (2015a)
17. Han, S., Pool, J., Tran, J., Dally, W.: Learning both weights and connections for efficient neural network. In: Advances in Neural Information Processing Systems, pp. 1135–1143 (2015)
18. Hassibi, B., Stork, D.G., Wolff, G.J.: Optimal brain surgeon and general network pruning. In: IEEE International Conference on Neural Networks, pp. 293–299. IEEE (1993)

19. He, Y., Kang, G., Dong, X., Fu, Y., Yang, Y.: Soft filter pruning for accelerating deep convolutional neural networks. In: Proceedings of the International Joint Conference on Artificial Intelligence, IJCAI 2018, pp. 2234–2240. AAAI Press (2018)
20. Hooker, S., Erhan, D., Kindermans, P.J., Kim, B.: A benchmark for interpretability methods in deep neural networks. In: Advances in Neural Information Processing Systems, pp. 9734–9745 (2019)
21. Huang, P.S., et al.: Achieving verified robustness to symbol substitutions via interval bound propagation. In: Proceedings of the 2019 Conference on Empirical Methods in Natural Language Processing and the 9th International Joint Conference on Natural Language Processing (EMNLP-IJCNLP), pp. 4083–4093 (2019)
22. Jordao, A., Yamada, F., Schwartz, W.R.: Deep network compression based on partial least squares. Neurocomputing **406**, 234–243 (2020)
23. Kingma, D., Ba, J.: Adam: a method for stochastic optimization. In: Proceedings of the 3rd International Conference for Learning Representations (ICLR 2015), San Diego (2015)
24. LeCun, Y., Bengio, Y., Hinton, G.: Deep learning. Nature **521**(7553), 436–444 (2015)
25. Krizhevsky, A.: Learning multiple layers of features from tiny images. Master's thesis, University of Toronto (2009)
26. LeCun, Y., Bottou, L., Bengio, Y., Haffner, P., et al.: Gradient-based learning applied to document recognition. Proc. IEEE **86**(11), 2278–2324 (1998)
27. LeCun, Y., Denker, J.S., Solla, S.A.: Optimal brain damage. In: Advances in Neural Information Processing Systems, pp. 598–605 (1990)
28. Lee, N., Ajanthan, T., Torr, P.H.S.: SNIP: single-shot network pruning based on connection sensitivity. In: International Conference on Learning Representations (ICLR) (2019)
29. Lei, W., Chen, H., Wu, Y.: Compressing deep convolutional networks using k-means based on weights distribution. In: Proceedings of the 2nd International Conference on Intelligent Information Processing, pp. 1–6 (2017)
30. Li, Y., Adamczewski, K., Li, W., Gu, S., Timofte, R., Van Gool, L.: Revisiting random channel pruning for neural network compression. In: Proceedings of the IEEE/CVF Conference on Computer Vision and Pattern Recognition, pp. 191–201 (2022)
31. Liang, T., Glossner, J., Wang, L., Shi, S., Zhang, X.: Pruning and quantization for deep neural network acceleration: a survey. Neurocomputing **461**, 370–403 (2021)
32. Liu, Z., Sun, M., Zhou, T., Huang, G., Darrell, T.: Rethinking the value of network pruning. In: International Conference on Learning Representations (2018)
33. Luo, J.H., Wu, J., Lin, W.: ThiNet: a filter level pruning method for deep neural network compression. In: Proceedings of the IEEE International Conference on Computer Vision, pp. 5058–5066 (2017)
34. Malach, E., Yehudai, G., Shalev-Schwartz, S., Shamir, O.: Proving the lottery ticket hypothesis: pruning is all you need. In: III, H.D., Singh, A. (eds.) International Conference on Machine Learning, Proceedings of Machine Learning Research, vol. 119, pp. 6682–6691. PMLR (2020)
35. Molchanov, P., Mallya, A., Tyree, S., Frosio, I., Kautz, J.: Importance estimation for neural network pruning. In: Proceedings of the IEEE/CVF Conference on Computer Vision and Pattern Recognition, pp. 11264–11272 (2019)
36. Molchanov, P., Tyree, S., Karras, T., Aila, T., Kautz, J.: Pruning convolutional neural networks for resource efficient inference. In: International Conference on Learning Representations (ICLR) (2017)

37. Morcos, A., Yu, H., Paganini, M., Tian, Y.: One ticket to win them all: generalizing lottery ticket initializations across datasets and optimizers. In: Advances in Neural Information Processing Systems, vol. 32. Curran Associates, Inc. (2019)
38. Mosek, A.: The mosek optimization software. **54**(2–1), 5 (2010). www.mosek.com
39. Neyshabur, B., Li, Z., Bhojanapalli, S., LeCun, Y., Srebro, N.: The role of over-parametrization in generalization of neural networks. In: 7th International Conference on Learning Representations, ICLR (2019)
40. Ramanujan, V., Wortsman, M., Kembhavi, A., Farhadi, A., Rastegari, M.: What's hidden in a randomly weighted neural network? In: Proceedings of the IEEE/CVF Conference on Computer Vision and Pattern Recognition, pp. 11893–11902 (2020)
41. Salama, A., Ostapenko, O., Klein, T., Nabi, M.: Pruning at a glance: global neural pruning for model compression. arXiv preprint arXiv:1912.00200 (2019)
42. Serra, T., Kumar, A., Ramalingam, S.: Lossless compression of deep neural networks. In: 2020 Fall Eastern Virtual Sectional Meeting, AMS (2020)
43. Shrikumar, A., Greenside, P., Kundaje, A.: Learning important features through propagating activation differences. In: International Conference on Machine Learning, pp. 3145–3153. PMLR (2017)
44. Simonyan, K., Zisserman, A.: Very deep convolutional networks for large-scale image recognition. In: Bengio, Y., LeCun, Y. (eds.) International Conference on Learning Representations (ICLR) (2015)
45. Srinivas, S., Babu, R.V.: Data-free parameter pruning for deep neural networks. arXiv preprint arXiv:1507.06149 (2015)
46. Tieleman, T., Hinton, G.: Lecture 6.5-rmsprop: divide the gradient by a running average of its recent magnitude. COURSERA: Neural Netw. Mach. Learn. **4**(2), 26–31 (2012)
47. Tjeng, V., Xiao, K.Y., Tedrake, R.: Evaluating robustness of neural networks with mixed integer programming. In: International Conference on Learning Representations (ICLR) (2019)
48. Tsay, C., Kronqvist, J., Thebelt, A., Misener, R.: Partition-based formulations for mixed-integer optimization of trained relu neural networks. Adv. Neural. Inf. Process. Syst. **34**, 3068–3080 (2021)
49. Verma, S., Pesquet, J.C.: Sparsifying networks via subdifferential inclusion. In: International Conference on Machine Learning, pp. 10542–10552. PMLR (2021)
50. Wang, C., Grosse, R., Fidler, S., Zhang, G.: EigenDamage: structured pruning in the kronecker-factored eigenbasis. In: International Conference on Machine Learning, pp. 6566–6575. PMLR (2019)
51. Wang, C., Zhang, G., Grosse, R.B.: Picking winning tickets before training by preserving gradient flow. In: International Conference on Learning Representations (ICLR) (2020)
52. Wang, Y., et al.: Pruning from scratch. In: AAAI, pp. 12273–12280 (2020)
53. Ye, M., Gong, C., Nie, L., Zhou, D., Klivans, A., Liu, Q.: Good subnetworks provably exist: pruning via greedy forward selection. In: International Conference on Machine Learning, pp. 10820–10830. PMLR (2020)
54. Yu, R., et al.: NISP: pruning networks using neuron importance score propagation. In: Proceedings of the IEEE Conference on Computer Vision and Pattern Recognition, pp. 9194–9203 (2018)
55. Yu, X., Serra, T., Ramalingam, S., Zhe, S.: The combinatorial brain surgeon: pruning weights that cancel one another in neural networks. In: International Conference on Machine Learning, pp. 25668–25683. PMLR (2022)

56. Zeng, W., Urtasun, R.: MLPrune: multi-layer pruning for automated neural network compression. In: International Conference on Learning Representations (ICLR) (2018)
57. Zhang, C., Bengio, S., Hardt, M., Recht, B., Vinyals, O.: Understanding deep learning requires rethinking generalization. In: International Conference on Learning Representations (ICLR) (2017)
58. Zhu, Z., et al.: A geometric analysis of neural collapse with unconstrained features. Adv. Neural. Inf. Process. Syst. **34**, 29820–29834 (2021)

Predicting the Optimal Period for Cyclic Hoist Scheduling Problems

Nikolaos Efthymiou and Neil Yorke-Smith

STAR Lab, Delft University of Technology, Delft, The Netherlands
nikolaos.efthymiou@epfl.ch, n.yorke-smith@tudelft.nl

Abstract. Since combinatorial scheduling problems are usually NP-hard, this paper investigates whether machine learning (ML) can accelerate exact solving of a problem instance. We adopt supervised learning on a corpus of problem instances, to acquire a function that predicts the optimal makespan for a given instance. The learned predictor is invariant to the instance size as it uses statistics of instance attributes. We provide this prediction to a solving algorithm in the form of bounds on the objective function. Specifically, this approach is applied to the well-studied Cyclic Hoist Scheduling Problem (CHSP). The goal for a CHSP instance is to find a feasible schedule for a hoist which moves objects between tanks with minimal cyclic period. Taking an existing Constraint Programming (CP) model for this problem, and an exact CP-SAT solver, we implement a Deep Neural Network, a Random Forest and a Gradient Boosting Tree in order to predict the optimal period p. Experimental results find that, first, ML models (in particular DNNs), can be good predictors of the optimal p; and, second, providing tight bounds for p around the predicted value to an exact solver significantly reduces the solving time without compromising the optimality of the solutions.

Keywords: combinatorial optimisation · supervised learning · cyclic hoist scheduling · constraint programming

1 Introduction

Computationally-challenging scheduling problems are common in industrial practice [8]. The hardness often arises from a combinatorial core in the problem, coupled by large size. Examples of such problems are satellite downlink scheduling [7], staff rostering [19], and hoist scheduling [11].

Contemporary machine learning (ML) methods are being exploited in the solving of large-scale combinatorial optimisation problems, including challenging scheduling problems [9]. Among the various approaches in the recent literature are learning problem-class-specific heuristics, end-to-end production of solutions, and warm-starting an optimisation solver. Bengio et al. [1], Kotary et al. [10] provide surveys. This paper explores the effectiveness of a loose coupling of the learning and optimisation. Thus we adopt the last of the above approaches:

© The Author(s), under exclusive license to Springer Nature Switzerland AG 2023
A. A. Cire (Ed.): CPAIOR 2023, LNCS 13884, pp. 238–253, 2023.
https://doi.org/10.1007/978-3-031-33271-5_16

warm-starting a solver using information provided for a given problem instance by a pre-trained ML model. Specifically, taking inspiration from Wang et al. [24], we acquire and leverage an oracle for the makespan of the scheduling problem.

In more detail, we perform supervised learning on a corpus of problem instances to acquire a function that predicts the optimal makespan for a given instance. We provide this prediction to a solving algorithm in the form of bounds on the objective function. In this way we study the effect on the solver's computation time and the solution quality.

In the current paper this approach is developed for the *Cyclic Hoist Scheduling Problem* (CHSP), an optimization problem of practical and theoretical importance [11,12]. The aim is to find a schedule for one or multiple industrial hoists on track(s) that move objects between tanks. Process constraints impose bounds for the processing time in each tank, while the time that a hoist needs to travel between different tanks depends on the tanks and whether the hoist is empty or loaded. An important characteristic of CHSP is that the fixed series of moves is repeatedly performed by the hoist(s). This repetitive – cyclic – hoist schedule introduces the notion of the *cycle period* p, which is defined as the difference between the start time of two consecutive objects. Note that minimising the period is the analogue of minimising the makespan in this cyclic setting.

The literature boasts a host of techniques for solving CHSP instances [6], including custom branch-and-bound, mixed integer linear programming (MIP), constraint programming (CP), and meta-heuristics and evolutionary algorithms. We study the idea of providing predicted bounds to a competitive CP model of the CHSP [23], using an exact CP-SAT solver as the backend. Results show that supervised learning can acquire a function that predicts the optimal CHSP period p, and that providing bounds of 10% of this predicted p value leads to a mean time decrease of 90% in finding a feasible solution on unseen industrial problem instances, and a mean time decrease in solving to optimality of 44%.

This accelerated solving is important for industrial practice, where hoist lines can be much longer than tackled in the bulk of the academic literature [21]. Further, when an event occurs that causes a deviation from the planned schedule, a rapid rescheduling is necessary. In addition, from an academic perspective, the CHSP is relatively simple in its essential form [23] – while remaining challenging – which suggests that the same kind of methodology used in this paper can benefit other combinatorial scheduling problems.

Summarising: 1) we show that supervised learning can effectively acquire a function that predicts the optimal CHSP makespan; 2) we provide evidence that feeding a solver with predicted bounds of the objective function can accelerate the solving process; and 3) through extensive computational experiments we provide the first demonstration of the value of ML in finding CHSP solutions.

The remainder of the paper is structured as follows. Section 2 introduces the CHSP. Section 3 explains our approach. Section 4 studies the approach empirically. Section 5 situates our work in the literature. Section 6 concludes with future directions.

2 Hoist Scheduling Problem

The hoist scheduling problem is to operate one or multiple hoists which move along a linear track above a set of tanks (Fig. 1). Among the many problem variants [2], the cyclic hoist scheduling problem assumes a fixed sequence of items to be processed. From a scheduling perspective, the challenge is to allow the processing of successive items to overlap, so that (different) tasks on different items may be carried out at the same time. The schedule of tasks for one item is repeated for subsequent items: the length of a cycle is the time between the start of processing for an item and that of its successor. The objective is to maximise throughput, i.e., to minimise the cycle period, p; this is equivalent to minimising the makespan. An efficient CP model for the generic CHSP problem is formulated by Wallace and Yorke-Smith [23]. At the centre is a three-variable disjunctive constraint, from which arises the hard combinatorial core of the CHSP. The CP approach is interesting because a single model can solve a set of CHSP problem variants, whereas solving is performed by any of a range of state-of-the-art backend solvers which can ingest the model.

Optimal p Predictor. The CP model uses static calculated lower and upper bounds of p, denoted B_{calc}, to specify the space of feasible solutions. Given that such computation reflects the theoretical maximum range of p, B_{calc} tend to be quite loose. This leads to the central hypothesis of the paper: predicting the optimal value of p – without solving the CSHP instance – and then restricting the range in which the solver is trying to find a solution could result in lower solving times (T). This is a form of predict-then-optimise in which the prediction is not *necessary* for the optimisation, but can serve as a catalyst in the solver's inference and search process for an optimal solution.

We would like the learned predictor to be invariant to the instance size. For a CHSP instance with n tanks, we have the following exhaustive set of possible raw features: number of hoists, number of tanks, minimum/maximum processing times of each tank, $(n+1)$-dimensional vector of loaded move times, $(n+1) \times (n+1)$ matrix with empty move times, and capacity of each tank. Considering all these features leads to a dimensionality of $(n+1)^2 + 3n + 4$. This is a large number of features, and also of varying dimension depending on the

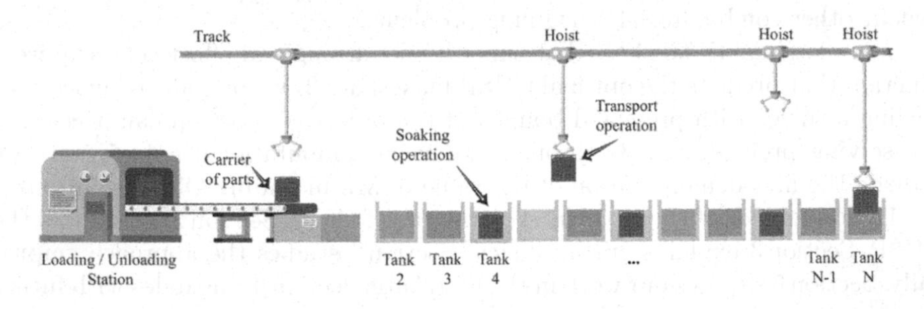

Fig. 1. Typical hoist scheduling line (from Laajili et al. [11]).

number of tanks. One approach would be to train a different ML model for each value of n [24]. Instead, we will study the possibility of having a fixed number of independent variables for the ML models, irrespective of the amount of tanks. To this end we will replace instances' attributes per tank with their descriptive statistics.

Hypothesis 1. *A universal regressor for the optimal value of p can be implemented that uses a fixed number of selected CHSP descriptive statistics.*

3 Methodology

In order to study accelerated solving of CHSP instances, we first perform supervised learning on a corpus of instances, to acquire a predictor of the optimal period for a given instance. We provide this prediction to a CP-SAT solver in the form of bounds on the objective function. This section explains the approach.

3.1 Data

Industry Instances. Seven sets of industry instances (PU [18], BO1, BO2, Zinc, Copper [13], Ligne1 and Ligne2 [15]) were used to: 1) analyse the various patterns that the values of instances' features follow, in order to implement the random generator (described below); 2) test the ML models; and 3) assess the performance of the CP solver when predicted bounds B_{pred} are used. Each instance has the following attributes: number of tanks n; minimum processing time for each tank $tmin$; maximum processing time for each tank $tmax$; vector of loaded move times from tank i to tank $i+1$, f; and 2D matrix of empty move times from tank i to tank j, e. Further, in order to extend the original instances to larger sizes, we considered values in $\{1, 2, 3, 4, 5\}$ for the tank capacity and the number of hoists (compare the 'multiplier' factor of Wallace and Yorke-Smith [23]). This yields a total of $7 \cdot 5 \cdot 5 = 175$ instances. We will denote this set with I_{ind}. We note that the values of 4 and 5 for the capacity and the number of hoists are not present in random instances used to train ML models, and thus we test the generalizability of the ML models. For comparison reasons, a subset of I_{ind} for which the number of hoists and the capacity take values in $\{1, 2, 3\}$ is also used (we will refer to this subset as $I_{ind[3]}$).

Random Instances. Since the number of industrial instances is limited, we generated random CHSP instances for training and testing the ML models. We implemented a random generator by making assumptions about the loading-unloading stations, the relative position of the tanks, and the times defining each instance. To this aim, we examined the patterns of the industry instances with respect to the following features and then uniform random values were chosen from a fixed range of each parameter: 1) Time window of treatments (minimum time, variability in minimum time, maximum to minimum time ratio) 2) Empty move times from tank i to tank j and their variability 3) Loaded move times from tank

i to tank $i+1$ and their variability 4) Number of tanks: $\{3, 4, \ldots, 24\}$ 5) Number of hoists: $\{1, 2, 3\}$ and 6) Tank capacity: $\{1, 2, 3\}$. These features are the ones mostly used in the literature and especially in the CP model that we utilise as a baseline. We assume the following: 1) treatment i occurs in tank i; 2) the loading and the unloading stations are the same; 3) there is one track; 4) the time to load/unload a job into/from a tank is 0; and 5) all tanks of an instance have the same capacity. These assumptions are made in most published works that include single-track industry hoist scheduling data. Regarding Hypothesis 1, we flattened each instance to a 19-dimensional vector (corresponding to the features of the ML models) consisting of: 1) the number of tanks; 2) the number of hoists; 3) the capacity of the tanks; and 4) the minimum, maximum, average and standard deviation of e, f, $tmin$ and $tmax$.

Four generators were implemented, each of which corresponds to a different topology. In particular, the following four spatial arrangements of the tanks were examined: 1) *linear* topology, where tank n is the farthest from the loading/unloading station (similar to Ligne2 [15]); 2) *reversed linear* topology with tank 1 being the farthest from the loading/unloading station (similar to BO1, BO2, Ligne1 [13,15]); 3) *ring* topology with tank $\frac{n}{2}$ being the farthest from the loading/unloading station (similar to Copper, Zinc, PU [13,18]); and 4) a transformed version of the linear topology with increased time for the loaded move from tank n to the loading/unloading station. In total we generated 166,320 random instances: 66,528 for the linear topology and 33,264 for each of the others. The observed variety in the descriptive statistics of the instances' e, f, $tmin$ and $tmax$ suggests the dataset captures many possible real life instances.

Solving Instances with Calculated Bounds. Of the generated instances, 98.6% (164,032) were solved with at least a feasible solution, within 6 min using the Google OR-Tools CP-SAT solver [17] ({*processes* = 8, *free search* = True, *optimisation level* = 1, *timeout* = 360 s}). We denote this set with I_{gen}; it was used to train and test ML models. For hypothesis testing purposes, a random sample of 4,000 instances was drawn from I_{gen} (1,000 for each topology). In addition, as we consider to be 'difficult' those instances that led to a Satisfied solution when B_{calc} were used, and in order to make our sample more demanding, we include the remaining 173 such instances to the sample. We denote this set of 4,173 instances with I_{sample}.

3.2 ML Model Training

We experimented with three ML models. The (optimal) cycle period p_{calc} found by using B_{calc} was used as a target value of the ML models to be trained. We split the data set I_{gen} into a train set I_{train} and a test set I_{test} (test size = 0.33).

First, we used the Keras library for training and testing a Deep Neural Network (DNN); we experimented with 3–8 hidden, non-linear, dense layers with various activation functions (ReLu, Leaky ReLu, PReLu, ELU). We normalized the input features and used a linear dense single-output layer for predicting the optimal p. Model compilation was performed with the MAPE loss function and the Adam optimizer (learning rate = 0.001). Fine-tuning was done with a validation split of 0.2. Six DNN variants were tested on I_{test}, I_{ind} and $I_{ind[3]}$.

Second, we also trained and tested a *Random Forest* (RF) regressor model using the Scikit-learn Python library. During the fine-tuning phase we tuned various values of the parameters *max depth, max features, min samples leaf, min samples split and n-estimators.* Third, we trained and tested a *Histogram-based Gradient Boosting Regression* (HGBR) model using Scikit-learn by tuning the following parameters: *l2 regularization, learning rate, loss, max iterations, max leaf nodes* and *min samples leaf.*

Lastly, to obtain a simpler and easier to interpret model, we performed feature extraction (by removing irrelevant features) on the flattened instances. For this we considered the *mean decrease in impurity* and the *permutation feature importance* [3], using the structure of the RF model. We give details below.

Solving Instances with Predicted Bounds. The best-performing ML model was applied to predict the optimal p value for the instances of I_{sample} and I_{ind}. This predicted value (p_{pred}) and its deviation from p_{calc} provide the basis for calculating the predicted bounds B_{pred}. We denote with $\hat{p}_l$ and $\hat{p}_u$ the lower and upper bounds derived accordingly. Note there is no guarantee that the true optimum is within these derived bounds. We denote with p_l and p_u the static lower and upper bounds from Wallace and Yorke-Smith [23], and write $B_{calc} = [p_l, p_u]$. These calculated bounds are conservative: they guarantee to contain the true optimum.

To solve problem instances from the sample dataset I_{sample} and the industrial dataset I_{ind}, we use the CP model unaltered, except that we give the following bounds on the objective function (note we always keep the tightest bounds):

$$p_l^* = \begin{cases} \hat{p}_l, & \text{if } \hat{p}_l \in (p_l, p_u) \\ p_l, & \text{otherwise} \end{cases} \qquad p_u^* = \begin{cases} \hat{p}_u, & \text{if } \hat{p}_u \in (p_l, p_u) \\ p_u, & \text{otherwise} \end{cases} \tag{1}$$

Given that in almost all cases the $\hat{p}_l, \hat{p}_u$ values were used, for simplicity we define the predicted bounds $B_{pred} = [p_l^*, p_u^*]$.

Throughout, the OR-Tools CP-SAT solver was used with the same machine configuration for experiments on each data set, to make the results comparable. We compare the solutions found by the solver when using B_{calc} and using B_{pred}, in terms of the following standard metrics: number of Optimal, Satisfied and Unsatisfiable cases; best found p value; number of solutions (N) found; and total solving time (T) to find and prove the optimal solution.

4 Experimental Results

With the pipeline described, this section reports an empirical study. Firstly in Sect. 4.1 we compare the various ML models to select the best performing model. Then in Sect. 4.2 we study the tightness of the predicted bounds versus their inclusion of the true optimal period. Thirdly in Sect. 4.3 we assess the performance of the whole pipeline in terms of solving time and optimal solution.

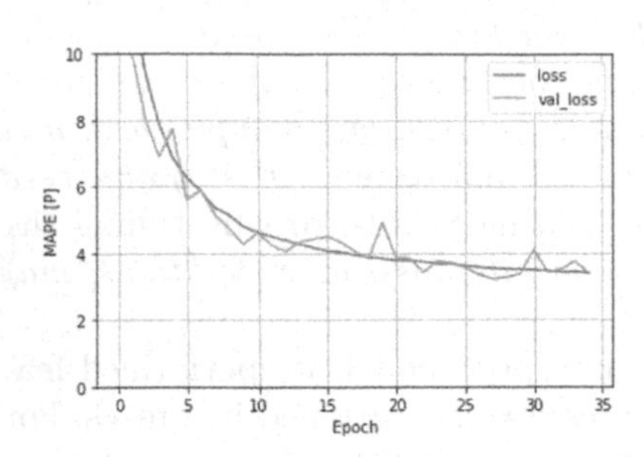

Fig. 2. DNN loss on training and validation sets

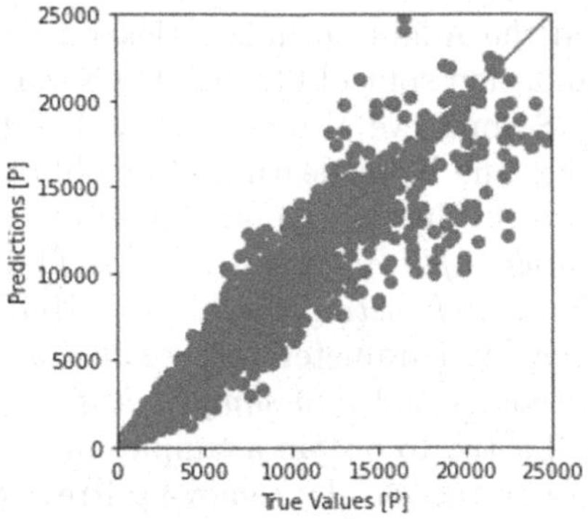

Fig. 3. Predicted vs. true values of p on dataset I_{test}

4.1 Experiment 1: ML Predictive Power and Model Selection

The DNN model that led to the best results (f^*) has 4 hidden dense layers with 92 neurons. The *Exponential Linear Units* activation function was used and the model was run for 35 epochs. Plots for the performance of this model are shown in Figs. 2 and 3. The graph of the training–validation loss reveals a rather normal learning progress over the number of epochs, and a good fit to the training data. The scatterplot shows a high correlation between predicted and true p values. The parameter values of the best RF model were the default values of the *Scikit-learn* library. As for the HGBR model, the strongest performance was obtained with the following parameter configuration: {*l2 regularization* = 0.22, *learning rate* = 0.065, *loss* = Poisson, *max iterations* = 1500, *max leaf nodes* = 100, *min samples leaf* = 160}.

Table 1 shows the seven most important features, as calculated by the RF regressor, with respect to two importance measures: *mean decrease in impurity* (MDI) and *feature permutation* (FP). These features were used to train and test a simpler DNN model f^- that receives 7-dimensional examples as input. Tables 2 and 3 report the MAPE values obtained with the best-performing models and the difference in performance between using all features and using only the important features. The best DNN model slightly outperforms the other ML methods used, on MAPE values on the test set I_{test}. The full model f^* achieved a MAPE of 3.38 on the test set I_{test}, while the value of 4.73 was reached using f^- on the same set. Further, the DNN model has significantly higher predictive power when testing on the industry sets I_{ind} and $I_{ind[3]}$. f^* performed adequately, even in the case of I_{ind} that contains instances with unseen attribute values. We note that the Ligne1 and PU industry sets have higher MAPE values compared to the other industry instances. One explanation might be that only these sets have tanks with infinite processing time (that is somehow treated by the CP solver) and there is no such case in the training dataset.

Table 1. Features sorted by importance

Feature	MDI	FP
$max(tmin)$	0.51	1.32
tank capacity	0.19	0.47
# hoists	0.09	0.31
$avg(f)$	0.05	0.23
# tanks	0.05	0.13
$max(f)$	0.04	0.07
$avg(tmax)$	0.02	0.04

Table 2. MAPE values (%) of f^* and f^- on the industry instances

	f^*		f^-	
	$I_{ind[3]}$	I_{ind}	$I_{ind[3]}$	I_{ind}
BO1	9	9	11	14
BO2	7	9	17	18
Copper	1	3	1	10
Ligne1	48	34	73	75
Ligne2	8	12	6	22
PU	13	19	22	23
Zinc	5	8	2	11
All	13	13	19	25

Table 3. MAPE values (%) of the ML models

Features	DNN			HGBR			Random Forest		
	$I_{ind[3]}$	I_{ind}	I_{test}	$I_{ind[3]}$	I_{ind}	I_{test}	$I_{ind[3]}$	I_{ind}	I_{test}
All	12.95	13.34	3.38	18.01	30.81	4.50	22.60	36.20	3.81
Important	18.85	24.74	4.73	27.61	37.18	4.89	26.61	37.16	3.98

Table 4. Cumulative relative frequency of instances per bound deviation d

d	f^*			f^-		
	$I_{ind[3]}$	I_{ind}	I_{test}	$I_{ind[3]}$	I_{ind}	I_{test}
5%	42.9	37.7	81.5	44.4	32.6	76.7
10%	66.7	61.7	90.2	57.1	42.3	86.7
15%	77.8	72.0	94.5	66.7	55.4	91.8
20%	84.13	77.7	96.6	71.4	60.6	94.7
>	100.0	100.0	100.0	100.0	100.0	100.0

Calculation of B_{pred} In order to minimise the possibility of p_{calc} being outside of the new predicted bounds of p, while keeping the bounds as narrow as possible, we calculated the cumulative relative frequency of instances per interval class of the *predicted to actual p deviation* ($d = |\frac{p_{pred}-p_{calc}}{p_{calc}}|\%$). Table 4 presents the percentage of instances with p_{pred} being at most $x\%$ away from p_{calc} (for $x \in \{5, 10, 15, 20\}$). The relative difference between p_{pred} and p_{calc} did not exceed 20% for 96.6% of random instances and 5% for 81.5% of instances. Thus we selected these *margins* for the further experiments. In case of I_{ind} these cumulative frequencies are lower and so we selected a margin of $\pm10\%$ instead of $\pm5\%$.

4.2 Experiment 2: Bounds and Solutions

Solutions on Random Instances. We examine first the results of the CP-SAT solver on the 4,173 instances of I_{sample}, when the predicted bounds B_{pred} (as predicted by f^*) were used. In most cases (98.3% for the ±5% margin and 90.2% for the ±20% margin), both the lower and upper predicted bounds of p (B_{pred}) are tighter than the calculated bounds (B_{calc}), and they were used by the solver. Further, in most cases (85.9% for the ±5% margin and 97.1% for the ±20% margin), B_{pred} contain the original p_{calc}. In the other 14.1% of cases, either the solver (incorrectly) found the instance unsatisfiable (e.g., $N = 421$ in case of ±5%), or the solver found a sub-optimal p (e.g., $N = 168$ in case of ±5%): usually only slightly sub-optimal.

A feasible solution was found for 89.9% of instances, and an optimal solution for 86.1% of instances, in the case of ±5% margin (respectively, 97.3% and 92.2%, in the case of ±20% margin). Hence only a relatively small number of instances have no solution ('unsatisfiable': there is no solution with p within B_{pred} or 'unknown'). As expected, this number reduces as the B_{pred} margin increases from 5% to 20%. The solver found the original p_{calc} in most cases: 85.0% and 96.1% of all instances or 94.6% and 98.8% of solved instances, per margin respectively. The percentage of satisfied cases is similar in all three scenarios: 5.4% for the original solver, 4.2% for the ±5% margin and 5.3% for the ±20% margin.

As the bound margin decreases, a very small number of instances have an optimal p greater (i.e., worse) than the original solver: $N = 147$ (plus $N = 18$ previously satisfied) for $B_{\text{pred\_5\%}}$ vs. B_{calc}, and $N = 9$ for $B_{\text{pred\_20\%}}$ vs. B_{calc}. This is so because the original optimal p is out of the predicted bounds used. Further, in case of satisfied solutions, there are only 13 and 20 cases respectively with p greater than the original solver and p_{calc} in B_{pred}. Table 5 reports in detail the solutions found with B_{pred} (in rows) in comparison with those found with B_{calc} (in columns) for each scenario.

Solutions on Industry Instances. Table 6 reports that, for most of the industry instances of I_{ind}, an (optimal) solution was found (163 out of 175, in case of ±10% margin and 168 in case of ±20%). However, especially in the case of ±10% margins, the B_{pred} of some instances do not contain the original p_{calc}. Thus the optimal p_{pred} found using B_{pred} is higher than p_{calc}. Table 6 presents the solutions per industry setting. We observe a variation in the performance of the solver that corresponds to the MAPE values presented in Table 2.

Table 5. Solutions for the set I_{sample}

	±5% *any p*		±5% p_{calc}		±20% *any p*		±20% p_{calc}	
	Opt	Sat	Opt	Sat	Opt	Sat	Opt	Sat
Optimal	3,546	46	3,399	23	3,834	14	3,825	14
Satisfied		159		126	1	213	1	172
Unknown		11						
Unsatisfiable	400	11			111			
Total	3,946	227	3,399	149	3,946	227	3,826	186

Table 6. Solutions per industry set

Industry instances	Unsatisfiable		p_{calc} found		$p \neq p_{calc}$ found	
	±10%	±20%	±10%	±20%	±10%	±20%
BO1	5		14	25	6	
BO2			17	24	8	1
Copper			25	25		
Ligne1	2	2	8	12	15	11
Ligne2	2	2	16	20	7	3
PU	3	3	10	16	12	6
Zinc			22	22	3	3
Total	12	7	112	144	51	24

Comparison with a Baseline Approach. We compare the above B_{pred}-based method with a simple bound estimation. First we compute the average r of $p_{calc}/\overline{B_{calc}}$ across all instances; then for each instance compute estimated bounds: $B_{est} = (lb_{est}, ub_{est}) = (r \cdot \overline{B_{calc}} \cdot 0.95, r \cdot \overline{B_{calc}} \cdot 1.05)$. B_{est} contain the original p_{calc} for 5% of the solved instances of I_{sample} and for 7% of the solved instances of I_{ind}. Hence if we apply these new estimated bounds to the CP solving process, for most instances either no solution will be found or a solution with a greater p.

4.3 Experiment 3: Solver Performance with Predicted Bounds

Our final experiment studies the impact of using B_{pred} in the CP-SAT solver. For this we used a subset of I_{sample} to only include cases for which an optimal solution was found, and we formulate three hypotheses:

Hypothesis 2. *A (CP) solver that uses B_{pred} instead of B_{calc} requires less time (T) to find an optimal solution.*

Hypothesis 3. *As B_{pred} become tighter, the solving time (T) decreases further.*

Table 7. Random Instances: Predicted vs. Calculated bounds

		Optimal solution				Satisfied			
		Inst.	Δp	ΔT	ΔN	Inst	Δp	ΔT	ΔN
Margin ±5%	Any p	3,546	0.3%	−70.7%	−56.6%	159	0.0%	−16.6%	−77.4%
	p_{calc}	3,399	—	−68.1%	−53.4%	126	—	−29.5%	−78.0%
Margin ±20%	Any p	3,834	0.1%	−33.1%	−27.1%	213	−0.1%	−9.5%	−36.3%
	p_{calc}	3,825	—	−33.0%	−26.9%	172	—	-22.9%	−40.2%

Table 8. Solving time difference (margin ±5%) – Negative vs. positive deviation

	T_{calc}	$T_{pred\_5\%}$	ΔT
$T_{pred\_5\%} - T_{calc} > 0$	0.18	0.22	0.04
$T_{pred\_5\%} - T_{calc} < 0$	2.31	0.66	−1.65
Total	1.91	0.58	−1.33

Hypothesis 4. *Using B_{pred} instead of B_{calc} does not increase the value of p of the optimal solution found.*

To test these hypotheses, we follow a repeated-measurements experimental design and thus only cases with a feasible solution in both conditions (calculated bounds vs. predicted bounds) are included. Note that cases with no solution are irrelevant here, for their failure relates to prediction error, i.e., Hypothesis 1. Comparisons are made independently for instances with an optimal solution in both conditions, and instances with a satisfied solution in both conditions.

Impact on Solving Time. Regarding Hypotheses 2 and 3, we can accept that the solving time is significantly lower when the predicted bounds of p (B_{pred}) are used, instead of the calculated bounds B_{calc} ($T_{pred\_5\%}$: $\overline{X} = 0.58$, $s = 4.93$; $T_{pred\_20\%}$: $\overline{X} = 1.27$, $s = 11.01$; T_{calc}: $\overline{X} = 1.91$, $s = 14.09$). Specifically, there is a decrease in time ($\Delta T = -1.3$) when $B_{pred\_5\%}$ is used instead of B_{calc}. There is a more modest decrease ($\Delta T = -0.6$) when $B_{pred\_20\%}$ is used instead of B_{calc}. Accordingly, there is a decrease in solving time when B_{pred} become narrower from ±20% to ±5% ($\Delta T = -0.7$).

As Table 7 shows, the decrease in the solving time for cases in which an optimal solution was found (−70.7%) is larger compared to cases with a satisfied solution (−16.6%). Figure 4 shows there was a decrease in the solving time for 81% of all solved instances. We also see that for a minority of some 60 instances, much more time was required (relative change higher than 90%). However, in most cases with $\Delta T > 0$, the solving time and its increase are very small in absolute terms. The corresponding means of ΔT and T are shown in Table 8.

Impact on p Value. Regarding Hypothesis 4, the impact of B_{pred} on the value of p of the optimal solution is negligible (p_{calc}: $\overline{X} = 3007$, $s = 3099$; $p_{pred\_5\%}$: $\overline{X} = 3017$, $s = 3117$; $p_{pred\_20\%}$: $\overline{X} = 3010$, $s = 3107$). We found a very small

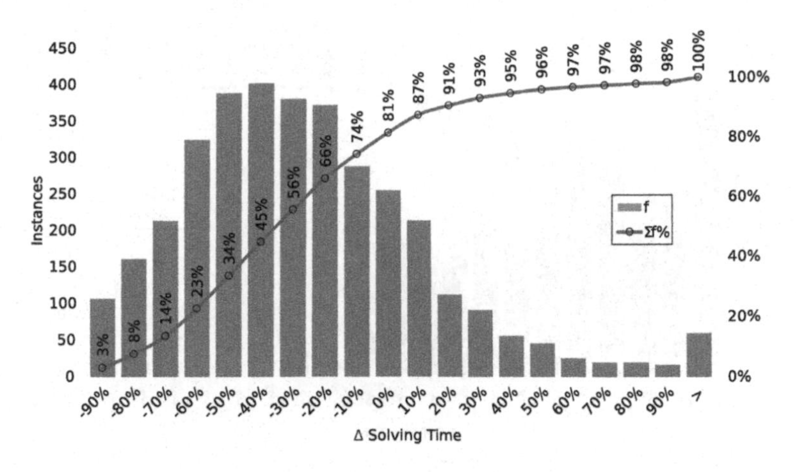

Fig. 4. Solving time difference with 5% margin on I_{sample} (optimal solution found)

Table 9. Industry Instances: Predicted vs. Calculated bounds

		Inst.	ΔP	ΔT	ΔN
Margin ±10%	Any p	163	1.2%	−90.3%	−32.6%
	p_{calc}	112	—	−43.5%	−18.8%
Margin ±20%	Any p	168	0.5%	−78.8%	−3.7%
	p_{calc}	144	—	−33.2%	2.0%

increase ($\Delta p = 9.4$) between $p_{pred\_5\%}$ and p_{calc}, an even smaller increase ($\Delta p = 2.2$) between $p_{pred\_20\%}$ and p_{calc}, and an analogous increase ($\Delta p = 7.2$) between $p_{pred\_5\%}$ and $p_{pred\_20\%}$. Table 7 reports the relative change in p.

Results on Industry Instances. Similar promising results were observed on the industry data set I_{ind} (Table 9). Using B_{pred} resulted in a large decrease in solving time with practically no increase in the value of p. Further, Fig. 5 shows that in 85% of all solved instances, solving time was reduced.

5 Related Work

The interplay between learning, search and inference for solving combinatorial optimisation problems was surveyed by Bengio et al. [1]. In CP, Cappart et al. [4] tightly hybridise reinforcement learning with a dynamic programming representation. Most works tend to emphasise one side or the other.

On the one hand, some authors emphasise the ML side followed by some search. In early work, Deudon et al. [5] for instance study the travelling salesperson problem, and improve a learned schedule with local search. In more recent work, Kool et al. [9] study vehicle routing problems with time windows and derive a schedule using neural-network guided dynamic programming.

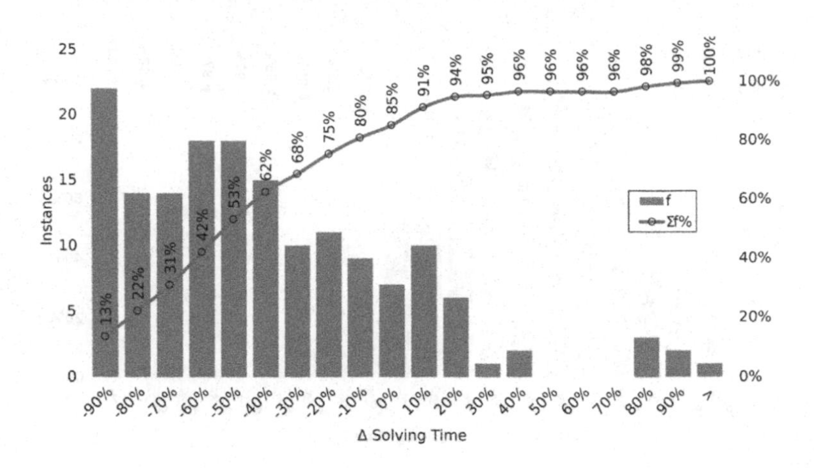

Fig. 5. Solving time difference with 10% margin on I_{ind} (all solved instances)

On the other hand – and closer to this paper – is the idea of pre-computing using ML and informing a 'standard' optimisation solver with this information. Osanlou et al. [16], for instance, study constrained path planning and use a graph convolution network to predict an optimal vertex order. They feed the resulting cost into a branch-and-bound search as an upper bound. Xu et al. [25] predict with supervised learning the satisfaction of a general (binary) CSP. Wang et al. [24] study job shop scheduling, and predict the optimal makespan of an instance. We follow the same approach, but on a more complex problem; we use aggregation functions in order to obtain a regressor independent of the problem size (number of tanks); we further simplify the ML models by performing feature extraction; and we use prediction bounding rather than the binary objective search. There are many recent similar works (Václavík et al. [26], Zhang et al. [22]); none to our knowledge consider the hoist scheduling problem.

Lastly we note the approach of decision-focussed learning (e.g., [14]), which offers a tighter integration between learning and optimisation than either the learning-to-search or the predict-then-optimise approaches.

6 Conclusion and Future Work

The growing research field at the intersection of machine learning and combinatorial optimisation leads in this paper to the question: how can a predict-then-optimise approach lead to faster solving of hard scheduling problems, without sacrificing solution quality?

This approach – in the form of offline supervised learning coupled with online exact solving – was applied to the CHSP, a hard combinatorial optimisation problem that involves search for a feasible hoist schedule that minimises the cycle period p. We tested various ML models for predicting the optimal p, in order to improve the effectiveness of a CP-SAT solver by providing tighter objective function bounds. Using a size-invariant representation, coupled with feature ablation, we found that a relatively simple dense neural network was effective in acquiring a predictor. Experiments on synthetic and industrial benchmarks showed that using tighter bounds derived from $\hat{p}$ leads to markedly lower solving times (mean 44% decrease on industrial instances), without increasing the value of p in the majority of cases. Only for a small fraction of instances is the solving time using the predicted bounds longer. This minority increase most commonly happens when the initial solving time is negligible (i.e., easy instances): thus the delta is small in absolute terms.

Our approach has intentionally emphasised simplicity and a loose coupling of the learning and optimisation. The highly-promising results suggest several future directions. First, it will be interesting to choose the relaxed bounds around $\hat{p}$ based not on static percentages, but according to the confidence of the ML model. Second, to further study the generalisation ability of the predictor: for instance, testing on instances obtained from a different generator, or exploring other backend solvers such as a MIP. Third, since the ML models implemented in this paper do not consider some CSHP instance attributes (e.g., number of tracks and loading/unloading times), their inclusion could be investigated in future work. Fourth, the ML model could be informed by a regret-based loss function, in the style of decision-focussed learning [20]. Besides these directions, we are exploring whether a more end-to-end ML approach could be promising. To this aim we currently study the use of a graph neural network embedding to try and predict the optimal solution (not just the optimal objective function value). Initial experiments indicate that predicting the whole solution is challenging; it seems more promising to predict only the removal time of the first item from its first tank, and we find this already is useful guidance for the solver.

Acknowledgements. We thank the anonymous reviewers. Thanks to K. van den Houten, S. van der Laan, M. Wallace and M. de Weerdt. Partially supported by TAILOR, funded by the EU Horizon 2020 programme under grant 952215.

References

1. Bengio, Y., Lodi, A., Prouvost, A.: Machine learning for combinatorial optimization: a methodological tour d'horizon. Eur. J. Oper. Res. **290**(2), 405–421 (2021). https://doi.org/10.1016/j.ejor.2020.07.063
2. Boysen, N., Briskorn, D., Meisel, F.: A generalized classification scheme for crane scheduling with interference. Eur. J. Oper. Res. **258**(1), 343–357 (2017). https://doi.org/10.1016/j.ejor.2016.08.041
3. Breiman, L.: Random forests. Mach. Learn. **45**(1), 5–32 (2001). https://doi.org/10.1023/A:1010933404324
4. Cappart, Q., Moisan, T., Rousseau, L.M., Prémont-Schwarz, I., Cire, A.A.: Combining reinforcement learning and constraint programming for combinatorial optimization. In: AAAI 2021, pp. 3677–3687 (2021). https://doi.org/10.1609/aaai.v35i5.16484
5. Deudon, M., Cournut, P., Lacoste, A., Adulyasak, Y., Rousseau, L.: Learning heuristics for the TSP by policy gradient. In: CPAIOR 2018, pp. 170–181 (2018). https://doi.org/10.1007/978-3-319-93031-2_12

6. Feng, J., Chu, C., Che, A.: Cyclic jobshop hoist scheduling with multi-capacity reentrant tanks and time-window constraints. Comput. Ind. Eng. **120**, 382–391 (2018). https://doi.org/10.1016/j.cie.2018.04.046
7. He, L., Guijt, A., de Weerdt, M., Xing, L., Yorke-Smith, N.: Order acceptance and scheduling with sequence-dependent setup times. Comput. Ind. Eng. **138**, 106102 (2019). https://doi.org/10.1016/j.cie.2019.106102
8. Kacem, I., Kellerer, H.: Foreword: combinatorial optimization for industrial engineering. Comput. Ind. Eng. **61**(2), 239–241 (2011). https://doi.org/10.1016/j.cie.2011.07.016
9. Kool, W., van Hoof, H., Gromicho, J.A.S., Welling, M.: Deep policy dynamic programming for vehicle routing problems. In: CPAIOR 2022, pp. 190–213 (2022). https://doi.org/10.1007/978-3-031-08011-1_14
10. Kotary, J., Fioretto, F., Hentenryck, P.V., Wilder, B.: End-to-end constrained optimization learning: a survey. In: IJCAI 2021, pp. 4475–4482 (2021). https://doi.org/10.24963/ijcai.2021/610
11. Laajili, E., Lamrous, S., Manier, M., Nicod, J.: An adapted variable neighborhood search based algorithm for the cyclic multi-hoist design and scheduling problem. Comput. Ind. Eng. **157**, 107225 (2021). https://doi.org/10.1016/j.cie.2021.107225
12. Lee, C.Y., Lei, L., Pinedo, M.: Current trends in deterministic scheduling. Ann. Oper. Res. **70**, 1–41 (1997). https://doi.org/10.1023/A:1018909801944
13. Leung, J.M., Zhang, G., Yang, X., Mak, R., Lam, K.: Optimal cyclic multi-hoist scheduling: a mixed integer programming approach. Oper. Res. **52**(6), 965–976 (2004). https://doi.org/10.1287/opre.1040.0144
14. Mandi, J., Bucarey, V., Mulamba Ke Tchomba, M., Guns, T.: Decision-focused learning: through the lens of learning to rank. In: ICML 2022, pp. 14935–14947 (2022). https://proceedings.mlr.press/v162/mandi22a.html
15. Manier, M.A., Lamrous, S.: An evolutionary approach for the design and scheduling of electroplating facilities. J. Math. Model. Algorithms **7**(2), 197–215 (2008). https://doi.org/10.1007/s10852-008-9083-z
16. Osanlou, K., Bursuc, A., Guettier, C., Cazenave, T., Jacopin, E.: Optimal solving of constrained path-planning problems with graph convolutional networks and optimized tree search. In: IROS 2019, pp. 3519–3525 (2019). https://doi.org/10.1109/IROS40897.2019.8968113
17. Perron, L., Furnon, V.: OR-Tools version 9.3 (2022). https://developers.google.com/optimization/
18. Phillips, L.W., Unger, P.S.: Mathematical programming solution of a hoist scheduling program. AIIE Trans. **8**(2), 219–225 (1976). https://doi.org/10.1080/05695557608975070
19. Quesnel, F., Wu, A., Desaulniers, G., Soumis, F.: Deep-learning-based partial pricing in a branch-and-price algorithm for personalized crew rostering. Comput. Oper. Res. **138**, 105554 (2022). https://doi.org/10.1016/j.cor.2021.105554
20. Teso, S., et al.: Machine learning for combinatorial optimisation of partially-specified problems. CoRR abs/2205.10157 (2022). https://doi.org/10.48550/arXiv.2205.10157
21. UTIKAL Automation: Private correspondence (2022). https://utikal-automation.com
22. Václavík, R., Novák, A., Sucha, P., Hanzálek, Z.: Accelerating the branch-and-price algorithm using machine learning. Eur. J. Oper. Res. **271**(3), 1055–1069 (2018). https://doi.org/10.1016/j.ejor.2018.05.046

23. Wallace, M., Yorke-Smith, N.: A new constraint programming model and solving for the cyclic hoist scheduling problem. Constraints **25**(3), 319–337 (2020). https://doi.org/10.1007/s10601-020-09316-z
24. Wang, T., Payberah, A.H., Vlassov, V.: CONVJSSP: convolutional learning for job-shop scheduling problems. In: ICMLA 2022, pp. 1483–1490 (2020). https://doi.org/10.1109/ICMLA51294.2020.00229
25. Xu, H., Koenig, S., Kumar, T.K.S.: Towards effective deep learning for constraint satisfaction problems. In: CP 2018, pp. 588–597 (2018). https://doi.org/10.1007/978-3-319-98334-9_38
26. Zhang, W., et al.: NLocalSAT: boosting local search with solution prediction. In: IJCAI 2020, pp. 1177–1183 (2020). https://doi.org/10.24963/ijcai.2020/164

Scalable and Near-Optimal ε-Tube Clusterwise Regression

Aravinth Chembu(✉), Scott Sanner, and Elias B. Khalil

Department of Mechanical and Industrial Engineering, University of Toronto, Toronto, Canada

aravinth.chembu@mail.utoronto.ca, {ssanner,khalil}@mie.utoronto.ca

Abstract. Clusterwise Regression (CLR) methods that jointly optimize clustering and regression tasks are useful for partitioning data into disjoint subsets with distinct regression trends. Due to the inherent difficulty in simultaneously optimizing clustering and regression objectives, it is not surprising that existing optimal CLR approaches do not scale beyond 100 s of data points. In an effort to provide more scalable and optimal CLR methods, we propose a novel formulation of the problem that takes inspiration from ε-tubes in Support Vector Regression (SVR). The advantage of this novel formulation, which aims to assign data points to clusters in order to minimize the largest ε-tube that encapsulates the regressed data, is that it admits an optimal MILP formulation. Furthermore, given that each constraint in our formulation corresponds to a single data point, we propose an efficient row generation solution that can optimally converge for the full dataset while only requiring optimization over a subset of the data. Our results on a variety of synthetic and benchmark real datasets show that our Clusterwise Regression MILP formulation provides near-optimal solutions up to 100,000 data points and the smallest data-encapsulating ε-tubes among CLR alternatives.

Keywords: Clusterwise Regression · Row Generation · Mixed-integer linear programming

1 Introduction

Clusterwise Regression (CLR) is a fundamental task in Machine Learning that jointly optimizes for clustering and regression tasks, where the data is partitioned into several clusters, each group fit by a regression plane, such that the overall regression error is minimized. CLR models find applications in a plethora of fields such as social science [17], marketing analysis [9], and climate modeling [1].

Traditionally, CLR models entailed jointly optimizing for clustering with the squared error objective for regression, as proposed in seminal work on CLR [20]. Existing greedy algorithms for CLR are sensitive to initialization and provide only locally optimal results, thus limiting clustering quality and reproducibility. Moreover, the classical CLR model [20] was recently shown to be NP-hard [18] and considered a very difficult problem to solve [14]. Thus, optimally solving for 100 s of data observations is challenging [4–7], even with synthetic examples.

© The Author(s), under exclusive license to Springer Nature Switzerland AG 2023
A. A. Cire (Ed.): CPAIOR 2023, LNCS 13884, pp. 254–263, 2023.
https://doi.org/10.1007/978-3-031-33271-5_17

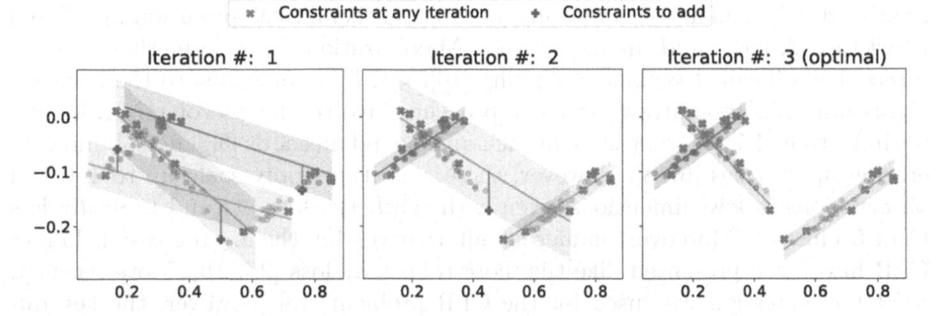

Fig. 1. We show the ε-tube CLR solution (right) with an illustrative example. Additionally, we demonstrate our row generation algorithm, where we start (left) with an initial set of points (denoted with $\times$) and run two iterations adding 3 constraints per iteration denoted with $+$ that are the farthest from the regression lines until we reach the optimal result in the third iteration. We observe that convergence to the optimal solution does not require ε to monotonically decrease.

In this work, we propose a novel approach to CLR that is inspired by the ε-tubes (or margins) that correspond to absolute values of the regression residuals in Support Vector Regression (SVR) [12,22]. In this formulation, we minimize the largest ε-tube across all clusters that encapsulates the regressed data. Such a formulation is inherently insensitive to cluster size imbalance since we only measure the worst-case residual. In addition, a core computational advantage of this formulation is that it can be expressed and optimally solved as a Mixed Integer Linear Program (MILP) that supports an efficient row (constraint) generation strategy. We illustrate this iterative row generation process in Fig. 1 demonstrating the evolution of data point (re)assignments to three clusters and their corresponding shaded ε-tube at each iteration until optimality. It is important to note that this solution only generated the most-violated constraints for all data points (most often near the ε-tube boundaries, by definition) since the remaining data lie within tube boundaries and automatically satisfy the optimality criteria.

Leveraging our novel MILP formulation and row generation solution can thus solve ε-tube CLR using a subset of the data (while guaranteeing optimality w.r.t. all data), hence providing near-optimal results for up to 100,000 data points in comparison to other CLR formulations and solutions that cannot scale optimally beyond 100 s of data points. We provide experiments on a variety of synthetic datasets (varying number of data points, dimensionality, clusters, and cluster imbalance) and 10 benchmark real datasets to demonstrate our algorithm's ability to reach the smallest ε-tube clusters when compared with several baselines.

2 Related Work

Several greedy algorithms have been proposed to solve the classical CLR problem, including exchange algorithms proposed in the pioneering works of

Späth [20,21], simulated annealing in [10], mathematical programming-based heuristics [2,3,13], and an Expectation-Maximization [8] type methodology in a recent work called k-plane clustering [16], which is analogous to the k-means algorithm [15]. In contrast, exact approaches involve the use of mixed-integer optimization [4,7,14], repetitive branch-and-bound methods [5], and column generation approaches [6,18]. However, these algorithms only scale up to 100 s of observations in low dimensions, even with synthetic datasets and typically less than 5 clusters. Moreover, numerous alternatives for the L_2 regression loss of CLR have been presented, like the more robust L_1 loss [2,4,19]. More recently, SVR for regression was used for the CLR problem [13]; however, the key difference with our approach is that we directly minimize the ε-tubes while they solve for pure SVRs in each cluster (by minimizing the slacks) with ε being a hyperparameter; further, they do not provide any optimality guarantees.

3 Optimal CLR with ε-Tube Objective

3.1 Reduction of ε-Tube CLR to a MILP

Our ε-tube objective for CLR minimizes the maximum regression residual for every point across all the clusters. More formally, consider that we have n observations $(\boldsymbol{x}_i, y_i)$ with d features in the dataset $(\boldsymbol{X}, y) \in \mathbb{R}^{n\times(d+1)}$ where $i \in N = \{1, ..., n\}$. The main goal in CLR is to find one regression plane for each of the k clusters (C_j), where the regression coefficients for the jth cluster are represented by weights $\boldsymbol{w}_j \in \mathbb{R}^d$ and bias $b_j \in \mathbb{R}$ for $j \in K = \{1, ..., k\}$. We will use $\boldsymbol{w}$ and b without cluster indices to refer to the collection of k regression plane parameters. Each data point is assigned to exactly one cluster, similar to a hard-partitioning setting in unsupervised clustering. We introduce binary variables c_{ij} that denote whether point i is assigned to cluster C_j ($c_{ij} = 1$) or not ($c_{ij} = 0$), thus enabling us to formulate a min-max mixed integer optimization problem. Using this notation, we define our **first key novel contribution of the ε-tube CLR objective**:

$$\min_{w,b,c} \max_{j\in K} \max_{i\in N} |y_i - \boldsymbol{w}_j^\intercal \boldsymbol{x}_i - b_j| \cdot c_{ij} \tag{1}$$

Here, we first observe that we can further reduce this ε-tube CLR objective to a bi-level optimization problem with the introduction of a new variable ε, which takes the value of the maximum residual from all points across the k regression planes (through the first constraint in (2)).

$$\begin{aligned}
&\min_{w,b,c} \ \varepsilon \\
&\text{s.t.} \ \ \varepsilon = \max_{j\in K} \max_{i\in N} |y_i - \boldsymbol{w}_j^\intercal \boldsymbol{x}_i - b_j| \cdot c_{ij} \\
&\qquad \sum_{j=1}^{k} c_{ij} = 1, \ \ i \in N; \ \ w_{j,1} < w_{j+1,1}, \ j \in K \backslash \{k\}; \\
&\qquad \boldsymbol{w}_j \in \mathbb{R}^d, \ \ b_j \in \mathbb{R}, \ \ j \in K; \ \ c_{ij} \in \{0,1\}, \ \ i \in N, \ j \in K;
\end{aligned} \tag{2}$$

In this bi-level problem, indicator variables c_{ij} ensure we only capture residuals for the regression plane (cluster) a point is assigned to. Moreover, constraints $\sum_{j=1}^{k} c_{ij} = 1$ guarantee every point in the dataset is assigned to exactly one cluster. We also add symmetry-breaking constraints of the form $w_{j,1} < w_{j+1,1}$ that enforce the first dimension of the regression coefficients across clusters to be in increasing order. This guarantees that we choose exactly one solution out of the $k!$ possible permutations with the same optimal value. A key observation is that we can *remove both* max*'s* from the first constraint and rewrite it to $\varepsilon \geq |y_i - \boldsymbol{w}_j^\intercal \boldsymbol{x}_i - b_j| \cdot c_{ij}, i \in N, j \in K$. While this constraint ensures that ε takes a value greater than or equal to the max prediction error, the minimization criteria in (2) enforces equality! Using this elegant transformation and indicator constraints to encode the product of regression residual and c_{ij} yields our **second key novel contribution of ε-tube CLR formulated as a MILP**:

$$
\begin{aligned}
\min_{\boldsymbol{w},b,c} \ & \varepsilon \\
\text{s.t.} \ & c_{ij} = 1 \implies \varepsilon \geq |y_i - \boldsymbol{w}_j^\intercal \boldsymbol{x}_i - b_j|, \ \ i \in N, \ j \in K \\
& \sum_{j=1}^{k} c_{ij} = 1, \ \ i \in N; \ \ w_{j,1} < w_{j+1,1}, \ j \in K\backslash\{k\}; \\
& \boldsymbol{w}_j \in \mathbb{R}^d, \ \ b_j \in \mathbb{R}, \ \ j \in K; \ \ c_{ij} \in \{0,1\}, \ \ i \in N, \ j \in K;
\end{aligned}
\tag{3}
$$

3.2 Row Generation Methodology

The final pure-MILP formulation for our ε-tube objective in (3) allows for the use of efficient branch-and-bound strategies through state-of-the-art MILP solvers. However, the large number of binary variables and constraints may present a challenge for MILP solvers. A possible solution would be to reduce the number of variables and constraints of the model without affecting its correctness.

In problem (3), we observe that we minimize a single variable ε whose value is governed through the $n \times k$ indicator constraints. If points that have large residuals w.r.t. the regression coefficients for the optimal result can be known a priori, the *indicator constraints corresponding to the points that have much smaller residuals can be neglected.* Neglecting these observations will not change the optimal value as ε is already larger than the residuals from these points.

This crucial insight can be leveraged in our **third novel contribution of an efficient row (constraint) generation MILP solution** by starting with a small subset of observations (with their associated variables and constraints) in a reduced version of (3) we term main problem (MP). Given an optimal solution to MP, we check whether it is optimal for the full problem (3). This check can be performed through a sub-problem (SP) that identifies points that have residuals larger than that of the current solution of the MP. In essence, the SP identifies the *most-violated constraints* corresponding to the points with largest residuals not yet included in the MP. These most-violated constraints can then be added to the MP. This procedure can be iteratively executed until the SP ensures that all observations have residuals smaller than that of the current optimal solution. In such a case, we have found an optimum for the full problem (3). Convergence to optimality is guaranteed in finite time since in the (unlikely) worst case this happens when we generate all rows (constraints) and recover the full problem (3).

Algorithm 1. Row generation

Input $(\boldsymbol{x}_i, y_i)$, k
1: $\varepsilon^* \leftarrow 0$, $\hat{\varepsilon} \leftarrow \infty$
2: $I \neq \emptyset$, CONS $\neq \emptyset$, $\boldsymbol{w}_j, b_j \leftarrow$ Initialize
▷ *Initial constraints* CONS *for points* I
3: $\varepsilon^*, \boldsymbol{w}_j^*, b_j^*, c_{ij}^* \leftarrow$ Solve MILP with CONS
4: $\hat{c_{ij}} \leftarrow \{\mathbb{1}_{j=\hat{j}} | \hat{j} = \arg\min_j |y_i - \boldsymbol{w}_j^{*\intercal}\boldsymbol{x}_i - b_j^*|\}$, ▷ *Assign points* $i \in N$
5: $\hat{\varepsilon}, I$, CONS $\leftarrow$ Add-Constraints()
6: **if** $\hat{\varepsilon} > \varepsilon^*$ **then**
7: **go to** line 3 ▷ *Re-solve MILP with augmented constraints set* CONS
8: **end if**
9: **return** $\varepsilon^*, \boldsymbol{w}_j^*, b_j^*, \hat{c_{ij}}$ ▷ *Optimal*

Algorithm 2. Add-Constraints

Input $(\boldsymbol{x}_i, y_i), \boldsymbol{w}_j^*, b_j^*, \hat{c_{ij}}$, CONS, ε^*
$\hat{\varepsilon} = \max_{i \in N,\ j \in K}\{|y_i - \boldsymbol{w}_j^{*\intercal}\boldsymbol{x}_i - b_j^*| \cdot \hat{c_{ij}}\}$
▷ *Check if* ε^* *is the max residual*
if $\hat{\varepsilon} > \varepsilon^*$ **then**
  for $j \in K$ **do**
    $I_{add} \leftarrow \{\arg\max_i |y_i - \boldsymbol{w}_j^{*\intercal}\boldsymbol{x}_i - b_j^*| \cdot \hat{c_{ij}}\}$ ▷ *Find largest residual*
    $I \leftarrow I \cup I_{add}$
  end for
  CONS $\leftarrow$ CONS $\cup$ $\{c_{ij} = 1 \implies \varepsilon \geq |y_i - \boldsymbol{w}_j^\intercal \boldsymbol{x}_i - b_j|, i \in I_{add}, j \in K\}$
end if
return $\hat{\varepsilon}, I$, CONS

We formalize the above row (or constraint) generation procedure through Algorithms 1 and 2 that primarily perform the MP and SP tasks, respectively. In Algorithm 1, we initialize our model with a small subset of observations in $I \subset N$, and their corresponding variables c_{ij} and constraints in C. We solve the MP with the reduced formulation of (3) using a MILP solver to obtain the optimal value ε^*. The coefficients for the k regression planes are used to assign a point $i \in N$ to the cluster to which it has the lowest prediction error (line 4 in Algorithm 1). This is a crucial step in our algorithm as we use the cluster assignment information to then compute the maximum residual stored in $\hat{\varepsilon}$ (line 1 in Algorithm 2) for all points w.r.t. the coefficients obtained from the MP. If $\hat{\varepsilon} > \varepsilon^*$, we are yet to reach the optimal solution. Hence, we identify the most violating constraints (if one exists) for each of the k clusters through the SP (line 7 in Algorithm 2) and add them to the MP. We stop iterating between the MP and SP when $\hat{\varepsilon} = \varepsilon^*$, i.e., when no more points $i \in N$ incur residuals larger than the current objective, implying optimality w.r.t. all constraints.

4 Empirical Evaluation

We first study the properties of our ε-tube CLR objective and comparatively evaluate our solution on synthetic and real datasets. We use a hyphenated three-part naming convention: (1) clustering criteria (k-means or direct point-to-cluster assignment), (2) regression loss (least squares, SVR, or ε-tube), and

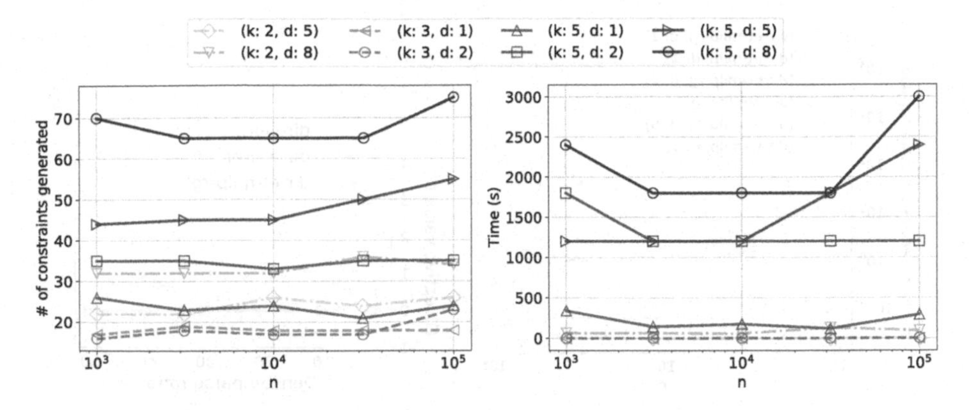

Fig. 2. # Constraints generated and solution time to 5% optimality gap for dir-et-milp-rg with number of clusters k and dimensionality d vs. amount of data n for the *klr-data*. # Constraints and time only increase marginally as n increases.

(3) optimization method (independent, iterative, or MILP). We compare the following methods: (i) **km-ls-indep**: k-means (km) followed by least-squares (ls) regression where the clustering and regression take place independently; (ii) **km-svr-indep**: k-means followed by SVR [22]; (iii) **km-et-indep**: k-means followed by optimizing our ε-tube objective; (iv) **dir-ls-iter**: k-planes [16] algorithm with least squares regression where the clustering is directly (dir) assigned by an iterative procedure (iter); (v) **dir-et-iter**: a novel k-means inspired approximate iterative algorithm to optimize for our ε-tube (et) objective — here we iterate between (a) finding the best set of hyperplanes (compute $\boldsymbol{w}_j, b_j, j \in K$) per cluster given a cluster assignment for all points and (b) re-assigning points to clusters (update c_{ij}) such that each point has the lowest prediction error when assigned to that cluster; (vi) **dir-et-milp**: our novel full MILP from Eq (3); (vii) dir-et-milp-rg: our novel full MILP with *row generation* (rg).

Across our experiments, similar to the approach followed in [5,6,16,18], we primarily focus on comparing the ε value, i.e., the maximum residual among all clusters, since providing an optimal algorithm for the ε-tube objective is our key contribution; we only include non-"et" methods for relative comparison with other common CLR methods. All experiments were run on Google Colab in the standard CPU setting (at 2.3 GHz and 32 GB memory) with Gurobi 9.5.2. All code is available at https://github.com/Aravinthck/CLR-epsTube.

4.1 Synthetic Dataset Experiments

Scalability. Our dir-et-milp-rg algorithm depends on solving a MILP at every iteration, which leads us to ask how well it scales vs. the dimensionality, number of clusters, and amount of data. To this end, we evaluate the scalability of dir-et-milp-rg when the ground truth clusters are recovered by constructing synthetic datasets (called *klr-data*) similar in spirit to [5,6,16] where we choose

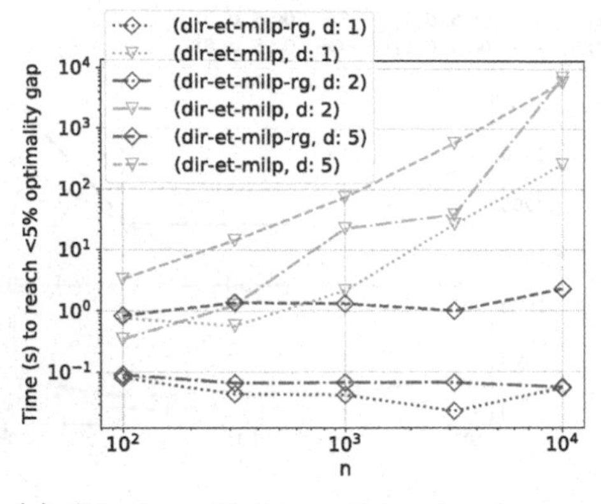

(a) We show that run-times for the row generation method are a fraction of dir-et-milp (as we increase n, d with $k = 2$) where at $n = 10^4$, the difference was 2s to 6030s.

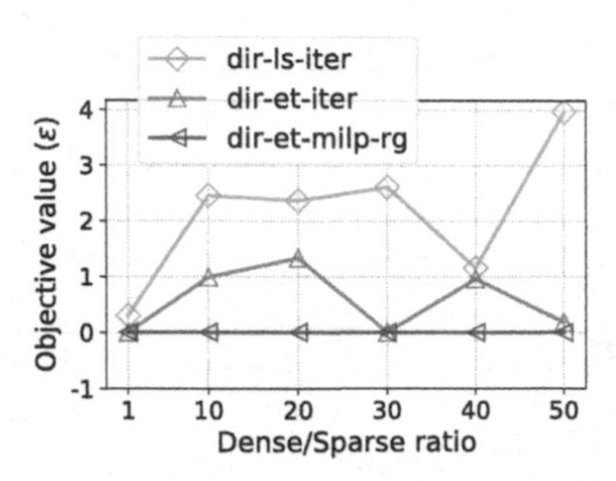

(b) We show that ε-tube objective is agnostic to cluster imbalance and finds the true ground-truth clusters, unlike the least squares objective.

Fig. 3. Comparative experiments of different algorithms for synthetic datasets.

$k \in \{2, 3, 5\}$ regression parameters uniformly at random with $d \in \{1, 2, 5, 8\}$ features and normal error in the regression. The feature vectors are extracted from Gaussian clusters with observations n varied from 10^3 to 10^5 points. Fig. 2 shows that the number of constraints generated to reach an optimality gap of 5% only increases marginally with the number of data points n. Similar trends are observed for the reported run-time for these experiments (cf. Fig. 2, right). These empirical results further suggest that only a small fraction of the observations are needed. For example, only ≈ 75 observations are needed to solve the problem to 5% optimality gap with 10^5 observations (top-right point in Fig. 2, left).

Performance Gain for Row Generation. We now compare the run-times of our row generation method dir-et-milp-rg and full-MILP solution **dir-et-milp** in Fig. 3a where we terminate on reaching the 5% optimality gap with both methods. With the *klr-data*, we experiment with n ranging from 100 to 10^4, $d \in \{1, 2, 5\}$ and $k = 2$ to ensure that **dir-et-milp** finishes in under 10^4 seconds. Here, dir-et-milp-rg strictly dominates its **dir-et-milp** counterpart in all cases and by more than 3 orders of magnitude for $n = 10^4$. What is more remarkable is that dir-et-milp-rg remains relatively flat as n increases in contrast to **dir-et-milp** that grows exponentially with n (i.e., as evidenced by the superlinear trend on this log-log plot).

Robustness to Cluster Imbalance. The optimal solution of our ε-tube CLR should be insensitive to imbalance in the number of points in the clusters because we only measure the worst-case residual. This alleviates the need for a higher concentration of clusters in areas where the data is denser. We validate this

Table 1. Comparative evaluation of objective values ε from (3) on 10 datasets shown in the columns and ordered by number of data points (n).

index	Iris	Autompg	Ceosalaries	Boston	Airfoil	Redwine	Abalone	Whitewine	Powerplant	Protein
# data (n)	150	392	500	506	1503	1599	4177	4898	9568	45730
# dimension (d)	4	7	1	13	5	11	7	11	4	9
# clusters (k)	3	3	6	2	4	3	3	3	5	2
km-ls-indep	0.8	1.5	10.64	2.34	2.77	3.04	4.56	3.98	2.71	3.94
km-svr-indep	0.94	1.38	5.69	2.02	2.3	2.81	3.57	3.39	1.99	1.72
km-et-indep	0.63	1.06	5.68	1.27	2.2	2.24	2.83	2.98	1.62	1.72
dir-ls-iter	0.48	1.0	2.5	1.76	1.21	2.34	2.82	3.05	1.76	2.16
dir-et-iter	0.32	0.59	0.7	0.99	0.86	1.55	1.75	1.86	0.53	1.67
dir-et-milp	0.24	0.42	0.64	0.78	0.77	1.13	1.75	1.82	1.45	1.72
dir-et-milp-rg	**0.22**	**0.38**	**0.48**	**0.67**	**0.59**	**0.62**	**0.88**	**1.03**	**0.3**	**0.82**

claim with imbalanced clusters with $n = 10^4$, $d = 1$, and $k = 3$. One of the clusters was designed to be dense, and the others were sparse, with the ratio of points given by *dense/sparse ratio.* From Fig. 3b, it is evident that all three methods came close to recovering the optimal *balanced* clusters (dense/sparse ratio = 1). However, for the *imbalanced* cases (ratio > 1), only dir-et-milp-rg identified the true ground truth cluster in all cases, while **dir-et-iter** struggled with local optimality of its greedy method (reaching optimality once) and **dir-ls-iter** appears to further suffer from the sensitivity of its least sum-of-squared loss to imbalanced clusters.

4.2 Real Dataset Experiments

We now benchmark our model with 10 commonly used CLR datasets found in [2,11,13,16]. In Table 1, we report the mean objective value ε obtained after 10 independent runs for the various greedy baselines compared with the full-MILP solution **dir-et-milp**, which is run for the same amount of time that it takes dir-et-milp-rg to reach the 5% optimality gap. We use the best value for k reported in the mentioned peer-reviewed works for each dataset. The ε values obtained for dir-et-milp-rg are better than all other methods, often substantially. Interestingly, we note that the greedy **dir-et-iter** that we have proposed outperforms **dir-et-milp** on several datasets. Moreover, we observed that the lower bound returned by Gurobi for **dir-et-milp** was zero and did not increase within the restricted time limit for these experiments. Improved formulation for **dir-et-milp** (and -rg) and tightening of lower bounds is an interesting direction for potential future work. However, the strictly dominant performance of dir-et-milp-rg underscores its ability to scale to large real datasets.

5 Conclusion

We provided a novel formulation for the ε-tube CLR problem that reduces to a MILP and further admits an efficient row generation solution. Our results on benchmark datasets make it evident that our row generation solution is much faster than solving the full MILP and that we outperform other CLR methods in terms of maximum ε-tube loss and different levels of cluster imbalance.

References

1. Bagirov, A.M., Mahmood, A., Barton, A.: Prediction of monthly rainfall in victoria, Australia: clusterwise linear regression approach. Atmos. Res. **188**, 20–29 (2017). https://doi.org/10.1016/j.atmosres.2017.01.003, https://www.sciencedirect.com/science/article/pii/S0169809517300285
2. Bagirov, A.M., Taheri, S.: Dc programming algorithm for clusterwise linear l_1 regression. J. Oper. Res. Soc. China **5**(2), 233–256 (2017)
3. Bagirov, A.M., Ugon, J., Mirzayeva, H.: Nonsmooth nonconvex optimization approach to clusterwise linear regression problems. Eur. J. Oper. Res. **229**(1), 132–142 (2013). https://doi.org/10.1016/j.ejor.2013.02.059, https://www.sciencedirect.com/science/article/pii/S0377221713002087
4. Bertsimas, D., Shioda, R.: Classification and regression via integer optimization. Oper. Res. **55**(2), 252–271 (2007). https://doi.org/10.1287/opre.1060.0360, https://doi.org/10.1287/opre.1060.0360
5. Carbonneau, R.A., Caporossi, G., Hansen, P.: Extensions to the repetitive branch and bound algorithm for globally optimal clusterwise regression. Comput. Oper. Res. **39**(11), 2748–2762 (2012). https://doi.org/10.1016/j.cor.2012.02.007
6. Carbonneau, R.A., Caporossi, G., Hansen, P.: Globally optimal clusterwise regression by column generation enhanced with heuristics, sequencing and ending subset optimization. J. Classif. **31**(2), 219–241 (2014). https://doi.org/10.1007/s00357-014-9155-x
7. Carbonneau, R.A., Caporossi, G., Hansen, P.: Globally optimal clusterwise regression by mixed logical-quadratic programming. Eur. J. Oper. Res. **212**(1), 213–222 (2011). https://doi.org/10.1016/j.ejor.2011.01.016, https://www.sciencedirect.com/science/article/pii/S0377221711000191
8. Dempster, A.P., Laird, N.M., Rubin, D.B.: Maximum likelihood from incomplete data via the em algorithm. J. R. Stat. Soc. Series B (Methodological) **39**(1), 1–38 (1977). http://www.jstor.org/stable/2984875
9. DeSarbo, W.S., Cron, W.L.: A maximum likelihood methodology for clusterwise linear regression. J. Classif. **5**(2), 249–282 (1988). https://doi.org/10.1007/BF01897167
10. DeSarbo, W.S., Oliver, R.L., Rangaswamy, A.: A simulated annealing methodology for clusterwise linear regression. Psychometrika **54**(4), 707–736 (1989)
11. Di Mari, R., Rocci, R., Gattone, S.A.: Clusterwise linear regression modeling with soft scale constraints. Int. J. Approximate Reasoning **91**, 160–178 (2017). https://doi.org/10.1016/j.ijar.2017.09.006, https://www.sciencedirect.com/science/article/pii/S0888613X17305686
12. Drucker, H., Burges, C.J.C., Kaufman, L., Smola, A., Vapnik, V.: Support vector regression machines. In: Mozer, M., Jordan, M., Petsche, T. (eds.) Advances in Neural Information Processing Systems, vol. 9. MIT Press (1996). https://proceedings.neurips.cc/paper/1996/file/d38901788c533e8286cb6400b40b386d-Paper.pdf
13. Joki, K., Bagirov, A.M., Karmitsa, N., Mäkelä, M.M., Taheri, S.: Clusterwise support vector linear regression. Eur. J. Oper. Res. **287**(1), 19–35 (2020). https://doi.org/10.1016/j.ejor.2020.04.032, https://www.sciencedirect.com/science/article/pii/S0377221720303830

14. Lau, K.N., Leung, P.l., Tse, K.K.: A mathematical programming approach to clusterwise regression model and its extensions. Eur. J. Oper. Res. **116**(3), 640–652 (1999). https://EconPapers.repec.org/RePEc:eee:ejores:v:116:y:1999:i:3:p:640-652
15. Lloyd, S.: Least squares quantization in PCM. IEEE Trans. Inf. Theory **28**(2), 129–137 (1982). https://doi.org/10.1109/TIT.1982.1056489
16. Manwani, N., Sastry, P.: K-plane regression. Inf. Sci. **292**(C), 39–56 (2015). https://doi.org/10.1016/j.ins.2014.08.058
17. Olson, A.W., Zhang, K., Calderon-Figueroa, F., Yakubov, R., Sanner, S., Silver, D., Arribas-Bel, D.: Classification and regression via integer optimization for neighborhood change. Geogr. Anal. **53**(2), 192–212 (2021)
18. Park, Y.W., Jiang, Y., Klabjan, D., Williams, L.: Algorithms for generalized clusterwise linear regression. INFORMS J. Comput. **29**(2), 301–317 (2017)
19. Späth, H.: Clusterwise linear least absolute deviations regression. Computing **37**(4), 371–377 (1986)
20. Späth, H.: Algorithm 39 clusterwise linear regression. Computing **22**(4), 367–373 (1979). https://doi.org/10.1007/BF02265317
21. Späth, H.: A fast algorithm for clusterwise linear regression. Computing **29**(2), 175–181 (1982). https://doi.org/10.1007/BF02249940
22. Vapnik, V.: The Nature of Statistical Learning Theory. Springer, New York (1999)

Branch & Learn with Post-hoc Correction for Predict+Optimize with Unknown Parameters in Constraints

Xinyi Hu[1(✉)], Jasper C. H. Lee[2(✉)], and Jimmy H. M. Lee[1(✉)]

[1] Department of Computer Science and Engineering, The Chinese University of Hong Kong, Shatin, N.T., Hong Kong
xyhu@cse.cuhk.edu.hk, jasper.lee@wisc.edu
[2] Department of Computer Sciences & Institute for Foundations of Data Science, University of Wisconsin-Madison, Madison, WI, USA
jlee@cse.cuhk.edu.hk

Abstract. Combining machine learning and constrained optimization, Predict+Optimize tackles optimization problems containing parameters that are unknown at the time of solving. Prior works focus on cases with unknowns only in the objectives. A new framework was recently proposed to cater for unknowns also in constraints by introducing a loss function, called Post-hoc Regret, that takes into account the cost of correcting an unsatisfiable prediction. Since Post-hoc Regret is non-differentiable, the previous work computes only its approximation. While the notion of Post-hoc Regret is general, its specific implementation is applicable to only packing and covering linear programming problems. In this paper, we first show how to compute Post-hoc Regret exactly for any optimization problem solvable by a recursive algorithm satisfying simple conditions. Experimentation demonstrates substantial improvement in the quality of solutions as compared to the earlier approximation approach. Furthermore, we show experimentally the empirical behavior of different combinations of correction and penalty functions used in the Post-hoc Regret of the same benchmarks. Results provide insights for defining the appropriate Post-hoc Regret in different application scenarios.

Keywords: Constraint Optimization · Machine Learning · Predict+Optimize

1 Introduction

Constraint optimization problems are ubiquitous and occur in many daily and industrial applications [2,7,8]. In practice, constraint optimization problems can contain certain parameters which are unknown at the time of solving and require prediction based on some historical records. For example, a train company needs to schedule a minimal number of trains while meeting the passenger demand, but the precise demand is unknown ahead of time and needs to be predicted. The task is to 1) predict the unknown parameters, then 2) solve the optimization problem using the predicted parameters, such that the resulting solutions are good even under true parameters. Traditionally, in the prediction stage, machine learning models are trained with error metrics independent of

© The Author(s), under exclusive license to Springer Nature Switzerland AG 2023
A. A. Cire (Ed.): CPAIOR 2023, LNCS 13884, pp. 264–280, 2023.
https://doi.org/10.1007/978-3-031-33271-5_18

optimization problems, such as mean squared error. However, this kind of error metric does not necessarily represent the performance of the resulted solutions. The predicted parameters may in fact lead to a low-quality solution for the (true) optimization problem despite being "high-quality" for the error metric. Predict+Optimize instead trains the prediction model with the more effective *regret* function, capturing the difference in objective between the estimated and true optimal solutions, both evaluated using the true parameters. The challenge is, regret, the new error metric, is piecewise constant and non-differentiable [4], thus gradient-based methods do not apply.

A number of prior works [3–5, 10, 20] propose methods to overcome the non-differentiability of regret, and they can be roughly divided into *approximation* and *exact* methods. Approximation methods [5, 20] compute the (approximate) gradients of (approximations of) the regret function. They work not with the regret loss itself, but an approximation of it. While novel, they are not always reliable. On the other hand, exact methods [3, 4, 10] work directly with the regret to find a good prediction model, even if the method cannot always find the global-optimum model for the training data (e.g. if the method uses a local optimization method to find the output model). To overcome the nondifferentiability of the regret, they exploit the structure of optimization problems to train models without computing gradients, and can be applied to dynamic programming solvable problems [4] and recursively solvable problems [10].

Despite the variety of approaches, most of the previous works [3–5, 10, 20] on Predict+Optimize handle problems with unknowns only in the objective. When constraints contain also unknown parameters, one major challenge is that the estimated solution may end up being *infeasible* under the true parameters—an issue inherent with uncertainty in constraints. The *regret* function designed for fixed solution space is not applicable in this situation. Hu *et al.* [9] propose a more general loss function called *post-hoc regret*, in which an infeasible estimated solution is first corrected into a feasible one (with respect to true parameters), and then the error of prediction is the sum of 1) the objective difference between the true optimal solution and the feasible solution, and 2) potential penalty incurred by correction. When unknown parameters only appear in the objective, the post-hoc regret degenerates into the regret. The post-hoc regret is also nondifferentiable, and Hu *et al.* further propose an approximation approach for packing and covering linear programs [9]. However, exact approaches considering the post-hoc regret remain uncovered.

The contributions of this paper are threefold. First, we propose the first exact approach for Predict+Optimize with unknown parameters in both the objective and constraints. The proposed method is extended from Branch & Learn [10] and handles recursively solvable problems with unknown constraints. Second, extensive experiments are conducted to investigate the performance of post-hoc regret. We experimentally compare the proposed method with the state-of-the-art approximation method [9] and investigate the performance of post-hoc regret on more general problems. Third, we empirically study different combinations of the two key components of post-hoc regret, i.e., the correction function and the penalty function to gain insights for defining post-hoc regret in different scenarios. Due to the page limit, we opted to present only part of the experiment results. For the complete experiment results and analysis, as well as future work, please refer to the extended version on arXiv with the same title.

2 Background

Without loss of generality, we define an *optimization problem* (OP) P as finding:

$$x^* = \arg\min_x obj(x) \quad \text{s.t. } C(x)$$

where $x \in \mathbb{R}^d$ is a vector of decision variables, $obj : \mathbb{R}^d \to \mathbb{R}$ is a real-valued objective function in x which is to be minimized, and C is a set of constraints over x. We say x^* is an *optimal solution* and $obj(x^*)$ is the *optimal value*. A *parameterized optimization problem (Para-OP)* $P(\theta)$ is an extension of an OP P:

$$x^*(\theta) = \arg\min_x obj(x, \theta) \quad \text{s.t. } C(x, \theta)$$

where $\theta \in \mathbb{R}^t$ is a vector of parameters. The objective $obj(x, \theta)$ and the constraints $C(x, \theta)$ all depend on θ. An OP is a degenerated case of a Para-OP when there are no unknowns.

The true parameters $\theta \in \mathbb{R}^t$ for a Para-OP are hidden at the time of solving in the *Predict+Optimize* (P+O) setting [3], and *estimated parameters* $\hat{\theta}$ are utilized in their places. Suppose each parameter is estimated by m features. The estimation will rely on a machine learning model trained over n observations of a training data set $\{(A^1, \theta^1), \ldots, (A^n, \theta^n)\}$ where $A^i \in \mathbb{R}^{t \times m}$ is a *feature matrix* for θ^i, in order to yield a *prediction function* $f : \mathbb{R}^{t \times m} \to \mathbb{R}^t$ predicting parameters $\hat{\theta} = f(A)$.

Solving the Para-OP under the estimated parameters, we can obtain an *estimated solution* $x^*(\hat{\theta})$. When constraints contain unknown parameters, a big challenge is that the feasible region is only approximated at solving time, and thus the estimated solution may be infeasible under the true parameters. Fortunately, some applications allow us to correct an infeasible solution into a feasible one, after the true parameters are revealed. Under these applications, Predict+Optimize can use a novel error measurement, called *Post-hoc Regret* [9], to evaluate the quality of the estimated parameters $\hat{\theta}$. The correction process can be formalized as a *correction function*, which takes an estimated solution $x^*(\hat{\theta})$ and true parameters θ and returns a *corrected solution* $x^*_{corr}(\hat{\theta}, \theta)$ that is feasible under θ. Although some scenarios may allow for post-hoc correct of an estimated solution, some penalties may incur from such correction. A *penalty function* $Pen(x^*(\hat{\theta}) \to x^*_{corr}(\hat{\theta}, \theta))$ takes an estimated solution $x^*(\hat{\theta})$ and the corrected solution $x^*_{corr}(\hat{\theta}, \theta)$ and returns a non-negative penalty. The choice of both the correction function and the penalty function are problem and application-specific.

With respect to the corrected solution $x^*_{corr}(\hat{\theta}, \theta)$ and penalty function Pen, we are now ready to define the *Post-hoc Regret*. The post-hoc regret contains two parts, one is the objective difference between the *true optimal solution* $x^*(\theta)$ and the corrected solution $x^*_{corr}(\hat{\theta}, \theta)$ under the true parameters θ, another one is the penalty that changing from the estimated solution $x^*(\hat{\theta})$ to the corrected solution $x^*_{corr}(\hat{\theta}, \theta)$ will incur. The *Post-hoc Regret* $PReg(\hat{\theta}, \theta)$ can be formally defined as:

$$PReg(\hat{\theta}, \theta) = obj(x^*_{corr}(\hat{\theta}, \theta), \theta) - obj(x^*(\theta), \theta) + Pen(x^*(\hat{\theta}) \to x^*_{corr}(\hat{\theta}, \theta)) \quad (1)$$

where $obj(x^*_{corr}(\hat{\theta}, \theta), \theta)$ is the *corrected optimal value* and $obj(x^*(\theta), \theta)$ is the *true optimal value*.

When only the objective contains unknown parameters, Post-hoc Regret degenerates into the *Regret* function [5], which compares the difference between the objective value of the *true optimal solution* $x^*(\theta)$ and the *estimated solution* $x^*(\hat{\theta})$ under true parameters θ. The regret function can be defined as: $Reg(\hat{\theta}, \theta) = obj(x^*(\hat{\theta}), \theta) - obj(x^*(\theta), \theta)$, where $obj(x^*(\hat{\theta}), \theta)$ is the *estimated optimal value*.

Following the empirical risk minimization principle, Hu *et al.* [9] choose the prediction function to be the function f from the set of models $\mathcal{F}$ attaining the smallest average post-hoc regret over the training data:

$$f^* = \arg\min_{f \in \mathcal{F}} \frac{1}{n} \sum_{i=1}^{n} PReg(f(A^i), \theta^i) \tag{2}$$

For discrete OPs and linear programs, the Post-hoc Regret is non-differentiable. Hence, traditional machine learning algorithms that rely on gradients are not applicable.

Branch & Learn (B&L) [10] is a Predict+Optimize framework for Para-OPs with unknown parameters only in the objective, which can compute the Regret exactly. B&L can handle optimization problems solvable with a recursive algorithm (under some restrictions). In B&L, Hu *et al.* study the class $\mathcal{F}$ of linear prediction functions and represent the solution structure of a Para-OP using (continuous) piecewise linear functions. A *piecewise linear function* h is a real-valued function defined on a finite set of (closed) intervals $\mathbb{I}(h)$ partitioning $\mathbb{R}$. Each interval $I \in \mathbb{I}(h)$ is associated with a linear function $h[I]$ of the form $h[I](r) = a_I r + b_I$, and the value of $h(r)$ for a real number $r \in \mathbb{R}$ is given by $h[I](r)$ where $r \in I$. An algebra can be canonically defined on piecewise linear functions [18]. For piecewise linear functions h and g, we define pointwise addition as $(h+g)(r) = h(r)+g(r)$ for all $r \in \mathbb{R}$. Pointwise subtraction, max/min and scalar products are similarly defined. All five operations can be computed efficiently by iterating over intervals of the operands [4].

This work is extended from B&L and in the rest of the paper, following the assumption in B&L, we assume that the prediction function f is a *linear mapping of the form* $f(A) = A\alpha$ for some m-dimensional vector of *coefficients* $\alpha \in \mathbb{R}^m$.

3 Branch & Learn with Post-hoc Correction

In this section, we extend B&L to cater for unknown parameters also in constraints, and call the extended framework *Branch & Learn with post-hoc correction (B&L-C)*. Post-hoc regret is used as the error metric.

To solve Problem 2, following the approach of Hu *et al.* [10], we update coefficients α of f iteratively via coordinate descent (Algorithm 1). The algorithm starts with an arbitrary initialization of α, and updates each coefficient in a round-robin fashion. Each iteration (Lines 3–11) contains four functions. `Construct` constructs a Para-OP as a function of the *free coefficient*, fixing the other coefficients in α, with an initial domain I_0. `Convert` returns a piecewise function of the free coefficient from the Para-OP, and each interval of the function corresponds to one or a set of estimated solution(s). `Correct` takes the returned piecewise function from `Convert` and the true parameters as inputs. Then makes the post-hoc correction, and returns a piecewise function of

Algorithm 1: Branch & Learn with Post-hoc Correction

Input: A Para-COP $P(\theta)$ and a training data set $\{(A^1,\theta^1),\ldots,(A^n,\theta^n)\}$
Output: a coefficient vector $\alpha \in \mathbb{R}^m$

1 Initialize α arbitrarily and $k \leftarrow 0$;
2 **while** *not converged* $\wedge$ *resources remain* **do**
3 $\quad k \leftarrow (k \bmod m) + 1$;
4 $\quad$ Initialize L to be the zero constant function;
5 $\quad$ **for** $i \in [1, 2, \ldots, n]$ **do**
6 $\quad\quad (P^i_\gamma, I_0) \leftarrow \texttt{Construct}(P(\theta), k, A^i)$;
7 $\quad\quad E^i(\gamma) \leftarrow \texttt{Convert}(P^i_\gamma, I_0)$;
8 $\quad\quad C^i(\gamma) \leftarrow \texttt{Correct}(E^i, \theta^i, I_0)$;
9 $\quad\quad L^i(\gamma) \leftarrow \texttt{Evaluate}(E^i, C^i, \theta^i, I_0)$;
10 $\quad\quad L(\gamma) \leftarrow L(\gamma) + L^i(\gamma)$;
11 $\quad \alpha_k \leftarrow \arg\min_{\gamma\in\mathbb{R}} L(\gamma)$*;
12 return α;

the free coefficient from the Para-OP. Each interval of the function corresponds to a data structure representing one or a set of corrected solution(s). `Evaluate` takes the two returned functions from `Convert` and `Correct`, and the true parameters as inputs. Then computes the corrected optimal value and the penalty, and obtains the post-hoc regret as a piecewise function of the free coefficient.

Let us describe lines 3–11 in Algorithm 1 in more detail. In each iteration (Lines 3–11), a coefficient α_k is updated. Iterating over index $k \in \{1, \ldots, m\}$, we replace α_k in α with a variable $\gamma \in \mathbb{R}$ by constructing $\alpha + (\gamma - \alpha_k)e_k$, where e_k is a unit vector for coordinate k. In Lines 5–11, we wish to update α_k as:

$$\alpha_k \leftarrow \arg\min_{\gamma\in\mathbb{R}} \sum_{i=1}^{n} \mathit{PReg}(A^i e_k \gamma + A^i(\alpha - \alpha_k e_k), \theta^i)$$

For notational convenience, let $a^i = A^i e_k \in \mathbb{R}^m$ and $b^i = A^i(\alpha - \alpha_k e_k) \in \mathbb{R}^m$, which are vectors independent of the free variable γ.

`Construct` synthesizes the Para-OP

$$P^i_\gamma \equiv x^*(a^i\gamma + b^i) = \arg\min_x \mathit{obj}(x, a^i\gamma + b^i) \text{ s.t. } C(x, a^i\gamma + b^i)$$

Sometimes, the Para-OP can also have an initial domain $I_0 \neq \mathbb{R}$ for γ.

`Convert` takes P^i_γ to create a function E^i mapping γ to the estimated objective $E^i(\gamma) = \mathit{obj}(x^*(a^i\gamma + b^i), a^i\gamma + b^i)$. Associated with each interval $I \in \mathbb{I}(E^i(\gamma))$, a linear function maps γ to the objective computed with the estimated parameters $a^i\gamma + b^i$. When the unknown parameters only appear in the objective, the estimated solution $x^*(a^i\gamma + b^i)$ remains the same in each interval I [3,4], i.e., each interval corresponds to one estimated solution. However, when the unknown parameters appear in constraints, the estimated solution may not remain the same in each interval I. If the estimated solution changes in one interval, one interval corresponds to a set of estimated solutions.

Whether the estimated solution will remain the same in each interval depends on the optimization problem and the positions of the unknown parameters. In Sect. 4, we show two examples that the estimated solution will remain the same in each interval and one example that the estimated solution will change in each interval.

`Correct` implements the correction function in the post-hoc regret. It takes the returned piecewise function $E^i(\gamma)$ from `Convert` and the true parameters θ^i as inputs. For each interval $I \in \mathbb{I}(E^i(\gamma))$, one or a set of estimated solution(s) $x^*(a^i\gamma + b^i)$ can be obtained. `Correct` makes the post-hoc correction with the true parameters θ^i, and creates a function C^i mapping γ to the corrected optimal value $C^i(\gamma) = obj(x^*_{corr}(a^i\gamma + b^i, \theta^i), \theta^i)$. Each interval $I \in \mathbb{I}(C^i(\gamma))$ corresponds to one or a set of corrected solution(s) $x^*_{corr}(a^i\gamma + b^i, \theta^i)$. Whether the corrected solution will remain the same in each interval I depends on the correction function. Besides, the form of the returned function $C^i(\gamma)$ depends on the correction function.

`Evaluate` takes the returned function from `Convert`, the returned function from `Correct`, and the true parameters as inputs. It computes the corrected optimal value and the penalty, and obtains the post-hoc regret L^i for each γ, i.e.

$$\begin{aligned} L^i[I] = PReg(a^i\gamma + b^i, \theta^i) &= obj(x^*_{corr}(a^i\gamma + b^i, \theta^i), \theta^i) - obj(x^*(\theta^i), \theta^i) \\ &+ Pen(x^*(a^i\gamma + b^i) \rightarrow x^*_{corr}(a^i\gamma + b^i, \theta)) \end{aligned}$$

When the unknown parameters only appear in objectives, the post-hoc regret function L^i returned from `Evaluate` is always a piecewise constant function of the free coefficient [3,4,10]. It is straightforward to compute the sum of two piecewise constant functions and update the coefficient (Lines 10–11 in Algorithm 1). However, when the unknown parameters appear in constraints, the form of the post-hoc regret function L^i returned from `Evaluate` depends on the correction function and the penalty function, which are both problem and application specific. Under different scenarios, the post-hoc regret function may even be a piecewise nonlinear function, which leads to a technical obstacle: how to sum up two piecewise nonlinear functions and find the minimum of the resulted function efficiently. We will discuss this obstacle in Sect. 4.

While coordinate descent is a standard technique, a challenge of using this framework is how to construct `Convert` for an algorithm. B&L presents a standard template for recursive algorithms and shows how to cleanly adapt a recursive algorithm to `Convert`. We use their template to construct `Convert` here. Therefore, the proposed B&L-C framework has the same restrictions on the optimization problems as the B&L.

4 Case Studies

4.1 Maximum Flow with Unknown Edge Capacities

We first demonstrate, using the example of the maximum flow problem (MFP), how our framework can solve problems solvable by a state-of-the-art approximation method (IntOpt-C) [9]. The problem aims to find the largest possible flow sent from a source s to a terminal t in a directed graph, under the constraints that the flow sent on each edge cannot exceed the edge capacity.

Using the template proposed in B&L [10], we adapt the Edmonds-Karp algorithm [19] to `Convert`, which recursively finds an unblocking path with remaining capacity and sends a flow such that at least one edge along the path is saturated. The estimated solution $x^*(a^i\gamma+b^i)$ of MFP is the flows sent through each path and therefore will change with the capacities of saturated edges. In each interval of $E^i(\gamma)$ returned by `Convert`, the saturated edges remain the same, but the estimated solution will change when γ changes. When the edge capacities are unknown, we need to consider a case where the flow computed with the estimated capacities might exceed the true capacities of some edges. We consider two possible correction functions:

- **Correction Function A**: given an infeasible estimated solution x^*, find the largest $\lambda \in [0, 1]$ such that λx^* satisfies the constraints under the true parameters.
  Note that Correction Function A is the same as the one used in IntOpt-C [9], which is designed for packing linear programs. By using the same correction function, we can investigate the performance difference between B&L-C and IntOpt-C.
- **Correction Function B**: re-compute the blocking flows of the chosen paths in the infeasible estimated solution with the true capacities, and then augment the paths one by one with the re-computed blocking flows. The ordering of path augmentation is important but computing the best order requires $O(n!)$ time. We can adopt an approximate method: the paths are augmented according to the order of the path augmentation of the Edmonds-Karp algorithm.

Using Correction Function B, augmenting the chosen paths by their blocking flows one by one may lead to a situation that, since some edges are shared by several paths, they may be blocked before some paths are used. Thus the blocking flows of some chosen paths in the estimated solution may be zero and these chosen paths are wasted. Therefore, we propose a penalty function:

- **Penalty Function I**: whenever a chosen path in the estimated solution is wasted, deduct K units of flow.

Penalty Function I is not needed if Correction Function A is used, since the true capacities are all positive, λ will not be zero in this problem and no chosen paths in the estimated solution will be wasted.

Using Correction Function A, the corrected solution $x^*_{corr}(a^i\gamma + b^i, \theta^i)$ will not remain the same in each interval $I \in \mathbb{I}(C^i(\gamma))$ either, where $C^i(\gamma)$ is the piecewise rational linear function returned from `Convert`. Since the true optimal value $obj(x^*(\theta^i), \theta^i)$ is a constant value, the post-hoc regret function $L(\gamma)$ returned from `Evaluate` is a piecewise rational linear function. This will lead to the technical obstacle mentioned in Sect. 3: how to sum up two piecewise rational linear functions and find the minimum of the resulting piecewise rational linear function efficiently. In this work, we deal with this obstacle by using grid search.

Using Correction Function B, the corrected solution $x^*_{corr}(a^i\gamma+b^i, \theta^i)$ remains the same in each interval $I \in \mathbb{I}(C^i(\gamma))$, where $C^i(\gamma)$ is the piecewise constant function returned from `Convert`. Using Penalty Function I, $Pen(x^*(a^i\gamma+b^i) \to x^*_{corr}(a^i\gamma+b^i, \theta^i))$ is also a piecewise constant function. Therefore, the post-hoc regret function $L(\gamma)$ returned from `Evaluate` is a piecewise constant function, and we can easily sum up two piecewise constant functions and minimizes $L(\gamma)$ in Lines 9 and 10 respectively in Algorithm 1.

4.2 0-1 Knapsack with Unknown Weights

In the second example, we showcase our framework on a packing integer programming problem, the 0-1 knapsack problem, which can be handled by our framework straightforwardly but not by IntOpt-C. Given a set of items, each with a weight w_i and a value v_i, and a knapsack with a maximum capacity C. The aim is to maximize the total value of the selected items under the constraint that the total weight of the selected items is less than or equal to the maximum capacity. Using the template proposed in B&L [10], we adapt the branching algorithm for the 0-1 knapsack problem to `Convert`. The estimated solution $x^*(a^i\gamma + b^i)$ is a set of the selected items. In each interval of $E^i(\gamma)$ returned by `Convert`, the set of the selected items, i.e., the estimated solution, remains the same. When the weights are unknown, we need to consider a case where the items are selected with the estimated weights, but the total true weights might exceed the capacity. We propose three correction functions here:

- **Correction Function A**: remove the selected items in the estimated solution one by one in increasing order of the ratios of value/weight until the capacity is sufficient.
- **Correction Function B**: remove the selected items in the estimated solution one by one in decreasing order of the weights until the capacity is sufficient.
- **Correction Function C**: remove all the selected items in the estimated solution when it is infeasible.

Removing selected items from the knapsack may incur some removal fees, which are formulated as the penalty function here. We consider two possible penalty functions:

- **Penalty Function I**: when the ith item is removed from the estimated solution, $\sigma_i v_i$ units of value is deducted, where $\sigma \geq 0$ is a non-negative tunable vector.
- **Penalty Function II**: whenever a selected item in the estimated solution is removed, K units of value is deducted, where K is a constant.

Since the solution set is discrete and finite, the estimated solution $x^*(a^i\gamma + b^i)$ remains the same in each interval $I \in \mathbb{I}(E^i(\gamma))$, where $E^i(\gamma)$ is the piecewise linear function returned from `Convert`. Using the above three correction functions, the corrected solution $x^*_{corr}(a^i\gamma + b^i, \theta^i)$ remains the same in each interval $I \in \mathbb{I}(C^i(\gamma))$, where $C^i(\gamma)$ is the piecewise constant function returned from `Convert`. Using the above two penalty functions, $Pen(x^*(a^i\gamma + b^i) \rightarrow x^*_{corr}(a^i\gamma + b^i, \theta^i))$ is also a piecewise constant function. Therefore, the post-hoc regret function $L(\gamma)$ returned from `Evaluate` is a piecewise constant function, and we can easily sum up two piecewise constant functions and minimizes $L(\gamma)$ in Lines 9 and 10 respectively in Algorithm 1.

4.3 Minimum Cost Vertex Cover with Unknown Costs and Edge Values

Our last example is a variant of the minimum cost vertex cover (MCVC) problem, where we show how to apply our framework to an optimization problem that has unknown parameters in both the objective and the constraints. This problem is also not solvable by IntOpt-C. Given a graph $G = (V, E)$, there is an associated *cost* $c \in \mathbb{R}^{|V|}$ denoting the cost of picking each vertex, as well as edge values $\ell \in \mathbb{R}^{|E|}$, one real value for each edge. Both the costs and edge values are unknown parameters. The goal is to pick a subset of vertices, minimizing the total cost, subject to the constraint that for all edges *except the one with the smallest edge value*, the edge needs to be covered, namely at least one of the two vertices on the edge needs to be picked. This problem is relevant in applications such as building public facilities. Consider, for example, the graph being a road network with edge values being traffic flow, and we wish to build speed cameras at intersections with minimum cost, while covering all the roads except the one with the least traffic.

Using the template proposed in B&L [10], we adapt the branching algorithm for the MCVC to `Convert`. The estimated solution $x^*(a^i\gamma+b^i)$ is a set of the picked vertices. In each interval of $E^i(\gamma)$ returned by `Convert`, the set of the picked vertices, i.e., the estimated solution, remains the same. When the edge values are unknown, the estimated edge values might cause an edge to be wrongly removed. The selected vertices might not cover all the edges that need to be covered. We thus propose one correction function:

- **Correction Function A**: if there is an edge not covered by the selected vertices, add both of the edge endpoints to the selection.

Since the solution set is discrete and finite, the estimated solution $x^*(a^i\gamma + b^i)$ remains the same in each interval $I \in \mathbb{I}(E^i(\gamma))$, where $E^i(\gamma)$ is the piecewise linear function returned from `Convert`. Using Correction Function A, the corrected solution $x^*_{corr}(a^i\gamma + b^i, \theta^i)$ remains the same in each interval $I \in \mathbb{I}(C^i(\gamma))$, where $C^i(\gamma)$ is the piecewise constant function returned from `Convert`. Since there is no penalty function in this example, $Pen(x^*(a^i\gamma + b^i) \rightarrow x^*_{corr}(a^i\gamma + b^i, \theta^i)) = 0$ is a constant function. Therefore, the post-hoc regret function $L(\gamma)$ returned from `Evaluate` is a piecewise constant function, and we can easily sum up two piecewise constant functions and minimizes $L(\gamma)$ in Lines 9 and 10 respectively in Algorithm 1.

5 Experimental Evaluation

In this section, we evaluate the proposed B&L-C framework and the post-hoc regret function on the three optimization problems mentioned in Sect. 4. We compare the proposed framework (B&L-C) with 7 different methods: the B&L framework [10], a state-of-the-art approximation method (IntOpt-C) [9], and 5 classical regression methods including linear regression (LR), k-nearest neighbors (k-NN), classification and regression tree (CART), random forest (RF) and neural network (NN) [6]. Below we briefly discuss the experiment setting of each problem:

MFP with Unknown Edge Capacities. Our aim is to use this problem to compare the proposed B&L-C framework with IntOpt-C [9]. Therefore, we use the same dataset and follow the experiment setting in the work of IntOpt-C. The real-life dataset [17] is used on three real-life graphs: POLSKA [15], with 12 vertices and 18 edges, USANet [13], with 24 vertices and 43 edges, and GÉANT [12], with 40 vertices and 61 edges. In this dataset, each unknown edge capacity is related to 8 features. Following the setting in IntOpt-C, we divide the dataset into two sets: training and test. For experiments on

POLSKA and USANet, 610 instances are used for training and 179 instances for testing the model performance, while for experiments on GÉANT, 490 instances are used for training and 130 instances for testing the model performance.

0-1 Knapsack with Unknown Weights. In this experiment, each instance consists of 10 items. The weights W will be predicted from data, while values V and capacity C are given. Given that we are unable to find datasets specifically for the 0-1 knapsack problem, we follow the experimental approach in the previous works of P+O [10, 14] and use real data from a different problem (the ICON scheduling competition) [17] as numerical values required for our experiment instances. In this dataset, each unknown weight is related to 8 features. We use a 70%/30% training/testing data split: 210 instances are used for training and 90 instances for testing the model performance.

We generate the values following the generation method proposed by Pisinger [16], which is widely used to generate knapsack data [1, 11]. Three groups of values, which are uncorrelated, weakly correlated, and almost strongly correlated with the weights, are considered. Suppose the value of the ith item is v_i, the weight of the ith item is w_i. These 3 groups of values are generated as: 1) uncorrelated: v_i is randomly chosen in $[1, R]$, 2) weakly correlated: v_i is randomly chosen in $[\max\{1, w_i - R/10\}, w_i + R/10]$, 3) almost strongly correlated: v_i is randomly chosen in $[w_i + R/10 - R/500, w_i + R/10 + R/500]$, where R is set to be 500 since the weights in the dataset are around 40 to 60. Since the average total weight of each instance is around 400, we conduct experiments on the 0-1 knapsack problem with 100, 200, and 300 capacities. The σ in Penalty Function I is set to be 0.1 and K in Penalty Function II is set to be 500.

MCVC with Unknown Costs and Edge Values. Since the MCVC is an NP-hard problem, we conduct experiments on two small graphs from the Survivable Network Design Library [15]: ABILENE, with 12 vertices and 15 edges, and PDH, with 11 vertices and 34 edges. Given that we are unable to find datasets specifically for the MCVC, we use the same real data from the ICON scheduling competition [17] as numerical values required for our experiment instances. We use a 70%/30% training/testing data split: 210 instances are used for training and 90 instances for testing the model performance.

5.1 B&L-C Versus IntOpt-C

In the first experiment, we compare our exact method (B&L-C) against an approximation method (IntOpt-C) in terms of solution quality and runtime. We conduct experiments on the MFP with unknown edge capacities, which can be solved by both the approximation method IntOpt-C [9] and the proposed exact method B&L-C. Following the experiment setting in IntOpt-C [9], Correction Function A is used and there is no penalty function here. We run 10 simulations on each graph and compare the solution quality and the runtime of each method.

Table 1 reports the mean post-hoc regrets and standard deviations for each approach. At the bottom of the table, we also report the *average True Optimal Values* (TOV) for reference. Note that B&L performs training with the regret but the testing is

Table 1. Mean post-hoc regrets and standard deviations for the MFP with unknown edge capacities using Correction Function A and no penalty function.

PReg	POLSKA	USANet	GÉANT
B&L-C	$\mathbf{9.07 \pm 0.67}$	$\mathbf{14.44 \pm 1.12}$	$\mathbf{10.18 \pm 1.02}$
B&L	17.01±2.00	21.79±1.53	17.04±2.11
IntOpt-C	10.00±0.67	16.64±1.34	10.84±1.10
Ridge	11.20±0.73	19.52±1.16	12.47±1.14
k-NN	14.39±0.83	22.89±1.58	15.13±1.08
CART	16.65±1.06	24.15±1.51	17.01±1.59
RF	12.30±0.90	22.27±1.34	12.52±1.19
NN	12.18±1.08	18.62±1.23	12.05±1.13
TOV	88.66±1.10	96.22±1.38	98.71±1.98

Table 2. Average runtimes (in seconds) for the MFP with unknown edge capacities.

Runtime(s)	POLSKA	USANet	GÉANT
B&L-C	66.54	411.67	48.32
B&L	40.30	288.43	29.90
IntOpt-C	18.65	132.22	15.48

with the post-hoc regret, while B&L-C and IntOpt-C use post-hoc regret in both training and testing. The results show that B&L-C always achieves the best performance, while IntOpt-C achieves the second-best performance in all cases. Compared with IntOpt-C, B&L-C obtains 9.29% smaller regret on POLSKA, 13.20% smaller regret on USANet, and 6.08% smaller regret on GÉANT. Considering the relative error, B&L-C achieves 10.23%, 15.01%, and 10.31% relative error on POLSKA, USANet, and GÉANT respectively. We also observe that using regret as the loss function, B&L does not have a better performance than the classical regression methods when the unknown parameters appear in constraints. Table 2 shows the average runtimes of B&L-C using Correction Function A, B&L, and IntOpt-C. Here, the runtime refers to the training time of the prediction model. The results show that the runtime of B&L-C using Correction Function A is larger than that of IntOpt-C, while the runtime of B&L is not that much larger than that of IntOpt-C. The reason is that, in this problem, the post-hoc regret function using Correction Function A is a piecewise rational linear function, while the regret function is a piecewise linear function. To compute the minimum of a piecewise rational linear function, grid search is used and is quite time-consuming. But the minimum of a piecewise linear function can be computed easily and grid search is not needed.

In conclusion, we observe that B&L-C can achieve much better solution quality but need longer runtime than IntOpt-C.

5.2 Post-hoc Regret on More General Problems

IntOpt-C is only applicable for packing and covering linear programming problems. We investigate the performance of post-hoc regret on two integer programming problems: 0–1 knapsack with unknown weights, and a variant of MCVC with unknown costs and edge values, both of which cannot be solved by IntOpt-C.

0-1 Knapsack with Unknown Weights. In this experiment, we use Correction Function A and Penalty Function I as an example, to show that the B&L-C framework can deal with 0-1 knapsack with unknown weights. Table 3 reports the solution qualities for

Table 3. Mean post-hoc regrets and standard deviations for the 0-1 knapsack problem with unknown weights and 3 groups of values using Correction Function A with Penalty Function I.

Preg		Cap=100	Cap=200	Cap=300
Uncorrelated	B&L-C	**112.66 ± 19.70**	**91.45 ± 9.43**	**53.90 ± 8.33**
	B&L	165.57±18.35	199.02±48.07	123.06±25.26
	Ridge	175.05±24.22	201.00±24.52	145.73±23.04
	k-NN	188.73±22.95	239.83±30.24	189.11±37.25
	CART	185.33±22.53	215.83±24.14	174.44±22.06
	RF	179.34±21.99	213.53±23.63	159.27±29.35
	NN	159.75±40.05	172.87±74.31	120.21±70.43
	TOV	942.76±37.06	1712.67±41.63	2174.50±42.63
Weakly Correlated	B&L-C	**20.81 ± 2.61**	**18.88 ± 1.56**	**12.91 ± 1.16**
	B&L	31.73±6.56	36.45±6.36	25.31±5.65
	Ridge	28.98±3.39	36.00±3.81	26.81±2.40
	k-NN	31.22±3.93	42.00±4.95	34.32±5.08
	CART	31.61±4.68	38.80±3.72	31.83±3.49
	RF	30.04±4.08	38.05±4.94	29.34±3.51
	NN	27.08±5.11	31.18±11.57	22.12±9.80
	TOV	165.96±4.62	309.28±4.69	397.94±5.77
Almost Strongly Correlated	B&L-C	**44.62 ± 4.01**	**59.77 ± 3.91**	**43.93 ± 3.41**
	B&L	51.95±5.17	84.11±13.64	97.72±22.87
	Ridge	49.98±3.47	82.92±8.01	96.53±10.61
	k-NN	51.73±3.38	87.15±7.68	110.98±12.66
	CART	52.40±4.48	81.02±6.73	101.58±8.95
	RF	50.75±4.29	80.72±9.87	98.31±12.44
	NN	50.11±3.63	75.34±18.42	85.56±32.39
	TOV	209.40±5.92	441.17±9.05	654.16±10.99

each approach across 10 runs on 0-1 knapsack problem with unknown weights and 3 different groups of values (uncorrelated, weakly correlated, and almost strongly correlated).

As shown in Table 3, B&L-C has the smallest mean post-hoc regrets in all cases. In the experiments on uncorrelated values, B&L-C obtains at least 29.48%, 47.10%, and 55.16% when the capacity is 100, 200, and 300 respectively. In the experiments on weakly correlated values, B&L-C obtains at least 23.13%, 39.45%, and 41.61% when the capacity is 100, 200, and 300 respectively. In the experiments on almost strongly correlated values, B&L-C obtains at least 10.72%, 20.67%, and 48.65% when the capacity is 100, 200, and 300 respectively. These results indicate that using Correction Function A, when the capacity grows larger, the advantage of B&L-C is more evident.

We also report the relative errors in these experiments. In the experiments on uncorrelated values, B&L-C achieves 11.95%, 5.34%, and 2.48% relative error when the capacity is 100, 200, and 300 respectively. In the experiments on weakly correlated values, B&L-C achieves 12.54%, 6.10%, and 3.25% relative error when the capacity is 100, 200, and 300 respectively. In the experiments on almost strongly correlated values, B&L-C achieves 21.31%, 13.55%, and 6.72% relative error when the capacity is 100, 200, and 300 respectively. These results indicate that when the values are more correlative with the weights, the relative error is larger. Besides, using Correction Function A, the relative error becomes smaller when the capacity grows larger.

Table 4. Mean post-hoc regrets and standard deviations for the MCVC with unknown costs and edge values.

PReg	ABILENE	PDH
B&L-C	**11.83 ± 2.79**	**55.94 ± 8.46**
B&L	15.26±3.56	73.6±8.55
Ridge	19.3±3.05	65.23±6.76
k-NN	33.08±4.55	70.52±6.72
CART	28.6±5.67	66.03±7.39
RF	27.91±4.25	65.29±8.01
NN	14.14±2.42	70.65±5.69
TOV	275.33±5.43	491.18±12.75

Table 5. Mean post-hoc regrets and standard deviations for the MFP with unknown edge capacities using Correction Function B without/with Penalty Function I.

PReg	Correction Function B			Correction Function B & Penalty Function I		
	POLSKA	USANet	GÉANT	POLSKA	USANet	GÉANT
B&L-C	**1.41 ± 0.26**	**4.49 ± 0.67**	**1.03 ± 0.24**	**6.14 ± 1.08**	**16.89 ± 1.10**	**2.04 ± 0.27**
B&L	1.51±0.30	10.09±1.37	5.82±1.68	8.59±0.45	20.89±1.55	7.00±2.01
Ridge	1.52±0.30	7.11±0.88	1.20±0.34	8.54±0.47	18.72±1.10	2.47±0.27
k-NN	2.42±0.36	8.03±0.86	1.47±0.52	8.22±0.53	23.57±1.04	3.10±0.46
CART	3.29±0.69	10.84±1.28	1.75±0.52	8.13±0.88	28.79±1.96	3.71±0.82
RF	1.81±0.33	8.72±1.19	1.27±0.41	7.27±0.52	21.30±1.36	2.54±0.44
NN	1.83±0.29	5.52±0.73	1.22±0.37	8.95±0.44	18.98±1.02	2.45±0.54
TOV	88.66±1.10	96.22±1.38	98.71±1.98	88.66±1.10	96.22±1.38	98.71±1.98

MCVC with Unknown Costs and Edge Values. Table 4 shows the solution qualities for each approach across 10 runs on the MCVC experiment. B&L-C achieves the best performance in both of the two graphs. B&L-C obtains at 16.13%-64.24% smaller post-hoc regret in ABILENE, and 14.24%-23.99% in PDH. Considering the relative error, B&L-C achieves 4.30% relative error in ABILENE, and 11.39% relative error in PDH.

5.3 Different Combinations of Correction Functions and Penalty Functions

The correction function and the penalty function are problem and application-specific. Even in the same problem but different scenarios, the correction function and the penalty function could be different. Here, we try out different combinations of correction functions and penalty functions in the same problem to provide insights for defining the appropriate post-hoc regret.

MFP with Unknown Edge Capacities. We conduct experiments on the MFP using Correction Function B with/without Penalty Function I. The experiment results are shown in Table 5. Since the correction function is changed, the gradient of the post-hoc regret with respect to edge capacities is also changed and thus IntOpt-C cannot be used.

First, to compare the performance of post-hoc regret with different correction functions, we compare the results of using Correction Function A and B in Tables 1 and 5 respectively. Table 5 shows that B&L-C always achieves the best performance. Compared with other methods, B&L-C obtains at least 7.03% smaller regret on POLSKA, 18.50% smaller regret on USANet, and 14.19% smaller regret on GÉANT. Considering the relative error, B&L-C achieves 1.59%, 4.67%, and 1.04% relative error on POLSKA, USANet, and GÉANT respectively. The results show that B&L-C using Correction Function B can achieve smaller mean post-hoc regret than B&L-C using Correction Function A, which indicates that using Correction Function B is more suitable.

Second, we investigate the performance of post-hoc regret when using penalty functions on the MFP. Due to the page limit, we opted to present only one value of K, which is the units of flow to be deducted. Here, K is set to be 10. For experiment results and analysis on $K = \{30, 50\}$, please refer to the extended version on arXiv with the same title. Results in Table 5 show that when using Correction Function B & Penalty Function I, B&L-C also achieves the best performance on all of the three graphs. Since the penalty term exists and is always non-negative, the mean post-hoc regrets here are larger than the mean post-hoc regrets of using Correction Function B without a penalty function. Compared with other methods, B&L-C obtains at least 15.46% smaller regret on POLSKA, 9.79% smaller regret on USANet, and 16.47% smaller regret on GÉANT. Considering the relative error, B&L-C achieves 6.93%, 17.53%, and 2.07% relative error on POLSKA, USANet, and GÉANT respectively.

Table 6. Mean post-hoc regrets and standard deviations for the 0-1 knapsack problem with weakly correlated values using different correction functions with Penalty Function I.

PReg	Corection Function B			Correction Function C		
	Cap=100	Cap=200	Cap=300	Cap=100	Cap=200	Cap=300
B&L-C	**24.25 ± 2.59**	**32.87 ± 3.55**	**31.70 ± 1.89**	**53.48 ± 5.27**	**116.56 ± 14.06**	**130.79 ± 9.80**
B&L	32.59±6.32	40.24±5.31	36.35±5.72	58.67±7.41	124.67±23.56	174.70±26.32
Ridge	29.88±3.22	39.62±3.60	36.27±2.33	55.31±7.29	124.23±14.73	160.04±20.54
k-NN	32.96±3.72	45.98±4.16	41.73±4.11	62.72±6.63	123.12±14.03	153.48±21.53
CART	35.14±4.71	45.79±4.37	42.28±2.79	70.97±11.57	141.80±19.47	171.11±23.31
RF	32.83±3.97	43.51±4.91	37.30±2.37	64.01±11.51	126.48±22.04	158.52±26.99
NN	28.60±4.58	39.05±8.02	38.52±11.05	74.05±28.73	163.71±64.50	213.17±80.20
TOV	165.96±4.62	309.28±4.69	397.94±5.77	165.96±4.62	309.28±4.69	397.94±5.77

0-1 Knapsack with Unknown Weights. Experiments on weakly correlated values are conducted as examples to show the performance difference of post-hoc regret when using different correction functions and penalty functions in the 0-1 knapsack problem.

First, to compare the performance when using different correction functions, we fix the penalty function as Penalty Function I. Experiments on 0-1 knapsack using Correction Function A & Penalty Function I are conducted in Sect. 5.1, and we only conduct experiments using Correction Function B and C here. The results are shown in Table 6. B&L-C outperforms other approaches. Besides, the mean post-hoc regrets achieved by B&L-C using Correction Functions B and C are both larger than that achieved by B&L-C using Correction Function A, which indicates that when using Penalty Function I, Correction Function A is more suitable than Correction Functions B and C.

Second, to compare the performance when using different penalty functions, we conduct experiments using Penalty Function II and Correction Function A, B, and C respectively. Experiment results are shown in Table 7. As Table 7 shows, B&L-C outperforms other approaches. The results show that when using Penalty Function II, B&L-C

Table 7. Mean post-hoc regrets and standard deviations for the 0-1 knapsack problem with unknown weights and weakly correlated values using Penalty Function II.

PReg		Cap=100	Cap=200	Cap=300
Correction Function A	B&L-C	**127.13 ± 27.52**	**202.99 ± 48.62**	**231.32 ± 58.74**
	B&L	280.65±73.44	482.37±134.13	668.78±167.01
	Ridge	263.55±68.79	439.81±101.30	608.07±121.73
	k-NN	323.26±67.78	445.81±119.63	566.84±159.66
	CART	566.15±188.48	687.51±184.25	697.52±176.48
	RF	378.68±131.24	536.21±170.43	613.09±153.58
	NN	357.23±121.00	614.05±191.50	608.49±200.71
Correction Function B	B&L-C	**127.24 ± 27.60**	**186.84 ± 37.45**	**193.92 ± 39.52**
	B&L	281.43±73.56	491.13±158.54	726.18±175.94
	Ridge	264.37±68.94	435.33±97.59	596.34±121.07
	k-NN	323.73±66.71	437.21±115.41	556.91±159.57
	CART	563.81±187.58	676.10±180.67	685.91±176.05
	RF	380.11±±132.23	527.84±163.75	595.89±151.56
	NN	356.05±119.96	552.03±217.25	592.49±190.67
Correction Function C	B&L-C	**173.12 ± 43.50**	**255.03 ± 73.81**	**335.59 ± 82.28**
	B&L	566.25±155.93	1546.77±518.46	2736.39±628.02
	Ridge	543.03±149.36	1472.25±283.24	2328.08±430.10
	k-NN	658.57±156.24	1376.22±322.83	2069.61±493.77
	CART	1030.83±320.56	1991.15±453.46	2581.91±509.94
	RF	775.12±270.11	1623.26±483.81	2322.75±582.54
	NN	697.35±269.30	1890.08±712.55	2485.80±892.67
TOV		165.96±4.62	309.28±4.69	397.94±5.77

using Correction Function B achieves smaller mean post-hoc regret than B&L-C using Correction Function A or C. This indicates that when using Penalty Function II, Correction Function B is more suitable to be used. The reason for this phenomenon lies in the definitions of Penalty Function II and Correction Function B. In Penalty Function II, removing more items leads to a larger penalty, while Correction Function B removes the selected items in the estimated solution one by one in decreasing order of the weights, thus will remove fewer items than Correction Functions A and C. We also notice that when the capacity is 100, B&L-C using Correction Function A outperforms B&L-C using Correction Function B. We give the explanation below. Since the average weight of the items is around 40 to 60, when the capacity is 100, the number of the selected items in the estimated solution and the number of the removal items from the post-hoc correction are both very small and the latter ones in Correction Function A and B may be almost the same. Under this situation, the penalty terms $(Pen(x^*(\hat{\theta}) \rightarrow x^*_{corr}(\hat{\theta}, \theta)))$ of using Correction Functions A and B in the post-hoc regret function are almost the same, then selecting items with higher value and has larger corrected optimal value $(obj(x^*_{corr}(\hat{\theta}, \theta), \theta))$ can achieve smaller post-hoc regret. Therefore, B&L-C using Correction Function A performs better.

6 Conclusion

We propose the first exact method for Predict+Optimize with unknown parameters in both the objective and constraints. The proposed framework is an extension of Branch & Learn, a framework for problems with only unknown objectives, and can handle recursively and iteratively solvable problems. Extensive experiments are conducted to compare the proposed method with the state-of-the-art Predict+Optimize approach and investigate the performance of the post-hoc regret on more general problems. Furthermore, we empirically study different combinations of correction functions and penalty functions to gain insights for defining post-hoc regret in different scenarios.

Acknowledgments. We thank the anonymous referees for their constructive comments. In addition, Xinyi Hu and Jimmy H.M. Lee acknowledge the financial support of a General Research Fund (RGC Ref. No. CUHK 14206321) by the University Grants Committee, Hong Kong. Jasper C.H. Lee was supported in part by the generous funding of a Croucher Fellowship for Postdoctoral Research, NSF award DMS-2023239, NSF Medium Award CCF-2107079 and NSF AiTF Award CCF-2006206.

References

1. Cappart, Q., Moisan, T., Rousseau, L.M., Prémont-Schwarz, I., Cire, A.A.: Combining reinforcement learning and constraint programming for combinatorial optimization. In: Proceedings of the Thirty-Fourth AAAI Conference on Artificial Intelligence, vol. 35, pp. 3677–3687 (2021)
2. Collet, M., Gotlieb, A., Lazaar, N., Carlsson, M., Marijan, D., Mossige, M.: RobTest: a CP approach to generate maximal test trajectories for industrial robots. In: Simonis, H. (ed.) CP 2020. LNCS, vol. 12333, pp. 707–723. Springer, Cham (2020). https://doi.org/10.1007/978-3-030-58475-7_41

3. Demirović, E., et al.: Predict+Optimise with ranking objectives: exhaustively learning linear functions. In: Proceedings of the Twenty-Eighth International Joint Conference on Artificial Intelligence, pp. 1078–1085 (2019)
4. Demirović, E., et al.: Dynamic programming for Predict+Optimise. In: Proceedings of the Thirty-Fourth AAAI Conference on Artificial Intelligence, pp. 1444–1451 (2020)
5. Elmachtoub, A.N., Grigas, P.: Smart predict, then optimize. Manag. Sci. **68**(1), 9–26 (2022)
6. Friedman, J., Hastie, T., Tibshirani, R.: The Elements of Statistical Learning, vol. 1, Number 10. Springer series in statistics, New York (2001). https://doi.org/10.1007/978-0-387-21606-5
7. Genc, B., O'Sullivan, B.: A two-phase constraint programming model for examination timetabling at university college cork. In: Simonis, H. (ed.) CP 2020. LNCS, vol. 12333, pp. 724–742. Springer, Cham (2020). https://doi.org/10.1007/978-3-030-58475-7_42
8. de Givry, S., Lee, J.H.M., Leung, K.L., Shum, Y.W.: Solving a judge assignment problem using conjunctions of global cost functions. In: O'Sullivan, B. (ed.) CP 2014. LNCS, vol. 8656, pp. 797–812. Springer, Cham (2014). https://doi.org/10.1007/978-3-319-10428-7_57
9. Hu, X., Lee, J.C., Lee, J.H.: Predict+Optimize for packing and covering LPs with unknown parameters in constraints. In: Proceedings of the Thirty-Second AAAI Conference on Artificial Intelligence (2022)
10. Hu, X., Lee, J.C., Lee, J.H., Zhong, A.Z.: Branch & learn for recursively and iteratively solvable problems in Predict+Optimize. Adv. Neural Inf. Process. Syst. **35** (2022)
11. Li, D., et al.: A novel method to solve neural knapsack problems. In: Proceedings of the Thirty-Eighth International Conference on Machine Learning, pp. 6414–6424. PMLR (2021)
12. LLC, M.: Geant topology map dec2018 copy (2018). https://www.geant.org/Resources/Documents/GEANT_Topology_Map_December_2018.pdf. Accessed 10 Sep 2020
13. Lucerna, D., Gatti, N., Maier, G., Pattavina, A.: On the efficiency of a game theoretic approach to sparse regenerator placement in WDM networks. In: GLOBECOM 2009–2009 IEEE Global Telecommunications Conference, pp. 1–6. IEEE (2009)
14. Mandi, J., Stuckey, P.J., Guns, T., et al.: Smart predict-and-optimize for hard combinatorial optimization problems. In: Proceedings of the Thirty-Fourth AAAI Conference on Artificial Intelligence, vol. 34, pp. 1603–1610 (2020)
15. Orlowski, S., Pióro, M., Tomaszewski, A., Wessäly, R.: SNDlib 1.0-Survivable network design library. In: Proceedings of the Third International Network Optimization Conference, April 2007. http://www.zib.de/orlowski/Paper/OrlowskiPioroTomaszewskiWessaely2007-SNDlib-INOC.pdf.gz, http://sndlib.zib.de, extended version accepted in Networks, 2009
16. Pisinger, D.: Where are the hard knapsack problems? Comput. Oper. Res. **32**(9), 2271–2284 (2005)
17. Simonis, H., O'Sullivan, B., Mehta, D., Hurley, B., Cauwer, M.D.: Energy-Cost Aware Scheduling/Forecasting Competition (2014). http://challenge.icon-fet.eu/sites/default/files/iconchallenge.pdf
18. Von Mohrenschildt, M.: A normal form for function rings of piecewise functions. J. Symb. Comput. **26**(5), 607–619 (1998)
19. Waissi, G.R.: Network flows: theory, algorithms, and applications (1994)
20. Wilder, B., Dilkina, B., Tambe, M.: Melding the data-decisions pipeline: decision-focused learning for combinatorial optimization. In: Proceedings of the Thirty-Third AAAI Conference on Artificial Intelligence, pp. 1658–1665 (2019)

Interpretable Clustering via Soft Clustering Trees

Eldan Cohen(✉)

University of Toronto, Toronto, Canada
ecohen@mie.utoronto.ca

Abstract. Clustering is a popular unsupervised learning task that consists of finding a partition of the data points that groups similar points together. Despite its popularity, most state-of-the-art algorithms do not provide any explanation of the obtained partition, making it hard to interpret. In recent years, several works have considered using decision trees to construct clusters that are inherently interpretable. However, these approaches do not scale to large datasets, do not account for uncertainty in results, and do not support advanced clustering objectives such as spectral clustering. In this work, we present soft clustering trees, an interpretable clustering approach that is based on soft decision trees that provide probabilistic cluster membership. We model soft clustering trees as continuous optimization problem that is amenable to efficient optimization techniques. Our approach is designed to output highly sparse decision trees to increase interpretability and to support tree-based *spectral* clustering. Extensive experiments show that our approach can produce clustering trees of significantly higher quality compared to the state-of-the-art and scale to large datasets.

1 Introduction

Clustering, an unsupervised learning task, typically consists of partitioning an unlabelled dataset into K groups of similar data points. Since most popular clustering algorithms do not provide any explanation or interpretation for the obtained partition, a post-hoc analysis is often required to characterize the groups. In recent years, different approaches for *interpretable* clustering aim to provide clustering together with explanations of the obtained groups. One of the most prominent directions for interpretable clustering is based on using decision trees to construct clusters [2,16,28,30]. However, existing approaches for clustering trees are not scalable to large datasets, do not account for uncertainty in results, and do not support advanced clustering objectives such as Spectral Clustering [44] and Kernel-PCA clustering [37].

Soft decision trees are decision trees where at each node a data point is directed left with some probability p and right with probability $1-p$. Soft decision trees have been used for classification and regression in a range of previous works [4,5,7,23,48]. Unlike *hard* decision trees that are typically optimized

© The Author(s), under exclusive license to Springer Nature Switzerland AG 2023
A. A. Cire (Ed.): CPAIOR 2023, LNCS 13884, pp. 281–298, 2023.
https://doi.org/10.1007/978-3-031-33271-5_19

using a specialized heuristic procedure, soft decision trees can be optimized using gradient-based continuous optimization techniques. However, soft decision trees have not been applied to clustering.

In this work we present soft clustering trees, the first approach for interpretable clustering via soft decision trees that provide probabilistic output on cluster membership. Our approach is scalable and supports advanced clustering objectives such as Spectral Clustering and Kernel-PCA clustering. Specifically, we make the following contributions:

1. We present a novel approach for interpretable clustering based on soft decision trees that provide probabilistic output on cluster membership.
2. We present a continuous optimization model for soft clustering trees that is designed to produce *fully sparse* trees and is amenable to efficient second-order continuious optimization algorithms as well as scalable, SGD-based optimization algorithms.
3. We extend our soft clustering trees model to support *spectral* and *Kernel-PCA* clustering objectives, while still using interpretable decision trees in the original feature space to construct clusters.
4. We run extensive experiments and show that: (1) our spectral clustering and Kernel-PCA clustering variants can significantly outperform the state-of-the-art clustering trees algorithm on small and medium datasets; (2) our scalable approach for training soft clustering trees can produce high-quality clustering trees for large datasets.

2 Soft Clustering Trees

2.1 Soft Decision Trees

A *tree* $\mathcal{T}$ is a tuple $(\mathcal{T}_B, \mathcal{T}_L, \delta, p, l, r)$ where $\mathcal{T}_B$ is the set of branching nodes and $\mathcal{T}_L$ is the set of leaf nodes. $\delta \in \mathcal{T}_B$ is the root node, $p : (\mathcal{T}_B \cup \mathcal{T}_L - \{\delta\}) \rightarrow \mathcal{T}_B$ is the parent function, and $l, r : \mathcal{T}_B \rightarrow (\mathcal{T}_B \cup \mathcal{T}_L)$ are the left and right child functions, respectively.

The *depth* of a node in the tree $t \in \mathcal{T}_B \cup \mathcal{T}_L$ is recursively defined as $depth(t) = depth(p(t)) + 1$ with $depth(\delta) = 0$. The depth of a tree $\mathcal{T}$ is defined as the maximum depth among its leaf nodes, $depth(\mathcal{T}) = \max_{t \in \mathcal{T}_L} depth(t)$. A tree is considered *complete* if all leaves have the same depth, $\forall t_1, t_2 \in \mathcal{T}_L : depth(t_1) = depth(t_2)$.

A *decision tree* maps each branching node $t \in \mathcal{T}_B$ with a feature $f_t \in F$ and a threshold value μ_t such that each data point $x_i \in \mathbb{R}^{|F|}$ is directed left if $x_i^{f_t} \leq \mu_t$. An oblique decision tree maps each branching node to an oblique cut $a_{.t}^T x_i - \mu_t$ such that x_i is directed left if $a_{.t}^T x_i \leq \mu_t$. In contrast, a *soft decision tree* is associated with a matrix $a \in \mathbb{R}^{|F| \times |\mathcal{T}_B|}$ and a vector $\mu \in \mathbb{R}^{|\mathcal{T}_B|}$ such that the *probability* of point x_i to be directed left at branching node $t \in \mathcal{T}_B$ is

$$P_{it} = Sigmoid(\Gamma_t \cdot (a_{.t}^T x_i - \mu_t)), \tag{1}$$

where $a_{\cdot t}$ is the column vector representing the coefficients of all features in branching node t, μ_t is the threshold value, and Γ_t controls the softness of the split at node t such that higher values leads to more deterministic decisions [4,27]. Therefore, P_{it} can be considered as a soft (probabilistic) version of the oblique cut $a_{\cdot t}^T x_i \leq \mu_t$. Note that the complement, $1 - P_{it}$, is the probability that point x_i is directed right at node $t \in \mathcal{T}_B$.

Finally, the probability that a data point x_i ends up at a leaf node $t \in \mathcal{T}_L$ is defined as

$$Q_{it} = \prod_{t' \in A_L(t)} P_{it'} \prod_{t' \in A_R(t)} (1 - P_{it'}), \tag{2}$$

where $A_L(t)$ (resp. $A_R(t)$) denotes the set of all ancestors of a leaf node $t \in \mathcal{T}_L$ such that t is a descendant of their left (resp. right) branch.[1]

2.2 Soft Clustering Trees

Let $X = \{x_i\}_{i=1}^n$ be a set of n data points with x_i being a finite-sized feature vector, $x_i \in \mathbb{R}^{|F|}$, and K be the number of clusters ($K < |X|$). To extend soft decision trees to perform clustering of X into K clusters, we consider the following objective function inspired by fuzzy clustering,

$$\min \sum_{i \in |X|} \sum_{k \in 1..K} w_{ik}^m \cdot \|x_i - z_k\|^2, \tag{3}$$

where w_{ik} (defined below) is the probability that data point x_i is in cluster $k \in 1..K$, z_k is the centroid of cluster k, $\|x_i - z_k\| = \sqrt{\sum_{f \in F}(x_i^f - z_k^f)^2}$ is the Euclidean distance between point x_i and the centroid z_k, and $m \geq 1.0$ is a hyper-parameter that controls the fuzziness of the clustering. Equation (3) is similar to the objective of Fuzzy C-Means (FCM) [3], however in our case w is defined based on our soft decision tree rather than being an unconstrained variable.

To define w_{ik}, we first define $c_{\cdot t}$ to represent the distribution over cluster labels, i.e., c_{kt} is the probability that data points reaching leaf node $t \in \mathcal{T}_L$ are in cluster k. Then, we define w_{ik}, the probability that point x_i is in cluster k as:

$$w_{ik} = \sum_{t \in \mathcal{T}_L} Q_{it} c_{kt}. \tag{4}$$

In Fig. 1, we present an example for a soft clustering tree of depth 2 for the Iris dataset.

2.3 Sparsity in Soft Clustering Trees

While the soft clustering trees in Sect. 2.2 provide inherent tree-based interpretation, oblique cuts can be hard to interpret as they may utilize many, or even all

[1] If $A_L(t) = \varnothing$ or $A_R(t) = \varnothing$ then the corresponding products in Eq. (2) are equal to 1.0.

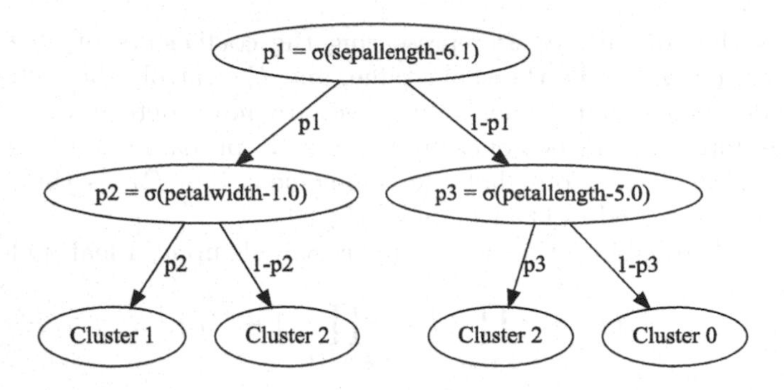

Fig. 1. Example soft clustering tree for the Iris dataset. We use σ to denote the Sigmoid function.

the features (i.e., cuts may have non-zero coefficients for many, or all, features). To obtain more interpretable trees, we would like to keep the number of non-zero coefficients in each branching node to a small number, ideally having only one non-zero coefficient similar to standard (non-oblique) decision trees. Previous works [4,21,27] on classification and regression have considered penalizing the ℓ_1-norm of the coefficient matrix a. However, this often results in some branching nodes having no non-zero coefficients while others having many non-zero coefficients, remaining difficult to interpret.

Instead of using oblique cuts, we consider the more restricted class of normalized cuts for branching nodes, i.e., all coefficients are non-negative ($\forall f \in F, t \in \mathcal{T}_B : a_{ft} > 0$) and the sum of coefficients in each branching node is equal to one ($\forall t \in \mathcal{T}_B : \sum_{f \in F} a_{ft} = 1$). This can be seen as a continuous relaxation of the typical univariate splits in standard decision trees by replacing the domain of each coefficient from the discrete set $\{0, 1\}$ to the continuous range $[0, 1]$ while keeping the sum of coefficients equal to one. Then, we introduce the following regularization term for each branching node $t \in \mathcal{T}_B$,

$$\phi_t = -\sum_{f \in F} a_{ft}^2. \tag{5}$$

The minimal value for each ϕ_t term is -1 which indicates a *fully sparse* cut, i.e., exactly one coefficient in the cut is equal to one and the rest are equal to zero.

2.4 Learning Sparse Soft Clustering Trees Using Continuous Optimization

We formulate the problem of learning soft clustering trees as a constrained continuous optimization problem. Given a dataset X and the number of clusters K, we search for an assignment of the variables a_{ft}, c_{kt}, μ_t, z_k, and Γ_t that minimizes our regularized clustering cost function.[2]

[2] Following [27], we keep Γ_t as variables rather than hyper-parameters.

In our continuous optimization model, we assume w.l.o.g that X is normalized in the range $[0, 1]$. We therefore bound the μ_t and z_k variables in the range $[0, 1]$. Further, to improve the optimization performance we consider $\mathbf{x}_i \in \mathbf{X}$ to be a transformation of the dataset $x_i \in X$ such that *each feature* is normalized within the range $[0, 1]$. In particular, we employed the quantile transformation following [27]. We redefine the cut in branching node $t \in \mathcal{T}_B$ to be $a_{.t}^T\mathbf{x}_i - \mu_t$, while keeping the clustering objective based on the original $x \in X$. As we focus on fully sparse trees, each cut can be converted back to its original values using the inverse transformation.

Constrained Continuous Optimization Model. Equation (6) presents the complete constrained optimization model for sparse soft clustering trees. Equation (6a) is the objective function that consists of the clustering cost and the sparsity regularization weighted by hyper-parameter Λ. Equations (6b)–(6e) are based on the definitions discussed in Sects. 2.1, 2.2 and 2.3. The constraints in Eq. (6f) and Eq. (6g) guarantee that label probabilities in leaf nodes and feature coefficients in branch nodes sum to one, respectively. Finally, Eqs. (6h)–(6l) define the bounds for each of the continuous decision variables.

$$\min \sum_{i \in |X|} \sum_{k \in 1..K} w_{ik}^m \cdot \|x_i - z_k\|^2 + \Lambda \sum_{t \in \mathcal{T}_B} \phi_t \tag{6a}$$

$$s.t.:$$

$$P_{it} := Sigmoid(\Gamma_t \cdot (a_{.t}^T\mathbf{x}_i - \mu_t)) \quad \forall \mathbf{x}_i \in \mathbf{X}, t \in \mathcal{T}_B \tag{6b}$$

$$Q_{it} := \prod_{t' \in A_L(t)} P_{it'} \prod_{t' \in A_R(t)} (1 - P_{it'}) \quad \forall x_i \in X, t \in \mathcal{T}_L \tag{6c}$$

$$w_{ik} := \sum_{t \in \mathcal{T}_L} Q_{it} c_{kt} \quad \forall x_i \in X, k \in 1..K \tag{6d}$$

$$\phi_t := -\sum_{f \in F} a_{ft}^2 \quad \forall t \in \mathcal{T}_B \tag{6e}$$

$$\sum_{k \in 1..K} c_{kt} = 1 \quad \forall t \in \mathcal{T}_L \tag{6f}$$

$$\sum_{f \in F} a_{ft} = 1 \quad \forall t \in \mathcal{T}_B \tag{6g}$$

$$0 \le a_{ft} \le 1 \quad \forall t \in \mathcal{T}_B, f \in F \tag{6h}$$

$$0 \le c_{kt} \le 1 \quad \forall t \in \mathcal{T}_L, k \in 1..K \tag{6i}$$

$$0 \le \mu_t \le 1 \quad \forall t \in \mathcal{T}_B \tag{6j}$$

$$0 \le z_k \le 1 \quad \forall k \in 1..K \tag{6k}$$

$$-1 \le \Gamma_t \le -128 \quad \forall t \in \mathcal{T}_B \tag{6l}$$

Constrained continuous optimization algorithms such as interior point optimization can be used to solve the optimization model in Eq. (6) for small and medium datasets, however they tend to be less efficient compared to unconstrained continuous optimization algorithms and are not amenable to scalable, mini-batch, stochastic gradient descent optimizers.

Unconstrained Optimization Model. We can reformulate the model in Eq. (6) to be an unconstrained optimization model by making the following changes.

Regularized Softmax Splits. To eliminate the constraints in Eqs. (6g)–(6h), we redefine a_{ft} based on Softmax normalization of the unnormalized variables $\hat{a}_{ft} \in \mathbb{R}^{|F|\times|\mathcal{T}_B|}$,

$$a_{ft} = \frac{\exp(\hat{a}_{ft})}{\sum_{f' \in F} \exp(\hat{a}_{f't}))}. \tag{7}$$

Similar to our constrained model, we use the regularization terms ϕ_t to guarantee sparse cuts. Due to the nature of Softmax, coefficients cannot be exactly zero, but could get very close to zero. However, since all features are scaled to the same range of [0, 1], features with near-zero coefficients will have negligible impact on the branching behavior and can be eliminated from the resultant clustering tree.[3] Although recent works have proposed several sparse variants of Softmax [10,29], we found that they can have negative impact on the optimization and are not needed in our case due to the feature-wise normalization.

Leaf Class Labels. To eliminate the constraints in Eq. (6f) and Eq. (6i), we redefine c_{kt} based on Softmax normalization of the unnormalized variables $\hat{c}_{kt} \in \mathbb{R}^{K\times|\mathcal{T}_L|}$,

$$c_{kt} = \frac{\exp(\hat{c}_{kt})}{\sum_{k' \in 1..K} \exp(\hat{c}_{k't}))}. \tag{8}$$

The bounds on the remaining variables, Eqs. (6j)–(6l), were used to improve the constrained model, but are not required and can be removed in our unconstrained model.

2.5 Interpretable Spectral and Kernel-PCA Clustering

One of the benefits of our formulation is that the feature representation used for constructing the decision tree and the feature representation used for the clustering do not have to be the same. Specifically, we can replace the objective in Eq. (3) with a more general objective,

$$\min \sum_{i \in |\bar{X}|} \sum_{k \in 1..K} w_{ik}^m \cdot \|\bar{x}_i - \bar{z}_k\|^2 \tag{9}$$

where $\bar{X} = \{\bar{x}_i\}_{i=1}^n$ is a (possibly) different representation of the dataset X based on feature set $\bar{F}$, $\bar{x}_i \in \mathbb{R}^{|\bar{F}|}$. We note that Eq. (3) is a special case of Eq. (9) with $\bar{F} = F$ and, consequently, $\bar{X} = X$. However, $\bar{X}$ can also be based on a different feature representation such as a spectral embedding or a PCA transformation of X. The decision tree is still constructed from the interpretable feature set X, i.e., branching node cuts based on $\mathbf{x} \in \mathbf{X}$, however the objective would be to optimize the clustering cost based on, for example, the spectral embedding or PCA transformation of the original dataset. In our experiment we consider two different objectives:

[3] For this purpose, we arbitrarily select 10^{-4} as the threshold for zeroing coefficients in the resultant clustering tree.

- Spectral clustering [44] where $\bar{X}$ is computed by applying spectral decomposition to the graph Laplacian of the k-nearest neighbors graph using the Laplacian Eigenmaps algorithm [1].
- Kernel-PCA (KPCA) clustering where $\bar{X}$ is computed using a non-linear dimensionality reduction through the use of kernels [37]. In our experiments, we use the radial basis function (RBF) kernel.

To our knowledge, this is the first approach for interpretable spectral and KPCA clustering that is based on decision trees in the original feature space.

2.6 Scalable Training of Soft Clustering Trees

Our unconstrained formulation in Sect. 2.4 is amenable for scalable training using mini-batch stochastic gradient descent algorithms to support interpretable clustering of large datasets using soft decision trees. However, consistent with work on soft classification trees [15], we found that training of soft clustering trees using first-order SGD optimizers can get stuck in poor solutions in which one or more of the branching nodes directs almost all the data points into one of the subtrees. We therefore introduce the following regularization term that encourages branching nodes to make equal use of both left and right branches, following [15],

$$\pi = -\sum_{t \in \mathcal{T}_B} \theta_t [0.5 \cdot \log(\alpha_t) + 0.5 \cdot \log(1 - \alpha_t)], \tag{10}$$

with α_t (resp. 1-α_t) being the fraction of probability mass directed to the left (resp. right) branch of branching node $t \in \mathcal{T}_B$ out of the probability mass directed to node t,

$$\alpha_t = \frac{\sum_{x_i \in X} Q_{i,l(t)}}{\sum_{x_i \in X} Q_{i,t}}, \tag{11}$$

and θ_t ensures the strength of the penalty decays exponentially with the depth of node t, $\theta_t = 2^{-depth(t)}$.

Training. The final objective function for our scalable training of soft clustering trees is

$$\min \sum_{i \in |X|} \sum_{k \in 1..K} w_{ik}^m \cdot \|x_i - z_k\|^2 + \Lambda \sum_{t \in \mathcal{T}_B} \phi_t + \omega\pi,$$

where ω is a hyper-parameter that controls the weight associated with the regularization term. To efficiently train clustering trees using mini-batch stochastic gradient descent algorithms, we start by training with no sparsity regularization, $\Lambda = 0$, for a fixed number of training steps. Then, we anneal Λ by increasing it every training step until we obtain a fully sparse tree.

3 Experiments

In this section, we perform extensive experiments with 18 datasets to evaluate the performance of soft clustering trees.

3.1 Implementation Details

Single-Batch Training of Soft Clustering Trees using Second-Order Optimizers. For our experiments with small and medium datasets, we implemented our constrained and unconstrained continuous optimization models for single-batch training using second-order optimizers. Our constrained optimization model was implemented in Julia using the JuMP library [13] and solved using IPOPT [45], a primal-dual interior-point algorithm with a filter line-search method for non-linear programming. As IPOPT converges to a local minimum on non-convex problems, we run the solver five times, starting from different random initializations, and select the lowest-cost solution.

Our unconstrained model was implemented in Python and solved using the Limited-memory BFGS with bounds (L-BFGS-B) solver and the Sequential Least Squares Programming (SLSQP) solver, both implemented in the Scipy library. We found the runtime for unconstrained optimization to be shorter compared to constrained optimization, however it requires more runs to converge to high-quality solutions. We therefore restart the solver 20 times using random initializations and select the lowest-cost solution.

Scalable Training of Soft Clustering Trees using SGD. For our experiments with large datasets, we implemented our scalable model for training soft clustering trees (Sect. 2.6) in Python using the PyTorch library [31]. We employ mini-batch stochastic optimization scheme using the RMSProp optimizer [20]. To obtain fully sparse trees, we train our model according to the training scheme described in Sect. 2.6: We first train the model for 25,000 steps with no sparsity regularization and then slowly anneal Λ by increasing it each step by 10^{-3}. In our experiments, we set the total number of training steps to be 50,000 and we employ cyclical learning rate schedule [39] in the range $[0.0005, 0.005]$.

Inference. Our decision trees are *soft* and represent probabilistic cluster membership. In our experiments, we obtain a *hard* clustering for each data point by selecting the cluster label for which the membership probability is the highest.

3.2 Datasets

To evaluate our single-batch approach for learning soft clustering trees, we use a set of 13 small- and medium-size real and synthetic datasets (Table 1). Seven real datasets were obtained from the UCI repository [12]: Glass, Ionosphere, Iris, TAE, Vertebral, Wine, Zoo. Four synthetic datasets representing challenging clustering problems in 2D and 3D were obtained from FCPS [43]. Finally, we generated two instances of the well-known clustering problems moons and circles.

Table 1. Details of small- and medium-size datasets used in the experiments. Synthetic datasets are marked "(s)".

Dataset	$\|X\|$	$\|F\|$	K
Atom (s)	800	3	2
Chainlink (s)	1,000	3	2
Circles (s)	500	2	2
Glass	214	9	6
Ionosphere	351	34	2
Iris	150	4	3
Moons (s)	500	2	2
TAE	151	5	3
Target (s)	770	2	6
Vertebral	310	6	3
Wine	178	12	3
Wingnut (s)	1,016	2	2
Zoo	101	16	7

Table 2. Details of large datasets used in the experiments.

Dataset	$\|X\|$	$\|F\|$	K
Adult†	48,842	105	2
Avila	20,867	10	12
Covertype	581,012	54	7
Pendigits	10,992	16	10
Shuttle	58,000	9	7

†Categorical features converted to one-hot encoding.

To evaluate our approach for scalable training of soft clustering trees, we run experiments on five large datasets obtained from UCI [12]: Adult, Avila, Covertype, Pendigits, and Shuttle, as described in Table 2. All datasets were standardized by removing the mean and scaling each feature to unit variance.

3.3 Evaluation

Since all datasets in Sect. 3.2 have ground-truth labels, we evaluate the quality of the obtained clusterings using the following external evaluation metrics. Note that we do not use internal evaluation metrics, such as the mean Silhouette Coefficient [36], as they depend on the feature representation and therefore are not comparable across different feature representations (such as the Spectral Embedding and the Kernel-PCA).

Adjusted Rand Index (ARI). Rand Index [35] measures agreement between two partitions of the same dataset, P_1 and P_2. Each partition represents $\binom{n}{2}$ decisions over all pairs, assigning them to the same or different clusters. Let a be the number of pairs assigned to the same cluster in both P_1 and P_2. Let b be the number of pairs assigned to different clusters. Rand Index is defined as follows:

$$RI(P_1, P_2) = \frac{a+b}{\binom{n}{2}}.$$

The Adjusted Rand Index (ARI) [22] is a correction for RI, based on its expected value:

$$ARI = \frac{RI - \mathbb{E}(RI)}{Max(RI) - \mathbb{E}(RI)}.$$

ARI score of zero indicates the partition is not better than a random assignment, while a score of one indicates a perfect match. We compute the ARI between the obtained clustering and the ground-truth labels.

Normalized Mutual Information (NMI). Mutual information quantifies the statistical information shared between two distributions [40]. $MI(P_1, P_2)$ denotes the mutual information between partitions P_1 and P_2, and $H(P_i)$ denotes the entropy of partition P_i. Normalized mutual information (NMI) [40] is normalized using the mean of $H(P_1)$ and $H(P_2)$:

$$NMI(P_1, P_2) = \frac{MI(P_1, P_2)}{Mean(H(P_1), H(P_2))}.$$

Values close to zero indicate independent partitions, while values close to one indicate a significant agreement between $P1$ and $P2$. We compute the NMI between the obtained clustering and the ground-truth labels.

Unsupervised Clustering Accuracy (ACC). The unsupervised clustering accuracy [46] is defined as:

$$ACC = \max_{map \in M} \frac{\sum_{i=1}^{n} \mathbb{1}\{l_i = map(c_i))\}}{n},$$

where $l(x_i)$ and $c(x_i)$ are the ground-truth label and the assigned cluster label for data point x_i, respectively, and M is the set of all possible one-to-one mappings from clusters to ground-truth labels.

3.4 Results

First, we compare our basic constrained and unconstrained optimization models to ExKMC [16], the state-of-the-art approach for interpretable clustering using decision trees. For our unconstrained model, we used both L-BFGS and SLSQP solvers. Each problem is solved 20 times starting from different initializations and the median runtime for one run was 1.81 s for L-BFGS and 2.01 s for SLSQP. As all runs are independent, they can be parallelized over multiple cores. As we are comparing our approaches that are probabilistic to the deterministic and fully sparse ExKMC that aims to optimize the standard K-means cost, we tuned the hyper-parameters of our approach over a small set of possible values, $\Lambda \in \{10^0, 10^1, 10^2\}$ and $m \in \{1.05, 1.1\}$, and select the ones that yielded the *hard* clustering with the lowest K-means cost while being fully sparse (all results presented for our approaches are therefore based on fully sparse trees). For our constrained model, solved using IPOPT, runs required longer runtime (median of 14.45 s) and we therefore opted for only five random initializations and considered only one value for m that was found to work well ($m = 1.0$). We note that the results for our approaches are not directly comparable in terms of optimization performance due to the large set of possible choices available for each solver (how many runs vs. how long each run, how many available

Table 3. Experimental Results on Soft Clustering Trees for Small and Medium Datasets.

dataset	Max $\lvert\mathcal{T}_L\rvert$	Adjusted Rand Index (ARI)				Clustering Accuracy (ACC)			
		BFGS	SLSQ	IPOP	ExKM	BFGS	SLSQ	IPOP	ExKM
atom	4	0.180	0.182	0.161	**0.186**	0.713	0.714	0.701	**0.716**
atom	8	0.149	**0.165**	0.161	0.159	0.694	**0.704**	0.701	0.700
atom	16	**0.189**	0.176	0.174	**0.189**	**0.718**	0.710	0.709	**0.718**
chainl	4	−0.001	−0.001	**0.183**	−0.001	0.500	0.500	**0.714**	0.504
chainl	8	−0.001	−0.001	**0.207**	−0.001	0.509	0.505	**0.728**	0.508
chainl	16	−0.001	−0.000	**0.107**	−0.001	0.505	0.514	**0.664**	0.509
circles	4	**−0.002**	**−0.002**	**−0.002**	**−0.002**	**0.502**	**0.502**	**0.502**	**0.502**
circles	8	**−0.002**	**−0.002**	**−0.002**	**−0.002**	**0.502**	**0.502**	**0.502**	**0.502**
circles	16	**−0.002**	**−0.002**	**−0.002**	**−0.002**	0.500	**0.502**	**0.502**	**0.502**
glass	8	**0.200**	0.188	0.168	0.148	**0.505**	0.458	0.481	0.425
glass	16	0.148	0.173	**0.230**	0.173	0.481	0.472	**0.519**	0.472
iono	4	0.158	0.145	0.149	**0.163**	0.701	0.692	0.695	**0.704**
iono	8	0.112	0.178	**0.183**	0.168	0.670	0.712	**0.715**	0.707
iono	16	**0.193**	0.178	0.168	0.168	**0.721**	0.712	0.707	0.707
iris	4	**0.759**	0.610	0.515	0.574	**0.907**	0.827	0.747	0.800
iris	8	**0.653**	0.620	0.574	0.601	**0.853**	0.833	0.800	0.820
iris	16	**0.642**	0.632	**0.642**	0.610	**0.847**	0.840	**0.847**	0.827
moons	4	**0.483**	**0.483**	**0.483**	0.456	**0.848**	**0.848**	**0.848**	0.838
moons	8	**0.472**	**0.472**	**0.472**	0.461	**0.844**	**0.844**	**0.844**	0.840
moons	16	0.472	0.472	0.472	**0.478**	0.844	0.844	0.844	**0.846**
tae	4	**0.064**	**0.064**	0.050	0.047	**0.510**	**0.510**	0.444	0.417
tae	8	0.047	0.047	0.047	**0.064**	0.417	0.417	0.417	**0.510**
tae	16	0.048	0.047	0.048	**0.064**	0.424	0.417	0.424	**0.510**
target	8	0.529	0.557	0.302	**0.636**	0.635	0.627	0.416	**0.638**
target	16	0.636	**0.637**	0.634	0.636	0.642	**0.652**	0.625	0.636
vert	4	0.163	**0.212**	0.175	0.165	0.465	**0.487**	0.471	0.452
vert	8	0.180	**0.221**	0.169	0.194	0.461	**0.516**	0.455	0.461
vert	16	**0.221**	0.210	0.219	0.196	0.506	**0.513**	0.490	0.461
wine	4	0.754	0.748	**0.848**	0.802	0.916	0.910	**0.949**	0.933
wine	8	0.732	0.757	0.741	**0.880**	0.904	0.916	0.910	**0.961**
wine	16	0.683	0.835	0.880	**0.897**	0.882	0.944	0.961	**0.966**
wingn	4	0.760	0.791	0.791	**0.930**	0.936	0.945	0.945	**0.982**
wingn	8	0.736	0.733	0.743	**0.764**	0.929	0.928	0.931	**0.937**
wingn	16	0.700	0.693	**0.730**	0.683	0.918	0.916	**0.927**	0.913
zoo	8	0.870	**0.871**	0.617	0.737	**0.871**	**0.871**	0.762	0.822
zoo	16	0.814	**0.815**	0.792	0.737	0.822	0.812	**0.832**	0.822
average		0.354	0.359	0.356	**0.360**	0.683	0.684	**0.687**	0.682

cores, how many hyper-parameters values to consider, etc.) We therefore simply demonstrate the performance of each approach with a reasonable set of choices.

For ExKMC, we run the algorithm from 100 different random initializations and choose the one with the lowest cost (experiments with additional runs did not lead to significant improvement). As ExKMC does not have a limit on the tree depth but on the maximum number of leaves, we compare the results for three different values of number of leaves, namely $4, 16, 32$. In our approach these values correspond to a maximum tree depth of $2, 3, 4$. For each dataset, we set K to be the number of ground-truth labels in the dataset. As the datasets Glass, Target, and Zoo have more than 4 clusters, we only run experiments for a maximum number of leaves of 16 and 32.

Table 3 shows the ARI and ACC scores obtained by each of the approaches for each of the datasets. It also reports the average scores across all datasets. Results on NMI exhibited similar trends and are omitted due to space. We can see that, in general, the different methods are relatively comparable. Each of the methods outperforms the other methods on some of the datasets, and we observe minor differences between the methods in the average scores. Specifically, ExKMC performed slightly better in terms of average ARI and IPOPT performs slightly better in terms of ACC as well as NMI (not presented).

In the next two sections we demonstrate the unique benefits of our approaches, namely that they can be extended to use Spectral and K-PCA objectives and that are amenable to scalable optimization procedures.

Spectral and K-PCA Clustering Trees. We present results for the extensions of our basic approach: our Kernel PCA model (*Ours-K*), and our Spectral Clustering model (*Ours-S*). For KPCA, we used 10 components. For the spectral embedding, we used k-nearest neighbors graph with $k = 10$ for all datasets and set the dimension of the projected subspace to be the number of clusters. Due to limited space, in this experiment we focus on our unconstrained model as it is the basis for our scalable model, and we present results for the L-BFGS solver. We compare our approaches to our basic model (*Ours*) and to ExKMC [16].

Table 4 shows the ARI and ACC scores obtained by each of the approaches for each of the datasets. Results on NMI exhibited similar patterns to ARI and are omited due to space. It also reports the average scores across all datasets. The best performing approach based on the average scores is *Ours-S* followed by *Ours-K*. Furthermore, we observe that in approximately 86% of the cases, for all evaluation metrics (including NMI), the top performing approach is one of our approaches. The results demonstrate the unique benefits of approaches like *Ours-S* in cases such as the datasets Atom, Chainlink, and Circles, where ExKMC and *Ours* find low-quality solutions compared to the high-quality solutions found by *Ours-S* due to the spectral embedding.

Results for Large Datasets. Next, we run experiments with our approach for scalable training of soft clustering trees (Sect. 2.6). As our approach is the first scalable approach for interpretable clustering based on decision trees, we

Table 4. Experimental Results on Soft Clustering Trees for Small and Medium Datasets. Our approaches are based on our unconstrained model solved by L-BFGS.

dataset	Max	Adjusted Rand Index (ARI)				Clustering Accuracy (ACC)			
	$\lvert\mathcal{T}_L\rvert$	Ours	Ours-K	Ours-S	ExKM	Ours	Ours-K	Ours-S	ExKM
atom	4	0.180	0.577	**0.779**	0.186	0.713	0.880	**0.941**	0.716
atom	8	0.149	0.865	**0.874**	0.159	0.694	0.965	**0.968**	0.700
atom	16	0.189	0.912	**0.970**	0.189	0.718	0.978	**0.993**	0.718
chain	4	−0.001	0.174	**0.861**	−0.001	0.500	0.709	**0.964**	0.504
chain	8	−0.001	0.178	**0.933**	−0.001	0.509	0.711	**0.983**	0.508
chain	16	−0.001	0.181	**0.941**	−0.001	0.505	0.713	**0.985**	0.509
circles	4	−0.002	−0.002	**0.369**	−0.002	0.502	0.504	**0.804**	0.502
circles	8	−0.002	−0.002	**0.639**	−0.002	0.502	0.502	**0.900**	0.502
circles	16	−0.002	−0.002	**1.000**	−0.002	0.500	0.504	**1.000**	0.502
glass	8	**0.200**	0.145	0.112	0.148	**0.505**	0.411	0.360	0.425
glass	16	0.148	**0.184**	0.133	0.173	**0.481**	0.439	0.379	0.472
iono	4	0.158	**0.203**	−0.028	0.163	0.701	**0.726**	0.538	0.704
iono	8	0.112	**0.208**	−0.034	0.168	0.670	**0.729**	0.504	0.707
iono	16	0.193	**0.224**	−0.034	0.168	0.721	**0.738**	0.501	0.707
iris	4	**0.759**	0.600	0.489	0.574	**0.907**	0.820	0.773	0.800
iris	8	0.653	**0.736**	0.394	0.601	0.853	**0.900**	0.700	0.820
iris	16	**0.642**	0.611	0.413	0.610	**0.847**	0.827	0.713	0.827
moons	4	0.483	0.512	**0.678**	0.456	0.848	0.858	**0.912**	0.838
moons	8	0.472	0.512	**0.853**	0.461	0.844	0.858	**0.962**	0.840
moons	16	0.472	0.512	**1.000**	0.478	0.844	0.858	**1.000**	0.846
tae	4	**0.064**	**0.064**	0.050	0.047	**0.510**	**0.510**	0.444	0.417
tae	8	0.047	**0.113**	0.047	0.064	0.417	**0.550**	0.417	0.510
tae	16	0.048	**0.113**	0.047	0.064	0.424	**0.550**	0.417	0.510
target	8	0.529	0.538	0.544	**0.636**	0.635	0.627	0.626	**0.638**
target	16	**0.636**	0.634	0.328	**0.636**	**0.642**	0.627	0.443	0.636
vert	4	0.163	0.169	**0.171**	0.165	**0.465**	0.458	0.461	0.452
vert	8	0.180	**0.254**	0.212	0.194	0.461	0.474	**0.500**	0.461
vert	16	0.221	**0.251**	0.219	0.196	0.506	**0.539**	0.474	0.461
wine	4	0.754	0.723	0.762	**0.802**	0.916	0.899	0.916	**0.933**
wine	8	0.732	0.642	0.754	**0.880**	0.904	0.871	0.916	**0.961**
wine	16	0.683	0.725	0.820	**0.897**	0.882	0.904	0.938	**0.966**
wingn	4	0.760	0.693	**1.000**	0.930	0.936	0.916	**1.000**	0.982
wingn	8	0.736	0.651	**1.000**	0.764	0.929	0.904	**1.000**	0.937
wingn	16	0.700	0.736	**0.984**	0.683	0.918	0.929	**0.996**	0.913
zoo	8	**0.870**	0.820	0.653	0.737	**0.871**	0.832	0.743	0.822
zoo	16	**0.814**	0.646	0.633	0.737	**0.822**	0.743	0.752	**0.822**
average		0.354	0.419	**0.544**	0.360	0.683	0.721	**0.748**	0.682

compare our approach to *non-interpretable* scalable clustering using Mini-Batch K-Means [38].

We run experiments for three tree depths: the minimum depth based on the number of ground-truth labels, as well as two levels deeper. We did not tune hyper-parameters for each dataset and instead fix $\omega = 0.1$ and $m = 1.05$ across datasets (hyper-parameter tuning per dataset may lead to further improvement). We run the training procedure five times, starting from different random initializations, using a batch size of 256. Similar to previous experiment, we select the one that yielded the hard clustering with the lowest K-means cost while being fully sparse. For Mini-Batch K-Means, we run the algorithm for 100 random initializations with a similar batch size of 256 and select the lowest cost solution.

Table 5 compares our approach for scalable training (*Ours*) to Mini-Batch K-Means (mKM) on the five large datasets. We note that the two methods are *not* directly comparable as Mini-Batch K-Means is not constrained to produce tree-based clusterings. The results show that for Adult, Covertype, and Shuttle, our approach can reach comparable results to mKM and even find higher-quality solutions according to some criteria. For Pendigits, we observe that as we increase the tree depth we are getting closer to mKM's performance however even a depth of 6 was not sufficient to reach the performance of mKM with a *fully-sparse* decision tree. For Avila, we interestingly find the best solution at the lowest tree depth. Overall, the results in Table 5 indicate that our scalable approach is able to produce high-quality, fully sparse clustering trees for large datasets.

Table 5. Experimental Results for Large Datasets.

X	Max	ARI		NMI		ACC	
	$\lvert T_L \rvert$	Ours	mKM	Ours	mKM	Ours	mKM
Adult	2	**0.184**	0.183	0.134	**0.136**	**0.719**	0.718
Adult	4	**0.184**		0.134		**0.719**	
Adult	8	0.180		0.133		0.717	
Avila	16	**0.064**	0.052	0.117	**0.136**	0.291	**0.292**
Avila	32	0.016		0.053		0.218	
Avila	64	0.055		0.108		0.232	
Cover	8	0.037	0.056	0.143	**0.150**	0.291	0.319
Cover	16	**0.057**		**0.150**		**0.329**	
Cover	32	0.031		0.145		0.309	
Pend.	16	0.403	**0.539**	0.554	**0.685**	0.590	**0.675**
Pend.	32	0.437		0.595		0.590	
Pend.	64	0.485		0.624		0.638	
Shut.	8	0.181	0.214	0.366	0.378	0.412	0.421
Shut.	16	0.196		0.329		0.444	
Shut.	32	**0.348**		**0.475**		**0.631**	

4 Related Work

Soft decision trees have been a popular choice in tasks such as classification and regression, solved using either constrained or unconstrained continuous optimization algorithms [4,5,23,27]. Some works have explored using soft decision trees together with learned representations by formulating the problem as a deep neural network [15,19,41,47]. To our knowledge, our work is the first approach that use soft decision trees for interpretable clustering.

Recent work on neural oblivious classification and regression trees has considered sparse alternatives of Softmax, such as entmax [33], to produce sparse trees [34], however we found it difficult to produce fully sparse trees without hurting the optimization performance.

Previous work on interpretable clustering primarily focused on using decision trees [2,14,16,18,26,30,42]. Other approaches also include polytope machines [6,25], rectangular rules [8,32], and layerwise relevance propagation [24]. To our knowledge, our work is the first to consider soft decision trees, to support scalable training, and to be extended to tree-based spectral and KPCA clustering.

Several works on interpretable clustering via decision trees focus on a setting in which each cluster corresponds to exactly one leaf, similar to hierarchical clustering [2,18,30,49]. This approach significantly restricts the expressive power of the decision trees and their ability to accurately match the observed clusters in the dataset. Similar to the recent ExKMC [16], our approach allows more than K leaves to support more expressive trees.

A very recent work has focused on clustering using hard, oblique decision trees via alternating optimization [17]. While their implementation is not available, their experiments show limited improvement over ExKMC for fully sparse (axis-aligned) trees. Different from our work, they focus on hard decision trees and their approach is not amenable to scalable, mini-batch, stochastic gradient descent optimization.

5 Conclusion

We present a novel approach for interpretable clustering based on soft clustering trees. We formulate the problem as a continuous optimization problem that can be efficiently solved by second-order optimizers, such as L-BFGS, as well as scalable SGD optimization. We extend our approach to support spectral and KPCA clustering trees. We conduct extensive experiments using 18 datasets and show that our spectral and KPCA approaches significantly outperform the state-of-the-art approach on small and medium datasets and our scalable training using SGD produces high quality clustering trees for large datasets.

Our work can be extended in a number of ways. Investigating approaches for joint construction of soft clustering trees where clustering is based on learned representations would be an interesting extension of our work. Investigating strategies to incorporate fairness considerations [9] is an important direction for future work. Finally, incorporating domain-specific knowledge in the form of constraints [11] could lead to higher-quality, yet interpretable, solutions.

References

1. Belkin, M., Niyogi, P.: Laplacian eigenmaps for dimensionality reduction and data representation. Neural Comput. **15**(6), 1373–1396 (2003)
2. Bertsimas, D., Orfanoudaki, A., Wiberg, H.: Interpretable clustering: an optimization approach. Mach. Learn. **110**(1), 89–138 (2021)
3. Bezdek, J.C., Ehrlich, R., Full, W.: FCM: the fuzzy C-means clustering algorithm. Comput. Geosci. **10**(2–3), 191–203 (1984)
4. Blanquero, R., Carrizosa, E., Molero-Río, C., Morales, D.R.: Sparsity in optimal randomized classification trees. Eur. J. Oper. Res. **284**(1), 255–272 (2020)
5. Blanquero, R., Carrizosa, E., Molero-Río, C., Morales, D.R.: Optimal randomized classification trees. Comput. Oper. Res. **132**, 105281 (2021)
6. Carrizosa, E., Kurishchenko, K., Marín, A., Morales, D.R.: Interpreting clusters via prototype optimization. Omega **107**, 102543 (2022)
7. Carrizosa, E., Molero-Río, C., Romero Morales, D.: Mathematical optimization in classification and regression trees. TOP **29**(1), 5–33 (2021). https://doi.org/10.1007/s11750-021-00594-1
8. Chen, J., et al.: Interpretable clustering via discriminative rectangle mixture model. In: 2016 IEEE 16th International Conference on Data Mining (ICDM), pp. 823–828. IEEE (2016)
9. Chhabra, A., Masalkovaitė, K., Mohapatra, P.: An overview of fairness in clustering. IEEE Access **9**, 130698–130720 (2021)
10. Correia, G.M., Niculae, V., Martins, A.F.: Adaptively sparse transformers. In: Proceedings of the EMNLP-IJCNLP (2019, to appear)
11. Dao, T.B.H., Vrain, C., Duong, K.C., Davidson, I.: A framework for actionable clustering using constraint programming. In: Proceedings of the Twenty-Second European Conference on Artificial Intelligence, pp. 453–461 (2016)
12. Dua, D., Graff, C.: UCI machine learning repository (2017). http://archive.ics.uci.edu/ml
13. Dunning, I., Huchette, J., Lubin, M.: JuMP: a modeling language for mathematical optimization. SIAM Rev. **59**(2), 295–320 (2017). https://doi.org/10.1137/15M1020575
14. Fraiman, R., Ghattas, B., Svarc, M.: Interpretable clustering using unsupervised binary trees. Adv. Data Anal. Classif. **7**(2), 125–145 (2013)
15. Frosst, N., Hinton, G.: Distilling a neural network into a soft decision tree. arXiv preprint arXiv:1711.09784 (2017)
16. Frost, N., Moshkovitz, M., Rashtchian, C.: ExKMC: expanding explainable k-means clustering. arXiv preprint arXiv:2006.02399 (2020)
17. Gabidolla, M., Carreira-Perpiñán, M.Á.: Optimal interpretable clustering using oblique decision trees. In: Proceedings of the 28th ACM SIGKDD Conference on Knowledge Discovery and Data Mining, pp. 400–410 (2022)
18. Gamlath, B., Jia, X., Polak, A., Svensson, O.: Nearly-tight and oblivious algorithms for explainable clustering. Adv. Neural. Inf. Process. Syst. **34**, 28929–28939 (2021)
19. Hazimeh, H., Ponomareva, N., Mol, P., Tan, Z., Mazumder, R.: The tree ensemble layer: Differentiability meets conditional computation. In: International Conference on Machine Learning, pp. 4138–4148. PMLR (2020)
20. Hinton, G., Srivastava, N., Swersky, K.: Neural networks for machine learning lecture 6a overview of mini-batch gradient descent. Cited on **14**(8), 2 (2012)
21. Hou, Q., Zhang, N., Kirschen, D.S., Du, E., Cheng, Y., Kang, C.: Sparse oblique decision tree for power system security rules extraction and embedding. IEEE Trans. Power Syst. **36**(2), 1605–1615 (2020)

22. Hubert, L., Arabie, P.: Comparing partitions. J. Classif. **2**(1), 193–218 (1985)
23. Irsoy, O., Yildiz, O.T., Alpaydin, E.: Soft decision trees. In: Proceedings of the 21st International Conference on Pattern Recognition (ICPR2012), pp. 1819–1822. IEEE (2012)
24. Kauffmann, J., Esders, M., Ruff, L., Montavon, G., Samek, W., Müller, K.R.: From clustering to cluster explanations via neural networks. IEEE Trans. Neural Netw. Learn. Syst. (2022)
25. Lawless, C., Kalagnanam, J., Nguyen, L.M., Phan, D., Reddy, C.: Interpretable clustering via multi-polytope machines. In: Proceedings of the AAAI Conference on Artificial Intelligence, pp. 7309–7316 (2022)
26. Liu, B., Xia, Y., Yu, P.S.: Clustering via decision tree construction. In: Chu, W., Young Lin, T. (eds.) Foundations and Advances in Data Mining. Studies in Fuzziness and Soft Computing, vol. 180, pp. 97–124. Springer, Heidelberg (2005). https://doi.org/10.1007/11362197_5
27. Luo, H., Cheng, F., Yu, H., Yi, Y.: SDTR: soft decision tree regressor for tabular data. IEEE Access **9**, 55999–56011 (2021)
28. Makarychev, K., Shan, L.: Near-optimal algorithms for explainable k-medians and k-means. In: International Conference on Machine Learning, pp. 7358–7367. PMLR (2021)
29. Martins, A., Astudillo, R.: From softmax to sparsemax: a sparse model of attention and multi-label classification. In: International Conference on Machine Learning, pp. 1614–1623. PMLR (2016)
30. Moshkovitz, M., Dasgupta, S., Rashtchian, C., Frost, N.: Explainable k-means and k-medians clustering. In: International Conference on Machine Learning, pp. 7055–7065. PMLR (2020)
31. Paszke, A., et al.: Pytorch: An imperative style, high-performance deep learning library. In: Wallach, H., Larochelle, H., Beygelzimer, A., d'Alché-Buc, F., Fox, E., Garnett, R. (eds.) Advances in Neural Information Processing Systems, vol. 32, pp. 8024–8035. Curran Associates, Inc. (2019). http://papers.neurips.cc/paper/9015-pytorch-an-imperative-style-high-performance-deep-learning-library.pdf
32. Pelleg, D., Moore, A.: Mixtures of rectangles: interpretable soft clustering. In: ICML, vol. 2001, pp. 401–408 (2001)
33. Peters, B., Niculae, V., Martins, A.F.: Sparse sequence-to-sequence models. arXiv preprint arXiv:1905.05702 (2019)
34. Popov, S., Morozov, S., Babenko, A.: Neural oblivious decision ensembles for deep learning on tabular data. arXiv preprint arXiv:1909.06312 (2019)
35. Rand, W.M.: Objective criteria for the evaluation of clustering methods. J. Am. Stat. Assoc. **66**(336), 846–850 (1971)
36. Rousseeuw, P.J.: Silhouettes: a graphical aid to the interpretation and validation of cluster analysis. J. Comput. Appl. Math. **20**, 53–65 (1987)
37. Schölkopf, B., Smola, A., Müller, K.-R.: Kernel principal component analysis. In: Gerstner, W., Germond, A., Hasler, M., Nicoud, J.-D. (eds.) ICANN 1997. LNCS, vol. 1327, pp. 583–588. Springer, Heidelberg (1997). https://doi.org/10.1007/BFb0020217
38. Sculley, D.: Web-scale k-means clustering. In: Proceedings of the 19th International Conference on World Wide Web, pp. 1177–1178 (2010)
39. Smith, L.N.: Cyclical learning rates for training neural networks. In: 2017 IEEE Winter Conference on Applications of Computer Vision (WACV), pp. 464–472. IEEE (2017)
40. Strehl, A., Ghosh, J.: Cluster ensembles–a knowledge reuse framework for combining multiple partitions. J. Mach. Learn. Res. **3**, 583–617 (2002)

41. Tanno, R., Arulkumaran, K., Alexander, D., Criminisi, A., Nori, A.: Adaptive neural trees. In: International Conference on Machine Learning, pp. 6166–6175. PMLR (2019)
42. Tavallali, P., Tavallali, P., Singhal, M.: K-means tree: an optimal clustering tree for unsupervised learning. J. Supercomput. **77**(5), 5239–5266 (2021)
43. Ultsch, A., Lötsch, J.: The fundamental clustering and projection suite (FCPS): a dataset collection to test the performance of clustering and data projection algorithms. Data **5**(1), 13 (2020)
44. Von Luxburg, U.: A tutorial on spectral clustering. Stat. Comput. **17**(4), 395–416 (2007)
45. Wächter, A., Biegler, L.T.: On the implementation of an interior-point filter line-search algorithm for large-scale nonlinear programming. Math. Program. **106**(1), 25–57 (2006)
46. Xie, J., Girshick, R., Farhadi, A.: Unsupervised deep embedding for clustering analysis. In: International Conference on Machine Learning, pp. 478–487. PMLR (2016)
47. Yang, Y., Morillo, I.G., Hospedales, T.M.: Deep neural decision trees. In: ICML Workshop on Human Interpretability in Machine Learning (WHI) (2018)
48. Yoo, J., Sael, L.: EDiT: interpreting ensemble models via compact soft decision trees. In: 2019 IEEE International Conference on Data Mining (ICDM), pp. 1438–1443. IEEE (2019)
49. Zantedeschi, V., Kusner, M., Niculae, V.: Learning binary decision trees by argmin differentiation. In: International Conference on Machine Learning, pp. 12298–12309. PMLR (2021)

NER4OPT: Named Entity Recognition for Optimization Modelling from Natural Language

Parag Pravin Dakle[1], Serdar Kadıoğlu[1,2](✉), Karthik Uppuluri[1], Regina Politi[1], Preethi Raghavan[1], SaiKrishna Rallabandi[1], and Ravisutha Srinivasamurthy[1]

[1] AI Center of Excellence, Fidelity Investments, Boston, USA
{paragpravin.dakle,serdar.kadoglu,karthik.uppuluri,regina.politi, preethi.raghavan,saikrishna.rallabandi,ravisutha.srinivasamurthy}@fmr.com
[2] Department of Computer Science, Brown University, Providence, USA

Abstract. Solving combinatorial optimization problems involves a two-stage process that follows the model-and-run approach. First, a user is responsible for formulating the problem at hand as an optimization model, and then, given the model, a solver is responsible for finding the solution. While optimization technology has enjoyed tremendous theoretical and practical advances, the overall process has remained the same for decades. To date, transforming problem descriptions into optimization models remains a barrier to entry. To alleviate users from the cognitive task of modeling, we study named entity recognition to capture components of optimization models such as the objective, variables, and constraints from free-form natural language text, and coin this problem as NER4OPT. We show how to solve NER4OPT using classical techniques based on morphological and grammatical properties and modern methods leveraging pre-trained large language models and fine-tuning transformers architecture with optimization-specific corpora. For best performance, we present their hybridization combined with feature engineering and data augmentation to exploit the language of optimization problems. We improve over the state-of-the-art for annotated linear programming word problems, identify several next steps and discuss important open problems toward automated modeling.

Keywords: Optimization Modeling · Named Entity Recognition · Natural Language Processing

1 Introduction

Optimization technology spans a wide range of applications, and over the years, combinatorial optimization solvers have enjoyed significant speed-ups [27]. In parallel, several high-level modeling languages (e.g., [16,36,56]) are designed to improve the accessibility of this powerful technology. Still, the overall process

© The Author(s), under exclusive license to Springer Nature Switzerland AG 2023
A. A. Cire (Ed.): CPAIOR 2023, LNCS 13884, pp. 299–319, 2023.
https://doi.org/10.1007/978-3-031-33271-5_20

of modeling and solving optimization problems remained the same for decades. The de facto approach is to follow the *model-and-run* strategy where the user is responsible for transforming the problem at hand as an optimization model, and then, given the model, a solver is responsible for finding the solution.

We envision an automated modeling assistant to help turn natural language into optimization formulations. To realize this future state, there is a necessary precursor: given the problem description of an optimization problem, finding key pieces of information to enable model formulation. This is exactly what we study in this paper; NER4OPT, the challenge of named entity recognition for extracting optimization-related information such as the objective, constraints, and variables from free-form natural language text. With this goal in mind, in this paper, we make the following contributions:

1. We formalize NER4OPT as an interdisciplinary problem at the intersection of Natural Language Processing and Combinatorial Optimization. We discuss how it differs from the standard named entity recognition (NER) (Sect. 2) in important ways stemming from the optimization context.
2. We start with baseline lexical solutions built on classical techniques commonly used in NER (Sect. 3.1) and then study the impact of recent advances in pre-trained large language models for building semantic solutions (Sect. 3.2).
3. We show how to combine the lexical and semantic models as a hybrid approach and propose several augmentation techniques including fine-tuning language models using optimization textbooks to achieve the best performance (Sect. 3.4).

Our key finding is that generalization for NER4OPT is possible. We learn from annotated optimization problems emerging in advertising, investment, and sales optimization as the source domain that is then tested on production, science, and transportation optimization as the target domain.

For computational experiments (Sect. 4), we consider the recently introduced linear programming word problems as the benchmark (Sect. 4.1). We improve over the best-known solutions on this dataset [48]. To foster further research, we release our code and demo that extracts optimization-related entities from input text[1]. More importantly, we show that it is possible to train effectively for generalization not only to new problem instances of the same domain but also to new applications. Our work is necessary but not sufficient for automated modeling assistants, and accordingly, we discuss important next steps and remaining open problems.

2 Problem Description

Let us start with a formal definition of the NER4OPT problem which can be viewed as an instantiation of the classical NER problem [10,15,55] with its particular challenges emerging from the optimization context.

[1] https://huggingface.co/spaces/skadio/ner4opt.

Definition 1 (Named Entity Recognition for Optimization (NER4OPT)). *Given a sequence of tokens* $s = \langle w_1, w_2, \cdots, w_n \rangle$*, the goal of* NER4OPT *is to output a list of tuples* $\langle I_s, I_e, t \rangle$ *each of which is a named entity specified in* s*. Here,* $I_s \in [1, n]$ *and* $I_e \in [1, n]$ *are the start and the end indexes of a named entity mention;* t *is the entity type from a predefined category set of constructs related to optimization.*

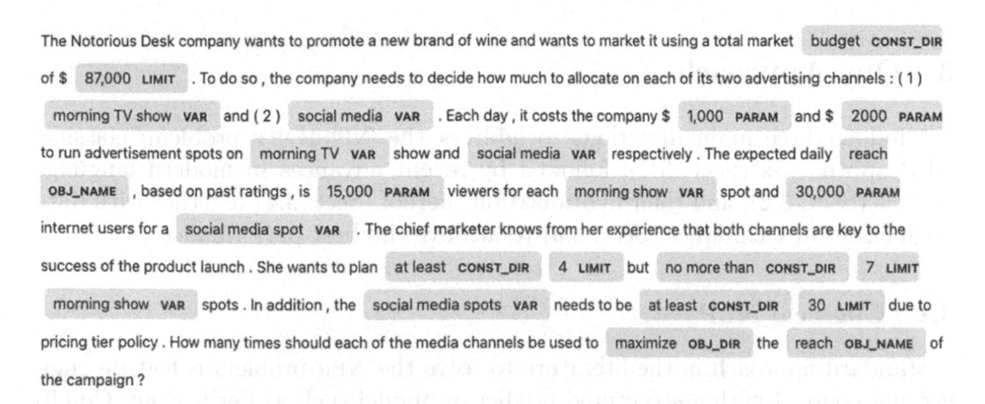

Fig. 1. NER4OPT Example: Given the problem description in free-form natural language text, the goal is to extract key information about the variables, parameters, constraint direction, limits, objective, and optimization direction.

Figure 1 illustrates our problem definition with an example. Given the problem description, the goal of NER4OPT is to extract constraints, parameters, variables, and the objective, among others.

Regarding entities, when NER was first defined in MUC-6 [15], the task was to recognize names of people, organizations, locations, time, currency, and percentage expressions in the text. As shown in our example, the predefined optimization entities in this paper are constraint direction (CONST_DIR), limits (LIMIT), objective direction (OBJ_DIR), objective name (OBJ_NAME), parameter (PARAM), and variable (VAR).

Regarding downstream applications, NER acts as an essential pre-processing step for information retrieval, question answering, and machine translation. Here, we leverage it as the precursor to automated modeling where NER4OPT treats the input description as a word problem that describes decision variables, the objective, and constraints. However, this multi-sentence word problem exhibits a high level of ambiguity due to the variability of the linguistic patterns, problem structure, and application domain.

Regarding NLP tasks, the NER4OPT differs from the NER problem in several challenging ways. First, optimization technology is a general-purpose tool that can tackle a wide range of applications. Accordingly, when parsing problem descriptions, the solution for NER4OPT must be domain-agnostic and generalize to new instances and applications. Second, given the complexity of building optimization models, we only have access to limited training data. Unlike many NLP tasks, we cannot even depend on human annotators since it requires modeling expertise. Therefore, we must rely on large-scale domain knowledge and data augmentation methods to train robust models in low-resource settings. Finally, while most existing parsers operate at the sentence level, optimization descriptions span long text inputs to describe variables and constraints with a high degree of compositionality and ambiguity.

3 Our Approach

We follow three main directions to address the NER4OPT problem: classical NLP approaches (Sect. 3.1), followed by recent advances in modern language models (Sect. 3.2), and their hybridization thereof (Sect. 3.4) together with data augmentation techniques (Sect. 3.3) to achieve the best performance.

3.1 Classical NLP

A standard approach in the literature to solve the NER problem is feature engineering coupled with a structured prediction model such as linear chain Conditional Random Field (CRF) [28,41]. This is what we start with as our baseline.

Overview of CRF. As shown in Fig. 2, given an input sequence of tokens x_i and a set of engineered feature extraction functions f_j at each token position, a conditional random field models a probability distribution of labels y_i that can be assigned to appropriate segments in x.

$$score\left(y\,|x\right) = \sum_{j=1}^{m}\sum_{i=1}^{n} w_j f_j\left(x, i, y_i, y_{i-1}\right) \tag{1}$$

$$p(y|x) \;=\; \frac{\exp^{score(y|x)}}{\sum_{y'} \exp^{score(y'|x)}} \tag{2}$$

Given a set of training examples D, a CRF finds an optimal label assignment using maximum likelihood. Here, w is the weight vector and C is the regularization parameter.

$$D = \left[\left(x_1, y_1\right), \left(x_2, y_2\right), \left(x_3, y_3\right), \ldots, \left(x_d, y_d\right)\right] \quad i.i.d\ training\ examples \tag{3}$$

$$L(w, D) \;=\; -\sum_{k=1}^{d} log\left[p(y^k|x^k)\right] \tag{4}$$

$$w^* \;=\; \operatorname*{arg\,min}_{w}\; L(w, D) \;+\; C\,\frac{1}{2}||w||^2 \tag{5}$$

Feature Extraction. In NLP, a feature extraction function explores the linguistic properties of a token or a group of tokens. For NER, different classes of properties, including grammatical (e.g., part-of-speech tagging and dependency relations), morphological (e.g., prefix, suffix, and word shape), vocabulary (e.g., gazetteers) and syntactic (noun phrases and prepositional phrases) are often used. For details on features used for NER, we refer to [49]. In addition to commonly used feature extraction functions, we engineered other features inspired by the structural characteristics of the optimization problems, as detailed next.

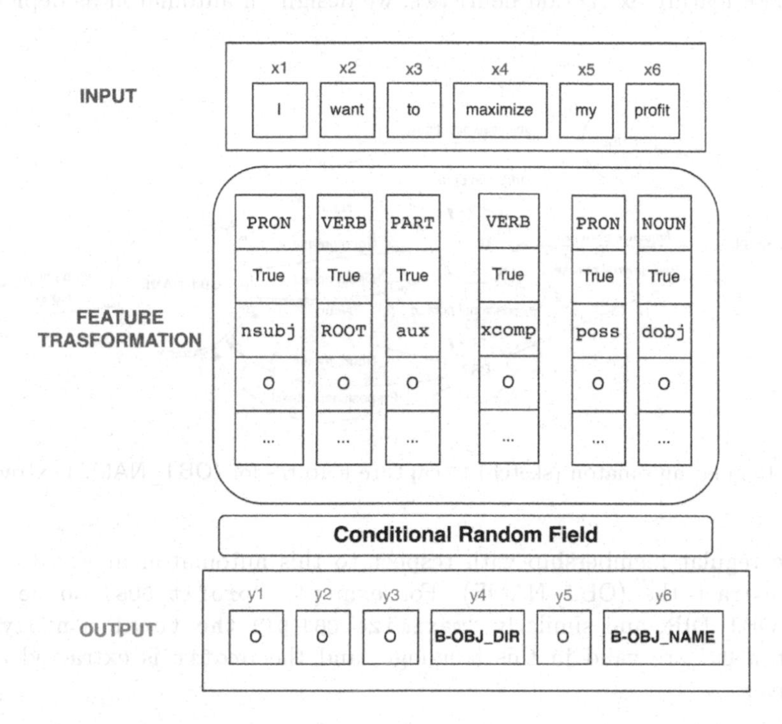

Fig. 2. NER4OPT CRF Example: Given the input sentence, feature extraction and transformation of each token is fed into the conditional random field to find the output of recognized entities.

Gazetteer Features: Gazetteers serve as lookup tables and are utilized as noisy priors to entity labels. These are especially useful when the entity class has frequent keywords and phrases. These key phrases are extracted from the training data. In our optimization setting, canonical keywords include `maximize OBJ_DIR` and `minimize OBJ_DIR`, and similarly, `at least CONST_DIR` and `at most CONST_DIR`.

Syntactic Features: In linguistics, a *conjunct* is a group of tokens joined together by conjunction or appropriate punctuation. The (VAR) and (OBJ_NAME) entities are associated with unique syntactical properties in the form of conjuncts. We observe four patterns for the (VAR) entity. First, these are often conjuncting noun chunks, e.g., "a factory produces rice VAR and corn VAR". These entities also appear as conjuncting prepositional chunks, e.g., "there are two types of cars: cars with automatic gear VAR and cars with manual gear VAR". The other patterns include conjuncts connected by a hyphen or a quote. For the ⟨OBJ_NAME⟩, we observe that ⟨OBJ_DIR⟩ appears in the context of defining the objective of the problem. Moreover, frequently, an ⟨OBJ_DIR⟩ is followed by a noun chunk denoting the ⟨OBJ_NAME⟩ often qualified by an adjective, verb, or prepositional phrase. To succinctly capture these feature extraction heuristics, we design an automaton as depicted in Fig. 3.

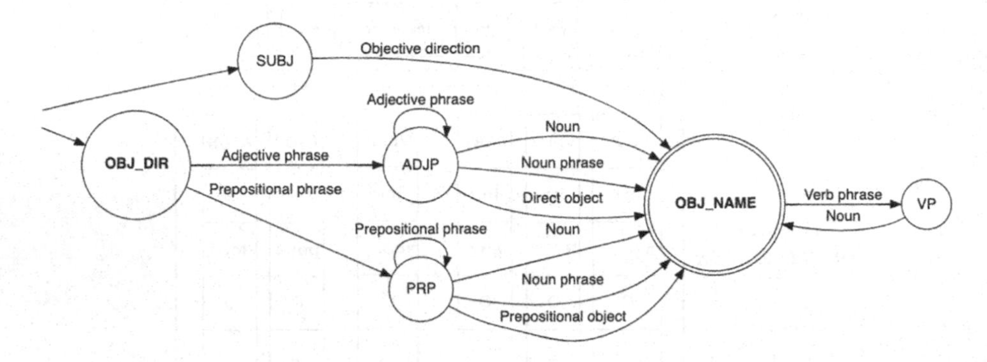

Fig. 3. Regular automaton (sketch) to capture features for ⟨OBJ_NAME⟩ extraction.

The regular membership with respect to this automaton in Fig. 3 enables us to extract the ⟨OBJ_NAME⟩. For example, "profit SUBJ to be maximized OBJ_DIR, and similarly, "maximize OBJ_DIR the total monthly ADJP profit NOUN" are valid in this language, and the profit is extracted as the objective.

Contextual Features: All the previous hand-crafted features operate at token level. In addition, we extract left and right contextual features around each token with window size, w. The parameter w is learned from the training data based on the longest entity phrase. These feature extraction functions act as noisy priors to each entity label. The CRF model relies on many such features and their respective contexts; hence, a few false positive features will not affect the model's overall performance. We studied other features engineering methods, including constituent parsing, quantized word embeddings, and word-frequency-based features that are omitted here for brevity.

3.2 Modern NLP

So far, our solution for NER4OPT only considered classical methods based on feature extraction and manual feature engineering. This helps us establish a baseline performance. The challenger to this baseline is motivated by the recent advances in NLP, offering advantages over traditional techniques. Specifically, deep neural networks alleviate the need for manual feature extraction. At a high level, Eq. 1 continues to apply whereby feature vectors $f(x)$ now correspond to dense embeddings retrieved from large language models. This not only saves a significant amount of time in creating features but offers more robust behavior. Moreover, the nonlinearity in the activation functions enables learning complex features and dependencies from the labeled training data.

In practice, NER problems require modeling long-range text dependencies. When operating on the long-range, recurrent architectures are known to struggle with vanishing and exploding gradients [39]. As a remedy, most recent works rely on the TRANSFORMERS architecture [57] that solve the long-range problem by replacing the recurrent component with the *attention* mechanism. There are many variants of this architecture, and here, we consider three distinct flavors based on RoBERTa [32] to generate the features vectors $f(x)$ used in CRF.

1. XLM-RB: The XLM-RoBERTa [11] is a self-supervised language model that follows the RoBERTa architecture with multilingual training. This is the state-of-the-art method [48] on the benchmark dataset we consider. One of our goals is to improve this existing approach and its large version, XLM-RL.
2. XLM-RL+: Our unique contribution that extends XLM-RL with fine-tuning over corpora related to optimization texts. We explain this in detail below.
3. ROBERTA: Another large language model that uses the same transformers architecture as BERT [12] and improves it with more robust training [32]. It achieves state-of-the-art results on well-known NLP benchmarks such as GLUE [58], RACE [29] and SQuAD [46,47]. As such, we consider it here and employ its large version.

[XLM-RL+] Fine-Tuning on Optimization Textbooks: Language models such as BERT, RoBERTa, and GPT [42] are pre-trained on non-domain specific texts where the goal of pre-training is to obtain good downstream performance on a diverse set of *language-oriented tasks* (e.g., sentiment analysis). The training is carried out in a self-supervised fashion via masked language modeling, next sentence prediction [12], and causal language modeling [42]. For *domain-specific* tasks (e.g., sentiment analysis in finance), performance can be improved further using domain-specific corpora to fine-tune pre-trained models [1,5,20,30].

We use a similar approach for fine-tuning XLM-RL leading to XLM-RL+. For that purpose, we source three publicly available optimization textbooks. The first is the well-known convex optimization book by S. Boyd [8]. The second one is about linear programming and game theory [54]. And finally, we consider the course notes on optimization [18] from the Open Optimization Platform[2] that shares educational content. We extract textual data from the PDFs of these textbooks, and then, fine-tune XLM-RL via masked language modeling. More precisely, we mask 15% of the words at random and replace 80% of the masked words with the MASK token, 10% with random words, and the remaining 10%

[2] https://github.com/open-optimization.

with the original word. Finally, the model is trained in self-supervised fashion to predict the masked words.

3.3 Data Augmentation

In parallel to modeling strategies, we also consider two methods for data augmentation to improve performance. First, we show how to find infrequent patterns to over-sample, and second, we introduce a simple yet effective technique, coined L2 augmentation, to disambiguate the objective variable from other variables.

Dealing with Infrequent Patterns: Over-sampling is an effective technique, especially when dealing with class imbalance. While the distribution of entity classes might not be imbalanced, the lexical features might exhibit popular traits with a few infrequent features. For example, the objective direction is almost always `maximize` or `minimize`. Yet, in a few cases, it is given as an adjective, e.g., "`cost to be` minimal ADJ". The challenge is to find a way to surface such infrequent cases without manual inspection so that we can over-sample the dataset.

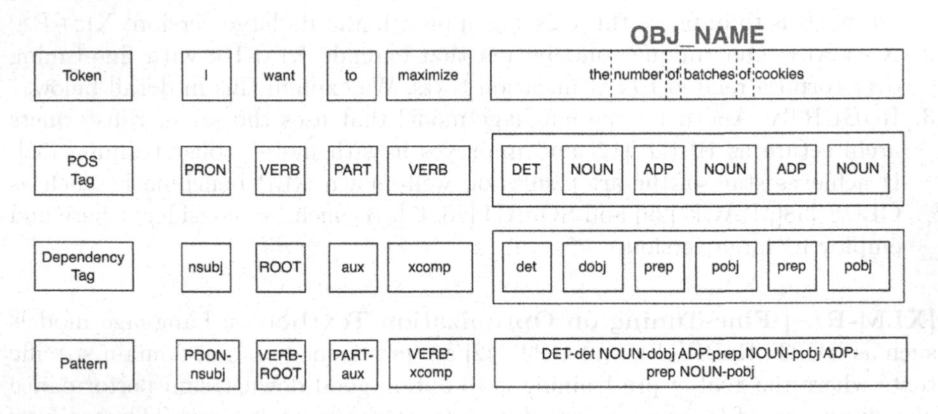

Fig. 4. Example pattern for the given objective name entity as the union of its part-of-speech and dependency tags.

We propose a simple approach as shown in Fig. 4. First, we extract part-of-speech and dependency tags for each token in a given sentence and consider their union as a *pattern*. We then build the set of unique patterns and map them to entity labels in the training data to find their occurrence counters. Consequently, problem descriptions with infrequent patterns are duplicated.

Dealing with Disambiguation: There is a severe ambiguity between the objective variable and other variables. After all, the objective is yet another variable, only with an optimization direction. The critical question is how to train a model to differentiate between the two effectively. Consider the scenario in Fig. 5. In this case, the objective is blood pressure reducing medicine, and there is no distinctive feature that separates it from other variables such as diabetic pill and diabetic shot. Even for the human annotator or the optimization expert, the objective remains unknown until the last sentence. Despite that, the model must label the objective correctly as early as the second sentence in its first occurrence. This is precisely the long-range text dependency aspect of NER4OPT. To combat this, we append the beginning of each problem description with the last two sentences and refer to this method as L2 augmentation.

3.4 Hybrid Modeling

Finally, we consider a hybrid approach that combines our proposed methods. Classical methods for feature engineering and modern techniques for feature learning have their strengths and weakness. While feature engineering can sometimes be brittle, feature learning struggles when there is long-range dependency and no semantic theme, as in the ⟨OBJ_NAME⟩. On the other hand, handcrafted features, such as our gazetteer, syntactic and contextual features, and knowledge injection, such as our purpose-built automaton, allow us to build apriori information. Our hybrid model uses the classical CRF approach boosted by feature engineering plus an additional feature provided by the prediction of a transformers-based model fine-tuned on optimization corpora with over-sampling and L2 data augmentation.

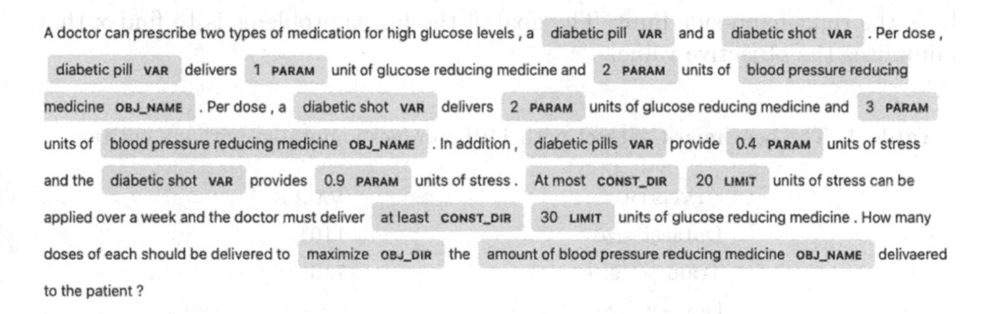

Fig. 5. The challenge of long-range text dependency in NER4OPT. Notice the disambiguation problem between the objective variable and other variables and the importance of the last sentences in capturing the goal of the problem description.

4 Experiments

To demonstrate the effectiveness of our approach when solving the NER4OPT problem in practice, we consider the following specific questions:

Q1: What is the baseline performance of classical methods (Sect. 3.1) and does feature engineering help?

Q2: How do modern NLP methods (Sect. 3.2) perform, how do we fare against the best-known solutions on the same dataset, and do we improve the state-of-the-art for NER4OPT?

Q3: Does the hybrid model (Sect. 3.4) that combines feature engineering with feature learning and augmentation perform better than its counterparts in isolation?

Let us start with an overview of the dataset, the experimental setup, and evaluation metrics and present numerical results with discussions and error analysis.

4.1 NER4OPT Dataset

We use linear programming word problems that are released as part of the NeurIPS'22 natural language for optimization challenge[3]. We are indebted to the organizers for contributing such a rich dataset to the community. Our formal definition of NER4OPT corresponds to Task–I from this challenge[4]. This dataset was first introduced in [48], which uses the XLM-RB model to solve the entity recognition problem. It contains 1101 linear programming word problems of the form:

$$\min_{\mathbf{x} \in \mathbb{R}^n} \quad \mathbf{c}^\top \mathbf{x} \quad \text{s.t.} \quad \mathbf{a}_i^\top \mathbf{x} \leq \mathbf{b}_i, \quad i = 1, ..., m \tag{6}$$

where $\mathbf{c}$ represents the parameters of the objective, $\mathbf{a}_i$ the i-th constraint, and $\mathbf{b}_i$ is the right-hand-side limit. The goal of the linear problems is to find $\mathbf{x}$ that minimizes the objective value.

Table 1. The benchmark dataset with 1101 samples annotated with six entities.

STATISTIC	VALUE
Dataset size	1101
Train set size	713
Dev set size	99
Test set size (not available)	289
Number of entity types	6
Number of VAR entities	5299
Number of PARAM entities	4113
Number of LIMIT entities	2064
Number of CONST_DIR entities	1877
Number of OBJ_DIR entities	813
Number of OBJ_NAME entities	2391

[3] https://github.com/nl4opt.

[4] https://github.com/nl4opt/nl4opt-subtask1-baseline.

Table 1 shows dataset statistics. The problems in the dataset belong to six domains grouped into two: the source domain comprising of problems from advertising, investment, and sales, and the target domain consists of problems from production, science, and transportation. As in our problem description (Sect. 2), it contains annotations for six entity types: variable (VAR), parameter (PARAM), limit (LIMIT), constraint direction (CONST_DIR), objective direction (OBJ_DIR) and objective name (OBJ_NAME). The test set is not public, hence we focus on train and dev sets and cannot directly compare with Task-I.

The training set consists of samples only from the source domain, whereas the dev and test sets consist of samples from both source and target domains in a 1:3 source-to-target domain ratio. According to [48], 15 annotators created the problem descriptions and the labels while 4 additional NLP/OR experts annotated more than 10% of the entire dataset to compute inter-annotator agreement. Average pairwise micro-averaged F1 score was used to measure the agreement and a score of 97.7% was reported. Figure 1 presents an annotated input sample.

4.2 Comparisons

We compare the classical, modern, and hybrid models with the following variants:

1. CLASSICAL: Our classical method (Sect. 3.1) based on grammatical and morphological features.
2. CLASSICAL+: Our classical method plus our hand-crafted gazetteer, syntactic, and contextual features.
3. XLM-RB: The state-of-the-art method on this dataset from [48] that we re-ran thanks to authors' code. We also consider its large version, XLM-RL.
4. ROBERTA: Transformers model with good default performance across several language tasks for comparison. We use its large model variant.
5. XLM-RL+: Our approach (Sect. 3.2) to fine-tune XLM-RL with optimization books.
6. HYBRID: Our hybrid approach (Sect. 3.4) that combines CLASSICAL+ with XLM-RL+ and data augmentation.

4.3 Experimental Setup

We use the train set for learning the model weights and the dev set for testing the performance for all the methods considered. We conduct limited parameter tuning to avoid overfitting. We leverage HuggingFace [59] and SimpleTransformers [45] for transformer models, spaCy [19] for part-of-speech and dependency tagging, and sklearn-crf[5] for the CRF model.

[5] https://github.com/TeamHG-Memex/sklearn-crfsuite.

Hyperparameters for CRF. There are four hyper-parameters for CRF: $c1$ & $c2$ controlling the amount of regularization, the context window size w, and the optimizer. We use gradient descent with the L-BFGS method [66] as our optimizer and random cross-validation search to find the best values for $c1$ & $c2$ from the continuous exponential distribution of scale 0.5 for c1 and 0.05 for c2. In addition, the window size is set to 6 tokens as the longest entity observed in the training data.

Hyperparameters for Transformers. We perform limited tuning for all the transformers models to avoid over-fitting. As mentioned in Sect. 4.1, dev and test sets have samples from both source and target domains. Therefore, any over-fitting of the training data will hurt the model's generalizability. For learning rate, we use the range {4E-5, ..., 1E-1} with a step size of 1E-2. For the maximum sequence length, we experiment with {256, 512}. In addition, for the l2 regularization coefficient, we use the range {1E-3, ..., 1E-1} with a step size of 1E-2. Finally, we run the training procedure for a maximum of 25 epochs, with an early stopping callback function set to stop training by monitoring the loss delta set to 1E-3 and patience of 5 epochs.

4.4 Evaluation Metrics

We evaluate all methods using the micro-averaged F1 score as suggested in the NeurIPS'22 competition and in the existing results [48]. The score is computed as follows:

$$\mathbf{F1} = \frac{2 \times \mathcal{P} \times \mathcal{R}}{\mathcal{P} + \mathcal{R}} \tag{7}$$

where $\mathcal{P}$ and $\mathcal{R}$ are the average precision and average recall of all entity types, respectively. The computation of true positives, false positives, and false negatives for precision and recall is done as follows:

- A predicted span is considered as *True Positive* if the span and the predicted entity type are present in the ground truth annotations.
- A predicted span is considered as *False Positive* if the span is present in the ground truth annotations but the predicted entity type is incorrect or the predicted span is not present in the ground truth annotations.
- A span is considered as *False Negative* if the span is present in the ground truth but is absent in the predicted spans.

Table 2. Numerical results that compare classical, modern, and hybrid models for precision, $\mathcal{P}$, and recall, $\mathcal{R}$ for each named entity together with average micro F1 score.

METHOD	CONST_DIR		LIMIT		OBJ_DIR		OBJ_NAME		PARAM		VAR		Average Micro F1
	$\mathcal{P}$	$\mathcal{R}$	$\mathcal{P}$	$\mathcal{R}$	$\mathcal{P}$	$\mathcal{R}$	$\mathcal{P}$	$\mathcal{R}$	$\mathcal{P}$	$\mathcal{R}$	$\mathcal{P}$	$\mathcal{R}$	
CLASSICAL	0.956	0.854	0.904	**0.954**	0.979	0.929	0.649	0.353	0.958	0.916	0.795	0.714	0.816
CLASSICAL+	**0.960**	0.858	0.931	0.942	**0.990**	0.970	0.726	0.544	0.953	0.935	0.823	0.787	0.853
XLM-RB [48]	0.887	0.897	0.965	0.950	0.949	0.999	0.617	0.469	0.960	0.969	0.909	0.932	0.888
XLM-RL	0.930	0.897	0.979	0.938	0.979	0.989	0.606	0.512	0.963	0.985	0.899	0.938	0.893
ROBERTA	0.895	**0.902**	0.984	0.950	**0.990**	**1.000**	0.668	0.597	0.965	0.983	0.916	0.940	0.904
XLM-RL+	0.901	0.897	**0.987**	0.953	0.989	0.999	0.665	0.583	**0.971**	**0.989**	0.918	0.946	0.907
HYBRID	0.946	0.890	0.980	0.942	**0.990**	**1.000**	**0.730**	**0.668**	0.957	0.983	**0.935**	**0.953**	**0.919**

4.5 Numerical Results

Table 2 presents our results that compare different classical, modern, and hybrid methods for solving the NER4OPT on the linear programming word problems dataset. We report the performance metrics for precision, $\mathcal{P}$, and recall, $\mathcal{R}$, for each entity class together with the micro F1 score averaged across classes.

[Q1] Performance of Classical NLP The CLASSICAL method based on grammatical and morphological features achieves an average micro F1 of 0.816. This result establishes our performance lower bound. From there, CLASSICAL+ jumps to 0.853 by leveraging our hand-crafted gazetteer, syntactic, and contextual features. The gazetteer features focus on (OBJ_DIR) and (CONST_DIR). Accordingly, for (OBJ_DIR), the $\mathcal{P}$ and $\mathcal{R}$ increase by 0.011 and 0.041, and for (CONST_DIR), both metrics increases slightly by 0.004. The syntactic features focus on (VAR) and (OBJ_NAME) entity types. Accordingly, for (VAR), $\mathcal{P}$ and $\mathcal{R}$ increase by 0.028 and 0.077, and for (OBJ_NAME), $\mathcal{P}$ and $\mathcal{R}$ increase by 0.077 and 0.191. While the F1 score 0.816 is relatively lower compared to other methods, both classical approaches report 0.90+ $\mathcal{P}$ and 0.85+ $\mathcal{R}$ on all entity types except (OBJ_NAME) and (VAR). These two classes stand out as the difficult labels, as we discussed earlier, due to ambiguity and long-range dependency.

[Q2] Performance of Modern NLP and the State-of-the-Art The state-of-the-art method for NER4OPT on this dataset from [48] is based on the modern transformers architecture, specifically XLM-RB. XLM-RB improves over the classical results from an F1 of 0.816 to 0.888. Switching the underlying transformer architecture from the base model to its large version, XLM-RL, or to a different model as ROBERTA improves the results further. Interestingly, ROBERTA outperforms XLM-RB, hinting that the multilingual pre-training objective of XLM-RB [11] is not beneficial in our NER4OPT task.

When comparing modern methods that use deep contextual embeddings with their classical counterpart, we observe that recent techniques perform better; 0.853 vs. 0.907. That said, it is worth noting that while it is possible to improve the overall average F1 score when compared to classical methods, modern methods do not improve $\mathcal{P}$ and $\mathcal{R}$ in every class.

Our approach XLM-RL+ that combines XLM-RL with fine-tuning on optimization corpora improves the F1 score of 0.893 to 0.907. It is encouraging to realize this performance improvement even when fine-tuning with only a few textbooks over the pre-trained model built with large corpora. Our XLM-RL+ stands out as the best-performing modern method in parts with small margins.

[Q3] Performance of Hybrid Modeling Finally, we consider the case for combining classical and modern techniques. Overall, the best performance is achieved with our HYBRID model with an F1 score of **0.919**. This result significantly outperforms the baseline classical approach from the F1 score of 0.816 and improves over the best-known results from the F1 score of 0.888. The HYBRID model benefits from data augmentation via over-sampling to address infrequent patterns and L2 augmentation to combat long-range dependency. Beyond average results, upon closer inspection of $\mathcal{P}$ and $\mathcal{R}$ for each entity class, we find that HYBRID offers the best result in half of the scores. The largest improvement is realized for the (OBJ_NAME) entity type, which is the most challenging label to predict. To summarize, our experiments demonstrate that integrating feature engineering with feature learning coupled with fine tuning and augmentation stands out as an attractive mechanism for NER4OPT.

4.6 Post-Mortem Analysis

A critical post-mortem error analysis is to inspect where the methods fail for NER4OPT. Our initial findings reveal conflicting token spans in the annotation of entities. Let's highlight a few examples to illustrate the issue:

⟹ `How many of each type of donut should be bought in order to maximize OBJ_DIR the total monthly profit OBJ_NAME?`

⟹ `How many of each type of transportation should the company schedule to move their lumber to minimize OBJ_DIR the total cost OBJ_NAME?`

⟹ `How many of each should the pharmaceutical manufacturing plant make to minimize OBJ_DIR the total number of minutes needed OBJ_NAME?`

In the first example from the training data, `profit` is annotated as the objective omitting the preceding adjective phrase `total monthly`. Contrarily, in the second example, this time from the dev data, `total cost` is annotated as the objective, considering the adjective as part of the entity span. On the other hand, in the last example, again from the dev data, `number of minutes` is annotated as the objective omitting the `total`.

This inconsistency is especially evident in (OBJ_NAME) entity, which turns out to be the hardest entity to predict. Similar inconsistencies exist in other classes as well. For example, in (VAR) entity, the prepositional phrase preceding a noun is sometimes tagged, sometimes ignored. Likewise, in (LIMIT) and (PARAM) non-alpha-numeric characters (e.g., $, %) preceding or succeeding the term is tagged inconsistently.

This is known as aleatoric uncertainty and is difficult to address [13]. We expect the performance of any method, classical, modern or hybrid, to saturate eventually due to inherent labeling issues. At times, even human annotators cannot agree on the exact span of annotations, making Ner4Opt challenging.

5 Related Work

The unique aspect of our paper is its interdisciplinary nature. At the intersection of NLP and Optimization, our main contribution is to formalize the Ner4Opt problem and offer an initial attempt at its solution. We consider classical, modern, and hybrid techniques commonly employed in the Ner literature.

Similar to our approach, early Ner systems were built via templates and hand-crafted rules. For example, in [50], corporate identity was extracted from financial texts using heuristics. As in our case, such rule-based approaches require domain experts to formulate templates. We design gazetteer, syntactic, and contextual features for optimization problems. It is also common to employ representative models such as Hidden Markov Models [34,63,65], and as in here, Conditional Random Fields [28,41]. A drawback of these methods is poor generalization, as observed in our experiments for the average scores.

Complementary to these are modern approaches, and in particular, masked language modeling employed in BERT [12], autoregressive training utilized in GPT [43], permutation-based training employed in XLNet [62]. We build on the idea of pre-trained language models to perform practical tasks without additional training [9,43]. We leverage pre-trained transformer models such as Xlm-Rb [11] and RoBERTa [32]. Additionally, we fine-tune these large language models using optimization textbooks. To the best of our knowledge, this is the first attempt to improve large language models with optimization verbiage. A recent survey of advances in modern Ner can be found in [60].

Apart from these, we exploit domain expertise in optimization and propose regular automaton to capture heuristics for the objective variable succinctly. This automaton can be further specialized for specific downstream optimization problems. Similarly, we propose the L2 data augmentation aimed at capturing the context of the objective earlier in the text.

As benchmark, we consider the linear programming word problems dataset from [48]. This work is the closest to our paper in spirit. Compared to our work, [48] goes a step further and attempts to build an interactive decision support system to assist modelers in formulating optimization problems from text. Unfortunately, [48] does not formally define the entity recognition problem. It is only mentioned briefly as one of the black boxes in the overall system architecture without details on how to solve it.

We introduce Ner4Opt as a standalone problem and an important building block of modeling assistants. We then provide an in-depth study of its solution approaches ranging from baselines to advanced hybrids. While [48] depends solely on off-the-shelf pre-trained models, we attempt domain-specific fine-tuning. Our results improve the best-known solutions from [48] considerably.

A growing body of research is dedicated to integrating machine learning and optimization. These include general algorithm configuration procedures [21,22,24], variable selection [6,22,23,31,33], branching constraint selection [61], cut selection [53], node selection [17,51], and theoretical results for tree-search configuration [2,3]. Compared to these efforts, the integration of natural language processing and optimization remains much more limited. Our paper is one of the first attempts in that direction. For optimization technology, the NER4OPT is immediately relevant to pave the road for modeling assistants.

For NLP, NER4OPT offers unique challenges as noted in Section (Sect. 2) such as multi-sentence dependency with high-level of ambiguity, low data regime with high-cost of annotation, and inherent aleatoric uncertainty. Linguistically, NER4OPT is somewhat counter-intuitive. In the classical NLP setting, entities in the same class refer to similar objects in the real world, e.g., person, place, and organization, and they share grammatical properties. Contrarily, in NER4OPT, the objects tagged in the same entity can wildly differ and be completely unrelated, e.g., the number of trucks, the number of coconuts, and the time spent brewing coffee might all be variables or even objectives. Given the challenging nature of NER4OPT, both communities would benefit from closer integration.

The need for significant expertise to formulate models as a barrier to entry is a common concern shared by many in the community. In that regard, our paper is closely related to learning constraint models. These include learning models using generate-and-test [26], from examples [44], from spreadsheets [25], from tables [38], from solutions as in model seeker [4], and from non-solutions [7,40]. Related to this are visualization frameworks, explanations, and user hints as part of human-computer interaction [14,35,37,52]. Our work differs substantially from all of these previous works, as we are working with free form natural language text. Let us note that analogous attempts have already succeeded in other domains, e.g., turning text into SQL queries [64].

6 Conclusions

We envision a future in which non-technical users are empowered with optimization techniques so that they can naturally interact in multi-modal settings via text, and even voice. This requires significant advances at the intersection of multiple domains, and our paper is an initial attempt toward automated modeling assistants. Still, more work is needed toward the integration of NER4OPT into high-level modeling frameworks. Our call to action for researchers and practitioners is to help us break the low annotated data regime to achieve revolutionary breakthroughs as realized in large language models. The optimization community is neatly suited for such success with a wide range of applications that have already equipped with exact model annotations in several problem domains.

References

1. Araci, D.: Finbert: financial sentiment analysis with pre-trained language models. arXiv preprint arXiv:1908.10063 (2019)
2. Balcan, M., Prasad, S., Sandholm, T., Vitercik, E.: Sample complexity of tree search configuration: cutting planes and beyond. In: Ranzato, M., Beygelzimer, A., Dauphin, Y.N., Liang, P., Vaughan, J.W. (eds.) Advances in Neural Information Processing Systems 34: Annual Conference on Neural Information Processing Systems 2021, NeurIPS 2021, 6–14 December 2021, pp. 4015–4027 (2021). https://proceedings.neurips.cc/paper/2021/hash/210b7ec74fc9cec6fb8388dbbdaf23f7-Abstract.html
3. Balcan, M., Prasad, S., Sandholm, T., Vitercik, E.: Improved sample complexity bounds for branch-and-cut. In: Solnon, C. (ed.) 28th International Conference on Principles and Practice of Constraint Programming, CP 2022, 31 July to 8 August 2022, Haifa, Israel. LIPIcs, vol. 235, pp. 3:1–3:19. Schloss Dagstuhl - Leibniz-Zentrum für Informatik (2022). https://doi.org/10.4230/LIPIcs.CP.2022.3
4. Beldiceanu, N., Simonis, H.: A model seeker: extracting global constraint models from positive examples. In: Milano, M. (ed.) CP 2012. LNCS, pp. 141–157. Springer, Heidelberg (2012). https://doi.org/10.1007/978-3-642-33558-7_13
5. Beltagy, I., Lo, K., Cohan, A.: SciBERT: a pretrained language model for scientific text. In: Proceedings of the 2019 Conference on Empirical Methods in Natural Language Processing and the 9th International Joint Conference on Natural Language Processing (EMNLP-IJCNLP), Hong Kong, China, pp. 3615–3620. Association for Computational Linguistics (2019). https://doi.org/10.18653/v1/D19-1371. https://aclanthology.org/D19-1371
6. Bengio, Y., Lodi, A., Prouvost, A.: Machine learning for combinatorial optimization: a methodological tour d'horizon. Eur. J. Oper. Res. **290**(2), 405–421 (2021). https://doi.org/10.1016/j.ejor.2020.07.063. https://www.sciencedirect.com/science/article/pii/S0377221720306895
7. Bessiere, C., Coletta, R., Freuder, E.C., O'Sullivan, B.: Leveraging the learning power of examples in automated constraint acquisition. In: Wallace, M. (ed.) CP 2004. LNCS, vol. 3258, pp. 123–137. Springer, Heidelberg (2004). https://doi.org/10.1007/978-3-540-30201-8_12
8. Boyd, S., Boyd, S.P., Vandenberghe, L.: Convex Optimization. Cambridge University Press, Cambridge (2004)
9. Brown, T., et al.: Language models are few-shot learners. Adv. Neural. Inf. Process. Syst. **33**, 1877–1901 (2020)
10. Chinchor, N., Robinson, P.: Appendix E: MUC-7 named entity task definition (version 3.5). In: Seventh Message Understanding Conference (MUC-7): Proceedings of a Conference Held in Fairfax, Virginia, 29 April–1 May 1998 (1998). https://aclanthology.org/M98-1028
11. Conneau, A., et al.: Unsupervised cross-lingual representation learning at scale. arXiv preprint arXiv:1911.02116 (2019)
12. Devlin, J., Chang, M.W., Lee, K., Toutanova, K.: Bert: pre-training of deep bidirectional transformers for language understanding. arXiv preprint arXiv:1810.04805 (2018)
13. Fisch, A., Jia, R., Schuster, T.: Uncertainty estimation for natural language processing. In: COLING (2022). https://sites.google.com/view/uncertainty-nlp
14. Goodwin, S., Mears, C., Dwyer, T., de la Banda, M.G., Tack, G., Wallace, M.: What do constraint programming users want to see? Exploring the role of visuali-

sation in profiling of models and search. IEEE Trans. Vis. Comput. Graph. **23**(1), 281–290 (2017). https://doi.org/10.1109/TVCG.2016.2598545
15. Grishman, R., Sundheim, B.: Message understanding conference- 6: a brief history. In: COLING 1996 Volume 1: The 16th International Conference on Computational Linguistics (1996). https://aclanthology.org/C96-1079
16. Guns, T.: On learning and branching: a survey. In: The 18th Workshop on Constraint Modelling and Reformulation (2019)
17. He, H., Daume III, H., Eisner, J.M.: Learning to search in branch and bound algorithms. In: Ghahramani, Z., Welling, M., Cortes, C., Lawrence, N., Weinberger, K. (eds.) Advances in Neural Information Processing Systems, vol. 27. Curran Associates, Inc. (2014). https://proceedings.neurips.cc/paper/2014/file/757f843a169cc678064d9530d12a1881-Paper.pdf
18. Hildebrand, R., Poirrier, L., Bish, D., Moran, D.: Mathematical programming and operations research (2022). https://github.com/open-optimization/open-optimization-or-book
19. Honnibal, M., Montani, I., Van Landeghem, S., Boyd, A.: Spacy: industrial-strength natural language processing in python (2020)
20. Howard, J., Ruder, S.: Universal language model fine-tuning for text classification. arXiv preprint arXiv:1801.06146 (2018)
21. Hutter, F., Hoos, H.H., Leyton-Brown, K., Stützle, T.: Paramils: an automatic algorithm configuration framework. J. Artif. Int. Res. **36**(1), 267–306 (2009)
22. Kadioglu, S., Malitsky, Y., Sabharwal, A., Samulowitz, H., Sellmann, M.: Algorithm selection and scheduling. In: Lee, J. (ed.) CP 2011. LNCS, vol. 6876, pp. 454–469. Springer, Heidelberg (2011). https://doi.org/10.1007/978-3-642-23786-7_35
23. Kadioglu, S., Malitsky, Y., Sellmann, M.: Non-model-based search guidance for set partitioning problems. In: Hoffmann, J., Selman, B. (eds.) Proceedings of the Twenty-Sixth AAAI Conference on Artificial Intelligence, 22–26 July 2012, Toronto, Ontario, Canada. AAAI Press (2012). http://www.aaai.org/ocs/index.php/AAAI/AAAI12/paper/view/5082
24. Kadioglu, S., Malitsky, Y., Sellmann, M., Tierney, K.: ISAC - instance-specific algorithm configuration. In: Coelho, H., Studer, R., Wooldridge, M.J. (eds.) ECAI 2010 - 19th European Conference on Artificial Intelligence, Lisbon, Portugal, 16–20 August 2010, Proceedings. Frontiers in Artificial Intelligence and Applications, vol. 215, pp. 751–756. IOS Press (2010). https://doi.org/10.3233/978-1-60750-606-5-751
25. Kolb, S., Paramonov, S., Guns, T., Raedt, L.D.: Learning constraints in spreadsheets and tabular data. Mach. Learn. **106**(9–10), 1441–1468 (2017). https://doi.org/10.1007/s10994-017-5640-x
26. Kumar, M., Kolb, S., Guns, T.: Learning constraint programming models from data using generate-and-aggregate. In: Solnon, C. (ed.) 28th International Conference on Principles and Practice of Constraint Programming, CP 2022, 31 July to 8 August 2022, Haifa, Israel. LIPIcs, vol. 235, pp. 29:1–29:16. Schloss Dagstuhl - Leibniz-Zentrum für Informatik (2022). https://doi.org/10.4230/LIPIcs.CP.2022.29
27. Laborie, P., Rogerie, J., Shaw, P., Vilím, P.: IBM ILOG CP optimizer for scheduling - 20+ years of scheduling with constraints at IBM/ILOG. Constraints **23**(2), 210–250 (2018). https://doi.org/10.1007/s10601-018-9281-x
28. Lafferty, J.D., McCallum, A., Pereira, F.C.N.: Conditional random fields: probabilistic models for segmenting and labeling sequence data. In: Brodley, C.E., Danyluk, A.P. (eds.) Proceedings of the Eighteenth International Conference on Machine

Learning (ICML 2001), Williams College, Williamstown, MA, USA, 28 June–1 July 2001, pp. 282–289. Morgan Kaufmann (2001)
29. Lai, G., Xie, Q., Liu, H., Yang, Y., Hovy, E.: Race: large-scale reading comprehension dataset from examinations. arXiv preprint arXiv:1704.04683 (2017)
30. Lee, J., Yoon, W., Kim, S., Kim, D., Kim, S., So, C.H., Kang, J.: BioBERT: a pre-trained biomedical language representation model for biomedical text mining. Bioinformatics (2019). https://doi.org/10.1093/bioinformatics/btz682
31. Liberto, G.M.D., Kadioglu, S., Leo, K., Malitsky, Y.: DASH: dynamic approach for switching heuristics. Eur. J. Oper. Res. **248**(3), 943–953 (2016). https://doi.org/10.1016/j.ejor.2015.08.018
32. Liu, Y., et al.: Roberta: a robustly optimized bert pretraining approach. arXiv preprint arXiv:1907.11692 (2019)
33. Lodi, A., Zarpellon, G.: On learning and branching: a survey. TOP **25**(2), 207–236 (2017). https://doi.org/10.1007/s11750-017-0451-6
34. Morwal, S., Jahan, N., Chopra, D.: Named entity recognition using hidden Markov model (HMM). Int. J. Nat. Lang. Comput. (IJNLC) **1** (2012)
35. do Nascimento, H.A.D., Eades, P.: User hints: a framework for interactive optimization. Future Gener. Comput. Syst. **21**(7), 1171–1191 (2005). https://doi.org/10.1016/j.future.2004.04.005
36. Nethercote, N., Stuckey, P.J., Becket, R., Brand, S., Duck, G.J., Tack, G.: MiniZinc: towards a standard CP modelling language. In: Bessière, C. (ed.) CP 2007. LNCS, vol. 4741, pp. 529–543. Springer, Heidelberg (2007). https://doi.org/10.1007/978-3-540-74970-7_38
37. O'Callaghan, B., O'Sullivan, B., Freuder, E.C.: Generating corrective explanations for interactive constraint satisfaction. In: van Beek, P. (ed.) CP 2005. LNCS, vol. 3709, pp. 445–459. Springer, Heidelberg (2005). https://doi.org/10.1007/11564751_34
38. Paramonov, S., Kolb, S., Guns, T., Raedt, L.D.: Tacle: learning constraints in tabular data. In: Lim, E., et al. (eds.) Proceedings of the 2017 ACM on Conference on Information and Knowledge Management, CIKM 2017, Singapore, 06–10 November 2017, pp. 2511–2514. ACM (2017). https://doi.org/10.1145/3132847.3133193
39. Pascanu, R., Mikolov, T., Bengio, Y.: On the difficulty of training recurrent neural networks. In: International Conference on Machine Learning, pp. 1310–1318. PMLR (2013)
40. Pawlak, T.P., Krawiec, K.: Automatic synthesis of constraints from examples using mixed integer linear programming. Eur. J. Oper. Res. **261**(3), 1141–1157 (2017). https://doi.org/10.1016/j.ejor.2017.02.034. https://www.sciencedirect.com/science/article/pii/S037722171730156X
41. Quattoni, A., Collins, M., Darrell, T.: Conditional random fields for object recognition. In: Advances in Neural Information Processing Systems, vol. 17 (2004)
42. Radford, A., Narasimhan, K., Salimans, T., Sutskever, I., et al.: Improving language understanding by generative pre-training (2018)
43. Radford, A., Wu, J., Child, R., Luan, D., Amodei, D., Sutskever, I.: Language models are unsupervised multitask learners (2019)
44. Raedt, L.D., Passerini, A., Teso, S.: Learning constraints from examples. In: AAAI Conference on Artificial Intelligence (2018)
45. Rajapakse, T.C.: Simple transformers (2019). https://github.com/ThilinaRajapakse/simpletransformers
46. Rajpurkar, P., Jia, R., Liang, P.: Know what you don't know: unanswerable questions for squad. arXiv preprint arXiv:1806.03822 (2018)

47. Rajpurkar, P., Zhang, J., Lopyrev, K., Liang, P.: Squad: 100,000+ questions for machine comprehension of text. arXiv preprint arXiv:1606.05250 (2016)
48. Ramamonjison, R., Li, H., et al.: Augmenting operations research with auto-formulation of optimization models from problem descriptions (2022). https://doi.org/10.48550/ARXIV.2209.15565. https://arxiv.org/abs/2209.15565
49. Ratinov, L., Roth, D.: Design challenges and misconceptions in named entity recognition. In: Proceedings of the Thirteenth Conference on Computational Natural Language Learning (CoNLL-2009), pp. 147–155 (2009)
50. Rau, L.F.: Extracting company names from text. In: Proceedings the Seventh IEEE Conference on Artificial Intelligence Application, pp. 29–30. IEEE Computer Society (1991)
51. Sabharwal, A., Samulowitz, H., Reddy, C.: Guiding combinatorial optimization with UCT. In: Beldiceanu, N., Jussien, N., Pinson, É. (eds.) CPAIOR 2012. LNCS, vol. 7298, pp. 356–361. Springer, Heidelberg (2012). https://doi.org/10.1007/978-3-642-29828-8_23
52. Simonis, H., Davern, P., Feldman, J., Mehta, D., Quesada, L., Carlsson, M.: A generic visualization platform for CP. In: Cohen, D. (ed.) CP 2010. LNCS, vol. 6308, pp. 460–474. Springer, Heidelberg (2010). https://doi.org/10.1007/978-3-642-15396-9_37
53. Tang, Y., Agrawal, S., Faenza, Y.: Reinforcement learning for integer programming: learning to cut. In: Daume III, H., Singh, A. (eds.) Proceedings of the 37th International Conference on Machine Learning. Proceedings of Machine Learning Research, vol. 119, pp. 9367–9376. PMLR (2020). https://proceedings.mlr.press/v119/tang20a.html
54. Thie, P.R., Keough, G.E.: An Introduction to Linear Programming and Game Theory. Wiley, Hoboken (2011)
55. Tjong Kim Sang, E.F.: Introduction to the CoNLL-2002 shared task: language-independent named entity recognition. In: COLING-02: The 6th Conference on Natural Language Learning 2002 (CoNLL-2002) (2002). https://aclanthology.org/W02-2024
56. Van Hentenryck, P.: The OPL Optimization Programming Language. MIT Press, Cambridge (1999)
57. Vaswani, A., et al.: Attention is all you need. In: Advances in Neural Information Processing Systems, vol. 30 (2017)
58. Wang, A., Singh, A., Michael, J., Hill, F., Levy, O., Bowman, S.R.: Glue: a multi-task benchmark and analysis platform for natural language understanding. arXiv preprint arXiv:1804.07461 (2018)
59. Wolf, T., et al.: Huggingface's transformers: state-of-the-art natural language processing. arXiv preprint arXiv:1910.03771 (2019)
60. Yadav, V., Bethard, S.: A survey on recent advances in named entity recognition from deep learning models. arXiv preprint arXiv:1910.11470 (2019)
61. Yang, Y., Boland, N., Dilkina, B., Savelsbergh, M.: Learning generalized strong branching for set covering, set packing, and 0-1 knapsack problems. Eur. J. Oper. Res. **301**(3), 828–840 (2022). https://doi.org/10.1016/j.ejor.2021.11.050. https://www.sciencedirect.com/science/article/pii/S0377221721010018
62. Yang, Z., Dai, Z., Yang, Y., Carbonell, J., Salakhutdinov, R.R., Le, Q.V.: Xlnet: generalized autoregressive pretraining for language understanding. In: Advances in Neural Information Processing Systems, vol. 32 (2019)
63. Zhao, S.: Named entity recognition in biomedical texts using an hmm model. In: Proceedings of the International Joint Workshop on Natural Language Processing in Biomedicine and its Applications (NLPBA/BioNLP), pp. 87–90 (2004)

64. Zhong, V., Xiong, C., Socher, R.: Seq2sql: generating structured queries from natural language using reinforcement learning. arXiv preprint arXiv:1709.00103 (2017)
65. Zhou, G., Su, J.: Named entity recognition using an HMM-based chunk tagger. In: Proceedings of the 40th Annual Meeting of the Association for Computational Linguistics, pp. 473–480 (2002)
66. Zhu, C., Byrd, R.H., Lu, P., Nocedal, J.: Algorithm 778: L-BFGS-B: fortran subroutines for large-scale bound-constrained optimization. ACM Trans. Math. Softw. (TOMS) **23**(4), 550–560 (1997)

Exploiting Entropy in Constraint Programming

Auguste Burlats[1(✉)] and Gilles Pesant[2(✉)]

[1] UCLouvain, Ottignies-Louvain-la-Neuve, Belgium
auguste.burlats@uclouvain.be
[2] Polytechnique Montréal, Montreal, Canada
gilles.pesant@polymtl.ca

Abstract. The introduction of Belief Propagation in Constraint Programming through the CP-BP framework makes possible the computation of an estimation of the probability that a given variable-value combination belongs to a solution. The availability of such marginal probability distributions, effectively ranking domain values, allows us to develop branching heuristics but also more generally to apply the concept of entropy to Constraint Programming. We explore how variable and problem entropy can improve how we solve combinatorial problems in the CP-BP framework. We evaluate our proposal on an extensive set of benchmark instances.

1 Introduction

Constraint Programming (CP) is a powerful approach to solve combinatorial problems. It can significantly reduce the search space by using constraints and their powerful inference algorithms to filter out infeasible variable-value combinations at each node of the search tree. The order in which variables are branched on has a significant impact on the shape of the tree and thus on search efficiency. This is why finding robust and generic variable ordering heuristics is crucial. The introduction of Belief Propagation (BP) in CP [6] makes possible the computation of an estimation of the probability that a given variable-value combination belongs to a solution. The availability of such marginal probabilities, effectively ranking domain values, allows us to develop variable ordering heuristics [1] but also more generally to apply the concept of entropy to CP, which is the subject of this paper.

A *Constraint Satisfaction Problem* (*CSP*) $P = \langle X, D, C\rangle$ is a combinatorial problem defined by a triplet where:

- $X = \{x_1, x_2, \ldots, x_n\}$ is a finite set of variables,
- $D = \{D(x_1), D(x_2), \ldots, D(x_n)\}$ is a finite set of finite domains,
- $C = \{c_1, c_2, \ldots, c_m\}$ is a finite set of constraints.

A. Burlats—Most of this work was carried out while the first author was at Polytechnique Montréal.

© The Author(s), under exclusive license to Springer Nature Switzerland AG 2023
A. A. Cire (Ed.): CPAIOR 2023, LNCS 13884, pp. 320–335, 2023.
https://doi.org/10.1007/978-3-031-33271-5_21

A *solution* $\mathbf{s} = (v_1, v_2, \ldots, v_n)$ to P assigns to each variable $x_i \in X$ a value v_i from its corresponding domain $D(x_i)$ such that all constraints in C are satisfied. Let S^P denote the set of all solutions to P and $\mathbf{s}[x]$ the value assigned to variable x in solution $\mathbf{s}$. Define

$$\theta_x^P(v) = \frac{|\{\mathbf{s} \in S^P : \mathbf{s}[x] = v\}|}{|S^P|}$$

as the proportion of solutions in which variable x takes value v.[1] We will call this quantity the *marginal* of variable-value pair (x, v) in reference to the marginal probability of x taking value v in a solution chosen uniformly at random from S. Note that we assume for the moment that S in nonempty i.e. that P is *satisfiable*: otherwise we will consider all marginals to be null. We define $H(x)$, the *entropy* of variable x using Shannon entropy [8]:

$$H(x) = \ -\textstyle\sum_{v \in D(x)} \theta_x(v) \log(\theta_x(v)).$$

This nonnegative quantity can be interpreted as the uncertainty about which value x should take in a solution: a null entropy corresponds to $\theta_x(v) = 1$ for some v, and so $\theta_x(v') = 0\ \forall v' \neq v$, i.e. $x = v$ in every solution (and x is thus a *backbone* variable); maximum entropy $\log(|D(x)|)$ is reached whenever $\theta_x(v) = \frac{1}{|D(x)|}\ \forall v \in D(x)$ i.e. its values are uniformly distributed among solutions. The normalized entropy (also called efficiency) of x divides its entropy by the logarithm of the cardinality of its domain, unless its domain is a singleton in which case the (normalized) entropy is null. We derive the entropy of problem P as

$$H(P) = \frac{\sum_{x \in X\,:\,|D(x)|>1} \frac{H(x)}{\log(|D(x)|)}}{|X|}.$$

It corresponds to the average normalized entropy of its variables and thus lies between 0 and 1 inclusive. We make two general observations about CSP entropy:

Observation 1. *A null CSP entropy only occurs when either it admits a single solution (including the special case where all variables are bound in a consistent assignment) or it has no solution. In the former case each variable has some unique value in its domain with a unit marginal whereas in the latter, each variable has all null marginals.*

Observation 2. *A CSP for which every assignment is a solution (or with uninformed marginals) will exhibit uniformly distributed marginals for each variable and an entropy equal to the proportion of its variables with non-singleton domains. In particular if all variables have non-singleton domains the CSP entropy is one.*

Of course our notion of entropy relies on marginals of which we typically do not know the exact value. This is where BP comes in to provide estimates of

[1] We will generally omit superscript P for ease of notation.

these marginals. In this paper we investigate several uses of entropy to help solve CSPs, particularly for branching heuristics.

We follow with a review of the CP-BP framework in Sect. 2 and then evaluate the accuracy of the marginals computed in this framework in Sect. 3. Section 4 presents different uses of entropy and follows with comparative experiments in Sect. 5. We then conclude in Sect. 6.

2 Belief Propagation for CSPs

Belief Propagation (BP) is an algorithm introduced by Pearl [5]. It is able to compute the marginal distribution for each non-observed node in a graphical model (e.g. a factor graph), conditioned by the value of the observed nodes.

Pesant [6] introduced a framework combining CP and BP in which beliefs about variable-value pairs are propagated as messages between variables and constraints, thus generalizing the simpler propagation of unsupported pairs. A CSP can be viewed as a factor graph where the constraints are the factor nodes and the variables are the variable nodes. We note $\mu_{c \to x}$ the message from constraint c to variable x, and $\mu_{x \to c}$ the message from variable x to constraint c. Their definition is

$$\begin{cases} \mu_{x \to c}(v) = \prod_{c' \in N(x) \setminus \{c\}} \mu_{c' \to x}(v) \\ \mu_{c \to x}(v) = \sum_{\mathbf{v}:\mathbf{v}[x]=v} f_c(\mathbf{v}) \prod_{x' \in N(c) \setminus \{x\}} \mu_{x' \to c}(\mathbf{v}[x']) \end{cases}$$

where $N(x)$ is the neighbourhood of variable x, i.e. the constraints applied to this variable, $N(c)$ is the neighbourhood of constraint c, i.e. its scope, $\mathbf{v}$ is a tuple from the Cartesian product of all the variables in the scope of c, $\mathbf{v}[x]$ is the value taken by x in $\mathbf{v}$ and $f_c(\mathbf{v})$ is a function that returns 1 if tuple $\mathbf{v}$ satisfies c and 0 otherwise. We are thus able to estimate the marginal of a variable x as

$$\hat{\theta}_x(v) = \prod_{c \in N(x)} \mu_{c \to x}(v) \qquad \forall v \in D(x).$$

Messages are sent iteratively: first, all variables send their messages (initially, uniform distributions over their domain); then all constraints send their messages. This cycle is repeated for a fixed number of iterations. Among other things, in this paper we offer a way to decide this number dynamically at each node of the search tree. Computing $\sum_{\mathbf{v}:\mathbf{v}[x]=v} f_c(\mathbf{v})$ is equivalent to counting solutions (local to c) where $\mathbf{v}[x] = v$. Therefore messages from constraints report the number of such solutions, each being weighted by the product of corresponding messages from variables. Pesant [6] provided efficient dedicated algorithms for weighted counting on several constraints. If there is no such algorithm for a given constraint it simply sends back a uniform distribution. Let's examine a small example from [6] to illustrate the behaviour of marginals.

Example 1. Consider variables a, b, c, and d with identical domains $\{1, 2, 3, 4\}$, and the following constraints:

$$\texttt{alldifferent}(a, b, c), \qquad a + b + c + d = 7, \qquad c \leq d.$$

Table 1. True marginals (a), initial estimated marginals (b), marginals after 1st iteration of BP (c) and after 10th iteration (d) for Example 1.

	1	2	3	4
θ_a	0	1/2	1/2	0
θ_b	0	1/2	1/2	0
θ_c	1	0	0	0
θ_d	1	0	0	0

(a) true marginals

	1	2	3	4
$\hat{\theta}_a$	.25	.25	.25	.25
$\hat{\theta}_b$	.25	.25	.25	.25
$\hat{\theta}_c$	.25	.25	.25	.25
$\hat{\theta}_d$	.25	.25	.25	.25

(b) initial marginals

	1	2	3	4
$\hat{\theta}_a$	.50	.30	.15	.05
$\hat{\theta}_b$	.50	.30	.15	.05
$\hat{\theta}_c$	.62	.28	.09	.01
$\hat{\theta}_d$	.29	.34	.26	.11

(c) 1st iteration

	1	2	3	4
$\hat{\theta}_a$	.01	.52	.46	.01
$\hat{\theta}_b$	.01	.52	.46	.01
$\hat{\theta}_c$	.98	.02	.00	.00
$\hat{\theta}_d$	.90	.10	.00	.00

(d) 10th iteration

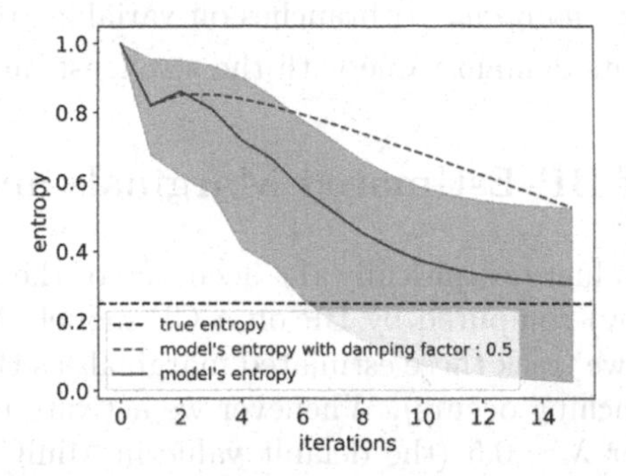

Fig. 1. Evolution of entropy during Belief Propagation for the CSP in Example 1.

This CSP has two solutions: $\langle a = 2, b = 3, c = 1, d = 1\rangle$ and $\langle a = 3, b = 2, c = 1, d = 1\rangle$. If we examine variable a, we observe that assignment $a = 2$ is present in one solution and that assignment $a = 3$ is present in the other one. There is no valid solution containing $a = 1$ or $a = 4$. Therefore its true marginal distribution is $\theta_a(1) = 0, \theta_a(2) = 1/2, \theta_a(3) = 1/2, \theta_a(4) = 0$. If we examine variable c, we can observe that only assignment $c = 1$ can be in a valid solution. Thus, its marginal distribution is $\theta_c(1) = 1, \theta_c(2) = 0, \theta_c(3) = 0, \theta_c(4) = 0$. BP starts from a uniform distribution for each variable: $\hat{\theta}_{x_i}(v) = 1/|D(x_i)|, \forall v \in D(x_i), \forall x_i \in X$. And, as we can see in Table 1, BP tends to converge to the true marginal distributions after a few iterations. Figure 1 also traces the evolution of the CSP entropy $H(P)$ (solid curve) with a lighter band showing the range of entropy for individual variables (normalized $H(x)$). For comparison the uninformed CSP entropy, i.e. considering domain values to be equally likely, corresponds to the initial value of the curve (1.0) and the true entropy is 0.25.

BP is assured to converge when there is no cycle in the graph [5] but the graphical representation of a CSP typically contains such cycles. However the large arity of global constraints, in addition to performing efficient inference, allows us to encapsulate some of those cycles and prevent the marginals from oscillating [6]. For instance in Example 1 marginals converge despite the remaining cycles in the model. In case marginals still oscillate, *message damping* has been known to help. Babaki et al. [1] propose using the weighted average of the current and previous messages from variables to constraints:

$$\mu_{x\to c}(v) = \lambda\mu_{x\to c}^{\text{current}}(v) + (1-\lambda)\mu_{x\to c}^{\text{previous}}(v)$$

where the *damping factor* ($0 \le \lambda \le 1$) balances the old and the new. Observe that for Example 1 damping is not needed and even slows down convergence (dashed curve in Fig. 1).

The estimated marginals $\hat{\theta}_x(v)$ can serve to inform dynamic search heuristics. For example *max-marginal* [1] branches on variable $\text{argmax}_{x\in X}\max_{v\in D(x)}(\hat{\theta}_x(v))$, assigning it its domain value with the strongest marginal.

3 Accuracy of BP-Estimated Marginals and Entropy

In this section we evaluate empirically the accuracy of the marginals (and ultimately of the entropy) computed by BP on a CP model. After the initial constraint propagation, we track these estimated marginals as BP iterations proceed (and before any branching occurs). Whenever we activate message damping we use a damping factor $\lambda = 0.5$ (the default value in MiniCPBP, the prototype solver implementing the CP-BP framework). We use several instances of combinatorial problems: some with a single solution (Sudoku and Nonogram) and others with a moderate number of solutions (n-queens and Feature model [9]). We enumerate the solutions to these instances and use them to compute the true marginals, against which we compare the estimated marginals using the *Kullback-Leibler divergence*, a measure of dissimilarity between two probability distributions:

$$\sum_{v\in D(x)} \theta_x(v)\cdot\log(\theta_x(v)/\hat{\theta}_x(v)).$$

For n-queens we use the instances ranging from $n = 5$ to 9 that have respectively 10, 4, 40, 92, and 352 solutions (we do not break symmetries). We observe at Fig. 2 that the KL-divergence stabilizes after a few iterations and to a low value. The exception is for 6-queens where the divergence is about one order of magnitude greater. That small instance has the fewest solutions: each variable has four domain values with identical true marginal (0.25) and the remaining two with a null true marginal. Upon inspection, no estimated marginal is null but the four outstanding values in each domain do have larger estimated marginals for all variables, meaning they are ranked correctly, though with less of a distinction for *q1* and *q4*. There are also a few misses: for example with 7-queens $\hat{\theta}_{q3}(3)$ is the lowest in the domain whereas it should be the highest. Damping does not make much of a difference here. Even though we typically cannot compute the KL-divergence because we do not know the true marginals, the observed entropy of the estimated marginals (Fig. 2, right column), which we can compute, will align very well with the unobserved divergence in terms of the iteration when they become stable, and this could be used to decide when to stop iterating BP. The difference between the estimated and true problem entropy is often under 1%.

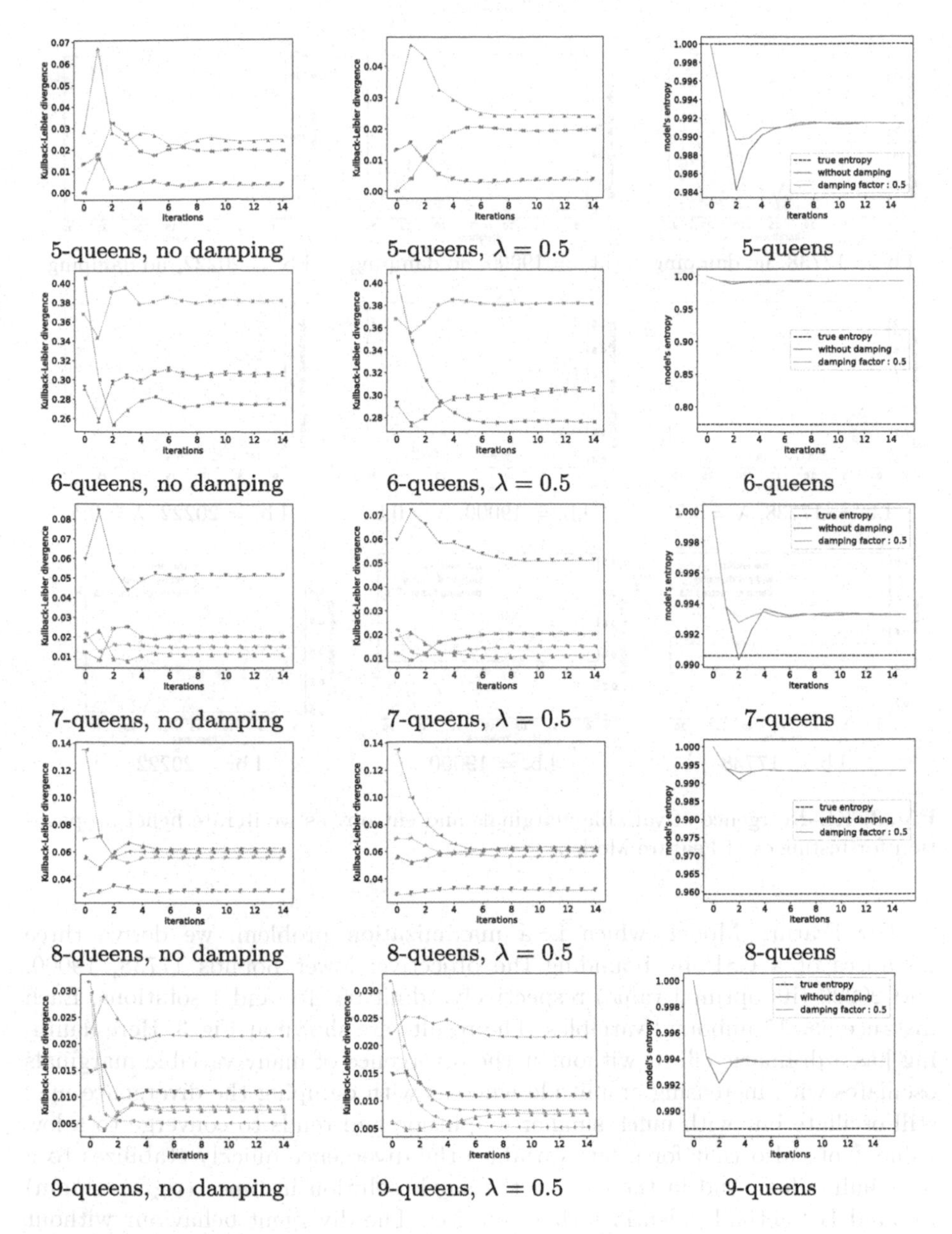

Fig. 2. KL-divergence of variable marginals and entropy as we iterate belief propagation for instances of n-queens.

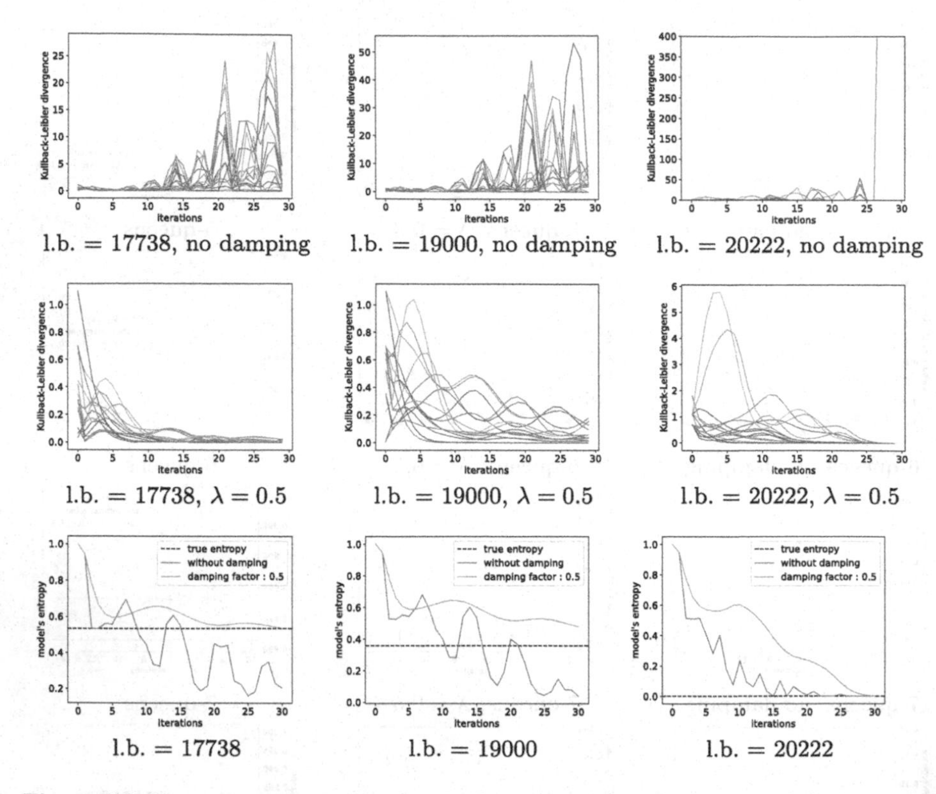

Fig. 3. KL-divergence of variable marginals and entropy as we iterate belief propagation for instances of Feature Model.

For Feature Model, which is a maximization problem, we derive three instances of a CSP by bounding the objective: lower bounds 17738, 19000, and 20222 (its optimal value) respectively admit 95, 15, and 1 solutions. Each instance has 15 unbound variables. The results are shown at Fig. 3. Here damping has a dramatic effect: without it the divergence of many variable marginals oscillates with increasing amplitude whereas with damping the divergence may still oscillate but with much smaller amplitude and tends to converge to a low value. Note also that for a few variables the divergence quickly stabilizes to a near null value. And in the case of the single-solution instance (right column) iterated BP actually identifies that solution. The divergent behaviour without damping appears as an oscillation in the observed problem entropy whereas the latter is smoother and even sometimes stable for the better-performing damping. Observe also how the estimated entropy with damping appears to converge to the true entropy.

Lastly we turn to instances with a larger number of variables and a single solution. The Sudoku instance we use at Fig. 4 features 33 unbound variables after constraint propagation. Without damping, severe oscillation occurs for many of

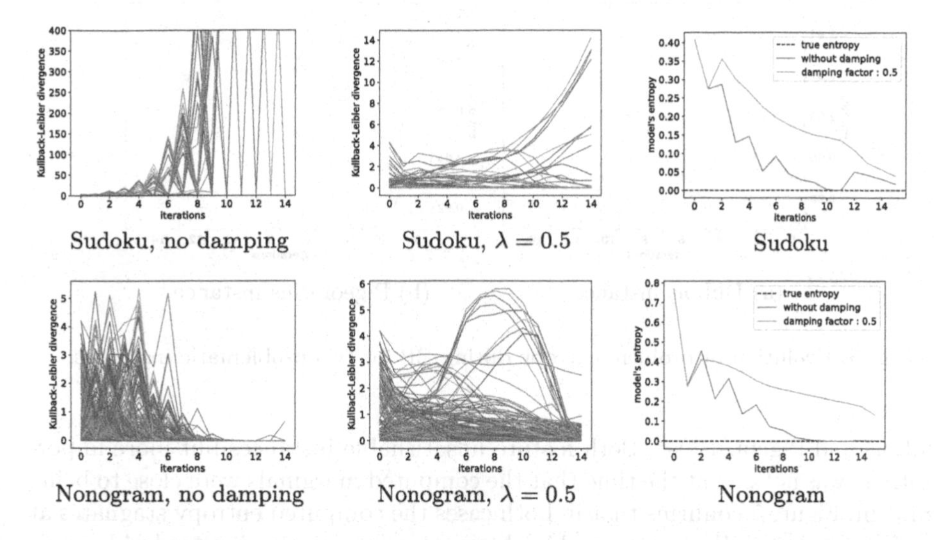

Fig. 4. KL-divergence of variable marginals and entropy as we iterate belief propagation for instances of Sudoku and Nonogram.

the variable marginals. It is again accompanied by an oscillation of the entropy, though less pronounced. Damping is very useful to keep the marginals under control except for a few which start to diverge around Iteration 10. The Nonogram instance has 444 unbound variables out of 576. In contrast with the previous instance no damping behaves better: it even momentarily stabilizes to the solution around Iteration 12. Another difference is that entropy without damping oscillates until Iteration 6, a behaviour that had coincided with increased divergence in the previous instances.

So, damping is not always better and entropy oscillation does not necessarily signal that we should use damping. But according to this limited empirical investigation damping generally helps much more than it hurts: for Feature Model and Sudoku it avoids very large (50) or even infinite divergence; for Nonogram the divergence with damping never strays above 6.

4 Exploiting Entropy

Now that we have empirical evidence for the accuracy of the computed marginals and entropy, we propose in this section several uses for such information.

4.1 Deciding When to Use BP

In their empirical evaluation of search based on BP for CSPs, Babaki et al. [1] reported two problems on which the approach performed particularly badly:

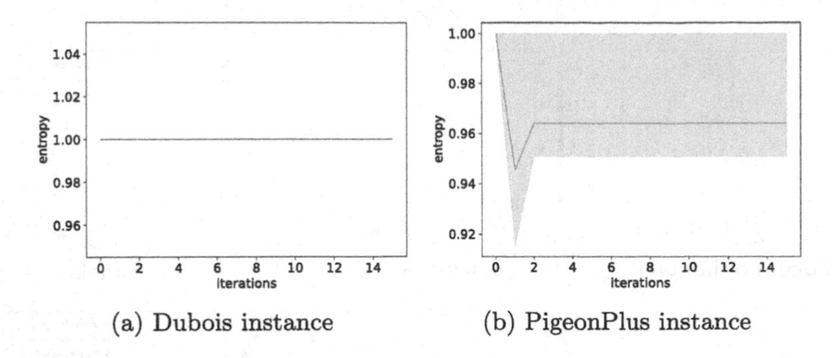

(a) Dubois instance (b) PigeonPlus instance

Fig. 5. Evolution of problem entropy during BP for two problematic instances.

Dubois and PigeonsPlus[2]. Both feature unsatisfiable instances but more importantly it was noticed at the time that the computed marginals were close to being uniform. Figure 5 confirms that in both cases the computed entropy stagnates at a value close to 1. We can turn this observation into a criterion to decide when problem entropy should be used to help solve an instance and when computationally expensive BP should be interrupted instead and replaced by a cheaper variable ordering heuristic, at least until useful information can be inferred again to guide search.

4.2 Deciding When to Stop BP Iterations

Variations of entropy give us information about the variations of marginals. If the entropy of a variable undergoes important variations along BP iterations, the marginals of this variable are varying too. It may mean that we shouldn't stop BP just yet. Based on this idea, we design a dynamic criterion to decide at each search-tree node when we should stop BP iterations. This criterion is based on the variations of the problem entropy $H(P)$. After iteration t, we compare the current entropy $H_t(P)$ to the entropy at the previous iteration $H_{t-1}(P)$. If $0 \leq H_{t-1}(P) - H_t(P) \leq \alpha$, for some threshold α, BP iterations are stopped and a branching decision is taken. This difference must be positive: otherwise it means that the entropy is increasing and that we shouldn't stop BP.

Another potential criterion is to look at the value of the smallest variable entropy. If this entropy becomes very low, we can consider this variable as almost decided, because we have strong knowledge about the value it should take. Thus, additional BP iterations are unnecessary and the variable should be branched on. Another incentive to do so is that in the next few iterations this variable will likely have all its marginals at zero except for the one value, which we observed will have a cascading effect on several other variables, dropping their entropy close to zero as well, which will make it harder to discriminate between the variable at the origin of this phenomenon and the other variables when deciding which one to branch on.

[2] http://www.xcsp.org/instances/.

4.3 Deciding When to Activate Damping

But perhaps the problem entropy never stabilizes and so does not meet our first stopping criterion. We saw in Sect. 3 that marginals may sometimes oscillate with increasing amplitude — which can be signaled by an oscillating entropy — and that damping can alleviate this issue. However damping is not necessarily desirable and can in some cases slow down convergence to the true marginals, as in the case of Example 1. As an alternative to activating damping by default, we will investigate starting BP without damping and switching it on whenever such entropy oscillation is observed.

4.4 Branching to Search for a Solution

The lower a variable's entropy, the stronger the information about which value the variable should take in a solution. Hence entropy is a powerful tool that we can exploit to make better branching decisions. We introduce variable ordering heuristic *min-entropy* that selects the variable with the lowest entropy and first tries fixing it to its domain value with the strongest marginal. Notice that, if the marginal distributions are uniform (i.e. we have no discriminating information between domain values), the variable with the lowest entropy will be the one with the smallest domain. Therefore, we can consider that *min-entropy* is a generalization of standard *smallest-domain* where we can discriminate between domain values based on the CSP.

5 Experimental Evaluation

In this section, we evaluate the quality of our resulting search strategy. In order to position our work with respect to the state of the art, we compare its performance to the *dom/wdeg* [2] and *IBS* [7] heuristics, and to another heuristic based on marginals and BP, *max-marginal*. Our metrics are the number of fails, which shows the accuracy of a heuristic, i.e. how good are the branching decisions, and the runtime, which indicates if the extra cost induced by our heuristics still makes them worthwhile.

5.1 Experimental Protocol

We ran our experiment on a set of 1319 instances from XCSP3[3] and the Minizinc Challenge[4]. One limitation of MiniCPBP is that it needs to store each value in the domain of each variable, and each corresponding marginal. Therefore when variables have very large domains this can be very space-consuming. We selected problems where variables have reasonable-size domains: summing over the variables, our largest instance has about 810 000 domain values. Our other criterion for problem selection was the constraints in the model. We chose problems where,

[3] Availables at http://www.xcsp.org/instances/.

[4] Available at https://www.minizinc.org/challenge2022/mznc2022_probs.tar.gz.

for the majority of the constraints present, our solver provides a weighted counting algorithm, in order to have a meaningful observation of the contribution of BP. The experiments were performed on a server with two Intel E5-2683 v4 Broadwell @ 2.1 GHz. We used the solver MiniCPBP[5], which is implemented over MiniCP [4] and is able to perform BP. Each run had a 20-min timeout and up to 12 GB of memory available.

Our results are presented as performance profiles: each point of a graph shows the proportion of instances (given on the y axis) that are solved with a number of failures or runtime less than or equal to the value on the x axis. We compare *min-entropy* with *max-marginal* during depth-first search (DFS), to see if entropy is a better exploitation of the marginals. Before each branching decision, unless indicated otherwise, five iterations of belief propagation are performed (the current default number in MiniCPBP). To avoid the oscillation of marginals, we apply damping on the messages sent during BP with a damping factor $\lambda = 0.5$.

As an attempt to improve basic *min-entropy* and as described in Sect. 4, we consider a dynamic configuration where the use of damping and the number of BP iterations are dynamically decided during the search. BP is stopped when the variation of the problem entropy is lower than threshold $\alpha = 0.1$ (see Sect. 4.2): experiments showed a strong variance on the best value for α depending on the problem but that parameter value generally performs well on our benchmark problems. In Sect. 4.2, we describe another stopping criterion which stops BP iterations when the smallest entropy among variables falls under a threshold. We tested it with different threshold values and it showed better performance than the criterion based on the entropy's variations on some problems, but the latter remained the best choice overall. At each search-tree node, BP is performed at first without damping and if oscillations in the problem entropy are detected, damping is activated (with $\lambda = 0.5$ as before). To detect oscillations, we count how many times we observe a decreasing entropy starting to increase: if it switches 3 times from a negative variation of the entropy to a positive variation, we activate damping for the rest of the propagation. After the branching decision, damping is deactivated again.

As a state-of-the-art reference, we use *dom/wdeg* with restarts (initial restart: $3n$ failures, where n is the number of variables in the problem; increased by a factor 1.4 after each restart) and Impact-Based-Search (*IBS*) [7], also with restarts (initial restart: $2n$ failures; increased by a factor 1.2 after each restart). For *IBS*, to initialize impacts before the search, we try each possible assignment and register its impact.

5.2 Evaluation

According to Fig. 6 *min-entropy* shows better performance than *max-marginal* on opt-cryptoanalysis, MagicSquare and MagicHexagon. For the other problems,

[5] Solver and used instances are available at https://github.com/PesantGilles/MiniCPBP.

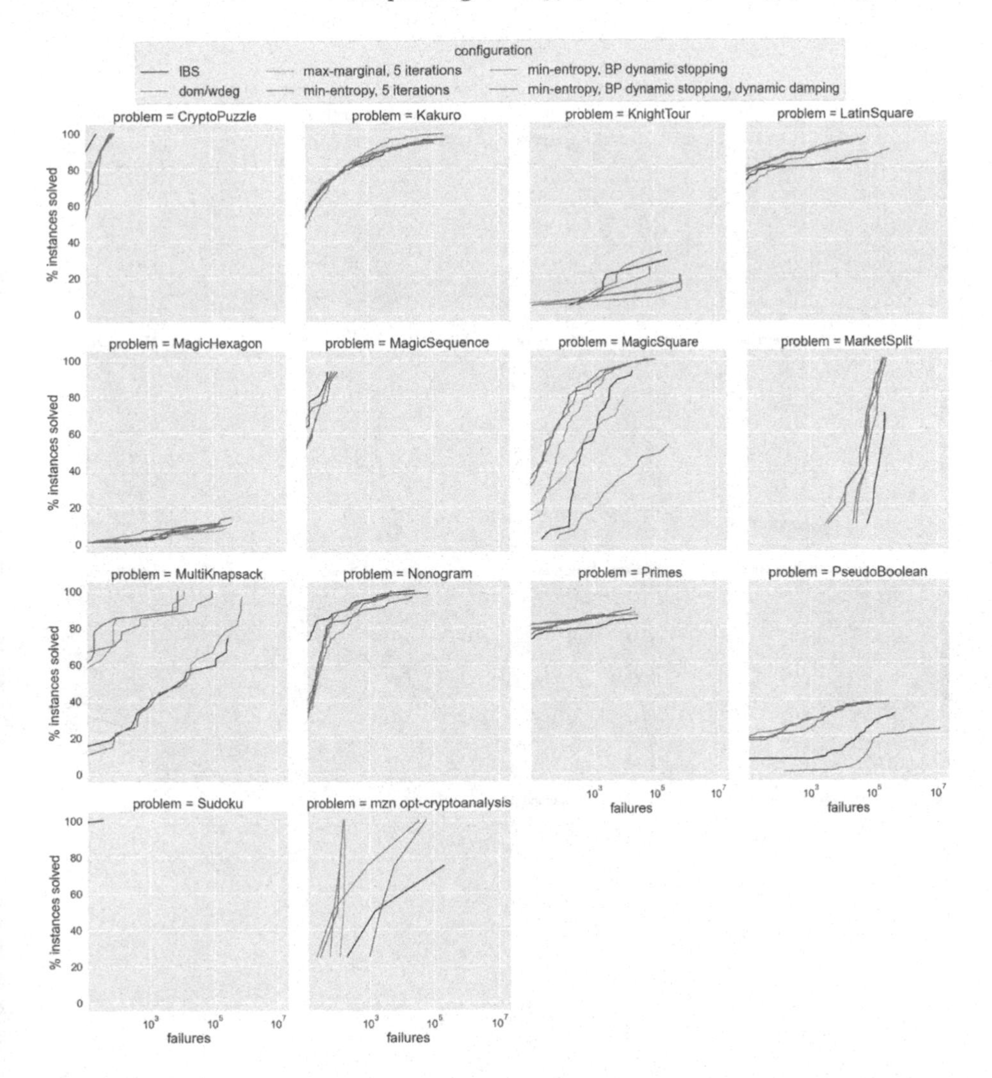

Fig. 6. % instances solved vs #fails for several branching heuristics

we observe similar performance for these two heuristics. The binary domains in Nonogram, PseudoBoolean, MultiKnapsack and MarketSplit explain that we observe identical performance between *max-marginal* and *min-entropy* (i.e. the orange and red curves coincide): with such domains, the variable that presents the strongest marginal is also the one with the lowest entropy. We can therefore conclude that *min-entropy* is a good improvement of marginal usage.

If we now compare our heuristics to *dom/wdeg* and *IBS*, we see that we outperform them on five problems in our dataset (LatinSquare, MagicSquare, MultiKnapsack, PseudoBoolean, and opt-cryptoanalysis). For three problems some of the state of the art is showing better performance: for KnightTour

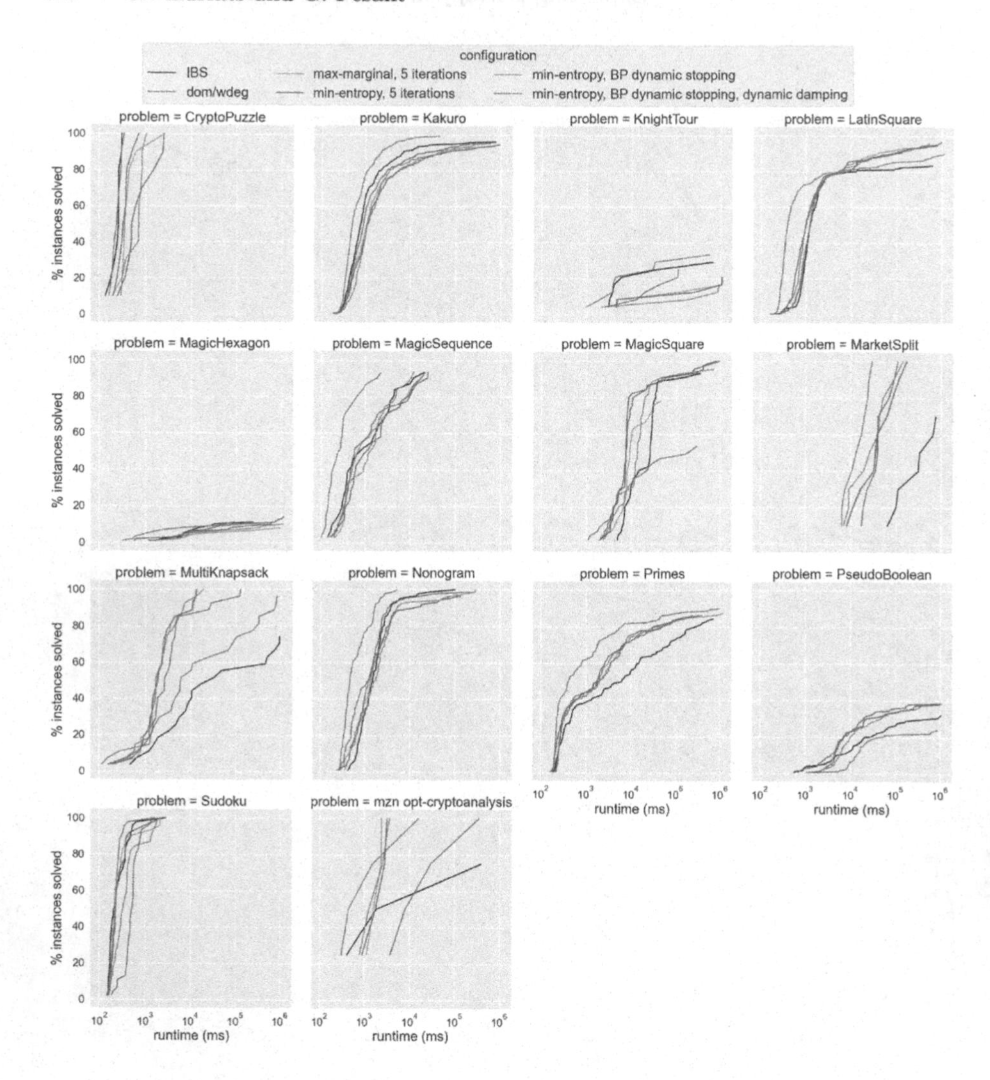

Fig. 7. % instances solved vs runtime (ms) for several branching heuristics

both *dom/wdeg* and *IBS* outperform *min-entropy*, for *Kakuro* only *dom/wdeg* is better, and for *CryptoPuzzle* only *IBS* is better. On the remaining six problems heuristics perform similarly. Based on these results, exploiting entropy to make branching decisions is a competitive approach.

Let's take a closer look at MarketSplit and MultiKnapsack. They are interesting because they present similar structure, i.e. their variables are binary and they only contains `sum` constraints applied on all variables. However, we observe a strong difference in the performance of our heuristic between these two problems. On MultiKnapsack, our heuristics clearly outperform *dom/wdeg* and *IBS*, whereas on MarketSplit the results are more mixed. If we compare the entropy of the variable in each problem, the reason is clear: if in MultiKnapsack they quickly decrease, in MarketSplit they often stay above 0.9. MarketSplit is thus a good example of a problem where BP is not as informed, as we discussed in Sect. 4.1, and typically we would detect this and then decide to use another branching heuristic.

We now turn to runtime to evaluate the cost of adding a second kind of propagation at each node of the search tree. If we look at Fig. 7 and compare with Fig. 6 we can observe this additional cost. If we encountered fewer failures with *min-entropy* and *max-marginal* than *dom/wdeg* for problems like MarketSplit and MagicSquare, we observe similar runtimes for these problems. For Kakuro, Nonogram and MagicSequence, the performance in terms of failures was similar, but *dom/wdeg* outperforms the other heuristics in terms of runtime. Sudoku is a particular problem because it is the only one for which we don't observe any additional cost. This is because the majority of instances are solved during the first constraint propagation, before the use of belief propagation. Despite this additional cost, our heuristics still outperform *dom/wdeg* and *IBS* on PseudoBoolean and MultiKnapsack. And for LatinSquare, *max-marginal* and *min-entropy* are able to solve more instances. With some optimization, this approach could be a good option for a wide spectrum of problems.

Speaking of BP optimization, we should now look at the performance of our *dynamic* strategies. From Fig. 6 we observe that the heuristic quality is not strongly impacted by the use of the dynamic parameters. We observe a small degradation of performance for PseudoBoolean. This deterioration is due to dynamic damping, because the configuration that is only using dynamic stopping for BP shows performance as good as static min-entropy. This degradation is stronger for MultiKnapsack and opt-cryptoanalysis and is once again mainly due to dynamic damping. On the contrary for KnightTour and Nonogram we observe a significant reduction of the number of failures. But the primary goal of the dynamic stopping criterion is to improve runtime by choosing the best moment to stop BP and thus spare useless iterations. At Fig. 7 we observe an improvement for CryptoPuzzle, Kakuro, PseudoBoolean, and KnightTour. Concerning the latter, the improvement is connected to the reduction of failures. For the other problems, the improvement is not as significant. By using this criterion, our goal is to spare a few iterations at each node of the search tree when it is adequate. Therefore, the reduction of runtime will be linear, which is less noticeable on a logarithmic scale. Finally we observe a small deterioration for Sudoku, for MultiKnapsack, which is linked to the increase of failures, and for MarketSplit. A limitation of our dynamic stopping criterion is the variability of the best value for α, as we mentioned in Sect. 4.2. We chose the value that showed the most stability in its performance, but it was not the best configuration for all problems, and for some problems, like MagicSquare, we observed significantly better performance by using our other stopping criterion, which is stopping BP when the smallest entropy falls under a threshold. In conclusion, using a dynamic number of iterations shows good potential to reduce the runtime, but it requires more work to find a stopping criterion that would show improvement consistently on a diverse set of problems. Concerning dynamic damping, we observe a slight improvement for *KnightTour* in Fig. 6, but otherwise using it shows similar or worse performance. Further work is required to find a better criterion for detecting the need of damping.

6 Conclusion

We investigated the entropy of CSPs and of their finite-domain variables, made possible by estimating marginal distributions using the CP-BP framework on constraint programming models. Our study showed that these estimated distributions can get quite close to the true distributions, and that message damping during BP may be necessary to obtain convergence. We proposed several ways to exploit entropy in order to help solve combinatorial problems. Our experiments on 1319 instances from 14 different problems showed that branching on the variable with the lowest entropy is an insightful variable-ordering strategy.

Because performing belief propagation does come with a computational cost, we considered two stopping criteria to decide dynamically when BP iterations should stop in an effort to avoid unproductive work. The experiments showed that such stopping criteria have potential but require further refinements in order to be robust across problems.

We also considered when message damping should be activated. Instead of turning it on with a fixed damping factor, we tried to adjust the use of damping dynamically according to the observed effect on entropy oscillation. Experimental results show that further work is required to find a more accurate criterion.

A possible pragmatic improvement could be not to use BP at each node of the search tree. If a branching decision has a very small impact on domains, we can consider that it would have a very small impact on the marginals. Therefore we could reuse the marginals computed at the previous node and spare a mostly redundant phase of belief propagation. A similar approach gave significant acceleration when applied to *maxSD* [3], a branching heuristic based on solution counting. Thus, it is a promising idea that could be explored in future work.

References

1. Babaki, B., Omrani, B., Pesant, G.: Combinatorial search in CP-based iterated belief propagation. In: Simonis, H. (ed.) CP 2020. LNCS, vol. 12333, pp. 21–36. Springer, Cham (2020). https://doi.org/10.1007/978-3-030-58475-7_2
2. Boussemart, F., Hemery, F., Lecoutre, C., Sais, L.: Boosting systematic search by weighting constraints. In: de Mántaras, R.L., Saitta, L. (eds.) Proceedings of the 16th European Conference on Artificial Intelligence, ECAI 2004, Including Prestigious Applicants of Intelligent Systems, PAIS 2004, Valencia, Spain, 22–27 August 2004, pp. 146–150. IOS Press (2004)
3. Gagnon, S., Pesant, G.: Accelerating counting-based search. In: van Hoeve, W.-J. (ed.) CPAIOR 2018. LNCS, vol. 10848, pp. 245–253. Springer, Cham (2018). https://doi.org/10.1007/978-3-319-93031-2_17

4. Michel, L., Schaus, P., Van Hentenryck, P.: MiniCP: a lightweight solver for constraint programming. Math. Program. Comput. **13**(1), 133–184 (2021). https://doi.org/10.1007/s12532-020-00190-7
5. Pearl, J.: Reverend Bayes on inference engines: a distributed hierarchical approach. In: Proceedings of the National Conference on Artificial Intelligence, Pittsburgh, PA, 18–20 August 1982, pp. 133–136 (1982). http://www.aaai.org/Library/AAAI/1982/aaai82-032.php
6. Pesant, G.: From support propagation to belief propagation in constraint programming. J. Artif. Intell. Res. **66**, 123–150 (2019). https://doi.org/10.1613/jair.1.11487
7. Refalo, P.: Impact-based search strategies for constraint programming. In: Wallace, M. (ed.) CP 2004. LNCS, vol. 3258, pp. 557–571. Springer, Heidelberg (2004). https://doi.org/10.1007/978-3-540-30201-8_41
8. Shannon, C.E.: A mathematical theory of communication. Bell Syst. Tech. J. **27**(3), 379–423 (1948). https://doi.org/10.1002/j.1538-7305.1948.tb01338.x
9. Vavrille, M., Truchet, C., Prud'homme, C.: Solution sampling with random table constraints. In: Michel, L.D. (ed.) 27th International Conference on Principles and Practice of Constraint Programming, CP 2021, Montpellier, France (Virtual Conference), 25–29 October 2021. LIPIcs, vol. 210, pp. 56:1–56:17. Schloss Dagstuhl - Leibniz-Zentrum für Informatik (2021). https://doi.org/10.4230/LIPIcs.CP.2021.56

Constraint Propagation on GPU: A Case Study for the Cumulative Constraint

Fabio Tardivo[1](✉), Agostino Dovier[2], Andrea Formisano[2], Laurent Michel[3], and Enrico Pontelli[1]

[1] New Mexico State University, Las Cruces, USA
{ftardivo,epontell}@nmsu.edu
[2] University of Udine, INdAM-GNCS, Udine, Italy
{agostino.dovier,andrea.formisano}@uniud.it
[3] Synchrony Chair in Cybersecurity, University of Connecticut, Storrs, USA
ldm@uconn.edu

Abstract. The *Cumulative* constraint is one of the most important global constraints, as it naturally arises in a variety of problems related to scheduling with limited resources. Devising fast propagation algorithms that run at every node of the search tree is critical to enable the resolution of a wide range of applications. Since its introduction, numerous propagation algorithms have been proposed, providing different tradeoffs between computational complexity and filtering power.

Motivated by the impressive computational power that modern GPUs provide, this paper explores the use of GPUs to speed up the propagation of the *Cumulative* constraint. The paper describes the development of a GPU-driven propagation algorithm, motivates the design choices, and provides solutions for several design challenges. The implementation is evaluated in comparison with state-of-the-art constraint solvers on different benchmarks from the literature. The results suggest that GPU-accelerated constraint propagators can be competitive by providing strong filtering in a reasonable amount of time.

Keywords: Constraint Propagation · Cumulative · Parallelism · GPU

1 Introduction

Industrial scheduling problems are derivatives of the so-called "Resource Constrained Project Scheduling Problem" (briefly, RCPSP) in which one must *order* non-preemptible activities of fixed duration to minimize the makespan, i.e., the project duration. Activities use a fixed amount of resources to execute and each resource has a fixed capacity. Unsurprisingly, industrial scheduling readily benefits from any improvements to solve the classic RCPSP problem.

The last three decades witnessed the development of multiple techniques to prune the search tree of such an NP-hard problem [3,16]. The most prominent techniques are Edge-Finding [29,33], Time-Tabling [26], Not-First/Not-Last [33, 40], and Energetic-Reasoning [28]. Edge-Finding was developed for cumulative

© The Author(s), under exclusive license to Springer Nature Switzerland AG 2023
A. A. Cire (Ed.): CPAIOR 2023, LNCS 13884, pp. 336–353, 2023.
https://doi.org/10.1007/978-3-031-33271-5_22

instances through a series of contributions and it deduces precedences between activities that must be satisfied in time $O(n^2k)$ [29] where n is the number of activities and k is the number of distinct capacity requirements of the activities. Time-Tabling focuses on resource usage profile. Several techniques based on *line-sweep* methods were proposed with an $O(n^2)$ [17] solution that separates profile building and inference phases. The core of the inference mechanism rests on the ability to deduce, from the mandatory part of the profile, whether an activity must be postponed or not. Not-First/Not-Last performs an orthogonal pruning with respect to the other approaches by deducing unfeasible precedences between activities in time $O(n^2 \log n)$ [39] using a Θ-tree data structure. Energetic-Reasoning calculates the resource usage in specific time intervals to check and adjust the activities so that there is no over-consumption. Its standard algorithm has $O(n^3)$ time complexity, that can be reduced to $O(n^2 \log n)$ [36] by using Monge matrices.

In practice, most CP solvers employ Edge-Finding or Time-Tabling techniques that exhibit a lower time complexity at each node of the search tree, despite the strength of the filtering one might benefit from with Energetic-Reasoning. This paper revisits this design decision and considers the use of an Energetic-Reasoning propagator in a CP solver. Specifically, the paper advocates that the high computational complexity cost at each fixpoint can be mitigated with the use of a Graphics Processing Unit (GPU) and deliver, overall, faster computation times, or better solutions within a given time horizon. The fundamental assumption is that the energetic filtering rule is easily parallelized on this class of hardware and that the benefits from the derived filtering can be significant.

This paper is organized as follows. Section 2 offers some general background. Section 3 discusses the design of the proposed solution. Section 4 discusses empirical results that pitch a GPU-based Energetic-Reasoning against multiple solvers using Edge-Finding and Time-Tabling techniques. Section 5 concludes the paper.

2 Background

This section establishes the required background knowledge on Constraint Satisfaction and Optimization [2,37], Cumulative Scheduling [3,6] and General-Purpose computing on Graphics Processing Units (GPGPU) [10,38].

2.1 Constraint Satisfaction/Optimization Problem

A *Constraint Satisfaction Problem* (CSP) is a triple $P = \langle V, D, C \rangle$, where $V = \{V_1, \dots, V_n\}$ is a finite set of *variables*, $D = \{D_1, \dots, D_n\}$ is the set of *finite domains*, and C is a collection of *constraints* on variables in V. Each constraint $c \in C$, defined over a set of variables $vars(c) = \{V_{i_1}, \dots, V_{i_m}\} \subseteq V$, defines a relation on $D_{i_1} \times \cdots \times D_{i_m}$, namely $c \subseteq D_{i_1} \times \cdots \times D_{i_m}$. A *solution* of $\langle V, D, C \rangle$ is an assignment $\sigma : V \to \bigcup_{i=1}^{n} D_i$ such that:

- $\sigma(V_i) \in D_i$ for each $i = 1, \dots, n$
- $\forall\, c \in C$ then $\langle \sigma(V_{i_1}), \dots, \sigma(V_{i_m}) \rangle \in c$, where $vars(c) = \{V_{i_1}, \dots, V_{i_m}\}$.

The set of solutions for the CSP $\langle V, D, C\rangle$ is denoted $S(\langle V, D, C\rangle)$. Given a CSP $\langle V, D, C\rangle$, a *constraint solver* searches for one or more solutions. A solver alternates two types of steps: (1) constraint propagation and (2) non-deterministic choice. The latter is used to select the next variable to be assigned and to select non-deterministically a value to be given to such a variable (drawn from its current domain). Constraint propagation uses the constraints to prune the domain of the variables, removing values that provably do not belong to a solution.

The propagation algorithm uses a queue to schedule the constraints that must be reconsidered when the domain of a variable changes. Namely, whenever the domain of a variable $x \in vars(c)$ for some $c \in C$ changes, the constraint c is added to the queue. The filtering algorithms of a constraint that shrinks domains enforces a level of consistency, such as *domain consistency* [37]. An m-ary constraint c on the variables $vars(c) = \{V_{i_1}, \dots, V_{i_m}\}$ is *domain consistent* if $\forall\, j \in \{1, \dots, m\}$ the following holds:

$$\forall a_j \in D_{i_j} : \exists a_1 \in D_{i_1} \cdots \exists a_{j-1} \in D_{i_{j-1}} \exists a_{j+1} \in D_{i_{j+1}} \cdots \exists a_m \in D_{i_m} : (a_1, \dots, a_m) \in c$$

A CSP is domain consistent if all constraints in $\mathcal{C}$ are domain consistent. For binary constraints (i.e., $m = 2$) domain consistency is known as *arc consistency*. Without loss of generality, a *Constraint Optimization Problem* is specified with $\langle V, D, C, f\rangle$ where $\langle V, D, C\rangle$ is a CSP and $f : D_1 \times \cdots \times D_n \to \mathbb{R}$ is an objective function to be minimized. The goal is to find a solution of $\langle V, D, C\rangle$

$$\sigma^* = \underset{\sigma \in S(\langle V,D,C\rangle)}{argmin} \quad f(\sigma)$$

that minimizes $f(\sigma(V_1), \cdots, \sigma(V_n))$.

2.2 Cumulative

The *Cumulative* constraint is one of the most used constraints in CP. It makes it easy to model and solve a variety of real-world problems, contributing to the success of CP in scheduling applications.

In detail, it models the Cumulative Scheduling Problem (CuSP): given a set A of activities that use a resource of capacity u and where the goal is to schedule the activities so that the last activity finishes as soon as possible and no more than u units of the resources are used at any time. Formally, each activity $a \in A$ is defined by its start time s_a, its processing time p_a and, its resource usage h_a. The end time of activity a is $e_a = s_a + p_a$ and the problem is defined as follows:

$$\begin{aligned} &\text{minimize} \quad \underset{a \in A}{max}(e_a) \\ &\text{subject to} \quad \sum_{\{a : a \in A,\, s_a \le t < e_a\}} h_a \le u \qquad t \in \mathbb{N} \end{aligned}$$

Since its introduction in [1], the *Cumulative* constraint has been the subject of many studies to improve its efficiency. The result is a collection of propagation algorithms with different trade-offs between filtering capability and computational complexity. Such algorithms are commonly classified by the filtering

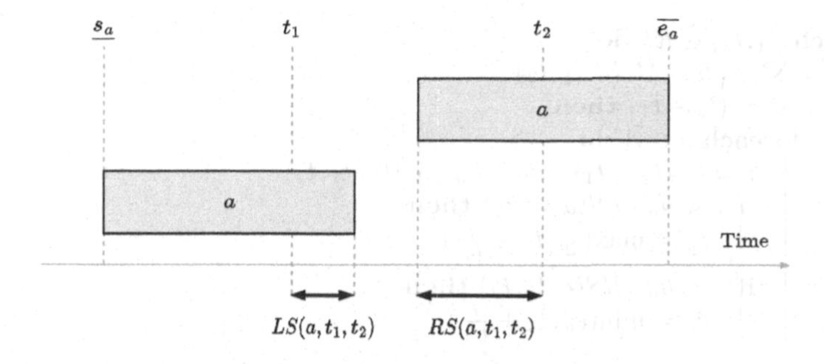

Fig. 1. Left shift and right shift for the activity a with respect to $[t_1, t_2)$.

technique they employ: Time-Tabling, Edge-Finding, Not-First/Not-Last, and Energetic-Reasoning. A complete description of these approaches is out of the scope of this work, interested readers can refer to [6]. This section describes the core ideas that characterize each method and points to the relevant literature for details.

Edge-Finding. This approach considers subsets of activities, determining if an activity must start before or end after the rest. It was introduced in [33], corrected in [29] and improved in different ways in [20,24,35,47,48].

Time-Table. This method consists of computing the minimal resource usage at every time and adjusting the starting time of the activities so that there is no over-consumption of the resource. It first appeared in [26] and was successively refined in [7,17,27].

Not-First/Not-Last. This approach considers subsets of activities and determines whenever an activity cannot be the first/last to be executed. Introduced in [33], it was corrected in [40] and improved in [22,23,39].

Energetic-Reasoning. This method checks some critical time intervals and adjusts the starting time of the activities so that there is no over-consumption of the resource. Introduced in [28], it was refined and improved in [4,13,36, 45,46].

Energetic-Reasoning is one of the strongest propagators for the *Cumulative* constraint, dominating both the Time-Table and Edge-Finding approaches [6]. Such filtering examines $O(n^2)$ time intervals for a total complexity of $O(n^3)$, too costly to be used in practice [13,36,46].

Preliminaries. Before proceeding to formalize the Energetic-Reasoning, we introduce some notation: $[t_1, t_2)$ denotes an open time interval. The lower and upper bounds of (the domain of) a variable x are denoted by $\underline{x}$ and $\overline{x}$, respectively. Given a time interval $[t_1, t_2)$ and an activity a, their *minimal intersection* is $MI(a, t_1, t_2) = \min(LS(a, t_1, t_2), RS(a, t_1, t_2))$ where LS and RS are the *left shift* and *right shift* (see Fig. 1):

$$LS(a, t_1, t_2) = \max(0, \min(\underline{e_a}, t_2) - \max(\underline{s_a}, t_1))$$
$$RS(a, t_1, t_2) = \max(0, \min(\overline{e_a}, t_2) - \max(\overline{s_a}, t_1))$$

```
foreach [t_1, t_2) ∈ RI do
    w = Σ_{a∈A} h_a · MI(a, t_1, t_2)
    if w < c · (t_2 − t_1) then
        foreach a ∈ A do
            r = c · (t_2 − t_1) − w + h_a · MI(a, t_1, t_2)
            if r < h_a · LS(a, t_1, t_2) then
                s̲_a = max(s̲_a, t_2 − r/h_a)
            if r < h_a · RS(a, t_1, t_2) then
                ē_a = min(ē_a, t_1 + r/h_a)
    else
        Fail
```

Algorithm 1: Energetic-Reasoning propagation algorithm.

The Energetic-Reasoning propagator considers the intervals whose extremes are related to the beginning/end of an action [4,13]. We define the set of *relevant intervals* as

$$RI = \bigcup_{(i,j)\in A\times A} O(i,j)$$

where:

$$\begin{aligned}O(i,j) = &\{[t_1,t_2) \ : t_1 < t_2, t_1 \in O_1(i), t_2 \in O_2(j)\} \cup \\ &\{[t_1,t_2) \ : t_1 < t_2, t_1 \in O_1(i), t_2 \in O_3(j, O_1(i))\} \cup \\ &\{[t_1,t_2) \ : t_1 < t_2, t_1 \in O_3(i, O_2(j)), t_2 \in O_2(j)\}\end{aligned}$$

and $O_1(a) = \{\underline{s_a}, \overline{s_a}\}$, $O_2(a) = \{\underline{e_a}, \overline{e_a}\}$, $O_3(a,T) = \{\underline{s_a} + \overline{e_a} - t : t \in T\}$. The consistency condition of is:

$$\forall [t_1,t_2) \in RI : \sum_{a\in A} h_a \cdot MI(a,t_1,t_2) \leq u \cdot (t_2 - t_1)$$

From such condition one adjusts the start time of activities to prevent over-usage as listed in Algorithm 1. Note that it could be worth to consider a set $O(i,j)$ only if s_i or s_j changed in the current propagation phase. The rational of this heuristics is to avoid to check consistent intervals and despite that it makes the filtering weaker, it proved to be effective for the sequential implementation (see Sect. 4).

2.3 GPUs and CUDA

Modern Graphical Processing Units (GPUs) are massively parallel architectures where thousands of computing units can process large amounts of data. Such power allows for the solution of problems that are out of reach with contemporary multi-core CPU technology. Recent research shows that the use of GPUs can

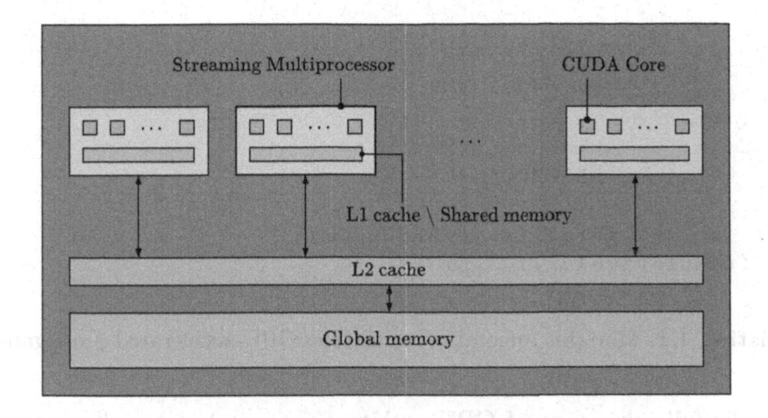

Fig. 2. High level GPU architecture.

be beneficial for speeding up basic Computational Logic tasks. See, among many, [11,12] for SAT, [14,15] for ASP, and [44] for CP. However, accessing the computational power offered by GPUs demands specific techniques and algorithms that proficiently exploit the peculiarities of the GPU architecture. To support developers and researchers, NVIDIA introduced *CUDA* (Computing Unified Device Architecture) [34], a C/C++ Application Programming Interface (API) that allows to ignore the underlying graphical concepts in favor of parallel computing concepts. A typical CUDA program is composed of parts executed by the CPU, the *host*, and parts designed to be executed on the GPU, the *device*. The host parts contain instructions for data movement and computation offloading, while the device parts contain the code that performs the computation.

A current NVIDIA GPU contains up to one hundred *Streaming Multiprocessors* (SM), each containing up to one hundred computational units called CUDA Cores (see Fig. 2). The main GPU memory is called *global memory*, and it can be tens of GB large. Between global memory and SMs there is a *L2 cache* of a few MB. Finally, each SM is equipped with some tens of KB of fast memory used as *L1 cache* and/or scratchpad memory, in which case it is referred to as *shared memory*.

The CUDA computational model is defined as *Single-Instruction Multiple-Thread* (SIMT). In this model, each thread executes the same C/C++ function, named *kernel*, and uses its unique index to identify the data fragments to fetch or the control flow. The case where different threads take different control flows is called *thread divergence*, and it is handled by executing the threads one after the other. Such a behavior may cause serious performance degradation. From a programmer's perspective, threads are logically grouped in *blocks* and blocks are organized in a *grid*. Blocks are dispatched to the Streaming Multiprocessors, that run the threads using their CUDA Cores. Threads in the same block can share data using the shared memory, while threads of different blocks can only share data through the global memory.

```
include "cumulative.mzn";
include "minicpp.mzn";
...
int: m; % Number of resources
set of int: RESOURCE = 1..m;
...
constraint forall(r in RESOURCE)
  (cumulative(...) ::gpu );
...
```

Listing 1.1. MiniZinc annotation to use the GPU-accelerated propagator.

To take full advantage of GPU architecture, one has to adhere to specific programming directives to distribute the workload among the cores, avoid thread divergence and optimize memory accesses. This usually involves exploiting the shared memory to reduce costly global memory accesses.

3 Design and Implementation

This section describes the process of developing a constraint solver which supports the GPU-accelerated propagation of *Cumulative*. The first part is about the constraint solver and can be used to estimate the effort necessary to integrate our ideas into an existing solver. The second part focuses on a GPU-accelerated propagator and can be useful to evaluate how effectively other propagators can be parallelized.

Solver. The exploitation of a GPU-based propagator within a solver has some caveats. The first is that the solver is *open-source* because intimate modifications of internals might be needed. Second, it is preferable that the solver is written in *C/C++* to facilitate the interaction with CUDA. Finally, it is convenient that the solver supports the *MiniZinc* language so that the GPU-accelerated propagator is easily accessible by the community.

We choose to work with MiniCP [30] because it is open-source and it is reasonably simple to modify thanks to the comprehensive documentation and the straightforward mapping between its architecture and the theory. In particular we used MiniCPP [19], a C++ implementation of MiniCP. The integration of the GPU-accelerated propagator is the same as any other propagator, but it requires modifying the build process to properly handle CUDA code. The addition of a FlatZinc frontend [42], few variable/value selection heuristics and some constraints were sufficient to obtain a solver compatible with MiniZinc [41]. To provide a simple mechanism to use the GPU-accelerated propagator, we introduced a new MiniZinc annotation. Specifically, a constraint annotated with `::gpu` is enforced using the GPU-accelerated propagator in place of the CPU implementation (see Listing 1.1).

Propagator. The GPU can be used to enhance constraints propagation according to two strategies: speed-up the fastest algorithms, or lower the computational price of strong filtering algorithms. The first strategy has different downsides: offloading to a GPU introduces an overhead that may overshadow the speed-up, and it may not be obvious to parallelize the best (sequential) algorithm because of its data structures. On the contrary, strong filtering algorithms may expose enough parallelizable work to make it convenient to offload the computation, but it may still be too slow to be beneficial.

Let us consider prior implementations proposed for Edge-Finding, Time-Tabling, etc. to single out the most promising one for GPU parallelization. We evaluate them based on the data structures they use, preferring plain data structures since *pointer chasing* (i.e., a sequence of irregular memory accesses following chains of pointers) is quite harmful on a GPU. The Time-Table propagator, as proposed in [17], seems to be a good candidate since other implementations are impeded by the use of heap data structures. Among the Edge-Finding propagators found in the literature, the most promising are those proposed in [24,29], as other approaches heavily rely on *linked* data-structures (trees, queues, lists) and involve pointer chasing. All the Not-First/Not-Last approaches are equally dependent on *linked* data-structures. The standard Energetic-Reasoning [4] is a strong filtering candidate which uses only array-like data structure. We decided to base our GPU-accelerated propagator on Energetic-Reasoning for both its GPU-friendly data structures, and because on typical instances and with "enough" GPU cores, it is possible to generate and check all the $O(n^2)$ intervals in parallel, reducing the running time from $O(n^3)$ to $O(n)$.

3.1 Parallelization

This section describes and motivates the developing of a parallel Energetic-Reasoning propagator. The first part introduces the notions of occupancy and latency, two fundamental concepts of GPU computing. The second part details how we parallelized the filtering algorithm, while the final part is about overhead reduction.

Performance Considerations. Propagators are called thousands of times and run for a few milliseconds. To derive a speedup, a GPU-accelerated propagator must *maximize occupancy* and *minimize latency.*

Occupancy refers to how many and how effectively GPU cores are used. A good algorithm uses fine-grain parallelism to engage many GPU cores, and relies on cache-friendly data structures to mitigate memory stalls. Latency refers to the time used to transfer data –and control– to the GPU as well as the time to retrieve results and recover control back to the CPU. Such operations are *very* expensive, so it is crucial to minimize both their duration and frequency.

Data Layout. It is convenient to specify how data is organized on the GPU. All the data are stored in the global memory using dynamically allocated and statically sized vectors. Such vectors are triples (p, s, c), where p is a *pointer* to the

```
propagationFailed = False
initStartingTimesFromDomains(S)
memcpyCpuToGpu([propagationFailed, S])                /* Asynchronous API */
calcIntervalsKernel(S, RI)                            /* Asynchronous launch */
updateBoundsKernel(S, RI, propagationFailed)          /* Asynchronous launch */
memcpyGpuToCpu([propagationFailed, S])                /* Asynchronous API */
waitGpu()
if ¬propagationFailed then
  | updateDomains(S)
else
  | fail()
```

Algorithm 2: Pseudocode of the parallel Energetic-Reasoning propagator.

allocated memory block, s is the current *size*, and c is the maximum *capacity*. This representation does not rely on links of pointers and can be allocated when the constraint is created. Specifically, four vectors are kept in the GPU memory: P containing the processing time of activities, H containing the resource usage of activities, RI[1] containing the relevant intervals (pairs of integers), and S containing pairs of integers representing the earlier/latest starting time of activities.

Parallel Algorithm. The parallelization of Algorithm 1 begins with the parallel computation of RI. A GPU kernel named *calcIntervalsKernel* calculates and merges the sets $O(i, j)$ of each $(i, j) \in A \times A$. Then, the outer loop is parallelized by a kernel named *updateBoundsKernel* that processes the intervals $[t_1, t_2) \in RI$. The resulting parallel propagator is listed in Algorithm 2 and available in the gpu branch of [41].

Let $\#SM$ be the number of Streaming Multiprocessors, and $\#CS$ be the number of CUDA Cores per Streaming Multiprocessors. We maximize the occupancy of *calcIntervalsKernel* by running it with $\#SM$ blocks, each of $\#CS$ threads so that each thread is responsible for about $\frac{|A \times A|}{\#SM \cdot \#CS}$ pairs of activities. In details, each thread generates some elements of RI in shared memory and stores them in RI, that is in global memory. To store the elements, each thread first reserves enough space in RI and then writes the elements. The reservation is done with a single atomic increment on the size of the vector. Such increment is the only point where threads might be serialized. The occupancy of *updateBoundsKernel* is maximized by launching it with $\#SM$ blocks, each of $\#CS$ threads so that each thread is responsible for about $\frac{|RI|}{\#SM \cdot \#CS}$ intervals (see Fig. 3). To retain correctness, each update of S must be *atomic*. Because of the massive number of threads concurrently accessing S, such atomic operations, if performed on global memory, would cause contention and slow down. Shared memory can be used instead by creating a copy S' of S for each block to reduce contention. Once all threads complete their computations, S is updated using S'. Naturally, updates

[1] From now on we will use RI to refer both to the set and to the relative vector.

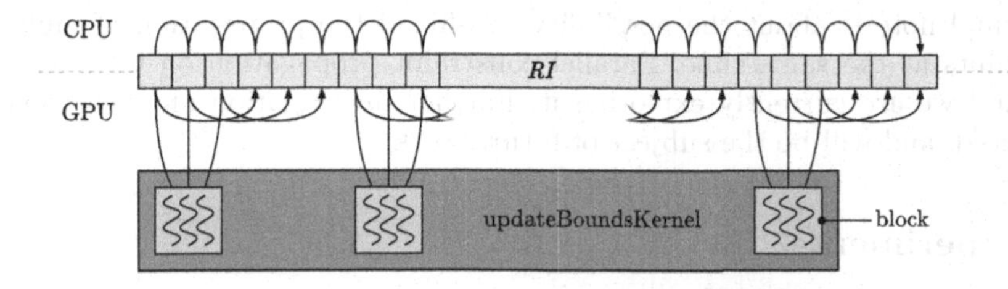

Fig. 3. Sequential (top) vs parallel (bottom) processing of RI.

of S's are still atomic, but since their scopes are single blocks, different blocks do not interfere.

Overhead Reduction. The first step to reduce the overhead is to minimize the volume of data transferred to/from the GPU. Since vectors P and H are constant, it suffices to copy them to the GPU when the constraint is posted. The only data that the host has to communicate to the GPU is the vector S, while it has to retrieve both the updated S and the Boolean *propagationFailed*. A possibility consists in using *CUDA Unified Memory* to exchange data between CPU and GPU. In this case, the CUDA runtime autonomously copies the data between host and device through a paging mechanism that, unfortunately, introduces a not negligible overhead. Hence, we packed S and *propagationFailed* into a structure and explicitly copy it to/from the GPU as a single block of data when needed. In Fig. 4 such transfers are represented in cyan and magenta.

Another source of overhead originates from CUDA asynchronous calls. The bottom part of Fig. 4 illustrates on a timeline the latency one experiences when multiple such calls occur. The alternative is to use *CUDA Graphs* to organize all kernel launches and memory operations in a dependency graph in such a way that they can be launched by means of a *single* API call.

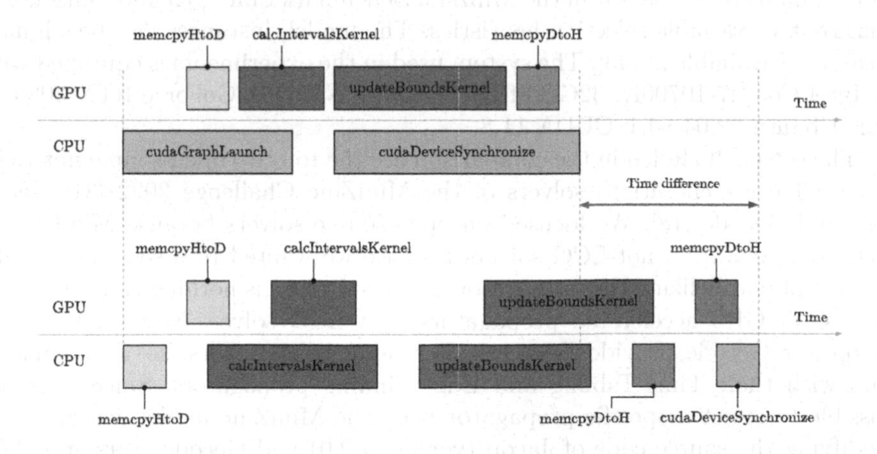

Fig. 4. Propagation with (top) and without (bottom) the use of CUDA Graph.

A final note is about the possibility to offload the propagation of multiple constraints at the same time. Parallel constraint propagation on GPU is possible and we are currently exploring it. Further investigation on such topic is warranted, and will be the subject of future work.

4 Experiments

This section presents the result of a comparison between the GPU-accelerated propagator, the CPU implementation and state-of-the-art solvers on different sets of instances from the literature. Moreover, it shows the benefits of the heuristics introduced in Sect. 2.2. We used the RCPSP as a benchmark. It is a generalization of CuPS with multiple resources and precedences between activities. In CP it is usually modeled with multiple *Cumulative* and linear constraints. Hence, it is particularly well-suited to evaluate the performance of a *Cumulative* propagator. The evaluation considered three established sets of instances, for a total of 299 instances:

PSPLib. Introduced in [25], it is the most popular benchmark for RCPSP. It contains synthetic instances of 30, 60, 90, and 120 activities. Instances are classified by their generation parameters, for a total of 204 classes, each of 10 instances. To have a reasonable benchmark time, we considered only the first instance of each class.

BL. Introduced in [5], it is part of a study about solving highly disjunctive and highly cumulative instances. It contains 40 highly cumulative synthetic instances, with 20 and 25 activities.

Pack. Introduced in [8], it is part of a study that uses sharp makespan's lower bounds to solve the RCPSP. It contains 55 highly cumulative synthetic instances, with 17 to 35 activities.

For a detailed description of the benchmarks, the reader can refer to [3]. Both the model and instances are from the MinZinc Benchmark Suite [32] and make use of `smallest` as variable selection heuristics. The model, instances and benchmark scripts are available at [43]. The system used in the experiments is equipped with an Intel Core i7-10700K, 32GB of RAM, and a NVIDIA GeForce RTX 3080. It runs Ubuntu 22.04 with CUDA 11.8.

The solvers included in the comparison are the top two open-source not-LCG (Lazy Clause Generation) solvers of the MiniZinc Challenge 2022 [31]: Jacop [21], and Gecode [18]. We focused on open-source solvers because MiniCPP is open-source, and on not-LCG solvers because we wanted to assess the specific benefits of parallelizing the propagator. However, there is nothing that precludes the use of GPU-accelerated propagators in a LCG solver. Note that neither Jacop nor Gecode provide Energetic-Reasoning propagators, so we compared ours with their Time-Tabling and Edge-Finding propagators. Since it is not possible to select a specific propagator from the MiniZinc model, we did it by modifying the source code of Jacop (version 4.9.0) and Gecode (version 6.3.0). We called such solvers `Jacop-TT`, `Jacop-EF`, `Gecode-TT`, and `Gecode-EF`, while

MiniCPP-ER and MiniCPP-ER-GPU stand for MiniCPP using the sequential and the GPU-accelerated Energetic-Reasoning propagators, respectively.

4.1 Results and Analysis

To give an effective and concise presentation, we focus on the instances for which a reasonable comparison of the solvers is possible. Namely, to be selected, an instance must satisfy one of the following criteria:

1. It has been solved by at least one solver, and at least half of the solvers had spent more than 10 s on the search. In this way, we rule out easier instances.
2. It was not solved by any solver and at least half of the solvers reported a solution after 10 s. This way we filter out instances for which there is not enough information on the progress of the search.

With 30 min timeout, these criteria select 31 instances in category 1 and 50 in category 2.

Category 1. The results of the instances in category 1 are illustrated in Fig. 5 and summarized in Table 2. The plots use logarithmic scale on the time axis, while each entry of the table is the sum of the corresponding statistic among all the instances. Overall, MiniCPP-ER-GPU results are compelling. It is the only approach that completed the search for all the instances and has the smallest total search time. The BL benchmark offers a direct comparison between MiniCPP-ER and MiniCPP-ER-GPU since both solved all its instances. In this case the GPU-accelerated solver is an order of magnitude faster. For the highly cumulative BL and Pack, Energetic-Reasoning leads to a smaller search tree that translates in a smaller search time only for MiniCPP-ER-GPU. For the PSPLib instances, Energetic-Reasoning leads to bigger search trees and Jacop-TT results the fastest solver. A similar outcome was observed in [9].

Table 1. Aggregate statistics for the proposed Energetic-Reasoning heuristics.

Benchmark	Solver	Optimal	Time (s)	Nodes (M)	Failures (M)	Depth
BL	MiniCPP-ER	9	3072	3.31	1.10	349
	MiniCPP-ER*	9	985	3.32	1.11	353
	MiniCPP-ER-GPU	9	228	3.31	1.10	349
	MiniCPP-ER*-GPU	9	237	3.32	1.11	353
Pack	MiniCPP-ER	3	2277	2.17	0.72	147
	MiniCPP-ER*	3	799	2.19	0.73	147
	MiniCPP-ER-GPU	3	148	2.17	0.72	147
	MiniCPP-ER*-GPU	3	159	2.19	0.73	147
PSPLib	MiniCPP-ER	6	2138	0.26	0.09	558
	MiniCPP-ER*	6	535	0.30	0.10	566
	MiniCPP-ER-GPU	6	28	0.26	0.09	558
	MiniCPP-ER*-GPU	6	30	0.30	0.10	566

Table 2. Aggregate statistics for the instances in category 1.

Benchmark	Solver	Optimal	Time (s)	Nodes (M)	Failures (M)	Depth
BL	`Gecode-TT`	15	8436	297.55	148.78	799
	`Gecode-EF`	0	30600	187.34	93.67	791
	`Jacop-TT`	13	11551	2307.47	1153.74	1123
	`Jacop-EF`	17	1022	22.76	11.38	787
	`MiniCPP-ER`	17	3163	3.46	1.15	605
	`MiniCPP-ER-GPU`	17	233	3.46	1.15	605
Pack	`Gecode-TT`	2	7812	222.34	111.17	308
	`Gecode-EF`	0	10800	40.62	20.31	272
	`Jacop-TT`	0	10782	2458.11	1229.05	814
	`Jacop-EF`	3	7219	177.93	88.97	463
	`MiniCPP-ER`	3	7674	5.37	1.79	286
	`MiniCPP-ER-GPU`	6	2509	23.76	7.92	307
PSPLib	`Gecode-TT`	4	8085	137.33	68.66	507
	`Gecode-EF`	0	14400	32.12	16.06	564
	`Jacop-TT`	8	404	34.01	17.00	502
	`Jacop-EF`	8	3506	33.40	16.70	496
	`MiniCPP-ER`	3	10049	11.97	3.99	485
	`MiniCPP-ER-GPU`	8	3142	43.88	14.63	492

Table 3. Aggregate statistics for the instances in category 2.

Benchmark	Solver	Solutions	AUC (K)	Nodes (M)	Failures (M)	Depth
Pack	`Gecode-TT`	41	16.99	493.47	246.74	520
	`Gecode-EF`	0	1157.40	60.17	30.09	458
	`Jacop-TT`	38	17.88	4112.02	2056.01	1646
	`Jacop-EF`	41	14.01	343.27	171.63	978
	`MiniCPP-ER`	28	45.66	14.22	4.74	669
	`MiniCPP-ER-GPU`	45	10.62	238.52	79.51	806
PSPLib	`Gecode-TT`	344	146.02	443.49	221.74	7613
	`Gecode-EF`	4	11176.49	25.44	12.72	9739
	`Jacop-TT`	412	22.70	4960.48	2480.23	6570
	`Jacop-EF`	382	84.48	861.81	430.90	5918
	`MiniCPP-ER`	312	213.83	7.15	2.38	4744
	`MiniCPP-ER-GPU`	389	76.45	354.61	118.20	5549

Category 2. The results of the instances in category 2 are reported in Table 3. The number of optimally solved instances is replaced with the number of solutions and the search time with the Area Under the Curve (AUC). There are no instances of BL for category 2. The numbers confirm what was observed for category 1: `MiniCPP-ER-GPU` is the best solver for the Pack benchmark, having the smaller AUC and the bigger number of solutions, while `Jacop-TT` is the best

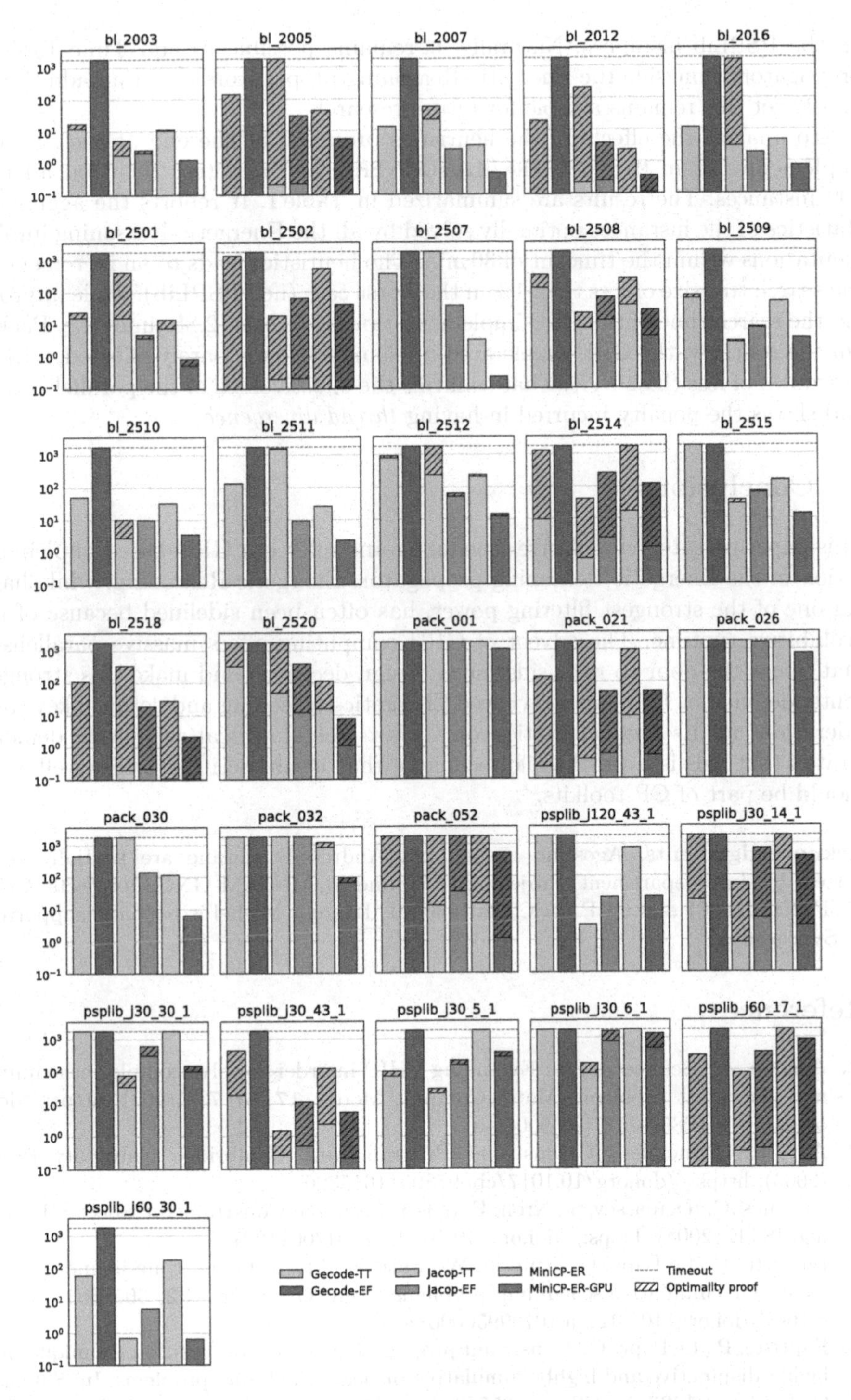

Fig. 5. Search time (in seconds) for all the instances in category 1.

on the PSPLib instances. Naturally, it remains possible to add Time-Tabling propagators alongside the Energetic-Reasoning propagators to get an additional boost, yet this remains a topic for future research.

To analyze the effects of the heuristics proposed at the end of Sect. 2.2 we implemented it in MiniCPP-ER*/MiniCPP-ER*-GPU, and test them on all the 299 instances. The results are summarized in Table 1. It reports the aggregate statistics of the instances optimally solved by all the Energetic-Reasoning implementations within the timeout of 30 min. The heuristics leads to an increment of the search tree size of less than 1% in the worst case (i.e., PSPLib), while improving the search time of the CPU implementation by at least 2.85 times (i.e. Pack). On the contrary, the GPU-accelerated implementation is barely affected, with a slowdown of less than 0.1%. That confirms the effectiveness of the parallelization and shows the penalty incurred in having *thread divergence*.

5 Conclusions

This paper revisited cumulative scheduling and offered a GPU-based implementation of the Energetic-Reasoning propagator. Energetic-Reasoning, while having one of the strongest filtering power, has often been sidelined because of its prohibitive runtime. The advent of GPU computing offers massive parallelism that opens the door to reconsider such design decisions and make this stronger contender viable. This paper reviewed Energetic-Reasoning and detailed key considerations for its implementation on GPUs. The empirical evaluation demonstrated that this is a worthwhile technique that is competitive, scales well and, should be part of CP toolkits.

Acknowledgements. Agostino Dovier and Andrea Formisano are partially supported by Interdepartment Project on AI and by INdAM-GNCS projects CUP E55F22000270001 and CUP E53C22001930001. Laurent Michel is partially supported by Synchrony.

References

1. Aggoun, A., Beldiceanu, N.: Extending CHIP in order to solve complex scheduling and placement problems. Math. Comput. Model. **17**, 57–73 (1993). https://doi.org/10.1016/0895-7177(93)90068-a
2. Apt, K.: Principles of Constraint Programming. Cambridge University Press (2003). https://doi.org/10.1017/cbo9780511615320
3. Artigues, C., Demassey, S., Nron, E. (eds.): Resource-Constrained Project Scheduling. ISTE (2008). https://doi.org/10.1002/9780470611227
4. Baptiste, P., Le Pape, C., Nuijten, W.: Satisfiability tests and time-bound adjustments for cumulative scheduling problems. Ann. Oper. Res. **92**, 305–333 (1999). https://doi.org/10.1023/a:1018995000688
5. Baptiste, P., Le Pape, C.: Constraint propagation and decomposition techniques for highly disjunctive and highly cumulative project scheduling problems. In: Smolka, G. (ed.) CP 1997. LNCS, pp. 375–389. Springer, Heidelberg (1997). https://doi.org/10.1007/bfb0017454

6. Baptiste, P., Le Pape, C., Nuijten, W.: Constraint-Based Scheduling. Springer, Heidelberg (2001). https://doi.org/10.1007/978-1-4615-1479-4
7. Beldiceanu, N., Carlsson, M.: A new multi-resource *cumulatives* constraint with negative heights. In: Van Hentenryck, P. (ed.) CP 2002. LNCS, vol. 2470, pp. 63–79. Springer, Heidelberg (2002). https://doi.org/10.1007/3-540-46135-3_5
8. Carlier, J., Néron, E.: On linear lower bounds for the resource constrained project scheduling problem. Eur. J. Oper. Res. **149**(2), 314–324 (2003). https://doi.org/10.1016/s0377-2217(02)00763-4
9. Van Cauwelaert, S., Lombardi, M., Schaus, P.: Understanding the potential of propagators. In: Michel, L. (ed.) CPAIOR 2015. LNCS, vol. 9075, pp. 427–436. Springer, Cham (2015). https://doi.org/10.1007/978-3-319-18008-3_29
10. Cheng, J., Grossman, M., McKercher, T.: Professional CUDA C Programming. EBL-Schweitzer. Wiley (2014). https://www.wiley.com/en-us/Professional+CUDA+C+Programming-p-9781118739310
11. Collevati, M., Dovier, A., Formisano, A.: GPU parallelism for SAT solving heuristics. In: Calegari, R., Ciatto, G., Omicini, A. (eds.) Proceedings of the CILC 2022. CEUR Workshop Proceedings, vol. 3204, pp. 17–31. CEUR-WS.org (2022)
12. Dal Palù, A., Dovier, A., Formisano, A., Pontelli, E.: CUD@SAT: SAT solving on GPUs. J. Exp. Theor. Artif. Intell. **27**(3), 293–316 (2015). https://doi.org/10.1080/0952813X.2014.954274
13. Derrien, A., Petit, T.: A new characterization of relevant intervals for energetic reasoning. In: O'Sullivan, B. (ed.) CP 2014. LNCS, vol. 8656, pp. 289–297. Springer, Cham (2014). https://doi.org/10.1007/978-3-319-10428-7_22
14. Dovier, A., Formisano, A., Pontelli, E.: Parallel answer set programming. In: Hamadi, Y., Sais, L. (eds.) Handbook of Parallel Constraint Reasoning, pp. 237–282. Springer, Cham (2018). https://doi.org/10.1007/978-3-319-63516-3_7
15. Dovier, A., Formisano, A., Vella, F.: GPU-based parallelism for ASP-solving. In: Hofstedt, P., Abreu, S., John, U., Kuchen, H., Seipel, D. (eds.) INAP/WLP/WFLP-2019. LNCS (LNAI), vol. 12057, pp. 3–23. Springer, Cham (2020). https://doi.org/10.1007/978-3-030-46714-2_1
16. Garey, M.R., Johnson, D.S.: Computers and Intractability: A Guide to the Theory of NP-Completeness. W. H. Freeman & Co., USA (1990)
17. Gay, S., Hartert, R., Schaus, P.: Simple and scalable time-table filtering for the cumulative constraint. In: Pesant, G. (ed.) CP 2015. LNCS, vol. 9255, pp. 149–157. Springer, Cham (2015). https://doi.org/10.1007/978-3-319-23219-5_11
18. Gecode Team: GECODE. https://github.com/Gecode/gecode
19. Gentzel, R., Michel, L., van Hoeve, W.-J.: HADDOCK: a language and architecture for decision diagram compilation. In: Simonis, H. (ed.) CP 2020. LNCS, vol. 12333, pp. 531–547. Springer, Cham (2020). https://doi.org/10.1007/978-3-030-58475-7_31
20. Gingras, V., Quimper, C.G.: Generalizing the edge-finder rule for the cumulative constraint. In: Kambhampati, S. (ed.) Proceedings IJCAI 2016, pp. 3103–3109. IJCAI/AAAI Press (2016)
21. JaCoP Team: JaCoP. https://github.com/radsz/jacop
22. Kameugne, R., Betmbe Fetgo, S., Gingras, V., Ouellet, Y., Quimper, C.-G.: Horizontally elastic not-first/not-last filtering algorithm for cumulative resource constraint. In: van Hoeve, W.-J. (ed.) CPAIOR 2018. LNCS, vol. 10848, pp. 316–332. Springer, Cham (2018). https://doi.org/10.1007/978-3-319-93031-2_23
23. Kameugne, R., Fotso, L.P.: A cumulative not-first/not-last filtering algorithm in $\mathcal{O}(n^2 \log n)$. Indian J. Pure Appl. Math. **44**(1), 95–115 (2013). https://doi.org/10.1007/s13226-013-0005-z

24. Kameugne, R., Fotso, L.P., Scott, J., Ngo-Kateu, Y.: A quadratic edge-finding filtering algorithm for cumulative resource constraints. In: Lee, J. (ed.) CP 2011. LNCS, vol. 6876, pp. 478–492. Springer, Heidelberg (2011). https://doi.org/10.1007/978-3-642-23786-7_37
25. Kolisch, R., Sprecher, A.: PSPLIB - a project scheduling problem library. Eur. J. Oper. Res. **96**(1), 205–216 (1997). https://doi.org/10.1016/s0377-2217(96)00170-1
26. Lahrichi, A.: Scheduling: the notions of hump, compulsory parts and their use in cumulative problems. Comptes Rendus De L Academie Des Sciences Serie I-mathematique **294**(6), 209–211 (1982)
27. Letort, A., Beldiceanu, N., Carlsson, M.: A scalable sweep algorithm for the *cumulative* constraint. In: Milano, M. (ed.) CP 2012. LNCS, pp. 439–454. Springer, Heidelberg (2012). https://doi.org/10.1007/978-3-642-33558-7_33
28. Lopez, P.: Energy-based approach for task scheduling under time and resource constraints. Ph.D. thesis, Université Paul Sabatier-Toulouse III (1991)
29. Mercier, L., Van Hentenryck, P.: Edge finding for cumulative scheduling. INFORMS J. Comput. **20**(1), 143–153 (2008). https://doi.org/10.1287/ijoc.1070.0226
30. Michel, L., Schaus, P., Van Hentenryck, P.: MiniCP: a lightweight solver for constraint programming. Math. Program. Comput. **13**(1), 133–184 (2021). https://doi.org/10.1007/s12532-020-00190-7
31. MiniZinc Team: MiniZinc Challenge 2022 Results. https://www.minizinc.org/challenge2022/results2022.html
32. MiniZinc Team: The MiniZinc Benchmark Suite. https://github.com/MiniZinc/minizinc-benchmarks
33. Nuijten, W.: Time and resource constrained scheduling: a constraint satisfaction approach. Ph.D. thesis, Eindhoven University of Technology (1994)
34. Nvidia Team: CUDA. https://developer.nvidia.com/cuda-toolkit
35. Ouellet, P., Quimper, C.-G.: Time-table extended-edge-finding for the cumulative constraint. In: Schulte, C. (ed.) CP 2013. LNCS, vol. 8124, pp. 562–577. Springer, Heidelberg (2013). https://doi.org/10.1007/978-3-642-40627-0_42
36. Ouellet, Y., Quimper, C.-G.: A $O(n \log^2 n)$ checker and $O(n^2 \log n)$ filtering algorithm for the energetic reasoning. In: van Hoeve, W.-J. (ed.) CPAIOR 2018. LNCS, vol. 10848, pp. 477–494. Springer, Cham (2018). https://doi.org/10.1007/978-3-319-93031-2_34
37. Rossi, F., van Beek, P., Walsh, T. (eds.): Handbook of Constraint Programming, Foundations of Artificial Intelligence, vol. 2. Elsevier (2006). https://www.sciencedirect.com/science/bookseries/15746526/2
38. Sanders, J., Kandrot, E.: CUDA by Example: An Introduction to General-Purpose GPU Programming. Pearson Education (2010). https://developer.nvidia.com/cuda-example
39. Schutt, A., Wolf, A.: A new $\mathcal{O}(n^2 \log n)$ not-first/not-last pruning algorithm for cumulative resource constraints. In: Cohen, D. (ed.) CP 2010. LNCS, vol. 6308, pp. 445–459. Springer, Heidelberg (2010). https://doi.org/10.1007/978-3-642-15396-9_36
40. Schutt, A., Wolf, A., Schrader, G.: Not-first and not-last detection for cumulative scheduling in $\mathcal{O}(n^3 \log n)$. In: Umeda, M., Wolf, A., Bartenstein, O., Geske, U., Seipel, D., Takata, O. (eds.) INAP 2005. LNCS (LNAI), vol. 4369, pp. 66–80. Springer, Heidelberg (2006). https://doi.org/10.1007/11963578_6
41. Tardivo, F.: Fzn-minicpp. https://bitbucket.org/constraint-programming/fzn-minicpp

42. Tardivo, F.: Libfzn. https://bitbucket.org/constraint-programming/libfzn
43. Tardivo, F.: MiniCPP-Benchmarks. https://bitbucket.org/constraint-programming/minicpp-benchmarks
44. Tardivo, F., Dovier, A., Formisano, A., Michel, L., Pontelli, E.: Constraints propagation on GPU: a case study for AllDifferent. In: Calegari, R., Ciatto, G., Omicini, A. (eds.) Proceedings of CILC 2022. CEUR Workshop Proceedings, vol. 3204, pp. 61–74. CEUR-WS.org (2022)
45. Tesch, A.: A nearly exact propagation algorithm for energetic reasoning in $\mathcal{O}(n^2 \log n)$. In: Rueher, M. (ed.) CP 2016. LNCS, vol. 9892, pp. 493–519. Springer, Cham (2016). https://doi.org/10.1007/978-3-319-44953-1_32
46. Tesch, A.: Improving energetic propagations for cumulative scheduling. In: Hooker, J. (ed.) CP 2018. LNCS, vol. 11008, pp. 629–645. Springer, Cham (2018). https://doi.org/10.1007/978-3-319-98334-9_41
47. Vilím, P.: Edge finding filtering algorithm for discrete cumulative resources in $\mathcal{O}(kn \log n)$. In: Gent, I.P. (ed.) CP 2009. LNCS, vol. 5732, pp. 802–816. Springer, Heidelberg (2009). https://doi.org/10.1007/978-3-642-04244-7_62
48. Vilím, P.: Timetable edge finding filtering algorithm for discrete cumulative resources. In: Achterberg, T., Beck, J.C. (eds.) CPAIOR 2011. LNCS, vol. 6697, pp. 230–245. Springer, Heidelberg (2011). https://doi.org/10.1007/978-3-642-21311-3_22

Constraint Programming for the Robust Two-Machine Flow-Shop Scheduling Problem with Budgeted Uncertainty

Carla Juvin[1(✉)], Laurent Houssin[1,2], and Pierre Lopez[1]

[1] LAAS-CNRS, Université de Toulouse, CNRS, Toulouse, France
{carla.juvin,laurent.houssin,pierre.lopez}@laas.fr
[2] ISAE-SUPAERO, Toulouse, France

Abstract. This paper addresses the robust two-machine permutation flow-shop scheduling problem considering non-deterministic operation processing times associated with an uncertainty budget. The objective is to minimize the makespan of the schedule.

Exact solution methods incorporated within the framework of a two-stage robust optimization are proposed to solve the problem. We first prove that under particular conditions the robust two-machine permutation flow-shop scheduling problem can be solved in polynomial time by the well-known Johnson's algorithm usually dedicated to the deterministic version. Then we tackle the general problem, for which we propose a column and constraint generation algorithm. We compare two versions of the algorithm. In the first version, a mixed-integer linear programming formulation is used for the master problem. In the second version, we use a constraint programming model for the master problem. To the best of our knowledge, the use of constraint programming for a master problem in a two-stage robust optimization problem is innovative.

The experimental results show the very good performance of the method based on the constraint programming formulation. We also notice that Johnson's algorithm is surprisingly efficient for the robust version of the general problem.

Keywords: Flow-Shop Scheduling · Robust Optimization · Uncertainty Budget · Mixed-Integer Linear Programming · Constraint Programming

1 Introduction

The permutation flow-shop scheduling is a well-studied problem where a set of jobs are to be processed on a set of machines. Each job must be processed on every machine, with given processing times, following the same given order of machines. Each machine can process only one job at a time and the processing sequence of jobs is the same on each machine. This problem is strongly NP-hard for three or more machines. However, the two-machine flow-shop scheduling

© The Author(s), under exclusive license to Springer Nature Switzerland AG 2023
A. A. Cire (Ed.): CPAIOR 2023, LNCS 13884, pp. 354–369, 2023.
https://doi.org/10.1007/978-3-031-33271-5_23

problem is solved by the well-known Johnson's algorithm [5] for the makespan minimization when operation processing times are assumed deterministic. In the real world, many sources of uncertainty (processing times variation, machine breakdown, addition of new operations, etc.) can affect the quality and even the feasibility of a schedule.

There exist two major approaches to deal with data uncertainty: stochastic optimization and robust optimization. While stochastic optimization considers probability distribution, robust optimization assumes that uncertain data belong to a given uncertainty set and aims to optimize performance considering the worst-case scenario within that set. The traditional robust optimization approach [8] consists in protecting against the case when all parameters can deviate at the same time, which makes the solution overly conservative. Indeed, there is a very low probability that all parameters take their worst value all together. To overcome this limitation, Bertsimas & Sim introduce an uncertainty budget approach that allows a restriction on the number of deviations that can occur simultaneously to a given budget [2]. In order to reach a trade-off between robustness and solution quality, we exploit this approach to define the uncertainty set. Multi-stage robust optimization has been introduced by Ben-Tal et al. [1]. In some optimization problems, only a part of the decision variables have to be determined before uncertainty is revealed, while the other variables can be chosen after the realization of the uncertainty and can thus be adjusted to the scenario. The authors introduce the *adjustable robust counterpart* where the set of decision variables is split into "here and now" decisions and "wait and see" decisions. Thus, the objective is to find a solution for the "here and now" decision variables such that there always exists "wait and see" variables meeting the constraints for all values of the uncertain parameters, and minimizing the objective value.

The purpose is to find the sequence on the machines (first stage decision), allowing to define a start time for each operation and each scenario (second stage decision), minimizing the makespan in the worst-case scenario considering the budget of uncertainty.

In this paper, we propose exact solution methods to solve the robust two-machine flow-shop scheduling problem. We first study a particular case of the problem. Next, we provide a robust counterpart model based on constraint programming formulation, that is embedded in a column and constraint generation framework. A discussion is conducted on the basis of an analysis of experimental results.

2 Problem Statement

An instance of the two-machine flow-shop scheduling problem implies a set of jobs $\mathcal{J}$ and two machines $\mathcal{M} = \{M_1, M_2\}$. Each job $i \in \mathcal{J}$ consists of two operations $O_{i,1}$ and $O_{i,2}$. The first one, $O_{i,1}$, must be processed by machine M_1 with a duration of $p_{i,1}$ and then, $O_{i,2}$ must be processed by machine M_2 with a duration of $p_{i,2}$. Each machine can process only one job at a time and each job can only be processed on one machine at a time. The objective is to find a permutation of jobs, denoted σ, minimizing the makespan.

When processing times are deterministically known, the problem can be solved in polynomial time by means of Johnson's rule [5], which states that job i must be processed before job j if $\min(p_{i,1}, p_{j,2}) < \min(p_{j,1}, p_{i,2})$.

2.1 Processing Times Uncertainty

Here, we consider that the processing times of operations are uncertain. Each processing time $p_{i,m}$ of an operation $O_{i,m}$, $i \in \mathcal{J}, m \in \{M_1, M_2\}$, belongs to the interval $[\bar{p}_{i,m}, \bar{p}_{i,m} + \hat{p}_{i,m}]$, where $\bar{p}_{i,m}$ is the nominal value and $\hat{p}_{i,m}$ the maximum deviation of the processing time from its nominal value.

Let Γ be the budget of uncertainty, that is the maximum number of operations whose processing time deviation can occur simultaneously. For each scenario ξ in the set of feasible scenarios $\mathcal{U}^{\Gamma}$, the processing time of operation $O_{i,m}$ is then given by:

$$p_{i,m}(\xi) = \bar{p}_{i,m} + \xi_{i,m} \cdot \hat{p}_{i,m}$$

where $\xi_{i,m}$ is equal to 1 if the processing time of the operation deviates, 0 otherwise.

In this study we consider two types of uncertainty budget:

1. A global budget Γ which denotes the number of operations that can deviate on both machines combined. In this case, the set of feasible scenarios is expressed as:

$$\mathcal{U}^{\Gamma} = \left\{ (\xi_{i,m})_{i \in \mathcal{J}, m \in \{M_1, M_2\}} \mid \sum_{i \in \mathcal{J}} \sum_{m=1}^{2} \xi_{i,m} \leq \Gamma,\ \xi_{i,m} \in \{0,1\} \right\}$$

2. A machine-dependent budget $\Gamma = (\Gamma_1, \Gamma_2)$ where Γ_1 and Γ_2 denote the number of operations whose processing time deviation can occur simultaneously on machines M_1 and M_2, respectively. In this case, the set of feasible scenarios is expressed as:

$$\mathcal{U}^{\Gamma} = \left\{ (\xi_{i,m})_{i \in \mathcal{J}, m \in \{M_1, M_2\}} \mid \sum_{i \in \mathcal{J}} \xi_{i,1} \leq \Gamma_1, \sum_{i \in \mathcal{J}} \xi_{i,2} \leq \Gamma_2,\ \xi_{i,m} \in \{0,1\} \right\}$$

Notations and Definitions

- $\mathcal{J}$: Set of jobs
- $\mathcal{M}$: Set of machines
- $O_{i,m}$: m^{th} operation of job i ($i \in \mathcal{J}$) to be executed on machine $m \in \mathcal{M}$
- σ : sequence of jobs
- $\sigma[i]$: i^{th} job of sequence σ
- Γ : uncertainty budget
- Γ_m : uncertainty budget on machine $m \in \mathcal{M}$
- $\mathcal{U}^{\Gamma}$: uncertainty set for a given uncertainty budget Γ ξ: scenario
- $C_{i,m}^{\gamma}(\sigma)$: maximum completion time of the ith job of sequence σ, on machine $m \in \mathcal{M}$, considering at most γ deviations

2.2 Worst-Case Evaluation

For a given sequence of jobs σ, there exists a worst-case scenario, maximizing the value of the makespan. Levorato et al. [6] developed a polynomial-time ($\mathcal{O}(n^2)$) worst-case determination procedure, using dynamic programming, for machine-dependent budget $\Gamma = (\Gamma_1, \Gamma_2)$. The same idea is now used when considering a global budget Γ.

In the deterministic case, the completion time $C_{i,1}(\sigma)$ of job $\sigma[i]$ on machine M_1 is the completion time $C_{i-1,1}(\sigma)$ of job $\sigma[i-1]$ plus the processing time $p_{\sigma[i],1}$ of job $\sigma[i]$ on machine M_1, i.e.:

$$C_{i,1}(\sigma) = C_{i-1,1}(\sigma) + p_{\sigma[i],1} \tag{1}$$

and the completion time $C_{i,2}(\sigma)$ of job $\sigma[i]$ on machine M_2 is the maximum between the completion time $C_{i-1,2}(\sigma)$ of job $\sigma[i-1]$ plus the processing time $p_{\sigma[i],2}$ of job $\sigma[i]$ on machine M_2 and the completion time $C_{i,1}(\sigma)$ of job $\sigma[i]$ on machine M_1 plus the processing time $p_{\sigma[i],2}$ of job $\sigma[i]$ on machine M_2 (see Fig. 1):

$$C_{i,2}(\sigma) = \max(C_{i-1,2}(\sigma), C_{i,1}(\sigma)) + p_{\sigma[i],2} \tag{2}$$

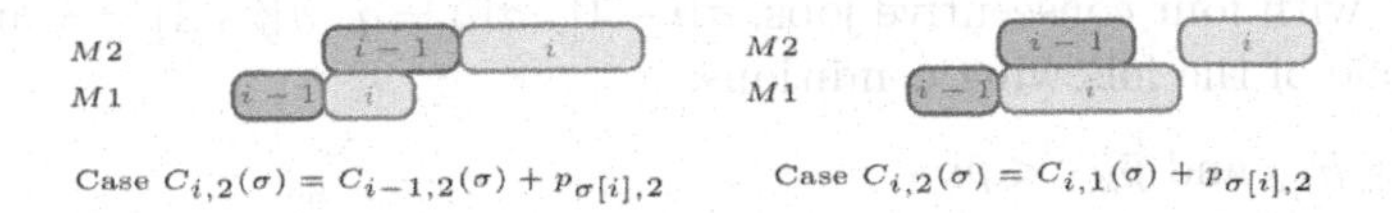

Fig. 1. Completion time on machine M_2

Given a machine $m \in \mathcal{M}$, and two integer numbers $i \in [1, |\mathcal{J}|]$ and $\gamma \in [0, \Gamma]$, let $C^{\gamma}_{i,m}(\sigma)$ be the maximum completion time of the ith job of sequence σ, on machine m, considering at most γ deviations. This value is defined by the following recurrence relations:

$$C^{\gamma}_{i,1}(\sigma) = \max(C^{\gamma}_{i-1,1}(\sigma) + \bar{p}_{\sigma[i],1},\ C^{\gamma-1}_{i-1,1}(\sigma) + \bar{p}_{\sigma[i],1} + \hat{p}_{\sigma[i],1}) \tag{3}$$

$$\begin{aligned} C^{\gamma}_{i,2}(\sigma) = \max(&C^{\gamma}_{i-1,2}(\sigma) + \bar{p}_{\sigma[i],2},\ C^{\gamma}_{i,1}(\sigma) + \bar{p}_{\sigma[i],2},\ C^{\gamma-1}_{i-1,2}(\sigma) \\ &+ \bar{p}_{\sigma[i],2} + \hat{p}_{\sigma[i],2},\ C^{\gamma-1}_{i,1}(\sigma) + \bar{p}_{\sigma[i],2} + \hat{p}_{\sigma[i],2}) \end{aligned} \tag{4}$$

with $C^{\gamma}_{i,m}(\sigma) = -\infty$ if $\gamma < 0$, $C^{\gamma}_{0,m}(\sigma) = 0$ if $\gamma \geq 0$ and $C^{0}_{i,2}(\sigma) = \bar{p}_{\sigma[i],2} + \max(C^{0}_{i-1,2}(\sigma),\ C^{0}_{i,1}(\sigma))$.

The worst-case makespan, under sequence σ, for a global uncertainty budget Γ, is given by $C^{\Gamma}_{|\mathcal{J}|,2}(\sigma)$.

3 Special Cases

In this section, we focus on the complexity of the robust two-machine permutation flow-shop scheduling problem with an uncertainty budget. In particular, we study the effect of two parameters, namely:

- the type of uncertainty budget, global budget or machine-dependent budget, as defined in Sect. 2.1;
- the type of processing time deviation,
  - with preserved order through deviation, i.e., $\forall(i,i') \in \mathcal{J}^2$, $\forall(m,m') \in \mathcal{M}^2$, $\bar{p}_{i,m} < \bar{p}_{i',m'} \Leftrightarrow \bar{p}_{i,m} + \hat{p}_{i,m} < \bar{p}_{i',m'} + \hat{p}_{i',m'}$, or
  - with unpreserved order, i.e., $\exists(i,i') \in \mathcal{J}^2$, $\exists(m,m') \in \mathcal{M}^2$, such that $\bar{p}_{i,m} < \bar{p}_{i',m'}$ and $\bar{p}_{i,m} + \hat{p}_{i,m} > \bar{p}_{i',m'} + \hat{p}_{i',m'}$.

3.1 Global Budget and Preserved Order of Processing Times

Proposition 1. *If the order of processing times is preserved through deviation, then a schedule following the Johnson's rule is optimal for any global uncertainty budget Γ.*

Proof. Suppose that σ is an optimal sequence for a given global uncertainty budget Γ, with four consecutive jobs, $\sigma[i-1]$, $\sigma[i] = j$, $\sigma[i+1] = k$ and $\sigma[i+2]$ meeting one of the following conditions:

(i) $\bar{p}_{j,1} > \bar{p}_{j,2}$ and $\bar{p}_{k,1} < \bar{p}_{k,2}$
(ii) $\bar{p}_{j,1} < \bar{p}_{j,2}$, $\bar{p}_{k,1} < \bar{p}_{k,2}$ and $\bar{p}_{j,1} > \bar{p}_{k,1}$
(iii) $\bar{p}_{j,1} > \bar{p}_{j,2}$, $\bar{p}_{k,1} > \bar{p}_{k,2}$ and $\bar{p}_{j,2} < \bar{p}_{k,2}$

That is, sequence σ such that j is before k does not respect Johnson's rule ($\min(p_{k,1}, p_{j,2}) < \min(p_{j,1}, p_{k,2}) \implies k$ before j).

Note σ' the sequence obtained from σ by pairwise interchanging jobs j and k. It suffices to show that under any of the three conditions the makespan of the schedule under σ' is not greater than under σ.

The maximum start time of job $\sigma[i+2] = \sigma'[i+2]$, on machine M_1, considering at most γ deviations, is the same under σ and σ', namely, the completion time of job $\sigma[i-1]$ plus the processing times of jobs j and k, on machine M_1, considering at most γ deviations, i.e., $C^{\gamma}_{i+1,1}(\sigma) = C^{\gamma}_{i+1,1}(\sigma') = \max(a_1, a_2, a_3)$ with:

$$a_1 = C^{\gamma}_{i-1,1}(\sigma) + \bar{p}_{j,1} + \bar{p}_{k,1}$$
$$a_2 = C^{\gamma-1}_{i-1,1}(\sigma) + \bar{p}_{j,1} + \bar{p}_{k,1} + \max(\hat{p}_{j,1}, \hat{p}_{k,1}) \text{ if } \gamma \geq 1, 0 \text{ otherwise}$$
$$a_3 = C^{\gamma-1}_{i-1,1}(\sigma) + \bar{p}_{j,1} + \hat{p}_{j,1} + \bar{p}_{k,1} + \hat{p}_{k,1} \text{ if } \gamma \geq 2, 0 \text{ otherwise}$$

Thus, the rest of the schedule on machine M_1 is not affected by the pairwise interchange. We now study the effect of the change on the schedule on machine M_2, in particular, the date of availability of machine M_2 to process job $\sigma[i+2]$, i.e., $C^{\gamma}_{i+1,2}(\sigma)$ under the original schedule and $C^{\gamma}_{i+1,2}(\sigma')$ after the interchange.

The worst-case completion time of job $\sigma[i+1] = k$ under original schedule σ, on machine M_2, depending on the operation in which the deviations occur, is $C^{\gamma}_{i+1,2}(\sigma) = \max(b_1, b_2, b_3, b_4, b_5, b_6, b_7, b_8, b_9, b_{10}, b_{11})$ with:

$b_1 = C^{\gamma}_{i-1,2} + \bar{p}_{j,2} + \bar{p}_{k,2}$
$b_2 = C^{\gamma-1}_{i-1,2} + \bar{p}_{j,2} + \bar{p}_{k,2} + \max(\hat{p}_{j,2}, \hat{p}_{k,2})$ if $\gamma \geq 1$, 0 otherwise
$b_3 = C^{\gamma-2}_{i-1,2} + \bar{p}_{j,2} + \hat{p}_{j,2} + \bar{p}_{k,2} + \hat{p}_{k,2}$ if $\gamma \geq 2$, 0 otherwise
$b_4 = C^{\gamma}_{i-1,1} + \bar{p}_{j,1} + \bar{p}_{j,2} + \bar{p}_{k,2}$
$b_5 = C^{\gamma-1}_{i-1,1} + \bar{p}_{j,1} + \bar{p}_{j,2} + \bar{p}_{k,2} + \max(\hat{p}_{j,1}, \hat{p}_{j,2}, \hat{p}_{k,2})$ if $\gamma \geq 1$, 0 otherwise
$b_6 = C^{\gamma-2}_{i-1,1} + \bar{p}_{j,1} + \bar{p}_{j,2} + \bar{p}_{k,2} + \max(\hat{p}_{j,1} + \hat{p}_{j,2}, \hat{p}_{j,1} + \hat{p}_{k,2}, \hat{p}_{j,2} + \hat{p}_{k,2})$ if $\gamma \geq 2$, 0 otherwise
$b_7 = C^{\gamma-3}_{i-1,1} + \bar{p}_{j,1} + \hat{p}_{j,1} + \bar{p}_{j,2} + \hat{p}_{j,2} + \bar{p}_{k,2} + \hat{p}_{k,2}$ if $\gamma \geq 3$, 0 otherwise
$b_8 = C^{\gamma}_{i-1,1} + \bar{p}_{j,1} + \bar{p}_{k,1} + \bar{p}_{k,2}$
$b_9 = C^{\gamma-1}_{i-1,1} + \bar{p}_{j,1} + \bar{p}_{k,1} + \bar{p}_{k,2} + \max(\hat{p}_{j,1}, \hat{p}_{k,1}, \hat{p}_{k,2})$ if $\gamma \geq 1$, 0 otherwise
$b_{10} = C^{\gamma-2}_{i-1,1} + \bar{p}_{j,1} + \bar{p}_{k,1} + \bar{p}_{k,2} + \max(\hat{p}_{j,1} + \hat{p}_{k,1}, \hat{p}_{j,1} + \hat{p}_{k,2}, \hat{p}_{k,1} + \hat{p}_{k,2})$ if $\gamma \geq 2$, 0 otherwise
$b_{11} = C^{\gamma-3}_{i-1,1} + \bar{p}_{j,1} + \hat{p}_{j,1} + \bar{p}_{k,1} + \hat{p}_{k,1} + \bar{p}_{k,2} + \hat{p}_{k,2}$ if $\gamma \geq 3$, 0 otherwise

while the completion time of job $\sigma'[i+1] = j$ on machine M_2 under sequence σ' is $C^{\gamma}_{i+1,2}(\sigma') = \max(c_1, c_2, c_3, c_4, c_5, c_6, c_7, c_8, c_9, c_{10}, c_{11})$, with:

$c_1 = C^{\gamma}_{i-1,2} + \bar{p}_{k,2} + \bar{p}_{j,2}$
$c_2 = C^{\gamma-1}_{i-1,2} + \bar{p}_{k,2} + \bar{p}_{j,2} + \max(\hat{p}_{k,2}, \hat{p}_{j,2})$ si $\gamma \geq 1$, 0 otherwise
$c_3 = C^{\gamma-2}_{i-1,2} + \bar{p}_{k,2} + \hat{p}_{k,2} + \bar{p}_{j,2} + \hat{p}_{j,2}$ if $\gamma \geq 2$, 0 otherwise
$c_4 = C^{\gamma}_{i-1,1} + \bar{p}_{k,1} + \bar{p}_{k,2} + \bar{p}_{j,2}$
$c_5 = C^{\gamma-1}_{i-1,1} + \bar{p}_{k,1} + \bar{p}_{k,2} + \bar{p}_{j,2} + \max(\hat{p}_{k,1}, \hat{p}_{k,2}, \hat{p}_{j,2})$ si $\gamma \geq 1$, 0 otherwise
$c_6 = C^{\gamma-2}_{i-1,1} + \bar{p}_{k,1} + \bar{p}_{k,2} + \bar{p}_{j,2} + \max(\hat{p}_{k,1} + \hat{p}_{k,2}, \hat{p}_{k,1} + \hat{p}_{j,2}, \hat{p}_{k,2} + \hat{p}_{j,2})$ if $\gamma \geq 2$, 0 otherwise
$c_7 = C^{\gamma-3}_{i-1,1} + \bar{p}_{k,1} + \hat{p}_{k,1} + \bar{p}_{k,2} + \hat{p}_{k,2} + \bar{p}_{j,2} + \hat{p}_{j,2}$ if $\gamma \geq 3$, 0 otherwise
$c_8 = C^{\gamma}_{i-1,1} + \bar{p}_{k,1} + \bar{p}_{j,1} + \bar{p}_{j,2}$
$c_9 = C^{\gamma-1}_{i-1,1} + \bar{p}_{k,1} + \bar{p}_{j,1} + \bar{p}_{j,2} + \max(\hat{p}_{k,1}, \hat{p}_{j,1}, \hat{p}_{j,2})$ if $\gamma \geq 1$, 0 otherwise
$c_{10} = C^{\gamma-2}_{i-1,1} + \bar{p}_{k,1} + \bar{p}_{j,1} + \bar{p}_{j,2} + \max(\hat{p}_{k,1} + \hat{p}_{j,1}, \hat{p}_{k,1} + \hat{p}_{j,2}, \hat{p}_{j,1} + \hat{p}_{j,2})$ if $\gamma \geq 2$, 0 otherwise
$c_{11} = C^{\gamma-3}_{i-1,1} + \bar{p}_{k,1} + \hat{p}_{k,1} + \bar{p}_{j,1} + \hat{p}_{j,1} + \bar{p}_{j,2} + \hat{p}_{j,2}$ if $\gamma \geq 3$, 0 otherwise

It is clear that $b_1 = c_1$, $b_2 = c_2$ and $b_3 = c_3$.
Under condition (i): with $\bar{p}_{j,1} > \bar{p}_{j,2}$ we get $c_4 \leq b_8$, $c_5 \leq b_9$, $c_6 \leq b_{10}$ and $c_7 \leq b_{11}$; with $\bar{p}_{k,1} < \bar{p}_{k,2}$ we get $c_8 \leq b_4$, $c_9 \leq b_5$, $c_{10} \leq b_6$ and $c_{11} \leq b_7$.
Under condition (ii): with $\bar{p}_{k,1} < \bar{p}_{j,1}$ we get $c_4 \leq b_4$, $c_5 \leq b_5$, $c_6 \leq b_6$ and $c_7 \leq b_7$; with $\bar{p}_{k,1} < \bar{p}_{k,2}$ we get $c_8 \leq b_4$, $c_9 \leq b_5$, $c_{10} \leq b_6$ and $c_{11} \leq b_7$.
Under condition (iii): with $\bar{p}_{j,1} > \bar{p}_{j,2}$ we get $c_4 \leq b_8$, $c_5 \leq b_9$, $c_6 \leq b_{10}$ and $c_7 \leq b_{11}$; with $\bar{p}_{k,2} > \bar{p}_{j,2}$ we get $c_8 \leq b_8$, $c_9 \leq b_9$, $c_{10} \leq b_{10}$ and $c_{11} \leq b_{11}$.

Thus under each condition, $C^{\gamma}_{i+1,2}(\sigma') \leq C^{\gamma}_{i+1,2}(\sigma)$, and therefore the makespan of the schedule under σ' is not greater than under σ. □

Note that this proof deals only with the relative position of jobs j and k; it is based on the general case where these jobs are surrounded by other jobs in the sequence. However, we can apply the same reasoning:

- **If jobs j and k are the first two jobs of the sequence.** In this case, the availability date of the machines to process the first job is zero ($C^{\gamma}_{i-1,m} = 0$ $\forall \gamma \leq \Gamma$, $\forall m \in \{M_1, M_2\}$, job $\sigma[i-1]$ can be seen as a virtual task with duration of 0).
- **If jobs j and k are the last two jobs of the sequence.** In this case, $C^{\gamma}_{i+1,2}(\sigma)$ under the original schedule and $C^{\gamma}_{i+1,2}(\sigma')$ after the interchange no longer represent the date of availability of machine M_2 to process job $\sigma[i+2]$, but the makespan of the solution.

Consequently, it is possible to find an optimal sequence for the robust two-machine permutation flow-shop scheduling problem, with a global uncertainty budget and preserved order of processing times, in polynomial time using Johnson's rule.

3.2 Machine-Dependent Budget $\Gamma = (\Gamma_1, \Gamma_2)$

In general, Johnson's rule does not lead to an optimal robust schedule when considering a machine-dependent uncertainty budget.

Example 1. Consider a robust two-machine flow-shop problem with 3 jobs with machine-dependent uncertainty budget $\Gamma = (1, 2)$. The intervals $[\bar{p}_{i,m}, \bar{p}_{i,m} + \hat{p}_{i,m}]$ of processing times $p_{i,m}$ of operations $O_{i,m}$, $i \in \mathcal{J}, m \in \mathcal{M}$, are given in Table 1.

Table 1. Numerical example of an instance of a two-machine flow-shop problem: preserved order of operation processing times.

	M_1	M_2
J_1	[6,9]	[8,12]
J_2	[10,15]	[4,6]
J_3	[4,6]	[3,5]

Applying Johnson's rule to this instance yields the sequence $\sigma = \{J_1, J_2, J_3\}$. Given sequence σ, and considering an uncertainty budget $\Gamma = (1, 2)$, the worst case, for this solution, is that the processing time of job J_2 on machine M_1 and jobs J_2 and J_3 on machine M_2 deviate and take their greatest value. Figure 2 depicts the Gantt chart in this case. The objective function value of this solution reaches a makespan equal to 32.

Another possible sequence is $\sigma' = \{J_3, J_1, J_2\}$. The worst case for this new solution is such that the processing time of job J_2 on machine M_1 and jobs J_1 and J_2 on machine M_2 deviate from their nominal value. Figure 3 depicts the Gantt chart in this case; it leads to a solution with a worst-case makespan equal to 31.

Although the order of processing times is preserved, Johnson's rule does not allow us to obtain the optimal sequence for this instance when considering a machine-dependent uncertainty budget $\Gamma = (1, 2)$.

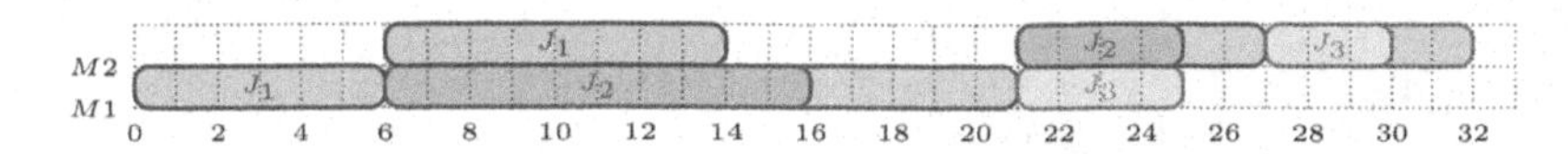

Fig. 2. Example 1 and sequence $\{J_1, J_2, J_3\}$: worst case under Johnson's schedule, $\Gamma_1 = 1, \Gamma_2 = 2$.

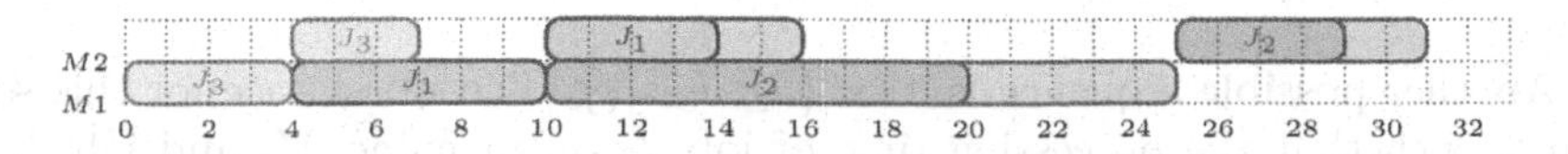

Fig. 3. Example 1 and sequence $\{J_3, J_1, J_2\}$: worst case under optimal robust schedule, $\Gamma_1 = 1, \Gamma_2 = 2$.

3.3 Unpreserved Order of Processing Times

In general, Johnson's rule does not lead to an optimal robust schedule when considering an instance with unpreserved order of processing times, even when we consider a global uncertainty budget.

Example 2. Consider a robust two-machine flow-shop problem with 3 jobs with the global uncertainty budget $\Gamma = 3$. The intervals $[\bar{p}_{i,m}, \bar{p}_{i,m}+\hat{p}_{i,m}]$ of processing times $p_{i,m}$ of operations $O_{i,m}$, $i \in \mathcal{J}, m \in \mathcal{M}$, are given in Table 2. The order of processing time is unpreserved, for example, $\bar{p}_{1,1} = 1 < \bar{p}_{2,1} = 2$ while $\bar{p}_{1,1} + \hat{p}_{1,1} = 5 > \bar{p}_{2,1} + \hat{p}_{2,1} = 3$.

Table 2. Numerical example of an instance of a two-machine flow-shop problem: unpreserved order of operation processing times.

	M_1	M_2
J_1	[1,5]	[2,3]
J_2	[2,3]	[1,5]
J_3	[2,20]	[4,5]

Applying Johnson's rule to this instance yields the sequence $\sigma = \{J_1, J_3, J_2\}$. Given sequence σ, and considering an uncertainty budget $\Gamma = 2$, the worst case, for this solution, is such that the processing time of job J_3 on machine M_1 and job J_2 on machine M_2 deviate and take their greatest value. Figure 4 depicts the Gantt chart in this case. The objective function value of this solution reaches a makespan equal to 30.

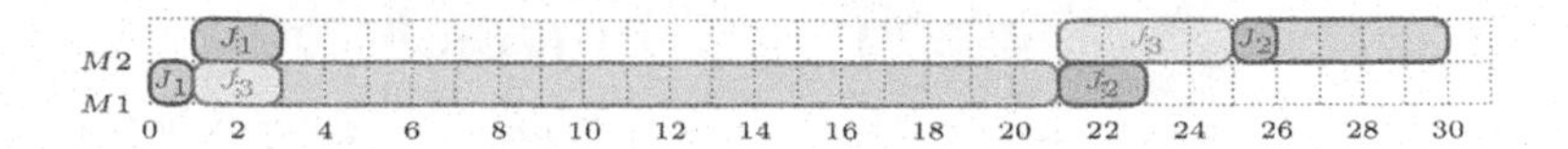

Fig. 4. Example 2 and sequence $\{J_1, J_3, J_2\}$: worst case under Johnson's schedule, $\Gamma = 2$.

Another possible sequence is $\sigma' = \{J_2, J_3, J_1\}$. The worst case for this solution is such that the processing time of job J_2 on machine M_1 and job J_1 on machine M_2 deviate from their nominal value. Figure 5 depicts the Gantt chart in this case; it leads to a solution with a worst-case makespan equal to 29.

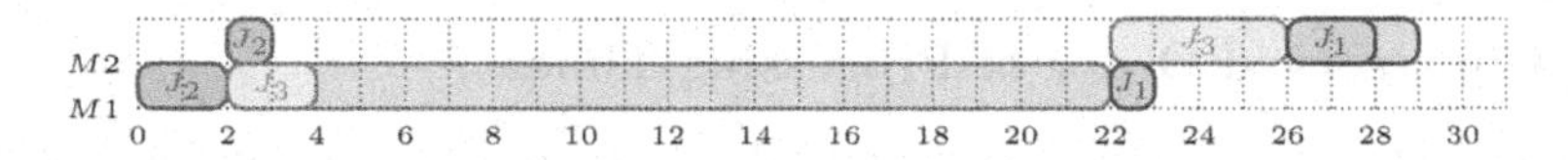

Fig. 5. Example 2 and sequence $\{J_2, J_3, J_1\}$: worst case under optimal robust schedule, $\Gamma = 2$.

Although the considered uncertainty budget is global, Johnson's rule does not allow us to obtain the optimal sequence for this instance whose order of processing times is not preserved.

4 General Case

As discussed in the previous section, in the general case, Johnson's rule is not guaranteed to find an optimal robust sequence. However, Mixed-Integer Linear Programming (MILP) or Constraint Programming (CP) allow the development of exact solution methods.

4.1 Mixed-Integer Linear Programming Robust Counterparts

Levorato et al. [6] proposed two mixed-integer linear programming robust counterparts for the two-machine permutation flow-shop problem.

The first one is adapted from the integer programming model for the three-machine deterministic flow-shop by Wagner [9]. It uses rank decision binary variables, which determine whether a job is placed at a given position in the sequence. It also uses two types of idle times variables. The first ones represent the time each job waits between the end of its execution on machine M_1 and its starting on machine M_2. The others represent the time machine M_2 idles between the execution of each pair of consecutive jobs. These idle times variables are duplicated for each considered scenario. Precedence constraints are addressed with *job-adjacency* and *machine-linkage* constraints, which exploit the special structure of the problem to describe the relation between idle times, both on machines and jobs, and processing times.

The second robust counterpart proposed by Levorato et al. is based on the formulation presented by Wilson [10]. It also uses rank decision binary variables to determine whether a job is placed at a given position in the sequence. Precedence constraints are based on start time variables defined for each job operation and each machine.

The numerical experiments in [6] highlight the superiority of Wagner's formulation over Wilson's method. Consequently, we only focus on Wagner's formulation and MILP always refers to this formulation in the following.

4.2 Constraint Programming Robust Counterparts

Another alternative is to use constraint programming. To present the CP model, we use the IBM CP Optimizer solver, which allows the use of specific decision variables and constraints. In particular, interval variables can be used to represent the time during which a task is processed and are defined by a starting value, an ending value and a size. Constraints such as `endBeforeStart()` allow us to constrain the relative positions of the interval variables. A sequence variable allows the representation of an order on a set of interval variables. Constraints can be applied on it such as `NoOverlap()` to prevent intervals from a sequence overlapping or `SameSequence()` which forces two sequences of intervals to follow the same order. We use interval and sequence variables defined as follows:

- $task_{i,m,\xi}$: interval variable between the start and the end of the processing of job $i \in \mathcal{J}$ on machine $m \in \{M_1, M_2\}$ in scenario $\xi \in \mathcal{U}^\Gamma$;
- $seqs_{m,\xi}$: sequence variable of operations scheduled on machine $m \in \{M_1, M_2\}$ in scenario $\xi \in \mathcal{U}^\Gamma$.

The CP model developed for the two-machine robust flow-shop problem is as follows:

$$\min C_{\max} \tag{5}$$

$$\text{s.t.} \quad C_{\max} \geq task_{i,2,\xi}.end \quad \forall i \in \mathcal{J},\ \xi \in \mathcal{U}^\Gamma \tag{6}$$

$$EndBeforeStart(task_{i,1,\xi}, task_{i,2,\xi}) \quad \forall i \in \mathcal{J},\ \xi \in \mathcal{U}^\Gamma \tag{7}$$

$$NoOverlap(seqs_{m,\xi}) \quad \forall m \in \{M_1, M_2\},\ \xi \in \mathcal{U}^\Gamma \tag{8}$$

$$SameSequence(seqs_{1,1}, seqs_{m,\xi}) \quad \forall m \in \{M_1, M_2\},\ \xi \in \mathcal{U}^\Gamma \tag{9}$$

Constraints (6) allow the determination of the makespan, which is equal to the end of the last job on machine M_2 in the worst-case scenario. Constraints (7) ensure the precedence relations between the two operations of a same job. Constraints (8) ensure that, in each scenario, each machine performs at most one operation at a time. Constraints (9) ensure that the sequence is the same on both machines, and the same for each scenario. The first scenario $\xi = 1$ is used as reference, and the constraint is duplicated for each scenario and each machine.

4.3 Column and Constraint Generation Algorithm

The column and constraint generation method has been introduced by Zeng and Zhao [12] to solve two-stage robust optimization problems. The procedure splits the problem into a master problem and an adversarial subproblem. The idea is to solve the robust counterpart problem (or master problem), for a limited subset of scenarios, that fixes the first stage variables, and then to identify which scenarios, if any, make the solution found in the master problem infeasible, using an adversarial subproblem. Then, these scenarios are included in the master problem by generating the corresponding recourse decision variables on the fly. This process repeats until a solution that is feasible for all scenarios is found [3,4,6,7]. Figure 6 depicts the scheme of the column and constraint generation algorithm.

Levorato et al. [6] propose a column and constraint generation framework for the two-machine permutation flow-shop problem where the "first-stage variables" are the sequence variables, that allow the determination of the order of the jobs on the machines and the makespan, and the "second-stage variables" are the start times for each operation and each scenario. It consists in relaxing one of the MILP formulations presented in Sect. 4.1 by considering only a subset of scenarios. Then, given a sequence, a polynomial time dynamic algorithm (see Sect. 2.2) is used to identify the worst-case makespan considering a given uncertainty budget.

Since constraint programming is often very efficient for scheduling problems, we try to improve this framework by replacing the master problem by a relaxed version of the constraint programming model presented in Sect. 4.2. For this purpose, each robust constraint (6–9) is defined only for a subset of scenarios. The rest of the algorithm remains identical to the version proposed by Levorato et al.

5 Experimental Results

We evaluate the performance of the column and constraint generation algorithm for both the MILP and the CP models. Experiments are performed on three cluster nodes with Intel Xeon E5-2695 v4 CPU at 2.1 GHz. The algorithms are implemented in C++, CPLEX 12.10 is used as the solver for the MILP master problem and CP Optimizer (CPO) 12.10 for the CP master problem. We limited time to 2 h, with 4 CPU and a total of 16 GB of RAM, per instance.

5.1 Instances from Literature

The instances we used in this section are the same as in [6], based on instances generated by Ying [11]. They are composed of six groups of instances of different size, where the number of jobs $|\mathcal{J}|$ belongs to $\{10, 20, 50, 100, 150, 200\}$. The nominal processing time $\bar{p}_{i,m}$, $i \in \mathcal{J}, m \in \{M_1, M_2\}$ is generated from the uniform distribution $U[10, 50]$ and the processing time deviation $\hat{p}_{i,m}$ is a ratio

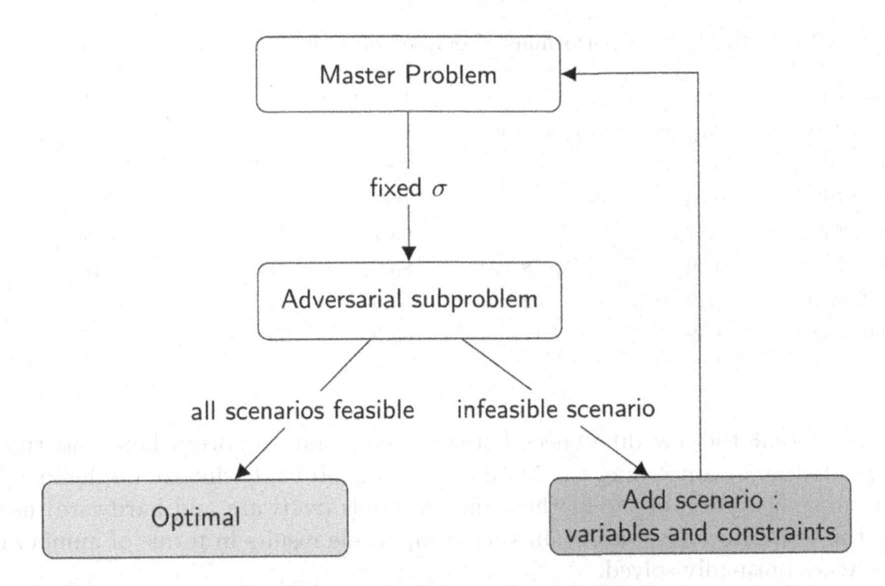

Fig. 6. Column and constraint algorithm.

of the nominal processing time $\alpha\bar{p}_{i,m}$ with $\alpha = 10, 20, 30, 40$ and 50%. Thus, the order of processing times is preserved through deviation. Ten sets of values for nominal duration were generated for each size $|\mathcal{J}|$, and all deviations ratios α were applied to each of them, giving a total of 300 test instances. The uncertainty budgets, Γ_1 and Γ_2, are set to 20%, 40%, 60%, 80% and 100% of $|\mathcal{J}|$.

We summarize the results in Table 3 where the performance over instances of different sizes is displayed. We report the percentage of instances solved to optimality before reaching the time limit (*Solved (%)*). For the instances solved to optimality, we display the average execution time, in seconds, to reach optimality (*Avg. time (sec)*). Lastly, we display the average percentage gap of non-optimally solved instances (*Avg. gap (%)*), where the gap is computed as follows:

$$gap^{method} = \frac{UB^{method} - LB^*}{UB^{method}} \tag{10}$$

with LB^* the best bound found among the two versions of the column and constraint generation algorithm.

We notice that the CP model outperforms the MILP one. Indeed, even for the largest instances (200 jobs) the CP-based method manages to solve almost all (98.83%) of the instances optimally, while the MILP-based method solves fewer and fewer instances as they grow (down to 68.92% for 200 jobs). We also observe that the time needed to reach the optimum is much lower for the CP method, whatever the size of the instances. Finally, we can see that for both methods, the gap is quite low (less than 1%, regardless of instance size).

Table 3. Methods performance comparison grouping by instance sizes.

$\lvert\mathcal{J}\rvert$	CP			MILP		
	Solved (%)	Avg. time (sec)	Avg. gap (%)	Solved (%)	Avg. time (sec)	Avg. gap (%)
10	100	0.42	–	100	14.7	–
20	100	0.76	–	100	191.59	–
50	100	1.47	–	98.75	194.9	0.23
100	99.58	0.99	0.08	85.5	319.17	0.31
150	98.92	3.12	0.02	74.67	341.65	0.47
200	98.83	8.49	0.1	68.92	390.73	0.4

Note that the few differences between the results reported here and those presented in [6] concerning the MILP method is probably due to a difference in the implementation of the method and the tools (software and hardware) used for the tests. However, we obtain very comparable results in terms of number of instances optimally solved.

We now examine the quality of the solution obtained by applying Johnson's rule on these instances. Table 4 presents the percentage of best known solution found (*Best known sol. (%)*) and the average percentage gap (*Avg. gap (%)*) of non-optimally solved instances ($UB^{method} > LB^*$).

We note that the polynomial time algorithm enables to find high quality solutions. Indeed, for almost all the instances (94.46%), Johnson's rule provides a robust solution with the same objective value as the best known solution, and the optimality gap is very low.

Table 4. Johnson's rule performance (instances from literature grouping by size).

$\lvert\mathcal{J}\rvert$	Best known sol. (%)	Avg. gap (%)
10	92.58	1.13
20	92.67	0.4
50	98.5	0.29
100	97.67	0.04
150	89.67	0.04
200	95.67	0.04

In view of these results, we notice that these instances are easy to solve, due to their particular structure that preserves the order of the processing times. To overcome this, we generated new instances and the results obtained are presented in the following section.

5.2 New Instances

In this section, we generated new instances which are also based on the ones from [11]. The nominal processing times $\bar{p}_{i,m}, i \in \mathcal{J}, m \in \{M_1, M_2\}$ remain the same as in the original instances [11]. However, the processing time deviations $\hat{p}_{i,m}$ are randomly generated to avoid the order preserving of processing times. Let $\bar{p}_{max}$ be the maximum nominal processing time for all operations. For each operation $O_{i,m}, i \in \mathcal{J}, m \in \{M_1, M_2\}$, we randomly generate a value for $\hat{p}_{i,m}$ within a range from 25% to 80% of the value of $\bar{p}_{max}$. We generate in total 60 instances that we use for our tests. Again, the uncertainty budgets, Γ_1 and Γ_2, are set to 20%, 40%, 60%, 80% and 100% of $|\mathcal{J}|$.

Table 5 presents the same performance indicators as Table 3 for the new generated instances.

By comparing these two tables, it can be seen that the new instances are more difficult to solve than the ones from literature. Indeed, there is a lower proportion of instances solved optimally, a higher average time needed to reach the optimum, as well as a higher average gap for the unsolved instances, for both methods. However, focusing on the information provided by Table 5, it is noticeable that the CP model still outperforms the MILP one, for all observed indicators.

Table 5. Methods performance comparison grouping by instance sizes (new instances).

$\lvert\mathcal{J}\rvert$	CP			MILP		
	Solved (%)	Avg. time (sec)	Avg. gap (%)	Solved (%)	Avg. time (sec)	Avg. gap (%)
10	100	12.36	–	100	46.77	–
20	80	167.11	2.04	75	1002.05	2.22
50	60	86.85	2.61	49	902.69	3.07
100	56	310.3	2.49	45	600.67	4.11
150	48	271.32	2.3	32	667.89	4.3
200	54	99.57	2.68	30	827.08	4.36

Tables 6 and 7 detail the percentage of solved instances according to uncertainty budget $\Gamma = (\Gamma_1, \Gamma_2)$ for the CP-based and the MILP-based column and constraint generation method, respectively.

By comparison of these two tables with each other, we notice that, for all combinations of Γ_1 and Γ_2, except one ($\Gamma_1 = 40\%$, $\Gamma_2 = 20\%$), the CP model outperforms the MILP one. We also see that the problem is more difficult to solve for medium uncertainty budgets (40% or 60%), for both methods. This can be explained by the fact that these uncertainty budgets generate a greater number of scenarios. However, the number of scenarios is not the only difficulty factor, as we can see, the methods are more efficient in solving instances with a large uncertainty budget. For example, instances with an uncertainty budget of 80% are better solved than those with a budget of 20%, while the number of possible scenarios are the same.

Table 6. Percentage of solved instances according to uncertainty budget $\Gamma = (\Gamma_1, \Gamma_2)$ for the CP-based column and constraint generation method.

Γ_2	Γ_1				
	20%	40%	60%	80%	100%
20%	28.33	25	41.67	70	100
40%	35	21.67	26.67	81.67	100
60%	36.67	28.33	28.33	83.33	100
80%	60	71.67	75	88.33	100
100%	100	100	100	100	–

Table 7. Percentage of solved instances according to uncertainty budget $\Gamma = (\Gamma_1, \Gamma_2)$ for the MILP-based column and constraint generation method.

Γ_2	Γ_1				
	20%	40%	60%	80%	100%
20%	26.67	26.67	33.33	58.33	100
40%	23.33	18.33	23.33	65	100
60%	23.33	21.67	26.67	65	98.33
80%	35	35	40	60	100
100%	76.67	83.33	91.67	98.33	–

6 Conclusion

In this paper, we investigate the robust two-machine flow-shop scheduling problem where the operation processing times are subject to uncertainty. A two-stage robust optimization is used to deal with this uncertainty, where the first stage is devoted to fixing the sequencing decisions whilst the second stage determines the start time of the operations. As a main contribution, we show that under specific conditions the problem can be solved in polynomial time. For the general case, we introduce a constraint programming formulation, which we embed in a column and constraint generation decomposition scheme. This method provides the best results compared to a literature algorithm based on a MILP formulation.

References

1. Ben-Tal, A., Goryashko, A., Guslitzer, E., Nemirovski, A.: Adjustable robust solutions of uncertain linear programs. Math. Program. **99**(2), 351–376 (2004)
2. Bertsimas, D., Sim, M.: The price of robustness. Oper. Res. **52**(1), 35–53 (2004)
3. Duarte, J.L.R., Fan, N., Jin, T.: Multi-process production scheduling with variable renewable integration and demand response. Eur. J. Oper. Res. **281**(1), 186–200 (2020)

4. Hamaz, I., Houssin, L., Cafieri, S.: The cyclic job shop problem with uncertain processing times. In: 16th International Conference on Project Management and Scheduling (PMS 2018), Rome, Italy, pp. 119–122 (2018)
5. Johnson, S.M.: Optimal two-and three-stage production schedules with setup times included. Nav. Res. Logist. Q. **1**(1), 61–68 (1954)
6. Levorato, M., Figueiredo, R., Frota, Y.: Exact solutions for the two-machine robust flow shop with budgeted uncertainty. Eur. J. Oper. Res. **300**(1), 46–57 (2022)
7. Silva, M., Poss, M., Maculan, N.: Solution algorithms for minimizing the total tardiness with budgeted processing time uncertainty. Eur. J. Oper. Res. **283**(1), 70–82 (2020)
8. Soyster, A.L.: Convex programming with set-inclusive constraints and applications to inexact linear programming. Oper. Res. **21**(5), 1154–1157 (1973)
9. Wagner, H.M.: An integer linear-programming model for machine scheduling. Nav. Res. Logist. Q. **6**(2), 131–140 (1959)
10. Wilson, J.: Alternative formulations of a flow-shop scheduling problem. J. Oper. Res. Soc. **40**(4), 395–399 (1989)
11. Ying, K.C.: Scheduling the two-machine flowshop to hedge against processing time uncertainty. J. Oper. Res. Soc. **66**(9), 1413–1425 (2015)
12. Zeng, B., Zhao, L.: Solving two-stage robust optimization problems using a column-and-constraint generation method. Oper. Res. Lett. **41**(5), 457–461 (2013)

A Weighted Counting Algorithm for the Circuit Constraint

Gauthier Pezzoli and Gilles Pesant(✉)

Polytechnique Montréal, Montreal, Canada
gilles.pesant@polymtl.ca

Abstract. The CIRCUIT constraint is useful to model many combinatorial problems in Constraint Programming. CP solvers extended with Belief Propagation, such as MiniCPBP, require that constraints be equipped with weighted counting algorithms in order to propagate probability mass functions over domains. This is not yet the case for CIRCUIT. To this purpose we introduce a probabilistic sampling algorithm to count Hamiltonian circuits in a weighted graph. We show that our resulting estimator is unbiased, measure its empirical accuracy, and evaluate its impact on search performance.

1 Introduction

Let $\mathcal{G} = (\mathcal{N}, \mathcal{A})$ be a directed graph. A Hamiltonian circuit of $\mathcal{G}$ is a path through arcs from $\mathcal{A}$ that visits exactly once all the nodes in $\mathcal{N}$ and ends where it started (Fig. 1). Let $\mathcal{H}(\mathcal{G})$ denote the set of all Hamiltonian circuits of $\mathcal{G}$. We are interested in counting the number of such circuits. For unweighted graphs, each Hamiltonian circuit counts for 1. In order to generalize the counting to weighted graphs, we can define the weight of a circuit as the product of the weights w_a of each arc a in the circuit. Thus an unweighted graph can be seen as a weighted graph with arcs of unit weight. If the weights are natural numbers we can make a parallel with multigraphs, with each weight representing the number of arcs between two nodes. Here the weights are instead interpreted as probabilities. We define the weighted count of all Hamiltonian circuits as

$$\mathbb{H}(\mathcal{G}) = \sum_{c \in \mathcal{H}(\mathcal{G})} \prod_{a \in c} w_a.$$

The CIRCUIT$(\{s_1, s_2, \ldots, s_{|\mathcal{N}|}\})$ constraint, defined over finite-domain variables $s_i \in \{j \in \mathcal{N} : (i,j) \in \mathcal{A}\}$, is useful to model many combinatorial problems in Constraint Programming (CP). It enforces that the set of arcs $\{(i, s_i) : i \in \mathcal{N}\}$ forms a Hamiltonian circuit of $\mathcal{G}$. Several domain-filtering algorithms have been proposed in the literature (e.g. [3]), including for the optimization version of the constraint (e.g. [2]), as well as some decompositions. CP solvers extended with Belief Propagation (BP), such as MiniCPBP [4], additionally require that constraints be equipped with weighted counting algorithms, which is not yet the case for CIRCUIT. This short paper aims to correct this.

In Sect. 2 we present our weighted counting algorithm and discuss its implementation into MiniCPBP in Sect. 3. Section 4 evaluates its usefulness to solve combinatorial problems. We conclude with Sect. 5.

© The Author(s), under exclusive license to Springer Nature Switzerland AG 2023
A. A. Cire (Ed.): CPAIOR 2023, LNCS 13884, pp. 370–377, 2023.
https://doi.org/10.1007/978-3-031-33271-5_24

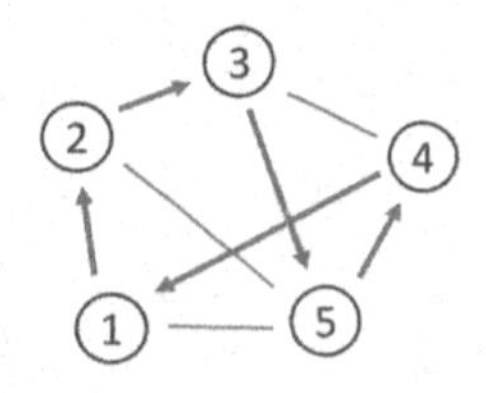

Fig. 1. A Hamiltonian circuit (shown in red) (Color figure online)

Algorithm 1: Sampling algorithm

Input: weighted digraph $\mathcal{G}(\mathcal{N}, \mathcal{A})$, with $|\mathcal{N}| = n$
Output: estimator $X_{\mathcal{G}}$

1 **if** $n = 1$ **then**
2 $\quad X_{\mathcal{G}} = \begin{cases} w_{(1,1)} & \text{if } (1,1) \in \mathcal{A} \\ 0 & \text{else} \end{cases}$
3 **else**
4 $\quad W = \{i > 1 : (1, i) \in \mathcal{A}\}$
5 $\quad$ **if** $W = \emptyset$ **then**
6 $\quad\quad X_{\mathcal{G}} \leftarrow 0$
7 $\quad$ **else**
8 $\quad\quad$ Choose $J \in W$ randomly according to the weight of each edge$(1, J)$
9 $\quad\quad X_{\mathcal{G}} \leftarrow X_{\mathcal{G}_{1,J}} \times \sum_{i \in W} w_{(1,i)}$

---

2 An Unbiased Estimator for the Weighted Count of Hamiltonian Circuits

Since deciding whether a graph admits a Hamiltonian circuit is already $\mathcal{NP}$-complete, we cannot hope to count them exactly in a reasonable amount of time.

2.1 A Sampling Algorithm

We turn to designing a probabilistic sampling algorithm adapted from Rasmussen's estimator for undirected, unweighted graphs [6].

Algorithm 1 starts from node 1 and tries to build a path neighbour after neighbour, being careful not to form a sub-circuit before encompassing all the nodes. At the end it checks whether the last node has an arc to the first one. The weighted count is accumulated as the path is built. $\mathcal{G}_{1,J}$ denotes graph $\mathcal{G}$ transformed in the following way: redirect node 1's incoming arcs to node J and node J's incoming arcs to node 1, delete node 1, renumber the nodes from 1 to $n - 1$ (no matter the order). Figure 2 illustrates one step of the algorithm. Changes with respect to the original algorithm of Rasmussen are shown in red.

We now proceed to show that our estimator is unbiased.

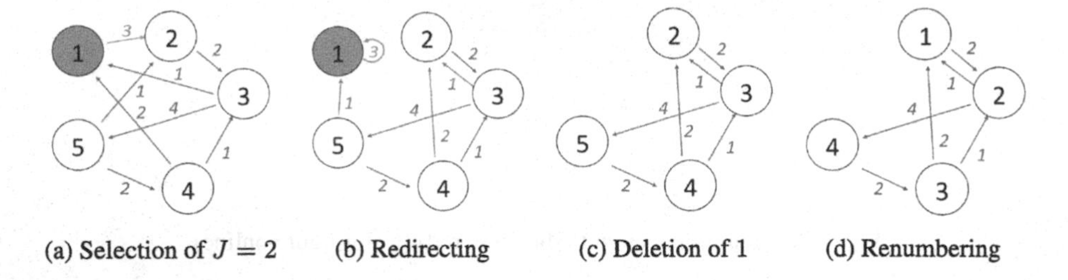

(a) Selection of $J = 2$ (b) Redirecting (c) Deletion of 1 (d) Renumbering

Fig. 2. One step of the estimator

Lemma 1. *Let $\mathcal{G}$ be a weighted directed graph and W the set of neighbours of node* 1. *Then* $\mathbb{H}(\mathcal{G}) = \sum_{i \in W} w_{(1,i)} \times \mathbb{H}(\mathcal{G}_{1,i})$

Proof. If W is empty then node 1 has no neighbour and $\mathbb{H}(\mathcal{G}) = 0$. Otherwise assume that W is not empty. Thus $\mathbb{H}(\mathcal{G}) = \sum_{i \in W} w_{(1,i)} \times C_{i,1}$ where $C_{k,\ell}$ is the sum of the costs (which is defined for a path here as the product of the weights of its arcs) of all Hamiltonian paths starting with k, ending with ℓ, and containing all the other nodes. We have $C_{k,\ell} = \mathbb{H}(\mathcal{G}_{k,\ell})$ because $\mathcal{G}_{k,\ell}$ corresponds to the graph $\mathcal{G}$ where we would have merged nodes k and ℓ into one new node that represents the link $k \rightarrow \ell$. Instead of actually merging the nodes, we consider ℓ as the node representing the link. It must therefore receive the incoming arcs of node k (since it is the beginning of the link) and delete the incoming arcs of node ℓ that no longer need to exist. This is exactly what we do with $\mathcal{G}_{k,\ell}$ by swapping the incoming arcs of k and ℓ, then deleting node k and all its arcs. It is the same for $C_{k,\ell}$ which represents the weight of all Hamiltonian circuits of $\mathcal{G}$ that include arc (k, ℓ) without counting the weight of it. □

Theorem 1. *Let $\mathcal{G}$ be a weighted directed graph. Then $\mathbb{E}(X_\mathcal{G}) = \mathbb{H}(\mathcal{G})$.*

Proof. We demonstrate this by induction on the number of nodes, n. If $n = 1$ then either there is a loop on node 1 and $\mathbb{H}(\mathcal{G}) = w_{(1,1)}$, or there is not and the algorithm returns 0. Assume the theorem holds for $n - 1$; we show it for n. If the set W is empty, node 1 has no neighbour and thus we won't be able to form a Hamiltonian circuit that way and we indeed return 0. Otherwise

$$
\begin{aligned}
\mathbb{E}(X_\mathcal{G}) &= \sum_{j \in W} \mathbb{E}(X_\mathcal{G} | J = j) \Pr(J = j), \quad \text{by definition of expectation} \\
&= \sum_{j \in W} \mathbb{E}\left(X_{\mathcal{G}_{1,j}} \times \sum_{i \in W} w_{(1,i)}\right) \frac{w_{(1,j)}}{\sum_{i \in W} w_{(1,i)}}, \quad \text{by construction} \\
&= \sum_{j \in W} w_{(1,j)} \times \mathbb{E}\left(X_{\mathcal{G}_{1,j}}\right) \\
&= \sum_{j \in W} w_{(1,j)} \times \mathbb{H}(\mathcal{G}_{1,j}), \quad \text{from the induction hypothesis} \\
&= \mathbb{H}(\mathcal{G}), \quad \text{from Lemma 1}
\end{aligned}
$$

□

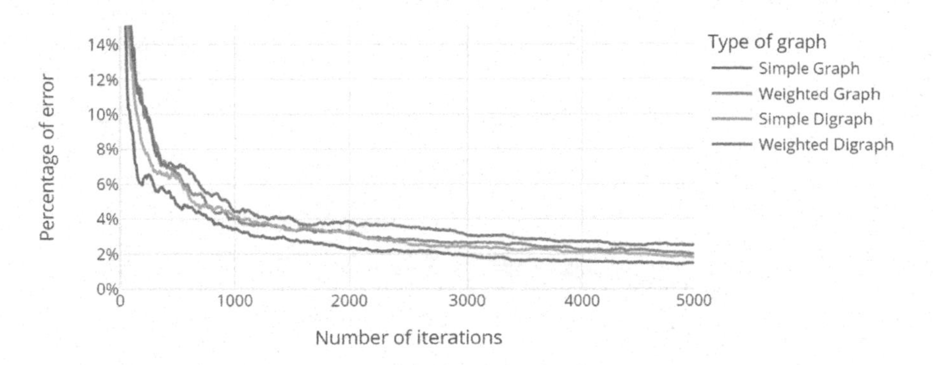

Fig. 3. Relative error versus number of iterations (graph with 10 nodes)

In practice we choose the starting node uniformly at random, run our algorithm several times, and take the average of the results. We have then a time complexity in $\mathcal{O}(kn^2)$ where k is the number of runs. Indeed for one step of the sampling algorithm, in the worst case, W isn't empty: we create W, pick up J, and construct $\mathcal{G}_{1,J}$ in $\mathcal{O}(n)$. Then we repeat it a maximum of n times for the n nodes of the graph. Hence in the worst case we have $\mathcal{O}(n^2)$ for one run of the sampling algorithm.

2.2 Empirical Accuracy of the Estimator

Despite the probabilistic guarantee from Theorem 1 that the expected value of our estimator is equal to the true value, we measure empirically its relative error and variance with respect to the number of runs and to characteristics of the graph. All the following tests have been run using randomly generated graphs on a small number of nodes, the limiting factor being the exact calculation by the naive algorithm whose runtime grows exponentially. The graphs we report on admit at least one Hamiltonian circuit: on an instance without a Hamiltonian circuit our estimator would necessarily return 0, which is the exact value.

Figure 3 shows the influence of the number of iterations. We set the arc density to 0.7 and each data point represents the average of 100 instances. Not surprisingly, the more iterations there are, the closer the relative error gets to 0. We also do not seem to observe any differences between the different types of graph. Figure 4 shows how the edge density acts on the relative error. Each data point represents the average of 10 instances. The algorithm performs well especially when the density is high, the relative error being fairly low and slowly increasing with graph size. However the estimator struggles around an edge density equal to 0.3. On sparser graphs which still admit Hamiltonian circuits, our algorithm may often fail (i.e. return 0) which increases the variance significantly.

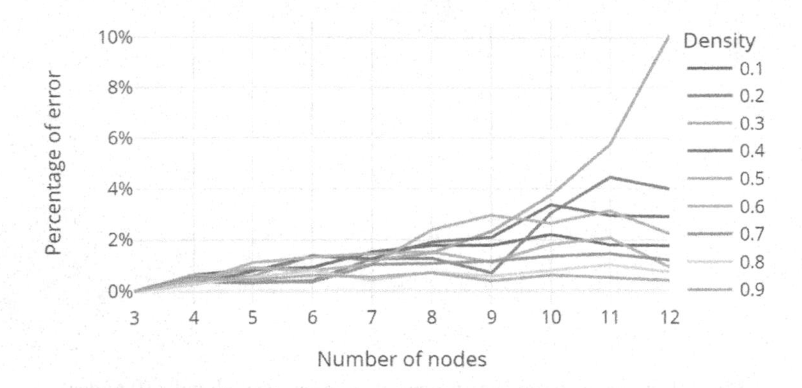

Fig. 4. Relative error versus number of nodes per density (nb iter = 10000)

3 Integration in the CP-BP Framework

3.1 CP-BP Framework

CP-based Belief Propagation, introduced in [4], offers a more informative propagation of constraints. In addition to removing unsupported values from the domain of variables, it computes the belief that each value will satisfy the constraint. Those beliefs for individual variables are then sent as messages to these variables, which combine them with messages from the other constraints and send back the resulting, more global, beliefs to each constraint. In the next round of messages, constraints use these updated beliefs to weight each of their solutions, hence the term *weighted counting*.

Specifically in our case, the (weighted) proportion of circuits that use a given arc $(i, j) \in \mathcal{A}$,

$$\theta_{\mathcal{G}}(s_i, j) = \frac{\mathbb{H}_{(i,j)}(\mathcal{G})}{\mathbb{H}(\mathcal{G})} \tag{1}$$

$$\text{where} \quad \mathbb{H}_{(i,j)}(\mathcal{G}) = \sum_{c \in \mathcal{H}(\mathcal{G}):(i,j) \in c} \left(\prod_{a \in c: a \neq (i,j)} w_a \right),$$

is called the *marginal* of assignment $s_i = j$ according to the Hamiltonian circuit constraint on $\mathcal{G}$.

This back-and-forth process is iterated and hopefully converges to the true marginal probabilities over the whole CSP: for each variable-value pair (x, v), the probability that x is assigned value v in a solution (i.e. where all the constraints are satisfied). *Message damping* [1], sending a convex combination of the newly computed and previous messages, may be used to help convergence. Branching heuristics have been designed based on marginals: for example *max-marginal* [1] selects a variable with the largest marginal probability for one of its domain values and assigns it to that value.

3.2 Implementation of Weighted Counting for CIRCUIT

For domain filtering, the current implementation of CIRCUIT relies on a standard decomposition into an ALLDIFFERENT constraint and subtour elimination constraints that keep track of the extremities `orig` and `dest` of current partial paths [5].

Algorithm 2: *updateBelief* algorithm

```
Input: beliefs from variables: b(i, j) for all variable indices i and values j
Output: unnormalized θ_G(s_i, j) for all unbound variables s_i
// Collect beliefs of partial paths: orig[i] ⇝ i, s_i unbound
1  B ← 1
2  foreach i ∈ Unbound do
3      j ← orig[i]
4      while j ≠ i do
5          B ← B × b(j, s_j)
6          j ← s_j
   // Clear frequency count of unbound variables
7  foreach i ∈ Unbound do
8      foreach j ∈ D(s_i) do
9          count[i][j] ← ε          // avoids null count if never sampled
   // Compute frequency counts
10 t ← 0
11 t* ← |Unbound|^2                // our target number of samples
12 for k ← 1 to t^max do
13     if findCircuit(B, b, count) > 0 then
14         t ← t + 1
15     if t = t* then
16         break
   // Set marginals
17 if t ≥ 1 then
18     foreach i ∈ Unbound do
19         foreach j ∈ D(s_i) do
20             θ_G(s_i, j) ← count[i][j]
21 else
       // no sample obtained; set default uniform marginals
22     foreach i ∈ Unbound do
23         foreach j ∈ D(s_i) do
24             θ_G(s_i, j) ← 1
```

In the CP-BP framework, messages from constraints to variables are assembled by calling the *updateBelief* method of each constraint. Algorithm 2 describes that method for the CIRCUIT constraint. Throughout the search we maintain a reversible set

`Unbound` of indices to the unbound successor variables. Array `count` accumulates the results of successful sampling trials in an effort to approximate the numerator in Eq. 1. Function *findCircuit* is basically our estimator described in Algorithm 1 which uses beliefs b as weights, starts with an initial weight B, and adds a contribution to `count` each time it succeeds.

4 Combinatorial Search Guidance

We now evaluate empirically the contribution of our weighted counting algorithm to solve combinatorial problems featuring CIRCUIT constraints using the CP-BP framework as implemented in the MiniCPBP solver[1]. We selected the Perfect 1-Factorization of a Complete Graph Problem[2] because it features several CIRCUIT constraints (and many INVERSE, ALLDIFFERENT, ELEMENT, LEXLESS, and SUM constraints) and because finding a feasible solution quickly becomes challenging for a combinatorial solver. According to the documentation:

> "A 1-factorization [of K_n, a complete graph on n vertices] is a partition of the edges of the graph into $n - 1$ complete matchings. For the 1-factorization to be perfect, every pair of matchings must form a Hamiltonian circuit of the graph."

Because a 1-factorization is only possible when n is even, instances alternate between being satisfiable and unsatisfiable as n increases.

We ran MiniCPBP with the following settings: depth-first search, ten iterations of belief propagation at each search-tree node, message damping with damping factor 0.75, *max-marginal* branching heuristic considering all model variables, maximum number of sampling trials $t^{\max} = 5000$ for CIRCUIT, and a one-hour time limit.

We compare solving instances of the above problem with and without the weighted counting algorithm for CIRCUIT. In the latter case, the CIRCUIT constraint simply returns uniform marginals in its messages to variables. All the other constraints in the model are already equipped with weighted counting algorithms. Table 1 reports the number of fails either to find a feasible solution or to show unsatisfiability for instances of increasing size. Since the weighted counting algorithm is not deterministic we report the median of five runs in that case. Looking first at satisfiable instances (i.e. even values of n) observe the marked difference in search guidance between the two (the "*no*" version times out on K_{10}). Remarkably, the addition of the weighted counting algorithm for the CIRCUIT constraint allows us to find a solution to both K_6 and K_8 without any backtracking. Turning to unsatisfiable instances (i.e. odd values of n) observe that informed marginals from the CIRCUIT constraint contribute to branching decisions yielding smaller (failed) search trees, especially for K_9 that times out without them.

[1] https://github.com/PesantGilles/MiniCPBP.
[2] https://github.com/MiniZinc/mzn-challenge/blob/develop/2021/p1f-pjs/p1f-pjs.mzn.

Table 1. Number of fails to find a feasible solution (n even) or show that none exists (n odd) for the Perfect 1-Factorization of K_n

weighted counting for CIRCUIT	n 6	8	10	5	7	9
no	0	2639	T.O.	5	45	T.O.
yes	0	0	335	4	40	5587

5 Conclusion

We designed a probabilistic algorithm to estimate the (weighted) number of Hamiltonian circuits in a directed weighted graph and proved our estimator to be unbiased. We then described how such an algorithm can be used to compute marginals for the CIRCUIT constraint in the CP-BP framework. Preliminary experiments suggest that it can greatly improve that framework's ability to solve combinatorial problems featuring that constraint.

References

1. Babaki, B., Omrani, B., Pesant, G.: Combinatorial search in CP-based iterated belief propagation. In: Simonis, H. (ed.) CP 2020. LNCS, vol. 12333, pp. 21–36. Springer, Cham (2020). https://doi.org/10.1007/978-3-030-58475-7_2
2. Benchimol, P., van Hoeve, W.J., Régin, J.C., Rousseau, L.M., Rueher, M.: Improved filtering for weighted circuit constraints. Constraints An Int. J. **17**(3), 205–233 (2012). https://doi.org/10.1007/s10601-012-9119-x
3. Isoart, N., Régin, J.C.: A linear time algorithm for the k-cutset constraint. In: Michel, L.D. (ed.) 27th International Conference on Principles and Practice of Constraint Programming (CP 2021). Leibniz International Proceedings in Informatics (LIPIcs), vol. 210, pp. 29:1–29:16. Schloss Dagstuhl - Leibniz-Zentrum für Informatik, Dagstuhl, Germany (2021). https://drops.dagstuhl.de/opus/volltexte/2021/15320
4. Pesant, G.: From support propagation to belief propagation in constraint programming. J. Artif. Intell. Res. **66**, 123–150 (2019). https://doi.org/10.1613/jair.1.11487
5. Pesant, G., Gendreau, M., Potvin, J.Y., Rousseau, J.M.: An exact constraint logic programming algorithm for the traveling salesman problem with time windows. Transp. Sci. **32**(1), 12–29 (1998). https://doi.org/10.1287/trsc.32.1.12
6. Rasmussen, L.E.: Approximating the permanent: a simple approach. Random Struct. Algorithms **5**(2), 349–362 (1994). https://doi.org/10.1002/rsa.3240050208

Boolean-Arithmetic Equations: Acquisition and Uses

R. Gindullin[1,2](✉), N. Beldiceanu[1,2], J. Cheukam Ngouonou[1,2,4], R. Douence[1,2,3], and C. -G. Quimper[4]

[1] IMT Atlantique, Nantes, France
[2] LS2N, Nantes, France
ramiz.gindullin@imt-atlantique.fr
[3] INRIA, Paris , France
[4] Université Laval, Quebec, Canada

Abstract. Motivated by identifying equations to automate the discovery of conjectures about sharp bounds on combinatorial objects, we introduce a CP model to acquire Boolean-arithmetic equations (BAE) from a table providing sharp bounds for various combinations of parameters.

Boolean-arithmetic expressions consist of simple arithmetic conditions (SAC) connected by a single commutative operator such as '∧', '∨', '⊕' or '+'. Each SAC can use up to three variables, two coefficients, and an arithmetic function such as '+', '−', '×', 'floor', 'mod' or 'min'. We enhance our CP model in the following way to limit the search space: (*i*) We break the symmetries linked to multiple instances of similar SACs in the same expression. (*ii*) We prevent the creation of SAC that could be simplified away. We identify several use cases of our CP model for acquiring BAE and show its applicability for learning sharp bounds for eight types of combinatorial objects as digraphs, forests, and partitions.

Keywords: Boolean-arithmetic equation · equation discovery · bounds

1 Introduction

In the context of finding conjectures about combinatorial objects, the relevance of Boolean and BAE has been noted but not fully developed. Larson and Cleemput describe in [21] the use of pure Boolean expressions to represent necessary or sufficient conditions for a graph property, while [8] depicts the application of BAE to express sharp bounds of graph characteristics. While the first work uses a systematic generate and test approach, the second does not describe how such BAE were produced. Our work is motivated by the following observations: (*i*) we want to go beyond a generate and test approach [21], and investigate how CP can be used to identify a wide range of concise BAE in the context of conjecture acquisition; (*ii*) while the experimental part of [8] indicates the relevance to use BAE to get sharp bounds for digraphs characteristics, it was still unclear whether this applies to other combinatorial objects. The contribution of the paper is threefold.

R. Gindulling is supported by the EU-funded ASSISTANT project no. 101000165.

© The Author(s), under exclusive license to Springer Nature Switzerland AG 2023
A. A. Cire (Ed.): CPAIOR 2023, LNCS 13884, pp. 378–394, 2023.
https://doi.org/10.1007/978-3-031-33271-5_25

1. First, it exhibits a variety of Boolean-arithmetic expressions $f(X_1, X_2, \ldots, X_n)$ which occur in practice when looking for sharp bounds.
2. Second, it provides CP models for acquiring Boolean arithmetic expressions. The main point of these models is to restrict the search space by breaking symmetries between similar simple arithmetic conditions (SACs) and avoiding generating simplifiable SACs.
3. Finally, it shows that Boolean-arithmetic expressions are not only relevant for expressing bounds for digraphs as mentioned in [8], but also for trees, forests, partitions, and some global constraints.

In Sect. 2, we describe four settings for the practical use of Boolean expressions that we observed in the context of sharp bound acquisition. In Sect. 3, we define the Boolean-arithmetic formulae that we consider throughout this paper. In Sect. 4, we provide a core CP model for learning a BAE that explains an output column of a table from a set of input columns. We show in Sect. 5 various extensions of the core model to restrict the search space of Boolean-arithmetic expressions. We evaluate the core model and its extensions in Sect. 6, discuss related work in Sect. 7, and conclude in Sect. 8.

2 The Relevance of Boolean-Arithmetic Equations

To show the expressive power of BAE, let us consider typical situations where they are relevant for acquiring sharp bounds, i.e. an inequality for which the equality holds for at least one example.

1. Using a BAE is a natural option when the codomain of $f(X_1, X_2, \ldots, X_n)$ is equal to $\{0, 1\}$ or more generally consists of only two distinct consecutive values v and $v+1$. For instance, let $v, a, \underline{os}$ and $\underline{s}$ be the number of vertices, of arcs, of smallest strongly connected components and the size of the smallest strongly connected component of a digraph. As found by the CP model of [8] when a is maximal, we have the relation $\underline{os} = \lfloor \frac{v}{\max(-\underline{s}+v,\underline{s})} \rfloor$, which is subsumed by the relation $\underline{os} = 1 + [v = 2 \cdot \underline{s}]$, where the Boolean expression $[v = 2 \cdot \underline{s}]$ is used as an integer, i.e. either 0 for false or 1 for true.
2. Even when the number of distinct values m of the codomain of $f(X_1, X_2, \ldots, X_n)$ is greater than two values, but still very small, we can use Boolean arithmetic expressions to capture concise formulae. This is done by summing up $m - 1$ Boolean-arithmetic conditions as illustrated now: e.g., let v, a, c_1 and $\underline{c}$ be the number of vertices, of arcs, of connected components having more than one vertex and the size of the smallest connected component of a digraph. As found by the CP model in [8], when a is maximal, we have the relation $c_1 = \lfloor \frac{(v+\max(-\underline{c}+v,\underline{c}))}{(2\cdot\max(\max(-\underline{c}+v,\underline{c}),2)-\max(-\underline{c}+v,\underline{c})+1)} \rfloor$, which is subsumed by the BAE $c_1 = 2 - ([\underline{c} = 1] + [(v - \underline{c}) \leq 1])$.
3. Quite often, using BAE allows one simplifying formulae with min and max as illustrated now. Let v, c, c_{23} and $\overline{s}$ be the number of vertices, connected components, connected components with two or three vertices where the size of each strongly connected component is equal to 1, and the size of the largest strongly connected component of a digraph: e.g for the graph $\bullet\!\to\!\bullet\!\to\!\bullet \quad \bullet\!\to\!\bullet \quad \bullet\!\leftrightarrow\!\bullet$

we have $v = 7, c = 3, c_{23} = 2, \overline{s} = 2$. As discovered by the CP model of [8], when c is minimal, we have $c_{23} = (v.\overline{s} \leq 3\,?\ \min(v-1,1) : 0)$,[1] which can be replaced by the Boolean relation $c_{23} = [\overline{s} = 1 \wedge v \in [2,3]]$.

4. It may occur that a formula can provide an approximate bound with an error of at most 1 on a parameter in $\mathbb{Z}$. Then, a way for getting a sharp bound is to find a Boolean formula which precisely describes the bound discrepancy. For instance, a non-sharp lower bound (with a deviation of at most 1) on the number of connected components c of a digraph $\mathcal{G}$ wrt the number of vertices v of $\mathcal{G}$, the size of the largest connected component $\overline{c}$ of $\mathcal{G}$, and the size of the smallest strongly connected component $\underline{s}$ of $\mathcal{G}$ is given by $\lceil \frac{v}{\overline{c}} \rceil$; but a sharp lower is given by $\lceil \frac{v}{\overline{c}} \rceil + [(\underline{s} < \text{fmod}(v, \overline{c})) \wedge (2 \cdot \underline{s} > \overline{c})]$, where $\text{fmod}(x, y)$ is defined by the conditional expression $(x \bmod y = 0\ ?\ y : x \bmod y)$.

The following points are specific to our equation discovery context [29]:

- As our samples are noise-free, we need to acquire formulae that *correctly represent all the samples* we have.
- As our samples correspond to instances of combinatorial objects reaching a sharp bound, this is why we search for equations rather than for inequalities.
- Simple conditions are not translated into a large set of features, which is the case for most decision tree approaches [3,4,19].
- We keep the original columns of the tables, as using one-hot encoding considerably increases the number of columns and affects the interpretability [28].

3 Describing Boolean-Arithmetic Expressions

The type of Boolean-Arithmetic expressions we consider is dictated by two opposite objectives. (*i*) On the one hand, we want to focus on concise expressions involving a limited number of variables and constants. This is motivated by the need to generate interpretable formulae and by the necessity to avoid a combinatorial explosion when searching for such formulae [12]. Consequently, we limit the number of variables and constants, as well as the number of subterms of Boolean-Arithmetic expressions. (*ii*) On the other hand, we aim at covering a variety of Boolean expressions which occurs in practice. This is done by allowing one to use a variety of arithmetic operators and Boolean functions.
To meet the above objectives, we use the following two-level description:

- First, we consider a Boolean-arithmetic condition (BAC) mentioning *no more than two variables and two constants*, the comparison operators $\leq$, $=$, $\geq$, $\in$, $\notin$ and a *variety of arithmetic operators* such as $+$, $-$, $\times$, $\lfloor \text{-} \rfloor$, $\lceil \text{-} \rceil$, mod, min, max. We currently have 53 elementary arithmetic conditions.
- Second, we build a Boolean-arithmetic term by feeding several arithmetic conditions, or their negation, to a *commutative and associative aggregation operator* such as $+$, $\vee$, $\wedge$, $\oplus$, eq, card1, voting, where:
  - $\oplus$ stands for XOR;
  - eq is equal to 1 iff all its conditions are evaluated to the same value;

[1] The expression $(cond\,?\,x : y)$ denotes x if condition $cond$ holds, y otherwise.

- card1 is equal to 1 iff only one of its conditions is evaluated to 1;
- voting is equal to 1 iff the majority of its conditions are evaluated to 1.

The use of a commutative and associative aggregation operator simplifies the interpretability of a formula and reduces the combinatorics, as the order of the BACs within a Boolean-arithmetic term is irrelevant. It allows for a compact representation of some Boolean expressions, that would otherwise be large when expressed in conjunctive or disjunctive normal form without introducing new variables. For instance, the n-ary XOR, i.e. $\oplus_{i=1}^{n} \ell_i$, is represented by a CNF consisting of 2^{n-1} clauses, where each clause mentions all literals $\ell_1, \ell_2, \ldots, \ell_n$. It also permits the use of the '+' operator in a natural way.

4 A Core Model for Acquiring BAE

This section introduces a CP-based core model for acquiring BAE. First, the model relies on soft constraints to represent learned Boolean expressions that mention a restricted number of arithmetic conditions taken from a large set of candidate conditions. Second, the model incorporates symmetry-breaking constraints resulting from the relaxation of arithmetic conditions. Section 5 will extend these symmetry-breaking constraints.

4.1 Problem Description

Given a two-dimensional table tab[1..r, 1..c] of integer values, consisting of r distinct rows and c distinct columns, where column c is functionally determined by columns $1, 2, \ldots, c-1$, the problem is to come up with a constraint model to acquire an equality constraint of the form

$$\forall j \in [1, r] : \text{tab}[j, c] = f(\text{tab}[j, 1], \text{tab}[j, 2], \ldots, \text{tab}[j, c-1]) \qquad (1)$$

i.e. a constraint that is valid for all rows of the table, where f is a Boolean-arithmetic expression mentioning $c-1$ parameters.

As we want to restrict the complexity of the acquired formulae, the expression f is limited to $n_{\text{AC}} \in \{1, 2, 3\}$ conditions chosen from $m = 53$ potential distinct BACs introduced in Sect. 3 (where a few conditions may be duplicated using different constants), and a single commutative and associative aggregation operator g selected from the set $\{\vee, \wedge, \oplus, +, \text{eq}, \text{card1}, \text{voting}\}$. As the acquisition system successively tries the different aggregation operators, we assume from now on that g is fixed. As we search for Boolean-arithmetic expressions by increasing number of BACs, we also assume that n_{AC} is fixed.

Each potential candidate BAC C_d of f (with $d \in [1, m]$) mentioning ℓ_d columns of the tab table (with $\ell_d \in [1, 3]$) and ℓ'_d coefficients (with $\ell'_d \in [0, 2]$) is represented by the term $C_d \begin{pmatrix} a_{d,1}, \ldots, a_{d,\ell_d}, \\ c_{d,1}, \ldots, c_{d,\ell'_d} \end{pmatrix}$, where:

- the variables $a_{d,1}, \ldots, a_{d,\ell_d}$ denote the indices of the distinct columns of the table tab[1..r, 1..c] mentioned by condition C_d,

– the variables $c_{d,1}, \ldots, c_{d,\ell'_d}$ represents the coefficients used in the arithmetic expression of condition C_d.

The problem is to come up with a CP-based model which, given (*i*) a commutative and associative Boolean operator $g \in \{\vee, \wedge, \oplus, +, \text{eq}, \text{card1}, \text{voting}\}$, and (*ii*) a fixed number of conditions n_{AC}, extracts the subset of relevant conditions for the expression f of Constraint (1), and finds for each used conditions all its parameters, i.e. which columns and which coefficient values it uses.

Example 1. On page 8, the left-hand side of Table 2 provides a table tab[1..9, 1..4] from which we acquire the following BAE $x_4 = [(x_1 - x_2) = 2] \vee [x_3 \leq 4]$. The acquisition process is now explained in Sect. 4.2.

4.2 A CP Core Model

Notation 1. *Given a table* tab[1..*r*, 1..*c*], *the j-th row of* tab[1..*r*, 1..*c*] *is called a* negative entry *if* tab[*j*, *c*] = 0, *and a* positive entry *otherwise.*

Selecting the BACs used in *f*. To each potential BAC C_d (with $d \in [1, m]$) of a Boolean-arithmetic expression f, we associate a variable b_d such that:

- $b_d = -1$ means that neither condition C_d, nor condition $\neg C_d$ are used in f,
- $b_d = -0$ indicates that the condition $\neg C_d$ is used in f, i.e. C_d is negated,
- $b_d = -1$ signifies that the condition C_d occurs in f.

As f should mention n_{AC} BACs, we set up the following AMONG constraint to specify that $m - n_{\text{AC}}$ conditions must be unused:

$$\text{AMONG}\,(m - n_{\text{AC}}, \langle b_1, b_2, \ldots, b_m \rangle, -1) \tag{2}$$

Selecting the Attributes Used in Each BAC. For each potential condition $C_d \begin{pmatrix} a_{d,1}, \ldots, a_{d,\ell_d}, \\ c_{d,1}, \ldots, c_{d,\ell'_d} \end{pmatrix}$ (with $d \in [1, m]$) we set all its variables $a_{d,1}, a_{d,2}, \ldots, a_{d,\ell_d}$ to 0 when the condition C_d is not used, i.e. when $b_d = -1$. We introduce the variables $a'_{d,1}, a'_{d,2}, \ldots, a'_{d,\ell_d}$ corresponding to $a_{d,1} + 1, a_{d,2} + 1, \ldots, a_{d,\ell_d} + 1$: we use the offset +1 as these variables will also be used in ELEMENT constraints whose index starts at 1.

For each potential condition C_d, its variables $a'_{d,1}, a'_{d,2}, \ldots, a'_{d,\ell_d}$ should either be all distinct and greater than or equal to 2, or be all equal to 1. This is expressed by the next GLOBAL_CARDINALITY (GCC) constraint.

$$\text{GCC}\begin{pmatrix} \left\langle a'_{d,1}, a'_{d,2}, \ldots, a'_{d,\ell_d} \right\rangle, \\ \langle 1 : \{0, \ell_d\},\ 2 : \{0, 1\},\ \ldots\ ,\ c : \{0, 1\} \rangle \end{pmatrix} \tag{3}$$

When the condition C_d is unused, i.e. $b_d = -1$, we set all its variables $a'_{d,1}, a'_{d,2}, \ldots, a'_{d,\ell_d}$ to 1 to break symmetry, i.e. to avoid enumerating over these variables:

$$\forall d \in [1, m], \forall k \in [1, \ell_d] : b_d = -1 \Leftrightarrow a'_{d,k} = 1 \tag{4}$$

To force the use of all attributes from 1 to $c-1$ of the table tab$[1..r, 1..c]$ across all selected conditions, i.e. those conditions C_d, (with $d \in [1,m]$) such that $b_d \neq -1$, we set up the following GCC constraint:

$$\text{GCC}\left(\left\langle \begin{array}{c} a'_{1,1}, a'_{1,2}, \dots, a'_{1,\ell_1}, \\ a'_{2,1}, a'_{2,2}, \dots, a'_{2,\ell_2}, \\ \dots\dots\dots\dots \\ a'_{m,1}, a'_{m,2}, \dots, a'_{m,\ell_m} \end{array} \right\rangle, \left\langle \begin{array}{c} 2 : o_2, \\ 3 : o_3, \\ \vdots \\ c : o_c \end{array} \right\rangle\right) \text{ with } o_i \in [1,m], \forall i \in [2,c] \quad (5)$$

Restricting the Coefficients of Each BAC. When the BAC $C_d\begin{pmatrix} a_{d,1}, \dots, a_{d,\ell_d}, \\ c_{d,1}, \dots, c_{d,\ell'_d} \end{pmatrix}$ is used, i.e. $b_d \neq -1$, the coefficient variables $c_{d,1}, \dots, c_{d,\ell'_d}$ denote the coefficients used in the arithmetic expression related to C_d. As we look for simple formulae, the initial domain of such variables is initialised to a small interval, e.g. $[-9, +9]$.

Note that we are not interested in acquiring conditions that can be substituted by `true` or `false` as they could be simplified away. For some types of conditions this would require additional constraints on the condition's coefficients, e.g. for $C_d(a_{d,1}, c_{d,1}, c_{d,2}) \equiv [(a_{d,1} \bmod c_{d,1}) = c_{d,2}]$ we post the additional constraints $b_d \neq -1 \Rightarrow c_{d,1} \geq 2$ and $b_d \neq -1 \Rightarrow c_{d,2} \in [0, c_{d,1} - 1]$:

- If $c_{d,1} = 1$ then the condition $C_d(a_{d,1}, c_{d,1}, c_{d,2})$ is either always `true`, when $c_{d,2} = 0$, or always `false` when $c_{d,2} \neq 0$.
- Otherwise, if $c_{d,1} \geq 2$ and $c_{d,2} \notin [0, c_{d,1} - 1]$ then the condition $C_d(a_{d,1}, c_{d,1}, c_{d,2})$ is always `false` as $(a_{d,1} \bmod c_{d,1}) \in [0, c_{d,1} - 1]$.

When the condition C_d is unused, we have $b_d = -1 \Rightarrow (c_{d,1} = \dots = c_{d,\ell'_d} = 0)$ to avoid multiple solutions stemming from the coefficients of an unused condition.

How to further restrict the initial domain of the coefficient variables wrt the entries of the table tab$[1..r, 1..c]$ will be explained in Sect. 5.3.

Setting Row Constraints. To evaluate each condition C_d wrt the j-th row of the table tab$[1..r, 1..c]$, we create the variables $v_{d,j,k}$ for the values of its k-th attributes and a variable $b_{d,j}$ for the value of C_d. This is described in the next two items:

- For each condition C_d (with $d \in [1,m]$), for each row j (with $j \in [1,r]$), and for each argument k (with $k \in [1, \ell_d]$) of condition C_d, we create a variable $v_{d,j,k}$ that gives, either the value of the k-th argument of condition C_d wrt the j-th row of the table tab$[1..r, 1..c]$, or 0 if the condition C_d is unused:
  - $\forall d \in [1,m]$, $\forall j \in [1,r]$, $\forall k \in [1, \ell_d]$:

    $$\text{ELEMENT}\left(a'_{d,k}, \langle 0, \text{tab}[j,1], \text{tab}[j,2], \dots, \text{tab}[j, c-1] \rangle, v_{d,j,k}\right).$$
- We also create a $0-1$ variable $b_{d,j}$ which will be set to true iff condition C_d holds for the j-th row of the table tab$[1..r, 1..c]$:
  - $\forall d \in [1,m]$, $\forall j \in [1,r] : b_{d,j} \Leftrightarrow C_d(v_{d,j,1}, v_{d,j,2}, \dots, v_{d,j,\ell_d})$.

Now, based on the aggregator g, we state a few row constraints for each used condition C_d (with $d \in [1, m]$) and wrt each row of the table tab$[1..r, 1..c]$. These row constraints are related to the type of aggregator g we are using. In this context, we distinguish the following types of aggregators I, II, and III:

I. Aggregators for which (*i*) positive and negative table entries have distinct row constraints and (*ii*) a single table entry may determine the Boolean arithmetic expression value. For instance, if g is the '$\wedge$' aggregator then on a positive entry, a condition C_d which is false (with $d \in [1, m]$) falsifies the Boolean arithmetic expression f. Aggregators '$\vee$' and '$\wedge$' belong to this class.
II. Aggregators for which (*i*) positive and negative table entries have distinct row constraints, and (*ii*) a single table entry cannot determine the Boolean arithmetic expression value. Aggregator 'eq' belongs to this class.
III. Aggregators for which (*i*) positive and negative table entries have the same row constraint, and (*ii*) a single table entry cannot determine the Boolean arithmetic expression value. Aggregators '+', '$\oplus$', 'card1', and 'voting' belong to this class.

Table 1 provides for each class in {I, II, III} of aggregator g the corresponding row constraints that determine the value of the Boolean arithmetic expression f. As mentioned earlier, for the first two classes, these row constraints depend on whether we have a positive or negative table entry; for the third class, the same constraint applies for both a positive and a negative table entry. We now explain the constraints stated in Table 1 for the first aggregator of each class.

[Case aggregator g is '$\vee$']

- For each positive row j (with $j \in [1, r]$), we post the constraint $\vee_{d=1}^{m} [b_d = b_{d,j}]$ to ensure that *at least one condition is true* so that the disjunction of conditions holds.
- For each condition C_d (with $d \in [1, m]$) and each negative row j (with $j \in [1, r]$), we post the constraint $\text{TABLE}(\langle (b_d, b_{d,j}) \rangle, \langle (-1, 0), (-1, 1), (0, 1), (1, 0) \rangle)$. When the condition C_d is not used, i.e. $b_d = -1$, there is no restriction on $b_{d,j}$, i.e. $b_{d,j} \in \{0, 1\}$; otherwise, *each condition must be falsified*, i.e. $b_{d,j} = 1 - b_d$, so that the corresponding disjunction of conditions is not true.

[Case aggregator g is 'eq']

- For each positive row j (with $j \in [1, r]$), we post the constraint:

$$\left[\sum_{d=1}^{m} [b_d = b_{d,j}] = n_{\text{AC}}\right] \vee \left[\sum_{d=1}^{m} [b_d = \neg b_{d,j}] = n_{\text{AC}}\right]$$

enforcing *either that all conditions hold or that all conditions are false.*
- For each negative row j (with $j \in [1, r]$) we post the constraint:

$$\left[\sum_{d=1}^{m} [b_d = b_{d,j}] < n_{\text{AC}}\right] \wedge \left[\sum_{d=1}^{m} [b_d = \neg b_{d,j}] < n_{\text{AC}}\right]$$

imposing that *at least one condition is false and at least one is true.*

Table 1. Row constraints which are posted on a positive or a negative table entry for computing the value of a Boolean arithmetic expression f, depending on the used aggregator g of classes I, II; for class III the same row constraint is posted for all entries.

Class	g	Positive entries ($\text{tab}[j,c]=1$)	Negative entries ($\text{tab}[j,c]=0$)
I	'$\vee$'	$\vee_{d=1}^{m}[b_d = b_{d,j}]$	$\text{TABLE}\left(\langle(b_d, b_{d,j})\rangle, \left\langle \begin{matrix}(-1,0),(-1,1),\\ (\ 0,1),(\ 1,0)\end{matrix} \right\rangle\right)$
	'$\wedge$'	$\text{TABLE}\left(\langle(b_d, b_{d,j})\rangle, \left\langle \begin{matrix}(-1,0),(-1,1),\\ (\ 0,0),(\ 1,1)\end{matrix} \right\rangle\right)$	$\vee_{d=1}^{m}[b_d = \neg b_{d,j}]$
II	'eq'	$\bigvee\left(\begin{matrix}\sum_{d=1}^{m}[b_d = \ b_{d,j}] = n_{\text{AC}},\\ \sum_{d=1}^{m}[b_d = \neg b_{d,j}] = n_{\text{AC}}\end{matrix}\right)$	$\bigwedge\left(\begin{matrix}\sum_{d=1}^{m}[b_d = \ b_{d,j}] < n_{\text{AC}},\\ \sum_{d=1}^{m}[b_d = \neg b_{d,j}] < n_{\text{AC}}\end{matrix}\right)$
III	'+'	$\text{tab}[j,c] = \sum_{d=1}^{m}[b_d = b_{d,j}]$	
	'$\oplus$'	$\text{tab}[j,c] = (\sum_{d=1}^{m}[b_d = b_{d,j}]) \bmod 2$	
	'card1'	$\text{tab}[j,c] = [(\sum_{d=1}^{m}[b_d = b_{d,j}]) = 1]$	
	'voting'	$\text{tab}[j,c] = [2 \cdot (\sum_{d=1}^{m}[b_d = b_{d,j}]) > n_{\text{AC}}]$	

Table 2. Illustrating the core model on the table tab[1..9, 1..4] (with columns x_1, x_2, x_3, x_4) for acquiring a Boolean-arithmetic expression explaining x_4 wrt x_1, x_2, x_3 using the '$\vee$' aggregator with two conditions C_1 and C_2 selected from the following potential candidate conditions $C_1 : x_i - x_j = cst$, $C_2 : x_i \leq cst$ and $C_3 : x_i = x_j$.

	j	table tab				$C_1 = [(x_1 - x_2) = 2]$			$C_2 = [x_3 \leq 4]$		$C_3 = [x_{k_1} = x_{k_2}]$			row constraint satisfaction
						$b_1{=}1$	$a'_{1,1}{=}2$	$a'_{1,2}{=}3$	$b_2{=}1$	$a'_{2,1}{=}4$	$b_3{=}{-1}$	$a'_{3,1}{=}1$	$a'_{3,2}{=}1$	
		x_1	x_2	x_3	x_4	$b_{1,j}$	$v_{1,j,1}$	$v_{1,j,2}$	$b_{2,j}$	$v_{2,j,1}$	$b_{3,j}$	$v_{3,j,1}$	$v_{3,j,2}$	
positive entries	1	4	2	5	**1**	1	4	2	0	5	1	0	0	`true`
	2	3	4	4	**1**	0	3	4	1	4	1	0	0	`true`
	3	1	1	3	**1**	0	1	1	1	3	1	0	0	`true`
	4	3	1	5	**1**	1	3	1	0	5	1	0	0	`true`
	5	4	1	2	**1**	0	4	1	1	2	1	0	0	`true`
negative entries	6	2	4	5	**0**	0	2	4	0	5	1	0	0	`true`
	7	4	1	5	**0**	0	4	1	0	5	1	0	0	`true`
	8	4	3	5	**0**	0	4	3	0	5	1	0	0	`true`
	9	3	5	5	**0**	0	3	5	0	5	1	0	0	`true`

[**Case aggregator** g is '+'] For each row j (with $j \in [1, r]$), we post the constraint $\text{tab}[j,c] = \sum_{d=1}^{m}[b_d = b_{d,j}]$ to ensure that *the appropriate number of conditions are satisfied.*

Example 2 (Continuation of Example 1). Table 2 summarises the acquisition of the BAE $x_4 = [(x_1 - x_2) = 2] \vee [x_3 \leq 4]$ from the table tab[1..9, 1..4]: it provides the main variables introduced by the core model. First, note that only conditions C_1 and C_2 are selected, as $b_3 = -1$. For the first positive entry (i.e. $j = 1$) and the first negative entry (i.e. $j = 6$), we now show that the corresponding row constraints described in Table 1 are true:

- As row 1 is a positive entry, we post the constraint $[b_1 = b_{1,1}] \vee [b_2 = b_{2,1}] \vee [b_3 = b_{3,1}]$ which is true as $b_1 = b_{1,1}$ holds.
- As row 6 is a negative entry, we post the constraint TABLE $(\langle(b_d, b_{d,6})\rangle, \mathcal{T})$, with $\mathcal{T} = \langle(-1,0), (-1,1), (0,1), (1,0)\rangle$, for each condition C_d ($d \in \{1,2,3\}$). All three constraints hold for the sixth row.

5 Enhancing the Core Model

5.1 Linking the Number of Conditions, Their Arity, and the Number of Attributes

We introduce the following constraints to explicitly restrict the potential combinations of unary, binary and ternary conditions to consider.

Notation 2. *Within the expression f formed by n_{AC} conditions, let $n_{\text{AC},k}$ denote the number of conditions mentioning k attributes.*

Since we restrict the Boolean-arithmetic expression f to at most three conditions, we state the constraints $n_{\text{AC}} = \sum_{k=1}^{3} n_{\text{AC},k}$ and $\sum_{k=1}^{3} k \cdot n_{\text{AC},k} = \sum_{i=2}^{c} o_i$, where o_i is the number of occurrences of value i in the variables $a'_{d,1}, a'_{d,2}, \ldots, a'_{d,\ell_d}$, as stated by the GCC constraint (3) of the core model. We now state the lower and upper bounds on the number of distinct attributes $c-1$ appearing in the expression f wrt $n_{\text{AC},k}$ (with $k \in [1,3]$):

$$\bullet\ c - 1 \geq \max_{k=1}^{3} (k \cdot \min(1, n_{\text{AC},k})), \quad \bullet\ c - 1 \leq \sum_{k=1}^{3} (k \cdot n_{\text{AC},k}).$$

5.2 Symmetry Breaking

As a formula may involve commutative arithmetic operators whose arguments can be interchanged, and mention several occurrences of the same condition which can be swapped, we show how to restrict the search space for formulae.

Commutative Arithmetic Operators. For each BAC $C_d\left(\frac{a_{d,1}, a_{d,2},}{c_{d,1}, \ldots, c_{d,\ell'_d}}\right)$ (with $d \in [1,m]$) mentioning two attributes $a_{d,1}$ and $a_{d,2}$, as well as a commutative arithmetic operator such as $+$, min, or max, we order its arguments only when the condition is used, by posting a constraint of the form $b_d \neq -1 \Rightarrow a'_{d,1} < a'_{d,2}$ on its variables $a'_{d,1}$ and $a'_{d,2}$.

Conditions Mentioning the Same Comparison and Arithmetic Operators. In case a same condition would occur several times in the expression f, positively or negatively, or with different attributes, we post symmetry-breaking constraints to prevent generating equivalent subexpressions. We order the list of potential BACs $C_1, C_2, \ldots, C_m$ so that conditions that use the same comparison operator $\leq$, $=$, $\geq$, $\in$, as well as the same arithmetic operator $+$, $-$, $\times$, $\lfloor - \rfloor$, $\lceil - \rceil$, mod, min, max are located consecutively. For each pair of consecutive conditions $C_d\begin{pmatrix} a_{d,1}, \ldots, a_{d,\ell_d}, \\ c_{d,1}, \ldots, c_{d,\ell'_d} \end{pmatrix}$, $C_{d+1}\begin{pmatrix} a_{d+1,1}, \ldots, a_{d+1,\ell_{d+1}}, \\ c_{d+1,1}, \ldots, c_{d+1,\ell'_{d+1}} \end{pmatrix}$ (with $d \in [1, m-1]$) using the same comparison and arithmetic operators, we enforce the following symmetry-breaking constraint.

The idea is to impose a strict lexicographic ordering constraint (SLOC) between the variables of such consecutive conditions C_d and C_{d+1}. However, we need to consider the cases where these conditions are unused ($b_d = -1$, $b_{d+1} = -1$), negated ($b_d = 0$, $b_{d+1} = 0$) or positively used ($b_d = 1$, $b_{d+1} = 1$). We use the following idea to adapt the SLOC to our context: a SLOC can be described as a finite automaton whose input alphabet consists of letters that pairwise compare the k-th components of two vectors [6]. We compare the vectors $\overrightarrow{U} = (b_d, a'_{d,1}, a'_{d,2}, \ldots, a'_{d,\ell_d}) = (u_1, u_2, \ldots, u_{\ell_d+1})$ and $\overrightarrow{V} = (b_{d+1}, a'_{d+1,1}, a'_{d+1,2}, \ldots, a'_{d+1,\ell_{d+1}}) = (v_1, v_2, \ldots, v_{\ell_d+1})$. Recall from Sect. 4.2 that (i) depending on whether condition C_d is unused, negated or used positively, b_d will be set to -1, 0 or 1 respectively, and that (ii) the variables $a'_{d,1}, a'_{d,2}, \ldots, a'_{d,\ell_d}$ are all in the range $[2, c]$ as we applied the offset $+1$. By pair-

Table 3. Definition of the input letters of the finite automaton depicted in Part (A) of Fig. 1 used for breaking symmetry between two consecutive conditions

Input letter	Corresponding condition	Comment
$w_k = 0$	$u_k = -1 \land v_k = -1$	Both conditions are unused.
$w_k = 1$	$(u_k = 0 \lor u_k = 1) \land v_k = -1$	Only one condition is used.
$w_k = 2$	$u_k = 1 \land v_k = 0$	The 1st condition is used positively, and the negation of the 2nd condition is used.
$w_k = 3$	$u_k = 0 \land v_k = 0$	The negation of the 1st condition is used, and the negation of the 2nd condition is used.
$w_k = 4$	$u_k = 1 \land v_k = 1$	$k = 1$: both conditions are used positively, $k > 1$: attributes of both conditions are unused.
$w_k = 5$	$u_k > 1 \land v_k = 1$	u_k is an attribute of the 1st condition, and v_k an unused attribute of the 2nd condition, as the 2nd condition is unused.
$w_k = 6$	$u_k > 1 \land v_k > 1 \land u_k = v_k$	u_k and v_k are attributes of the two used conditions, such that $u_k = v_k$.
$w_k = 7$	$u_k > 1 \land v_k > 1 \land u_k > v_k$	u_k and v_k are attributes of the two used conditions, such that $u_k > v_k$.
$w_k = 8$	$u_k > 1 \land v_k > 1 \land u_k < v_k$	u_k and v_k are attributes of the two used conditions, such that $u_k < v_k$.

wise comparing the k-th components of vectors $\overrightarrow{U}$ and $\overrightarrow{V}$ (with $k \in [1, \ell_d + 1]$) we create the following vector $\overrightarrow{W} = (w_1, w_2, \ldots w_{\ell_d+1})$, where each component is defined by one of the nine letters $0, 1, \ldots, 8$ described in Table 3.

We then force the components of vector $\overrightarrow{W}$ to be accepted by the finite automaton given in Fig. 1. The three accepting states labelled by **n**, **o**, and **t** respectively correspond to the fact that (*i*) none of the conditions C_d, C_{d+1} is used, (*ii*) only the first condition C_d is used, and (*iii*) the two conditions C_d, C_{d+1} are both used. The outgoing transitions from state ϵ to states $\overset{\mathbf{t}}{\neq}$ and $\overset{\mathbf{t}}{>}$ enforce that, when using a condition and its negated form, the negated form is located in the second position. The two outgoing transitions of state $\overset{\mathbf{t}}{>}$ ensure that the arguments of the first used condition are lexicographically strictly greater than the arguments of the second condition, while the two outgoing transitions of state $\overset{\mathbf{t}}{\neq}$ force the two conditions to not use the same arguments.

5.3 Pre-computing the Combinations of Possible Values of the Coefficients of a Condition

Most BACs $C_d\begin{pmatrix} a_{d,1}, \ldots, a_{d,\ell_d}, \\ c_{d,1}, \ldots, c_{d,\ell'_d} \end{pmatrix}$ can be presented as a comparison of the form $C'_d(P) \diamond c_{d,\ell'_d}$ (with $\diamond \in \{\leq, =, \geq\}$), where $C'_d(P)$ is an arithmetic expression parameterised by $P = \begin{pmatrix} a'_{d,1}, \ldots, a'_{d,\ell_d}, \\ c_{d,1}, \ldots, c_{d,\ell'_d-1} \end{pmatrix}$. Such BACs in a Boolean formula f must not be equivalent to `true` or `false`, as otherwise they could be simplified away from f. We also want to avoid generating a condition involving an inequality when an equality would suffice. For this purpose we proceed as follows.

- For each possible combination of values p of parameter P wrt the potential values of $a'_{d,1}, \ldots, a'_{d,\ell_d}, c_{d,1}, \ldots, c_{d,\ell'_d-1}$, we compute the feasible values of $C'_d(p)$ wrt all the table entries of tab[1..r, 1..c]. We denote by $\mathcal{V}_{d,p}$ such sets.

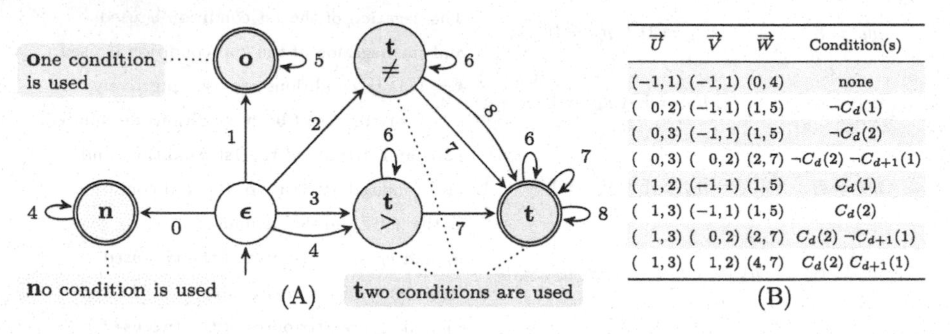

$\overrightarrow{U}$	$\overrightarrow{V}$	$\overrightarrow{W}$	Condition(s)
(−1, 1)	(−1, 1)	(0, 4)	none
(0, 2)	(−1, 1)	(1, 5)	$\neg C_d(1)$
(0, 3)	(−1, 1)	(1, 5)	$\neg C_d(2)$
(0, 3)	(0, 2)	(2, 7)	$\neg C_d(2)$ $\neg C_{d+1}(1)$
(1, 2)	(−1, 1)	(1, 5)	$C_d(1)$
(1, 3)	(−1, 1)	(1, 5)	$C_d(2)$
(1, 3)	(0, 2)	(2, 7)	$C_d(2)$ $\neg C_{d+1}(1)$
(1, 3)	(1, 2)	(4, 7)	$C_d(2)$ $C_{d+1}(1)$

Fig. 1. (A) Automaton for breaking symmetries between two consecutive conditions C_d, C_{d+1} sharing the same comparison and arithmetic operators, where accepting states are denoted by a double circle; (B) Examples of vectors $\overrightarrow{U}$, $\overrightarrow{V}$, $\overrightarrow{W}$ and corresponding used conditions with their arguments (each condition mentions one single attribute).

Table 4. Example table for pre-computing the possible values of the coefficients for conditions C_1 and C_2

x_1	x_2	x_c	$[x_1 - x_2]$	$[x_2 - x_1]$	$[x_1 \bmod 3]$	$[x_2 \bmod 3]$
1	2	0	−1	1	1	2
2	1	0	1	−1	2	1
1	2	1	−1	1	1	2
1	3	1	−2	2	1	0
1	4	1	−3	3	1	1

- Then, depending on the comparison operator $\diamondsuit$ used in condition C_d, we derive for each combination of values p of parameter P, the set of values of coefficient c_{d,ℓ'_d} which does not make condition C_d always true or always false. We denote such sets as $\mathcal{V}_{d,p}^{\diamondsuit}$. They are obtained from the sets $\mathcal{V}_{d,p}$ in the following way.
  - [$\diamondsuit$ is '=']: when the coefficient c_{d,ℓ'_d} is assigned a value outside $\mathcal{V}_{d,p}$ the condition C_d would always be false; if the cardinality of $\mathcal{V}_{d,p}$ is 1 then $\mathcal{V}_{d,p}^{\diamondsuit} = \emptyset$ (i.e. if there is only one value the condition would always be true), otherwise $\mathcal{V}_{d,p}^{\diamondsuit} = \mathcal{V}_{d,p}$.
  - [$\diamondsuit$ is '$\leq$' or '$\geq$']: let α and ω respectively be the smallest and the largest value of the set $\mathcal{V}_{d,p}$; then $\mathcal{V}_{d,p}^{\diamondsuit} = \mathcal{V}_{d,p} \setminus \{\alpha, \omega\}$. The intuition for $\diamondsuit =$'$\leq$' is as follows: if we keep α then $\diamondsuit =$'$\leq$' is equivalent to $\diamondsuit =$'$=$' ; if we keep ω the condition will always be true. For '$\geq$' the intuition is symmetrical.
- We may further reduce the set $\mathcal{V}_{d,p}^{\diamondsuit}$ by considering the aggregator g. First, for each possible combination of values p of parameter P, we compute the feasible values of $C'_d(p)$ wrt all the positive (resp. negative) table entries of tab[1..r, 1..c]. We denote by $\mathcal{V}_{d,p}^{\text{pos}}$ (resp. $\mathcal{V}_{d,p}^{\text{neg}}$) such sets. From these sets, we compute the further restricted set $\mathcal{V}_{d,p}^{\diamondsuit,g}$ as follows:
  - [g is '$\wedge$']: if $\diamondsuit$ is '=' then $\mathcal{V}_{d,p}^{\diamondsuit,g} = \mathcal{V}_{d,p}^{\text{pos}}$ else $\mathcal{V}_{d,p}^{\diamondsuit,g} = \mathcal{V}_{d,p}^{\text{pos}} \cap \mathcal{V}_{d,p}^{\diamondsuit}$,
  - [$g \in \{`\vee', `+'\}$]: $\mathcal{V}_{d,p}^{\diamondsuit,g} = \mathcal{V}_{d,p}^{\diamondsuit} \setminus \mathcal{V}_{d,p}^{\text{neg}}$,
  - [$g \notin \{`\wedge', `\vee', `+'\}$]: $\mathcal{V}_{d,p}^{\diamondsuit,g} = \mathcal{V}_{d,p}^{\diamondsuit}$.
- Finally, we set up the table constraint TABLE $\left(\left\langle \begin{matrix} a'_{d,1}, \ldots, a'_{d,\ell_d}, \\ c_{d,1}, \ldots, c_{d,\ell'_d} \end{matrix} \right\rangle, \mathcal{S}\right)$ where $\mathcal{S}$ corresponds to the union of cartesian products $\cup_{p \in P}(p \times \mathcal{V}_{d,p}^{\diamondsuit,g})$.

Example 3. To illustrate the process, consider Table 4. There are two input columns 1 and 2 and the output column c. Consider the two conditions $C_1 = [a_{1,1} - a_{1,2} = c_{1,1}]$ and $C_2 = [a_{2,1} \bmod c_{2,1} \geq c_{2,2}]$.

- For C_1 we have only two options for $p = \{a_{1,1}, a_{1,2}\}$, namely:

  1) $p = \{1, 2\}$: $\begin{cases} \mathcal{V}_{1,p} = \{-3, -2, -1, 1\}, \ \mathcal{V}_{1,p}^{=} = \mathcal{V}_{1,p}, \\ \mathcal{V}_{1,p}^{=,`\wedge'} = \{-3, -2, -1\}, \ \mathcal{V}_{1,p}^{=,`\vee'} = \mathcal{V}_{1,p}^{=} \setminus \{-1, 1\} = \{-3, -2\}. \end{cases}$

2) $p = \{2,1\}$: $\begin{cases} \mathcal{V}_{1,p} = \{-1,1,2,3\}, \ \mathcal{V}^{=}_{1,p} = \mathcal{V}_{1,p}, \\ \mathcal{V}^{=,`\wedge'}_{1,p} = \{1,2,3\}, \quad \mathcal{V}^{=,`\vee'}_{1,p} = \mathcal{V}^{=}_{1,p} \setminus \{-1,1\} = \{2,3\}. \end{cases}$

- For C_2 we need to enumerate on $c_{2,1}$. Wlog, we only consider the case $c_{2,1} = 3$. In this context, the options for $p = \{a_{2,1}, c_{2,1}\}$ are:
  1) $p = \{1,3\}$: $\mathcal{V}_{2,p} = \{1,2\}$, $\alpha = 1$, $\omega = 2$, $\mathcal{V}^{\geq}_{2,p} = \mathcal{V}_{2,p} \setminus \alpha, \omega = \emptyset$, i.e. this set of options for this condition is not considered any further.
  2) $p = \{2,3\}$: $\begin{cases} \mathcal{V}_{2,p} = \{0,1,2\}, \ \alpha = 0, \omega = 2, \ \mathcal{V}^{\geq}_{1,p} = \mathcal{V}_{2,p} \setminus \{\alpha,\omega\} = \{1\}, \\ \mathcal{V}^{\text{pos}}_{2,p} = \{1\}, \ \mathcal{V}^{\text{neg}}_{2,p} = \{1,2\}, \\ \mathcal{V}^{\geq,`\wedge'}_{2,p} = \mathcal{V}^{\text{pos}}_{2,p} \cap \mathcal{V}^{\geq}_{2,p} = \{1\}, \mathcal{V}^{\geq,`\vee'}_{2,p} = \mathcal{V}^{\geq}_{2,p} \setminus \mathcal{V}^{\text{neg}}_{2,p} = \emptyset. \end{cases}$

6 Evaluation

The CP core model introduced in Sect. 4 and its extension described in Sect. 5 were evaluated in the context of the search of conjectures on sharp bounds on characteristics of several combinatorial objects, which we now describe.

- **digraph (without isolated vertex):** a set of vertices $\mathcal{V}$ and a set of ordered pairs of vertices $\mathcal{A}$ with the restriction that each vertex of $\mathcal{V}$ occurs in at least one pair of $\mathcal{A}$ [5].
- **rooted tree:** a connected acyclic undirected graph where a vertex is designed as the "root" of the tree [18].
- **rooted forest:** a disjoint union of rooted trees [18]; we also consider a variant, **rooted forest2**, where all rooted trees have at least two vertices.
- **partition:** a partition of a set $\mathcal{S}$ is a collection of possibly empty subsets of $\mathcal{S}$ such that every element of $\mathcal{S}$ is in exactly one of the subsets of the collection. The use of a partition was motivated the by fact that a partition can be interpreted as a solution to the conjunction of the NVALUE (i.e. the number of partition subsets, see [26]) and the BALANCE (i.e. the difference between the cardinalities of the largest and smallest subsets of the partition, see [7]) constraints. Motivated by the extension of the BALANCE constraint, i.e. ALL_BALANCE [9], we also consider a version of partition named **partition0** where all subsets of $\mathcal{S}$ are non-empty.
- **stretch:** a solution of a stretch constraint on 0–1 variables, where a subsequence of 1 immediately preceded and followed by a 0 is called a *stretch* [27]; we also consider the variant named **cyclic stretch** where, when the sequence begins and terminates by 1, those two 1 belong to the same stretch.

Table 5 shows for each combinatorial object some characteristics we consider, and some conjectures found using the Boolean model described in this paper. Those conjectures are equalities which express (*i*) either the value of a characteristic when another characteristic is reaching its sharp bound, (*ii*) either a sharp bound formulated wrt other characteristics [8]. We evaluate the CP models of Sect. 4 and 5 wrt to the following two aspects:

Table 5. Examples of characteristics (char.) of combinatorial objects and corresponding conjectures: (*i*) $c, s, oc, \underline{c}$ and $\overline{s}$: *number of connected components (cc), strongly connected components (scc), connected components with at least two vertices, size of the smallest cc* and *size of the largest scc* of a **digraph**; (*ii*) c_O: *denote* 0 *if all the cc have same maximal size, and* $\underline{c}$ *otherwise*, for a **digraph**; (*iii*) v and f: *number of vertices* and *leaves* in a **rooted tree**; (*iv*) $\overline{d}$: *largest degree of a parent node* in a **rooted tree** or a **rooted forest**; (*v*) $\underline{p}$ and $\underline{t}$: *minimum depth* and *size of the smallest tree* in a **rooted forest**; (*vi*) n, $nval$, and $\underline{m}$: *number of elements, number of subsets,* and *cardinality of the smallest subset* in a **partition**; (*vii*) sr, dr, and $\underline{dm}$: *difference between the number of elements of the largest and smallest stretches, difference between the maximum and minimum distance of consecutive stretches,* and *minimum distance between consecutive stretches* in **stretch**; (*viii*) n, ng, and osc: *number of elements, total number of stretches, and number of stretches which have more than one element when the number of element of the largest stretch is maximal* in **cyclic stretch**.

Combinatorial object	Number of char.	Some of the used char.	Examples of discovered conjectures
digraph	20	$c, \underline{c}, s, oc$	$c = 1 + [\neg(s \leq \underline{c} \wedge oc \leq 1)]$
digraph	20	$c_O, c, s, \overline{s}$	$c_O = [\neg\,\text{voting}(c = s, c = 1, \min(c, \overline{s}) = 1)]$
rooted tree	6	$\overline{d}, v, f$	$\overline{d} = (v = f \vee v = 1\,?\,0 : f)$
rooted forest	11	$\underline{p}, \overline{d}, \underline{t}$	$\overline{d} = 2 - ([\underline{p} = 0] + [(\underline{t} - \underline{p}) = 1])$
partition	14	$n, nval, \underline{m}$	$nval = 1 + [2 \cdot \underline{m} \leq n]$
stretch	26	$sr, dr, \underline{dm}$	$\underline{dm} = [(sr + dr) \geq 1]$
cyclic stretch	26	osc, n, ng	$osc = [\neg\,\text{card1}(n = 2 \cdot ng, n \cdot ng = 3, n \cdot ng \leq 3)]$

Table 6. Computing time for the Core and the Enhanced models wrt aggregators (left side) and combinatorial objects (right side)

g	n_{AC}	Number of conjectures	Average time per conjecture C. Model	E. Model
$\wedge$	1	2311	0.36s	0.35s
$\wedge$	2	341	25.5s	11s
$\wedge$	3	5	2591s	542s
$\vee$	2	190	18.9s	6.1s
$\vee$	3	1	3062s	292s
eq	2	143	35.7s	14.7s
eq	3	47	309s	50s
+	2	662	2s	1.3s
card1	3	2	1003s	27.5s
voting	3	4	345s	66s

Combinatorial object	Number of conjectures	Average time per conjecture C. Model	E. Model
digraph	546	43s	10s
rooted tree	56	1.5s	1s
rooted forest	229	3.4s	1.3s
rooted forest2	586	18.2s	5.9s
partition	24	2.7s	1.7s
partition0	24	0.8s	0.5s
stretch	1059	10s	2.7s
cyclic stretch	1182	6.4s	1.8s
Total	3706	14.4h	3.9h

1. The computing time spent by the core model of Sect. 4 (i.e. C. Model) and by its enhanced version of Sect. 5 (i.e. E. Model) wrt (*i*) the type of aggregator used in a BAE, and wrt (*ii*) the kind of combinatorial object. For this aspect, we test 3706 examples of acquired BAE on a MacBookPro with a 2.6 GHz Core i7 and 16 Gb of memory using SICStus 4.6.0. Table 6 shows that the E. model acquires a BAE with, on average, 73% less time than the C. Model. Additional tests showed that using just the constraints from Sect. 5.1 increases the speed of the C. Model by ≈5%, just the constraints from Sect. 5.2 - by ≈63% and just the constraints from Sect. 5.3 - by ≈48%.

2. Using the enhanced model together with the model acquiring polynomial of [8] (i.e. the EP. Model), Table 7 gives (*i*) the number of Boolean formulae found, i.e. 642 (sum of second columns), replacing a formula with a polynomial, and (*ii*) the number of new Boolean formulae, i.e. 56 (sum of third columns) discovered compared to the model described in [8], which only looked for formulae with polynomials and arithmetic functions involving two polynomials (i.e. the P. Model).

7 Related Work

Learning purely Boolean expressions from data is widely reported in the literature. A significant number of papers explore the acquisition of relevant features, often called the "*relevant features problem*" (RFP). Blum formalises the RFP in [10], and provides a survey of various algorithms in [11]. The RFP can be applied to features that are Boolean, integer or continuous, each of which requires its own approach [15, chapter 1.2]. Some of the works focusing on purely Boolean RFP are described in [13,23,25]. In [24], Mutlu and Oghaz provide a taxonomy of Boolean and non-Boolean feature extraction techniques applied to graphs. Other works present the acquisition of Boolean expressions as a part of the Boolean rules extraction process for classification problems using SAT [31] or neural networks [22]. Lastly, there are papers [14,16] focusing on the construction and the simplification of Boolean functions. The acquisition of Boolean-arithmetic expressions is often used in the context of classification problems. Random forest [17], decision trees [3,4,19], Bayesian rule lists [30], fuzzy association rules [2] and rough sets [20] approaches are used. Most of the work considers the acquisition of relatively simple Boolean-arithmetic expressions of the type "*attribute has a value of*". The SEEN system [19] extracts more complex Boolean-arithmetic expressions that contain the $+$, $\times$ and / arithmetic operators: it calls such domain "*logical-arithmetic expression mining*".

Beyond the domain of Boolean formulae, synthesising formulae from data [1] mostly relies on a generate and test approach to produce candidate formulae of increasing complexity for a fixed grammar. In our context, applying techniques that minimise an error function produces complicated formulae that are not verified wrt all input data. In [8] we compared our approach to methods used for symbolic regression such as GPlearn and ffx: GPlearn generally found no formulae, while ffx discovered formulae with a large number of terms.

Table 7. Contribution of the EP. Model that both acquires Boolean formulae (BF) and polynomials compared to searching only formulae with polynomials with the P. Model

Combi-natorial object	Number of (BF) which are: found	replacing polynomials	new
digraph	164	118	46
rooted tree	26	26	0
rooted forest	91	86	5
rooted forest2	149	145	4

Combi-natorial object	Number of (BF) which are: found	replacing polynomials	new
partition	20	20	0
partition0	10	10	0
stretch	93	93	0
cyclic stretch	145	144	1

8 Conclusion

The paper presents a CP model for learning BAE that can cope with a variety of expressions involving the most common Boolean aggregators and arithmetic operators. In the context of sharp bound acquisition, this complements the model introduced in [8] for learning equations whose right-hand sides are polynomials. The model is relevant not only in the context of digraphs, but also for other combinatorial objects such as rooted trees or partitions.

References

1. Alur, R., Singh, R., Fisman, D., Solar-Lezama, A.: Search-based program synthesis. Commun. ACM **61**(12), 84–93 (2018). https://doi.org/10.1145/3208071
2. Au, W.H., Chan, K.C.: Mining fuzzy association rules in a bank-account database. IEEE Trans. Fuzzy Syst. **11**(2), 238–248 (2003)
3. Aung, M.S.H., et al.: Comparing analytical decision support models through Boolean rule extraction: a case study of ovarian Tumour malignancy. In: Liu, D., Fei, S., Hou, Z., Zhang, H., Sun, C. (eds.) ISNN 2007. LNCS, vol. 4492, pp. 1177–1186. Springer, Heidelberg (2007). https://doi.org/10.1007/978-3-540-72393-6_139
4. Barbareschi, M., Barone, S., Mazzocca, N.: Advancing synthesis of decision tree-based multiple classifier systems: an approximate computing case study. Knowl. Inf. Syst. **63**(6), 1577–1596 (2021). https://doi.org/10.1007/s10115-021-01565-5
5. Beldiceanu, N.: Global constraints as graph properties on a structured network of elementary constraints of the same type. In: Dechter, R. (ed.) CP 2000. LNCS, vol. 1894, pp. 52–66. Springer, Heidelberg (2000). https://doi.org/10.1007/3-540-45349-0_6
6. Beldiceanu, N., Carlsson, M., Petit, T.: Deriving filtering algorithms from constraint checkers. In: Wallace, M. (ed.) CP 2004. LNCS, vol. 3258, pp. 107–122. Springer, Heidelberg (2004). https://doi.org/10.1007/978-3-540-30201-8_11
7. Beldiceanu, N., Carlsson, M., Rampon, J.X.: Global Constraint Catalog, 2nd Edition (revision a). Technical report T2012–03, Swedish Institute of Computer Science (2012). http://ri.diva-portal.org/smash/get/diva2:1043063/FULLTEXT01.pdf
8. Beldiceanu, N., Cheukam-Ngouonou, J., Douence, R., Gindullin, R., Quimper, C.G.: Acquiring maps of interrelated conjectures on sharp bounds. In: 28th International Conference on Principles and Practice of Constraint Programming (CP 2022). Schloss Dagstuhl-Leibniz-Zentrum für Informatik (2022)
9. Bessiere, C., et al.: The balance constraint family. In: O'Sullivan, B. (ed.) CP 2014. LNCS, vol. 8656, pp. 174–189. Springer, Cham (2014). https://doi.org/10.1007/978-3-319-10428-7_15
10. Blum, A.: Relevant examples and relevant features: thoughts from computational learning theory. In: AAAI Fall Symposium on Relevance, vol. 5, p. 1 (1994)
11. Blum, A.L., Langley, P.: Selection of relevant features and examples in machine learning. Artif. Intell. **97**(1–2), 245–271 (1997)
12. Brence, J., Todorovski, L., Džeroski, S.: Probabilistic grammars for equation discovery. Knowl.-Based Syst. **224** (2021). https://doi.org/10.1016/j.knosys.2021.107077

13. Forman, G., Kirshenbaum, E.: Extremely fast text feature extraction for classification and indexing. In: Proceedings of the 17th ACM Conference on Information and Knowledge Management, pp. 1221–1230 (2008)
14. Golia, P., Slivovsky, F., Roy, S., Meel, K.S.: Engineering an efficient Boolean functional synthesis engine. In: 2021 IEEE/ACM International Conference On Computer Aided Design (ICCAD), pp. 1–9. IEEE (2021)
15. Guyon, I., Elisseeff, A.: An introduction to feature extraction. In: Guyon, I., Nikravesh, M., Gunn, S., Zadeh, L.A. (eds.) Feature Extraction. Studies in Fuzziness and Soft Computing, vol. 207, pp. 1–25. Springer, Berlin, Heidelberg (2006). https://doi.org/10.1007/978-3-540-35488-8_1
16. Jakobovic, D., Picek, S., Martins, M.S., Wagner, M.: Toward more efficient heuristic construction of Boolean functions. Appl. Soft Comput. **107**, 107327 (2021)
17. Jun, S., Lee, S., Chun, H.: Learning dispatching rules using random forest in flexible job shop scheduling problems. Int. J. Prod. Res. **57**(10), 3290–3310 (2019)
18. Knuth, D.: Art of Computer Programming, Volume 4, Generating All Trees, pp. 461–462. Addison-Wesley, Boston (2006)
19. Kosman, E., Kolchinsky, I., Schuster, A.: Mining logical arithmetic expressions from proper representations. In: Proceedings of the 2022 SIAM International Conference on Data Mining (SDM), pp. 621–629. SIAM (2022)
20. Lambert-Torres, G.: Application of rough sets in power system control center data mining. In: 2002 IEEE Power Engineering Society Winter Meeting. Conference Proceedings (Cat. No. 02CH37309). vol. 1, pp. 627–631. IEEE (2002)
21. Larson, C.E., Van Cleemput, N.: Automated conjecturing iii. Ann. Math. Artif. Intell. **81**(3), 315–327 (2017)
22. Mereani, F., Howe, J.M.: Exact and approximate rule extraction from neural networks with Boolean features. In: Proceedings of the 11th International Joint Conference on Computational Intelligence, pp. 424–433. SCITEPRESS (2019)
23. Mossel, E., O'Donnell, R., Servedio, R.A.: Learning functions of k relevant variables. J. Comput. Syst. Sci. **69**(3), 421–434 (2004)
24. Mutlu, E.C., Oghaz, T.A.: Review on graph feature learning and feature extraction techniques for link prediction. arXiv preprint arXiv:1901.03425 (2019)
25. Nguifo, E.M., Njiwoua, P.: Using lattice-based framework as a tool for feature extraction. In: Nédellec, C., Rouveirol, C. (eds.) ECML 1998. LNCS, vol. 1398, pp. 304–309. Springer, Heidelberg (1998). https://doi.org/10.1007/BFb0026700
26. Pachet, F., Roy, P.: Automatic generation of music programs. In: Jaffar, J. (ed.) CP 1999. LNCS, vol. 1713, pp. 331–345. Springer, Heidelberg (1999). https://doi.org/10.1007/978-3-540-48085-3_24
27. Pesant, G.: A filtering algorithm for the stretch constraint. In: Walsh, T. (ed.) CP 2001. LNCS, vol. 2239, pp. 183–195. Springer, Heidelberg (2001). https://doi.org/10.1007/3-540-45578-7_13
28. Schelldorfer, J., Wuthrich, M.V.: Nesting classical actuarial models into neural networks (2019). Available at SSRN 3320525
29. Todorovski, L.: Equation discovery. In: Sammut, C., Webb, G.I. (eds.) Encyclopedia of Machine Learning, pp. 327–330. Springer, Boston, MA (2011). https://doi.org/10.1007/978-0-387-30164-8_258
30. Yang, H., Rudin, C., Seltzer, M.: Scalable Bayesian rule lists. In: International Conference on Machine Learning, pp. 3921–3930. PMLR (2017)
31. Yu, J., Ignatiev, A., Stuckey, P.J., Le Bodic, P.: Computing optimal decision sets with SAT. In: Simonis, H. (ed.) CP 2020. LNCS, vol. 12333, pp. 952–970. Springer, Cham (2020). https://doi.org/10.1007/978-3-030-58475-7_55

Generating Random Instances of Weighted Model Counting

An Empirical Analysis with Varying Primal Treewidth

Paulius Dilkas(✉)

National University of Singapore, Singapore, Singapore
paulius.dilkas@nus.edu.sg

Abstract. Weighted model counting (WMC) is an extension of propositional model counting with applications to probabilistic inference and other areas of artificial intelligence. In recent experiments, WMC algorithms perform similarly overall but with significant differences on specific subsets of benchmarks. A good understanding of the differences in the performance of algorithms requires identifying key characteristics that favour some algorithms over others. In this paper, we introduce a random model for WMC instances with a parameter that influences primal treewidth—the parameter most commonly used to characterise the difficulty of an instance. We then use this model to experimentally compare the performance of WMC algorithms c2D, CACHET, D4, DPMC, and MINIC2D. Using these random instances, we show that the easy-hard-easy pattern is different for algorithms based on dynamic programming and algebraic decision diagrams than for all other solvers. We also show how all WMC algorithms scale exponentially with respect to primal treewidth and how this scalability varies across algorithms and densities. Finally, we combine insights from experiments involving both random and competition instances to determine how the best-performing WMC algorithm varies depending on clause density and primal treewidth.

Keywords: Weighted model counting · Random model · Parameterised complexity

1 Introduction

Weighted model counting (WMC)—a weighted generalisation of propositional model counting (#SAT) [19]—has emerged as a powerful computational framework for problems in a variety of domains. In particular, WMC has been used to perform probabilistic inference for graphical models [8,16,17,29,71], probabilistic programs [55], and probabilistic logic programs [45]. More recently, WMC was used in the context of neural-symbolic artificial intelligence as well [75]. Extensions of WMC add support for continuous variables [11], infinite domains [10], and first-order logic [51,73] and generalise the definition to

The work was done while the author was a PhD student at the University of Edinburgh.

© The Author(s), under exclusive license to Springer Nature Switzerland AG 2023
A. A. Cire (Ed.): CPAIOR 2023, LNCS 13884, pp. 395–416, 2023.
https://doi.org/10.1007/978-3-031-33271-5_26

support arbitrary pseudo-Boolean functions instead of clauses [35]. Exact WMC algorithms can be broadly classified as based on search [69,72], knowledge compilation [30,58,61], and dynamic programming [38,39]. Other alternatives include approximate [15,65] and parallel algorithms [24,44], hybrid approaches [53], quantum computing [66], and reduction to model counting [14].

Recent papers that include experimental comparisons of WMC algorithms show many of them performing very similarly overall [38,39] but with overwhelming differences when run on specific subsets of data [34,35,58]. Examples of such segregating data sets include bipartite Bayesian networks by Sang et al. [71] and relational Bayesian networks by Chavira et al. [20] that encode reachability in graphs under node deletion. So far, such performance differences remain unexplained. However, knowledge about the nature of these differences can inform our choices and aid in further algorithmic developments. Moreover, identifying performance predictors of algorithms is often an important step in developing a portfolio approach to the problem [76]. Lastly, if new algorithms are always tested on the same set of benchmarks, eventually they may become somewhat fitted to the particular characteristics of those instances, leading to algorithms that may perform worse when run on new types of data [56].

Both theoretical and experimental analysis of SAT algorithms on random instances is a rich area of research spanning almost forty years. Variations of some of the first random models ever proposed [46,63] continue to be instrumental up to this day for, e.g., establishing the location of the threshold between satisfiable and unsatisfiable instances [2] and approximating #SAT [47]. Other random models consider non-uniform variable frequencies [3], fixing the number of times each variable occurs both positively and negatively [22], and adding other constraints such as cardinality and 'exclusive or' [62]. Experimental work investigating how SAT algorithms behave on random instances typically focuses on parameters that describe each instance independently of its size. The most common parameter is the ratio of clauses to variables, i.e., *(clause) density*. Early work in the area showed random 3-SAT instances to be at their hardest when density is around 4.25 [59]. Later work revealed that the interaction between density and empirical hardness is much more solver-dependent [21]. Many other parameters such as heterogeneity, locality, and modularity have emerged from attempts to generate random instances similar to industry benchmarks [3,13,48,49].

In contrast, the analysis of WMC algorithms on random instances is only beginning to be developed. Early on, Sang et al. [69,70] ran one of the WMC algorithms on random 3-CNF formulas and observed an easy-hard-easy pattern with respect to (w.r.t.) clause density. Recently, Gupta et al. [52] included some WMC algorithms in their study on phase transitions in knowledge compilation. Two phase transitions were identified: one w.r.t. clause density and another w.r.t. a new parameter called *solution density*. There is also a recent attempt [33] to compare WMC algorithms on random instances of a particular application of WMC, i.e., probabilistic logic programs. However, it finds no meaningful differences among the algorithms in that context. Our work complements previous results by including WMC algorithms of various kinds (i.e., not just those based on knowledge compilation) and introducing another parameter of interest.

What parameters are most appropriate to study WMC? Like SAT [4], WMC is known to be fixed-parameter tractable w.r.t. primal treewidth (or a closely related notion) [5,26,30,69]. However—as we show in Sect. 4—instances generated by a standard random model for k-CNF formulas fail to exhibit enough variance in primal treewidth for us to infer its effect on the behaviour of the algorithms. Therefore, we present an extension of this model with a parameter that influences primal treewidth. The performance of WMC algorithms that use data structures called *algebraic decision diagrams* (ADDs) [6] is also known to depend on the numerical values of weights [38,39]. Thus, our random model also includes two parameters that control redundancies in these values.

In addition to introducing a new random model for WMC instances, the contributions of this paper include several findings about the behaviour of WMC algorithms on instances generated by our model. First, we show that the easy-hard-easy pattern w.r.t. density is different for dynamic programming algorithms than for all other algorithms. Second, we present statistical evidence that all the algorithms scale exponentially w.r.t. primal treewidth and estimate how the base of that exponential changes w.r.t. density. Third, we show how the performance of ADD-based algorithms gradually improves w.r.t. the proportion of weights that have repeating values. Fourth, we complement our findings on random instances with an experimental study on WMC competition benchmarks, showing how the best-performing algorithm changes depending on density and primal treewidth.

2 Preliminaries

Notation. For any graph G, we write $\mathcal{V}(G)$ for its set of nodes and $\mathcal{E}(G)$ for its set of edges. Let S be a finite set. We write 2^S to denote the powerset of S and $\mathcal{U}S$ for the discrete uniform probability distribution on S. We represent any other probability distribution as a pair (S, p) where $p\colon S \to [0,1]$ is a probability mass function. For any probability distribution $\mathcal{P}$, we write $x \leftsquigarrow \mathcal{P}$ to denote the act of sampling x from $\mathcal{P}$. For instance, $x \leftsquigarrow (\{1,2\}, \{1 \mapsto 0.1, 2 \mapsto 0.9\})$ means that x becomes equal to 1 with probability 0.1 or to 2 with probability 0.9.

By *variable*, we always mean a Boolean variable. A *literal* is either a variable (say, v) or its negation (denoted $\neg v$), respectively called *positive* and *negative* literal. A *clause* is a disjunction of literals. A *formula* is any well-formed expression consisting of variables, negation, conjunction, and disjunction. A formula is in *conjunctive normal form* (CNF) if it is a conjunction of clauses, and it is in k-CNF if every clause has exactly k literals. While we use the set-theoretic notation for CNF formulas (e.g., writing $c \in \phi$ to mean that clause c is one of the clauses in formula ϕ), duplicate clauses are still allowed. The *primal graph* of a CNF formula is a graph that has a node for every variable, and there is an edge between two variables if they coappear in some clause. The *primal treewidth* of a formula is the treewidth of its primal graph, where treewidth is as in Definition 1.

Definition 1 ([67]). *A* tree decomposition *of a graph G is a pair (T, χ), where T is a tree and $\chi\colon \mathcal{V}(T) \to 2^{\mathcal{V}(G)}$ is a labelling function, with the following properties:*

- $\bigcup_{t \in \mathcal{V}(T)} \chi(t) = \mathcal{V}(G)$;
- *for every $e \in \mathcal{E}(G)$, there is $t \in \mathcal{V}(T)$ such that e has both endpoints in $\chi(t)$;*
- *for all $t, t', t'' \in \mathcal{V}(T)$, if t' is on the path between t and t'', then $\chi(t) \cap \chi(t'') \subseteq \chi(t')$.*

The width *of tree decomposition (T, χ) is $\max_{t \in \mathcal{V}(T)} |\chi(t)| - 1$. The* treewidth *of graph G is the smallest w such that G has a tree decomposition of width w.*

Given a CNF formula ϕ, SAT is a decision problem that asks whether there exists a way to assign values to all variables in ϕ such that ϕ evaluates to true. Such a formula is said to be *satisfiable*; otherwise, it is *unsatisfiable*. #SAT is a problem that asks to count the number of such assignments. WMC extends #SAT with a weight function on literals and asks to compute the sum of the weights of the models of the given formula, where the weight of a model is the product of the weights of the literals in it [19]. For example, the WMC of the formula $x \lor y$ with a weight function $w\colon \{ x, y, \neg x, \neg y \} \to \mathbb{R}_{\geq 0}$ defined as $w(x)=0.3$, $w(y) = 0.2$, $w(\neg x) = 0.7$, $w(\neg y) = 0.8$ is $w(x)w(y) + w(x)w(\neg y) + w(\neg x)w(y) = 0.3 \times 0.2 + 0.3 \times 0.8 + 0.7 \times 0.2 = 0.44$.

3 Background on WMC Algorithms

In this section, we briefly review the three major approaches to WMC—search, knowledge compilation, and dynamic programming—and their corresponding algorithms. The main search-based WMC algorithm CACHET[1] [69] is based on a conflict-driven clause learning SAT solver [60], which is then extended with a component caching scheme and adapted to counting.

Knowledge compilation refers to transformations of propositional formulas into more restrictive formats that make various operations (such as model counting) tractable in the size of the representation [32]. C2D[2] [30], D4[3] [58], and MINIC2D[4] [61] are all algorithms of this type. C2D compiles to deterministic decomposable negation normal form (d-DNNF) [27]. Similarly, D4 compiles to decision-DNNF (also known as decomposable decision graphs) [42]. The only difference between d-DNNF and decision-DNNF is that decision-DNNF has if-then-else constructions instead of disjunctions [58]. Finally, MINIC2D compiles

[1] https://henrykautz.com/Cachet/index.htm.
[2] http://reasoning.cs.ucla.edu/c2d/.
[3] https://www.cril.univ-artois.fr/KC/d4.html.
[4] http://reasoning.cs.ucla.edu/minic2d/.

to decision-SDDs—a subset of sentential decision diagrams (SDDs) that form a subset of d-DNNF [31].

All of the algorithms mentioned above run the same way regardless of whether computing WMC or #SAT. Two recent WMC algorithms instead use data structures whose size (and thus the runtime of the algorithm) depends on the numerical values of weights. These data structures represent *pseudo-Boolean functions*, i.e., functions of the form $f\colon 2^X \to \mathbb{R}_{\geq 0}$, where X is a set. ADDMC is the first such algorithm [38]. It uses ADDs to represent pseudo-Boolean functions, combining and simplifying them in a bottom-up dynamic programming fashion. Since the size of an ADD for f depends on the cardinality of the range of f [6], the performance of the algorithm is sensitive to the numerical values of weights, e.g., to how frequently they repeat. DPMC[5] extends ADDMC in two ways [39]. First, DPMC allows for the order and nesting of operations on ADDs to be determined from an approximately-minimal-width tree decomposition rather than by heuristics.[6] Second, tensors are offered as an alternative to ADDs.

In all known parameterised complexities of WMC algorithms, the exponential factor is a function of primal treewidth or a closely related parameter. Interestingly, C2D is specifically designed to handle high primal treewidth (which the author refers to as *connectivity* [25]) and improves upon an earlier algorithm that has $\mathcal{O}(mw2^w)$ time complexity, where m is the number of clauses, and w is the width of the decomposition tree which is known to be at most primal treewidth [26,30]. While the complexity of CACHET was not analysed directly, the algorithm is based on component caching which is known to have a $2^{\mathcal{O}(w)}n^{\mathcal{O}(1)}$ time complexity, where n is the number of variables, and w is the branchwidth of the underlying hypergraph [5,69], which is known to be within a constant factor of primal treewidth [68]. Similarly, the complexity of DPMC is not described in the paper, although the authors define a notion of width w that is at most primal treewidth plus one and estimate the runtime of the (execution part of the) algorithm to be proportional to 2^w [39].

4 Random k-CNF Formulas with Varying Primal Treewidth

Our random model is based on the following parameters: (a) the number of variables $\nu \in \mathbb{N}^+$, (b) density $\mu \in \mathbb{R}_{>0}$, (c) clause width $\kappa \in \mathbb{N}^+$ (for k-CNF formulas, $\kappa = k$), (d) a parameter $\rho \in [0, 1]$ that influences the primal treewidth of the formula, (e) the proportion $\delta \in [0, 1]$ of variables x such that $w(x) = 1$ and $w(\neg x) = 0$ or $w(x) = 0$ and $w(\neg x) = 1$, (f) and the proportion $\epsilon \in [0, 1 - \delta]$ of variables x such that $w(x) = w(\neg x) = 0.5$. The first three parameters are the standard parameters used to generate random κ-CNF formulas with $\nu\mu$ clauses (up to rounding). Parameters δ and ϵ control the numerical values of weights similarly to

[5] https://github.com/vardigroup/dpmc.

[6] There is also a recent line of work in using tree decompositions to guide the heuristics of search-based model counters [57].

Algorithm 1: Generating a random formula

Input: $\nu, \kappa \in \mathbb{N}^+$ (such that $\kappa < \nu$), $\mu \in \mathbb{R}_{>0}$, $\rho \in [0, 1]$.
Output: A k-CNF formula ϕ.

```
1  φ ← empty CNF formula;
2  G ← empty graph;
3  for i ← 1 to ⌊νμ⌋ do
4      X ← ∅;
5      for j ← 1 to κ do
6          x ← NewVariable(X, G);
7          V(G) ← V(G) ∪ {x};
8          E(G) ← E(G) ∪ {{x, y} | y ∈ X};
9          X ← X ∪ {x};
10     φ ← φ ∪ {l ↜ U{x, ¬x} | x ∈ X};
11 return φ;
12 Function NewVariable(X, G):
13     N ← {e ∈ E(G) | |e ∩ X| = 1};
14     if N = ∅ then
15         return x ↜ U({x_1, x_2, ..., x_ν} \ X);
16     return x ↜ ({x_1, x_2, ..., x_ν} \ X,
17         y ↦ (1−ρ)/(ν−|X|) + ρ |{z∈X | {y,z} ∈ E(G)}| / |N|);
```

determinism and parameter equality—facets of local structure considered in the literature on probabilistic models [74]. While all other WMC algorithms disregard the weights, DPMC [39] can exploit both determinism and equal weights to solve the problem faster. Indeed, higher values of both δ and ϵ result in ADDs having fewer real-numbered values they need to represent. Thus, the ADDs are smaller and can be handled more efficiently.

The process for generating random k-CNF formulas is summarized as Algorithm 1. The idea behind the algorithm is to reduce the density of the primal graph (via having some overlapping edges) while: (a) avoiding having many variables that do not occur in any clause and (b) promoting tree-like subgraphs that are likely to have low treewidth. For the rest of this section, let $\{x_i\}_{i=1}^{\nu}$ be the variables of the formula under construction. We simultaneously construct both formula ϕ and its primal graph G.[7] Each iteration of the first for-loop adds a clause to ϕ. This is done by constructing a set X of variables to be included in the clause, and then randomly adding either x or $\neg x$ to the clause for each $x \in X$ on line 10. Function `NewVariable` randomly selects each new variable x, and lines 7 to 9 add x to the graph and the formula while also adding edges between x and all the other variables in the clause. To select each variable, line 13 defines set N to contain all edges with exactly one endpoint in X. The edges added to

[7] The idea to directly take the primal graph into consideration while generating the formula is new—cf. random SAT instance generators based on, e.g., adversarial evolution [56] and community structure [48].

G by line 8 form a subset of N. If $N = \varnothing$, we select the variable uniformly at random (u.a.r.) from all viable candidates. Otherwise, ρ determines how much we bias the uniform distribution towards variables that would introduce fewer new edges to G.

When $\rho = 0$, Algorithm 1 reduces to what has become the standard random model for k-CNF formulas. Equivalently to Franco and Paull [46], we independently sample a fixed number of clauses, each clause has no duplicate variables, and each variable becomes either a positive or a negative literal with equal probabilities. At the other extreme, when $\rho = 1$, the first variable of a clause is still chosen u.a.r., but all other variables are chosen from those that already coappear in a clause (if possible). The probability that a variable is selected to be included in a clause scales linearly w.r.t. the proportion of edges in N that would be repeatedly added to G if the variable y was added to the clause. This is an arbitrary choice (which appears to work well, see Sect. 4.1) although alternatives (e.g., exponential scaling) could be considered. As long as $\rho < 1$, every k-CNF formula retains a positive probability of being generated by the algorithm.

To transform the generated formula into a WMC instance, we need to define weights on literals.[8] We want to partition all variables into three groups: those with weights equal to zero and one, those with weights equal to 0.5, and those with arbitrary weights, where the size of each group is determined by δ and ϵ. To do this, we sample a permutation $\pi \leftsquigarrow \mathcal{U}S_\nu$ (where S_ν is the permutation group on $\{1, 2, \dots, \nu\}$), and assign to each *variable* x_n a weight drawn u.a.r. from (a) $\mathcal{U}\{0, 1\}$ if $\pi(n) \le \nu\delta$, (b) $\mathcal{U}\{0.5\}$ if $\nu\delta < \pi(n) \le \nu\delta + \nu\epsilon$, and (c) $\mathcal{U}\{0.01, 0.02, \dots, 0.99\}$[9] if $\pi(n) > \nu\delta + \nu\epsilon$. We extend these weights to weights on *literals* by choosing the weight of each positive literal to be equal to the weight of its variable, and the weight of each negative literal to be such that $w(x) + w(\neg x) = 1$ for all variables x. This restriction is to ensure consistent answers among the algorithms.

Example 1. Let $\nu = 5$, $\mu = 0.6$, $\kappa = 3$, $\rho = 0.3$, $\delta = 0.4$, and $\epsilon = 0.2$ and consider how Algorithm 1 generates a random instance. Since $\kappa = 3$, and $\lfloor \nu\mu \rfloor = 3$, the algorithm will generate a 3-CNF formula with three clauses.

For the first variable of the first clause, we are choosing u.a.r. from $\{x_1, x_2, \dots, x_5\}$. Suppose the algorithm chooses x_5. Graph G then gets its first node but no edges. The second variable is chosen u.a.r. from $\{x_1, x_2, x_3, x_4\}$. Suppose the second variable is x_2. Then G gets another node and its first edge between x_2 and x_5. The third variable in the first clause is similarly chosen u.a.r. from $\{x_1, x_3, x_4\}$ because the only edge in G has both endpoints in $X = \{x_2, x_5\}$, and so $N = \varnothing$. Suppose the third variable is x_1. Graph G becomes a triangle connecting x_1, x_2, and x_5. Each of the three variables is then added to the clause as either a positive or a negative literal (with equal probabilities). Thus, the first clause becomes, e.g., $\neg x_5 \lor x_2 \lor x_1$.

---

[8] Algorithms such as DPMC and ADDMC [38,39] support a more flexible way of assigning weights that can lead to significant performance improvements [34,35].

[9] For convenience, we represent $(0, 1)$ as 99 discrete values.

The first variable of the second clause is chosen u.a.r. from $\{x_1, x_2, \ldots, x_5\}$. Suppose it is x_5 again. When the function `NewVariable` tries to choose the second variable, $X = \{x_5\}$, and so $N = \{\{x_1, x_5\}, \{x_2, x_5\}\}$. The second variable is chosen from the discrete probability distribution $\Pr(x_1) = \Pr(x_2) = \frac{1-0.3}{5-1} + 0.3 \times \frac{1}{2} = 0.325$ and $\Pr(x_3) = \Pr(x_4) = \frac{1-0.3}{5-1} = 0.175$.

We skip the details of how all remaining variables and clauses are selected and consider the weight assignment. First, we shuffle the list of variables and get, e.g., $L = (x_4, x_3, x_2, x_1, x_5)$. This means that the first $\nu\delta = 5 \times 0.4 = 2$ variables of L get weights u.a.r. from $\{0, 1\}$, the next $\nu\epsilon = 5 \times 0.2 = 1$ variable gets a weight of 0.5, and the remaining two variables get weights u.a.r. from $\{0.01, 0.02, \ldots, 0.99\}$. The weight function $w\colon \{x_1, x_2, \ldots, x_5, \neg x_1, \neg x_2, \ldots, \neg x_5\} \to [0, 1]$ can then be defined as, e.g., $w(x_4) = w(\neg x_3) = 0$, $w(x_3) = w(\neg x_4) = 1$, $w(x_2) = w(\neg x_2) = 0.5$, $w(x_1) = 0.23$, $w(\neg x_1) = 0.77$, $w(x_5) = 0.18$, and $w(\neg x_5) = 0.82$.

4.1 Validating the Model

The idea behind our model is that manipulating the value of ρ should allow us to generate instances of varying primal treewidth. Is this effect observable in practice? In addition, as WMC instances are mostly used for probabilistic inference, they tend to be satisfiable. Therefore, we want to filter out unsatisfiable instances from those generated by the model and need to ensure that the proportion of satisfiable instances remains sufficiently high. Given that higher values of ρ can result in constraints on variables being more localised and concentrated, we ask: are instances generated with higher values of ρ less likely to be satisfiable? To answer both questions, we run the following experiment.

Experiment 1. We fix $\nu = 100, \delta = \epsilon = 0$, and consider random instances with $\mu = 2.5 \times \sqrt{2}^{-5}, 2.5 \times \sqrt{2}^{-4}, \ldots, 2.5 \times \sqrt{2}^{5}$, $\kappa = 2, 3, 4, 5$, and ρ going from 0 to 1 in steps of 0.01. For each combination of parameters, we generate ten instances.[10] We check if each instance is satisfiable using MiniSat[11] 2.2.0 [41] and calculate its (approximate) primal treewidth using htd[12] [1].

Remark 1. Here and henceforth, we use htd to provide heuristic upper bounds on true treewidth as exact computation would make the experiments significantly more time-consuming. However, we compared the accuracy of htd with exact treewidth algorithm Jdrasil[13] [7] on 3% of our random instances. The difference between the upper bound produced by htd and the exact value was never higher than four and up to two in 85% of all cases. Since the difference is small enough to not have a qualitative effect, hereafter we write '(primal) treewidth' to mean 'the heuristic upper bound on treewidth found by htd'.

[10] Since one expects similar values of ρ to produce instances with similar properties, and ρ's are enumerate quite densely, generating only ten instances is sufficient.

[11] http://minisat.se/MiniSat.html.

[12] https://github.com/mabseher/htd.

[13] https://maxbannach.github.io/Jdrasil/.

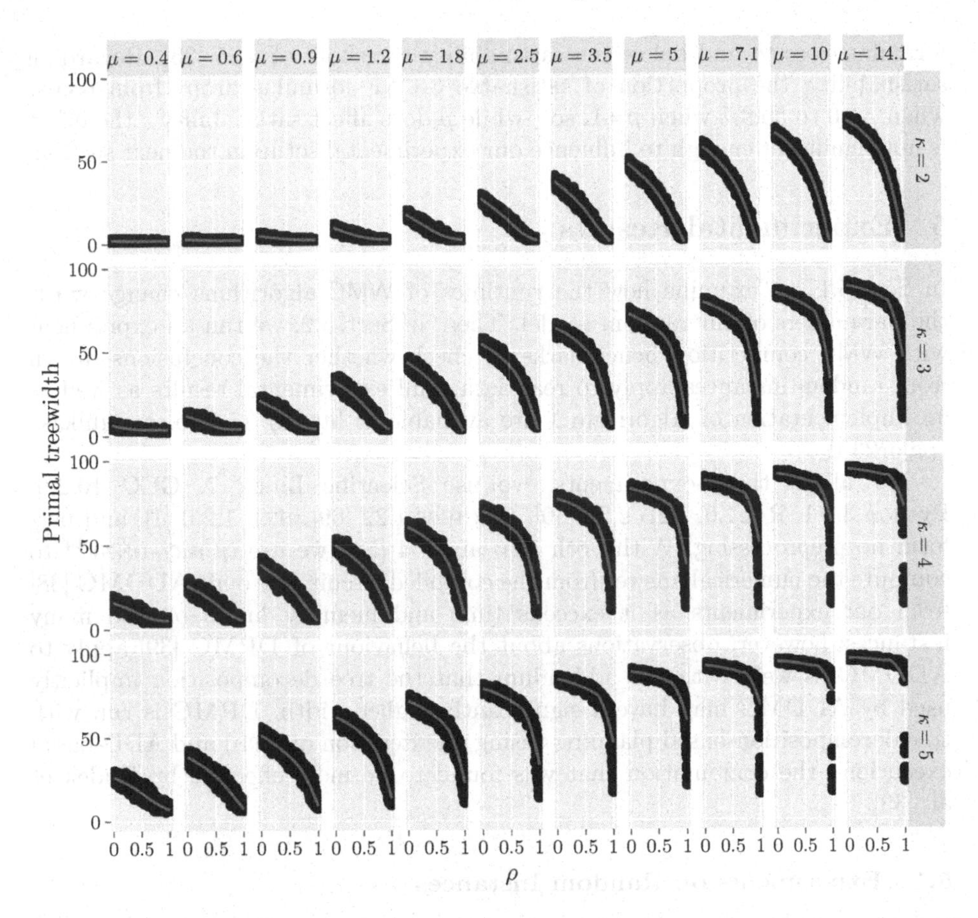

Fig. 1. The relationship between ρ and primal treewidth for various values of μ and κ for k-CNF formulas from Experiment 1. Black points represent individual instances, and blue lines are smoothed means computed using locally weighted smoothing. The values of μ are rounded to one decimal place.

Figure 1 shows the relationship between ρ and primal treewidth. Except for when both μ and κ are very low (i.e., the formulas are small in both clause width and the number of clauses), primal treewidth decreases as ρ increases. This downward trend becomes sharper as μ increases, however, not uniformly: it splits into a roughly linear segment that approaches a horizontal line (for most values of ρ) and a sharply-decreasing segment that approaches a vertical line (when ρ is close to one). Higher values of κ seem to expedite this transition, i.e., with a higher value of κ, a lower value of μ is sufficient for a smooth downward curve between ρ and primal treewidth to turn into a combination of a horizontal and a vertical line. While this behaviour may be troublesome when generating formulas with higher values of μ (almost all of which would be unsatisfiable), the relationship between ρ and primal treewidth is excellent for generating 3-CNF formulas close to and below the satisfiability threshold of 4.25 [23]. Regarding satisfiability, the proportion of satisfiable 3-CNF formulas drops from 63.6% when $\rho = 0$ to 50.9% when $\rho = 1$, so—while ρ does affect satisfiability—the effect is not significant enough to influence our experimental setup in the next section.

5 Experimental Results

In Sect. 5.1, we examine how the runtimes of WMC algorithms change w.r.t. the parameters of our random model. Then, in Sect. 5.2, we run an experiment with WMC competition benchmarks to check whether the conclusions drawn from random instances apply to real data. Full experimental results as well as an implementation of Algorithm 1 are available at https://github.com/dilkas/cpaior23-d.

For all of these experiments, we use Scientific Linux 7, GCC 10.2.0, Python 3.8.1, R 4.1.0, C2D 2.20 [30], CACHET 1.22 [69], HTD 1.2.0 [1], and perform no preprocessing. With both C2D and D4 [58], we use QUERY-DNNF[14] to compute the numerical answer from the compiled circuit. We omit ADDMC [38] from our experiments as it exceeds time and memory limits on too many instances; however, observations about the behaviour of DPMC [39] apply to ADDMC as well, with the addendum that the tree decomposition implicitly used by ADDMC may have a significantly higher width. DPMC is run with tree decomposition-based planning (using one iteration of HTD) and ADD-based execution—the combination that was found to be most effective by Dudek et al. [39].

5.1 Experiments on Random Instances

We restrict our attention to 3-CNF formulas, generate 100 satisfiable instances for each *combination* of parameters, and run each of the five algorithms with a 500 s time limit and an 8 GiB memory limit on Intel Xeon E5–2630. While both limits are somewhat low, we prioritise large numbers of instances to increase the accuracy and reliability of our results. Unless stated otherwise, in each plot of this section, lines denote median values, and shaded areas show interquartile ranges. We run the following three experiments, setting $\nu = 70$ in all of them as we found that this produces instances of suitable difficulty.

Experiment 2 (Density and Primal Treewidth). Let $\nu = 70$, μ go from 1 to 4.3 in steps of 0.3, ρ go from 0 to 0.5 in steps of 0.01, and $\delta = \epsilon = 0$.

Experiment 3 (δ). Let $\nu = 70$, $\mu = 2.2$[15], $\rho = 0$, δ go from 0 to 1 in steps of 0.01, and $\epsilon = 0$.

Experiment 4 (ϵ). Same as Experiment 3 but with $\delta = 0$ and ϵ going from 0 to 1 in steps of 0.01.

[14] http://www.cril.univ-artois.fr/kc/d-DNNF-reasoner.html.
[15] Experiment 2 shows this density to be the most challenging for DPMC.

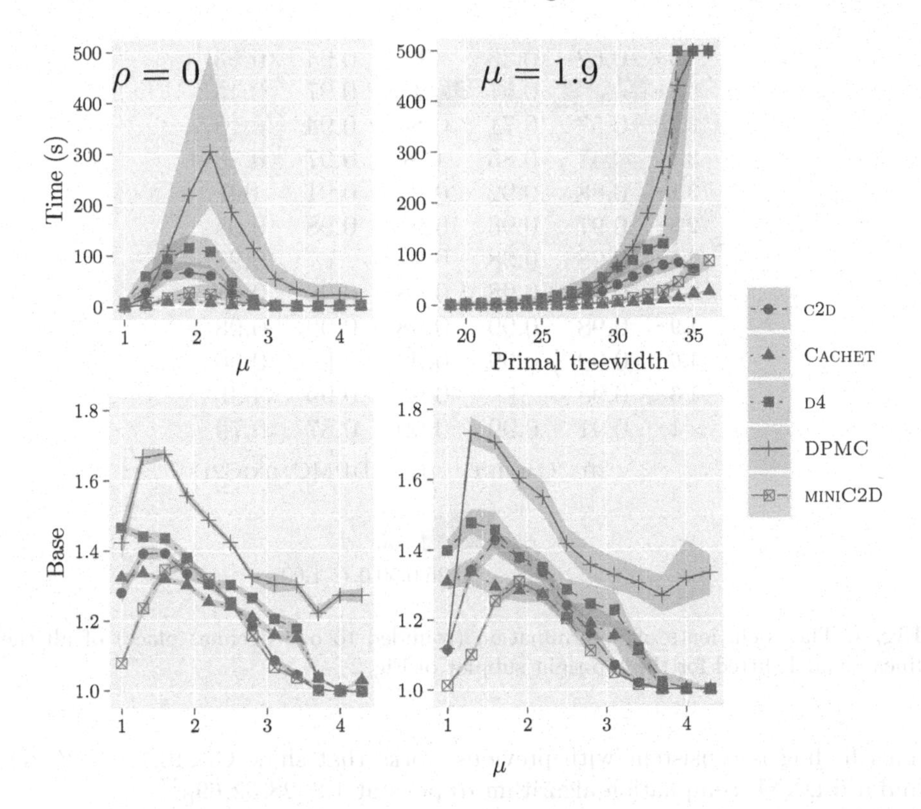

Fig. 2. Visualisations of the data from Experiment 2. The top-left plot shows how the runtime of each algorithm changes w.r.t. density when $\rho = 0$. The top-right plot shows changes in the runtime of each algorithm w.r.t. primal treewidth with μ fixed at 1.9. The plots at the bottom show how the estimated base of the exponential relationship between primal treewidth and the runtime of each algorithm depends on μ. The bottom-left plot is for the simple linear model (with shaded areas showing standard error), and the bottom-right plot uses the estimates provided by ESA [64] (with shaded areas showing 95% confidence intervals).

In each experiment, the proportion of algorithm runs that timed out never exceeded 3.8%. While in Experiment 2 only 1% of experimental runs ran out of memory, the same percentage was higher in Experiment 3 and 4—10 and 12%, respectively. D4 [58] and C2D are the algorithms that experienced the most issues fitting within the memory limit, accounting for 66–72% and 28–33% of such instances, respectively. We exclude the runs that terminated early due to running out of memory from the rest of our analysis.

In Experiment 2, we investigate how the runtime of each algorithm depends on the density and primal treewidth by varying both μ and ρ. The results are in Fig. 2. The first thing to note is that the peak hardness w.r.t. density occurs at around 1.9 for all algorithms except for DPMC, which peaks at 2.2 instead.[16]

[16] The exact values—while illegible from the plot—can be confirmed by numerical data.

μ	C2D	CACHET	D4	DPMC	MINIC2D
4.3	0.62	0.33	1	0.94	0.53
4	0.19	0.51	0	0.97	0.43
3.7	0.57	0.71	0.83	0.94	0.18
3.4	0.47	0.85	0.8	0.97	0.53
3.1	0.88	0.92	0.91	0.91	0.9
2.8	0.97	0.96	0.98	0.98	0.95
2.5	0.98	0.98	0.97	1	0.98
2.2	0.99	0.98	0.98	0.99	0.98
1.9	0.98	0.99	0.98	0.99	0.98
1.6	0.99	0.99	0.98	1	0.96
1.3	0.98	1	0.99	0.99	0.9
1	0.91	0.99	0.99	0.87	0.79

R^2 0.25 0.50 0.75 1.00

Fig. 3. The coefficients of determination (rounded to one decimal place) of all the linear models fitted for the top-right subplot of Fig. 2

This finding is consistent with previous works that show CACHET, MINIC2D, and a d-DNNF compilation algorithm to peak at 1.8 [28,52,69].[17]

The other question we want to investigate is how each algorithm scales w.r.t. primal treewidth. The top-right plot in Fig. 2 shows this relationship for a fixed value of μ, and one can see some evidence that the runtime of DPMC grows faster w.r.t. primal treewidth compared to the other algorithms. We use two statistical techniques to quantify this growth: a simple linear regression model and the empirical scaling analyzer (ESA) v2[18] [64]. In both cases, for each algorithm and value of μ in Experiment 2, we select the median runtime for all available primal treewidth values. In the former case, we fit the model $\ln t \sim \alpha w + \beta$, where t is the median runtime of the algorithm, w is the primal treewidth, and α and β are parameters.[19] In other words, we express median runtime as $e^{\beta}(e^{\alpha})^{w}$. In the latter case, we run ESA with 1001 bootstrap samples, a window of 101, and use the first 30% of the data for training.

The results of both models are qualitatively the same (except for DPMC run on instances with $\mu = 1$) and are displayed at the bottom of Fig. 2. We find that DPMC scales worse w.r.t. primal treewidth than any other algorithm across all values of μ and is the only algorithm that does not become indifferent to primal treewidth when faced with high-density formulas. A second look at the

[17] For comparison, #SAT algorithms are known to peak at densities 1.2 and 1.5 [9,12].
[18] https://github.com/YashaPushak/ESA.
[19] Similar analyses have been used to investigate polynomial-to-exponential phase transitions in SAT [21] and the behaviour of SAT solvers on CNF-XOR formulas [37].

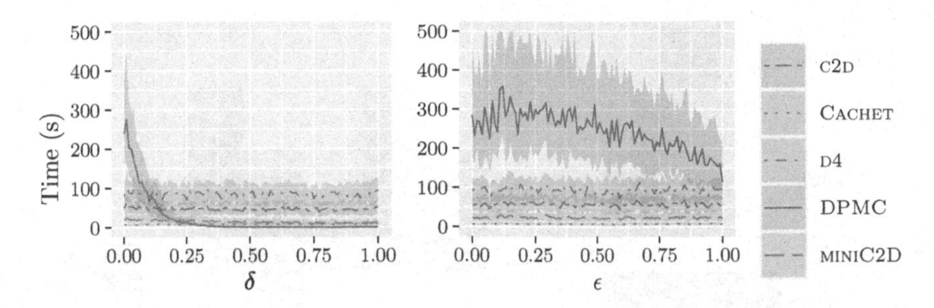

Fig. 4. Changes in the runtime of each algorithm as a result of changing δ (on the left-hand side) and ϵ (on the right-hand side) as in Experiments 3 and 4

top-left subplot of Fig. 2 suggests an explanation for the latter observation. The runtimes of all algorithms except for DPMC approach zero when $\mu > 3$ while the median runtime of DPMC approaches a small non-zero constant instead. This observation also explains why Fig. 3 shows that the fitted models fail to explain the data for non-ADD algorithms running on high-density instances—the runtimes are too small to be meaningful. In all other cases, an exponential relationship between primal treewidth and runtime fits the experimental data remarkably well.

Another thing to note is that MINIC2D [61] is the only algorithm that exhibits a clear low-high-low pattern in the bottom subplots of Fig. 2. To a smaller extent, the same may apply to C2D and DPMC, although the evidence for this is limited due to relatively large gaps between different values of μ. In contrast, the runtimes of CACHET and D4 remain dependent on primal treewidth even when the density of the WMC instance is very low, suggesting that MINIC2D should have an advantage on low-density high-primal-treewidth instances.

Finally, Experiments 3 and 4 investigate how changing the numerical values of weights can simplify a WMC instance. The results are in Fig. 4. As expected, the runtimes of all algorithms other than DPMC stay the same regardless of the value of δ or ϵ. The runtime of DPMC, however, experiences a sharp (exponential?) decline with increasing δ. The decline w.r.t. ϵ is also present, although significantly less pronounced and with high variance.

To sum, we found that C2D and D4 are the most memory-intensive algorithms, CACHET is great on random instances in general, MINIC2D excels on low-density high-primal-treewidth instances, and DPMC is at its best on low-density low-primal-treewidth instances. Furthermore, a median instance with all weights equal to each other is about three times easier for DPMC than a median instance with random weights. Another important observation is about how peak hardness w.r.t. density depends on the algorithm: DPMC peaks at a higher density than all other WMC algorithms, which peak at a higher density than (some) #SAT algorithms.

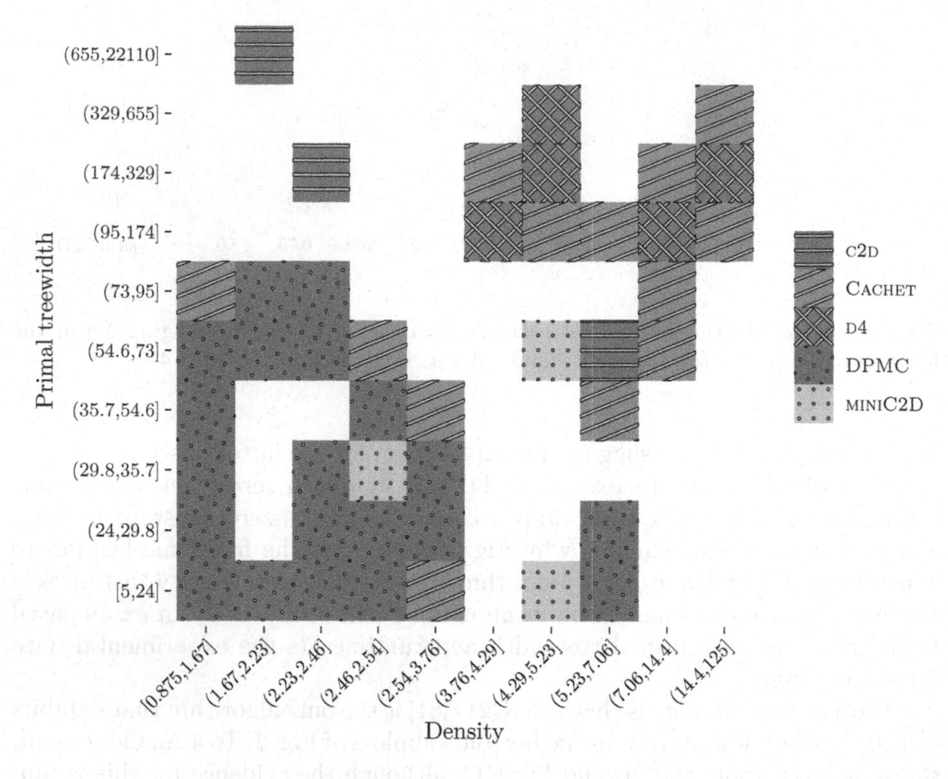

Fig. 5. The best-performing algorithm for each combination of density and primal treewidth according to the experiments on competition benchmarks. Both ranges of values are divided into ten bins so that there are ten instances in each bin. The best-performing algorithm for each combination of bins is the algorithm that solved the largest number of instances, with ties broken by minimising total runtime. An empty cell means that either no benchmark had this combination of density and primal treewidth or all algorithms failed on all such instances.

5.2 Experiments on Competition Benchmarks

To check whether our observations on random instances are accurate on real data, we use the 100 public instances from track 2 of the 2022 model counting competition[20]—an annual competition that has been running since 2020 [43]. This time, we run the algorithms on Intel Xeon Gold 6138 with 32 GiB of memory and a one hour time limit. As in Sect. 5.1, we compute the density and the primal treewidth of each instance.

Figure 5 shows the best-performing algorithm for various combinations of the parameters. We observe that: (a) DPMC [39] is best on most instances with low primal treewidth, (b) C2D [30] can handle some low-density high-primal-treewidth instances that all the other algorithms fail on, (c) CACHET [69] (as well

[20] https://mccompetition.org/2022/mc_description.

as D4 [58] to some extent) excel when both density and primal treewidth are quite high, (d) and MINIC2D [61] does not have a clear niche. Hence, the observation in Sect. 5.1 that DPMC is good on low-density low-primal-treewidth instances is confirmed by the experiments on real data. Moreover, higher density instances can also favour DPMC as long as primal treewidth is sufficiently low. On the other hand, while the experiments on random instances suggested that MINIC2D might excel at low-density high-primal-treewidth instances, our experiments on competition benchmarks suggest otherwise. Instead, C2D, CACHET or DPMC could be the right choice depending on the exact values.

6 Conclusions and Future Work

In this paper, we studied the behaviour of and differences among WMC algorithms on random instances generated by a standard model for k-CNF formulas extended with parameters that control primal treewidth and literal weights. Among other things, we established statistical evidence for the existence of an exponential relationship between primal treewidth and the runtimes of all WMC algorithms on instances generated by our model. The runtime of the ADD-based algorithm was observed to peak at a higher density, scale worse w.r.t. primal treewidth, and depend negatively on repeating weight values compared to algorithms based on search or knowledge compilation. These observations can, to some degree, be extended to a closely related weighted projected model counting algorithm [40] as well as to other applications of ADDs more generally, e.g., probabilistic inference [18,50] and stochastic planning [54].

One limitation of our work is that variability in primal treewidth was achieved via a parameter, and this could bias randomness in some unexpected way (although it is encouraging that there is only a slight decrease in the proportion of satisfiable instances between $\rho = 0$ and $\rho = 1$). Perhaps a theoretical investigation of the proposed model is warranted, including a characterisation of how ρ influences primal treewidth and the structure of the primal graph more generally. Since treewidth is widely used in parameterised complexity [36], formally establishing a connection with ρ could make our random model useful for a variety of other hard computational problems.

To keep the number of experiments feasible, we restricted our attention to 3-CNF formulas, although, of course, this is not very representative of real-world WMC instances. The model could be adapted to generate non-k-CNF formulas, and perhaps a more representative structure could be achieved by introducing new variables that clauses define to be equivalent to select conjunctions of literals as is done in one of the WMC encodings for Bayesian networks [29].

Acknowledgements. The author would like to thank Vaishak Belle and the anonymous reviewers for their feedback on earlier versions of this work. The author was supported by the EPSRC Centre for Doctoral Training in Robotics and Autonomous Systems, funded by the UK Engineering and Physical Sciences Research Council (grant EP/L016834/1). This work has made use of the resources provided by the Edinburgh Compute and Data Facility (ECDF) (http://www.ecdf.ed.ac.uk/). For the purpose of open access, the author has applied a Creative Commons Attribution (CC BY) licence to any Author Accepted Manuscript version arising from this submission.

References

1. Abseher, M., Musliu, N., Woltran, S.: htd – a free, open-source framework for (customized) tree decompositions and beyond. In: Salvagnin, D., Lombardi, M. (eds.) CPAIOR 2017. LNCS, vol. 10335, pp. 376–386. Springer, Cham (2017). https://doi.org/10.1007/978-3-319-59776-8_30
2. Achlioptas, D., Moore, C.: The asymptotic order of the random k-SAT threshold. In: 43rd Symposium on Foundations of Computer Science (FOCS 2002), 16–19 November 2002, Vancouver, BC, Canada, Proceedings, pp. 779–788. IEEE Computer Society (2002). https://doi.org/10.1109/SFCS.2002.1182003
3. Ansótegui, C., Bonet, M.L., Levy, J.: Towards industrial-like random SAT instances. In: Boutilier, C. (ed.) IJCAI 2009, Proceedings of the 21st International Joint Conference on Artificial Intelligence, Pasadena, California, USA, 11–17 July 2009, pp. 387–392 (2009). http://ijcai.org/Proceedings/09/Papers/072.pdf
4. Atserias, A., Fichte, J.K., Thurley, M.: Clause-learning algorithms with many restarts and bounded-width resolution. J. Artif. Intell. Res. **40**, 353–373 (2011). https://doi.org/10.1613/jair.3152
5. Bacchus, F., Dalmao, S., Pitassi, T.: Solving #SAT and Bayesian inference with backtracking search. J. Artif. Intell. Res. **34**, 391–442 (2009). https://doi.org/10.1613/jair.2648
6. Bahar, R.I., et al.: Algebraic decision diagrams and their applications. Formal Meth. Syst. Des. **10**(2/3), 171–206 (1997). https://doi.org/10.1023/A:1008699807402
7. Bannach, M., Berndt, S., Ehlers, T.: Jdrasil: a modular library for computing tree decompositions. In: Iliopoulos, C.S., Pissis, S.P., Puglisi, S.J., Raman, R. (eds.) 16th International Symposium on Experimental Algorithms, SEA 2017, 21–23 June 2017, London, UK. LIPIcs, vol. 75, pp. 28:1–28:21. Schloss Dagstuhl - Leibniz-Zentrum für Informatik (2017). https://doi.org/10.4230/LIPIcs.SEA.2017.28
8. Bart, A., Koriche, F., Lagniez, J., Marquis, P.: An improved CNF encoding scheme for probabilistic inference. In: Kaminka, G.A., et al. (eds.) ECAI 2016–22nd European Conference on Artificial Intelligence, 29 August–2 September 2016, The Hague, The Netherlands - Including Prestigious Applications of Artificial Intelligence (PAIS 2016). Frontiers in Artificial Intelligence and Applications, vol. 285, pp. 613–621. IOS Press (2016). https://doi.org/10.3233/978-1-61499-672-9-613
9. Bayardo Jr., R.J., Pehoushek, J.D.: Counting models using connected components. In: Kautz, H.A., Porter, B.W. (eds.) Proceedings of the Seventeenth National Conference on Artificial Intelligence and Twelfth Conference on on Innovative Applications of Artificial Intelligence, 30 July–3 August 2000, Austin, Texas, USA, pp. 157–162. AAAI Press/The MIT Press (2000). http://www.aaai.org/Library/AAAI/2000/aaai00-024.php
10. Belle, V.: Open-universe weighted model counting. In: Singh, S., Markovitch, S. (eds.) Proceedings of the Thirty-First AAAI Conference on Artificial Intelligence, 4–9 February 2017, San Francisco, California, USA, pp. 3701–3708. AAAI Press (2017). http://aaai.org/ocs/index.php/AAAI/AAAI17/paper/view/15008

11. Belle, V., Passerini, A., Van den Broeck, G.: Probabilistic inference in hybrid domains by weighted model integration. In: Yang, Q., Wooldridge, M.J. (eds.) Proceedings of the Twenty-Fourth International Joint Conference on Artificial Intelligence, IJCAI 2015, Buenos Aires, Argentina, 25–31 July 2015, pp. 2770–2776. AAAI Press (2015). http://ijcai.org/Abstract/15/392
12. Birnbaum, E., Lozinskii, E.L.: The good old Davis-Putnam procedure helps counting models. J. Artif. Intell. Res. **10**, 457–477 (1999). https://doi.org/10.1613/jair.601
13. Bläsius, T., Friedrich, T., Sutton, A.M.: On the empirical time complexity of scale-free 3-SAT at the phase transition. In: Vojnar, T., Zhang, L. (eds.) TACAS 2019. LNCS, vol. 11427, pp. 117–134. Springer, Cham (2019). https://doi.org/10.1007/978-3-030-17462-0_7
14. Chakraborty, S., Fried, D., Meel, K.S., Vardi, M.Y.: From weighted to unweighted model counting. In: Yang, Q., Wooldridge, M.J. (eds.) Proceedings of the Twenty-Fourth International Joint Conference on Artificial Intelligence, IJCAI 2015, Buenos Aires, Argentina, 25–31 July 2015, pp. 689–695. AAAI Press (2015). http://ijcai.org/Abstract/15/103
15. Chakraborty, S., Meel, K.S., Vardi, M.Y.: Approximate model counting. In: Biere, A., Heule, M., van Maaren, H., Walsh, T. (eds.) Handbook of Satisfiability - Second Edition, Frontiers in Artificial Intelligence and Applications, vol. 336, pp. 1015–1045. IOS Press (2021). https://doi.org/10.3233/FAIA201010
16. Chavira, M., Darwiche, A.: Compiling Bayesian networks with local structure. In: Kaelbling, L.P., Saffiotti, A. (eds.) IJCAI-05, Proceedings of the Nineteenth International Joint Conference on Artificial Intelligence, Edinburgh, Scotland, UK, 30 July–5 August 2005, pp. 1306–1312. Professional Book Center (2005). http://ijcai.org/Proceedings/05/Papers/0931.pdf
17. Chavira, M., Darwiche, A.: Encoding CNFs to empower component analysis. In: Biere, A., Gomes, C.P. (eds.) SAT 2006. LNCS, vol. 4121, pp. 61–74. Springer, Heidelberg (2006). https://doi.org/10.1007/11814948_9
18. Chavira, M., Darwiche, A.: Compiling Bayesian networks using variable elimination. In: Veloso, M.M. (ed.) IJCAI 2007, Proceedings of the 20th International Joint Conference on Artificial Intelligence, Hyderabad, India, 6–12 January 2007, pp. 2443–2449 (2007). http://ijcai.org/Proceedings/07/Papers/393.pdf
19. Chavira, M., Darwiche, A.: On probabilistic inference by weighted model counting. Artif. Intell. **172**(6–7), 772–799 (2008). https://doi.org/10.1016/j.artint.2007.11.002
20. Chavira, M., Darwiche, A., Jaeger, M.: Compiling relational Bayesian networks for exact inference. Int. J. Approx. Reason. **42**(1–2), 4–20 (2006). https://doi.org/10.1016/j.ijar.2005.10.001
21. Coarfa, C., Demopoulos, D.D., Aguirre, A.S.M., Subramanian, D., Vardi, M.Y.: Random 3-SAT: the plot thickens. Constraints **8**(3), 243–261 (2003). https://doi.org/10.1023/A:1025671026963
22. Coja-Oghlan, A., Wormald, N.: The number of satisfying assignments of random regular k-SAT formulas. Comb. Probab. Comput. **27**(4), 496–530 (2018). https://doi.org/10.1017/S0963548318000263
23. Crawford, J.M., Auton, L.D.: Experimental results on the crossover point in random 3-SAT. Artif. Intell. **81**(1–2), 31–57 (1996). https://doi.org/10.1016/0004-3702(95)00046-1
24. Dal, G.H., Laarman, A.W., Lucas, P.J.F.: Parallel probabilistic inference by weighted model counting. In: Studený, M., Kratochvíl, V. (eds.) International

Conference on Probabilistic Graphical Models, PGM 2018, 11–14 September 2018, Prague, Czech Republic. Proceedings of Machine Learning Research, vol. 72, pp. 97–108. PMLR (2018). http://proceedings.mlr.press/v72/dal18a.html
25. Darwiche, A.: Compiling knowledge into decomposable negation normal form. In: Dean, T. (ed.) Proceedings of the Sixteenth International Joint Conference on Artificial Intelligence, IJCAI 99, Stockholm, Sweden, 31 July–6 August 1999, 2 Volumes, 1450 pages, pp. 284–289. Morgan Kaufmann (1999). http://ijcai.org/Proceedings/99-1/Papers/042.pdf
26. Darwiche, A.: Decomposable negation normal form. J. ACM **48**(4), 608–647 (2001). https://doi.org/10.1145/502090.502091
27. Darwiche, A.: On the tractable counting of theory models and its application to truth maintenance and belief revision. J. Appl. Non Class. Logics **11**(1–2), 11–34 (2001). https://doi.org/10.3166/jancl.11.11-34
28. Darwiche, A.: A compiler for deterministic, decomposable negation normal form. In: Dechter, R., Kearns, M.J., Sutton, R.S. (eds.) Proceedings of the Eighteenth National Conference on Artificial Intelligence and Fourteenth Conference on Innovative Applications of Artificial Intelligence, 28 July–1 August 2002, Edmonton, Alberta, Canada, pp. 627–634. AAAI Press/The MIT Press (2002). http://www.aaai.org/Library/AAAI/2002/aaai02-094.php
29. Darwiche, A.: A logical approach to factoring belief networks. In: Fensel, D., Giunchiglia, F., McGuinness, D.L., Williams, M. (eds.) Proceedings of the Eights International Conference on Principles and Knowledge Representation and Reasoning (KR-02), Toulouse, France, 22–25 April 2002, pp. 409–420. Morgan Kaufmann (2002)
30. Darwiche, A.: New advances in compiling CNF into decomposable negation normal form. In: de Mántaras, R.L., Saitta, L. (eds.) Proceedings of the 16th Eureopean Conference on Artificial Intelligence, ECAI'2004, including Prestigious Applicants of Intelligent Systems, PAIS 2004, Valencia, Spain, 22–27 August 2004, pp. 328–332. IOS Press (2004)
31. Darwiche, A.: SDD: A new canonical representation of propositional knowledge bases. In: Walsh, T. (ed.) IJCAI 2011, Proceedings of the 22nd International Joint Conference on Artificial Intelligence, Barcelona, Catalonia, Spain, 16–22 July 2011, pp. 819–826. IJCAI/AAAI (2011). https://doi.org/10.5591/978-1-57735-516-8/IJCAI11-143
32. Darwiche, A., Marquis, P.: A knowledge compilation map. J. Artif. Intell. Res. **17**, 229–264 (2002). https://doi.org/10.1613/jair.989
33. Dilkas, P., Belle, V.: Generating random logic programs using constraint programming. In: Simonis, H. (ed.) CP 2020. LNCS, vol. 12333, pp. 828–845. Springer, Cham (2020). https://doi.org/10.1007/978-3-030-58475-7_48
34. Dilkas, P., Belle, V.: Weighted model counting with conditional weights for Bayesian networks. In: de Campos, C.P., Maathuis, M.H., Quaeghebeur, E. (eds.) Proceedings of the Thirty-Seventh Conference on Uncertainty in Artificial Intelligence, UAI 2021, Virtual Event, 27–30 July 2021. Proceedings of Machine Learning Research, vol. 161, pp. 386–396. AUAI Press (2021). https://proceedings.mlr.press/v161/dilkas21a.html
35. Dilkas, P., Belle, V.: Weighted model counting without parameter variables. In: Li, C.-M., Manyà, F. (eds.) SAT 2021. LNCS, vol. 12831, pp. 134–151. Springer, Cham (2021). https://doi.org/10.1007/978-3-030-80223-3_10
36. Downey, R.G., Fellows, M.R.: Fundamentals of Parameterized Complexity. TCS, Springer, London (2013). https://doi.org/10.1007/978-1-4471-5559-1

37. Dudek, J.M., Meel, K.S., Vardi, M.Y.: The hard problems are almost everywhere for random CNF-XOR formulas. In: Sierra, C. (ed.) Proceedings of the Twenty-Sixth International Joint Conference on Artificial Intelligence, IJCAI 2017, Melbourne, Australia, 19–25 August 2017, pp. 600–606. ijcai.org (2017). https://doi.org/10.24963/ijcai.2017/84
38. Dudek, J.M., Phan, V., Vardi, M.Y.: ADDMC: weighted model counting with algebraic decision diagrams. In: The Thirty-Fourth AAAI Conference on Artificial Intelligence, AAAI 2020, The Thirty-Second Innovative Applications of Artificial Intelligence Conference, IAAI 2020, The Tenth AAAI Symposium on Educational Advances in Artificial Intelligence, EAAI 2020, New York, NY, USA, 7–12 February 2020, pp. 1468–1476. AAAI Press (2020). https://ojs.aaai.org/index.php/AAAI/article/view/5505
39. Dudek, J.M., Phan, V.H.N., Vardi, M.Y.: DPMC: weighted model counting by dynamic programming on project-join trees. In: Simonis, H. (ed.) CP 2020. LNCS, vol. 12333, pp. 211–230. Springer, Cham (2020). https://doi.org/10.1007/978-3-030-58475-7_13
40. Dudek, J.M., Phan, V.H.N., Vardi, M.Y.: ProCount: weighted projected model counting with graded project-join trees. In: Li, C.-M., Manyà, F. (eds.) SAT 2021. LNCS, vol. 12831, pp. 152–170. Springer, Cham (2021). https://doi.org/10.1007/978-3-030-80223-3_11
41. Eén, N., Sörensson, N.: An extensible SAT-solver. In: Giunchiglia, E., Tacchella, A. (eds.) SAT 2003. LNCS, vol. 2919, pp. 502–518. Springer, Heidelberg (2004). https://doi.org/10.1007/978-3-540-24605-3_37
42. Fargier, H., Marquis, P.: On the use of partially ordered decision graphs in knowledge compilation and quantified Boolean formulae. In: Proceedings, the Twenty-First National Conference on Artificial Intelligence and the Eighteenth Innovative Applications of Artificial Intelligence Conference, 16–20 July 2006, Boston, Massachusetts, USA, pp. 42–47. AAAI Press (2006). http://www.aaai.org/Library/AAAI/2006/aaai06-007.php
43. Fichte, J.K., Hecher, M., Hamiti, F.: The model counting competition 2020. ACM J. Exp. Algorithmics **26**, 13:1–13:26 (2021). https://doi.org/10.1145/3459080
44. Fichte, J.K., Hecher, M., Woltran, S., Zisser, M.: Weighted model counting on the GPU by exploiting small treewidth. In: Azar, Y., Bast, H., Herman, G. (eds.) 26th Annual European Symposium on Algorithms, ESA 2018, 20–22 August 2018, Helsinki, Finland. LIPIcs, vol. 112, pp. 28:1–28:16. Schloss Dagstuhl - Leibniz-Zentrum für Informatik (2018). https://doi.org/10.4230/LIPIcs.ESA.2018.28
45. Fierens, D., et al.: Inference and learning in probabilistic logic programs using weighted Boolean formulas. Theor. Pract. Log. Program. **15**(3), 358–401 (2015). https://doi.org/10.1017/S1471068414000076
46. Franco, J., Paull, M.C.: Probabilistic analysis of the Davis Putnam procedure for solving the satisfiability problem. Discret. Appl. Math. **5**(1), 77–87 (1983). https://doi.org/10.1016/0166-218X(83)90017-3
47. Galanis, A., Goldberg, L.A., Guo, H., Yang, K.: Counting solutions to random CNF formulas. SIAM J. Comput. **50**(6), 1701–1738 (2021). https://doi.org/10.1137/20M1351527
48. Giráldez-Cru, J., Levy, J.: Generating SAT instances with community structure. Artif. Intell. **238**, 119–134 (2016). https://doi.org/10.1016/j.artint.2016.06.001
49. Giráldez-Cru, J., Levy, J.: Locality in random SAT instances. In: Sierra, C. (ed.) Proceedings of the Twenty-Sixth International Joint Conference on Artificial Intelligence, IJCAI 2017, Melbourne, Australia, 19–25 August 2017, pp. 638–644. ijcai.org (2017). https://doi.org/10.24963/ijcai.2017/89

50. Gogate, V., Domingos, P.M.: Approximation by quantization. In: Cozman, F.G., Pfeffer, A. (eds.) UAI 2011, Proceedings of the Twenty-Seventh Conference on Uncertainty in Artificial Intelligence, Barcelona, Spain, 14–17 July 2011, pp. 247–255. AUAI Press (2011)
51. Gogate, V., Domingos, P.M.: Probabilistic theorem proving. Commun. ACM **59**(7), 107–115 (2016). https://doi.org/10.1145/2936726
52. Gupta, R., Roy, S., Meel, K.S.: Phase transition behavior in knowledge compilation. In: Simonis, H. (ed.) CP 2020. LNCS, vol. 12333, pp. 358–374. Springer, Cham (2020). https://doi.org/10.1007/978-3-030-58475-7_21
53. Hecher, M., Thier, P., Woltran, S.: Taming high treewidth with abstraction, nested dynamic programming, and database technology. In: Pulina, L., Seidl, M. (eds.) SAT 2020. LNCS, vol. 12178, pp. 343–360. Springer, Cham (2020). https://doi.org/10.1007/978-3-030-51825-7_25
54. Hoey, J., St-Aubin, R., Hu, A.J., Boutilier, C.: SPUDD: Stochastic planning using decision diagrams. In: Laskey, K.B., Prade, H. (eds.) UAI 1999: Proceedings of the Fifteenth Conference on Uncertainty in Artificial Intelligence, Stockholm, Sweden, 30 July–1 August 1999, pp. 279–288. Morgan Kaufmann (1999)
55. Holtzen, S., Van den Broeck, G., Millstein, T.D.: Scaling exact inference for discrete probabilistic programs. Proc. ACM Program. Lang. **4**(OOPSLA), 140:1–140:31 (2020). https://doi.org/10.1145/3428208
56. Hossain, M.M., Abbass, H.A., Lokan, C., Alam, S.: Adversarial evolution: phase transition in non-uniform hard satisfiability problems. In: Proceedings of the IEEE Congress on Evolutionary Computation, CEC 2010, Barcelona, Spain, 18–23 July 2010, pp. 1–8. IEEE (2010). https://doi.org/10.1109/CEC.2010.5586506
57. Korhonen, T., Järvisalo, M.: Integrating tree decompositions into decision heuristics of propositional model counters (short paper). In: Michel, L.D. (ed.) 27th International Conference on Principles and Practice of Constraint Programming, CP 2021, Montpellier, France (Virtual Conference), 25–29 October 2021. LIPIcs, vol. 210, pp. 8:1–8:11. Schloss Dagstuhl - Leibniz-Zentrum für Informatik (2021). https://doi.org/10.4230/LIPIcs.CP.2021.8
58. Lagniez, J., Marquis, P.: An improved decision-DNNF compiler. In: Sierra, C. (ed.) Proceedings of the Twenty-Sixth International Joint Conference on Artificial Intelligence, IJCAI 2017, Melbourne, Australia, 19–25 August 2017, pp. 667–673. ijcai.org (2017). https://doi.org/10.24963/ijcai.2017/93
59. Mitchell, D.G., Selman, B., Levesque, H.J.: Hard and easy distributions of SAT problems. In: Swartout, W.R. (ed.) Proceedings of the 10th National Conference on Artificial Intelligence, San Jose, CA, USA, 12–16 July 1992, pp. 459–465. AAAI Press/The MIT Press (1992). http://www.aaai.org/Library/AAAI/1992/aaai92-071.php
60. Moskewicz, M.W., Madigan, C.F., Zhao, Y., Zhang, L., Malik, S.: Chaff: Engineering an efficient SAT solver. In: Proceedings of the 38th Design Automation Conference, DAC 2001, Las Vegas, NV, USA, 18–22 June 2001, pp. 530–535. ACM (2001). https://doi.org/10.1145/378239.379017
61. Oztok, U., Darwiche, A.: A top-down compiler for sentential decision diagrams. In: Yang, Q., Wooldridge, M.J. (eds.) Proceedings of the Twenty-Fourth International Joint Conference on Artificial Intelligence, IJCAI 2015, Buenos Aires, Argentina, 25–31 July 2015, pp. 3141–3148. AAAI Press (2015). http://ijcai.org/Abstract/15/443
62. Pote, Y., Joshi, S., Meel, K.S.: Phase transition behavior of cardinality and XOR constraints. In: Kraus, S. (ed.) Proceedings of the Twenty-Eighth International

Joint Conference on Artificial Intelligence, IJCAI 2019, Macao, China, 10–16 August 2019, pp. 1162–1168. ijcai.org (2019). https://doi.org/10.24963/ijcai.2019/162
63. Purdom, P.W., Jr., Brown, C.A.: An analysis of backtracking with search rearrangement. SIAM J. Comput. **12**(4), 717–733 (1983). https://doi.org/10.1137/0212049
64. Pushak, Y., Hoos, H.H.: Advanced statistical analysis of empirical performance scaling. In: Coello, C.A.C. (ed.) GECCO 2020: Genetic and Evolutionary Computation Conference, Cancún Mexico, 8–12 July 2020, pp. 236–244. ACM (2020). https://doi.org/10.1145/3377930.3390210
65. Renkens, J., Kimmig, A., Van den Broeck, G., De Raedt, L.: Explanation-based approximate weighted model counting for probabilistic logics. In: Brodley, C.E., Stone, P. (eds.) Proceedings of the Twenty-Eighth AAAI Conference on Artificial Intelligence, 27–31 July 2014, Québec City, Québec, Canada, pp. 2490–2496. AAAI Press (2014). http://www.aaai.org/ocs/index.php/AAAI/AAAI14/paper/view/8484
66. Riguzzi, F.: Quantum weighted model counting. In: Giacomo, G.D., et al. (eds.) ECAI 2020–24th European Conference on Artificial Intelligence, 29 August-8 September 2020, Santiago de Compostela, Spain, 29 August–8 September 2020 - Including 10th Conference on Prestigious Applications of Artificial Intelligence (PAIS 2020). Frontiers in Artificial Intelligence and Applications, vol. 325, pp. 2640–2647. IOS Press (2020). https://doi.org/10.3233/FAIA200401
67. Robertson, N., Seymour, P.D.: Graph minors. III. Planar tree-width. J. Comb. Theor. Ser. B **36**(1), 49–64 (1984). https://doi.org/10.1016/0095-8956(84)90013-3
68. Robertson, N., Seymour, P.D.: Graph minors. X. Obstructions to tree-decomposition. J. Comb. Theor. Ser. B **52**(2), 153–190 (1991). https://doi.org/10.1016/0095-8956(91)90061-N
69. Sang, T., Bacchus, F., Beame, P., Kautz, H.A., Pitassi, T.: Combining component caching and clause learning for effective model counting. In: SAT 2004 - The Seventh International Conference on Theory and Applications of Satisfiability Testing, 10–13 May 2004, Vancouver, BC, Canada, Online Proceedings (2004). http://www.satisfiability.org/SAT04/programme/21.pdf
70. Sang, T., Beame, P., Kautz, H.: Heuristics for fast exact model counting. In: Bacchus, F., Walsh, T. (eds.) SAT 2005. LNCS, vol. 3569, pp. 226–240. Springer, Heidelberg (2005). https://doi.org/10.1007/11499107_17
71. Sang, T., Beame, P., Kautz, H.A.: Performing Bayesian inference by weighted model counting. In: Veloso, M.M., Kambhampati, S. (eds.) Proceedings, the Twentieth National Conference on Artificial Intelligence and the Seventeenth Innovative Applications of Artificial Intelligence Conference, 9–13 July 2005, Pittsburgh, Pennsylvania, USA, pp. 475–482. AAAI Press/The MIT Press (2005). http://www.aaai.org/Library/AAAI/2005/aaai05-075.php
72. Sharma, S., Roy, S., Soos, M., Meel, K.S.: GANAK: A scalable probabilistic exact model counter. In: Kraus, S. (ed.) Proceedings of the Twenty-Eighth International Joint Conference on Artificial Intelligence, IJCAI 2019, Macao, China, 10–16 August 2019, pp. 1169–1176. ijcai.org (2019). https://doi.org/10.24963/ijcai.2019/163
73. Van den Broeck, G., Taghipour, N., Meert, W., Davis, J., De Raedt, L.: Lifted probabilistic inference by first-order knowledge compilation. In: Walsh, T. (ed.) IJCAI 2011, Proceedings of the 22nd International Joint Conference on Artificial Intelligence, Barcelona, Catalonia, Spain, 16–22 July 2011, pp. 2178–2185. IJCAI/AAAI (2011). https://doi.org/10.5591/978-1-57735-516-8/IJCAI11-363

74. Vlasselaer, J., Meert, W., Van den Broeck, G., De Raedt, L.: Exploiting local and repeated structure in dynamic Bayesian networks. Artif. Intell. **232**, 43–53 (2016). https://doi.org/10.1016/j.artint.2015.12.001
75. Xu, J., Zhang, Z., Friedman, T., Liang, Y., Van den Broeck, G.: A semantic loss function for deep learning with symbolic knowledge. In: Dy, J.G., Krause, A. (eds.) Proceedings of the 35th International Conference on Machine Learning, ICML 2018, Stockholmsmässan, Stockholm, Sweden, 10–15 July 2018. Proceedings of Machine Learning Research, vol. 80, pp. 5498–5507. PMLR (2018). http://proceedings.mlr.press/v80/xu18h.html
76. Xu, L., Hutter, F., Hoos, H.H., Leyton-Brown, K.: SATzilla: portfolio-based algorithm selection for SAT. J. Artif. Intell. Res. **32**, 565–606 (2008). https://doi.org/10.1613/jair.2490

Virtual Pairwise Consistency in Cost Function Networks

Pierre Montalbano[1], David Allouche[1], Simon de Givry[1](✉), George Katsirelos[2], and Tomáš Werner[3]

[1] Université Fédérale de Toulouse, ANITI, INRAE, UR 875, 31326 Toulouse, France
{pierre.montalbano,david.allouche,simon.de-givry}@inrae.fr

[2] Université Fédérale de Toulouse, ANITI, INRAE, MIA Paris, AgroParisTech, 75231 Paris, France

[3] Department of Cybernetics, Faculty of Electrical Engineering, Czech Technical University in Prague, Prague, Czech Republic
werner@fel.cvut.cz

Abstract. In constraint satisfaction, pairwise consistency (PWC) is a well-known local consistency improving generalized arc consistency in theory but not often in practice. A popular approach to enforcing PWC enforces arc consistency on the dual encoding of the problem, allowing to reuse existing AC algorithms. In this paper, we explore the benefit of this simple approach in the optimization context of cost function networks and soft local consistencies. Using a dual encoding, we obtain an equivalent binary cost function network where enforcing virtual arc consistency achieves virtual PWC on the original problem. We experimentally observed that adding extra non-binary cost functions before the dual encoding results in even stronger bounds. Such supplementary cost functions may be produced by bounded variable elimination or by adding ternary zero-cost functions. Experiments on (probabilistic) graphical models, from the UAI 2022 competition benchmark, show a clear improvement when using our approach inside a branch-and-bound solver compared to the state-of-the-art.

Keywords: dual encoding · non-binary cost function network · soft local consistency · branch-and-bound · graphical model · discrete optimization

1 Introduction

Cost Function Networks (CFNs) can represent many combinatorial problems in a compact way as a sum of local functions over discrete variables. They have been used in bioinformatics [1,27], resource allocation [2,5], and elsewhere [9]. They can model probabilistic graphical models such as Bayesian networks and

This research was funded by the grants ANR-18-EURE-0021 and ANR-19-P3IA-0004. It receives support from the Genotoul (Toulouse) Bioinformatic platform.

© The Author(s), under exclusive license to Springer Nature Switzerland AG 2023
A. A. Cire (Ed.): CPAIOR 2023, LNCS 13884, pp. 417–426, 2023.
https://doi.org/10.1007/978-3-031-33271-5_27

Markov random fields [8], which find many applications in artificial intelligence [17,28,31]. We focus here on the minimization task, *a.k.a.* Weighted Constraint Satisfaction Problem (WCSP), where exact methods mostly rely on a branch-and-bound procedure. Its efficiency depends on the compromise between the quality of its lower bound and the time to construct it. Several directions have been studied, inspired by Arc Consistency (AC) in CSP [6]. Stronger *soft* local consistencies were rarely considered, except in [7,13,22]. Pairwise Consistency (PWC) is known as the strongest consistency that can be enforced without introducing new cost functions when computing the lower bound [34]. It was never implemented nor tested in a branch-and-bound WCSP solver. In constraint programming, PWC was compared to generalized AC for solving non-binary CSPs given in extension [26,29,32,33]. These approaches rely on a dual encoding into a binary CSP. We explore a similar idea in the CFN framework.

2 Background

2.1 Weighted Constraint Satisfaction Problem

A *Cost Function Network (CFN)* is a quadruplet (V, D, S, f) where V is a set of variable indices (or *variables* in short), $D = (D_i)_{i \in V}$ is the list of *finite domains* for all the variables, S is a set of subsets of V, and $f = (f_A)_{A \in S}$ is the list of all the cost functions (defined below). A *value* of variable $i \in V$ is denoted by $x_i \in D_i$. By $D_A = \prod_{i \in A} D_i$ we denote the Cartesian product of the domains of variables $A \subseteq V$, and by $x = (x_i)_{i \in A} \in D_A$ an *assignment* to variables A. For $B \subseteq A \subseteq V$, $x|_B = (x_i)_{i \in B}$ denotes the projection of $x \in D_A$ to variables B. A *cost function* f_A is a function of a set of variables A taking non-negative values or possibly infinity (representing forbidden assignments), *i.e.*, it is a function $D_A \to \mathbb{R}_+ \cup \{\infty\}$, where $A \subseteq V$ is the *scope* of the function and $|A|$ its *arity*. A nullary cost function $f_\emptyset$ with an empty scope is just a constant. We assume that the network is normalized (with one cost function per scope), $\emptyset \in S$ ($f_\emptyset$ will be used as a problem lower bound) and $\{i\} \in S \ \forall i \in V$ (the network contains all unary functions). Given a CFN (V, D, S, f), the *Weighted Constraint Satisfaction Problem (WCSP)* is to find a complete non-forbidden assignment $x \in D_V$ minimizing the function $\sum_{A \in S} f_A(x|_A)$. This problem is NP-hard.

Example 1. Let $V = \{1, 2, 3, 4, 5\}$, $S = \{\emptyset, \{1\}, \{1, 2, 3\}, \{1, 4\}, \{2\}, \{2, 3, 4\}, \{2, 3, 5\}, \{3\}, \{4\}, \{5\}\}$, $D_2 = D_3 = D_5 = \{a, b\}$, and $D_1 = D_4 = \{a, b, c\}$. The WCSP aims to minimize the objective function $f_\emptyset + f_1(x_1) + f_{123}(x_1, x_2, x_3) + f_{14}(x_1, x_4) + f_2(x_2) + f_{234}(x_2, x_3, x_4) + f_{235}(x_2, x_3, x_5) + f_3(x_3) + f_4(x_4) + f_5(x_5)$ (where we abbreviated $f_{\{1,2,3\}}$ by f_{123}, etc.) over all assignments $(x_1, x_2, x_3, x_4, x_5) \in D_V$.

2.2 Constraint Satisfaction Problem and Local Consistencies

If all cost functions in a CFN take only values 0 or ∞, the cost functions are called *constraints* and the WCSP reduces to the *Constraint Satisfaction Problem (CSP)*. In this case, we denote the cost functions by r_A rather than f_A, so the

Constraint Network (CN) is defined by (V, D, S, r). The values 0 and ∞ act as the logical values *true* and *false*, respectively. For $u, v \in \{0, \infty\}$, we will denote logical conjuction by $u \wedge v = u + v$ and the disjunction by $u \vee v = \min\{u, v\}$. As in CFN, we assume a CN contains all unary constraints, i.e., $\{i\} \in S \; \forall i \in V$.

For any $B \subseteq A \subseteq V$, we define the *projection* of a constraint r_A: $D_A \to \{0, \infty\}$ onto variables B to be the constraint $r_A|_B$: $D_B \to \{0, \infty\}$ given by

$$r_A|_B(x) = \bigvee_{x' \in D_A:\, x'|_B = x} r_A(x') \qquad \forall x \in D_B. \tag{1}$$

We say that a pair of constraints $\{r_A, r_B\}$ is *Pairwise Consistent (PWC)* if they admit the same set of assignments to their shared variables, i.e.,

$$r_A|_{A \cap B} = r_B|_{A \cap B} \tag{2}$$

where '=' denotes here equality of functions. A CN is PWC if all possible pairs of its constraints are PWC.[1] If we restrict PWC to pairs of constraints where one constraint is unary, we get (generalized) arc consistency (GAC, AC for binary CNs). PWC or GAC can be enforced on a CN P by iteratively forbidding assignments that violate (2). The minimal set of changes required to do this is unique and the resulting CN is called the PWC (or GAC) *closure* of P.

We say that a local consistency ψ' is *not weaker* than a local consistency ψ if for every CN instance for which the ψ-consistency closure is empty, the ψ'-consistency closure is also empty. We say that ψ and ψ' are *equally strong* if ψ' is not weaker than ψ and *vice versa*. We say that ψ' is *strictly* stronger than ψ if ψ' is not weaker than ψ but they are not equally strong. It can be shown that: (i) for binary CNs, AC is equally strong as PWC; (ii) for non-binary CNs, PWC is strictly stronger than GAC.

The PWC relation of constraints is clearly reflexive and symmetric. It is in general not transitive but it satisfies the following weaker condition:

Theorem 1. *[16] Let $C_1, \dots, C_n \in S$ be such that for every $i = 1, \dots, n$, we have $C_1 \cap C_n \subseteq C_i$. Let for every $i = 1, \dots, n-1$, constraint r_{C_i} be PWC with $r_{C_{i+1}}$. Then r_{C_1} is PWC with r_{C_n}.*

Thus, enforcing PWC for some constraint pairs implies that the PWC condition holds also for some other pairs, which can simplify algorithms [16, 29].

2.3 Soft Local Consistencies

To solve a WCSP to optimality, most methods rely on a branch-and-bound algorithm. At each node, the solver computes a bound using either static memory-intensive bounds [11] or memory-light ones [6] better suited to dynamic variable orderings. We focus on the latter, called *Soft Arc Consistencies (SAC)*, because they reason on each non-unary cost function one by one, in a generalization of

[1] This corresponds to *full* PWC because unary constraints may appear in these pairs.

propagation in CSPs. In particular, *Virtual Arc Consistency (VAC)* is characterized by AC of a CN derived from the CFN. To any CFN $P = (V, D, S, f)$ we associate the CN $\text{Bool}(P) = (V, D, S, r)$ where $r_\emptyset = 0$ and

$$r_A(x) = \begin{cases} 0 & \text{if } f_A(x) = 0 \\ \infty & \text{if } f_A(x) > 0 \end{cases} \qquad \forall A \in S \setminus \{\emptyset\},\ x \in D_A. \tag{3}$$

Definition 1. *[6] A CFN P is VAC if the GAC closure of* Bool(P) *is non-empty.*

Algorithms enforcing SACs apply a sequence of *Equivalence-Preserving Transformations (EPTs)* to the CFN. An EPT moves finite costs between two cost functions. That is, for some $A, B \in S$ we add a function $\varphi_{AB}\colon D_{A \cap B} \to \mathbb{R}$ with scope $A \cap B$ to cost function f_A and subtract it from function f_B:

$$f_A(x) := f_A(x) + \varphi_{AB}(x|_{A \cap B}) \qquad \forall x \in D_A, \tag{4a}$$

$$f_B(x) := f_B(x) - \varphi_{AB}(x|_{A \cap B}) \qquad \forall x \in D_B. \tag{4b}$$

This operation can be seen as moving a set of costs (stored as the values of the φ_{AB}) from f_B to f_A. The values of φ_{AB} cannot be arbitrary because we require the resulting cost functions to have non-negative values. EPTs preserve the WCSP objective function because the terms φ_{AB} and $-\varphi_{AB}$ cancel out in the sum $\sum_{A \in S} f_A(x|_A)$. CFNs (V, D, S, f) and (V, D, S, f') are *equivalent* if one can be obtained from the other by a sequence of EPTs. SAC algorithms aim to derive an equivalent CFN where $f'_\emptyset > f_\emptyset$.

2.4 Dual Encoding of a Cost Function Network

An encoding into a binary CFN is a way to get better bounds. The *dual encoding* of a CFN $P = (V, D, S, f)$ is a CFN $\text{Dual}(P) = (S \setminus \{\emptyset\}, \bar{D}, \bar{S}, \bar{f})$ where:

- The variables of the dual problem are the scopes $S \setminus \{\emptyset\}$ of P.
- The domain of variable $A \in S \setminus \{\emptyset\}$ of the dual problem is $\bar{D}_A = D_A$.
- The scopes are $\bar{S} = \{\emptyset\} \cup \{\, \{A\} \mid A \in S \,\} \cup \{\, \{A, B\} \mid A, B \in S,\ A \cap B \neq \emptyset \,\}$.
- The dual nullary cost function is unchanged: $\bar{f}_\emptyset = f_\emptyset$.
- The dual unary cost function with scope $\{A\} \in \bar{S}$ is the function $\bar{f}_A = f_A$.
- The dual binary cost function with scope $\{A, B\} \in \bar{S}$ is the *channeling constraint* $\bar{f}_{AB}\colon D_A \times D_B \to \{0, \infty\}$ with values:

$$\bar{f}_{AB}(y, y') = \begin{cases} 0 & \text{if } y_i = y'_i \forall i \in A \cap B \\ \infty & \text{otherwise} \end{cases} \qquad \forall y = (y_i)_{i \in A} \in D_A,\ y' = (y'_i)_{i \in B} \in D_B.$$

Example 2. Let P be the CFN described in Example 1, represented by the hypergraph in Fig. 1(a). Then, Dual(P) has 9 dual variables, $y_1, y_{123}, y_{14}, \ldots, y_5$, and 16 binary channeling constraints, as shown by the constraint graph in Fig. 1(b). Using Theorem 1, a minimal dual graph can be produced with only 9 binary constraints (Fig. 1(c)).

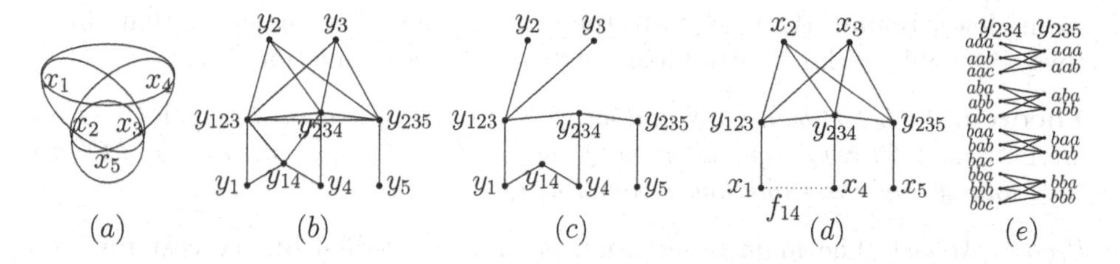

Fig. 1. (a) Hypergraph of a CFN, (b) its dual graph, (c) a minimal dual graph, (d) the partial dual graph used in the experiments, (e) a binary channeling constraint created by the dual encoding (an edge depicts a 0-cost assignment).

Table 1. (a) Original CFN. (b) dual unary cost functions (missing tuples have 0 cost).

(a)	f_{123}	x_1	x_2	x_3	Cost	f_{234}	x_2	x_3	x_4	Cost	f_{14}	x_1	x_4	Cost	(b)	y_{123}	Cost	y_{234}	Cost	y_{14}	Cost
		a	a	a	0		a	a	a	1		a	a	2		aaa	0	aaa	1	aa	2
		a	a	b	1		a	a	b	1		a	b	2		aab	1	aab	1	ab	2
		b	a	a	1		a	a	c	1		a	c	2		baa	1	aac	1	ac	2
		b	a	b	1	f_1			x_1		f_2		x_2			bab	1	y_1		y_2	
		c	a	a	1				b	2			a	0		caa	1	b	2	a	0
		c	a	b	1				c	2			b	2		cab	1	c	2	b	1

3 Virtual Pairwise Consistency

Following the idea of VAC, we introduce *Virtual Pairwise Consistency (VPWC)*, a stronger soft local consistency than VAC.

Definition 2. *A CFN P is VPWC if the PWC closure of* Bool(P) *is non-empty.*

Combining Definition 2 and previous results [16], we get that enforcing VPWC is possible using existing algorithms.

Theorem 2. *Let P be a CFN. P is VPWC if and only if* Dual(P) *is VAC.*

Proof. It is known that a CN has a non-empty PWC closure if and only if its dual has a non-empty AC closure [16]. Clearly, for any CFN P we have Dual(Bool(P)) = Bool(Dual(P)). Therefore, P is VPWC iff Dual(Bool(P)) = Bool(Dual(P)) has a non-empty AC closure, which means Dual(P) is VAC. □

Example 3. Following Example 2, we give the costs for each cost function in Table 1(a). VAC on this problem derives a lower bound of 2, since $x_1 = a$ is not consistent with r_{14}. VAC on the dual (Table 1(b)) derives a lower bound of 3, because (a) all values in y_{14} compatible with $y_1 = a$ (*i.e., aa, ab, ac*) have cost 2, and (b) all values compatible with $y_{123} = aaa$ in y_{234} (*i.e., aaa, aab, aac*) have a cost of 1, therefore they do not support $y_2 = a$, making it inconsistent in y_{123}. This leads to a lower bound of 3.

The dual can help derive better lower bounds, as we show in the next section, but introduces a possibly large number of variables with large domains which may slow down search. We propose to first dualize the problem and get a first strong lower bound, then return to the primal. The following shows that this is always possible without introducing higher order cost functions[2].

Theorem 3. *Let P be a CFN and let Q be a CFN equivalent to* Dual(P). *Then there exists a CFN Q' equivalent to Q such that all binary constraints of Q' are hard and Q' has the same lower bound as Q.*

Proof (Sketch). The main observation is that every dual binary cost function (the channeling constraint) r_{ij} has a block structure (see Fig. 1(e)): there exists a partition $H_i = \{s_1, ..., s_m\}$ of the domain D_i and a partition $H_j = \{s'_1, ..., s'_m\}$ of D_j such that for each $x_i \in s_k$ and $x_j \in s'_l$ we have $r_{ij}(x_i, x_j) = 0$ whenever $k = l$ and $r_{ij}(x_i, x_j) = \infty$ whenever $k \neq l$. This implies that every EPT that moves cost into r_{ij} can be matched with another EPT that moves cost out of it without affecting the lower bound. □

We can now summarize the base version of our approach. Given a CFN P, we apply EPTs to its dual encoding Dual(P) (using a VAC algorithm) to obtain a CFN Q with an increased lower bound. Theorem 3 lets us obtain from Q another CFN Q' in which all channeling cost functions are constraints. We can thus undo the dual encoding, i.e., obtain a CFN P', equivalent to P, such that $Q' = \text{Dual}(P')$. If Q was VAC then, by Theorem 2, P' is VPWC.

4 Experimental Results on UAI 2022 Competition

We won a recent competition on probabilistic graphical models.[3] We present results on a set of 120 tuning instances where 63 have maximum arity of 3.

We evaluate three solvers: daoopt (version from UAI 2012 competition with 1-h settings as given in [23]), cplex (version 20.1.0.0, forcing completeness with zero absolute and relative gaps, translating CFN to 0–1 LP by the tuple encoding [15]), and toulbar2 (version 1.2.0) using two state-of-the-art methods, Variable Neighborhood Search (VNS) [24] winner of UAI 2014 competition,[4] and Hybrid Best-First Search with VAC in preprocessing (VACpre-HBFS), including VAC integrality heuristics [30].[5] We implemented VPWC in the latest version of toulbar2. It is either enforced in preprocessing (and then converted back to the primal, see Theorem 3) (VPWCpre-HBFS) or maintained during search (HBFS-VPWC). EDAC is always enforced [19,27], providing a default value ordering heuristic when no solution is found for solution-based heuristics [12]. The branching heuristic is *dom/wdeg* [4] combined with *last conflict* [20].

We use a slightly different binary encoding, a hybrid between the dual and *hidden variable encoding* [25].[6] We keep the original variables and the original

[2] This is unsurprising because the strongest bound that can be derived using EPTs is obtained using a linear program which includes pairwise consistency constraints [34].
[3] https://uaicompetition.github.io/uci-2022, see MPE and MMAP entries.
[4] http://auai.org/uai2014/competition.shtml, http://miat.inrae.fr/toulbar2.
[5] Options *-A -P=1000 -T=1000 -vacint -vacthr -rasps -raspsini* in toulbar2-vacint.
[6] Called *double encoding* in [26], it allows more flexibility to enforce various levels of consistency from GAC to PWC depending on the selected channeling constraints.

Table 2. UAI 2022 detailed results on a selection of four instances for HBFS methods. '-' means the instance is unsolved in 1h. (in parentheses, remaining optimality gap).

instance	(n, d, e, a)	(n', e', a')	(n'', d'', e'')	VACpre-HBFS time *(gap)*	VPWCpre-HBFS time *(gap)*	HBFS-VPWC time *(gap)*
Grids21	(1600,2,4800,2)	(799,2810,4)	(1628,16,4675)	- *(42.4%)*	- *(3%)*	1216.83
Promedas12	(1766,2,1766,3)	(826,1884,4)	(1373,16,2223)	5.17	6.34	7.7
ProteinFold11	(400,2,1160,2)	(190,604,4)	(381,16,1005)	- *(16.1%)*	8.48	12.43
wcsp12	(311,4,5732,3)	(305,5887,3)	(12708,64,70959)	- *(49.9%)*	- *(19.3%)*	- *(54.8%)*

binary cost functions unchanged, and only dualize the original non-binary cost functions. We add channeling constraints between those pairs of dual variables that are not redundant by Theorem 1 and with intersecting scopes strictly greater than 1. We also add channeling constraints between dual and primal variables.[7] Note that Theorem 2 and 3 remain valid. Moreover, we apply this encoding only partially, indeed for high-arity constraints, a full dual encoding might mean prohibitive amount of memory to store the dual domains. Hence, only non-binary cost functions of arity less than 10 and fewer than 2^{15} non-forbidden tuples are dualized. Those remaining are lazily propagated by VAC/EDAC when they have less than three unassigned variables in their scope. The memory used by each channeling constraint between a pair of dual variables is restricted to at most 1 MB (arbitrarily chosen). Larger channeling constraints are ignored.

Additional preprocessing is performed beforehand for all the HBFS methods in order to find better bounds. An initial upper bound is found by local search [3, 21] and VAC-based heuristics [30]. To reduce the problem size and improve lower bounds, we apply bounded variable elimination with a min-fill ordering [10,14, 18][8] and add ternary zero-cost functions on the *most-preferred triangles* (total memory space of extra ternary functions limited to 1 MB).[9] It results in at most 6-ary (resp. zero-cost ternary) cost functions for 84 (resp. 81) instances, making our encoding applicable to 85 instances rather than 63. Finding a (quasi-)minimal dual graph (see Theorem 1) yielded 700.3 channeling constraints on average, a 4.5% savings compared to the complete dual graph.

The experiments were run on a single core of Intel Xeon E5-2683 2.1 GHz processors with 1-h CPU-time and 8 GB memory limit. toulbar2 was able to solve optimally 86 instances using VACpre-HBFS or VNS. daoopt solved 92 instances and cplex 95 instances. Using our partial dual encoding with VPWC applied in preprocessing, VPWCpre-HBFS solved 95 instances, and when applied during search, HBFS-VPWC solved 99 instances, 15% above VACpre-HBFS, being the best exact method for this benchmark.

[7] The resulting non-minimal graph for Example 2 is shown in Fig. 1(d).

[8] It is done only if the median degree in the original problem is less than 8, eliminating variables with a current degree less than or equal to the original median degree.

[9] With additional options *-i -pils -p* $=$ *-8 -O* $= -3$ *-t* $=$ *1*. A triangle is defined by three variables involved in three binary cost functions. The score of a triangle is given by the average cost in the three functions. Triangles with the largest score are selected first. This approach allows to simulate soft path inverse consistency [22].

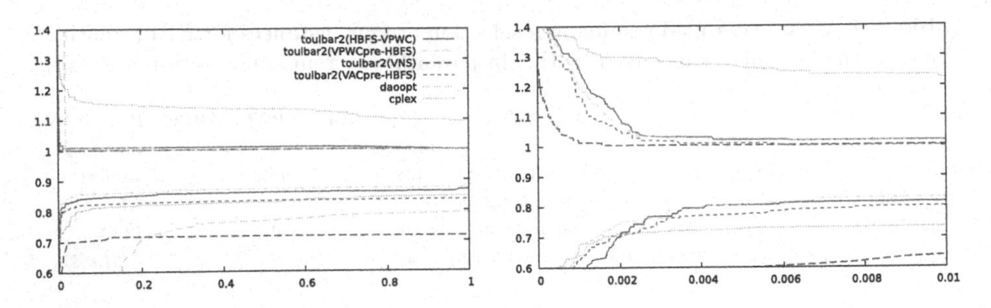

Fig. 2. Normalized lower and upper bounds (y-axis) as time passes (x-axis in hour, zoomed on the right fig.) for cplex, daoopt, and toulbar2 on UAI 2022 tuning benchmark.

Table 2 shows for a selection of UAI 2022 instances their size in terms of number of variables n, maximum domain size d, number of cost functions e, maximum arity a of the original problem, after preprocessing it with bounded variable elimination and adding triangles (n', d', e', a' with $d' = d$), and after applying our partial dual encoding (n'', d'', e'', a'' with $a'' = 2$). It gives also the CPU-time in seconds to solve an instance using HBFS methods or the remaining optimality gap if unsolved after 1 h. On Grids21, only HBFS-VPWC solves the instance. Notice the large improvement on the optimality gap by VPWCpre compared to VACpre. On Promedas12 and ProteinFolding11, VPWCpre-HBFS develops 13 and 32250 nodes, respectively, and takes about the same time as HBFS-VPWC which develops 4 and 992 nodes, respectively. For wcsp12, the size of the encoding slows down the search too much, suggesting harder limits for our partial dual encoding.

We report in Fig. 2 the average normalized lower and upper bounds as time passes (computed as in [30]). Here VNS provides the best upper bounds in limited time whereas HBFS-VPWC is slightly slower than VPWCpre-HBFS, VACpre-HBFS, and VNS, but still faster than daoopt and cplex. Both VPWCpre-HBFS and HBFS-VPWC offer the best average lower bounds in less than 1 h. HBFS-VPWC found 117 best solutions, VPWCpre-HBFS 112, VACpre-HBFS 106, VNS 105, daoopt 99, and cplex 95. VNS found 2 single-best solutions (wcsp11, wcsp12). For the competition, we combined VNS and HBFS-VPWC sequentially.[10]

5 Conclusion

We have defined virtual pairwise consistency and shown how it can efficiently be used in preprocessing or during search by applying the existing VAC algorithm to a dual encoding of the problem. In the future we will explore the benefit of other binary encodings [33] and adapt the VAC algorithm to the specific constraints

[10] See toulbar2-ipr results on the UAI 2022 Tuning Leader Board. Multiple runs of VNS with increasing floating-point precision were done with a total amount of time of $\frac{1}{2}$h. The remaining time is allocated to HBFS-VPWC. Each search procedure gives its best solution found to the next search procedure. On UAI 2022 tuning instances, this approach found 119 best solutions, ranking first among our 7 tested methods.

of the encoding as it is done in CSPs [29,32]. Finding good heuristics to exploit a partial dual encoding in conjunction with bounded variable elimination and zero-cost function addition is also an interesting question.

References

1. Allouche, D., et al.: Computational protein design as an optimization problem. Artif. Intell. **212**, 59–79 (2014)
2. Bensana, E., Lemaître, M., Verfaillie, G.: Earth observation satellite management. Constraints **4**(3), 293–299 (1999)
3. Beuvin, F., de Givry, S., Schiex, T., Verel, S., Simoncini, D.: Iterated local search with partition crossover for computational protein design. Proteins Struct. Funct. Bioinf. **87**, 1522–1529 (2021)
4. Boussemart, F., Hemery, F., Lecoutre, C., Sais, L.: Boosting systematic search by weighting constraints. In: ECAI. vol. 16, p. 146 (2004)
5. Cabon, B., de Givry, S., Lobjois, L., Schiex, T., Warners, J.: Radio link frequency assignment. Constraints J. **4**, 79–89 (1999)
6. Cooper, M., de Givry, S., Sanchez, M., Schiex, T., Zytnicki, M., Werner, T.: Soft arc consistency revisited. Artif. Intell. **174**(7–8), 449–478 (2010)
7. Cooper, M.C.: High-order consistency in valued constraint satisfaction. Constraints **10**, 283–305 (2005)
8. Cooper, M.C., de Givry, S., Schiex, T.: Graphical models: queries, complexity, algorithms (tutorial). In: 37th International Symposium on Theoretical Aspects of Computer Science (STACS-20). LIPIcs, vol. 154, pp. 4:1–4:22. Montpellier, France (2020)
9. Cooper, M.C., de Givry, S., Schiex, T.: Valued constraint satisfaction problems. In: Marquis, P., Papini, O., Prade, H. (eds.) A Guided Tour of Artificial Intelligence Research, pp. 185–207. Springer, Cham (2020). https://doi.org/10.1007/978-3-030-06167-8_7
10. Dechter, R.: Bucket elimination: a unifying framework for reasoning. Artif. Intell. **113**(1–2), 41–85 (1999)
11. Dechter, R., Rish, I.: Mini-buckets: a general scheme for bounded inference. J. ACM (JACM) **50**(2), 107–153 (2003)
12. Demirović, E., Chu, G., Stuckey, P.J.: Solution-based phase saving for CP: a value-selection heuristic to simulate local search behavior in complete solvers. In: Hooker, J. (ed.) CP 2018. LNCS, vol. 11008, pp. 99–108. Springer, Cham (2018). https://doi.org/10.1007/978-3-319-98334-9_7
13. Dlask, T., Werner, T., de Givry, S.: Bounds on weighted CSPs using constraint propagation and super-reparametrizations. In: Proceedings of CP-21. Montpellier, France (2021)
14. Favier, A., de Givry, S., Legarra, A., Schiex, T.: Pairwise decomposition for combinatorial optimization in graphical models. In: Proceedings of IJCAI-11. Barcelona, Spain (2011). http://www.inra.fr/mia/T/degivry/Favier11.mov
15. Hurley, B., et al.: Multi-language evaluation of exact solvers in graphical model discrete optimization. Constraints **21**(3), 413–434 (2016)
16. Janssen, P., Jégou, P., Nouguier, B., Vilarem, M.C.: A filtering process for general constraint-satisfaction problems: achieving pairwise-consistency using an associated binary representation. In: IEEE International Workshop on Tools for Artificial Intelligence, pp. 420–421. IEEE Computer Society (1989)

17. Kappes, J.H., et al.: A comparative study of modern inference techniques for structured discrete energy minimization problems. Intl. J. of Comput. Vis. **115**(2), 155–184 (2015)
18. Larrosa, J.: Boosting search with variable elimination. In: Dechter, R. (ed.) CP 2000. LNCS, vol. 1894, pp. 291–305. Springer, Heidelberg (2000). https://doi.org/10.1007/3-540-45349-0_22
19. Larrosa, J., Heras, F.: Resolution in Max-SAT and its relation to local consistency in weighted CSPs. In: Proceedings of IJCAI 2005, pp. 193–198. Edinburgh, Scotland (2005)
20. Lecoutre, C., Saïs, L., Tabary, S., Vidal, V.: Reasoning from last conflict(s) in constraint programming. Artif. Intell. **173**, 1592–1614 (2009)
21. Neveu, B., Trombettoni, G., Glover, F.: ID Walk: a candidate list strategy with a simple diversification device. In: Proceedings of CP, pp. 423–437 (2004)
22. Nguyen, H., Bessiere, C., de Givry, S., Schiex, T.: Triangle-based consistencies for cost function networks. Constraints **22**(2), 230–264 (2017)
23. Otten, L., Ihler, A., Kask, K., Dechter, R.: Winning the pascal 2011 map challenge with enhanced AND/OR branch-and-bound. In: DISCML 2012 Workshop, at NIPS 2012. Lake Tahoe, NV (2012)
24. Quali, A.: Variable neighborhood search for graphical model energy minimization. Artif. Intell. **278**(103194), 22p (2020)
25. Rossi, F., Petrie, C.J., Dhar, V.: On the equivalence of constraint satisfaction problems. In: ECAI, vol. 90, pp. 550–556 (1990)
26. Samaras, N., Stergiou, K.: Binary encodings of non-binary constraint satisfaction problems: algorithms and experimental results. J. Artif. Intell. Res. **24**, 641–684 (2005)
27. Sánchez, M., de Givry, S., Schiex, T.: Mendelian error detection in complex pedigrees using weighted constraint satisfaction techniques. Constraints **13**(1), 130–154 (2008)
28. Savchynskyy, B.: Discrete graphical models - an optimization perspective. Found. Trends Comput. Graph. Vis. **11**(3–4), 160–429 (2019)
29. Schneider, A., Choueiry, B.Y.: PW-AC: extending compact-table to enforce pairwise consistency on table constraints. In: Hooker, J. (ed.) CP 2018. LNCS, vol. 11008, pp. 345–361. Springer, Cham (2018). https://doi.org/10.1007/978-3-319-98334-9_23
30. Trösser, F., de Givry, S., Katsirelos, G.: Relaxation-aware heuristics for exact optimization in graphical models. In: Proceedings of CP-AI-OR'2020, pp. 475–491. Vienna, Austria (2020)
31. Wainwright, M.J., Jordan, M.I., et al.: Graphical models, exponential families, and variational inference. Found. Trends Mach. Learn. **1**(1–2), 1–305 (2008)
32. Wang, R., Yap, R.H.C.: Arc consistency revisited. In: Rousseau, L.-M., Stergiou, K. (eds.) CPAIOR 2019. LNCS, vol. 11494, pp. 599–615. Springer, Cham (2019). https://doi.org/10.1007/978-3-030-19212-9_40
33. Wang, R., Yap, R.H.: Bipartite encoding: a new binary encoding for solving non-binary CSPs. In: Proceedings of the Twenty-Ninth International Conference on International Joint Conferences on Artificial Intelligence, pp. 1184–1191 (2021)
34. Werner, T.: Revisiting the linear programming relaxation approach to Gibbs energy minimization and weighted constraint satisfaction. IEEE Trans. Pattern Anal. Mach. Intell. **32**(8), 1474–1488 (2010)

Multi-objective Optimization for the Design of Salary Structures

François-Alexandre Tremblay(✉), Dominique Piché-Meunier, and Louis Dubois

Department of Computer Science and Software Engineering, Université Laval, Québec, QC, Canada
{francois-alexandre.tremblay.1,dominique.piche-meunier.1, louis.dubois.2}@ulaval.ca

Abstract. In a context of labor shortage and strong global competition for talent, salary management is becoming a critical issue for companies wishing to attract, engage and retain qualified employees. This paper presents a multi-objective optimization model to balance the internal equity, external equity and costs trade-offs associated with the design of salary structures. Solutions are generated to estimate and explore the Pareto frontier using real compensation data from a unionized establishment in the public sector. Our work shows the interest of using combinatorial optimization techniques in the design of salary structures.

Keywords: Multi-objective optimization · constraint programming · salary structure · decision support system

1 Introduction

In a context of global labor shortage, companies must offer competitive wages to attract and retain qualified employees while minimizing their payroll to remain competitive. Managing this trade-off is no small task, especially for organizations that heavily rely on human capital. In North America, salary structures are the most common approach to determine and manage salaries fairly [13]. Fair and aligned organizational pay policies are important since they can significantly improve individual and organizational performance [1,6]. As illustrated by Fig. 1, these structures allow to integrate various jobs in a coherent manner according to their relative value within the company (i.e., internal equity) and to align internal wages with those on the market (i.e., external equity) [12].

However, establishing or updating salary structures usually requires a considerable amount of time and resources for companies. A decision maker (DM) must also consider several different scenarios simultaneously, and the set of constraints makes it difficult, without an analytical approach, to ensure that the structure obtained is close to an optimal one. For some companies, the law requires to follow some guidelines when establishing job grades [3], increasing the process complexity.

© The Author(s), under exclusive license to Springer Nature Switzerland AG 2023
A. A. Cire (Ed.): CPAIOR 2023, LNCS 13884, pp. 427–442, 2023.
https://doi.org/10.1007/978-3-031-33271-5_28

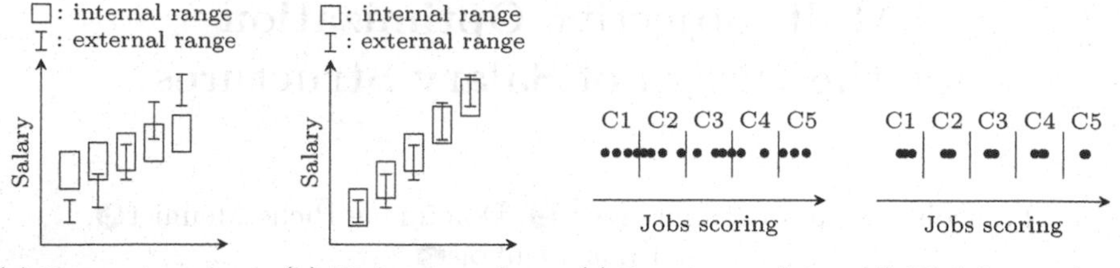

(a) Low external eq. (b) High external eq. (c) Low internal eq. (d) High internal eq.

Fig. 1. Concept of equity in the design of salary structures: (a) low external equity: internal pay ranges are misaligned with the market; (b) high external equity: internal pay ranges are aligned with external ranges on the market; (c) low internal equity: some jobs could belong to a different grade; (d) high internal equity: jobs likely belong to the right grade.

As depicted by Table 1, the literature addressing the subject analytically is surprisingly sparse. Bruno [4] proposes a linear program that consider individual performance (i.e., individual equity) and job evaluation. An employee's salary is determined by a weighted sum of the relative value of his job and his personal contribution. Wallace and Steuer [14] design a linear multi-objective program that reconciles internal and external equity issues by constraining the available budget. The user must iteratively modify the variables and constraints until a satisfactory salary structure is obtained. Finally, Kassa [7] develops a goal programming model to generate the minimum and maximum wage of grades by minimizing structure average wages with the external market and costs. However, all previous models are interactive or *a priori* approaches and partially take into account the organizational equity dimensions (i.e., individual, internal and external). Authors [7,14] suggest that subsequent models should include variables, such as the number of grades, the salary range spreads, the level of overlap between ranges, as well as wages offered per job, instead of per grade, on the market. In our work, we propose a multi-objective *a posteriori* optimization model to simultaneously maximize all equity dimensions and minimize costs while including the variables suggested by previous authors.

Table 1. Summary of the existing approaches in the literature for the design of salary structure based on the whether they consider a single objective (SO) or multiple objectives (MO), and based on the optimization techniques (LP: linear programming, GP: goal programming, CP: constraint programming). For internal equity, – means a less rigorous definition of the concept is used.

	Cost	Internal Eq	External Eq	Individual eq	Approach	Method
Bruno [4]	x	x		x	SO	LP
Wallace [14]	x	–	x		MO iterative	LP
Kassa [7]	x	–	x		MO *a priori*	GP
Ours	x	x	x	x	MO *a posteriori*	CP

This paper is organized as follows: in Sect. 2, we review the necessary concepts behind salary structures. Section 3 details the optimization model. In Sect. 4, we explain the experiments we perform to evaluate our model. The experimental results are discussed in Sect. 5 and we conclude in Sect. 6.

2 Preliminary Concepts

The process of designing salary structure can be summarized by the three following steps [2,12,13]: **(1)** The employer reviews and analyzes jobs to identify key characteristics and requirements. The end result is usually job descriptions. **(2)** Based on the previous analysis, managers undertake a job evaluation method to determine the relative importance of jobs within the organization. Even though several methods can be used to respect equity principles, the *points and factors* evaluation method, sometimes implicitly suggested by the law (e.g., Pay Equity Act [3]), is often preferred because of its neutral, analytical and systematic nature. Specifically, this method consists of associating points to each job according to several sub-factors (e.g. physical effort, decision-making, job complexity) established by the employer. These points come from a weighted scoring table which shows the sub-factor and their respective levels of points. The total points assigned to a job thus represent its relative value in the company. At this point, since the evaluation process could be affected by human judgment, the method requires determining ranges of points, called grades, to group jobs of similar value together. Therefore, a difference of 20 points does not necessarily mean that one job is more valued than another. **(3)** Finally, the employer assigns a salary range to each grade consisting of a minimum and maximum salary between which employees wages are positioned on. Higher grades imply higher salary ranges. Sometimes the ranges integrate steps, which refers to discrete salary levels within the salary range. At each salary increase period, an employee's wage will increase by one salary step, and consequently one wage rate, until the scale's maximum salary is reached.

3 Optimization Model of Salary Structure Design

The optimization model is built around three components presented previously in steps 2 and 3 of the salary structure design process: the scoring table, the score ranges and the salary structure.

3.1 Scoring Table

The scoring table $T_{s,n}$, illustrated by Fig. 2, represents the score of each level $n \in \mathcal{N}$ for each sub-factor $s \in \mathcal{S}$ where $\mathcal{N}$ is the set of possible levels and $\mathcal{S}$ is the set of sub-factors previously determined by the employer. Since the sub-factors do not necessarily possess the same number of levels, the constant N_s, which is the number of levels per sub-factor s, forces the variables of the

Table 2. Structure of a scoring table. As an example, if the third sub-factor (SF3) referred to physical effort, a desk job could be assigned to level one ($T_{3,1} = 5$), and a warehouse job to level 4 ($T_{3,4} = 20$).

Sub-factors		$SF1$	$SF2$	$SF3$	...
Min scoring		$T_1^{min} = 15$	$T_2^{min} = 10$	$T_3^{min} = 5$	...
Levels	$n = 1$	$T_{1,1} = 15$	$T_{2,1} = 10$	$T_{3,1} = 5$	...
	$n = 2$	$T_{1,2} = 45$	$T_{2,2} = 20$	$T_{3,2} = 10$	...
	$n = 3$	$T_{1,3} = 75$	$T_{2,3} = 30$	$T_{3,3} = 15$	...
	$n = 4$	$T_{1,4} = 105$		$T_{3,4} = 20$	...
Max scoring		$T_1^{max} = 105$	$T_2^{max} = 30$	$T_3^{max} = 20$	...
Max levels		$N_1 = 4$	$N_2 = 3$	$N_3 = 4$	...
Coefficients		$C_1 = 30$	$C_2 = 10$	$C_3 = 5$	...

table $T_{s,n}$ to remain null when levels are not used (1). The variables T_s^{min} and T_s^{max}, which indicate the score of the first and last levels of each sub-factor s respectively, allow for a more flexible manipulation of the variables related to the scoring table (2, 3). Since sub-factors are not all equally important for the company and must remain within a reasonable range, the constants L_s and U_s are used as lower and upper bounds respectively for the levels' maximum and minimum scores (4). To compute scores of intermediate levels, we introduce the variable C_s, which represents the gap between two levels of a sub-factor (5). This coefficient thus allows intermediate score values to be calculated by an arithmetic progression (6) [13]. Finally, we constrain the sum of maximum scores T_s^{max} to be 1000, as is often the case in practice to facilitate the communication of sub-factors impact with employees (7).

Parameters		**Variables**	
N_s	# of levels of sub-factor s	$T_{s,n}$	Score of the sub-factor
$\mathcal{N} = \{1, ..., \max(N_s)\}$	Set of levels	T_s^{min}	Min. score of the sub-factor
$\mathcal{S}$	Set of sub-factors	T_s^{max}	Max. score of the sub-factor
		C_s	Score between sub-factor levels

Constraints

$$T_{s,n} = 0 \qquad \forall s \in \mathcal{S} \mid n > N_s \qquad (1)$$

$$T_{s,1} = T_s^{min} \qquad \forall s \in \mathcal{S} \qquad (2)$$

$$T_{s,N_s} = T_s^{max} \qquad \forall s \in \mathcal{S} \qquad (3)$$

$$T_s^{min} \geq L_s \wedge T_s^{max} \leq U_s \qquad \forall s \in \mathcal{S} \qquad (4)$$

$$C_s = (T_s^{max} - T_s^{min})/N_s \qquad \forall s \in \mathcal{S} \qquad (5)$$

$$T_{s,n} = T_{s,n-1} + C_s \qquad \forall n \in \{2, ..., N_s\}, s \in \mathcal{S} \tag{6}$$

$$\sum_{s} T_s^{max} = 1000 \qquad \forall s \in \mathcal{S} \tag{7}$$

From the scoring table $T_{s,n}$ and the job evaluation ratings $R_{s,j}$ predetermined by the employer, it is possible to calculate, for each job $j \in \mathcal{J}$, it's total relative value V_j within the organization (8).

$\mathcal{J}$ Set of jobs

$$V_j = \sum_{s \in \mathcal{S}} T_{s,R_{s,j}} \qquad j \in \mathcal{J} \tag{8}$$

3.2 Score Ranges

Figure 2a illustrates an example of score ranges. The variables B_c^{inf} and B_c^{sup} express respectively the lower and upper bounds of each scoring range of grade $c \in \mathcal{C} = \{1, ..., C\}$ where C is the largest possible number of grades of the optimized salary structure. However, the number of grades is unknown and potentially less than C. We thus introduce the variable k, the number of grades in the structure, which strongly influences the separation of scores, and consequently, of jobs.

Several constraints force the score bounds to take a valid form. The lower bounds B_c^{inf} and upper bounds B_c^{sup} are null when the grade c is not used (9). To remain consistent, the first lower bound B_1^{inf} starts from the sum of the first levels of the weighting table T (10). The lower intermediate bound B_c^{inf} starts one point over the upper bound of the previous grade B_{c-1}^{sup} (11) while the upper bound B_c^{sup} of grade $c \in \{1, ..., k\}$ is p points over the minimum bound B_c^{inf} (12). This gap p is obtained via the number of grades k and the minimum bound B_1^{inf} (13).

Smaller number of grades compared to the number of grades in an original structure (i.e., actual structure on which the employees' salaries are positioned) systematically causes a compression in the classification of jobs and rise in costs. On the opposite, a number of grades that is too high could potentially lead to unbalanced score ranges and internal inequity problems. Thus, the number of

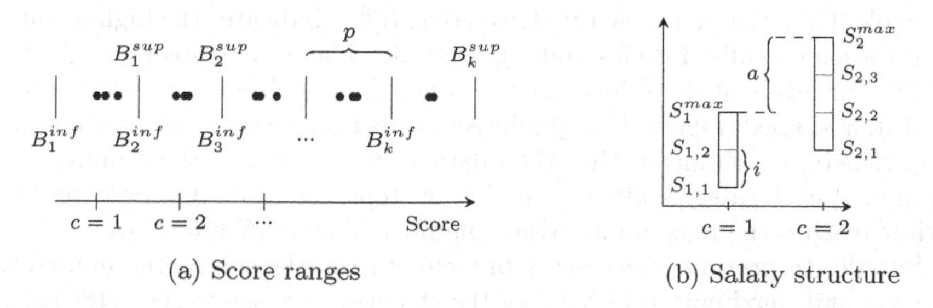

(a) Score ranges (b) Salary structure

Fig. 2. Variables related to (a) the score ranges, and (b) the salary structure (b).

grades k of the modeled structure is constrained to remain close to the number of grades K of the original structure (14). Finally, the variable D_j allows us to directly obtain the new grade of job j without iterating over the vector of bounds B^{sup} (15).

Parameters		**Variables**	
$C = 15$	Max number of grades	B_c^{inf}	Lower bound of grade c
$\mathcal{C} = \{1, ..., C\}$	Set of possible grades	B_c^{sup}	Upper bound of grade c
K	# of grades of original struct	k	# of grades of optimized struct
		p	Points gap between two bounds
		D_j	New grade of job j

Constraints

$$B_c^{inf} = 0 \wedge B_c^{sup} = 0 \quad \forall c \in \mathcal{C} | c > k \tag{9}$$

$$B_1^{inf} = \sum_{s \in \mathcal{S}} T_{1,s} \tag{10}$$

$$B_c^{inf} = B_{c-1}^{sup} + 1 \quad \forall c \in \{2, ..., k\} \tag{11}$$

$$B_c^{sup} = B_c^{inf} + p \quad \forall c \in \{1, ..., k\} \tag{12}$$

$$p = \left\lfloor \frac{\max_j V_j - B_1^{inf}}{k} \right\rfloor + 1 \tag{13}$$

$$k \geq K - 1 \wedge k \leq K + 3 \tag{14}$$

$$D_j = \left\lfloor \frac{(V_j - B_1^{inf})}{p + 1} \right\rfloor + 1 \quad \forall j \in \mathcal{J} \tag{15}$$

3.3 Salary Structure

Figure 2b is an example of the salary structure S. The variable $S_{c,e}$ denotes the salary of grade $c \in \mathcal{C}$ at step $e \in \mathcal{E} = \{1, ..., E\}$ where E represents the highest possible number of steps. Each grade c has a number of steps $M_c \in \mathcal{E}$, a minimum annual wage rate $S_{c,1} \in \mathcal{R}$ and a maximum wage rate $S_c^{max} \in \mathcal{R}$ where $\mathcal{R}$ is the set of possible wage values of the optimized structure. The minimum wage rate of grade $S_{c,1}$ is the wage rate offered to employees with no previous experience in the job while the maximum wage rate for a grade S_c^{max} indicates the highest rate the organization can offer for jobs under that grade. The set of steps $\{1, ..., M_c\}$ refer to different levels of the salary scale. In unionized environments, structures are often stair-shaped, especially in public service collective agreements according to practitioners, which means that the salary scale starts with the minimum steps M_1 and, at each salary scale, the number of steps per grade M_c increases by one until it reaches the maximum possible number of steps E (16, 17).

In order to prevent unnecessary branching from the solver, the unused wage rate $S_{c,e}$ and maximum rate S_c^{max} of the structure are set to zero (18, 19). The maximum wage rate S_c^{max} of grade c, associated with the maximum step M_c, is equivalent to the maximum grade's rate of the structure S_{c,M_c} (20). Since it is not realistic in practice to lower the organization's pay scales too much, a lower bound constrains the minimum rate $S_{1,1}$ to remain greater than or equal to 90% of the minimum rate of the original salary structure O (21). The optimized structure must also take into account employees' years of experience by integrating salary progression mechanisms and reasonably distinct salary gaps between maximum ranges [12]. To incorporate these aspects, we add the following constraints: the wage gap is i between two consecutive rates $S_{c,e-1}$ and $S_{c,e}$ (22); the wage gap is a between the previous scale maximum rate S_{c-1}^{max} and the next scale maximum rate S_c^{max} (23); the minimum wage rate of a grade $S_{c,1}$ remains higher than the previous minimum grade rate $S_{c-1,1}$ (24); the salary scales must overlap in accordance to the stair-shaped principle (25). In this work, the wage gap a between the scales maximums' rates is constant (i.e., arithmetic progression), as is often the case in practice. Other types of progression (e.g., geometric) could be integrated by modifying the constraint (23).

The variables F_j and H_j, which allow us to calculate the external equity objective function later, correspond to the new minimum and maximum wage rate of job $j \in \mathcal{J}$ and are respectively equal to the minimum $S_{c,1}$ and maximum S_c^{max} annual wage rate of their new grade D_j (26, 27).

Parameters		**Variables**	
$\mathcal{E}$	Set of possible steps	$S_{c,e}$	Salary of grade c at step e
$\mathcal{R}$	Set of possible steps c	M_c	# of steps in grade c
$\mathcal{R}$	Set of possible wages	X_c^{max}	Max salary in grade c
O	Original salary structure	i	Salary gap between cons. steps
		a	Salary gap between two scales
		F_j	Minimum wage of job j

Constraints

$$M_c = \min(M_{c-1} + 1, E) \quad \forall c \in \{2, ..., k\} \tag{16}$$

$$M_c = E \quad \forall c \in \{k+1, ..., C\} \tag{17}$$

$$S_{c,e} = 0 \quad \forall c \in \{k+1, ..., C\}, e \in \{M_c + 1, ..., E\} \tag{18}$$

$$S_c^{max} = 0 \quad \forall c \in \{k+1, ..., C\} \tag{19}$$

$$S_{c,M_c} = S_c^{max} \quad \forall c \in \{1, ..., k\} \tag{20}$$

$$S_{1,1} \geq \lfloor O_{1,1} \times 0.9 \rfloor \tag{21}$$

$$S_{c,e} = S_{c,e-1} + i \quad \forall c \in \{1, ..., k\}, e \in \{2, ..., M_c\} \tag{22}$$

$$S_c^{max} = S_{c-1}^{max} + a \quad \forall c \in \{2, ..., k\} \tag{23}$$

$$S_{c-1,1} < S_{c,1} \quad \forall c \in \{2, ..., k\} \tag{24}$$

$$S_{c,1} < S_{c-1}^{max} \quad \forall c \in \{2, ..., k\} \tag{25}$$

$$F_j = S_{D_j,1} \quad \forall j \in \mathcal{J} \tag{26}$$

$$H_j = S_{D_j}^{max} \quad \forall j \in \mathcal{J} \tag{27}$$

Since many employees share the same job and step in the original structure, it is possible to speed up certain calculation operations (e.g., total cost, step assignment) by arranging the employees into separate groups. Indeed, staff members are necessarily positioned at the same wage rate in the optimized structure if they hold the same wage rate in the original one.

For this purpose, we include the constants V_g, W_g, X_g, Y_g and Z_g which correspond respectively to the job, step, number of employees, salary and maximum possible step of a distinct group of employees $g \in \mathcal{G} = \{1, ..., G\}$ in the original salary structure where G denotes the number of distinct groups of employees. Before assigning a new salary to each group g, it is first necessary to obtain the new grade Q_g and the new step A_g. The first variable is obtained simply by taking the jobs' new grade (28). As for A_g, it is more difficult to obtain since the new wage's rate must be positioned on a step that is different from the one in the original structure. In other words, the salary in the current structure Y_g, which is tied to the step W_g, must now be associated with a new salary R_g and paired with the new step A_g. In practice, the salary Y_g is integrated at the step in the new structure (i.e., optimized structure) which salary is equal to or immediately greater than the salary the employee is currently earning. To respect this principle, we first convert, via the temporary variable Θ_g, the group salary into a step from a base change (29). The mathematical formulation of this constraint allows to avoid iterating through the grades and steps of the optimized structure to find the corresponding wage rate. Since the temporary variable Θ_g can take a value less than one or greater than the number of steps M_c of the related grade, we constrain A_g to remain within the range of valid steps (30).

Despite the determination of the group's new grade and step, it is still possible that some salaries Y_g of group g evolve outside the maximum of the new grade salary range. Employees in this situation are referred to as red circles [2]. Since the employer must at least maintain the wages of its employees, a group's new wage R_g is maintained if it exceeds the wage rate S_{Q_g,A_g} of the optimized structure (31). In addition, the concept of red circle causes employee dissatisfaction and demotivation [13]. The employer and union representatives (if any) thus have a common interest in limiting the number of employees in this situation. Consequently, we constrain the number of red circles r to remain less than or equal to 10% of the number of employees as the general rule of thumb used by practitioners (32, 33).

Parameters		**Variables**	
$\mathcal{G}$	Job of employees in group g	Q_g	New grade of group g
V_g	Job of employees in group g	Θ_g	Temporary variable
W_g	Step of employees group g	A_g	New step of group g
X_g	# of employees in group g	r	# of red circles
Y_g	Salary of group g	R_g	New salary of group g
Z_g	Highest step in group g		

Constraints

$$Q_g = D_{V_g} \qquad \forall g \in \mathcal{G} \quad (28)$$

$$\Theta_g = 2 + \left\lfloor \frac{(Y_g - S_{Q_g,1})(M_{Q_g} - 1)}{S_{Q_g}^{max} - S_{Q_g,1}} \right\rfloor \qquad \forall g \in \mathcal{G} \quad (29)$$

$$A_g = \min(\max(\Theta_g, 1), M_{Q_g}) \qquad \forall g \in \mathcal{G} \quad (30)$$

$$R_g = \max(S_{Q_g,A_g}, Y_g) \qquad \forall g \in \mathcal{G} \quad (31)$$

$$r = \sum_g X_g [S_{Q_g}^{max} < Y_g] \quad (32)$$

$$r \leq \lfloor 0.1 \times P \rfloor \quad (33)$$

3.4 Objective Functions

Our model has three objective functions: additional cost, internal equity and external equity. The additional cost objective α is defined as the extra cost an employer must pay over a given number of years y_{max} ($y_{max} = 3$ in our experiments) to implement the optimized structure.

Specifically, it is obtained by calculating the salary progression of employees over y_{max} years for each structure (i.e., optimized and initial) and summing the cost differences (34). Considering the evolution of the cost over several years allow to avoid solutions where the payroll increase is low the first year (when employees are integrated into the optimized structure) but high in subsequent years due to significant salary differences between steps. In Eq. (34), $O_{y,g}$ correspond to the initial structure cost at year $y \in \{0, ..., y_{max}\}$ of group g.

$$\alpha = \sum_{y=0}^{y_{max}-1} \sum_{g \in \mathcal{G}} \overbrace{X_g}^{\#\text{ employees}} \Big[\overbrace{\max(S_{Q_g,\min(y+A_g,M_{Q_g})}, Y_g)}^{\text{Optimized structure cost}} - \overbrace{O_{y,g}}^{\text{Initial structure cost}} \Big] \quad (34)$$

There are several ways to define an internal equity objective. Kassa [7] models this objective using the differentials between the grade's mid wages and target wages specified in advance by the DM, whereas Bruno [4] considers internal equity and individual equity under the same concept. In our work, we propose a distinct and more rigorous definition of internal equity that does not depend on parameters chosen by the user. Concretely, the internal equity objective β represents the sum of squared errors of jobs' scores from the center of their associated score ranges (35). Squaring errors gives quadratically more weight to scores that are further away from their grade center. Intuitively, a low value indicates that similar jobs are closer from the center of their grade, while a high value implies that they are nearer their bounds. The fact that job scores must stay close to each other reinforces the idea that each job is paid its fair share.

$$\beta = \sum_{j \in \mathcal{J}} \left[\left\lfloor \frac{B_{D_j}^{inf} + B_{D_j}^{sup}}{2} \right\rfloor - V_j \right]^2 \quad (35)$$

The external equity objective δ represents the sum of absolute errors between the new scale's minimum and maximum rates (F_j and H_j) and the average external minimum and maximum rates (λ_j^{min} and λ_j^{max}) (36). A low value means that the scale's minimum and maximum rates are close to those of the market, while high value signifies that they are misaligned with the market's realities.

$$\delta = \sum_{j \in \mathcal{J}} |F_j - \lambda_j^{min}| + |H_j - \lambda_j^{max}| \tag{36}$$

The objective functions are optimized using the ϵ-constraint method (see Sect. 4.2).

3.5 Model Complexity

Let N be the number of levels, S the number of sub-factors, J the number of jobs, C the maximum number of grades, E_o the maximum number of steps in the optimized structure, E_a the maximum number of steps in the current structure, and G the number of employee groups. The model has

- $\Theta(NS)$ variables and $\Theta(NS + J)$ constraints related to the scoring table (Sect. 3.1).
- $\Theta(J + C)$ variables and constraints related to the score ranges (Sect. 3.2)
- $\Theta(CE_o{+}G)$ variables and constraints linked to the salary structure (Sect. 3.3).

In total, the optimization model has $\Theta(NS{+}CE_o{+}G)$ variables and constraints. Since organizations typically have more distinct employee groups than other parameters, it is reasonable to assume that the model is upper bounded by G. Nevertheless, to reach the worst case of G, it is necessary that at least one employee covers each step of each job, which is very unlikely. The model is therefore suitable for large organizational workforce.

4 Experiments

4.1 Data

The optimization model is evaluated using compensation data from a unionized institution with a total of 127 employees that hold one of 20 distinct jobs in the organization. The employer has previously evaluated its job using a 13 sub-factors evaluation plan. The database reports, for each employee, the job title, the salary, the current salary grade and step, as well as the external market salary ranges. To incorporate this notion of external markets, the organization's jobs are matched by compensation experts when external data was available. This procedure, known as *job matching* [12], links the unionized institution's jobs with those in the market to obtain the external compensation components.

4.2 Optimization Method

In general, multi-objective optimization problems take the following form:

$$\min_{\mathbf{x}} \mathbf{J}(\mathbf{x}) = [J_1(\mathbf{x}), J_2(\mathbf{x}), ..., J_k(\mathbf{x})]^T \quad \text{s.t. } \mathbf{g}(\mathbf{x}) \leq 0 \tag{37}$$

where $\mathbf{J}$ is the objective functions vector and $\mathbf{x}$ is a feasible solution with inequality constraints $\mathbf{g}$. The optimal space of solutions is called the Pareto frontier (i.e., the set of non-dominated points).

To estimate the Pareto frontier, we use the ϵ-constraint optimization method, which consists in optimizing a single objective by transforming the others into bound constraints. The optimization is thus performed by successively changing the ϵ bounds values to explore different regions of the Pareto frontier. The algorithm generally achieves a good distribution of solutions, handles non-convex solution spaces, explores the Pareto frontier without *a priori* preferences and can solve problems with three or fewer objectives relatively quickly [8,10]. Since the solution space is non-convex, the salary structure design is an exploratory process and our model incorporates three objective functions, ϵ-constraint is the chosen optimization technique. In our work, we choose to minimize the additional cost and transform the internal equity and external equity objectives into bound constraints.

Considering the large solution space of the problem, it is necessary to make a compromise between the computation time and the quality of the solution in order to avoid, on the one hand, a loss of time without significant improvement of the solution, and on the other, a significant degradation of quality. As shown by Fig. 3, we found that 30 min was sufficient to bring the cost within 2.5% of the optimal cost obtained from optimizing over a 120-min period (using a M1 MacBook Pro).

Moreover, the ϵ-constraint algorithm requires to find the ranges of the ϵ-bounds on the objectives to constrain (i.e., external and external equity). To find the lower ϵ-bound, we separately minimize, over a one-hour interval, the internal and external equity objectives. To find the upper ϵ-bound of the internal equity, we use the internal equity value obtained by minimizing the external equity (and vice versa for the upper ϵ-bound of the external equity).

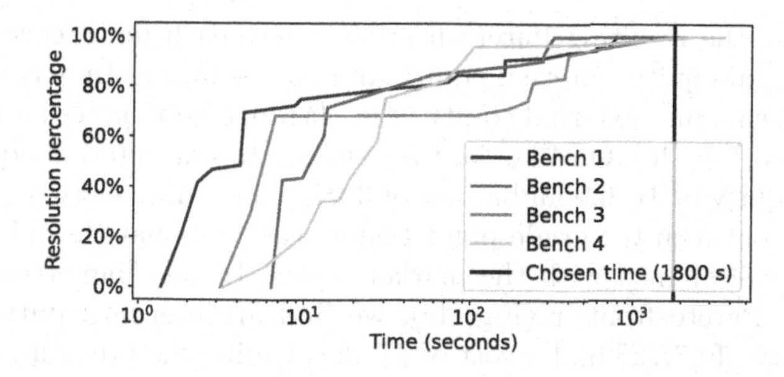

Fig. 3. Resolution of the cost objective (in %) compared to the minimum cost found (after a 120-min period) for each instance.

In our experiments, we optimize the additional cost objective over the ϵ equity parameter space for three years (i.e., $y_{max} = 3$), according to the time and bounds found in the previous step, on a 6X6 grid search. This optimization is run repeatedly, for each point of the grid search, from a Python script calling the Gecode solver in backend.

4.3 Heuristic Search

The solver chooses, in order, the number of grades k, the scoring table minimums T_s^{min} and maximums T_s^{max}, the highest step of the first salary grade M_1, the wage gap between two steps i, the wage gap between scale's maximum a, and the first grade maximum wage rate S_1^{max}. The first three variables are used to ensure a valid job classification and are therefore concerned with internal equity, while the last four are more concerned with wage differentials, and therefore external equity and cost. Thus, we first generate valid rankings before building the salary structure. Whereas it is straightforward to define a variable selection heuristic that helps guide the exploration of the search tree, it is more difficult to define a value selection heuristic that could work with all ϵ-constraint bounds. Consequently, we assign the values randomly, and we use a linear restart strategy with one unit increments. This restart allows to do a breadth-first search and gradually explore deeper levels of the search tree. The strategy used allows to quickly obtain good solutions while globally exploring the search tree.

To optimize the model, we use the Gecode solver in MiniZinc [11] as the Chuffed solver, which learns clauses from implication graphs that led to failures, was empirically found to be slower to converge. This outcome probably comes from the fact that domains are very poorly filtered (e.g., wage gap between scale maximums a) or never filtered at all (e.g., the number of grades k). Clause generation thus appears to be of little help compared to random sequence generation as permitted by Gecode.

5 Results and Discussion

5.1 Pareto Frontier

Figure 4 shows the resulting Pareto frontier where each point corresponds to a different internal equity, external equity and cost solution. In general, solutions with high internal and external equity (Fig. 4a) involve more costs than the less equitable ones (Fig. 4c). On Fig. 4a, the solution has an internal equity of 1267, an external equity of 41.8k$ and a cost of 368k$. This implies that most jobs are well centered between the grade point boundaries and that the salary structure is relatively well aligned with the market scales. In the diametrically opposite corner of the Pareto frontier (Fig. 4c), we find an internal equity of 2962, an external equity of 67.7k$ and a cost of 77.4k$. Unlike the previous solution, this solution has more job scores near the class boundaries as well as higher wages compared to the market.

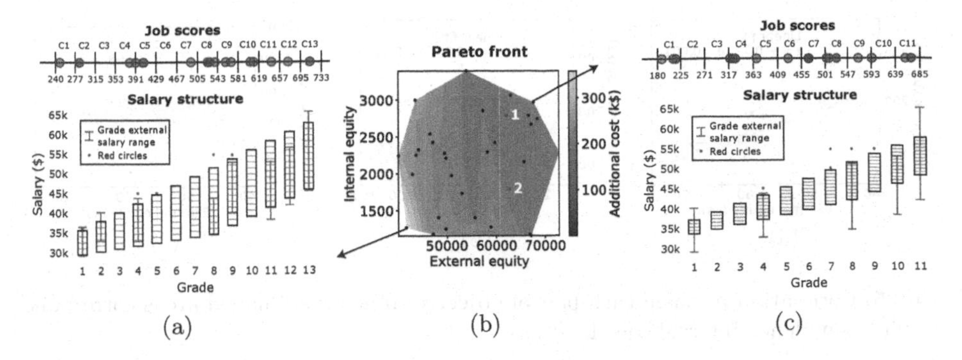

Fig. 4. Pareto frontier approximation using $6 \times 6 = 36$ values. Lower values of external and internal equities are better. Point 1 is an example of a weakly Pareto solution as there exists another solution (i.e., point 2) with a better internal equity keeping the external equity and cost objectives constant.

The obtained results suggest different strategies of the solver depending on the importance given to objective functions. On the one hand, to obtain a structure with strong equity (Fig. 4a), the number of grades takes the maximum value of the domain (i.e., 13). This is consistent with the fact that an increase of job rankings decreases the score gap, which in turn reduces the squared error of internal equity. The same is true for the absolute error of the external equity objective since it becomes easier to align structure rates with market rates when jobs are less clustered. However, the salary ranges are becoming more spread out (ranging from 29.5k$ to 62.9k$) to align the structure with the external market. This leads to an additional cost increase via higher wage progression and wage rates. On the other hand, to obtain a low-cost salary structure (Fig. 4c), the solver focuses on variables that lower the payroll such as the wage gap between scale maximums (861$) and the wage gap between steps (2048$). While there are solutions with even less internal and external equity on the Pareto frontier, the solver gets a relatively compressed structure (salaries from 33.8k$ to 57.7k$) with closer steps. This limits progression of employee salaries over time and reduces the additional cost accordingly.

Figure 5 illustrates the correlation between the external equity, internal equity and cost objectives. Figure 5a shows a statistically negative relationship ($r = -0.78, p < 0.001$) between external equity and cost; that is, as the structure becomes more aligned with the external market, the wage bill increases. In Fig. 5b, the internal equity and additional cost objectives are not statistically linked ($r = -0.25, p > 0.01$), and the same is true in Fig. 5c between the internal equity and external equity ($r = 0.09, p > 0.01$). Thus, employers cannot improve their external salary gaps without generating additional costs but, they can maintain fairness without compromising cost increases. In our case, this result seems to suggest that the internal equity objective could be replaced by a bound constraint, as there it incurs little trade-off with the other objectives. However, this

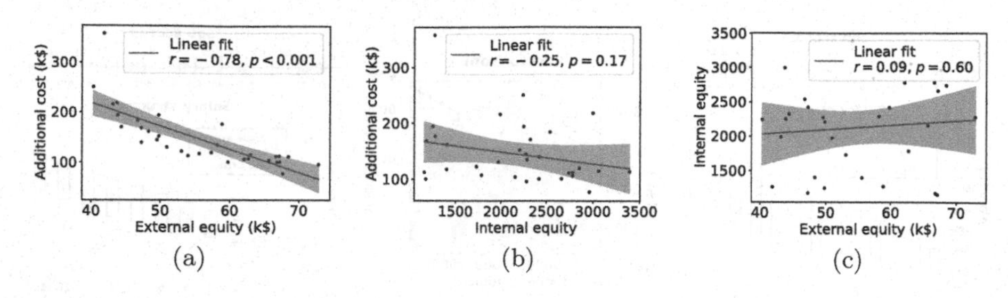

Fig. 5. Correlation between each pair of objective functions. Shaded areas correspond to 95% confidence intervals on the fit.

might not be the case for all organizations, as the internal equity objective is influenced by multiple variables (e.g., the scoring the table).

The Pareto frontier on Fig. 4b also suggests weakly Pareto optimal solutions (e.g., point 1). The presence of these points likely comes from the fact that there is no apparent relationship between internal equity and the other two objective functions (Fig. 5b and Fig. 5c). When this situation occurs, only the lowest internal equity value is relevant. Unfortunately, the ϵ-constraint method does not allow to avoid weakly Pareto optimal solutions and, consequently, results in wasted computation time.

5.2 Comparison with the Negotiated Structure

Since we have access to real data from a unionized establishment, we compare the solution obtained from an actual negotiated salary structure to our closest solution on the Pareto frontier in Fig. 6. The internal equity, external equity, and additional cost objectives of the negotiated solution are computed following Eqs. (34) to (36).

Looking at the job rankings, it is interesting to note that both solutions have 11 grades and the difference between the score boundaries remains comparable (49 points versus 43 points). On the other hand, the optimized structure has two more steps per grade, up to a maximum of 12 steps, so the maximum wage rates

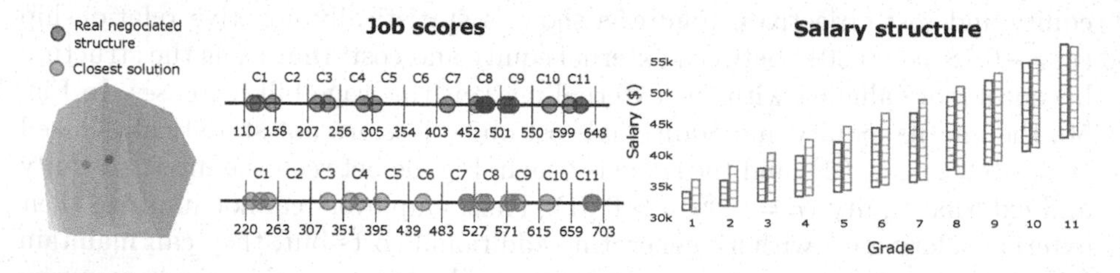

Fig. 6. Negotiated salary structure (internal equity: 2287, external equity: 54,577$ and cost: 106,890$) compared with the closest solution on the Pareto frontier (internal equity: 2208, external equity: 49,802$ and cost: 152,187$.

for the first grades are higher than those in the negotiated structure. This results, not surprisingly, in a slightly higher payroll cost. Moreover, despite the proximity of the internal and external equity values for the negotiated and optimized solution, the optimized one remains more expensive (152k$ versus 106k$). We interpret this result by the fact that the maximum rates in the negotiated structure are based on a geometrical growth (constant percentage gaps) in contrast to our model where it is linear (constant gaps). The negotiated structure thus benefits from more flexibility by allowing salary ranges to start at a lower rate and end at approximately the same rate as in our model. With lower maximum rates, the resulting solution necessarily becomes less expensive.

While the negotiated structure required nearly a month of work and meetings, our approach finds a similar solution (as well as multiple other ones) in only a few hours of unattended time. The generation of solutions on the Pareto frontier thus highlights the time and resource savings of using, in whole or in part, combinatorial optimization techniques in the design of salary structures.

6 Conclusion

Our work shows that designing salary structures with multi-objective optimization can definitely benefit companies that wish to obtain fair pay policies while reducing their payroll operating costs. The resulting Pareto frontier allows to explore and compare different solutions and trade-offs when implementing or updating salary structures. A decision maker could use it in real time on negotiation tables when renewing collective agreements. Finally, the efficiency of the model makes it an interesting solution for both small and large organizations.

Future work may address some limitations of our approach. On the model side, it would be interesting to validate the model on bigger instances to empirically test its scalability potential. Linearizing the problem (with some model simplifications) or using parallel computing (since each point on the Pareto front is independent) would be interesting options to decrease running time. Using reinforcement learning to learn value selection heuristics [5] or perform a lexicographic optimization to find the ϵ bounds values [9] could also be considered to further reduce computation time. Nevertheless, this paper is an additional step towards addressing the challenges of designing salary structures and hopefully pave the way for more research on the subject.

References

1. Armstrong, M., Chapman, A.: The Reward Management Toolkit: A Step-By-Step Guide to Designing and Delivering Pay and Benefits. Kogan Page, London (2011)
2. Barry, G., Newman, J.: Compensation, 13th edn. McGraw-Hill, New York (2019)
3. Branch, L.S.: Consolidated federal laws of Canada, Pay Equity Act (2021). https://laws-lois.justice.gc.ca/eng/acts/p-4.2/FullText.html. Accessed 31 Aug 2021
4. Bruno, J.E.: Compensation of school district personnel. Manage. Sci. **17**(10), B569–B587 (1971). http://www.jstor.org/stable/2628995. INFORMS

5. Chalumeau, F., Coulon, I., Cappart, Q., Rousseau, L.-M.: SeaPearl: a constraint programming solver guided by reinforcement learning. In: Stuckey, P.J. (ed.) CPAIOR 2021. LNCS, vol. 12735, pp. 392–409. Springer, Cham (2021). https://doi.org/10.1007/978-3-030-78230-6_25
6. Downes, P.E., Choi, D.: Employee reactions to pay dispersion: a typology of existing research. Hum. Resour. Manag. Rev. **24**(1), 53–66 (2014). https://doi.org/10.1016/j.hrmr.2013.08.009
7. Kassa, B.A.: A decision support model for salary structure design. Compensation Benefits Rev. **52**(3), 109–120 (2020). https://doi.org/10.1177/0886368720905696. SAGE Publications Inc
8. Laumanns, M., Thiele, L., Zitzler, E.: An adaptive scheme to generate the pareto front based on the epsilon-constraint method. In: Practical Approaches to Multi-Objective Optimization, 7–12 November 2004. Dagstuhl Seminar Proceedings, vol. 04461. IBFI, Schloss Dagstuhl, Germany (2005). https://doi.org/10.4230/DagSemProc.04461.6
9. Laumanns, M., Thiele, L., Zitzler, E.: An efficient, adaptive parameter variation scheme for metaheuristics based on the epsilon-constraint method. Eur. J. Oper. Res. **169**(3), 932–942 (2006). https://doi.org/10.1016/j.ejor.2004.08.029
10. Mavrotas, G.: Effective implementation of the epsilon-constraint method in multi-objective mathematical programming problems. Appl. Math. Comput. **213**(2), 455–465 (2009). https://doi.org/10.1016/j.amc.2009.03.037
11. Nethercote, N., Stuckey, P.J., Becket, R., Brand, S., Duck, G.J., Tack, G.: MiniZinc: towards a standard CP modelling language. In: Bessière, C. (ed.) CP 2007. LNCS, vol. 4741, pp. 529–543. Springer, Heidelberg (2007). https://doi.org/10.1007/978-3-540-74970-7_38
12. Singh, P., Long, R.J.: Strategic compensation in Canada, 6th edn. Nelson Education Ltd., Ontario (2018). oCLC: 1292020761
13. St-Onge, S., Morin, G.: Gestion de la rémunération: théorie et pratique, 4e édition edn. Chenelière éducation, Montréal (2020). https://doi.org/10.7202/000155ar
14. Wallace, M.J., Steuer, R.E.: Multiple objective linear programming in the design of internal wage structures. Acad. Manage. Proc. **1**, 251–255 (1979). https://doi.org/10.5465/ambpp.1979.4977109

Scheduling Complex Observation Requests for a Constellation of Satellites: Large Neighborhood Search Approaches

Samuel Squillaci(✉), Cédric Pralet, and Stéphanie Roussel

ONERA/DTIS, Université de Toulouse, Toulouse, France
{samuel.squillaci,cedric.pralet,stephanie.roussel}@onera.fr

Abstract. Nowadays, constellations of satellites have to deal with heterogeneous and complex observation requests, such as one-shot, video, stereoscopic, and periodic requests. In this paper, we consider the problem of scheduling these requests in order to maximize a measure of global utility. To solve this problem, we propose two Large Neighborhood Search algorithms that exploit problem decompositions. These algorithms explore large neighborhoods respectively based on heuristic search and Constraint Programming. The experiments performed on instances generated from real constellation features and weather data show that the approaches improve the state of the art.

Keywords: Satellites constellation · Large Neighborhood Search · Heuristic Search · Constraint Programming

1 Introduction

In this paper, we consider a future Earth observation constellation composed of 16 low-orbit satellites that has to fulfill heterogeneous observation requests on specific targets on Earth. Requests can consist in taking one picture of a target area (*one-shot* requests), a video of a target (*video* requests), two temporally close pictures of a target using two different observation angles (*stereoscopic* requests), or one picture of a target every X hours (*periodic* requests). In this paper, to deal with such heterogeneous requests in a unified way, we consider a set of *modes* for each observation request, *i.e.* alternatives composed of elementary observations that allow one to satisfy the request, even partially. Each mode has a utility that represents not only the fulfillment of the pattern required by the request but also the ability to get valid images depending on the cloud cover forecast for each elementary observation in the mode.

Then, we consider the following problem: "given a set of candidate requests along with a cloud cover forecast, choose a mode for each request and schedule the associated observations over each satellite, in order to maximize the global utility collected". Note that despite the increasing size of today's constellations, the number of requests posted by end-users makes it impossible to satisfy them

© The Author(s), under exclusive license to Springer Nature Switzerland AG 2023
A. A. Cire (Ed.): CPAIOR 2023, LNCS 13884, pp. 443–459, 2023.
https://doi.org/10.1007/978-3-031-33271-5_29

all, hence there is an actual need to solve an optimization problem to get the best global performance for the system. Moreover, to be used in an operational context, a tool for this problem must compute plans within a few minutes since in practice, it is called several times per day so as to exploit the most recent weather forecast and be reactive to urgent observation requests.

To create such a tool, we introduce two Large Neighborhood Search (LNS) algorithms that exploit a temporal decomposition of the problem. As introduced in [20], one basic idea is that the candidate observations can be partitioned into temporal clusters. The first LNS algorithm consists in iteratively removing requests from the current plans of the satellites and reinserting new ones based on heuristic search. The second LNS algorithm exploits the Constraint Programming (CP) model exposed in [20] to explore a neighborhood where the content of a subset of the temporal clusters is revised at each step, one objective being to exploit powerful CP techniques available for scheduling problems. Experiments have been performed on a large set of instances generated from real constellation features and cloud cover forecast that we have made publicly available. The results show that the two LNS approaches outperform the state of the art.

The paper is organized as follows. Section 2 presents related work. Section 3 defines the Earth Observation Scheduling problem we consider. Section 4 introduces the two LNS algorithms proposed. Section 5 describes the experiments and analyzes the results. Finally, Sect. 6 concludes on future works.

2 Related Work

In the last decade, several optimization techniques were defined to manage constellations of satellites, which offer the possibility to answer new user needs for Earth observation compared to single-satellite systems. Thereupon, [19,20] deal with heterogeneous observation requests such as one-shot requests or periodic requests, respectively requiring one picture and one picture for each period in a day. In [19], a Large Neighborhood Search is proposed to solve the associated observation scheduling problem and in [20], a greedy algorithm is designed to work on different parts of the problem simultaneously. In parallel to heterogeneous observation requests, some authors studied multi-instrument observations [11,12]. For instance, in [11], the authors consider the choice of different observation instruments providing different error measures for a soil moisture estimation model, while in [12], an adaptive tabu search algorithm is proposed for solving a scheduling problem involving a choice among different parameters of satellite cameras. Meanwhile, other authors considered constellation scheduling problems involving multiple users and proposed an auction-based algorithm to optimize the sharing of the constellation resources among the users [15]. Features of constraint programming solvers are also likely to be used to model and solve satellite scheduling problems and other problems, as suggested in [9,10].

In a different direction, some authors exploited the fact that satellite constellation planning problems are amenable to problem decomposition [7,8,20]. In [20], the authors cluster the candidate observations according to their time windows and a greedy algorithm exploits this decomposition to perform several insertions of observation tasks in parallel. In [8], a two-step algorithm first creates sub-problems by planning activities in overloaded parts of the time horizon, and then uses MILP to optimize the plans in each sub-problem independently. Similarly, parts of the problem that are temporally independent can be identified as in [7] to apply a column generation process that exploits this decomposition.

Various other aspects were investigated in the literature related to satellite constellation planning. For instance, the authors of [23] merge observations beforehand using their time windows and the required observation angles. In [4], the authors exploit a graph of infeasible sets of observations, i.e. observations that cannot be inserted in the plan all together. A dispatch system that split the scheduling problem of a constellation into multiple mono-satellite scheduling problems is introduced in [3], using a neural network to predict the probability to actually perform a task if the latter is allocated to a given satellite.

Also, several works reuse ideas developed for mono-satellite scheduling, such as the Adaptive Large Neighborhood Search (ALNS) introduced in [13], or the dynamic programming approach implemented in [14]. Finally, several references take into account time-dependent effects [5,6,22], considering that the duration of a satellite maneuver between two successive observations requiring different pointing angles actually depends on the time at which the maneuver is triggered.

To finish this literature review, an analogy can be made between the satellite constellation scheduling problem and the Vehicle Routing Problem (VRP), especially its variant called the Team Orienteering problem with Time Windows (TOPTW) where a fleet of vehicles must visit a set of (optional) customers. For this problem, in [18], a large neighborhood search that iteratively removes and inserts customers based on very efficient heuristics is implemented. A greedy randomized adaptive search procedure to build good quality solutions is exploited in [17]. Authors of [21] explore infeasible solutions through tabu search and compute high-quality solutions by recombining the best solutions found. Finally, a parallel can be made between references using problem decomposition for satellite constellation scheduling and clustered VRPTW (CluVRPTW) [1,2]. However, while observations are usually clustered with regards to their time windows in the satellite context, they are usually clustered according to their geographical positions for CluVRPTW.

With regards to these existing works, our contribution corresponds to a significant improvement of the techniques for managing heterogeneous requests (strong need to satisfy the end-users) and extends the possible scope of application of Earth observation systems.

3 Earth Observation Scheduling Problem

3.1 Problem Modeling

The Earth Observation Scheduling Problem (EOSP) we consider is defined by a set of satellites $\mathcal{S}$, a set of observation requests $\mathcal{R}$ and a set of download requests $\mathcal{D}$. In the following, we recall from [20] the definitions of these elements.

Time Windows. The satellites we consider are on low Earth orbit. This means that each target t on the Earth surface is visible only during some time windows, *i.e.* temporal intervals during which a satellite passes over t and is able to take a picture of it with respect to a maximum zenith-angle (angle between the target-satellite direction and the vertical at the level of the target). A temporal window w has an earliest start date $Earliest_w$, a latest end date $Latest_w$, and an associated satellite sat_w in $\mathcal{S}$. Each target is viewed several times per day by the satellites of the constellation (the number of time windows is a function of the altitudes of the satellites).

Observation Requests and Modes. The set of observation requests is denoted $\mathcal{R}$. An observation request $r \in \mathcal{R}$ is defined by a target t_r on the Earth surface and an observation duration δ_r. Each request r has an associated set of time windows, denoted $\mathcal{W}_r$. There are several ways to satisfy (or partially satisfy) a request from an end-user point of view. The ways to satisfy a request r are called its *modes*, denoted $\mathcal{M}_r$, and depend of the type of r. In this paper, we consider the following request types:

- *one-shot request*, requiring one observation of duration δ_r in one time window of $\mathcal{W}_r$. The set of modes $\mathcal{M}_r$ is a subset of $\{\{w\}, w \in \mathcal{W}_r\}$;
- *video request*, requiring one video of the target, *i.e.* several pictures that altogether require 40 s to 1 min. It can be seen as a one-shot request with a much larger duration;
- *stereoscopic request*, requiring two observations of duration δ_r from different angles with a single satellite. Formally, a mode m in $\mathcal{M}_r$ is a set $\{w_1, w_2\}$ containing two time windows in $\mathcal{W}_r$ such that $sat_{w_1} = sat_{w_2}$ and the time windows are close to each other;
- *periodic request*, requiring repetitive observations of duration δ_r, *e.g.* in three periods [11am, 1pm], [3pm, 5pm] and [7pm, 9pm]. Each mode m in $\mathcal{M}_r$ is a subset of $\mathcal{W}_r$ such that two windows w_1 and w_2 in m are not associated with the same period. Set $\mathcal{M}_r$ can contain many modes for a periodic request, however it does not need to be listed explicitly for the algorithms we propose.

To model the preferences of end-users, we define, for each request r and each mode m in $\mathcal{M}_r$, a utility $u_m \in \mathbb{R}$. Such a utility can depend on the cloud cover forecast associated with each time window, or on the satisfaction of the periodicity in the case of periodic requests. More details on a possible way to obtain these utilities are given in Sect. 5. Finally, the set of all windows associated with the requests in $\mathcal{R}$ is denoted $\mathcal{W}$. Formally, $\mathcal{W} = \bigcup_{r \in \mathcal{R}} \mathcal{W}_r$.

Download Request. For each communication window booked between a satellite and a ground station, we consider that a download activity of duration Δ must be performed (i.e. a data transfer from the satellite to the ground station) without overlapping any observation. For simplicity, such download activities are modeled within the observation request and mode framework defined previously. More precisely, the set of download requests $\mathcal{D}$ contains one request r_d per ground station d. Its associated duration is $\delta_{r_d} = \Delta$, and the set of time windows in $\mathcal{W}_{r_d}$

contains one element per communication window booked for ground station d. Moreover, there is a unique mode for r_d that contains all elements in $\mathcal{W}_{r_d}$. We can define a null utility for this mode as it is mandatory anyway, and to get more compact notations all download requests in $\mathcal{D}$ are added to $\mathcal{R}$.

Maneuvers. The satellites we consider in this problem are agile, *i.e.* they have to maneuver between successive activities in order to point to the right targets. The satellites cannot observe targets or download data during maneuvers. For two time windows w_1 and w_2 in $\mathcal{W}$ that respectively have $target_1$ and $target_2$ as targets, the time for any satellite of the constellation to go from a pointing attitude to $target_1$ to a pointing attitude to $target_2$ is denoted τ_{w_1,w_2}. Taking into account the time-dependent effects mentioned before is left for future work.

Solution. The EOSP is generally over-constrained since it may not be possible to satisfy all the observation requests. Therefore, some requests must be selected, and the associated observations must be scheduled on each satellite while respecting the temporal constraints. Formally, a solution plan σ is defined by:

- a subset of $\mathcal{R}$, denoted $\mathcal{R}(\sigma)$, that represents the selected requests;
- for each selected request $r \in \mathcal{R}(\sigma)$, its selected mode $m(r) \in \mathcal{M}_r$;
- for each satellite $s \in \mathcal{S}$, a sequence $seq(\sigma, s)$ that contains all windows w that concern s and belong to the selected modes, that is all windows in $\{w \in \mathcal{W} \mid sat_w = s\} \cap \left(\bigcup_{r\in\mathcal{R}(\sigma)} m(r)\right)$;
- a utility $u(\sigma) = \sum_{r\in\mathcal{R}(\sigma)} u_{m(r)}$ (sum of the utilities of the selected modes).

A solution σ is *feasible* if all download requests belong to $\mathcal{R}(\sigma)$ and for each satellite $s \in \mathcal{S}$ and each window $w \in seq(\sigma, s)$, it is possible to assign a start date $start_w$ in $[Earliest_w, Latest_w - \delta_r]$ (with r the request associated with w) such that there is no overlap between the successive activities of satellite s (including maneuvers).[1] The goal is to find a feasible solution having the maximum utility.

Example. Figure 1 presents a toy example involving four observation requests r_1, r_2, r_3, r_4 that are respectively *one-shot*, *video*, *stereoscopic*, and *periodic* requests, and a download request r_d. Figure 1a gives the time windows associated with each request for a constellation of two satellites ($\mathcal{S} = \{s_1, s_2\}$). For instance, $\mathcal{W}_{r_1} = \{w_1, w_2\}$, with $sat_{w_1} = s_1$ and $sat_{w_2} = s_2$. Figure 1b lists modes associated with the requests. One-shot request r_1 and video request r_2 have one mode for each of their time windows. Stereoscopic request r_3 has two modes. Periodic request r_4 aims at observing a target at dates t_1, t_2, and t_3 which are the midpoints of three periods. Suitable time windows for each of these dates are respectively $\{w_9\}$, $\{w_{10}\}$, and $\{w_{11}\}$. Therefore, the only mode that makes r_4 fully satisfied is mode $\{w_9, w_{10}, w_{11}\}$. In this work, we allow a partial satisfaction of periodic requests. In fact, end-users that ask for n observations might prefer having $k <$

[1] Memory and energy aspects are ignored, as it is assumed that they are not restrictive.

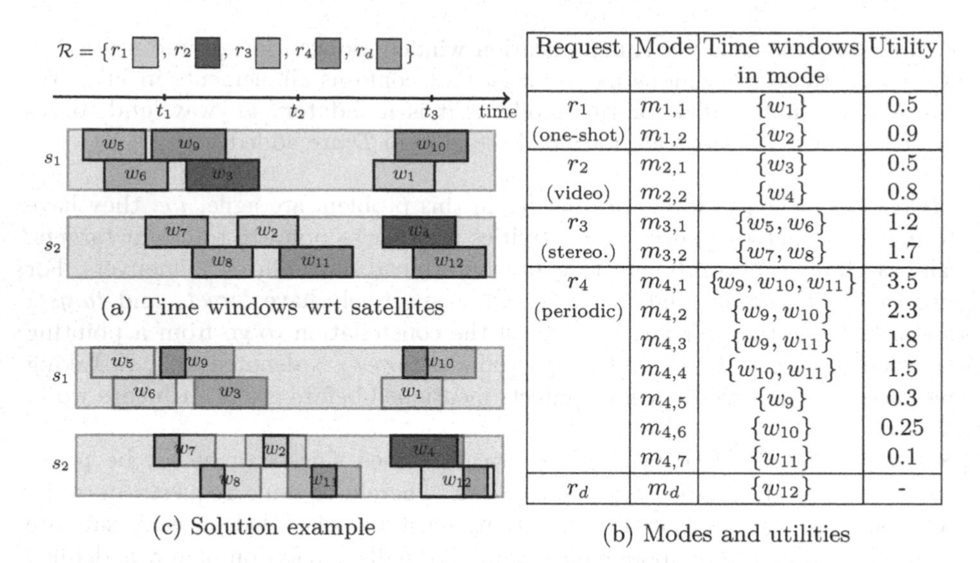

(a) Time windows wrt satellites

(c) Solution example

Request	Mode	Time windows in mode	Utility
r_1	$m_{1,1}$	$\{w_1\}$	0.5
(one-shot)	$m_{1,2}$	$\{w_2\}$	0.9
r_2	$m_{2,1}$	$\{w_3\}$	0.5
(video)	$m_{2,2}$	$\{w_4\}$	0.8
r_3	$m_{3,1}$	$\{w_5, w_6\}$	1.2
(stereo.)	$m_{3,2}$	$\{w_7, w_8\}$	1.7
r_4	$m_{4,1}$	$\{w_9, w_{10}, w_{11}\}$	3.5
(periodic)	$m_{4,2}$	$\{w_9, w_{10}\}$	2.3
	$m_{4,3}$	$\{w_9, w_{11}\}$	1.8
	$m_{4,4}$	$\{w_{10}, w_{11}\}$	1.5
	$m_{4,5}$	$\{w_9\}$	0.3
	$m_{4,6}$	$\{w_{10}\}$	0.25
	$m_{4,7}$	$\{w_{11}\}$	0.1
r_d	m_d	$\{w_{12}\}$	-

(b) Modes and utilities

Fig. 1. Example of an EOSP involving four observation requests

n observations than nothing. In this example, we consider all combinations of windows that contain at least one element, resulting in 7 modes.

Mode utilities aim to reflect end-users preferences. For instance, the fact that $u_{m_{1,1}}$ is less than $u_{m_{1,2}}$ could indicate that the cloud coverage is better for window w_2 in $m_{1,2}$. In this example, for request r_4, we consider that the more windows in the mode, the greater the associated utility, but such a property is not always true in the general case. Moreover, mode utilities are not additive: $u_{m_{4,1}}$ is greater than the sum of $u_{m_{4,5}}$, $u_{m_{4,6}}$ and $u_{m_{4,7}}$.

Figure 1c gives a solution σ where all requests are selected ($\mathcal{R}(\sigma) = \mathcal{R}$). The selected modes are $m(r_1) = m_{1,2}$, $m(r_2) = m_{2,2}$, $m(r_3) = m_{3,2}$, $m(r_4) = m_{4,1}$, and $m(r_d) = m_d$. The sequences for satellites s_1 and s_2 are respectively $[w_9, w_{10}]$ and $[w_7, w_8, w_2, w_{11}, w_4, w_{12}]$. The utility of σ is equal to $u_{m_{1,2}} + u_{m_{2,2}} + u_{m_{3,2}} + u_{m_{4,1}}$, that is to 6.9.

3.2 Connected Components

As introduced in [20], a connected component is a cluster of windows that are temporally linked to each other, meaning that inserting one activity in one time window of a component may impact the insertion of activities in other time windows of that component. Formally, we define an undirected graph G containing one node per window w in $\mathcal{W}$. Then, we connect two nodes of G if and only if their associated time windows w_1 and w_2 are on the same satellite ($sat_{w_1} = sat_{w_2}$) and overlap when considering maneuvers (neither $Latest_{w_1} + \tau_{w_1,w_2} \leq Earliest_{w_2}$ nor $Latest_{w_2} + \tau_{w_2,w_1} \leq Earliest_{w_1}$ holds). The connected components of G represent the clusters of independent time windows that can be exploited by the algorithms. The set of these connected components is denoted $\mathcal{C}$.

Example. Figure 2 presents the connected components associated with the example of Fig. 1. The set of connected components is $\mathcal{C} = \{c_1, c_2, c_3, c_4\}$, with $c_1 = \{w_5, w_6, w_9, w_3\}$, $c_2 = \{w_7, w_8, w_2, w_{11}\}$, $c_3 = \{w_{10}, w_1\}$ and $c_4 = \{w_4, w_{12}\}$.

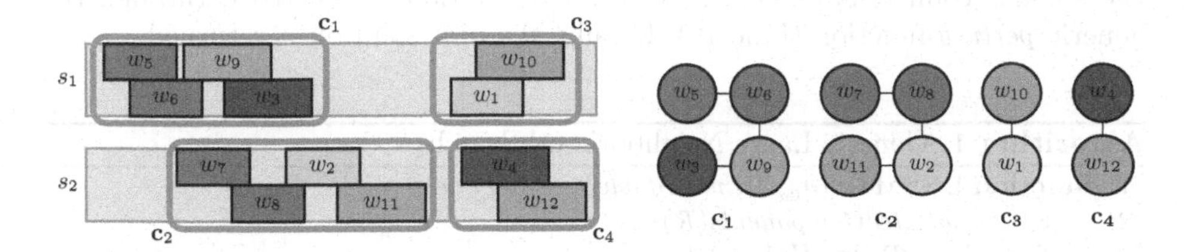

Fig. 2. Connected components obtained for the example of Fig. 1

4 Large Neighborhood Search Algorithms

In this section, we present two LNS approaches for solving EOSPs. We first describe a generic LNS scheme and then detail the two algorithms based on it. Without loss of generality, we assume that for every satellite s, there is no conflict between two ground station communication windows booked for s. As a result, the algorithms can take as an input an initial solution σ_d where all download requests are already planned, and in the following we only consider the decisions concerning the "actual" observation requests in $\mathcal{R}$.

4.1 Generic Large Neighbourhood Search

The generic LNS scheme is presented in Algorithm 1. It starts with an initialization step and then goes on with several neighborhood exploration steps until a maximum computation time is reached.

Initialization. The algorithm first computes the connected components (Line 2). It also initializes, for each request r, the set of windows $\mathcal{A}(r)$ that are usable to build a mode for r. Set $\mathcal{A}(r)$ initially contains all windows associated with r. Starting from solution σ_d that contains all the download activities, solution σ is initialized by calling function *greedyFill* that iteratively inserts new observations into the plan (Line 4, more details later on this point). This solution becomes the best known solution σ_{best}, and a counter called *stableIt*, that counts the number of LNS iterations without improvement in the global utility, is initialized.

Neighborhood Exploration. While the maximum computation time given in the input is not reached, we repetitively call the *destroyAndRepair* method (Line 8). The latter explores the current solution neighborhood based on destroy and repair operations. This function is generic and two versions will be defined in Sects. 4.3 and 4.4. If the solution returned (σ') is strictly better than the current solution, then the number of stable iterations is reset, σ' becomes the new current solution, and the best-known solution is updated if needed (Lines 10-12). Otherwise, the number of stable iterations is incremented and σ' becomes the new current solution only if it does not decrease the global utility (Line 15). After that, if the maximum number of stable iterations (*maxStableIt*) is reached, the *stableIt* counter is reset and the current solution is perturbed through the generic *perturb* function (Line 18). Finally, the best solution is returned.

Algorithm 1. Generic Large Neighborhood Search

```
1: function LNSMain(σ_d, R, maxStableIt, maxTime)
2:     C ← connectedComponents(R)
3:     for all r ∈ R do A(r) ← W_r
4:     σ ← greedyFill(σ_d, R, A, C, False)
5:     σ_best ← σ
6:     stableIt ← 0
7:     while timeSpent() < maxTime do
8:         σ' ← destroyAndRepair(σ, R, A, C)
9:         if u(σ') > u(σ) then
10:            stableIt ← 0
11:            σ ← σ'
12:            if u(σ) > u(σ_best) then σ_best ← σ
13:        else
14:            stableIt ← stableIt + 1
15:            if u(σ') = u(σ) then σ ← σ'
16:            if stableIt = maxStableIt then
17:                stableIt ← 0
18:                σ ← perturb(σ, R, A, C)
19:    return σ_best
```

4.2 Greedy Fill Method

Algorithm 2 details the *greedyFill* procedure. It takes as an input the current solution σ, the set of observation requests $\mathcal{R}$, the time windows allowed for the requests $\mathcal{A}$, the set of connected components $\mathcal{C}$, and a Boolean *reorder* expressing whether the observations already planned in σ can be reordered for the sake of other observation insertions. Procedure *greedyFill* aims at inserting the best modes of all requests that are unsatisfied so far (set $\mathcal{R}'$ initialized at Line 2). For each unsatisfied request r, its best possible mode $m(r)$ is computed through function *getBestMode*, which takes as an input the set of windows $\mathcal{A}(r)$ that are still usable to build a mode for r (Line 3). Function *getBestMode* returns *nil* if the request has no more feasible modes given set $\mathcal{A}(r)$.

Then, the algorithm tries to insert the best mode of all requests. In the main loop of the greedy filling procedure (Lines 4–11), we select a request r^* whose best mode has the highest utility among all unsatisfied requests (Line 5). We then try to insert mode $m(r^*)$ in the solution based on function *tryInsert* (Line 6). The latter tries to insert every time window associated with $m(r^*)$

into its connected component. Following these insertion attempts, two cases can occur:

1. every insertion attempt is accepted in its corresponding connected component; in this case, the mode is accepted and the function returns the new solution as well as an empty set *fails*;
2. or at least one connected component rejects the insertion; in this case, solution σ is unchanged and set *fails* contains a set of time windows whose insertion has been rejected.

In the first case, request r^* is removed from the set of unsatisfied requests (Line 8), while in the second case, we update the set of windows allowed for r^* and generate a new mode for this request (Lines 10–11).

Algorithm 2. Greedy filling

1: **function** $greedyFill(\sigma, \mathcal{R}, \mathcal{A}, \mathcal{C}, reorder)$
2: $\mathcal{R}' \leftarrow \mathcal{R} \backslash \mathcal{R}(\sigma)$
3: **for all** $r \in \mathcal{R}'$ **do** $m(r) \leftarrow getBestMode(r, \mathcal{A}(r))$
4: **while** $\exists r \in \mathcal{R}' \mid m(r) \neq nil$ **do**
5: $r^* \leftarrow \arg\max_{r \in \mathcal{R}' | m(r) \neq nil} u_{m(r)}$
6: $\sigma, fails \leftarrow tryInsert(\sigma, m(r^*), \mathcal{C}, reorder)$
7: **if** $fails = \varnothing$ **then**
8: $\mathcal{R}' \leftarrow \mathcal{R}' \setminus \{r^*\}$
9: **else**
10: $\mathcal{A}(r^*) \leftarrow \mathcal{A}(r^*) \setminus fails$
11: $m(r^*) \leftarrow getBestMode(r^*, \mathcal{A}(r^*))$
12: **return** σ

To detail, procedure *tryInsert* tries to insert every time window in $m(r^*)$ into the current plan based on two possible mechanisms. The first one is fast and used in the initialization procedure while the second one is slower but more efficient. The latter is used later in the algorithm. Both are described in the following:

- If flag *reorder* is set to *False*, then for each window $w \in m(r^*)$, the procedure simply searches for an insertion position for w that leads to a feasible plan. More precisely, if $[w_1, \ldots, w_p]$ denotes the current sequence of windows to use in the connected component of w, the procedure searches for a position i such that sequence $[w_1, \ldots, w_i, w, w_{i+1}, \ldots, w_p]$ is still feasible from a temporal point of view, i.e. starting dates can be computed. If several feasible positions exist, the algorithm chooses a position such that $\tau_{w_i,w} + \tau_{w,w_{i+1}}$ is minimum. If no feasible insertion position is found, window w is rejected and is added to set *fails*, hence using window w will not be allowed for the future steps.
- If flag *reorder* is set to *True* and if the greedy insertion seen before does not work for a given window w, more effort is made to insert w into the plan. More precisely, in this case, we call a recent state-of-the-art incomplete solver [16] developed for solving *Traveling Salesman Problem with Time Windows*. This

solver is requested to quickly find, for the connected component c associated with w, a feasible solution performing all activities in the current plan of c, plus the observation to be performed in w. This solver can reorder the current sequence of c, contrarily to the greedy insertion method.

4.3 Greedy LNS Destroying Requests

In the first version of the LNS introduced in Algorithm 1, the *destroyAndRepair* method works with the following settings.

For the destroy step, the algorithm first randomly selects a set of planned requests and removes the mode of each of these requests from the current solution. This step also updates the sets of allowed time windows $\mathcal{A}(r)$ that is passed as a parameter. More precisely, for every connected component c such that there exists a removed request r impacting c (i.e. a request r such that $m(r) \cap c \neq \emptyset$), all time windows contained in c for all requests r' are added to $\mathcal{A}(r')$ again, since the removal of r might have freed some space to insert new time windows.

For the repair step, the plan is refilled using the *greedyFill* method seen before, but with the extended insertion process (parameter *reorder* set to *True*).

Finally, the perturbation method is identical to the *destroyAndRepair* procedure: it randomly removes some requests and fills the solution again with the best modes (parameter *reorder* is also set to *True*). Note that the global LNS algorithm accepts perturbations leading to a lower global utility.

4.4 Hybrid LNS Destroying Connected Components

We describe the second implementation of the generic LNS search scheme, which revises the content of a subset of the components at each step.

Destroy Procedure. It consists in selecting a set $\mathcal{C}'$ of K connected components whose content is revised and a subset of requests $\mathcal{R}'$ whose mode can change.

To select the first component c_1 in $\mathcal{C}'$, we set a probability to pick a component c as proportional to the number of observations planned in c. Then, at each step $i \in [2..K]$, given the subset of components $\mathcal{C}' = \{c_1, \ldots, c_{i-1}\}$ selected so far, we proceed as follows to select the ith component to revise. We compute for any candidate $c \in \mathcal{C} \backslash \mathcal{C}'$ a measure $Common(c, \mathcal{C}')$ that counts the requests involved both in c and in the components of $\mathcal{C}'$. Formally, we first define such a measure for all pairs (c, c') such that c' belongs to $\mathcal{C}'$ by $Common(c, c') = |\{r \in \mathcal{R} | (\mathcal{W}_r \cap c \neq \varnothing) \wedge (\mathcal{W}_r \cap c' \neq \varnothing)\}|$, and then we define $Common(c, \mathcal{C}') = \sum_{c' \in \mathcal{C}'} Common(c, c')$. Intuitively, this measure reflects the possibility to relocate requests between the components in $\mathcal{C}'$. Then, we pick the next component c_i to add to $\mathcal{C}'$ according to a probability proportional to $Common(c_i, \mathcal{C}')$.

To restrict the size of the neighborhood, we define a subset of requests $\mathcal{R}'$ whose mode is permitted to change wrt the current solution σ. This means that for every request $r \in \mathcal{R} \backslash \mathcal{R}'$, the current mode $m(r)$ is mandatory for all solutions in the neighborhood. Mandatory download requests never belong to $\mathcal{R}'$.

CP-Based Repair Method. Following the destroy step, the goal is to determine the new modes to use in order to maximize the total utility. To do this, for each request $r \in \mathcal{R}$, we compute in a preprocessing phase all modes that can be considered for r, denoted as $\mathcal{M}_r(\sigma, \mathcal{C}', \mathcal{R}')$. If r is a fixed request in $\mathcal{R} \setminus \mathcal{R}'$, then we simply have $\mathcal{M}_r(\sigma, \mathcal{C}', \mathcal{R}') = \{m(r)\}$. If $r \in \mathcal{R}'$, set $\mathcal{M}_r(\sigma, \mathcal{C}', \mathcal{R}')$ is typically much more compact than the set of all modes of r in $\mathcal{M}_r$, especially if set $\mathcal{C}'$ contains just one or two connected components. As an illustration, if σ is the solution given in Fig. 1c, $\mathcal{C}' = \{c_1, c_2\}$ and $\mathcal{R}' = \{r_3, r_4\}$, then:

- $\mathcal{M}_{r_1}(\sigma, \mathcal{C}', \mathcal{R}') = \{\{w_2\}\}$ and $\mathcal{M}_{r_2}(\sigma, \mathcal{C}', \mathcal{R}') = \{\{w_4\}\}$ (fixed mode for requests r_1 and r_2 that are not in $\mathcal{R}'$);
- $\mathcal{M}_{r_3}(\sigma, \mathcal{C}', \mathcal{R}') = \{\{w_5, w_6\}, \{w_7, w_8\}\}$ (two possible modes for request r_3 when revising the content of components c_1, c_2);
- $\mathcal{M}_{r_4}(\sigma, \mathcal{C}', \mathcal{R}') = \{\{w_9, w_{10}, w_{11}\}, \{w_9, w_{10}\}, \{w_{10}, w_{11}\}, \{w_{10}\}\}$ (four possible modes where the selected window w_{10} in component c_3 is still present);
- $\mathcal{M}_{r_d}(\sigma, \mathcal{C}', \mathcal{R}') = \{\{w_{12}\}\}$ (fixed mode for the download request).

We also define the set of windows $\mathcal{W}_r(\sigma, \mathcal{C}', \mathcal{R}')$ to consider for request $r \in \mathcal{R}$ for the repair phase. This set contains the windows involved both in the modes belonging to $\mathcal{M}_r(\sigma, \mathcal{C}', \mathcal{R}')$ and in the connected components in $\mathcal{C}'$, that is $\mathcal{W}_r(\sigma, \mathcal{C}', \mathcal{R}') = \{w \in \mathcal{W}_r \mid (w \in \cup_{c \in \mathcal{C}'} c) \wedge (w \in \cup_{m \in \mathcal{M}_r(\sigma, \mathcal{C}', \mathcal{R}')} m)\}$.

Then, we introduce the CP model given in Eqs. 1–5. For each request r whose mode is not fixed ($r \in \mathcal{R}'$), this CP model contains one decision variable $\mathbf{y_m} \in \{0, 1\}$ per mode in $\mathcal{M}_r(\sigma, \mathcal{C}', \mathcal{R}')$. Variable $\mathbf{y_m}$ takes value 1 if mode m is selected, and value 0 otherwise. Equation (1) expresses that the objective is to maximize the sum of the utilities of all the selected modes, while Eq. (2) imposes that at most one mode is selected for each request. Note that choosing $\mathbf{y_m} = 0$ for each candidate mode m in $\mathcal{M}_r(\sigma, \mathcal{C}', \mathcal{R}')$ is equivalent to not planning r and in this case, all windows of r are removed from the solution.

The CP model models the temporal intervals over which activities are performed through *interval variables* (examples of *CP* features are available in [9,10]). For each request $r \in \mathcal{R}$, we introduce one interval variable Itv_w per time window $w \in \mathcal{W}_r(\sigma, \mathcal{C}', \mathcal{R}')$, *i.e.* involved in a component over which the repair phase works. Such an interval has a duration δ_r and must be inside time interval $[Earliest_w, Latest_w]$. Several constraints are defined over these variables. The first one imposes that intervals Itv_w must be present in the solution for time windows w belonging to the mode of a fixed request (Eq. 3). The second constraint imposes that an interval Itv_w associated with a non-fixed request r is present if and only if it belongs to the mode chosen for r (Eq. 4). Finally, for every component $c \in \mathcal{C}'$, there must be no temporal overlap between the selected intervals Itv_w associated with c, given the transition duration matrix τ (Eq. 5).

$$\max \textstyle\sum_{r \in \mathcal{R}', m \in \mathcal{M}_r(\sigma, \mathcal{C}', \mathcal{R}')} u_m \mathbf{y_m} \quad (1)$$

$$\forall r \in \mathcal{R}', \textstyle\sum_{m \in \mathcal{M}_r(\sigma, \mathcal{C}', \mathcal{R}')} \mathbf{y_m} \leq 1 \quad (2)$$

$$\forall r \in \mathcal{R} \setminus \mathcal{R}', \forall w \in \mathcal{W}_r(\sigma, \mathcal{C}', \mathcal{R}'),\ presenceOf(Itv_w) = 1 \quad (3)$$

$$\forall r \in \mathcal{R}', \forall w \in \mathcal{W}_r(\sigma, \mathcal{C}', \mathcal{R}'),\ presenceOf(Itv_w) = \textstyle\sum_{m \in \mathcal{M}_r(\sigma, \mathcal{C}', \mathcal{R}') \mid w \in m} \mathbf{y_m} \quad (4)$$

$$\forall c \in \mathcal{C}', noOverlap(\{Itv_w \mid r \in \mathcal{R}, w \in \mathcal{W}_r(\sigma, \mathcal{C}', \mathcal{R}') \cap c\}, \tau) \quad (5)$$

The repair method aims to provide a new better solution by solving the CP model given a maximum CPU time limit. At the next iterations, other parts of the current solution can be considered by choosing different sets $\mathcal{C}'$ and $\mathcal{R}'$.

5 Experiments

5.1 Instances

Constellation and Downloads. We consider a Walker constellation composed of 16 satellites dispatched equally on two distinct orbit planes at 800 km (low Earth orbits). We consider a unique ground station in Toulouse, France. The duration of download activities is set to $\Delta = 180$ s. The time to maneuver between two targets is proportional to the euclidean distance between them (from a few seconds to several tens of seconds). The scheduling horizon is 24 h.

Targets. All targets are located in West Europe. We consider two types of instances depending on the precise location of the targets. In the first type of instances, namely *spread target instances*, we define latitude and longitude bounds (respectively $[-10°, 10°]$ and $[40°, 55°]$) and choose, for each target, a random coordinate using a uniform probability inside these bounds. In the second type of instances, namely *concentrated target instances*, we randomly choose 50 airports among West Europe ones. Then, for each request, we randomly choose one of these airports and add a noise with a maximum amplitude of 1° in latitude and longitude. Figure 3 illustrates both configurations for 1000 requests.

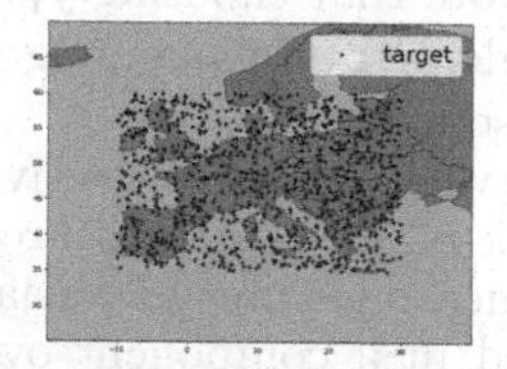

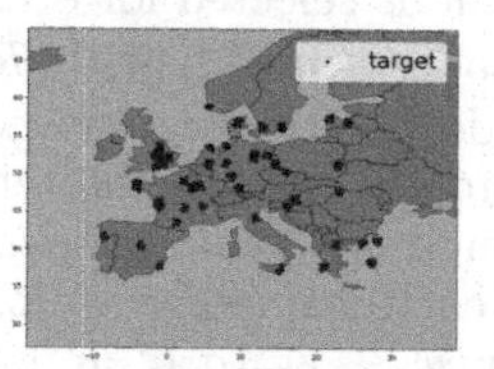

Fig. 3. Example of a spread instance (left) and a concentrated instance (right)

Requests. The instances contain from 50 to 1140 requests and vary in the number of requests of each type. Duration of observations for each request are randomly chosen following a truncated normal law $\mathcal{N}(5, 59, 1, 60)^2$ for one-shot requests, $\mathcal{N}(50, 20, 40, 60)$ for video requests, $\mathcal{N}(5, 9, 1, 10)$ for stereoscopic requests, and $\mathcal{N}(5, 19, 1, 20)$ for periodic requests. We generated 20 instance configurations in which the number of requests and their types vary. Then, for each configuration, we considered the spread target and concentrated target instances and 5 different seeds for target generation. For all these instances, the number of connected components is around 125. In fact, such a number depends more on the constellation features than on the considered number of requests.

[2] $\mathcal{N}(\mu, v, a, b)$: truncated normal law with mean μ, variance v, bounded by a and b.

Mode Utility. For each request r, all modes that contain at least one observation are considered. Hence, for each periodic request, all modes containing at least one period are taken into account. For each time window w in $\mathcal{W}$, we associate a utility u_w that depends on the real cloud cover percentage cc_w of that window, obtained from real weather data. More precisely, if $1 - cc_w$ belongs to range $[0, 0.5)$ (resp. $[0.5, 0.6)$, $[0.6, 0.7)$, $[0.7, 0.8)$, $[0.8, 0.9)$, $[0.9, 1)$), then $u_w = 0.05$ (resp. $0.1, 0.4, 0.7, 0.95, 1$). Then, for each request r that has type one-shot, video, or stereoscopic, for each mode $m \in \mathcal{M}_r$, $u_m = \sum_{w \in m} u_w$. For each periodic request r and for each mode $m \in \mathcal{M}_r$, utility is computed through the following non-additive formula : $u_m = \sum_{w \in m} u_w + 0.5\,(nPeriods - maxGapPeriods)$, where *nPeriods* is the number of periods in r and *maxGapPeriods* is the maximum number of consecutive periods of r that do not have a corresponding time window in m.

All instances are available at https://github.com/ssquilla/Earth_Observing_Satellites_benchmarks.git.

5.2 Experimental Setup

All the algorithms are implemented in Python 3.8.5. Experiments are run on a 20-core Intel(R) Xeon(R) CPU E5-2660 v3 @ 2.60 GHz, 62 GB RAM, with a time out of 5 min per run. Each run is allowed to use up to 10 cores. The *maxStableIt* parameter is set to 15. The following algorithms are compared.

CP is the Constraint Programming approach presented in [20]. It amounts to solving Eq. 1–5 with $\mathcal{C}' = \mathcal{C}$ and $\mathcal{R}' = \mathcal{R} \setminus \mathcal{D}$ (full revision of all the components).

BPCCAS (Batch Parallel Connected Component of Activities based Search) is the state-of-the-art approach defined in [20] for EOSPs.

GLNS (Greedy LNS) is the LNS approach defined in Sect. 4.3, that uses a request-based neighborhood explored by greedy search. The *destroyAndRepair* method destroys 20% of the observation requests in the solution. One LNS process is run on each of the 10 available cores, and the best solution is returned.

CPLNS (CP LNS) and **MCPLNS** (Multi-Core CP LNS) are the LNS approaches using a component-based neighborhood explored with Constraint Programming (Cplex Studio 20.1 used through docplex 2.22.213). We impose a time limit equal to 3 s for each call to the CP solver and $\mathcal{R}'$ is set to $\mathcal{R} \setminus \mathcal{D}$. For *CPLNS* (resp. *MCPLNS*), we consider a neighborhood with 1 (resp. 2) component(s) revised at each step. Moreover, for *CPLNS*, we run 10 LNS threads in parallel (one per core) and return the best solution found, while for *MCPLNS*, we use only one LNS thread but permit the CP solver to use the 10 cores.

5.3 Results

Table 1 gives representative results. Each line of the table is the mean value obtained over three runs on the same instance. For all instances (concentrated and spread) involving 250 requests or less, all the algorithms find the same utility except for *CP*. Moreover, the gaps associated with *BPCCAS* and *CP* increase as the number of requests grows. Globally, all the LNS algorithms find the same best utility except for instances involving a large number of periodic requests. In that case, *CPLNS* and *MCPLNS* outperform *GLNS*. Last, the results provided by *CPLNS* and *MCPLNS* are very close (if not equal) to the best solutions on all instances, and they lead to gaps that are always less than 2%.

Table 1. Representative results (one line per instance). The columns are: number of requests, per type (OS: One-Shot, V: Video, S: Stereo, P: Periodic), cardinality of $\mathcal{W}$, best utility obtained by the solvers, gaps (in percent) wrt the best utility for each method (Let b the best reward and a the score provided by the algorithm. $gap = 100(b-a)/b$), along with the number of calls to *destroyAndRepair* (nDR).

	Instance					$\|\mathcal{W}\|$	best	*CP*	*BPCCAS*	*GLNS*		*CPLNS*		*MCPLNS*	
	$\|\mathcal{R}\|$	OS	V	S	P		utility	gap	gap	gap	nDR	gap	nDR	gap	nDR
concentrated targets	50	50	0	0	0	2607	3.9	**0.0**	**0.0**	**0.0**	2620	**0.0**	9790	**0.0**	6778
	50	0	50	0	0	2436	40.6	**0.0**	**0.0**	**0.0**	2755	**0.0**	11128	**0.0**	7539
	50	0	0	0	50	568	82.8	**0.0**	**0.0**	**0.0**	4002	**0.0**	337	**0.0**	150
	57	12	15	27	3	2125	24.6	**0.0**	**0.0**	**0.0**	2893	**0.0**	10329	**0.0**	6684
	250	250	0	0	0	13149	16.6	1.81	**0.0**	**0.0**	275	**0.0**	4267	**0.0**	3015
	250	0	250	0	0	12162	74.4	24.19	**0.0**	**0.0**	79	**0.0**	4165	**0.0**	3021
	250	0	0	0	250	2359	400.0	31.15	6.72	5.95	217	**0.0**	107	0.38	94
	285	60	75	135	15	10327	85.2	**0.0**	0.12	**0.0**	316	**0.0**	4372	**0.0**	1418
	500	500	0	0	0	25646	50.7	12.23	**0.0**	**0.0**	25	**0.0**	1398	**0.0**	1064
	500	0	500	0	0	25554	209.6	24.9	**0.0**	**0.0**	12	**0.0**	771	**0.0**	608
	500	0	0	0	500	4786	754.4	49.96	14.16	12.43	44	**0.0**	92	0.16	84
	570	120	150	270	30	19203	269.7	3.11	**0.0**	0.37	92	1.22	1767	1.0	399
	1000	1000	0	0	0	51189	114.7	18.22	9.33	**0.0**	5	**0.0**	193	0.26	461
	1000	0	1000	0	0	49149	530.4	23.06	21.95	**0.0**	2	0.19	56	0.41	21
	1000	0	0	0	1000	9769	1157.3	60.69	80.8	30.24	3	**0.0**	73	1.24	49
	1140	240	300	540	60	45985	252.3	39.67	28.02	0.04	4	**0.0**	277	0.04	108
spread targets	50	50	0	0	0	2438	7.2	**0.0**	**0.0**	**0.0**	2838	**0.0**	6790	**0.0**	7014
	50	0	50	0	0	2427	28.8	**0.0**	**0.0**	**0.0**	2785	**0.0**	6816	**0.0**	7029
	50	0	0	0	50	560	85.6	**0.0**	**0.0**	**0.0**	4239	**0.0**	194	**0.0**	181
	57	12	15	27	3	2025	22.2	0.45	**0.0**	**0.0**	3215	**0.0**	7137	**0.0**	7275
	250	250	0	0	0	12269	46.3	0.22	0.22	**0.0**	332	**0.0**	3887	**0.0**	3504
	250	0	250	0	0	12190	184.9	**0.0**	0.59	1.14	226	1.51	4027	1.19	3611
	250	0	0	0	250	2340	439.2	12.89	0.61	0.96	270	**0.0**	96	0.14	97
	285	60	75	135	15	10467	145.1	**0.0**	**0.0**	**0.0**	365	0.07	3974	0.07	2170
	500	500	0	0	0	24475	92.5	**0.0**	0.97	1.08	71	1.62	2494	0.97	2018
	500	0	500	0	0	24326	358.8	5.71	**0.0**	0.53	28	1.39	2426	**0.0**	1906
	500	0	0	0	500	4586	775.9	41.17	9.76	8.84	46	**0.0**	88	0.08	86
	570	120	150	270	30	20820	266.6	3.49	0.08	**0.0**	75	0.98	962	0.3	499
	1000	1000	0	0	0	49035	177.8	10.74	0.11	**0.0**	7	1.12	1094	0.79	674
	1000	0	1000	0	0	48686	676.0	19.56	2.51	**0.0**	3	0.24	650	0.22	383
	1000	0	0	0	1000	9154	1226.0	53.0	43.59	25.46	5	0.55	70	**0.0**	58
	1140	240	300	540	60	41693	525.8	26.13	0.55	**0.0**	10	1.24	379	1.07	196

Figure 4 presents the value of the global utility wrt time for solvers *GLNS*, *BPCCAS* and *CPLNS* on two large instances. As detailed in [20], *BPCCAS* performs several iterations, where each iteration starts with an empty schedule and consists in inserting modes with the highest utilities first until all requests have been considered. At each iteration, the order of modes is modified. For instances that do not contain periodic requests (left plot), *BPCCAS* performs almost two iterations, which allows one to get a high utility solution. However, it is far from finishing the first iteration for instances involving a large number of periodic requests (right plot), which results in a low global utility in such cases.

For instances with few periodic requests, the starting solution of LNS approaches has a higher utility. With more periodic requests, *GLNS* slowly improves the solution because each insertion of a periodic request involves computation on multiple components, impacting the number of destroys and repairs that can be performed (nDR in Table 1). On such instances, the *CPLNS* neighborhood exploration allows one to significantly improve the global utility. In fact, the destroy phase partially destroys periodic requests, and the repair phase allows one to assign modes that are not necessarily the best ones for each individual request but give a higher global utility. This phenomenon is more noticeable for concentrated target instances, where the higher targets concentration causes more conflicts between requests. Therefore, in order to get good quality solutions, one has to perform trade-offs between requests (i.e. assigning partial modes instead of complete ones), which is particularly suited for *CPLNS* and *MCPLNS* that optimize a global utility through their calls to the CP solver.

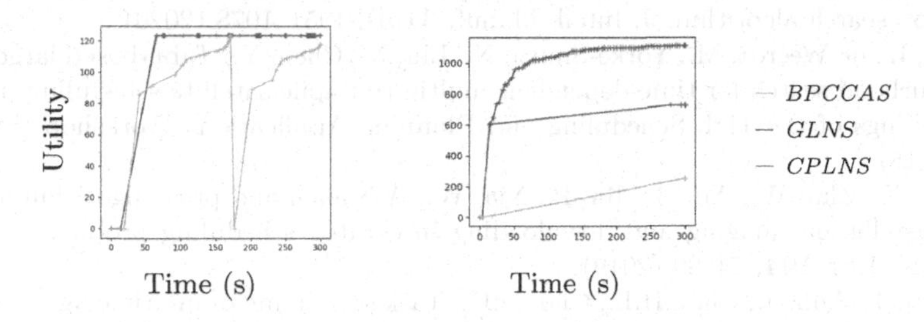

Fig. 4. Utility wrt time for instances 500 OS 500 S (left), and 1000 P (right).

6 Conclusion

In this paper, we proposed two LNS approaches outperforming the state of the art for planning complex observation requests for a constellation of Earth observing satellites. The first approach, which uses a destroy method centered on Request Removals (destroyRR) and a repair method based on Heuristic Search (repairHS), could be referred to as destroyRR-repairHS. The second approach, which uses a destroy method centered on Connected Components (destroyCC) and a repair method using CP (repairCP), could be referred to as destroyCC-repairCP, and it leads to high quality solutions whatever the instance configuration (number and location of targets or requests types). There exist other possible variants, such as "destroyCC-repairHS" (content of some components revised by heuristic search) or "destroyRR-repairCP" (request removals and CP-based reinsertions). Variant destroyCC-repairHS has been tested but is dominated by destroyCC-repairCP, and variant destroyRR-repairCP is left for future works.

Acknowledgments. This work has been performed with the support of BPI through PSPC project "LiChIE" of the "Programme d'Investissements d'Avenir".

References

1. Abbatecola, L., Fanti, M.P., Pedroncelli, G., Ukovich, W.: A new cluster-based approach for the vehicle routing problem with time windows. In: 2018 IEEE 14th International Conference on Automation Science and Engineering (CASE), pp. 744–749 (2018)
2. Dondo, R., Cerdá, J.: A cluster-based optimization approach for the multi-depot heterogeneous fleet vehicle routing problem with time windows. Eur. J. Oper. Res. **176**(3), 1478–1507 (2007)
3. Du, Y., Wang, T., Xin, B., Wang, L., Chen, Y., Xing, L.: A data-driven parallel scheduling approach for multiple agile earth observation satellites. IEEE Trans. Evol. Comput. **24**(4), 679–693 (2020)
4. Eddy, D., Kochenderfer, M.: A maximum independent set method for scheduling earth-observing satellite constellations. J. Spacecr. Rocket. **58**, 1–14 (2021)
5. He, L., de Weerdt, M., Yorke-Smith, N.: Time/sequence-dependent scheduling: the design and evaluation of a general purpose tabu-based adaptive large neighbourhood search algorithm. J. Intell. Manuf. **31**(4), 1051–1078 (2020)
6. He, L., de Weerdt, M., Yorke-Smith, N., Liu, X., Chen, Y.: Tabu-based large neighbourhood search for time-dependent multi-orbit agile satellite scheduling. In: Proceedings of the 11th Scheduling and Planning Applications Workshop (SPARK) (2018)
7. Hu, X., Zhu, W., An, B., Jin, P., Xia, W.: A branch and price algorithm for EOS constellation imaging and downloading integrated scheduling problem. Comput. Oper. Res. **104**, 74–89 (2019)
8. Kim, J., Ahn, J., Choi, H.L., Cho, D.H.: Task scheduling of multiple agile satellites with transition time and stereo imaging constraints. J. Aerosp. Inf. Syst. **17**(6) (2020)
9. Laborie, P.: IBM ILOG CP optimizer for detailed scheduling illustrated on three problems. In: van Hoeve, W.-J., Hooker, J.N. (eds.) CPAIOR 2009. LNCS, vol. 5547, pp. 148–162. Springer, Heidelberg (2009). https://doi.org/10.1007/978-3-642-01929-6_12
10. Laborie, P., Rogerie, J., Shaw, P., Vilím, P.: IBM ILOG CP optimizer for scheduling: 20+ years of scheduling with constraints at IBM/ILOG. Constraints **23**(2), 210–250 (2018)
11. Levinson, R., Nag, S., Ravindra, V.: Agile satellite planning for multi-payload observations for earth science. CoRR abs/2111.07042 (2021)
12. Liu, L., Dong, Z., Su, H., Yu, D., Lin, Y.: Research on a heterogeneous multi-satellite mission scheduling model for earth observation based on adaptive genetic-tabu hybrid search algorithm. In: IEEE 5th Advanced Information Technology, Electronic and Automation Control Conference (IAEAC), pp. 1684–1690 (2021)
13. Liu, X., Laporte, G., Chen, Y., He, R.: An adaptive large neighborhood search metaheuristic for agile satellite scheduling with time-dependent transition time. Comput. Oper. Res. **86**, 41–53 (2017)
14. Peng, G., Dewil, R., Verbeeck, C., Gunawan, A., Xing, L., Vansteenwegen, P.: Agile earth observation satellite scheduling: an orienteering problem with time-dependent profits and travel times. Comput. Oper. Res. **111**, 84–98 (2019)

15. Picard, G.: Auction-based and distributed optimization approaches for scheduling observations in satellite constellations with exclusive orbit portions. In: International Conference on Autonomous Agents and Multiagent Systems (2021)
16. Pralet, C.: Iterated maximum large neighborhood search for the traveling salesman problem with time windows and its time-dependent version. Comput. Oper. Res. **150**, 106078 (2022)
17. Ruiz-Meza, J., Brito, J., Montoya-Torres, J.R.: A GRASP to solve the multi-constraints multi-modal team orienteering problem with time windows for groups with heterogeneous preferences. Comput. Ind. Eng. **162**, 107776 (2021)
18. Schmid, V., Ehmke, J.F.: An effective large neighborhood search for the team orienteering problem with time windows. In: ICCL 2017. LNCS, vol. 10572, pp. 3–18. Springer, Cham (2017). https://doi.org/10.1007/978-3-319-68496-3_1
19. Squillaci, S., Roussel, S., Pralet, C.: Managing complex requests for a constellation of earth observing satellites. In: International Workshop on Planning and Scheduling for Space (2021)
20. Squillaci, S., Roussel, S., Pralet, C.: Parallel scheduling of complex requests for a constellation of earth observing satellites. In: Passerini, A., Schiex, T. (eds.) Frontiers in Artificial Intelligence and Applications. IOS Press (2022)
21. Su, X., Nan, H.: An enhanced heuristic for the team orienteering problem with time windows considering multiple deliverymen. Soft Comput. **27**(6), 2853–2872 (2022)
22. Wei, L., Xing, L., Wan, Q., Song, Y., Chen, Y.: A multi-objective memetic approach for time-dependent agile earth observation satellite scheduling problem. Comput. Ind. Eng. **159**, 107530 (2021)
23. Wu, G., Wang, H., Pedrycz, W., Li, H., Wang, L.: Satellite observation scheduling with a novel adaptive simulated annealing algorithm and a dynamic task clustering strategy. Comput. Ind. Eng. **113**, 576–588 (2017)

Predicting Wildlife Trafficking Routes with Differentiable Shortest Paths

Aaron Ferber[1(✉)], Emily Griffin[2], Bistra Dilkina[1], Burcu Keskin[3], and Meredith Gore[4]

[1] University of Southern California, Los Angeles, USA
{aferber,dilkina}@usc.edu
[2] Babson College, Babson Park, USA
egriffin@babson.edu
[3] University of Alabama, Tuscaloosa, USA
bkeskin@cba.ua.edu
[4] University of Maryland, College Park, USA
gorem@umd.edu

Abstract. Wildlife trafficking (WT), the illegal trade of wild fauna, flora, and their parts, directly threatens biodiversity and conservation of trafficked species, while also negatively impacting human health, national security, and economic development. Wildlife traffickers obfuscate their activities in plain sight, leveraging legal, large, and globally linked transportation networks. To complicate matters, defensive interdiction resources are limited, datasets are fragmented and rarely interoperable, and interventions like setting checkpoints place a burden on legal transportation. As a result, interpretable predictions of which routes wildlife traffickers are likely to take can help target defensive efforts and understand what wildlife traffickers may be considering when selecting routes. We propose a data-driven model for predicting trafficking routes on the global commercial flight network, a transportation network for which we have some historical seizure data and a specification of the possible routes that traffickers may take. While seizure data has limitations such as data bias and dependence on the deployed defensive resources, this is a first step towards predicting wildlife trafficking routes on real-world data. Our seizure data documents the planned commercial flight itinerary of trafficked and successfully interdicted wildlife. We aim to provide predictions of highly-trafficked flight paths for known origin-destination pairs with plausible explanations that illuminate how traffickers make decisions based on the presence of criminal actors, markets, and resilience systems. We propose a model that first predicts likelihoods of which commercial flights will be taken out of a given airport given input features, and then subsequently finds the highest-likelihood flight path from origin to destination using a differentiable shortest path solver, allowing us to automatically align our model's loss with the overall goal of correctly predicting the full flight itinerary from a given source to a destination. We evaluate the proposed model's predictions and interpretations both quantitatively and qualitatively, showing that the predicted paths are aligned with observed held-out seizures, and can be interpreted by policy-makers.

© The Author(s), under exclusive license to Springer Nature Switzerland AG 2023
A. A. Cire (Ed.): CPAIOR 2023, LNCS 13884, pp. 460–476, 2023.
https://doi.org/10.1007/978-3-031-33271-5_30

1 Introduction

Wildlife Trafficking (WT) broadly impacts biodiversity, human health, economic development, and national security [37]. It encompasses a wide array of species that originate from, and are transported to, supply and demand markets around the world. WT spans over 150 countries and includes more than 37,000 species of fauna and flora [37]. Transnational criminal organizations are known to leverage the increasingly interconnected air transportation network to move illegal wildlife products from source to destination locations, generating $19 billion annually in black market proceeds [18,28,38]. The massive scope, scale, and diversity of wildlife trafficking networks present a complex and dynamic challenge for authorities and researchers trying to understand and interrupt the transiting of illegal wildlife products using detection, interdiction, deterrence, education, or other activities. Stakeholders working to combat wildlife trafficking also face limited social, physical, and financial capital compared to other illicit activities such as drug trafficking. Current practice is to rely heavily on trusted and established personal relationships, "tip-offs" about specific flights, use of specially trained sniffer dogs, and education of airport personnel; these practices can be successful in one-off contexts but lack a desired deterrent effect. Network interdiction models can assist in determining the optimal allocation of scarce resources along known trafficking networks but have yet to be systematically applied to the transiting stage of wildlife trafficking supply chains [17,31]. Data-driven methods for understanding underlying wildlife trafficking patterns could help advance on the ground practice and expand modeling techniques to a novel domain space and are a necessary first step before targeted interdiction allocation can be applied effectively and efficiently.

Recognizing the potential for data-driven methods to dramatically enhance solutions to the problem of wildlife trafficking, multiple sectors have increased their data collection activities. For example, The Convention on International Trade in Endangered Species of Wild Fauna and Flora (CITES) is a global agreement among governments to regulate international trade in species under threat that was established in 1976 and is currently signed by 183 countries and the European Union. TRAFFIC is an organization that was established in 1976 by The World Wide Fund for Nature (WWF) and International Union for Conservation of Nature (IUCN) as a wildlife trade monitoring network to undertake data collection, analysis, and provision of recommendations to inform decision making on wildlife trade. In 2015, the U.S. Agency for International Development (IUCN) established the Reducing Opportunities for Unlawful Transport of Endangered Species (ROUTES) Partnership to bring together transport and logistics companies, government agencies, law enforcement, and conservation organizations to eliminate wildlife trafficking from the air transport supply chain. Importantly, these efforts have contributed to collection and synthesis of a limited but growing global database of illegal wildlife trade seizure data.

Overall, the flight network's widespread use for moving illegal goods, as well as the presence of structured data make it a promising setting for data analysis to help inform defensive measures. Center for Advanced Defense Studies (C4ADS), a nonprofit that is a member of ROUTES, produced in-depth summary analysis of the global wildlife trade flight seizure data from 2009–2017 [38] and 2016–2018 [39] and derived insights based on observed concentration of illegal activity and outliers. Some studies and reports describe traffickers' modus operandi, or factors that may influence their decisions to traffic products through certain ports over others [3,34]. Factors, such as larger airports with higher volume, prevalence of corruption, lower financial costs, and smaller legal penalties, have been shown to possibly be beneficial for traffickers [16]. However, there is limited quantitative research into the factors that impact traffickers' transit choices and their relative importance [22,33,35]. In fact, to our knowledge, predictive models have not been applied to the wildlife trafficking domain. Machine learning models can be instrumental in extrapolating the patterns from the limited seizure data to other airports and routes. They can highlight important factors and their weights to provide insight into traffickers' objectives that can be utilized when making interdiction decisions and predicting trafficker responses.

To this end, in this paper, we formulate wildlife trafficking across the global flight network as a route prediction problem on a graph, synthesize historical seizure data with data that describes airport nodes and flight edges, and propose a maximum likelihood machine learning model that exploits recent developments in differentiable optimization. In particular, we model probabilities of trafficking on each edge in the transportation network as a function of node and edge features, and train the model by comparing the maximum likelihood path (identified by computing the shortest path in log space) to the ground truth paths. We demonstrate the predictive power of our model. We analyze our model's results to understand the discrepancies between our predictions and the ground truth seizure data. By utilizing an interpretable linear model with respect to input features, we are also able to provide feature importance insights.

A key area of concern in combating WT is the convergence of multiple forms of illicit trade [14,35]. Convergence can take a variety of forms. For instance, revenue from WT activities can fund arms trafficking. Additionally, the people, countries, and transit routes used for various forms of trafficking can substantially overlap due to factors that are mutually beneficial. Convergence has long been an area of concern but the amount of scientific, quantitative, evidence for convergence is still limited [15]. Our work makes a step towards quantifying the scale and impact of convergence by directly incorporating measures of other illicit activities at given locations as features when predicting wildlife trafficking paths. Understanding the impact of other illicit activities on the path probabilities of traffickers provides a quantitative measure of geographic convergence.

2 Related Work

The overall problem of learning route choices may be considered an inverse optimization problem, where we are given "solutions" to optimization problems and we want to identify what optimization parameters yields those observed solutions as optimal [1]. Indeed, previous work in trajectory prediction has modeled hidden

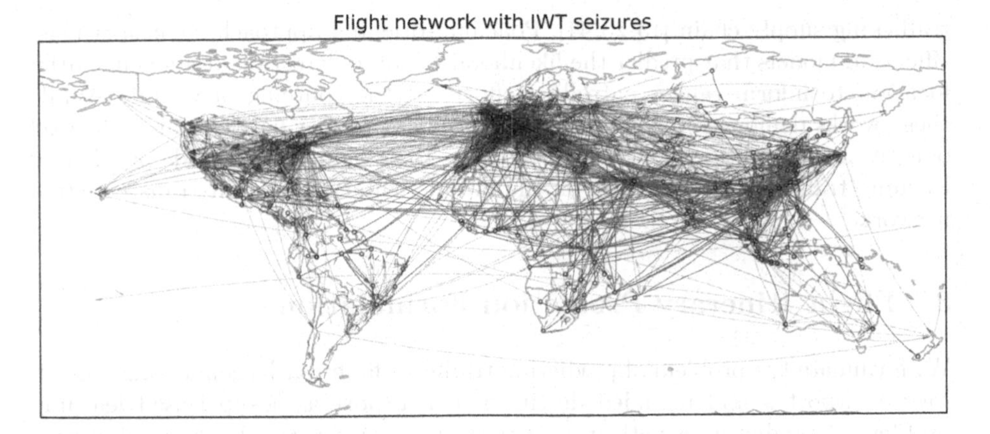

Fig. 1. Visualization of itineraries with historical seizures in red as well as a subset of the global flight network in grey. (Color figure online)

latencies for travel networks by solving an inverse shortest path problem [42], or learning transportation preferences for a road network which results in a given traffic flow on the network [11]. The area of trajectory prediction [10,12,29,43] aims to predict paths for individuals and thus tend to assume access to the start location, or continually updating sequence of locations, and try to predict the rest of the trajectory that the person will take. However, in our case, we have generally-known source and destination pairs and try to understand what are the most likely paths that traffickers will take without continuously updating information.

Recent work in the machine learning literature has investigated how to integrate optimization solvers as differentiable components in machine learning pipelines. This effectively allows the model designer to state that the model predictions will be used downstream by a structured optimization problem which will output an optimal solution to a problem with given predicted inputs. The seminal OptNet paper [2] introduces the quadratic optimization program as a differentiable layer for use in deep learning pipelines, by implicitly differentiating through the KKT optimality conditions, with follow-up work extending the approach to linear programs [40]. In a different vein, researchers investigated differentiating through blackbox optimizers [26] and differentiating through maximum likelihood estimation which can represent the optimal solution to a mathematical program [24]. Our approach directly builds off of [26] and leverages empirical insights in order to speed up gradient computation. Lastly, several approaches for smart predict then optimize have been proposed which compute subgradients of the optimal solution with respect to the inputs in order to train the predictive model [9]. This smart predict then optimize area has work on applicable theoretical guarantees and integration with decision trees [4,8].

Prior work has successfully used machine learning in the context of wildlife poaching in conservation areas, but poaching is only the "first" step in the wildlife trafficking supply chain [13,23,41]. Poaching-oriented approaches consider classification models that predict the likelihood of snare detection at a given spatial location to inform ranger patrolling efforts at the sourcing of wildlife. While these works demonstrated the ability to predict poacher behavior at each pixel of a given conservation area, here we address the global wildlife trade problem of learning trafficker route choices on the broader international air transportation network.

3 Flight Itinerary Prediction Formulation

We formulate the problem of predicting trafficker flight paths connecting a given source airport s and intended destination d airport as a supervised learning problem of predicting a path from s to d on a flight network represented as a directed graph G. The flight network G represents airports as nodes and the flights between them as directed edges. We augment the flight network with WT-related features ϕ on both nodes ϕ^v and edges ϕ^e. We collect N groundtruthed trafficker paths $\mathcal{D}^\pi = \{\pi_{s_i,d_i}\}_{i=1}^N$ from centralized databases of seizure reports. These reports contain the traffickers' intended itineraries between fixed source s_i and destination d_i. We encode these WT itineraries π_i as paths in the flight network, representing them as either a sequence of airport nodes or flight edges as needed.

Our data sources, collection, and synthesis are described in the section "Data Sources". To get a sense of the magnitude of the problem at hand, we visualize the observed trafficker paths as well as 20% of the full flight network in Fig 1. We subsample due to the density of the global flight network consisting of 14,118 flight edges connecting 1,933 airport nodes, rendering the image unreadable otherwise.

Formally, we aim to train a model that correctly predicts the observed structured path π_i given the input source s_i, destination d_i, flight network G, and features ϕ.

3.1 Predictive Model: Edge Transition Estimator

In order to predict full flight paths from features on just edges and nodes, we cannot simply predict how likely any individual path is, as the number of possible simple paths is exponentially large in the size of the network. Instead, we consider predicting a probability for each edge which then can be used to compute path likelihood.

We propose an approach for modeling the path prediction problem by predicting "transition" probabilities, or probabilities on which flight edges a trafficker may take to exit a given "current" node. Since our setting requires a simple model that can be easily handed off to domain experts and deliver actionable insights for interdiction, we forego complex architectures in favor of a simplistic predictive model. This models the trafficker as taking a biased random walk from the source airport to the destination airport on the flight network where our model learns the biased probabilities given edge and node features. With this transition probability modeling approach, we can compute the probability of taking any given source-destination path as being the product of individual edge probabilities.

Formally, we model the problem as finding the probability $P(i,j)$ of using a directed edge (i,j) to leave a starting node i. Here, probabilities on all edges leaving a given node i sum to 1. We use a parametrized model m, with parameters θ, to obtain probability estimates given the relevant features i.e. $\hat{P}(i,j) = m\left(\phi^e_{i,j}, \phi^v_i, \phi^v_j; \theta\right)$. For notational simplicity, we consider the feature vector for a given edge to be the concatenation of edge-specific features, origin features, and destination features $\phi_{i,j} = \left[\phi^e_{i,j}, \phi^v_i, \phi^v_j\right]$. The edge probability prediction model limits the number of trained parameters to prevent overfitting. This parameter sharing means that the same model is used to predict which flights will be taken out of an airport whether it is Addis Ababa or Charles de Gaulle. Furthermore, by predicting edge probabilities from edge and node features, we can understand how these features impact our model's estimates and thus better understand what factors may be driving wildlife trafficking. Hence, in our experiments we use a linear model relating the features to the predicted probabilities to ensure that the resulting model is interpretable.

We denote the set of edges leaving i as $\delta(i)$, and fully specify our linear model as making predictions on each edge as computing logits with a linear model, and using a softmax to normalize the edge logits based on the flight origin node to ensure that the outgoing probabilities sum to one. Mathematically our probability prediction model is described in Eq. 1.

$$\hat{P}(i,j) = m\left(\phi^e_{i,j}, \phi^v_i, \phi^v_j; \theta\right) = \frac{\exp\left(\theta^T \phi_{i,j}\right)}{\sum\limits_{j' \in \delta(i)} \exp\left(\theta^T \phi_{i,j'}\right)} \tag{1}$$

Our formulation ensures that the output probability estimates are a differentiable function of the parameters θ to be trained using standard deep learning libraries like pytorch [25]. Additionally, we experimented using a 3-layer multilayer perceptron (MLP) as well as gradient-boosted decision trees but found poor generalization of the MLP and the gradient-boosted decision trees performed on par with our linear model so we opted for the linear model as it was interpretable with no drawback in performance.

With the given formulation, the probability of a path $P(\pi)$ is the product of individual edge probabilities $\Pi_{(i,j)\in\pi}\hat{P}((i,j)|i)$. Furthermore, we can identify the model's highest-likelihood path by finding a shortest path with edge weights corresponding to the negative log probability. A path minimizing the sum of negative log probabilities is a path that maximizes the sum of log probabilities which, due to the logarithm's product rule and monotonicity, is a maximum likelihood path. The goal now is to find model parameters θ such that the observed trafficking paths π have the highest likelihood.

At deployment time, this edge transition model will enable us to identify easily the highest-likelihood path by solving a shortest path problem in log probability space, obtain other highly-likely paths by identifying other near-optimal solutions, and allows us to easily evaluate the likelihood of any other alternative path.

3.2 Model Training: Path-Integrated Learning

Given that we want to predict full paths in the flight network, we propose training the parameters θ to directly minimize differences between the predicted highest-probability path and observed trafficking paths. We consider a differentiable pipeline and loss function that directly aligns model training with the problem of recovering the ground truth path, and can be optimized using gradient descent.

Using the above definition of our edge transition probability estimator, we express model training as solving the optimization problem in Eq. 2 which minimizes the expected Hamming loss between a given ground-truth path $\pi_{s,d}$ with corresponding source s and destination d against the highest-likelihood path $\hat{\pi}_{s,d}$ predicted by the model connecting that source to that destination. The highest-likelihood path is computed by Single Source Single Destination shortest path solver (SSSDSolver) over the negative log of predicted transition probabilities $\hat{P}$. Transition probabilities $\hat{P}$ are computed according to Eq. (1).

Ultimately, to train the model we compute gradients for the model parameters via backpropagation of the hamming loss to the predicted highest-probability path $\hat{\pi}$, back to the predicted transition probabilities $\hat{P}$, and then to the model parameters θ.

$$\min_{\theta} E_{\pi_{s,d}} \left[H\left(\pi_{s,d}, \text{SSSDSolver}\left(-log\left(\hat{P}\right); s, d\right)\right)\right] \tag{2}$$

For completeness, we can define the single source single destination shortest path solver in Eq. (3) as finding the path minimizing the sum of weights on edges used in the path π, which in our case are negative log probabilities.

$$\text{SSSDSolver}(w; s, d) = \underset{\pi_{s,d}}{\arg\min} \left(\sum_{(i,j)\in\pi_{s,d}} w_{i,j}\right) \tag{3}$$

Here we can use any off-the-shelf shortest path solver without worrying about negative edge weights since the probabilities are all between 0 and 1 (exclusive), so the negative log of the probabilities are all positive values. In practice, we use Dijkstra's shortest path algorithm. Note that the forward pass to get predicted path $\hat{\pi}$ is the same approach we use for determining the highest-likelihood path, thus aligning our model's training with the overall deployment pipeline of correctly identifying the full path.

In order for us to use gradient descent to train our model parameters, we need to ensure that all steps from the model predictions to the loss evaluation are differentiable so that gradients may be easily computed using chain rule. All of the components except for the SSSDSolver are readily differentiable functions available in Pytorch [25], as a result we need to define a backward pass for the shortest path solver to enable model training.

Using the formulation enabling differentiation of blackbox solvers proposed in [26], we make our forward and gradient update explicit below. In the forward pass, we simply solve the shortest path problem and cache the solution $\hat{\pi} := \text{SSSDSolver}(w; s, d)$. The backward pass itself expects incoming gradients from the loss layer, and returns outgoing gradients with respect to the input edge costs w. Overall, the intention of the gradient is to give an indication of what changes in the edge costs w will produce the desired change in the returned path to minimize the loss and better align the path with the ground truth solution. The method for differentiating blackbox solvers introduced in [26] essentially perturbs the input objective coefficients w in the direction of the gradient to find a "locally-improved" solution. It then computes the gradients as the difference between the resulting "locally-improved" solution and the previously predicted solution. When used in conjunction with the hamming loss, the "locally-improved" objective coefficients are simply the input objective coefficients with a given amount increased or decreased depending on whether the decision component, such as the edge usage, should be used or not. In order to specify the degree that the input costs should be perturbed, the authors use a hyperparameter λ which determines the degree to which the weights w should be perturbed in the desired direction. Formally, in the backward pass, we are given input gradients $\nabla_{\hat{\pi}} L$ of the loss with respect to the shortest path $\hat{\pi}$. We compute improved edge weights $w' = w + \lambda \nabla_{\hat{\pi}} L$. Then we re-solve the problem with improved edge weights to find a better solution $\pi' = \text{SSSDSolver}(w'; s, d)$. Finally, we compute gradients of this layer as $-\frac{1}{\lambda}(\hat{\pi} - \pi')$.

In our setting, this method corresponds to solving the shortest path problem with perturbed weights where weight is slightly decreased on edges that should appear in the ground truth solution and slightly increased on edges that aren't in the ground truth solution. The gradient that is passed back to the edge costs is the difference between the predicted path and the "locally-improved" path. Intuitively, the approach aims to decrease cost on edges that should be in the locally-improved short path but aren't in the predicted path, and increase cost on edges that are in the outputted shortest path but don't appear in the locally-improved path. Additionally, in our initial experiments, we found that performant values of λ were large enough so that the weight perturbation eclipsed the initial weights themselves, meaning that overall the "locally-improved" solution was simply the ground truth solution. As such, to cut the number of solves down by half, we simply used the ground truth solution path π as the "locally-improved" solution.

Note that this approach is akin to updating the gradients such that it scores the ground truth solution π to have better objective value than the predicted solution $\hat{\pi}$. Additionally, in this scenario we consider that the path π is encoded as a 0–1 vector with a given entry indicating whether edge (i, j) is used in the path or not. As such, the weight vector is updated by the difference between path solutions.

3.3 Model Training: Edge-Myopic Learning

We compare our path-integrated learning method with an approach that is trained to minimize the Kullback-Leibler (KL) divergence [19] between the edge probabilities computed directly from training data P' to the edge probabilities predicted as a function of features $\hat{P}$. This approach focuses on correctly predicting rates at which different edges are used for trafficking in the ground truth rather than looking at full paths, and is a slight variant of baseline approaches in previous work [5] that is adapted for our setting where we have known source and destination locations, as well as network features. Previous work estimates the transition probabilities between different locations, and here we estimate these transitions with a logistic regression model to obtain a model of how the features are related to the observed transitions. Using raw training data, we estimate transition probabilities $P'(e)$ as the number of times that a given flight e is used for trafficking divided by the number of times that the source airport is used for trafficking. The predictive model's parameters are then trained to closely match these transition probabilities based on the given features. Given probability predictions $\hat{P}$ and data-driven estimates $P'(e)$, the KL divergence is $KL\left(\hat{P}||P'(e)\right) = \sum_e \hat{P}_e \log \frac{\hat{P}_e}{P'(e)}$. Overall, the Edge-Myopic learning trains the parameters θ to minimize this edge-level KL divergence.

4 Data Sources

Centralized and comprehensive data sources are critical for combating wildlife trafficking [16]; however, they are often lacking in practice, complicating the application of models to different domains. In our experiments, we leverage data regarding wildlife trafficking seizures, flight pricing, available flights, and indices of general crime prevalence and resilience infrastructure. The global flight network was collected from OpenFlights.org which hosts open-source information about airports, routes, and flights. The data was last updated on January 2017, and we have manually added several airports and routes to ensure that we can place as many seizure records on the flight network as possible. Overall, this dataset allows us to construct a flight network of 1,933 airport nodes connected by 14,118 flight edges.

For each flight edge, we record the distance and collect flight pricing estimates using the Skyscanner API [30]. Since flight pricing depends on several components such as the amount of time before the flight, we collect prices for all flight routes one month in advance. For each flight edge, we used the API on October 14, 2021 to request flight quotes for November 2021. The API did not return valid responses for several airport pairs due to no valid flight plans existing in the database accessed by the API which we determined manually from searching google flights. Additionally, we note that data was collected during the coronavirus pandemic impacting flight availability, as historical data was not available.

Each airport is associated with its country's metrics reported in the Global Organized Crime Index for 2021 [7], the first year the indices were published by

Table 1. Node and edge features of the flight network. Features in Bold were selected by recursive feature elimination.

NODE FEATURES	
METADATA	
Population	CITES membership
Flight Count	
GITOC - CRIMINAL MARKETS	
Criminal Markets (Average)	**Fauna Crimes**
Human Trafficking	**Heroin Trade**
Human Smuggling	Cocaine Trade
Arms Trafficking	Cannabis Trade
Flora Crimes	Synthetic Drug Trade
Non-Renewable Resource Crimes	
GITOC - RESILIENCE	
Anti-Money Laundering Systems	Resilience (Average)
Political Leadership And Governance	Territorial Integrity
Govt. Transparency & Accountability	Law Enforcement
Economic Regulatory Capacity	International Cooperation
Victim & Witness Support	National Policies & Laws
Judicial System And Detention	Prevention
GITOC - CRIMINAL ACTOR	
Criminal Actors (Average)	State-Embedded Actors
Mafia-Style Groups	Foreign Actors
Criminal Networks	Non-State Actors
EDGE FEATURES	
Price	Distance

the Global Initiative Against Transnational Organized Crime (GITOC). These indices represent expert opinion of a country's relationship with various forms of organized crime, including the prevalence of different criminal actors, strength of resilience resources, and presence of criminal markets. These indices score countries from 1 to 10 based on 5 rounds of anonymous and independent expert reviews in 2020. We also add information about whether the airport's country is a member of CITES, the city's population, and the number of flights that serve the given airport. The node and edge features we collected are summarized in Table 1.

We obtained seizure data from the Wildlife Trade Portal (WTP) [36] through which TRAFFIC provides historical seizure data with detailed records like intended itinerary (source, destination, transit points), trafficked wildlife, trafficker details, and legal outcomes. In total, we accessed 1,067 records between

2017 and 2021 to synthesize a dataset of 454 itineraries of wildlife trafficking. Only 362 of the 1,933 airport nodes in the global flight network are used by traffickers in the historical seizure data, highlighting the data sparsity. Furthermore, in terms of the paths themselves, the data is biased towards shorter paths, having 1-hop, 2-hop, 3-hop, and 4-hop paths making up 60.6%, 24.2%, 15%, and 0.2% of the data respectively.

Seizure data provides a glimpse of how WT networks operate, alert experts to trends in supply and demand for different species, and point to key locations for deterring wildlife crime [20]. However, it is important to understand the biases in seizure data due to being collected by different law enforcement agencies, using several means of detection, against various criminal agents [6,15]. As a result, seizure data not only reflects the criminal network, but also the defensive resources. Nevertheless, seizure data is one of the few tools we have available to peer into WT networks in a scalable manner.

5 Experiments

Table 2. Summary statistics from 10-fold cross-validation of models using either the full set of features or an algorithmically-selected subset. We evaluate two training methods, edge-myopic learning which aims to correctly predict how often individual edges are used, and path-integrated learning which aims to identify the complete intended source-destination path. We report the average performance across folds with 95% confidence intervals.

Training Method	Features	Path recall ↑	Edge recall ↑	Edge precision ↑	Edit distance ↓
Edge-Myopic	Selected	89.6% ± 1.2	83.1% ± 2.6	86.1% ± 1.0	0.115 ± 0.04
Path-Integrated	Selected	92.4% ± 2.7	88.4% ± 4.3	90.5% ± 2.6	0.088 ± 0.03
Edge-Myopic	All	89.2% ± 2.5	82.8% ± 3.5	85.5% ± 3.1	0.113 ± 0.03
Path-Integrated	All	89.1% ± 3.1	82.6% ± 4.0	85.4% ± 3.4	0.113 ± 0.03

5.1 Feature Selection

Feature selection identifies the highest-impact features, limits overfitting, and avoids correlated features. We use recursive feature elimination to iteratively remove the least-useful feature from the current feature set by testing each of them and evaluating the change in 10-fold path recall. Since node features appear twice for a given edge, once for the edge's head and again for the tail, we drop both as needed. The full set and selected features in bold are in Table 1.

5.2 Metrics

We train the models using the Adam optimizer with amsgrad [27], and evaluate the models using 10-fold cross-validation, splitting the dataset by source-destination pair. This produces a prediction for every source-destination path using a model that wasn't trained on information from the given source-destination pair. We evaluate using several metrics at the path and edge level.

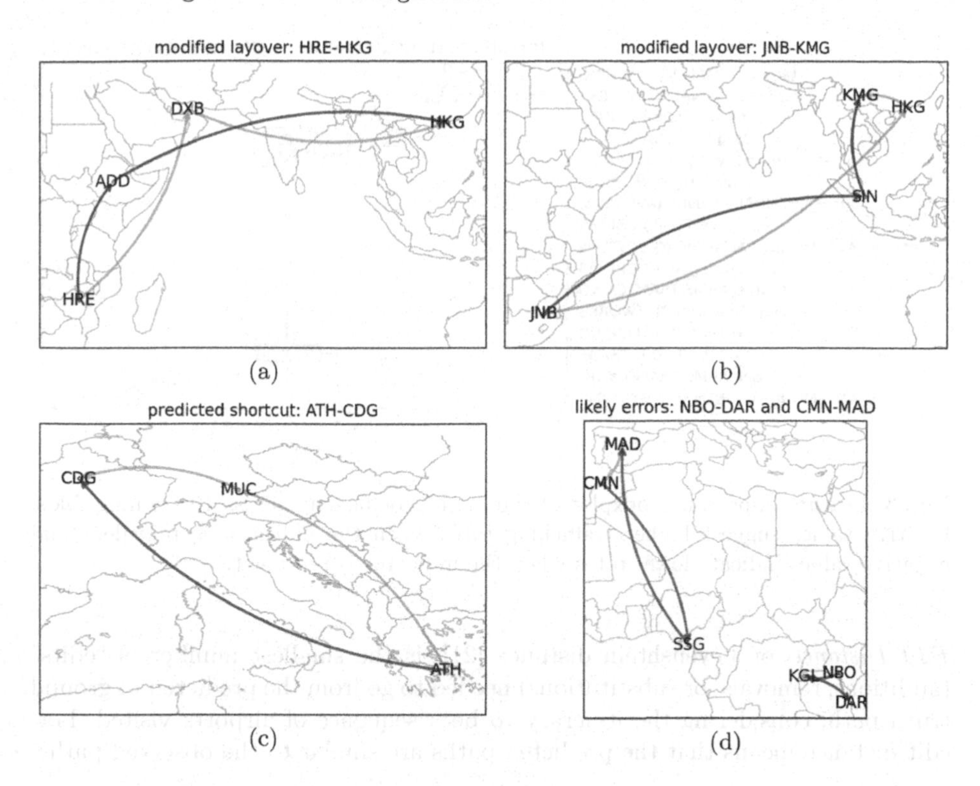

Fig. 2. Visualization of discrepancies between Path-Integrated predicted itineraries in blue and observed itineraries in red. Additionally, domain experts identified two likely errors in Fig. d where our model's predictions are unrealistic. (Color figure online)

Path Recall is the percent of the ground truth paths the model completely predicted correctly. Given the N ground truth paths in the dataset $\mathcal{D}^{\pi}$, the path recall is $\frac{1}{N}\sum_{\pi_{s,d}\in\mathcal{D}^{\pi}}\delta\left(\pi_{s,d}=\hat{\pi}_{s,d}\right)$. Here δ is just a 1 if the paths are completely equal (taking the same sequence of edges) and 0 otherwise. Higher values here mean that our model is not likely to miss out on trafficked paths.

Edge Recall is the percent of trafficked edges that our model predicts to have trafficking. Mathematically this is $\left(\sum_{\pi_{s,d}\in\mathcal{D}^{\pi}}\sum_{e\in\pi_{s,d}}\delta\left(e\in\hat{\pi}_{s,d}\right)\right)/\sum_{\pi_{s,d}\in\mathcal{D}^{\pi}}|\pi_{s,d}|$. High values here mean that a large proportion of observed trafficked edges are picked up by our model.

Edge Precision measures the percent of edges that our model predicts to have trafficking which did in fact exhibit trafficking in the seizure data. Mathematically this is $\left(\sum_{\pi_{s,d}\in\mathcal{D}^{\pi}}\sum_{e\in\hat{\pi}_{s,d}}\delta\left(e\in\pi_{s,d}\right)\right)/\sum_{\pi_{s,d}\in\mathcal{D}^{\pi}}|\hat{\pi}_{s,d}|$. High values here mean that our model's predictions are trustworthy and that domain users can expect that the model's predictions will likely contain trafficking.

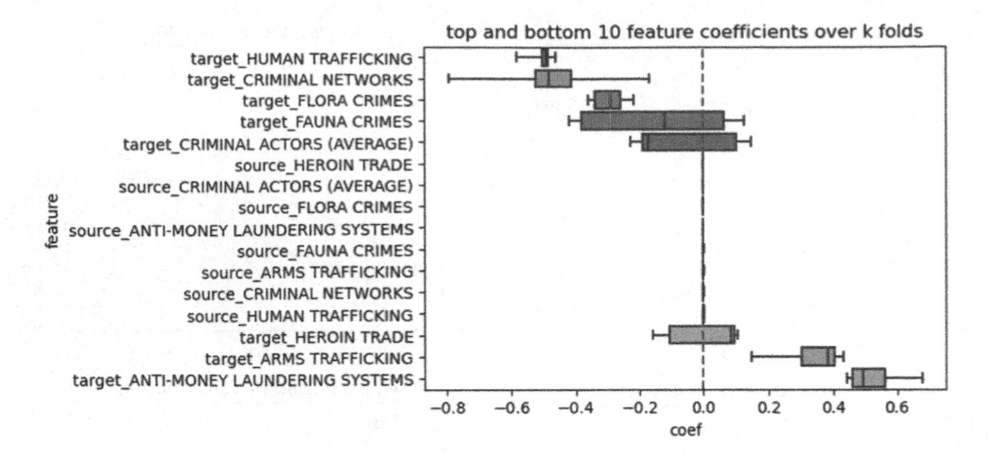

Fig. 3. Feature importance boxplot of the model coefficients across 10 training folds. Positive values suggest higher trafficking rates when the indicator is prevalent and negative values indicate lower rates when the indicator is prevalent.

Edit Distance or Levenshtein distance [21], is the smallest number of "edits" (additions, removals, or substitutions) needed to go from the predicted to ground truth path, considering the itinerary to be a sequence of airports visited. Low edit distance means that the predicted paths are similar to the observed paths.

5.3 Results Discussion

We present numerical results in Table 2, computing the average and standard deviation in performance with 10-fold cross-validation. Given the data size of only 454 itineraries, the differences in performance come from only a few predicted paths. Additionally, both models benefit from feature selection, with feature-selected path-integrated learning improving over edge-myopic learning. The performance of path-integrated learning with feature selection is high in that the models are able to recall 92.4% of the paths completely, 88.4% of the edges, and the predicted edges contain trafficking at a rate of 90.5%. Additionally, breaking the results down by path length, we find that on the 1-hop, 2-hop, 3-hop, and 4-hop paths, path-integrated learning with selected features gets path recalls of 98.9%, 83.1%, 80.7%, and 100%, respectively, and average Levenstein distances of 0.031, 0.169, 0.192, and 0 respectively. On the other hand, across 1-hop, 2-hop, 3-hop, and 4-hop paths, the edge-myopic learning with selected features has path recall of 100%, 83.1%, 58.6%, and 0%, respectively, with Levenstein distances of 0.0, 0.261, 0.314, 2.0, respectively. Across all folds, our model differs with 31 ground truth paths between 25 origin-destination pairs. We note that our performance improvements aren't statistically significant for most metrics, except for edge precision, due to our small sample size. However, given that path-integrated learning gives fewer errors in our low data regime and are promising for future work when more data is available for evaluation. Given the data bias, the predicted alternate routes may contain wildlife trafficking even though it is not present in the ground-truth data. We visualize representative discrepancies between predicted and observed paths in Fig. 2 with predicted paths in blue and observed paths in red. We categorize the discrepancies into 10 origin-destination pairs where our predictions shortcut the observed itinerary by removing stops (Fig. 2a), and 13 cases where our model predicts different layovers than the observed (Fig. 2b, 2c), identified as highly plausible in informal consultations with experts. The two cases where our model predicted additional layovers are visualized in Fig. 2d and are likely errors.

We visualize the path-integrated learning model's feature importance in Fig. 3. Here, positive values mean that high feature values induce high estimated probability, whereas negative values mean that high feature values induce low estimated probability. Overall, the model considers that traffickers are likely to travel to locations with high arms trafficking as well as resilience against money laundering. The convergence between wildlife trafficking and arms trafficking has been documented and has broad implications for interdiction [32]. Additionally, the model predicts that traffickers are less likely to enter regions with high flora crime, criminal networks, or human trafficking. The negative value for the flight destination's flora crimes is interesting and warrants further investigation, and may reflect seizure data bias, or traffickers wanting to flee suspicion. Some features have 0 weight from selecting features based on edge-myopic learning, as well as having correlated features. Ultimately, we propose a model and a training approach, presenting promising results with the best data available so far. More in-depth and robust conclusions about wildlife trafficking route prediction can be made in the future as more complete seizure data and richer feature sets become available, which can leverage our modeling work.

5.4 Conclusion

We approach the problem of predicting wildlife trafficking on the flight transportation network with differentiable optimization. To align our network training with the goal of correctly identifying full paths, we train with a differentiable highest-probability path solver We show that a path-integrated learning model learns over the available airport and flight features with limited training data, slightly improving over edge-myopic learning, and can likely further improve as more seizure data is collected. Lastly, we identify several features that may contribute to traffickers being more likely to take a given path. We hope that our method will help inform interdiction efforts and the study of wildlife trafficking networks, and we intend to use our predictions in conjunction with combinatorial interdiction in future work.

Acknowledgments. All authors were supported by U.S. NSF awards CMMI-1935451; Gore was also supported by IIS-2039951. The information contained herein does not represent the opinions of the U.S. Government or any author affiliations.

References

1. Ahuja, R.K., Orlin, J.B.: Inverse optimization. Oper. Res. **49**, 771–783 (2001). https://doi.org/10.1287/opre.49.5.771.10607
2. Amos, B., Kolter, J.Z.: OptNet: differentiable optimization as a layer in neural networks. In: Proceedings of the 34th International Conference on Machine Learning. Proceedings of Machine Learning Research, vol. 70, pp. 136–145. PMLR (2017)
3. Arroyave, F.J., Petersen, A.M., Jenkins, J., Hurtado, R.: Multiplex networks reveal geographic constraints on illicit wildlife trafficking. Appl. Netw. Sci. **5**(1), 1–20 (2020). https://doi.org/10.1007/s41109-020-00262-6
4. Balghiti, O.E., Elmachtoub, A.N., Grigas, P., Tewari, A.: Generalization bounds in the predict-then-optimize framework. In: Advances in Neural Information Processing Systems, vol. 32 (2019)
5. Choi, S., Yeo, H., Kim, J.: Network-wide vehicle trajectory prediction in urban traffic networks using deep learning. Transp. Res. Rec. **2672**(45), 173–184 (2018)
6. CITES: https://cites.org/eng ()
7. Global Initiative Against Transnational Organized Crime. The global organized crime index (2021). https://globalinitiative.net/analysis/ocindex-2021/
8. Elmachtoub, A., Liang, J.C.N., McNellis, R.: Decision trees for decision-making under the predict-then-optimize framework. In: International Conference on Machine Learning, pp. 2858–2867. PMLR (2020). https://github.com/rtm2130/SPOTree
9. Elmachtoub, A.N., Grigas, P.: Smart "predict, then optimize". Manage. Sci. **68**(1), 9–26 (2021). https://doi.org/10.1287/mnsc.2020.3922
10. Feng, J., et al.: DeepMove: predicting human mobility with attentional recurrent networks. In: Proceedings of the 2018 World Wide Web Conference, pp. 1459–1468 (2018)
11. Fosgerau, M., Paulsen, M., Rasmussen, T.K.: A perturbed utility route choice model. Transp. Res. Part C Emerg. Technol. **136**, 103514 (2022). https://doi.org/10.1016/j.trc.2021.103514, https://www.sciencedirect.com/science/article/pii/S0968090X21004976
12. Gambs, S., Killijian, M.O., del Prado Cortez, M.N.: Next place prediction using mobility Markov chains. In: Proceedings of the First Workshop on Measurement, Privacy, and Mobility, pp. 1–6 (2012)
13. Gholami, S., et al.: Adversary models account for imperfect crime data: forecasting and planning against real-world poachers (corrected version). In: 17th International Conference on Autonomous Agents and Multiagent Systems (2018)
14. Gore, M.L., et al.: Transnational environmental crime threatens sustainable development. Nat. Sustain. **2**(9), 784–786 (2019)
15. Gore, M.L., Mwinyihali, R., Mayet, L., Baku-Bumb, G.D.M., Plowman, C., Wieland, M.: Typologies of urban wildlife traffickers and sellers. Glob. Ecol. Conserv. **27**, e01557 (2021)
16. Gore, M.L., et al.: Voluntary consensus based geospatial data standards for the global illegal trade in wild fauna and flora. Sci. Data **9**(1), 1–8 (2022)
17. Haas, T.C., Ferreira, S.M.: Finding politically feasible conservation policies: the case of wildlife trafficking. Ecol. Appl. **28**(2), 473–494 (2018)
18. IATA: Combating wildlife trafficking. https://www.iata.org/en/programs/environment/wildlife-trafficking/. Accessed 15 Aug 2022
19. Kullback, S.: Information theory and statistics. Courier Corporation (1997)

20. Kurland, J., Pires, S.F.: Assessing us wildlife trafficking patterns: how criminology and conservation science can guide strategies to reduce the illegal wildlife trade. Deviant Behav. **38**(4), 375–391 (2017)
21. Levenshtein, V.I., et al.: Binary codes capable of correcting deletions, insertions, and reversals. In: Soviet Physics Doklady, vol. 10, pp. 707–710. Soviet Union (1966)
22. Magliocca, N., et al.: Comparative analysis of illicit supply network structure and operations: cocaine, wildlife, and sand. J. Illicit Econ. Dev. **3**(1), 50–73 (2021)
23. Nguyen, T.H., et al.: Capture: a new predictive anti-poaching tool for wildlife protection. In: Proceedings of the 2016 International Conference on Autonomous Agents & Multiagent Systems, pp. 767–775 (2016)
24. Niepert, M., Minervini, P., Franceschi, L.: Implicit MLE: backpropagating through discrete exponential family distributions. In: Advances in Neural Information Processing Systems, vol. 34 (2021). https://github.com/nec-research/tf-imle
25. Paszke, A., et al.: PyTorch: an imperative style, high-performance deep learning library. In: Wallach, H., Larochelle, H., Beygelzimer, A., d'Alché-Buc, F., Fox, E., Garnett, R. (eds.) Advances in Neural Information Processing Systems, vol. 32, pp. 8024–8035. Curran Associates, Inc. (2019). http://papers.neurips.cc/paper/9015-pytorch-an-imperative-style-high-performance-deep-learning-library.pdf
26. Pogančić, M.V., Paulus, A., Musil, V., Martius, G., Rolinek, M.: Differentiation of blackbox combinatorial solvers. In: International Conference on Learning Representations (2020). https://openreview.net/forum?id=BkevoJSYPB
27. Reddi, S.J., Kale, S., Kumar, S.: On the convergence of Adam and Beyond. In: International Conference on Learning Representations (2018). https://openreview.net/forum?id=ryQu7f-RZ
28. ROUTES: How the aviation industry transformed to combat wildlife trafficking (2022). https://www.internationalairportreview.com/article/173456/how-the-aviation-industry-transformed-to-combat-wildlife-trafficking/. Accessed 15 Aug 2022
29. Rudenko, A., Palmieri, L., Herman, M., Kitani, K.M., Gavrila, D.M., Arras, K.O.: Human motion trajectory prediction: a survey. Int. J. Robot. Res. **39**(8), 895–935 (2020)
30. skyscanner (2020). https://skyscanner.github.io/slate/
31. Smith, J.C., Song, Y.: A survey of network interdiction models and algorithms. Eur. J. Oper. Res. **283**(3), 797–811 (2020)
32. Spevack, B.: Shared skies convergence of wildlife trafficking with other illicit activities in the aviation industry. In: C4ADS (2021). www.routespartnership.org
33. Stringham, O.C.: Text classification to streamline online wildlife trade analyses. PLoS ONE **16**(7), e0254007 (2021)
34. Stringham, O.C., et al.: Dataset of seized wildlife and their intended uses. Data Brief **39**, 107531 (2021). https://doi.org/10.1016/j.dib.2021.107531. https://www.sciencedirect.com/science/article/pii/S2352340921008076
35. Stringham, O.C., et al.: A guide to using the internet to monitor and quantify the wildlife trade. Conserv. Biol. **35**(4), 1130–1139 (2021)
36. TRAFFIC: Wildlife trade portal (2021). www.wildlifetradeportal.org
37. UNODC: Enhancing the Detection, Investigation, and Disruption of Illicit Financial Flows from Wildlife Crime (2017). http://www.unodc.org/unodc/en/data-and-analysis/wildlife.html
38. Utermohlen, M., Baine, P.: In plane sight: wildlife trafficking in the air transport sector (2018). https://www.traffic.org/publications/reports/in-plane-sight/. Accessed 15 Aug 2022

39. Utermohlen, M.: Runway to extinction: wildlife trafficking in the air transport sector (2020). https://routespartnership.org/industry-resources/publications/runway-to-extinction-report. Accessed 15 Aug 2022
40. Wilder, B., Dilkina, B., Tambe, M.: Melding the data-decisions pipeline: decision-focused learning for combinatorial optimization. In: Proceedings of the AAAI Conference on Artificial Intelligence, vol. 33, 1658–1665 (2019). https://doi.org/10.1609/aaai.v33i01.33011658
41. Xu, L., Bondi, E., Fang, F., Perrault, A., Wang, K., Tambe, M.: Dual-mandate patrols: multi-armed bandits for green security. In: Thirty-Fifth AAAI Conference on Artificial Intelligence (AAAI-21) (2021)
42. Zhang, J., Paschalidis, I.C.: Data-driven estimation of travel latency cost functions via inverse optimization in multi-class transportation networks. In: 2017 IEEE 56th Annual Conference on Decision and Control, CDC 2017, pp. 6295–6300 (2018). https://doi.org/10.1109/CDC.2017.8264608
43. Ziebart, B.D., et al.: Planning-based prediction for pedestrians. In: 2009 IEEE/RSJ International Conference on Intelligent Robots and Systems, pp. 3931–3936. IEEE (2009)

Iterated Greedy Constraint Programming for Scheduling Steelmaking Continuous Casting

Dongyun Kim[1], Yeonjun Choi[1], Kyungduk Moon[1], Myungho Lee[1], Kangbok Lee[1(✉)], and Michael L. Pinedo[2]

[1] Pohang University of Science and Technology, Pohang, South Korea
{dykim97,choichoi,kaleb.moon,hojoung10,kblee}@postech.ac.kr
[2] Stern School of Business, New York University, New York, NY, USA
mlp5@stern.nyu.edu

Abstract. We consider a steelmaking-continuous casting (SCC) scheduling problem in the steel industry, which is a variant of the hybrid flow shop scheduling problem subject to practical constraints. Recently, Hong et al. [Hong, J., Moon, K., Lee, K., Lee, K., Pinedo, M.L., International Journal of Production Research **60**(2), 623-643 (2022)] developed an algorithm, called Iterated Greedy Matheuristic (IGM), in which a Mixed Integer Programming (MIP) model was proposed and its subproblems are iteratively solved to improve the solution. We propose a new constraint programming (CP) formulation for the SCC scheduling problem and develop an algorithm, called Iterated Greedy CP (IGC), which uses the framework of IGM but replaces the MIP model with our CP model. When we solve the CP subproblems iteratively, we also refine them by adding appropriate constraints, reducing the domains of the variables, and giving the variables hints derived from the current solution. From computational experiments in various settings, we show that IGC implemented with an open-source CP solver can be competitive with IGM running on a commercial MIP solver.

Keywords: Scheduling · steelmaking-continuous casting process · hybrid flow shop · constraint programming · computational experiments

1 Introduction

Constraint Programming (CP) has been frequently used to solve classical machine scheduling problems along with other methods such as exact algorithms (e.g., Mixed Integer Programming (MIP) formulation with a solver, dynamic programming (DP)), heuristics (e.g., dispatching, decomposition), and metaheuristics (e.g., genetic algorithm, artificial bee colony). Since CP recently has improved benchmark results for many scheduling problems such as the classical job shop [2], the open shop [6] and other complex shop scheduling problems [7,8], it has received attention from the academic community. Although CP has the

© The Author(s), under exclusive license to Springer Nature Switzerland AG 2023
A. A. Cire (Ed.): CPAIOR 2023, LNCS 13884, pp. 477–492, 2023.
https://doi.org/10.1007/978-3-031-33271-5_31

potential to represent complicated constraints of practical scheduling problems, metaheuristics are often preferred due to their capability of finding a fairly good solution in a decent amount of time.

Recently, an algorithm called *Iterated Greedy Matheuristic* (IGM) was developed by Hong et al. [5] to solve a practical scheduling problem in the steel industry. They proposed a MIP model and solved its subproblems iteratively to improve the solution. IGM reportedly outperforms pure MIP and genetic algorithms.

IGM can be regarded as a *Large Neighborhood Search* (LNS) algorithm, which dynamically relaxes a part of the variables of an incumbent solution (and fixes the remaining part to a certain extent) and reoptimizes by solving a smaller subproblem. First introduced by Shaw [13], LNS has been used to solve scheduling problems in job shop environments [1,3] and in flexible job shop environments [10]. In scheduling problems, LNS can be implemented by imposing precedence relationships to the operations except the ones to be relaxed. The relaxed operations can be selected randomly, from a subset of machines, or within a time window. We refer to [12] as a survey of LNS for various combinatorial optimization problems.

In this paper, we propose *Iterated Greedy CP* (IGC), a revised version of IGM by replacing a MIP model with a CP model. Our computational results show that IGC implemented with an open-source CP solver can be competitive with IGM using a commercial MIP solver. Hence, IGC seems to have the potential to solve hard scheduling problems. We also compare the performances of CP and MIP in the Iterated Greedy (IG) framework for practical SCC scheduling problems, which has not often been done in the literature.

The comparison between IGC and IGM can be summarized as follows.

	IGC	IGM
Modeling CP or MIP		
Main variable types	interval, integer, binary	continuous, binary
Logical implication	optional & present vars	big-M & binary vars
Tightening formulation	domain reduction	valid inequalities
Iteration in LNS		
From previous iteration	hints for some vars	initial values for all vars
Fixing machine assignment	fixing present vars	fixing binary vars
Fixing precedence	adding constraints	fixing binary vars

The organization of the paper is as follows. In Sect. 2, we describe the problem as a variant of the hybrid flow shop, and present a CP formulation. In Sect. 3, the IGC algorithm is presented. Sect. 4 shows experimental settings, computational results and analysis. Sect. 5 provides concluding remarks.

2 Problem Description

We consider a practical scheduling problem for the steel-making and continuous casting (SCC) process, which is known as a bottleneck process. The steel industry is one of the largest energy-consuming and pollution-generating industries in the world. Due to its importance, scheduling for the SCC process has been studied extensively in the last two decades [9,11,14]. We describe the SCC scheduling problem considered in [5] as a variant of the hybrid flow shop scheduling problem with additional constraints.

In the SCC scheduling problem, we consider a production unit called a *charge* that corresponds to a job. The stages consisting of the SCC process have a fixed order. A charge must be processed in the first and last stages, whereas it may not be processed in some or all of the stages in between. Each charge has its own predetermined route defined as a sequence of stages it has to go through. There are unrelated parallel machines in each stage, and when a charge is brought to the next stage, transportation time is required. If a charge is transported to the next stage but cannot start its processing immediately, a waiting time occurs and it cannot exceed the maximum waiting time limit. At the last stage, we consider another production unit called a *cast* which is defined as a predetermined sequence (serial batch) of charges, and each charge belongs to exactly one cast. Charges in a cast must be processed at the same machine in the last stage in a predetermined order without any idle time in between. A machine-dependent setup time takes place before processing the first charge of every cast. As objectives, we consider the minimization of a linear combination of the total waiting time, earliness, and tardiness. The waiting time is computed for each transportation of a charge between stages. Earliness and tardiness are defined for the last stage because each charge has a due date for its last operation. There should not be any idle times between charges in a cast; we incorporate these constraints as another objective term with a very large penalty called *cast break penalty*, which is common in many papers dealing with the SCC scheduling problem. As stated in [5], even the single-cast version of the described SCC scheduling problem is more general than the hybrid flow shop scheduling problem, and thus is NP-hard.

Table 1 shows the notations of the parameters and variables in the CP model. We assume that all parameters and variables have integral values. All parameters are adapted from [5] except the scheduling horizon H for the CP model. The formula of H in Table 1 expresses the makespan of a schedule where the operations of all charges are assigned to a common machine after time $\max_{k \in \Omega} d_k$ with their largest processing times and the transportation times plus the maximum waiting times. We assume that the setup times in the last stage are sufficiently small enough compared with the processing times. Preliminary experiments showed that using a small H value to reduce the domain of the variables significantly enhances the performance of the CP solver.

We also further elaborate on the notations used in our CP model and in our algorithm. An interval variable x is defined by three integer variables: starting time s, duration d, and ending time e. Hereafter, s and e are denoted as `startOf`(x) and `endOf`(x), respectively. An optional interval variable has an additional Boolean variable, denoted as `presenceOf`(x) that represents if the corresponding interval variable is present (`True`) or absent (`False`) within the

Table 1. Notations used in the constraint programming model

Parameters	
$\mathcal{S}$	The sequence of all stages, $\mathcal{S} = \{1, ..., l, ..., L\}$ where L is the last stage for continuous casting
J	The set of all casts, $J = \{1, ..., j, ..., m\}$ where m is the number of casts
Ω	The set of all charges, $\Omega = \{1, ..., k, ..., n\}$ where n is the number of charges
Ω_j	The sequence of charges in cast j, $\Omega_j := \{\Omega_j[1], \Omega_j[2], ..., \Omega_j[n_j]\}$ where n_j is the number of charges in cast j $(\forall j \in J)$
$\mathcal{S}_k$	The sequence of stages in charge k's route, $\mathcal{S}_k := \{\mathcal{S}_k[1], \mathcal{S}_k[2], ..., \mathcal{S}_k[c_k]\}$ where c_k is the number of stages in charge k's route $(\forall k \in \Omega)$ and $\mathcal{S}_k[1] = 1, \mathcal{S}_k[c_k] = L$
$\hat{\mathcal{S}}_k$	The set of pairs of two consecutive stages in the route of charge k, $\hat{\mathcal{S}}_k := \{(\mathcal{S}_k[\rho], \mathcal{S}_k[\rho+1]) : \rho \in \{1, 2, ..., c_k - 1\}\}$ $(\forall k \in \Omega)$
M_l	The set of machines at stage l $(\forall l \in \mathcal{S})$
p_{ik}	The processing time of charge k on machine i $(\forall k \in \Omega, i \in \bigcup_{l \in \mathcal{S}_k} M_l)$
$\tau_{ii'}$	The transportation time from machine i to i' $(\forall i, i' \in \bigcup_{l \in \mathcal{S}} M_l)$
r_{kl}	The earliest release time of charge k at stage l given as $r_{k1} := 0$ and $r_{kl'} := r_{kl} + \min_{i \in M_l, i' \in M_{l'}} \{p_{ik} + \tau_{ii'}\}$ $(\forall k \in \Omega, (l, l') \in \hat{\mathcal{S}}_k)$
s_{ij}	The setup time of cast j on machine i at stage L $(\forall j \in J, i \in M_L)$
d_k	The due date of charge k at stage L $(\forall k \in \Omega)$
$W_{\max}$	The maximum waiting time
π_1-π_4	Coefficients of penalty for (cast break / waiting time / earliness / tardiness)
H	An upper bound of the time horizon defined as follows $H := \max_{k \in \Omega} d_k + (\lvert\mathcal{S}\rvert - 1) \cdot \left(\max_{i, i' \in \bigcup_{l \in \mathcal{S}} M_l} \tau_{ii'} + W_{\max} \right) + \sum_{k \in \Omega} \sum_{l \in \mathcal{S}_k} \max_{i \in M_l} p_{ik}$
Variables and domains	
Chg_{ilk}	Optional interval variable with fixed duration p_{ik} representing charge k assigned to machine i at stage l $(\forall k \in \Omega, l \in \mathcal{S}_k, i \in M_l)$; $Chg_{ilk} \in [r_{kl}, H)$
Chg_{lk}	Interval variable for charge k assigned at stage l $(\forall k \in \Omega, l \in \mathcal{S}_k)$; $Chg_{lk} \in [r_{kl}, H)$
Cst_{ij}	Optional interval variable for cast j assigned to machine i at stage L $(\forall j \in J, i \in M_L)$; $Cst_{ij} \in [\max\{r_{\Omega_j[1]L} - s_{ij}, 0\}, H)$
Cst_j	Interval variable for cast j assigned at stage L $(\forall j \in J)$; $Cst_j \in [\max\{r_{\Omega_j[1]L} - \max_{i \in L} s_{ij}, 0\}, H)$
U_k	The idle time between charge k and its following charge in the same cast at stage L $(\forall k \in \Omega \setminus \cup_{j \in J}\{\Omega_j[n_j]\})$; $0 \leq U_k \leq H$
W_{kl}	The waiting time of charge k between stage l and its next stage l' in its route $(\forall k \in \Omega, (l, l') \in \hat{\mathcal{S}}_k)$; $0 \leq W_{kl} \leq W_{\max}$
$Tr_{kll'}$	The transportation time of charge k from stage l to the next stage l' in its route $(\forall k \in \Omega, (l, l') \in \hat{\mathcal{S}}_k)$; $\min_{i \in M_l, i' \in M_{l'}} \tau_{ii'} \leq Tr_{kll'} \leq \max_{i \in M_l, i' \in M_{l'}} \tau_{ii'}$
E_k	The earliness of charge k $(\forall k \in \Omega)$; $0 \leq E_k \leq \max\{d_k - r_{kL} - \min_{i \in M_L} p_{ik}, 0\}$
T_k	The tardiness of charge k $(\forall k \in \Omega)$; $0 \leq T_k \leq H - d_k$

schedule. For two Boolean variables b_1 and b_2, $b_1 \wedge b_2$ denotes the logical AND operator applied to b_1 and b_2. Function $\texttt{max}\{u, v\}$ denotes the maximum of two integer values u and v. For two constraints C_1 and C_2, we denote the conditional constraint "if C_1 (is satisfied) then C_2" as $C_1 \Rightarrow C_2$. We use constraint $\texttt{alternative}(x, \{y_1, ..., y_n\})$ with interval variables $x, y_1, ..., y_n$ in our model to ensure that exactly one interval out of $\{y_1, ..., y_n\}$ is present and has the same starting time and ending time as interval x. Constraint $\texttt{noOverlap}(\{y_1, ..., y_n\})$ ensures that the present intervals of $\{y_1, ..., y_n\}$ do not mutually overlap. Using Google OR-Tools CP-SAT solver, we can give *hints* when solving a CP. A *hint* is a (partial) solution that the solver uses as a warm start to create its initial feasible solution. It is allowed to give hints to some variables prior to solving a CP problem; $\texttt{addHint}(x, \hat{x})$ gives a variable x a value $\hat{x}$ as a hint.

The CP model is described with an objective (1) and constraints (2)–(14).

The CP model

minimize

$$\pi_1 \cdot \sum_{j \in J} \sum_{\kappa=1}^{n_j-1} U_{\Omega_j[\kappa]} + \pi_2 \cdot \sum_{k \in \Omega} \sum_{\rho=1}^{c_k-1} W_{k,\mathcal{S}_k[\rho]} + \pi_3 \cdot \sum_{k \in \Omega} E_k + \pi_4 \cdot \sum_{k \in \Omega} T_k \quad (1)$$

subject to

$$\texttt{alternative}(Chg_{lk}, \{Chg_{ilk}\}_{i \in M_l}) \qquad \forall k \in \Omega, l \in \mathcal{S}_k \quad (2)$$

$$\texttt{noOverlap}(\{Chg_{ilk}\}_{k \in \{k' :\ l \in \mathcal{S}_{k'}\}}) \qquad \forall l \in \mathcal{S}, i \in M_l \quad (3)$$

$$\texttt{alternative}(Cst_j, \{Cst_{ij}\}_{i \in M_L}) \qquad \forall j \in J \quad (4)$$

$$\texttt{noOverlap}(\{Cst_{ij}\}_{j \in J}) \qquad \forall i \in M_L \quad (5)$$

$$\texttt{presenceOf}(Chg_{iLk}) = \texttt{presenceOf}(Cst_{ij}) \qquad \forall j \in J, k \in \{\Omega_j[1], ..., \Omega_j[n_j]\}, i \in M_L \quad (6)$$

$$\texttt{startOf}(Chg_{l'k}) = \texttt{endOf}(Chg_{lk}) + W_{kl} + Tr_{kll'} \qquad \forall k \in \Omega, (l, l') \in \hat{\mathcal{S}}_k \quad (7)$$

$$\big(\texttt{presenceOf}(Chg_{ilk}) \wedge \texttt{presenceOf}(Chg_{il'k})\big) \Rightarrow (Tr_{kll'} = \tau_{ii'}) \qquad \forall k \in \Omega, (l, l') \in \hat{\mathcal{S}}_k, i \in M_l, i' \in M_{l'} \quad (8)$$

$$\texttt{presenceOf}(Cst_{ij}) \Rightarrow \big(\texttt{startOf}(Cst_j) = \texttt{startOf}(Chg_{L\Omega_j[1]}) - s_{ij}\big) \qquad \forall j \in J, i \in M_L \quad (9)$$

$$\texttt{presenceOf}(Cst_{ij}) \Rightarrow \big(\texttt{endOf}(Cst_j) = \texttt{endOf}(Chg_{L\Omega_j[n_j]})\big) \qquad \forall j \in J, i \in M_L \quad (10)$$

$$\texttt{endOf}(Chg_{L\Omega_j[\kappa]}) + U_{\Omega_j[\kappa]} = \texttt{startOf}(Chg_{L\Omega_j[\kappa+1]}) \qquad \forall j \in J, \kappa \in \{1, ..., n_j - 1\} \quad (11)$$

$$E_k = \texttt{max}\{d_k - \texttt{endOf}(Chg_{Lk}), 0\} \qquad \forall k \in \Omega \quad (12)$$

$$T_k = \texttt{max}\{\texttt{endOf}(Chg_{Lk}) - d_k, 0\} \qquad \forall k \in \Omega \quad (13)$$

$$T_k - E_k = \texttt{endOf}(Chg_{Lk}) - d_k \qquad \forall k \in \Omega \quad (14)$$

Objective (1) is the weighted sum of the total cast break penalty, total waiting time, earliness, and tardiness of charges. Constraints (2) ensure that a charge is present at exactly one machine in each stage along its route, and constraints (3) mean that the present charge intervals at each machine must not overlap. Constraints (4) and (5) are the analogies for each cast in the last stage. Constraints (6) ensure that a cast and its charges must be assigned to the same machine in the last stage. Constraints (7) define the waiting time and transportation time between two consecutive stages in a charge's route. Constraints (8) determine the machine-dependent transportation time of a charge between two stages in its route. Constraints (9) and (10) define the starting time and the ending time of a cast interval Cst_j as the starting time of the setup for cast j on the assigned machine and the ending time of the last charge in cast j, respectively. Constraints (11) define the idle times between consecutive charges of the same cast in the last stage. Fig. 1 illustrates an example of a cast interval and its charge intervals. Constraints (12) and (13) define the earliness and tardiness of each present charge, respectively. In addition, constraints (14) are valid equalities involving the tardiness and earliness of each present charge.

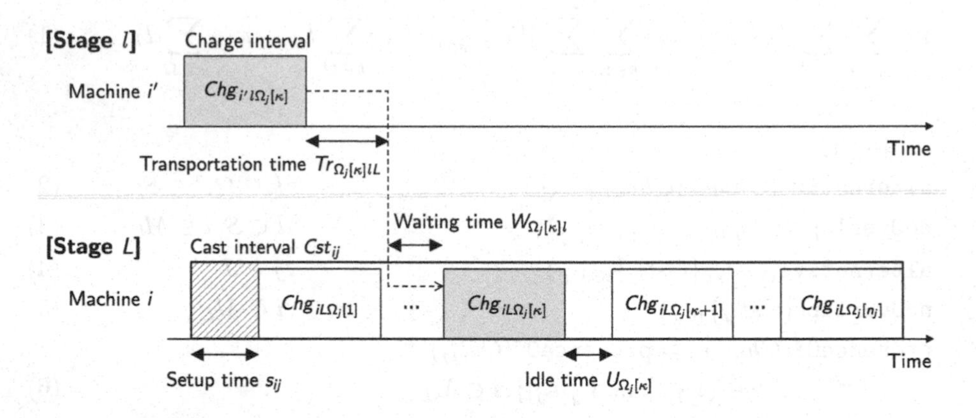

Fig. 1. Notations of a cast interval and its charge intervals

3 Iterated Greedy CP Algorithm

In this section, we describe our Iterated Greedy CP algorithm (IGC), a CP-based IG framework for solving the SCC problem in detail. Our IG framework consists of four procedures: LC, IH, DC, and CI. The LC procedure computes lower bounds for each single cast scheduling problem. The IH procedure finds an initial solution by a greedy heuristic. The DC procedure has two heuristics to iteratively improve the solution. Lastly, the CI procedure solves the entire CP model in search of a better solution than the incumbent solution. We refer to [5] for the original MIP version of the Iterated Greedy Matheuristic (IGM). The following notations are used throughout this section.

Notations used in IGC	
$\mathrm{CP}(I)$	The CP model in Sect. 2 with restricted set of casts $I \subseteq J$
master CP	The CP problem containing all casts, i.e., $\mathrm{CP}(J)$
σ	A partial or feasible solution (schedule) of a CP (sub)problem
$Z(\sigma)$	The objective function value of solution σ to a CP (sub)problem
$\mathcal{C}^{\mathrm{LB}}$	An ordered list of LB constraints for the objective terms in $\mathrm{CP}(J)$
$\hat{(\cdot)}^{\sigma}$	The value of a variable determined by a solution σ
$\mathcal{H}^{\sigma}$	A set of hints that are derived from solution σ
$\langle \mathcal{C}, \mathcal{H}, \overline{T} \rangle$	Control parameters in solving a CP (sub)problem; $\mathcal{C}$: a set of additional constraints, $\mathcal{H}$: a set of hints, and $\overline{T}$: a time limit

3.1 Lower Bound Computation

The first step of IGC is the Lower bound Computation (LC). We consider a single-cast scheduling problem for charges in a single cast, and the optimal objective value serves as a lower bound of the contribution of that single cast to the master problem's objective. Denoting σ_j as an optimal solution of $\mathrm{CP}(\{j\})$, the following constraint is valid.

$$\pi_2 \cdot \sum_{k \in \Omega_j} \sum_{\rho=1}^{c_k - 1} W_{k, \mathcal{S}_k[\rho]} + \pi_3 \cdot \sum_{k \in \Omega_j} E_k + \pi_4 \cdot \sum_{k \in \Omega_j} T_k \geq Z(\sigma_j) \tag{15}$$

This constraint for cast $j \in J$ is saved and used as an additional constraint throughout all procedures of IGC when solving a CP (sub)problem involving cast j. Since the starting time of cast j in σ_j may represent the desired starting time of cast j in the master CP, the casts are rearranged in nondecreasing order of $\texttt{startOf}(\widehat{Cst_j})^{\sigma_j}$ values. Algorithm 1 summarizes the LC procedure.

When we solve $\mathrm{CP}(\{j\})$, the upper bound of the time horizon H in the CP model is computed by considering only the charges in cast j as the entire charge set. That is, denoting H_j^0 as the H value for $\mathrm{CP}(\{j\})$, we have

$$H_j^0 := \max_{k \in \Omega_j} d_k + (|\mathcal{S}| - 1) \cdot \left(\max_{i,i' \in \cup_{l \in \mathcal{S}} M_l} \tau_{ii'} + W_{\max} \right) + \sum_{k \in \Omega_j} \sum_{l \in \mathcal{S}_k} \max_{i \in M_l} p_{ik} \tag{16}$$

In order to use a better upper bound on the time horizon for the subsequent procedures, we use the solutions σ_j ($\forall j \in J$) obtained in the LC procedure as follows. We define $R_j := \min_{k \in \Omega_j} \{\texttt{startOf}(\widehat{Chg_{lk}})^{\sigma_j} : l \in \mathcal{S}\}$ and $H_j := \max_{k \in \Omega_j} \{\texttt{endOf}(\widehat{Chg_{lk}})^{\sigma_j} : l \in \mathcal{S}\}$. Then, when we solve a $\mathrm{CP}(I)$ where $I \subseteq J$ (including the master CP) during the subsequent procedures, we take the H value as the minimum of its original value and $\max_{j \in I}\{R_j\} + \sum_{j \in I}(H_j - R_j)$.

Algorithm 1: Lower bound Computation (LC)

Input : A set of casts J
Output: $\mathcal{C}^{\text{LB}}$, a rearranged sequence of casts J'
begin
  $\mathcal{C}^{\text{LB}} \leftarrow [\ ]$ (an empty list);
  for j **in** J **do** $\{ \sigma_j \leftarrow$ Solve CP($\{j\}$); $\mathcal{C}^{\text{LB}}[j] \leftarrow (15);$ $\}$
  $J' \leftarrow$ Sort J in the increasing order of $\texttt{startOf}(\widehat{Cst_j})^{\sigma_j}$ for $j \in J$;
return $\mathcal{C}^{\text{LB}}, J'$

3.2 Initial Heuristic

Before describing the Initial Heuristic (IH) procedure, we define function RELOCATE described in Algorithm 2 that contains a key idea of IG framework. We will use RELOCATE in IH and DC heuristics described in the next subsection. Let σ be an input solution and Ω' be a set of charges which we desire to relocate (destruct and construct). RELOCATE(σ, Ω') returns the set of constraints that fix the machine assignment of each present charge in $\Omega \setminus \Omega'$ and the sequences of present charges in $\Omega \setminus \Omega'$ on each machine in solution σ. In Algorithm 2, constraint $\texttt{presenceOf}(Chg_{ilk}) = \texttt{True}$ fixes the machine assignment of a charge interval Chg_{ilk} before solving a CP problem, and constraint $\texttt{endOf}(Chg_{ilk}) \leq \texttt{startOf}(Chg_{ilk'})$ fixes the relative positions of two consecutive charge intervals Chg_{ilk} and $Chg_{ilk'}$ that are assigned to the same machine in σ. Note that IGC adds corresponding constraints using RELOCATE, while IGM fixes the corresponding variables.

We can also provide hints based on the current solution. For a given solution σ, we give the values $\texttt{startOf}(\widehat{Chg_{lk}})^{\sigma}$ and $\texttt{endOf}(\widehat{Chg_{lk}})^{\sigma}$ as hints for variables $\texttt{startOf}(Chg_{lk})$ and $\texttt{endOf}(Chg_{lk})$, respectively for all $k \in \Omega, l \in \mathcal{S}_k$. In addition, we give values $\texttt{presenceOf}(\widehat{Chg_{ilk}})^{\sigma}$ as hints to variables $\texttt{presenceOf}(Chg_{ilk})$ for all optional charge intervals. If $\texttt{presenceOf}(\widehat{Chg_{ilk}})^{\sigma}$ equals $\texttt{True}$, we also give the values $\texttt{startOf}(\widehat{Chg_{ilk}})^{\sigma}$ and $\texttt{endOf}(\widehat{Chg_{ilk}})^{\sigma}$ as hints for variables $\texttt{startOf}(Chg_{ilk})$ and $\texttt{endOf}(Chg_{ilk})$, respectively. For the rest of this paper, we denote the set of these hints as $\mathcal{H}^{\sigma}$.

In the Initial Heuristic (IH) procedure with a given ordered cast set J', we compute an initial schedule by solving a CP subproblem for the charges of the first cast in J'. Then, we add the charges in the next cast of J' to the output schedule by solving a corresponding CP subproblem and repeat the same with the next cast until we schedule all charges. Algorithm 3 describes the entire IH procedure.

When we schedule the charges in cast j, the input schedule σ is a partial schedule for charges of casts $1, \ldots, (j-1)$. Since we do not relax any charges in σ and maintain their machine assignments and relative positions, we use RELOCATE($\sigma, \varnothing$) to derive the constraints.

Algorithm 2: RELOCATE

Input : A partial solution σ, a set of charges $\Omega' \subseteq \Omega$
Output: A set of constraints $\mathcal{C}^{\text{fix}}$
begin
  $\mathcal{C}^{\text{fix}} \leftarrow \varnothing$;
  $\sigma \leftarrow$ the partial solution constructed from σ by removing the charges in Ω';
  for charge k **in** Ω **do**
    for stage l **in** $\mathcal{S}_k$ **do**
      for machine i **in** M_l **do**
        if $\texttt{presenceOf}(\widehat{Chg_{ilk}})^{\sigma} = \texttt{True}$ **then**
          $\mathcal{C}^{\text{fix}} \leftarrow \mathcal{C}^{\text{fix}} \cup \{\texttt{presenceOf}(Chg_{ilk}) = \texttt{True}\}$;
  for stage l **in** $\mathcal{S}$ **do**
    for machine i **in** M_l **do**
      $\{k_1, \ldots, k_\beta\} \leftarrow$ the sequence of charges on machine i in σ;
      for $\alpha = 1$ **to** $\beta - 1$ **do**
        $\mathcal{C}^{\text{fix}} \leftarrow \mathcal{C}^{\text{fix}} \cup \{\texttt{endOf}(Chg_{ilk_\alpha}) \leq \texttt{startOf}(Chg_{ilk_{\alpha+1}})\}$;
  return $\mathcal{C}^{\text{fix}}$

Algorithm 3: Initial Heuristic (IH)

Input : A sorted list of casts J', a sorted list of lower bound constraints $\mathcal{C}^{\text{LB}}$, a time limit $\overline{T}^{\text{IH}}$
Output: A feasible solution σ of the master CP
begin
  $\mathcal{C} \leftarrow \varnothing$, $\sigma \leftarrow \varnothing$, $\mathcal{H}^{\sigma} \leftarrow \varnothing$;
  for j **in** J' **do**
    $\mathcal{C} \leftarrow \mathcal{C} \cup \{\mathcal{C}^{\text{LB}}[j]\}$;
    $\mathcal{C}^{\text{fix}} \leftarrow$ RELOCATE$(\sigma, \varnothing)$;
    $\sigma \leftarrow$ Solve CP$(\{1, ..., j\})$ with $\langle \mathcal{C} \cup \mathcal{C}^{\text{fix}}, \mathcal{H}^{\sigma}, \overline{T}^{\text{IH}} \rangle$;
  return σ

3.3 Destruction and Construction Heuristics

The Destruction and Construction (DC) heuristics attempt to improve a feasible solution σ by relocating certain charges to better positions in the subsequent solution. This is implemented by iteratively solving CP subproblems subject to additional constraints.

Depending on how we define the set of charges to relocate, we use two types of DC heuristics: DC-cast (DA) and DC-charge (DH). DA selects the charges in a cast and does loops over the casts. On the other hand, DH chooses the charges completed within a time window and does loops over sliding windows. DH sequentially relocates the charges in the set of charge intervals that overlap with a time window in some stages. We repeatedly move the time windows forward by a given step size and redefine the set of charge intervals accordingly. Algorithm 4 and 5 illustrate the DA and DH, respectively.

Algorithm 4: DC-cast heuristic (DA)

Input : A set of casts J, a feasible solution σ, a sorted list of lower bound constraints $\mathcal{C}^{\mathrm{LB}}$, a time limit $\overline{T}^{\mathrm{DA}}$
Output: An improved solution σ^*
begin
  $J' \leftarrow$ Sort J in the increasing order of $\texttt{startOf}(\widehat{Cst_j})^{\sigma}$;
  $\sigma^* \leftarrow \sigma$;
  for j **in** J' **do**
    $\mathcal{C}^{\mathrm{fix}} \leftarrow \textsc{Relocate}(\sigma^*, \Omega_j)$;
    $\hat{\sigma} \leftarrow$ Solve master CP with $\langle \mathcal{C}^{\mathrm{LB}} \cup \mathcal{C}^{\mathrm{fix}}, \mathcal{H}^{\sigma^*}, \overline{T}^{\mathrm{DA}} \rangle$;
    if $\hat{\sigma}$ improves σ^* **then** $\sigma^* \leftarrow \hat{\sigma}$;
return σ^*

3.4 CP Improvement

The last procedure of IGC is the CP Improvement (CI). This is done by solving the master CP with the incumbent solution σ obtained so far and control parameters $\langle \mathcal{C}^{\mathrm{LB}}, \mathcal{H}^{\sigma}, \overline{T} \rangle$. We denote this procedure as $\mathrm{CI}(\sigma, \mathcal{C}^{\mathrm{LB}}, \overline{T})$.

3.5 Iterated Greedy CP

The previous procedures define the steps in the Iterated Greedy CP (IGC). Algorithm 6 describes the entire IGC algorithm. Parameters R^{DC}, R^{DA}, R^{DH} determine how many times we repeat the DC heuristics. Notice that when applying DA (or DH), if we do not obtain an improved solution within a given time limit, we stop repeating DA (or DH) and move to the next procedure.

Algorithm 5: DC-charge heuristic (DH)

Input : A feasible solution σ, a window duration D, a step size Δ, a sorted list of lower bound constraints $\mathcal{C}^{\mathrm{LB}}$, a time limit $\overline{T}^{\mathrm{DH}}$

Output: An improved solution σ^*

begin

$\sigma^* \leftarrow \sigma$;

for l **in** $\{1, L\}$ **do**

$\bar{S}_l(\sigma) \leftarrow \min_{k \in \Omega} \texttt{startOf}(Chg_{lk})$;

$\bar{C}_l(\sigma) \leftarrow \max_{k \in \Omega} \texttt{endOf}(Chg_{lk})$;

$\delta \leftarrow \min\left\{\frac{\bar{S}_L(\sigma)-\bar{S}_1(\sigma)}{L-1}, \frac{\bar{C}_L(\sigma)-\bar{C}_1(\sigma)}{L-1}\right\}$;

for l **in** $\mathcal{S}$ **do** $\{t^s_l \leftarrow (l-1)\delta + \bar{S}_1(\sigma); \quad t^e_l \leftarrow t^s_l + D;\}$

do

$\Omega^D \leftarrow \{k : k \in \Omega, \exists l \in S_k \text{ such that } \texttt{endOf}(\widehat{Chg_{lk}})^{\sigma^*} \in [t^s_l, t^e_l]\}$;

$\mathcal{C}^{\mathrm{fix}} \leftarrow \textsc{Relocate}(\sigma^*, \Omega^D)$;

$\hat{\sigma} \leftarrow$ Solve master CP with $\langle \mathcal{C}^{\mathrm{LB}} \cup \mathcal{C}^{\mathrm{fix}}, \mathcal{H}^{\sigma^*}, \overline{T}^{\mathrm{DH}} \rangle$;

if $\hat{\sigma}$ improves σ^* **then** $\sigma^* \leftarrow \hat{\sigma}$;

for l **in** $\mathcal{S}$ **do** $\{ t^s_l \leftarrow t^s_l + \Delta; \quad t^e_l \leftarrow t^e_l + \Delta; \}$

while Ω^D is not empty;

return σ^*

Algorithm 6: Iterated Greedy CP (IGC)

Input : A set of jobs J, a window duration D, a step size Δ, time limits $\overline{T}^{\mathrm{IH}}, \overline{T}^{\mathrm{DA}}, \overline{T}^{\mathrm{DH}}, \overline{T}^{\mathrm{IGC}}$, number of iterations $R^{\mathrm{DC}}, R^{\mathrm{DA}}, R^{\mathrm{DH}}$

Output: A feasible solution σ

begin

$\mathcal{C}^{\mathrm{LB}}, J' \leftarrow \mathrm{LC}(J)$;

$\sigma \leftarrow \mathrm{IH}(J', \mathcal{C}^{\mathrm{LB}}, \overline{T}^{\mathrm{IH}})$;

repeat R^{DC} **times**

repeat R^{DA} **times** $\sigma \leftarrow \mathrm{DA}(J, \sigma, \mathcal{C}^{\mathrm{LB}}, \overline{T}^{\mathrm{DA}})$ **until** *not improved*;

repeat R^{DH} **times** $\sigma \leftarrow \mathrm{DH}(\sigma, D, \Delta, \mathcal{C}^{\mathrm{LB}}, \overline{T}^{\mathrm{DH}})$ **until** *not improved*;

until *not improved*;

$\sigma \leftarrow \mathrm{CI}(\sigma, \mathcal{C}^{\mathrm{LB}}, \overline{T}^{\mathrm{IGC}} - \mathrm{ElapsedTime})$;

return σ

4 Experiments

4.1 Experimental Setting

Problem Instances. We use open problem instances which can be found from [4], which are classified into three groups according to the numbers of casts and charges: small, medium, and practical. Each group consists of 30 instances.

Size	No. of casts($\|J\|$)	No. of charges in a cast($\|\Omega_j\|$)	No. of charges($\|\Omega\|$)
Small	2–3	3–4	6–12
Medium	3–4	5–6	15–24
Practical	4–7	3–9	30–36

Regardless of the size, the same environment data are used as follows.

Type	Data	Values or distributions
Machine	L, M_l	$L = 5$, $(\|M_1\|, \|M_2\|, \|M_3\|, \|M_4\|, \|M_5\|) = (4, 2, 2, 2, 4)$
Charge	$\mathcal{S}_k$	The probability of skipping stages $= 2/3$
	p_{ik}	$U(45, 55)$ $i \in M_1$, $U(35, 45)$ $i \in M_L$, $U(30, 40)$ otherwise
	d_k	$U(100, 100 + 20n)$ where $n = \|\Omega\|$
Time		$\tau_{ii'} = 10$ min, $W_{\max} = 30$ min, and $s_{ij} = 30$ min
Objective	$\pi_{1\text{-}4}$	$(\pi_1, \pi_2, \pi_3, \pi_4) = (10^5, 1.5, 1, 1)$

Algorithm Parameters. The parameters of IGC and IGM are adopted from [5] as follows:

Module	Parameters
IH	$\overline{T}^{\text{IH}} = 60$ sec
DC	$R^{\text{DC}} = 4$, $R^{\text{DA}} = 2$, $R^{\text{DH}} = 1$, $\overline{T}^{\text{DA}} = 60$ sec, $\overline{T}^{\text{DH}} = 60$ sec
	$D = 90$ min, $\Delta = 45$ min
IGC (IGM)	$\overline{T}^{\text{IGC}} = \overline{T}^{\text{IGM}} = 600$ sec

Computation Environment. We compare six algorithms: IGC, the master CP (:= CP), IGM, the master MIP (:= MIP), the Non-dominated Sorting Genetic Algorithm II (:= NSGA-II), and a regular GA (:= GA). We reused the codes for MIP, IGM, NSGA-II, and GA developed by Hong et al. [5]; we refer to the paper for the details of these algorithms. For a fair comparison, all algorithms were tested with a single core in the same computation of environment with the following specifications.

Type	Computation environment
Software	Programming language: Python 3.10
	Solver: Gurobi 9.5.2 (MIP solver), OR-Tools 9.4 (CP solver)
Hardware	Desktop computer with Windows 10 OS,
	CPU: Intel Core i9-12900K @ 3.19GHz processor, RAM: 32GB

4.2 Experimental Results

Table 2 summarizes the computational results. We computed the optimality gap as $(Z - LB)/LB$ where Z is the best objective value found within the time limit, and LB is the best-known lower bound from IGC, CP, IGM, and MIP. Both IGC and CP obtained the optimal solutions within a very short time (2.0 sec., 1.1 sec. each) for all small instances while IGM and MIP took 73.2 sec., and 68.4 sec., respectively. In fact, IGC and CP optimally solved all 30 instances while IGM and MIP optimally solved 28 instances. NSGA-II and GA recorded the average gaps of 2.80% and 2.86% each. For the medium size problems, CP

Table 2. Performance comparison of algorithms

Size		IGC	IGM	CP	MIP	NSGA-II	GA
Small	Gap	**0.00%**	**0.00%**	**0.00%**	**0.00%**	2.80%	2.86%
		(0.00%)	(0.00%)	(0.00%)	(0.00%)	(2.91%)	(2.97%)
	Time	2.02	73.2	**1.10**	68.4	384.6	287.9
		(4.31)	(152.9)	(1.84)	(162.5)	(209.4)	(181.6)
	#opt	**30**	28	**30**	28	-	-
Medium	Gap	2.38%	2.87%	**2.22%**	5.75%	13.8%	13.0%
		(4.06%)	(4.20%)	(4.13%)	(6.61%)	(12.0%)	(10.6%)
	Time	281.4	600.1	**253.7**	600.2	600.2	600.2
		(274.7)	(0.1)	(272.5)	(0.1)	(0.1)	(0.1)
	#opt	18	0	**20**	0	-	-
Practical	Gap	4.87%	**4.84%**	5.16%	16.0%	34.3%†	37.1%†
		(3.57%)	(3.56%)	(3.98%)	(9.31%)	(11.9%)	(14.5%)
	Time	566.7	600.4	**562.1**	600.4	600.4	600.3
		(111.7)	(0.1)	(123.1)	(0.2)	(0.2)	(0.3)
	#opt	**4**	0	3	0	-	-

Gap: the average optimality gap
Time: the average elapsed time under 600 sec. time limit
#opt: the number of instances proved to be optimally solved
(stdev): the value in parenthesis corresponds to the standard deviation
†Excluded one instance that resulted a cast break

based algorithms (IGC and CP) outperform the other four algorithms in terms of the optimality gap as well as the computational time. Moreover, IGC and CP proved the optimality for 18, 20 instances each while IGM and MIP could not prove any. The performance of the genetic algorithms deteriorates significantly as the problem size becomes larger. For practical size problems, the algorithms with the IG framework (IGC and IGM) perform better than solving the master problem directly with either CP or MIP. We can conclude that IGC could be a practical alternative because the optimality gap of IGC is comparable to that of IGM and it proved the optimality for four instances while IGM could not prove any. Figure 2 shows optimality gaps of four algorithms (IGC, IGM, CP, and MIP) for all instances of medium and practical size. These results imply that IGC implemented with an open-source solver can be as powerful as IGM with a commercial solver even for large instances.

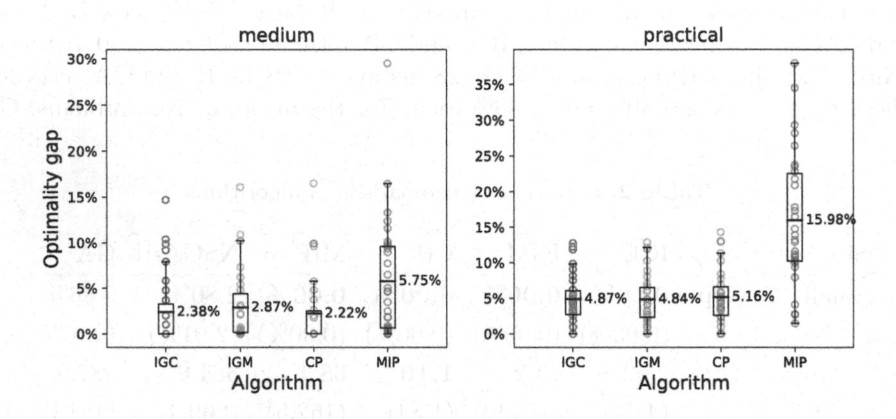

Fig. 2. The optimality gaps of four algorithms

In order to analyze the performance of each component of the IG framework, we plotted the average optimality gap and total elapsed time at each step in Fig. 3. 'DAθ' and 'DHθ' indicate the results obtained after the θ-th iteration of DA and DH. 'MI' or 'CI' shows the result obtained after MIP or CP improvement. Notably, the performance of IH under 50 sec. (see Fig. 3) is better than the final result of MIP (see Table 2), which shows the excellence of the IG framework. Compared to IGM, IH seems to run slightly slower in IGC, and DC heuristics are terminated quicker with less improvement. Therefore, the optimality gap of IGC is slightly higher than that of IGM before entering the CI step. On the other hand, IGC showed more improvement than IGM in the CI step while IGM showed little improvement in the MI step.

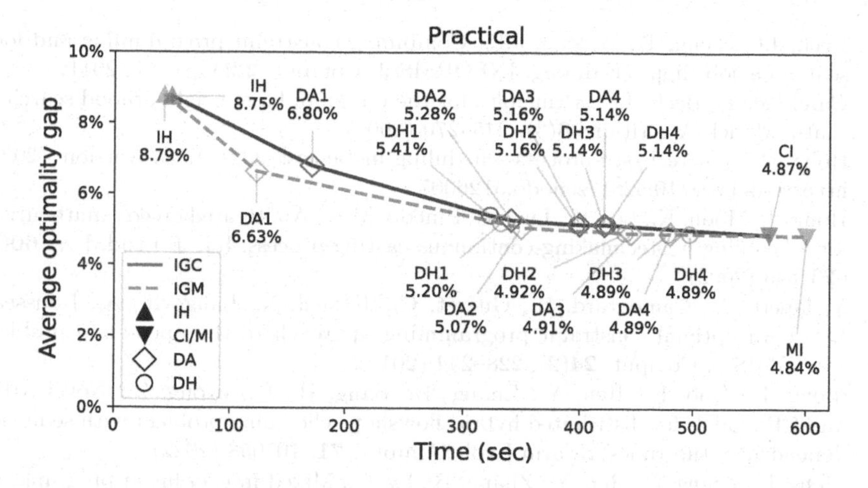

Fig. 3. The average optimality gap of IGC and IGM over steps and times

5 Concluding Remark

In this paper, we proposed a CP model for a practical scheduling problem of the steelmaking-continuous casting process. Based on our CP model, we proposed an iterative algorithm called IGC by adapting the IG framework of [5]. Our algorithm with an open-source CP solver outperforms the original IGM with a commercial MIP solver for small and medium size instances and shows a comparable performance for practical instances.

From the observation, we can consider various ways for improving IGC. Reducing the domain range of each variable in the CP model can be extremely helpful in the context of constraint programming. Hence, we can reduce the horizon using information achieved by solving subproblems in future works. We can also make our own propagation rules to expedite the solution process for the CP (sub)problems.

In the current experiments, we did not change any parameters regarding the algorithm design such as time limit and the number of loops in IGM. There are chances to improve the performance of IGC by tuning parameters. For instance, since the CI step in IGC showed more improvement than MI step in IGM, adjusting the parameters like $R^{DC}, R^{DA}, \overline{T}^{DA}, \overline{T}^{DH}$ would be helpful. A collaborative framework of IGC and IGM may be a promising future research direction and applications of our approach to new problems will also be interesting.

References

1. Abderrazzak, S., Hamid, A., Omar, S.: Adaptive large neighborhood search for the just-in-time job-shop scheduling problem. In: 2022 International Conference on Control, Automation and Diagnosis (ICCAD), pp. 1–6 (2022)

2. Beck, J.C., Feng, T., Watson, J.P.: Combining constraint programming and local search for job-shop scheduling. INFORMS J. Comput. **23**(1), 1–14 (2011)
3. Carchrae, T., Beck, J.C.: Principles for the design of large neighborhood search. J. Math. Model. Algorithms **8**(3), 245–270 (2009)
4. Hong, J.: junetech/scc-process-scheduling-instances: 2021-07-23 version (2021). https://doi.org/10.5281/zenodo.5126007
5. Hong, J., Moon, K., Lee, K., Lee, K., Pinedo, M.L.: An iterated greedy matheuristic for scheduling in steelmaking-continuous casting process. Int. J. Prod. Res. **60**(2), 623–643 (2022)
6. Malapert, A., Cambazard, H., Guéret, C., Jussien, N., Langevin, A., Rousseau, L.M.: An optimal constraint programming approach to the open-shop problem. INFORMS J. Comput. **24**(2), 228–244 (2012)
7. Meng, L., Gao, K., Ren, Y., Zhang, B., Sang, H., Chaoyong, Z.: Novel MILP and CP models for distributed hybrid flowshop scheduling problem with sequence-dependent setup times. Swarm Evol. Comput. **71**, 101058 (2022)
8. Meng, L., Zhang, C., Ren, Y., Zhang, B., Lv, C.: Mixed-integer linear programming and constraint programming formulations for solving distributed flexible job shop scheduling problem. Comput. Ind. Eng. **142**, 106347 (2020)
9. Missbauer, H., Hauber, W., Stadler, W.: A scheduling system for the steelmaking-continuous casting process. A case study from the steel-making industry. Int. J. Prod. Res. **47**(15), 4147–4172 (2009)
10. Pacino, D., Van Hentenryck, P.: Large neighborhood search and adaptive randomized decompositions for flexible jobshop scheduling. In: Proceedings of the Twenty-Second International Joint Conference on Artificial Intelligence - Volume Three, Barcelona, Catalonia, Spain, pp. 1997–2002. AAAI Press (2011)
11. Pan, Q.K., Wang, L., Mao, K., Zhao, J., Zhang, M.: An effective artificial bee colony algorithm for a real-world hybrid flowshop problem in steelmaking process. IEEE Trans. Autom. Sci. Eng. **10**(2), 307–322 (2013)
12. Pisinger, D., Ropke, S.: Large neighborhood search. In: Gendreau, M., Potvin, J.Y. (eds.) Handbook of Metaheuristics, pp. 399–419. Springer, Boston (2010). https://doi.org/10.1007/978-1-4419-1665-5_13
13. Shaw, P.: Using constraint programming and local search methods to solve vehicle routing problems. In: Maher, M., Puget, J.-F. (eds.) CP 1998. LNCS, vol. 1520, pp. 417–431. Springer, Heidelberg (1998). https://doi.org/10.1007/3-540-49481-2_30
14. Tang, L., Luh, P.B., Liu, J., Fang, L.: Steel-making process scheduling using Lagrangian relaxation. Int. J. Prod. Res. **40**(1), 55–70 (2002)

Combining Incomplete Search and Clause Generation: An Application to the Orienteering Problems with Time Windows

Trong-Hieu Tran[1,2,3](✉), Cédric Pralet[2,3], and Hélène Fargier[1,3]

[1] IRIT-CNRS, Toulouse, France
[2] ONERA/DTIS, Toulouse, France
trantronghieu97@gmail.com
[3] Université de Toulouse, Toulouse, France

Abstract. In this paper, we present a hybrid optimization architecture which combines on one side incomplete search processes that are often used to quickly find good-quality solutions to large-size problems, and on the other side clause generation techniques that are known to be efficient to boost systematic search. In this architecture, clauses are generated once a locally optimal solution is found. We introduce a generic component to store these clauses generated step-by-step. This component is able to prune neighborhoods by answering queries formulated by the incomplete search process. We define three versions of this clause basis manager and then experiment with an Operations Research problem known as the Orienteering Problem with Time Windows (OPTW) to show the efficiency of the approach.

Keywords: Incomplete search · Clause generation · Orienteering Problem with Time Windows

1 Introduction

Incomplete search methods are often used on large-size problems to quickly produce good-quality solutions. Such methods include *heuristic search*, where a solution is progressively built based on efficient heuristics, *local search*, where various neighborhoods help improve the current solution, and *metaheuristics* like tabu search, genetic algorithms, or iterated local search, to name just a few. To increase the performance of these incomplete methods, several hybridizations with complete search techniques developed for SAT and Constraint Programming (CP) have been proposed in the past [24], and there is rich literature on the topic both in terms of methods and applications.

In this paper, we study a new architecture combining incomplete search and SAT techniques. This architecture, which is given in Fig. 1, is inspired by the efficient complete search methods based on clause generation, namely CDCL [1,18]

© The Author(s), under exclusive license to Springer Nature Switzerland AG 2023
A. A. Cire (Ed.): CPAIOR 2023, LNCS 13884, pp. 493–509, 2023.
https://doi.org/10.1007/978-3-031-33271-5_32

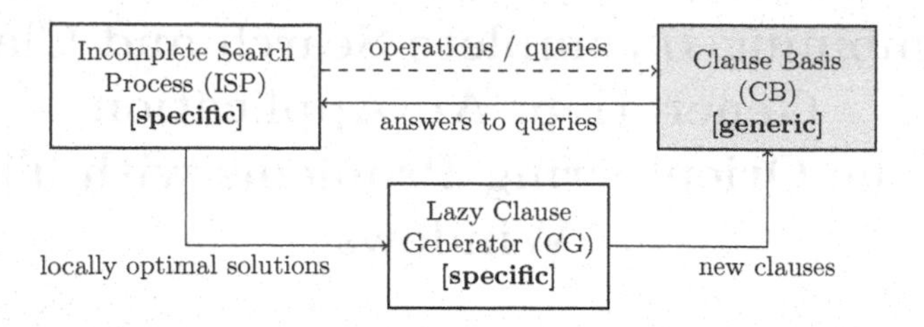

Fig. 1. Incomplete search combined with a clause basis

and Lazy Clause Generation (LCG [28]). The global search scheme works as follows. Each time the Incomplete Search Process (ISP) converges to a locally optimal solution, a *Clause Generator* (CG) analyzes this solution and produces clauses holding on Boolean decision variables of the problem. The clauses generated represent either the reasons why the current solution cannot be improved or conditions forbidding the local optimum or regions around to be reached again in the future. The clauses generated are then sent to a *Clause Basis* (CB). The latter is responsible for storing the clauses and answering various queries that are relevant for the main ISP to prune or to guide the neighborhood exploration. In this architecture, the clauses are generated in a lazy way, only for the parts of the search space that the ISP decides to explore. By doing so, the architecture involves a tight interaction between ISP and CB as well as a less frequent clause generation phase.

With regards to tabu search [9], the architecture obtained memorizes clauses instead of just storing recent local moves or recent solutions in a tabu list. One impact is that the clause manager must be able to quickly reason about the clauses collected, instead of just reading explicitly forbidden configurations. Concerning CDCL or LCG, one key difference is that the ISP is free to assign or unassign variables *in any order*, while the standard *implication graph* data structure used by CDCL or LCG relies on the assumption that the variable ordering in different layers of the graph is consistent with the order used for assigning and unassigning the decision variables. All these points raise several basic research questions:

- Which generic clause basis data structure should be used to be able to follow the decisions made in any order by an incomplete search process and to quickly reason about the set of clauses memorized?
- What is the effort required to integrate an existing specific ISP within such a generic architecture, and which key functions should the clause basis offer?
- What is the content of the clause generation module that analyzes the locally optimal solutions?
- From a practical point of view, is clause generation beneficial for an incomplete search that might have to explore thousands of successive neighborhoods per second?

To answer these questions, all along the paper, we take as an example a problem known as the Orienteering Problem with Time Windows (OPTW [32]). The paper is organized as follows: first, we recall the definition of OPTW and present a hybrid algorithm that combines a state-of-the-art search algorithm for OPTW with a clause basis. Following this, we introduce clause generation mechanisms and three data structures for the clause basis. We present experimental results obtained on OPTW benchmarks to demonstrate that the architecture proposed allows boosting the baseline ISP, while the integration effort required is rather small. Finally, we compare our approach with relevant works in the literature.

2 Orienteering Problem with Time Windows

The OPTW belongs to the class of vehicle routing problems with profits, where a vehicle has a limited time budget to visit a subset of nodes. Formally, we consider a set of nodes $i \in \{0, 1, \ldots, N+1\}$, each with a reward R_i and a predefined time window $[e_i, l_i]$. Nodes 0 and $N+1$ correspond to the start and end depots, with $R_0 = R_{N+1} = 0$ and a time window $[0, T_{max}]$ where T_{max} is the limited time budget. A non-negative travel time t_{ij} is associated with each pair of nodes $i \neq j$. A visit duration d_i can also be considered for each node, but to keep it simple, we assume that the visit duration is already included in the travel time. A solution is a sequence $\sigma = [\sigma_0, \sigma_1, \ldots, \sigma_K, \sigma_{K+1}]$ that starts at node $\sigma_0 = 0$, visits a set of distinct nodes $\sigma_i \in \{1, \ldots, N\}$, and returns to node $\sigma_{K+1} = N+1$. Early arrival to a particular node leads to a waiting time, and a solution is feasible when it visits each selected node before its latest arrival time. More precisely, the visit start time of node 0 is $s_0 = 0$, and for two consecutive nodes i, j in σ the visit start time of node j is $s_j = \max(s_i + t_{ij}, e_j)$, and solution σ is feasible if and only if $s_j \leq l_j$ for every node j in σ. Next, an optimal solution is a feasible solution σ that maximizes the total reward ($\sum_{i \in \sigma} R_i$). Basically, an OPTW involves both selection and sequencing decisions, i.e. selection of a subset of nodes S and search for a feasible visit order σ for these nodes. Regarding the selection aspect, we introduce one Boolean decision variable $x_i \in \{0, 1\}$ per node i, where $x_i = 1$ means that node i is visited.

Challenging applications were modeled as OPTW in the past, such as delivery problems, satellite planning problems, or tourist trip design problems [11,32]. Since OPTW is proven as NP-hard [10], many researches on the topic rely on incomplete search. One first investigation in this direction was a tree heuristic for building a single tour without violating the time windows constraints [16]. Incremental neighborhood evaluation methods were also introduced to quickly determine the feasible node insertion moves given a current solution [29,31]. Later, [27] proposed an effective Large Neighborhood Search (LNS) strategy that was shown to outperform the previous approaches. The basic idea is to iteratively remove and reinsert nodes based on well-tuned removal and insertion heuristics, and to use restarts from a pool of elite solutions.

3 Incomplete Search Using a Clause Basis

In the rest of the article, we integrate the state-of-the-art LNS algorithm for OPTW defined in [27] within the hybrid architecture proposed. The enhanced version, called *LNS-CB* for *LNS with a Clause Basis*, is depicted in Algorithm 1, where the few changes made on the baseline LNS version are highlighted in gray. Starting from an initial solution (Line 1), it iteratively destroys and repairs the current solution following the standard concept of LNS (Lines 5–6). It also uses an elite pool to record the best solutions obtained so far. This pool is reset whenever a better solution is found, and extended when a new equivalent solution is obtained (Line 9). When no improvement is found after R iterations, a restart is performed by picking a random solution in the elite pool (Line 11). The differences compared to the classical LNS algorithm are (a) the call to the clause generation function each time a full solution is produced (Lines 2, 7), and (b) the use of the CB as an argument to the repair function, the objective being to improve the repair phase (Line 6).

Algorithm 1. LNS-CB

```
1: σ ← CONSTRUCT()
2: CLAUSEGENERATION(σ, CB, maxConfSize)
3: σ* ← σ;  elitePool ← {σ}
4: while time limit is not reached do
5:     σ ← DESTROY(σ);
6:     σ ← REPAIR(σ, CB)
7:     CLAUSEGENERATION(σ, CB, maxConfSize)
8:     if σ better than σ* then
9:         σ* ← σ;  update elitePool
10:    else if no improvement after R iterations then
11:        σ ← a random solution in elitePool
12:    end if
13: end while
14: return σ*
```

The new repair phase is detailed in Algorithm 2. It takes as an input the current solution σ and the CB. We denote U as the set of unvisited nodes, and F as the set of feasible insertion moves (n, p) defined by a node $n \in U$ and a position p in σ. All insertion alternatives for each unvisited node are explored by EVALNEIGHBORHOOD(σ,U, CB) (Lines 2, 7). In this procedure, CB is used to prune neighbors that are invalid according to the clauses registered. Node insertions are iterated by selecting at each step a move that has the best evaluation according to the well-tuned heuristics of the original LNS method (Line 4), and they are performed until there is no more feasible move (Line 3).

The neighborhood evaluation function corresponds to Algorithm 3. It first determines the unvisited nodes that *must* be visited according to CB (Line 1), and if there is no such mandatory node, it determines the unvisited nodes that *can* be visited according to CB (Line 3). Then, for each node selected, the algorithm determines its best insertion position according to tuned insertion heuristics and the algorithm returns all pairs made by a node and its best insertion position.

Algorithm 2. REPAIR(σ,CB)

1: $U \leftarrow$ nodes that are not in σ
2: $F \leftarrow$ EVALNEIGHBORHOOD(σ, U, CB)
3: **while** $F \neq \emptyset$ **do**
4: $(n^*, p^*) \leftarrow$ SELECT(F)
5: Insert node n^* at position p^* in σ
6: $U \leftarrow U \setminus \{n^*\}$
7: $F \leftarrow$ EVALNEIGHBORHOOD(σ, U, CB)
8: **end while**
9: **return** σ

Algorithm 3. EVALNEIGHBORHOOD(σ,U,CB)

1: $U' \leftarrow \{n \in U \mid CB$ allows decision $[x_n = 1]$ and forbids decision $[x_n = 0]\}$
2: **if** $U' = \emptyset$ **then**
3: $U' \leftarrow \{n \in U \mid CB$ allows decision $[x_n = 1]\}$
4: **end if**
5: $F \leftarrow \emptyset$
6: **for each** $n \in U'$ **do**
7: $p \leftarrow$ best feasible insertion position for n in σ
8: **if** $p \neq nil$ **then**
9: $F \leftarrow F \cup \{(n, p)\}$
10: **end if**
11: **end for**
12: **return** F

4 Lazy Clause Generation Module

Several kinds of clauses are generated during the search, and the generation of these clauses exploits problem-dependent techniques, as for cuts generated in Logic-Based Benders decomposition [14]. Note that for OPTW, we consider only clauses holding over the selection decisions, and not clauses related to the detailed sequencing decisions defining the order of the visits.

4.1 Clauses Generated from Time-Window Conflicts

A Time-Window conflict (TW-conflict) is a subset $S_c \subseteq [1..N]$ such that there is no feasible solution visiting all nodes in S_c. In terms of clause generation, a TW-conflict S_c corresponds to clause $\vee_{i \in S_c} \neg x_i$. Due to the exponential number of possible sets S_c, we generate TW-conflicts in a lazy way i.e. only when a local optimum is reached. Moreover, determining whether S_c defines a TW-conflict is NP-hard [26], but it is an easy problem if $|S_c|$ is bounded. This is why we consider a predefined maximum TW-conflict size referred to as *maxConfSize*.

Technically, whenever a locally optimal sequence σ^* is found over nodes in S^*, we seek TW-conflicts preventing the other nodes from being added to σ^*. In Algorithm 4, we try to find explanations for every unvisited node i (Line 1). With a predefined *maxConfSize*, the algorithm first heuristically selects a

Algorithm 4. CLAUSEGENERATION(σ^*,CB,*maxConfSize*)

1: **for** $i \notin \sigma^*$ **do**
2: $S_c \leftarrow$ SELECT(σ^*, i, *maxConfSize*)
3: $\mathcal{C} \leftarrow$ EXTRACTMINTWCONFLICTS($S_c \cup \{i\}$)
4: **for** each TW-conflict $C \in \mathcal{C}$ **do**
5: Generate clause $\bigvee_{j \in C} \neg x_j$
6: **end for**
7: **end for**
8: [optional] Generate a temporary clause $\bigvee_{j \in Y} x_j$ with Y a subset of the nodes that are not selected in σ^*

set $S_c \subset S^* \setminus \{0, N+1\}$ containing *maxConfSize* $- 1$ nodes in σ^* that *might* prevent node i from being visited. Then, in function EXTRACTMINTWCONFLICTS, a dynamic programming (DP) procedure determines whether $S_c \cup \{i\}$ is truly a TW-conflict. If so, it also extracts TW-conflicts of minimal cardinality (Line 3). Indeed, the smaller the clauses the better, since smaller clauses prune larger parts of the search space. Function EXTRACTMINTWCONFLICTS takes as an input a set of nodes $S \subseteq [1..N]$ and determines all minimal sets $S' \subseteq S$ (minimal in terms of cardinality) such that there is no feasible solution visiting all nodes in S'. For space limitation reasons, we do not detail the pseudo-code of EXTRACTMINTWCONFLICTS, but the key idea is to compute, for each set $C \subseteq S$, quantities of the form $a(C, i)$ representing the earliest arrival time at node $i \in C$ for a path starting at node 0, visiting all nodes in $C \setminus \{i\}$, and ending at node i. As in existing methods for Traveling Salesman Problems with Time Windows [3], these quantities are computed by increasing the size of C following a recursive formula. Then, $C \subseteq S$ is a TW-conflict when for every $i \in C$, either $a(C, i) > l_i$ or $\max\{a(C, i), e_i\} + t_{i,N+1} > T_{max}$. The first condition corresponds to late arrivals for every candidate last node i, while the second one corresponds to the violation of the time limit when returning to node $N + 1$.

4.2 Clauses Related to Local Optima: Lopt-Conflicts

To avoid revisiting again and again the same solution, whenever reaching a locally optimal solution σ^*, it is possible to generate clause $\bigvee_{j \notin \sigma^*} x_j$ to force that at least one node unvisited in σ^* must be selected in the future. Such a clause is called a local optimum conflict or *Lopt-conflict*. To get small clauses that have a higher pruning power, we consider a maximum clause size *approxSize* and derive an *approximate* Lopt-conflict corresponding to a smaller clause $\bigvee_{j \in Y} x_j$ where Y contains at most *approxSize* nodes that are not involved in σ^* and that are chosen in function of their rewards (Algorithm 4, Line 8). To avoid pruning optimal solutions, such approximate clauses are used by CB only during a certain number of steps called *tabuSize*, similarly to a tabu search procedure, the main objective being to diversify search. We could also generate Pseudo-Boolean constraint $\sum_{i \in \{1,\ldots,N\}} R_i x_i \geq LB + 1$ whenever a new best total reward LB is found, but we focus here on clause generation.

5 Clause Basis Data Structures

The CB part is responsible for storing the clauses generated during search. Besides, the ISP needs to frequently query the clause basis, meaning that there is a need for *continuous* and *incremental* interactions between these two components. This raises many challenging questions about the choice of a specific data structure for CB. In principle, a clause basis manager must be able to:

- quickly integrate all the clauses generated step-by-step and compactly represent them (possibly with some trashing when the size of CB becomes too large);
- frequently update the partial assignment of the decision variables over which the clauses hold, to keep up-to-date knowledge of the content of the current solution considered by the main ISP. For LNS-CB, this occurs whenever a node is selected or removed, and these assign/unassign decisions can be sent to CB in any order;
- quickly answer to queries formulated by ISP, such as "evaluate whether decision $[x_i = 1]$ is feasible". For OPTW, if CB proves that this decision is infeasible given the current assignment and the clauses generated, then testing the insertion of node i in the current solution σ is unnecessary (neighborhood pruning). Another example is: "evaluate whether decision $[x_i = 1]$ is mandatory". If so, node i *must* be inserted into σ.

In the following we study three generic versions for CB:

- CB-UnitPropagation, where CB stores a list of clauses and performs incremental and decremental unit propagation to *evaluate* the consistency of the clause store for a given partial assignment of the x_i variables;
- CB-IncrementalSAT, where CB stores a list of clauses and employs powerful modern SAT solvers supporting incremental or assumption-based solving [2,7,21];
- CB-OBDD, where the clauses are stored in an Ordered Binary Decision Diagram (OBDD), a data structure defined in the field of knowledge compilation that has good compactness and efficiency properties [4,6].

5.1 CB-UnitPropagation

For this version of CB, unit propagation is used to prune infeasible values for the decision variables. In SAT, unit propagation can be achieved based on the *two-watched literals* technique, which consists in maintaining, in each clause, two distinct literals that can take value *true* [20]. In case there is no valid watched literal for a clause c, an inconsistency is detected. If only a single valid watched literal l is found, then clause c becomes unit and l must necessarily be *true* to satisfy the clause. In this case, literal $\neg l$ takes value *false* and unit propagation is applied to other clauses to further detect other propagated decisions.

In SAT, one advantage of the watched literals is that no literal reference needs to be updated when chronological backtracking occurs. But during incomplete

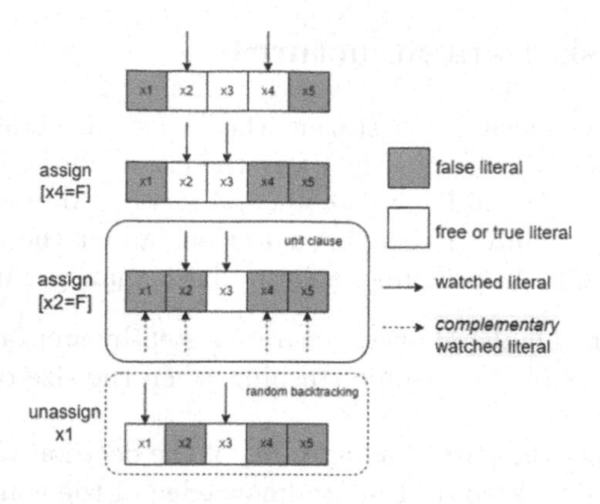

Fig. 2. Incremental and decremental unit propagation

search, variables can be assigned or unassigned in any order and some adaptations are required to maintain the watched literals. Precisely, to handle random variable unassignments and perform *decremental* unit propagation, we maintain a list of *complementary* watched literals *for each unit clause* c (see Fig. 2). Clause c is revised whenever one complementary watched literal l' becomes free due to unassignment decisions, and in this case l' can directly become a watched literal for c.

To answer the queries formulated by the ISP, we record a *justification* $justif(l)$ for each literal l. Basically, $justif(l) = \top$ means that literal l takes value true because of a decision received from the ISP, $justif(l) = c$ means that literal l is propagated by unit clause c, and $justif(l) = nil$ means that there is no clue about the truth value of l. Then, a decision like $[x = 1]$ is allowed if and only if literal $\neg x$ is not propagated or decided yet, i.e. $justif(\neg x) = nil$. The justification of each literal is updated during incremental and decremental unit propagation. Obviously, as unit propagation is incomplete, CB-UnitPropagation may not detect some infeasible or mandatory node selections. For example, let us consider four clauses $c_1 : \neg x_1 \lor \neg x_2$, $c_2 : \neg x_4 \lor \neg x_5$, $c_3 : x_2 \lor x_3 \lor x_4$, $c_4 : x_2 \lor x_3 \lor x_5$. If decision $[x_1 = 1]$ is made, clause c_1 becomes unit and we have $justif(x_1) = \top$ and $justif(\neg x_2) = c_1$. The other justifications take value *nil*. This implies that decision $[x_3 = 0]$ is still evaluated as possible, even if it would lead to a dead-end.

5.2 CB-IncrementalSAT

The idea of using incremental SAT solving was first proposed to improve the efficiency of the search for Minimal Unsatisfiable Sets [2]. In this case, the goal is to reuse as much information as possible between the successive resolutions of similar SAT problems. This is done by working with *assumptions*. Basically, an assumption $\mathcal{A}$ is a set of literals $\{l_1, \ldots, l_k\}$ where each literal is considered as an additional (unit) clause by the solver, but this unit clause is not permanently added to the original CNF formula $\mathcal{F}$ defining the problem to be solved. For

OPTW, the assumptions are exactly the node selection decisions. Then, a call $solve(\mathcal{F}, \mathcal{A})$ to an incremental SAT solver tries to find a model of $\mathcal{F}$ that satisfies all the assumptions in $\mathcal{A}$. Doing this, the incremental solver can reuse some previous knowledge and learn new clauses that will potentially be reused for future calls $solve(\mathcal{F}', \mathcal{A}')$ using an updated CNF formula $\mathcal{F}'$ or an updated set of assumptions $\mathcal{A}'$.

At the level of CB, to determine whether literal $l : [x_i = a]$ can still be assigned value true, it suffices to call $solve(\mathcal{F}, \mathcal{A} \cup \{l\})$ where $\mathcal{A}$ is the set of assumptions representing the selection decisions made so far by the search engine. Then, decision $[x_i = a]$ is forbidden by CB if and only if this call returns UNSAT. Contrarily to CB-UnitPropagation, the CB-IncrementalSat method is complete (it performs a full look ahead). One optimization allows us to reduce the number of calls to the *solve* function: when searching for the possible values of variable x_i given a set of assumptions $\mathcal{A}$, if $solve(\mathcal{F}, \mathcal{A} \cup \{x_i\})$ or $solve(\mathcal{F}, \mathcal{A} \cup \{\neg x_i\})$ returns a solution where another variable x_j takes value 1, then $x_j = 1$ is allowed and calling $solve(\mathcal{F}, \mathcal{A} \cup \{x_j\})$ is unnecessary.

5.3 CB-OBDD

Storing conflict clauses in an OBDD during a systematic search process has been explored in the past, e.g. for a search process based on DPLL [15]. We extend such an approach to deal with an incomplete search engine that again can assign/unassign the decision variables of the problem in any order.

As illustrated in Fig. 3, an OBDD defined over a set of Boolean variables X is a directed acyclic graph composed of one root node, two leaf nodes labeled by $\top$ and $\bot$, and non-leaf nodes labeled by a decision variable $x_i \in X$. Each node associated with variable x_i has two outgoing arcs corresponding to assignments $[x_i = 0]$ and $[x_i = 1]$ respectively (dotted and plain arcs in the figure). The paths from the root node to leaf node $\top$ correspond to the assignments that satisfy the logical formula represented by the OBDD, while the paths leading to leaf node $\bot$ correspond to the inconsistent assignments. Additionally, OBDDs are *ordered*, meaning that the variables always appear in the same order in any path from the root to the leaves. In practice, they are also *reduced*, meaning that redundant nodes (that have the same children) are recursively merged to save some space. Such a data structure offers several advantages, including the capacity to be exponentially more compact than an explicit representation of all models of a logical formula, and the capacity to perform several operations and answer several queries in polynomial time. For instance, given two OBDDs $\mathcal{O}_F$ and $\mathcal{O}_G$ representing logical formulas f and g and that use the same variable ordering, operation "$\mathcal{O}_F \wedge \mathcal{O}_G$" computes an OBDD representing $f \wedge g$ in polynomial time in the number of nodes in $\mathcal{O}_F$ and $\mathcal{O}_G$.

In CB-OBDD, one OBDD referred to as $\mathcal{O}_{CB}$ stores the clauses learned during search. Initially, $\mathcal{O}_{CB}$ only contains the leaf node $\top$ since all models are accepted. Each generated clause c_k can be transformed into an OBDD $\mathcal{O}_{c_k}$, and a set of new clauses $\{c_1, \ldots, c_n\}$ is added to $\mathcal{O}_{CB}$ by $\mathcal{O}_{CB} \leftarrow [\mathcal{O}_{c_1} \wedge \ldots \wedge \mathcal{O}_{c_n}] \wedge \mathcal{O}_{CB}$ (conjunction of the elementary OBDDs associated with the new clauses

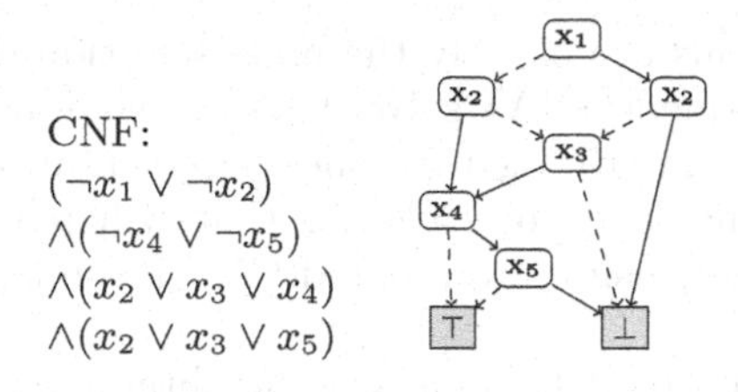

Fig. 3. A conjunction of clauses and an equivalent OBDD

followed by a batch addition into $\mathcal{O}_{CB}$). During search, CB-OBDD records the current list of assignments A_{CB} made by the incomplete search process (the assumptions). To determine whether a decision $[x = 1]$ is allowed, it suffices to *condition* $\mathcal{O}_{CB}$ by A_{CB}, and then to check that assignment $x = 0$ is not *essential* (not mandatory) for the resulting OBDD. The *conditioning primitive* and the search for *essential variables* are standard operations in OBDD packages. Their time complexity is linear in the number of OBDD nodes.

6 Computational Study

We carried out experiments on standard OPTW benchmarks[1] whose features are summarized in Table 1. The best known total reward for each instance is retrieved from [27]. All the experiments are performed on Intel(R) Core(TM) i5-8265U 1.60 GHz processors with 32 GB RAM. All implementations[2] are in C++ and compiled in a Linux environment with g++17.

Table 1. Features of the OPTW benchmarks

Instance Set	#instances	#nodes	remark
Solomon 1 (c1*, r1*, rc1*)	29	100	–
Solomon 2 (c2*, r2*, rc2*)	27	100	wider TW
Cordeau 1 (pr01-pr10)	10	48–288	–
Cordeau 2 (pr11-pr20)	10	48–288	wider TW

As the implementation of the state-of-the-art LNS algorithm [27] is not available online, we re-implemented it. We recover a similar performance even if there are some differences wrt. the results provided in the original paper, possibly due to random seeds or to a lack of information concerning a reset parameter R (we set $R = 50$ in our LNS implementation). Anyway, our primary objective was to determine whether conflict generation can help a baseline algorithm, therefore the slight differences in performance are not a real issue. The three CB proposed were implemented as follows:

[1] https://www.mech.kuleuven.be/en/cib/op.
[2] Github URL of the source code: https://github.com/thtran97/kb_ls_cpp.

- The CB-UnitPropagation data structure was implemented from scratch.
- For CB-IncrementalSat, we reused CryptoMiniSat[3] [30] that won the Incremental Track in the SAT competition 2020.
- For CB-OBDD, we reused the CUDD library that offers many functions to manage OBDDs.[4] CB-OBDD uses the dynamic reordering operations of CUDD [25]. Dynamic reordering can take some time but reducing the size of OBDDs can pay off in the long term.

6.1 Parameter Settings for clauseGeneration

In the hybrid optimization architecture proposed, the clauseGeneration procedure is problem-specific. For OPTW, we observed that the length of time windows has a large impact on the number of TW-conflicts generated for a given value of *maxConfSize*: many TW-conflicts are generated for the Solomon 1 & Cordeau 1 instances, contrarily to the Solomon 2 & Cordeau 2 instances that involve longer time windows. This is reasonable since longer time windows make the problem less constrained when considering only a few nodes. Besides, the complexity of the dynamic programming algorithm producing the TW-conflicts is exponential in *maxConfSize*. Thus, we decided to set $maxConfSize = 4$ after the analysis of the global search efficiency.

Another parameter is the heuristic according to which, given a locally optimal solution σ^* visiting a set of nodes S^*, we choose a subset $S_c \subseteq S^*$ for trying to explain why a customer $i \notin S^*$ cannot be inserted into σ^* (TW-conflicts). For this, we use the *NearestTimeWindow* heuristic: to define S_c, we choose $maxConfSize - 1$ nodes $j \in S^*$ such that the distance between the midpoint of the time window of j and the midpoint of the time window of i is as small as possible. However, generating TW-conflicts all the time can slow down the global search. Therefore, we define an *explanation quota* *xpQuota* for every node to reduce the workload of function extractMinTWconflicts. This quota is decreased by one unit each time a TW-conflict explaining the absence of i in a locally optimal solution is looked for. When the quota of i becomes 0 after *xpQuota* searches for TW-conflicts related to i, the absence of i in a locally optimal solution is not explained anymore. With such an approach, there is somehow a warm-up phase where TW-conflicts are learned, followed by an exploitation phase of these conflicts. After performing tests with different values of $xpQuota \in \{20, 60, 100\}$, we decided to set $xpQuota = 20$.

Last, concerning the generation of Lopt-conflicts to diversify search, we need to forbid during *tabuSize* iterations a region around a locally optimal solution, where the region size is controlled by the *approxSize* parameter which defines the maximum size of the approximate Lopt-conflicts. After several tests performed with $approxSize \in \{3, 5, 7\}$ and $tabuSize \in \{10, 50, 100, 200\}$, we set $approxSize = 7$ and $tabuSize = 50$ for the experiments.

---

[3] https://github.com/msoos/cryptominisat.

[4] https://github.com/ivmai/cudd.

6.2 Performance of the Versions of CB

Experiments are performed for the three CB data structures presented before. For LNS-CB-UnitPropagation (or shortly LNS-CB-UP), we actually consider two versions: one called LNS-CB-UP where no Lopt-conflict is generated, and another called LNS-CB-UP-Lopt where Lopt-conflicts are generated. For LNS-CB-IncrementalSAT (or shortly LNS-CB-Sat), we do not present the results obtained with the Lopt-conflicts due to space limitation reasons. For LNS-CB-OBDD, we do not use the temporary Lopt-conflicts as it would require (a) maintaining an OBDD containing only permanent TW-conflicts, and (b) making time-consuming conjunctions with the temporary Lopt-clauses that are still active at the current iteration.

Overall Performance. To quickly compare the baseline incomplete search algorithm (called LNS-noCB) and the versions using a CB, we first measured, for each solver and each instance, the average gap to the best known solution after five runs, each within 1 min. This gap g_s for solver s is defined by $g_s = 100 * (bk - bf_s)/bk$ where bf_s is the total reward of the best feasible solution found by s and bk is the best known objective value. Table 2 shows that for 1-minute time limit, using CB-UP globally improves the gaps (0.851% compared to 0.886% when using noCB), while using CB-UP-Lopt also generates competitive results. On the contrary, CB-Sat and CB-OBDD deteriorate the average gap (mean gaps equal to 1.739% and 1.418% respectively). Moreover, we also implemented a simple tabu list that prevents the algorithm from inserting (or removing) customers that were removed (or inserted) during the last k iterations. This tabu list made the LNS method highly effective for instances in the Solomon1 set, with an average gap of 0.083%. However, the average gaps obtained on other three sets are much larger, leading to a higher grand mean of the average gaps (2.332% for LNS-SimpleTabu, compared to 0.851% for LNS-UP).

To further analyze the results, each version of the solver is executed during 10 000 LNS iterations and the total time elapsed over each set is measured. Then, a speed-up rate compared to the noCB version is computed by $speedUp_s = 100 * (timeNoCB - timeWithCB_s)/timeNoCB$. Table 3 shows that the search process is accelerated with CB-UP and CB-UP-Lopt almost all the time, especially on the Cordeau instances where the speed-up reaches almost 50%. On the contrary, the search process is drastically slowed down with CB-Sat and CB-OBDD.

Table 2. Average gap (%) over 5 runs (maxCPUtime=60s, best average gaps in **bold**)

Instance set	Variants of CB in LNS					
	noCB	UP	UP-Lopt	SAT	OBDD	simpleTabu
Solomon1	1.093	1.093	1.304	1.492	1.315	**0.083**
Solomon2	0.416	0.387	**0.345**	0.607	0.497	4.097
Cordeau1	0.139	**0.078**	0.351	1.125	0.903	1.540
Cordeau2	1.898	**1.846**	1.900	3.729	2.958	2.119
Grand mean	0.886	**0.851**	0.977	1.739	1.418	2.332

Table 3. Speed-up (%) when solving during 10 000 LNS iterations

Instance set	Variant of CB in LNS			
	UP	UP-Lopt	SAT	OBDD
Solomon1	−8.83	−18.66	−2517.14	−646.14
Solomon2	**25.17**	**25.15**	−492.75	−163.62
Cordeau1	**48.66**	**47.04**	−2779.32	−2446.31
Cordeau2	**45.96**	**47.83**	−2092.35	−610.95

Slow Convergence with CB-SAT *and* CB-OBDD. Despite the rapidity of incremental solving with CryptoMiniSat, the results obtained show that the search process is slower for the LNS-CB-SAT version. The main reason for this is that there are numerous calls to $solve(\mathcal{F}, \mathcal{A} \cup \{l\})$, and each call must either find a full solution or prove that none exists.

As for CB-OBDD, while querying in OBDD is fast, the results are not as good as expected. Table 4 shows that the OBDDs obtained are globally compact given the number of conflicts. But the reordering operations performed to get such a compactness can take a lot of time: on some instances, CB-OBDD spends more than 60% of the CPU time for reordering the variables. Alternatively, it is challenging to heuristically compute in advance a good static variable ordering for the OBDDs, since we do not have the entire information about the conflicts when a static ordering must be defined. Meanwhile, we tested eight problem-dependent heuristics (e.g. ordering the selection variables depending on the rewards, the time windows, etc.), and as shown in Table 5, the best heuristics give poor results on some instances.

Better and Faster Search with CB-UP. Figure 4 details the evolution of the mean gap over each set of instances. Globally, we observe that LNS is boosted by CB-UP. In particular, for set Cordeau 1 involving many TW-conflicts, the search process converges much more quickly with the support of CB-UP. This is because more LNS iterations are performed thanks to the effectiveness of neighborhood pruning through CB-UP. The strength of CB-UP-LOPT is particularly visible over instance sets Cordeau 2 and Solomon 2. In these cases, even with very few TW-conflicts, the approximate Lopt-conflicts help guide the search towards other interesting search regions.

Table 4. Size of CB for each instance group (CPU time: 10 s)

Instance set	#OPTW nodes	#conflicts (average)	#OBDDnodes (average)	reorderingtime (%)
Solomon1	100	509.66	257.59	34.15
Solomon2	100	19.19	11.19	6.91
Cordeau1	48–288	109.10	303.50	67.73
Cordeau2	48–288	0.40	1.70	2.15

Table 5. Performance of the static and dynamic ordering strategies for OBDDs on two instances (pr01: 48 variables, best static order found = "increasing opening time"; pr06: 288 variables, best static order found = "decreasing rewards")

instance	LNS iteration	#conflicts	best-static-ordering		dynamic-ordering	
			#nodes	time(s)	#nodes	time(s)
pr01	1	0	1	0.0006	1	0.0012
	2	2	7	0.0013	5	0.0026
	3	8	36	0.0021	13	0.0041
	4	8	36	0.0030	12	0.0054
pr06	1	0	1	0.0885	1	0.4886
	2	55	69219	0.1803	477	2.5110
	3	80	6342191	19.1407	533	2.6108
	4	94	38250383	367.198	833	2.6422

7 Related Works

Incomplete search and SAT/CP were combined in Large Neighborhood Search [23], where a sequence of destroy-repair operations is performed on an incumbent solution. The destroy phase unassigns a subset S of the decision variables, while the repair phase can be delegated to a SAT/CP engine capable of quickly exploring all possible reassignments of S given the current partial assignment. Some authors also proposed to represent specific neighborhood structures using a tailored CP model and to translate the solutions found for this model into changes at the level of the global solution [22]. Others propose an efficient neighborhood exploration algorithm with the help of restricted decision diagrams [8]. In the same spirit, our CB is built to quickly detect inconsistent assignments at the selection level, therefore it can significantly reduce the neighborhood size to explore in the repair phase, but one difference is that we generate new conflicts during search and for the incomplete search process, CB only acts as a constraint propagation engine.

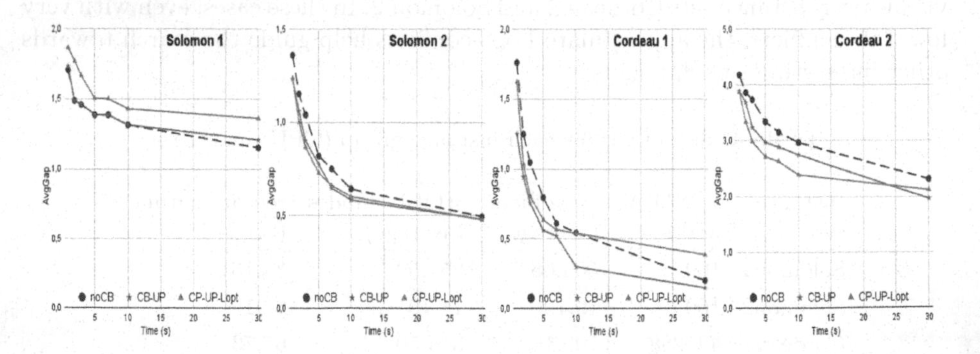

Fig. 4. Evolution of the average gaps for CB-UP and CB-UP-LOPT

Other hybrid approaches exploit the strengths of incomplete search and complete SAT/CP techniques at different search phases. As an illustration, in SAT, *Stochastic Local Search* (SLS) has been combined with DPLL or *Conflict Directed Clause Learning* (CDCL) [1,5,19]. For the SLS-CDCL version, the idea is that on one side, SLS can be run first to help CDCL have a heuristic for choosing variable values or to help CDCL update the activities of the variables, and on the other side CDCL can help SLS espace local optima. Another example is the composition of traditional CP search and Constraint-Based Local Search (CBLS [12]), where the two search approaches can exchange bounds, solutions, etc. In line with previous studies, inconsistency explanations generated at each iteration are stored in CB and then reused to help the search engine escape explored or invalid regions. In our case, by taking into account the current search state along with the clauses learned in the past iterations, CB may suggest mandatory assignments to quickly lead the search to promising regions.

Another technique uses inference methods such as unit propagation or constraint propagation, initially developed for complete search strategies, to speed up the neighborhood exploration during local search. One example following this line for SAT is the *unitWalk* algorithm [13,17]. At each iteration, it considers a complete variable assignment and performs a pass over this assignment to iteratively update the values of the variables with unit propagation. Compared to this work, one of the novelties in CB-UP is the decremental propagation aspect.

Last, the use of an external CB coupled with incomplete search can be compared with the use of a memory data structure in tabu search. On this point, instead of a simple list of forbidden local moves or forbidden variable assignments as in tabu search [9], CB memorizes logical formulas about the selection of nodes in a long-term way (possibly with some trashing when the size of CB becomes too large). CB is also equipped with efficient mechanisms to quickly reason about the formulas collected, instead of just reading explicit forbidden configurations. Another remark is that traditional tabu search is usually not recyclable i.e. the memory is reset at each resolution, while the time window conflicts stored in CB are easily recyclable for dynamic OPTWs where the reward associated with each node can change.

8 Conclusion and Perspectives

This paper presented a new hybrid optimization architecture combining an incomplete search process with clause generation techniques. Three generic clause basis managers were studied instead of just arbitrarily choosing a unique option, and the efficiency of the approach using unit propagation was demonstrated. One next step is to apply the approach to other problems like Team OPTW or flexible scheduling problems. Now that the generic clause bases are defined, the main effort to tackle a new problem is the definition of the problem-dependent clause generation procedure. Another perspective is to explore other clause basis managers (e.g. based on 0/1 linear programming and reduced-cost filtering), or knowledge bases covering pseudo-boolean constraints or cardinality constraints.

References

1. Audemard, G., Lagniez, J.M., Mazure, B., Saïs, L.: Integrating conflict driven clause learning to local search. In: 6th International Workshop on Local Search Techniques in Constraint Satisfaction (LSCS 2009) (2009)
2. Audemard, G., Lagniez, J.-M., Simon, L.: Improving glucose for incremental SAT solving with assumptions: application to MUS extraction. In: Järvisalo, M., Van Gelder, A. (eds.) SAT 2013. LNCS, vol. 7962, pp. 309–317. Springer, Heidelberg (2013). https://doi.org/10.1007/978-3-642-39071-5_23
3. Bellman, R.: Dynamic programming treatment of the travelling salesman problem. J. ACM (JACM) **9**(1), 61–63 (1962)
4. Bryant, R.E.: Graph-based algorithms for Boolean function manipulation. Comput. IEEE Trans. **100**(8), 677–691 (1986)
5. Crawford, J.: Solving satisfiability problems using a combination of systematic and local search. In: Second Challenge on Satisfiability Testing organized by Center for Discrete Mathematics and Computer Science of Rutgers University (1996)
6. Darwiche, A., Marquis, P.: A knowledge compilation map. J. Artif. Intell. Res. **17**, 229–264 (2002)
7. Eén, N., Sörensson, N.: Temporal induction by incremental SAT solving. Electron. Notes Theor. Comput. Sci. **89**(4), 543–560 (2003)
8. Gillard, X., Schaus, P.: Large neighborhood search with decision diagrams. In: International Joint Conference on Artificial Intelligence (2022)
9. Glover, F., Laguna, M.: Tabu search. In: Du, D.Z., Pardalos, P.M. (eds.) Handbook of Combinatorial Optimization, pp. 2093–2229. Springer, Boston (1998). https://doi.org/10.1007/978-1-4613-0303-9_33
10. Golden, B.L., Levy, L., Vohra, R.: The orienteering problem. Naval Res. Logistics (NRL) **34**(3), 307–318 (1987)
11. Gunawan, A., Lau, H.C., Vansteenwegen, P.: Orienteering problem: a survey of recent variants, solution approaches and applications. Eur. J. Oper. Res. **255**(2), 315–332 (2016)
12. Hentenryck, P.V., Michel, L.: Constraint-Based Local Search. The MIT Press, Cambridge (2005)
13. Hirsch, E., Kojevnikov, A.: UnitWalk: a new SAT solver that uses local search guided by unit clause elimination. Ann. Math. Artif. Intell. **43**, 91–111 (2002)
14. Hooker, J., Ottosson, G.: Logic-based Benders' decomposition. Math. Program. Ser. B **96**, 33–60 (2003)
15. Ignatiev, A., Semenov, A.: DPLL+ROBDD derivation applied to inversion of some cryptographic functions. In: Sakallah, K.A., Simon, L. (eds.) SAT 2011. LNCS, vol. 6695, pp. 76–89. Springer, Heidelberg (2011). https://doi.org/10.1007/978-3-642-21581-0_8
16. Kantor, M.G., Rosenwein, M.B.: The orienteering problem with time windows. J. Oper. Res. Soc. **43**(6), 629–635 (1992)
17. Li, X.Y., Stallmann, M.F., Brglez, F.: QingTing: a fast SAT solver using local search and efficient unit propagation. In: Proceedings of the Sixth International Conference on Theory and Applications of Satisfiability Testing (SAT 2003) (2003)
18. Marques-Silva, J., Lynce, I., Malik, S.: Conflict-driven clause learning SAT solvers. In: Handbook of Satisfiability, pp. 133–182. IOS Press (2021)
19. Mazure, B., Sais, L., Grégoire, É.: Boosting complete techniques thanks to local search methods. Ann. Math. Artif. Intell. **22**(3), 319–331 (1998)

20. Moskewicz, M.W., Madigan, C.F., Zhao, Y., Zhang, L., Malik, S.: Chaff: Engineering an efficient SAT solver. In: Proceedings of the 38th annual Design Automation Conference, pp. 530–535 (2001)
21. Nadel, A., Ryvchin, V.: Efficient SAT solving under assumptions. In: Cimatti, A., Sebastiani, R. (eds.) SAT 2012. LNCS, vol. 7317, pp. 242–255. Springer, Heidelberg (2012). https://doi.org/10.1007/978-3-642-31612-8_19
22. Pesant, G., Gendreau, M.: A constraint programming framework for local search methods. J. Heuristics **5**(3), 255–279 (1999)
23. Pisinger, D., Ropke, S.: Large neighborhood search. In: Gendreau, M., Potvin, J.-Y. (eds.) Handbook of Metaheuristics. ISORMS, vol. 272, pp. 99–127. Springer, Cham (2019). https://doi.org/10.1007/978-3-319-91086-4_4
24. Prestwich, S.: The relation between complete and incomplete search. In: Blum, C., Aguilera, M.J.B., Roli, A., Sampels, M. (eds.) Hybrid Metaheuristics. Studies in Computational Intelligence, vol. 114, pp. 63–83. Springer, Heidelberg (2008). https://doi.org/10.1007/978-3-540-78295-7_3
25. Rudell, R.: Dynamic variable ordering for ordered binary decision diagrams. In: Proceedings of 1993 International Conference on Computer Aided Design (ICCAD), pp. 42–47. IEEE (1993)
26. Savelsbergh, M.W.: Local search in routing problems with time windows. Ann. Oper. Res. **4**(1), 285–305 (1985)
27. Schmid, V., Ehmke, J.F.: An effective large neighborhood search for the team orienteering problem with time windows. In: ICCL 2017. LNCS, vol. 10572, pp. 3–18. Springer, Cham (2017). https://doi.org/10.1007/978-3-319-68496-3_1
28. Schutt, A., Feydy, T., Stuckey, P., Wallace, M.: Solving RCPSP/max by lazy clause generation. J. Sched. **16**(3), 273–289 (2013)
29. Solomon, M.M.: Algorithms for the vehicle routing and scheduling problems with time window constraints. Oper. Res. **35**(2), 254–265 (1987)
30. Soos, M., Nohl, K., Castelluccia, C.: Extending SAT solvers to cryptographic problems. In: 12th International Conference on Theory and Applications of Satisfiability Testing (SAT), pp. 244–257 (2009)
31. Vansteenwegen, P., Souffriau, W., Berghe, G.V., Van Oudheusden, D.: Iterated local search for the team orienteering problem with time windows. Comput. Oper. Res. **36**(12), 3281–3290 (2009)
32. Vansteenwegen, P., Souffriau, W., Van Oudheusden, D.: The orienteering problem: a survey. Eur. J. Oper. Res. **209**(1), 1–10 (2011)

Author Index

© The Editor(s) (if applicable) and The Author(s), under exclusive license
to Springer Nature Switzerland AG 2023
A. A. Cire (Ed.): CPAIOR 2023, LNCS 13884, pp. 511–512, 2023.
https://doi.org/10.1007/978-3-031-33271-5

Printed in the United States
by Baker & Taylor Publisher Services